Petroleum Engineering

DRILLING AND WELL COMPLETIONS

Petroleum Engineering

DRILLING AND WELL COMPLETIONS

CARL GATLIN

Department of Petroleum Engineering, The University of Texas

PRENTICE-HALL, INC. | Englewood Cliffs, N.J.

Dedicated to
LILA, AMY, *and* JEFF

Library of Congress Catalog Card Number: 60-6874

Current printing (last digit):
13 12 11 10

PRINTED IN THE UNITED STATES OF AMERICA
66215

Preface

In this book I have attempted to present an integrated picture of drilling and well completion operations as they are normally encountered by the petroleum engineer. In order to do this without assuming prior knowledge in the field, it was necessary to introduce a number of rather general topics. The chapters on reservoir fluid properties, reservoir rock properties, exploration and leasing practices, core analysis, well logging and formation damage fall into this category. The coverage in these sections is, of course, limited, and the emphasis is on the problem at hand rather than an over-all implications. It is hoped, however, that these treatments will form a sound basis for later, more detailed study.

Petroleum engineering curricula vary widely as to the level at which the drilling course(s) is taught. By including the necessary background material this text may be used in a first course. Similarly, by proper selection and deletion of chapters it can serve the needs of a more advanced course. The numerous references cited form adequate outside reading for a course at any level. It is also hoped that the many charts and example problems will make the book valuable as a reference for those practicing in these areas.

It is my opinion that for the most part petroleum engineers are best utilized in drilling operations rather than as designers of equipment. Hence, this text is primarily concerned with operational procedures and not with detailed descriptions and analysis of equipment. The latter coverage is, therefore, restricted to a level necessary for formulation and understanding of the problems. I feel this much is desirable.

During the writing of this book it was necessary to ask permission from numerous organizations and individuals for use of various material. It was gratifying to me that in no case was permission denied. In fact, in most cases much more was offered than was asked for. I have attempted throughout to acknowledge credit for this aid, and I hope no one has been overlooked. Similarly, I have tried to be scrupulously honest in the numerous references cited for it is these authors who have written this book. I merely put it together. I am sure oversights must exist; however, I hope they are few and excusable.

I wish to make several specific acknowledgements. First, I express my gratitude to the Society of Petroleum Engineers of AIME and to the American Petroleum Institute, from whose transactions I borrowed heavily.

I also wish to thank the following individuals for their counsel and assistance at various stages of the writing: John A. Casner, Dick Cavnar, Gerald L. Farrar, B. E. Groenewold, Robt. E. Hensley, Frank W. Jessen, G. W. (Sandy) McGaha, Phil C. Montgomery, Edward E. Runyan, Carroll V. Sidwell, Dwight K. Smith, Gould Whaley, Jr., and the late A. W. Walker.

I am greatly indebted to the following for their corrections and criticisms of specific chapters: Robert P. Alger, Fred W. Chisholm, Arthur Lubinski, E. A. Morlan, R. H. Nolley, Robt. L. Slobod, C. Drew Stahl, and Henry B. Woods. I also express my thanks to Joseph J. Cosgrove, Donald H. Crago, and Kenneth E. Gray for checking numerous derivations and problems.

I shall welcome at any time correspondence concerning errors, suggestions for improvement, or criticisms of the text. Indeed, if I were to start over, I would change a great many things myself. I am informed, however, that he who demands perfection never finishes his book. No perfection exists here for I have finished.

Carl Gatlin

Contents

Petroleum

Engineering

DRILLING AND WELL COMPLETIONS

Chapter 1 . . .

The Nature of Petroleum

There is probably no other technical field in which so many major questions remain unanswered and yet which functions as efficiently as the oil industry. For example, geologists have yet to agree completely on the origin and accumulation of petroleum, geophysicists have no tool which searches directly for oil, petroleum engineers are still leaving unrecoverable oil in the ground, and chemists and chemical engineers must still evaluate crude oils on the basis of empirical tests rather than by precise analyses. In short, we don't know what it is, how it originates and accumulates, how to find it, or how to get it all out of the ground.

Fortunately, there are a few things we do know and others about which we have workable ideas. The purpose of this book is to present a fundamental treatment of the techniques employed in the drilling phase of the oil industry. We will first consider the basic physical and chemical properties of petroleum.

1.1 Chemical Composition

Petroleum may be defined as a naturally occurring, complex mixture of hydrocarbons which may be either gas, liquid, or solid, depending upon its own unique composition and the pressure and temperature at which it is confined. The principal hydrocarbon series found in petroleum are

1. Paraffins (also called saturated hydrocarbons or alkanes) which have the general formula C_nH_{2n+2}. These compounds are chemically stable and have either straight or branched chains. The branched chain members are called *isomers* and exhibit somewhat different properties than their straight chain counterparts. All crude oils contain some paraffins, particularly as the more volatile (low boiling point) constituents. The first few members of this series are:

```
Abbreviation  Formula   Chemical structure        Name

                              H
                              |
C1            CH4           H—C—H                 Methane
                              |
                              H
                             H H
                             | |
C2            C2H6         H—C—C—H                Ethane
                             | |
                             H H
                            H H H
                            | | |
C3            C3H8        H—C—C—C—H               Propane
                            | | |
                            H H H
                           H H H H
                           | | | |
C4            C4H10      H—C—C—C—C—H              Normal Butane
                           | | | |
                           H H H H
                            H H H
                            | | |
                          H—C—C—C—H
iC4           C4H10         | | |                 iso-Butane
                            H | H
                            H—C—H
                              |
                              H
```

Note that *iso*-butane differs from *normal* by the manner in which the carbon atoms are arranged. This is the only isomer possible with C_4; the possible atomic combinations increase, however, with the length of the chain, as illustrated in Table 1.1.

TABLE 1.1

POSSIBLE ISOMERS OF VARIOUS PARAFFINS[1]

Carbon content	*Number of isomers*
C_8	18
C_{10}	75
C_{20}	366, 319
C_{40}	62, 491, 178, 805, 831
etc.	

[1]All references are listed at end of chapter.

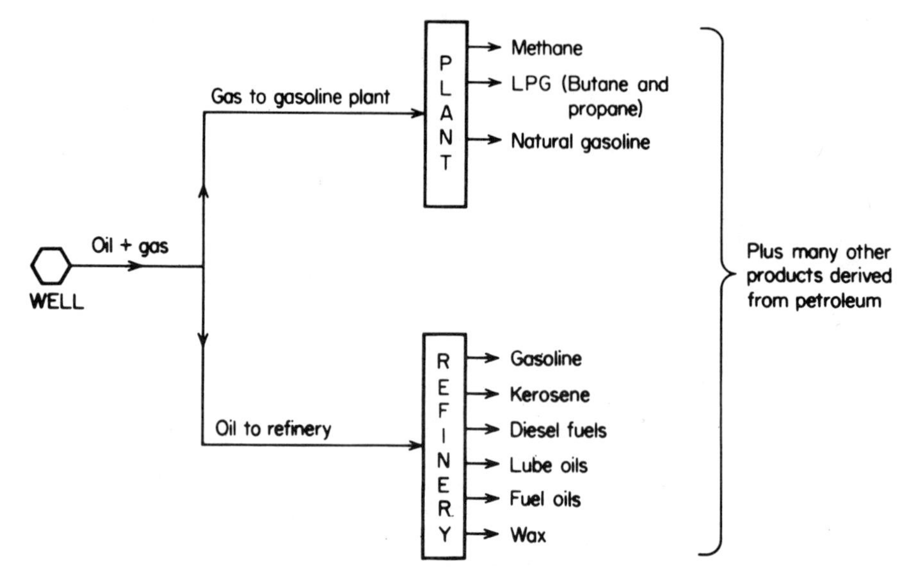

Fig. 1.1 Schematic flow diagram illustrating process route and ultimate products of produced oil and gas.

2. Cycloparaffins (naphthenes) having the general formula C_nH_{2n}. These compounds have a ring structure, the simplest member being cyclopropane. Typical members of this group are

C_3H_6 Cyclopropane

C_4H_8 Cyclobutane

3. Aromatics (benzene series) having the general formula C_nH_{2n-6}. These compounds are chemically active and contain the benzene ring. The simplest member is

C_6H_6 Benzene

In addition to hydrocarbons, petroleum may contain numerous impurities such as carbon dioxide (CO_2), hydrogen sulfide (H_2S), and other complex compounds of nitrogen, sulphur, and oxygen. When it is considered that any particular crude oil may contain a portion of each of the listed hydrocarbon series plus numerous impurities, and that its composition therefore is unique, it becomes obvious that precise chemical analysis is impossible. Consequently, petroleum is often classified by a *base* designation — as either a paraffin base, asphalt base, or mixed base crude.

A paraffin base crude is an oil whose chief components are paraffins, and which, when completely distilled, leaves a solid residue of wax. An asphalt base crude is an oil composed primarily of cyclic compounds (mostly naphthenes) which, when distilled, leaves a solid residue of asphalt. Oils which fall in the middle of these categories are classified as mixed base. In general, the crudes of California and the Gulf Coast are asphaltic, the Pennsylvanian and other Appalachian crudes are paraffinic, and the bulk of mid-continent crudes are mixed base. It should be recognized that these composition-by-locality statements are general and that any type crude oil may be found in some areas.

Crude petroleum yields a large number of products. The simplest refining process is fractional distillation whereby the constituents in the oil are separated by utilizing their differences in boiling points. A few of these products are shown schematically in Figure 1.1.

1.2 Properties of Liquid Petroleum

The most widely used indicator of a crude oil's worth to the producer is its API (American Petroleum Institute) gravity. This value is actually a measure of an oil's density, and is related to specific gravity by the following formula:

$$\text{API Gravity (degrees)} = \frac{141.5}{\text{sp. gr.}} - 131.5$$

Note that an API Gravity of 10° is equivalent to a specific gravity of 1.

Normally, the price which a producer receives for his oil depends on its gravity, the less dense oils (higher API gravity) being the most valuable. This price schedule is based on the premise that the lighter oils contain higher percentages of the more valuable products such as gasoline. It is quite possible, however, that a particular 30° oil may be more valuable than some 40° oil, due to a particularly high yield of a desirable product. Evidently, the refiners feel that these discrepancies average out; it appears probable that crude oil will always be sold on a gravity basis. Current price-gravity schedules may be found in a number of the trade journals.

The surface or tank oil as finally sold by the producer is not the same liquid which existed underground. The differences between tank oil and reservoir oil are of fundamental importance, and to cover them completely is beyond the present treatment. There are, however, certain basic concepts which can be discussed at this time.

A reservoir oil always contains in solution some components which would be gases at standard temperature and pressure. Their solubility is due to the elevated pressure and temperature existing underground.

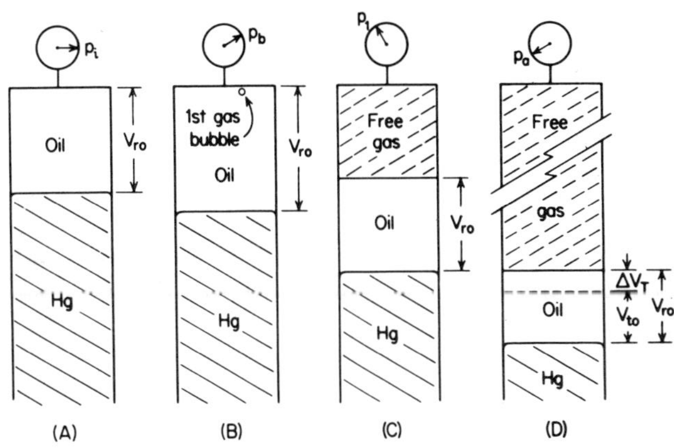

Fig. 1.2. Behavior of typical reservoir oil sample on isothermal pressure reduction. **(A)** Oil sample at original reservoir conditions. All gas is in solution and oil is undersaturated since $p_i > p_b$. **(B)** Pressure is reduced to p_b by removing mercury from the cell. First bubble of gas escapes from solution, hence p_b = bubble point or saturation pressure of the oil. Liquid volume has expanded slightly. **(C)** Pressure is reduced to p_i and considerable free gas has evolved. Liquid volume has shrunk due to loss of volatile fractions. **(D)** Pressure is now atmospheric. Liquid volume has shrunk to V_{ro}, the oil volume at the reservoir temperature, and 14.7 psia. Cooling this oil to standard temperature (60°F) results in its shrinking by an amount ΔV_T to the tank oil volume, V_{to}.

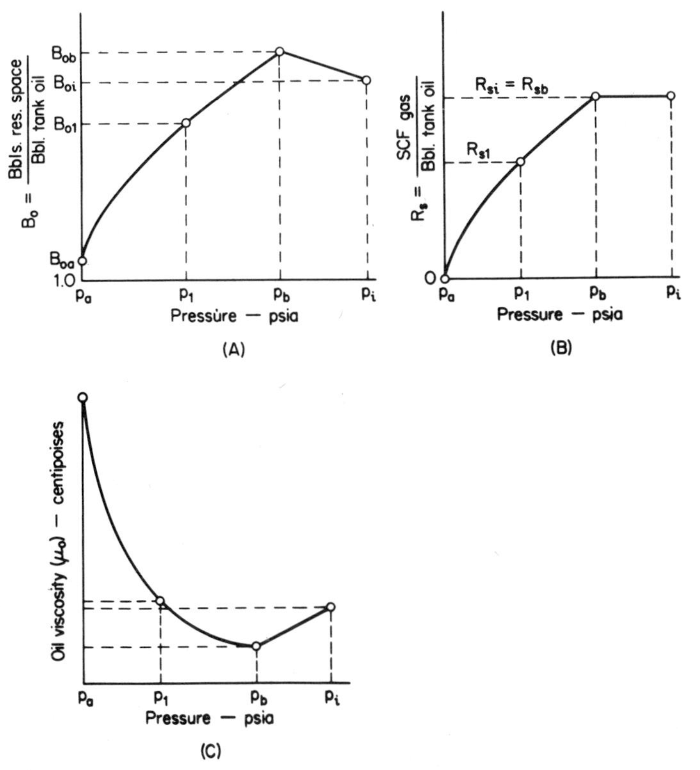

Fig. 1.3. Graphical representation of fluid properties depicted in Fig. 1.2. **(A)** Formation volume factor for oil as a function of pressure.

$$B_o = \frac{V_{ro} \text{ (at each pressure)}}{V_{to}}$$

(B) Solution gas-oil ratio as a function of pressure. **(C)** Oil viscosity as a function of pressure.

As oil is produced (brought from underground to the surface), the pressure is decreased until it reaches substantially atmospheric conditions in the stock tanks. This pressure reduction causes certain changes in the reservoir fluid properties:

a. Some of the volatile fractions vaporize, causing
b. the liquid volume to shrink, and
c. the liquid viscosity to increase.

These effects are shown in Figures 1.2 and 1.3. Also illustrated in these are a number of fundamental ideas which must be thoroughly defined:

1. *Bubble Point Pressure:* The pressure shown as p_b is the bubble point or saturation pressure. It is the pressure at which the first gas is liberated from the reservoir oil upon isothermal pressure reduction at the reservoir temperature. A more rigorous definition commonly used is, "the pressure at which an infinitesimal amount of gas is in equilibrium with an infinite quantity of liquid."
2. *Formation Volume Factor:* This is the quantity denoted by B_o, the reservoir volume occupied per volume of tank oil (oil reduced to standard conditions — 14.7 psia and 60°F) and its dissolved gas. As defined, this quantity is always greater than 1.0 because of (a) thermal expansion [shown as ΔV_T in Figure 1.2(*D*)] and (b) swelling as gas is dissolved at the higher pressures [shown by increasing values of V_{ro} in Figure 1.2 (A), (B), (C), (D)]. Note that B_o increases as the pressure is decreased from p_i to p_b, due to the liquid's expansion. This effect is exaggerated for clarity.
3. The *solution gas-oil ratio,* denoted by R_s, is the number of standard cubic feet of gas dissolved per barrel of tank oil.
4. The *oil viscosity* (μ_o) behavior shown in Figure 1.3 (C) is typical and is found by stepwise determinations in a high pressure viscosimeter. This behavior may be explained as follows:
   (a) Viscosity decreases as pressure is reduced from p_i to p_b due to the liquid's expansion; greater intermolecular freedom of motion is possible, and internal friction is reduced.
   (b) Viscosity increases with pressure reduction below p_b because the low viscosity fractions are lost; this more than compensates for the effect of liquid expansion.

The process depicted in Figure 1.2 is defined as a *flash vaporization,* since the composition of the system remained constant. If the cell pressure had been reduced by removing liberated gas while holding cell volume constant, the process would have been a *differential vaporization.* The actual process taking place in the underground petroleum reservoir more nearly fits the differential process, and it is commonly used in laboratory analyses.

The various fluid properties just discussed are commonly spoken of as *PVT* (Pressure-Volume-Temperature) relationships and are of fundamental importance to the solution of many petroleum engineering problems. These values are normally obtained from analyses of subsurface or recombined surface fluid samples. However, there are correlations in the literature which enable these factors to be estimated from readily available field data.[2,3]

1.3 Gaseous Petroleum (Natural Gas)

It is only in comparatively recent years that natural gas has come into its own as a highly valuable product. Prior to the development of our present extensive transcontinental gas transmission lines, gas produced with oil was sold primarily on a local scale and any excess was flared. Certainly these practices were wasteful, but, at the time, necessary. As the natural gasoline and liquefied petroleum gas (butane and propane) industry developed, the utilization of the residue gas (dry gas remaining after liquids removed) also increased. The recent growth of this phase of the oil industry has been very rapid, and today natural gas and its associated liquid products are virtually as much in demand as oil.

Natural gas is produced from three classes of wells:

1. From wells where the dominant product is oil (oil wells).
2. From wells where the gas itself is the principal product (gas wells).
3. As gas from condensate wells. Condensate wells produce from reservoirs in which the hydrocarbons (gas and liquid) originally existed as a single fluid (or phase), the reservoir temperature and pressure being above the critical point of the hydrocarbon mixture.

Each natural gas, like each crude oil, is a unique mixture of hydrocarbons. All are, however, composed primarily of the light members of the paraffin series and are predominantly methane. Numerous impurities are found in petroleum gases, some of the more common being carbon dioxide (CO_2), hydrogen sulfide (H_2S), water vapor, nitrogen, and helium. These impurities detract from the value of a natural gas by raising the costs of processing it to pipe line and consumer standards. The field of gas processing is extensive and complicated and is itself the subject of all or parts of several books.

There are a number of basic definitions which can be presented here.

1. *Wet gas:* A natural gas is said to be wet if it contains an appreciable natural gasoline content as determined by standard tests.[4]

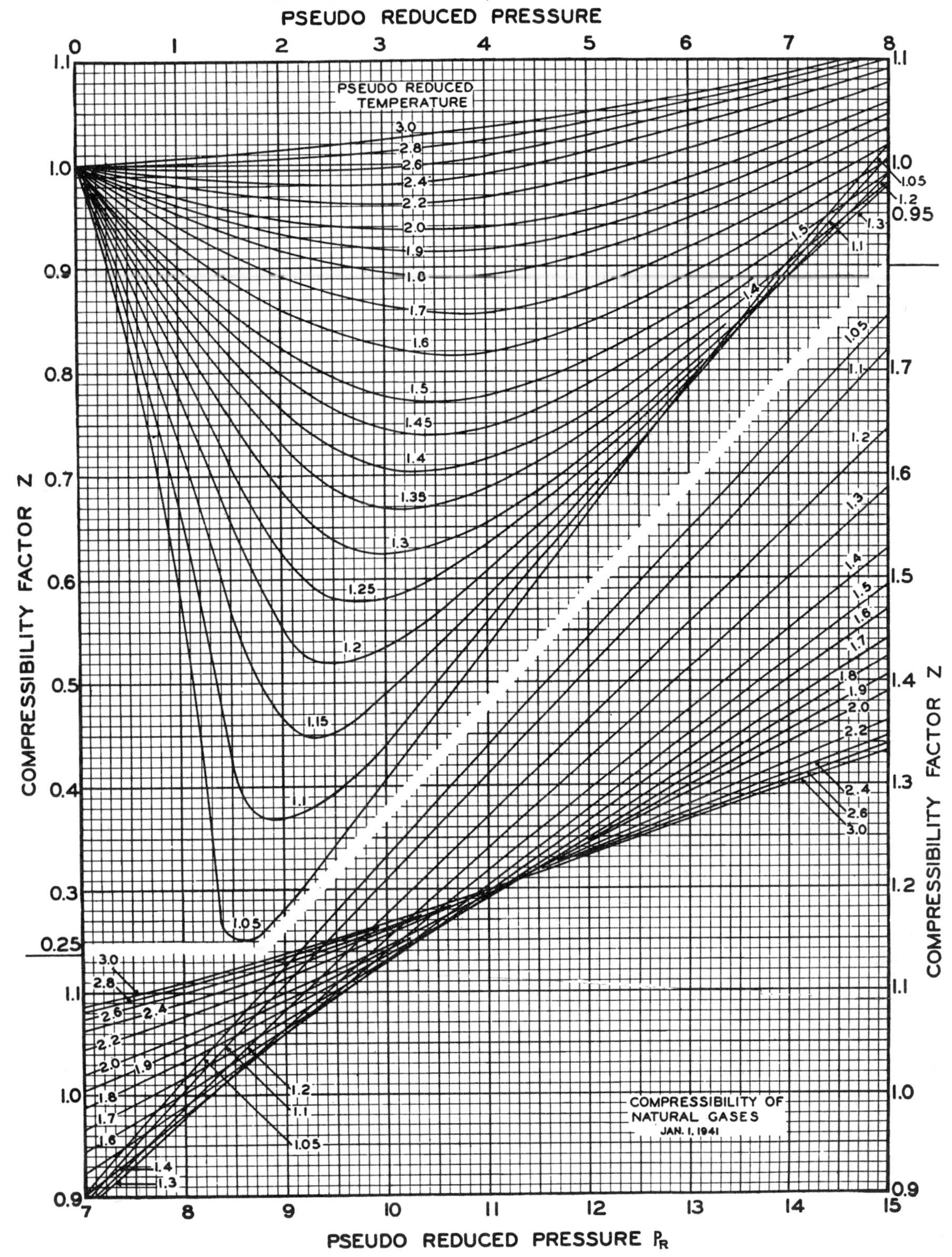

Fig. 1.4. Compressibility of natural gases as a function of reduced pressure and temperature. After Standing and Katz,[8] courtesy AIME.

2. *GPM:* The natural gasoline content of a gas expressed in gallons per thousand standard cubic feet (MCF). Gases having a GPM of 1 to 2 are wet, while gas with a GPM of 0.2 would be considered somewhat dry.
3. *Sour gas:* Natural gas containing hydrogen sulfide.
4. *Sweet gas:* Natural gas containing no hydrogen sulfide.
5. *Gas gravity:* The ratio of the density of a gas to the density of air at standard conditions. In other words, a specific gravity scale based on air.
6. *Standard conditions:* 14.7 psia and 60°F.

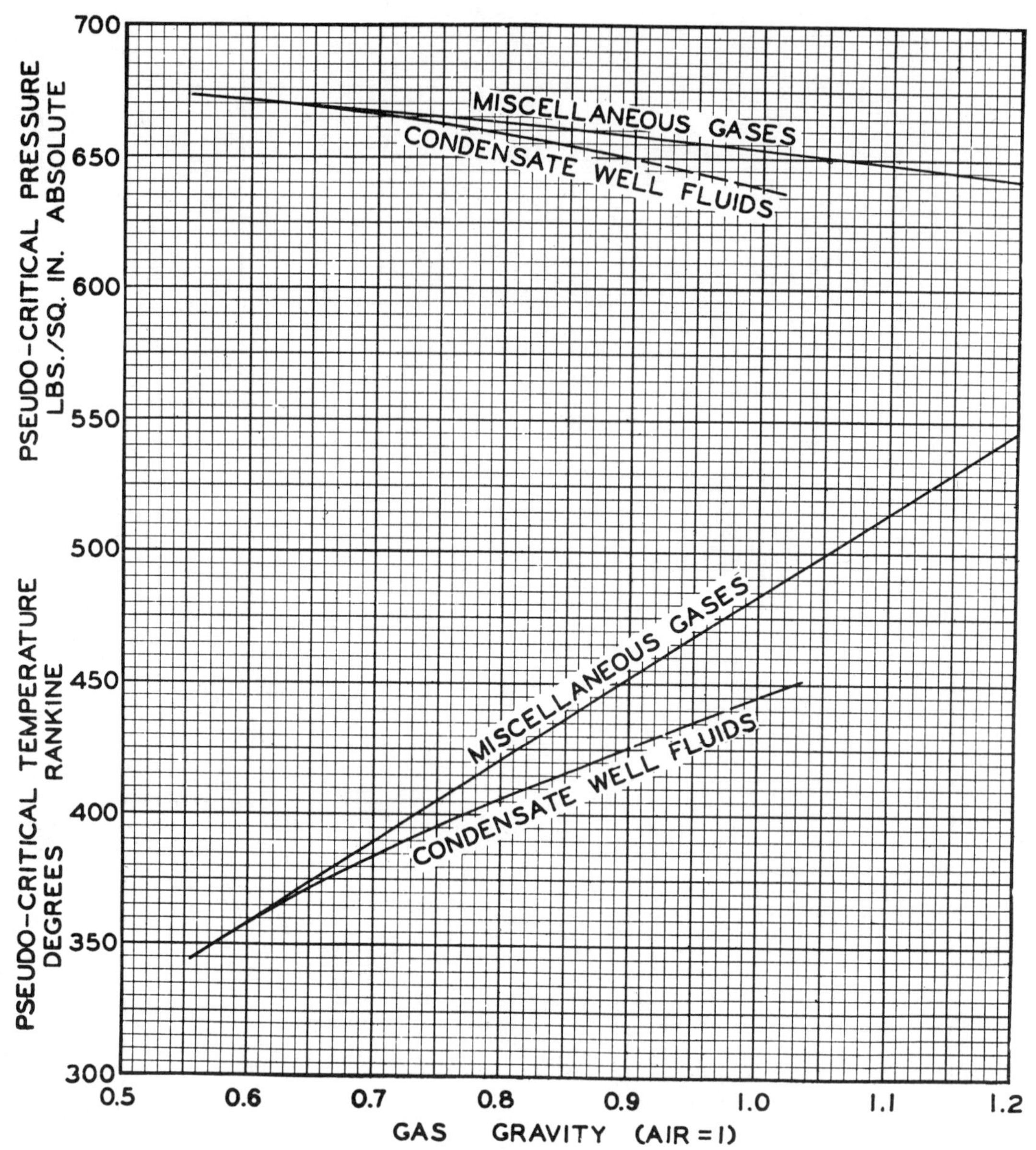

Fig. 1.5. Pseudo-critical properties of natural gases as functions of gas gravity. Courtesy G. G. Brown, *et al.*[5]

1.4 The Gas Law

The gas law as applied to the behavior of natural gas is most commonly stated as

$$pV = znRT \tag{1.1}$$

where p = pressure, absolute
V = volume
n = number of mols
R = gas constant
T = absolute temperature
z = deviation (also called compressibility) factor to account for the difference between actual and ideal gas volumes.

The value of R is dependent on the system of units used, as given in Table 1.2.

TABLE 1.2

VALUES OF THE GAS CONSTANT R FOR DIFFERENT VALUES OF p, V, AND T

p	V	T	R
atmospheres	cc	°K	82.1
atmospheres	liters	°K	0.0821
mm mercury	cc	°K	62369.
gm per sq cm	cc	°K	8.315
lb per sq in.	cu ft	°R	10.7
lb per sq ft	cu ft	°R	1545.
atmospheres	cu ft	°R	0.730

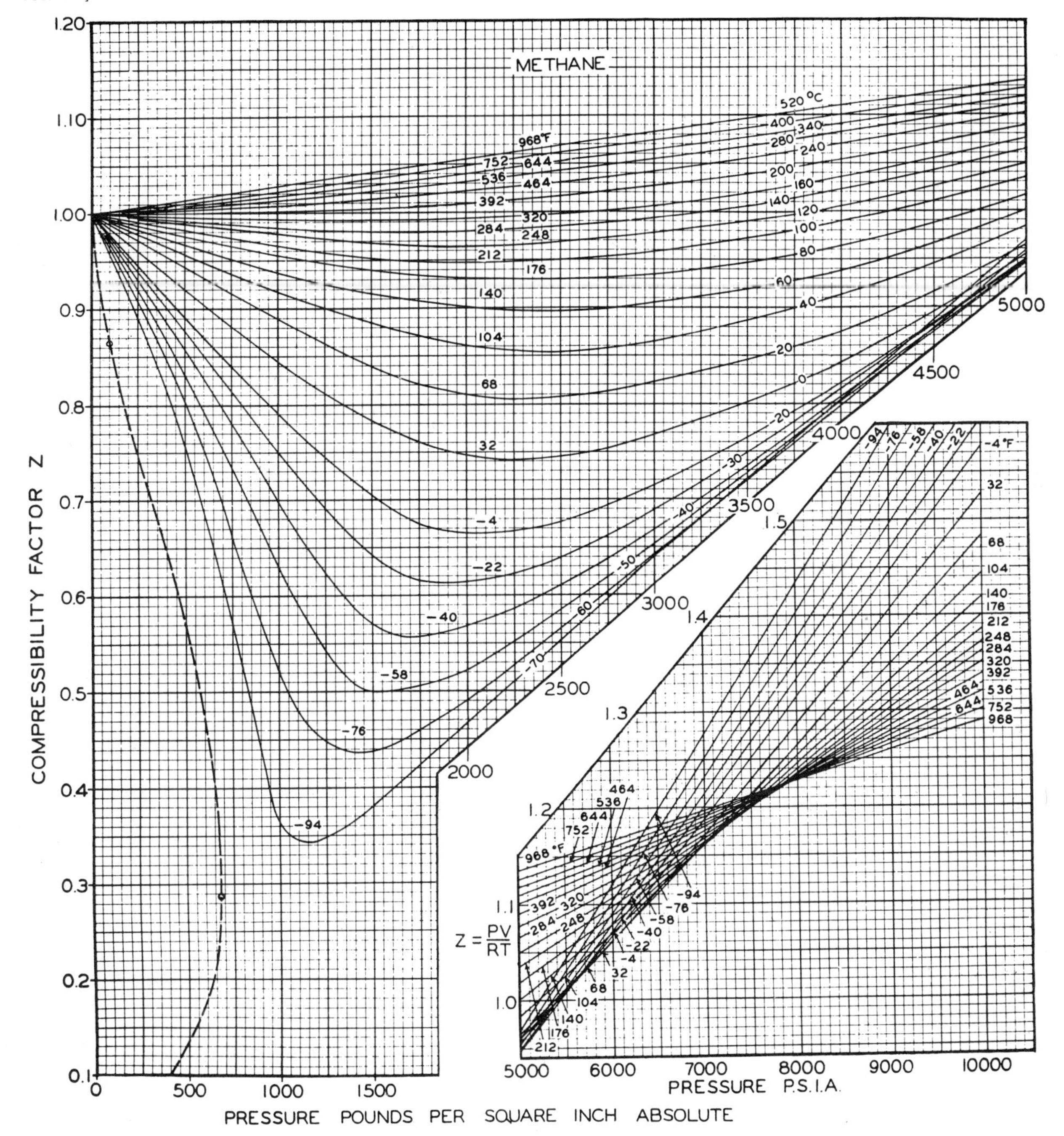

Fig. 1.6. Compressibility factors of methane. Courtesy G. G. Brown, *et al.*[5]

Equation (1.1) may be rewritten:

$$pV = z\frac{W}{M}RT \tag{1.2}$$

where $W/M = n$

W = total wt. of gas

M = molecular wt. of gas

Hence

$$p\frac{V}{W} = \frac{zRT}{M} \tag{1.3}$$

or

$$v = \frac{zRT}{pM} \tag{1.3a}$$

where $v = V/W$ = specific volume of the gas.

Also,

$$\frac{1}{v} = \rho = \frac{pM}{zRT} \tag{1.4}$$

where ρ = gas density.

Another useful expression relating the pVT behavior of a constant number of mols of gas is

$$\frac{p_1V_1}{z_1T_1} = \frac{p_2V_2}{z_2T_2} = nR = \text{Constant} \tag{1.5}$$

Determination of z

To cover completely the theory and background of z determinations is beyond this treatment and the reader is referred to other texts for a complete discussion.[5,6,7] Briefly, the values of z for natural gas mixtures have been experimentally correlated as functions of pressure, temperature, and composition. This correlation is based on the well known Theorem of Corresponding States which proposes that the ratio of the volume of a particular substance to its volume at its critical point is the same for all substances at the same ratio of absolute pressure to critical pressure, and absolute temperature to critical temperature. This theorem is not completely true but may satisfactorily be applied to compounds of similar molecular structure such as the light paraffins and natural gases. In preparing a correlation for hydrocarbon mixtures, the ratios of actual pressure and temperature to the molal average critical or *pseudo-critical* pressure and temperature have been used. These ratios are called reduced pressures and reduced temperatures. Figure 1.4 is a correlation of z as a function of these quantities.[8]

The chemical analysis of a natural gas is not always available and an alternate method is needed for determination of the pseudo-critical properties. Correlations of these as functions of gas gravity, (a factor which is always readily obtainable), have been found

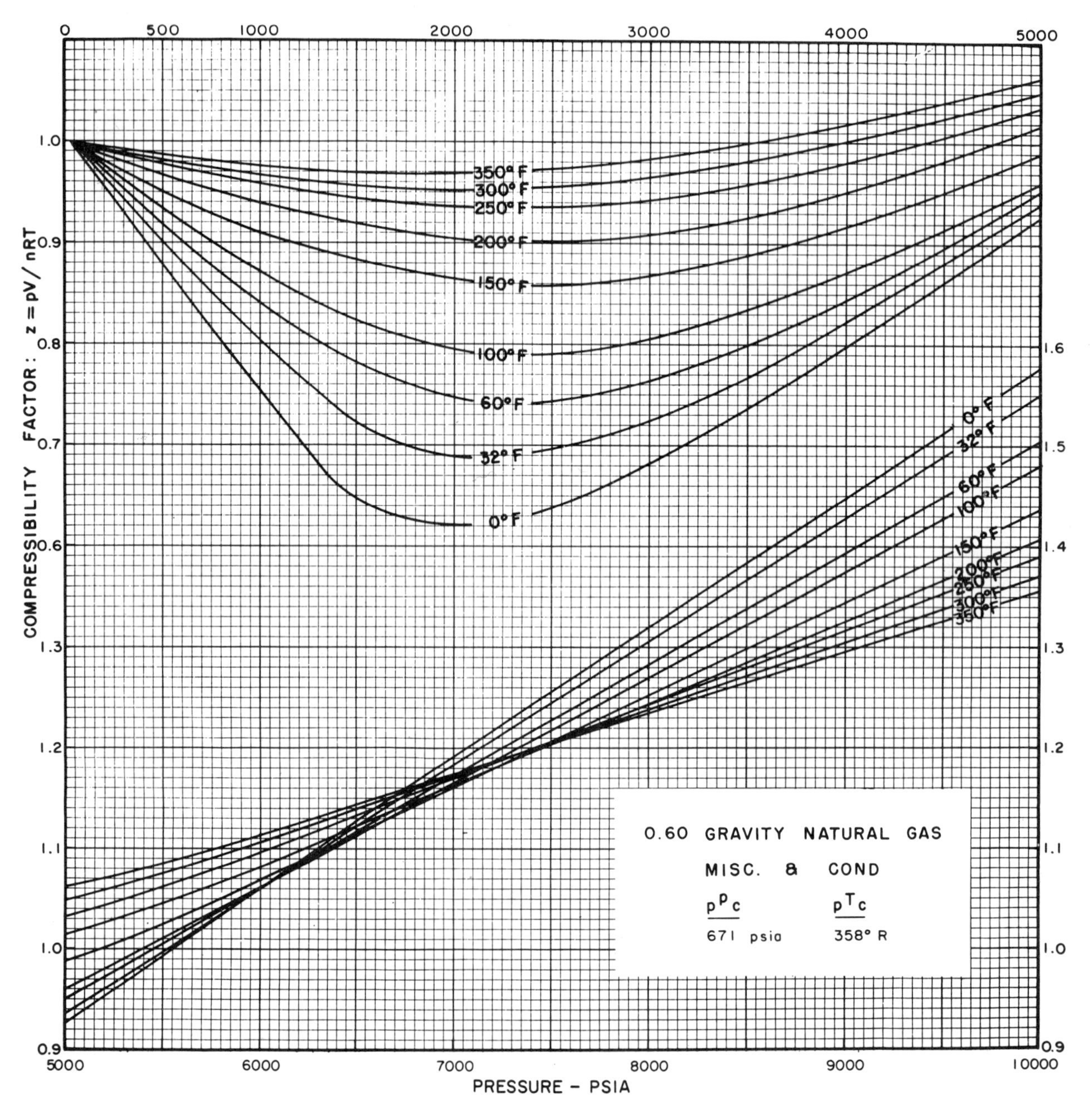

Fig. 1.7. 0.60 gravity natural gas.

to be sufficiently accurate for engineering purposes. Figure 1.5 shows this correlation as presented by Brown, et al.[5] This approach is recommended only if a chemical analysis is not available.

Rearrangement of the variables involved allows construction of the convenient curves, Figures 1.7 to 1.14, which give z as a direct function of pressure, temperature and gravity.[9] Curves of this type for natural gases were first presented by Buthod[10] using a different gravity correlation. The curves shown here are plotted from the data of Figures 1.4 and 1.5. Note the separate curves for miscellaneous and condensate systems above 0.60 gravity. The main disadvantage of this method is that values for intermediate gravities must be interpolated. Figure 1.6 is a similar presentation for methane.[5]

Various gas law calculations will be demonstrated in the following section.

TABLE 1.3

PHYSICAL PROPERTIES OF LIGHT PARAFFIN HYDROCARBONS AND MISCELLANEOUS COMPOUNDS

Compound: *Abbreviation of formula*	Compound: *Name*	*Molecular weight*	*Critical pressure, psia*	*Critical temperature, °Rankine*
C_1	Methane	16.04	673	344
C_2	Ethane	30.07	709	550
C_3	Propane	44.09	618	666
iC_4	*iso*-Butane	58.12	530	733
nC_4	*normal*-Butane	58.12	551	766
iC_5	*iso*-Pentane	72.15	482	830
nC_5	*n*-Pentane	72.15	485	847
nC_6	*n*-Hexane	86.17	434	915
nC_7	*n*-Heptane	100.2	397	973
nC_8	*n*-Octane	114.2	370	1025
nC_9	*n*-Nonane	128.3	335	1073
nC_{10}	*n*-Decane	142.3	312	1115
—	Air	28.97	547	239
N_2	Nitrogen	28.02	492	227
O_2	Oxygen	32.00	732	278
CO_2	Carbon Dioxide	44.01	1072	548
H_2S	Hydrogen Sulfide	34.08	1306	673
H_2O	Water	18.02	3206	1165

1.5 Sample Problems

1. Given the analysis in the table below of a natural gas produced from an oil well, compute, (a) the gas gravity, and (b) the pseudo-critical temperature and pressure.

$$\text{mol. wt. } C_7+ = 140$$

$$\text{Sp. Gravity of } C_7+ = 0.85$$

Solution:

(a) Column (1) is given. Column (2) is obtained from Table 1.3, and the molecular wt. of the mixture is the total of column (3).

$$\text{Gas gravity} = G_g = \frac{\text{density of gas at Std. Cond.}}{\text{density of air at Std. Cond.}}$$

Since one mol of any gas occupies the same volume at std. conditions (379 ft³),

$$G_g = \frac{22.13/379}{29/379} = \frac{22.13}{29} = 0.76$$

where 29 = mol wt of air.

(b) The C_7+ fraction is itself a mixture, and its pseudo-critical properties must be obtained from Figure 1.15. The other values in columns (4) and (5) are from Table 1.3. *Ans.* ${}_pp_c$ = 664 psia, ${}_pT_c$ = 409°R.

2. What volume will 100 lb of the above gas occupy at $p = 3000$ psig, $T = 170$°F?

Solution:

$$V = \frac{znRT}{p} \qquad n = \frac{100}{22.1} = 4.52 \text{ moles}$$

$$T = 460 + 170 = 630°\text{R}$$

$$p = 3000 + 14.7 \cong 3015 \text{ psia}$$

$$R = 10.7$$

(a) To obtain z from Figure 1.4.

$$\text{Reduced pressure, } p_r = \frac{p}{{}_pp_c} = \frac{3015}{664} = 4.54$$

$$\text{Reduced temperature, } T_r = \frac{T}{{}_pT_c} = \frac{630}{409} = 1.54$$

From this, $z = 0.81$, and

$$V = \frac{(0.81)(4.52)(10.7)(630)}{3015} = 8.2 \text{ ft}^3$$

(b) To obtain z from Figures 1.9 and 1.10.

G_g	z
0.70	0.84
0.80	0.79

$$z \text{ for } 0.76 \text{ gas} = 0.79 + \left(\frac{4}{10} \times .05\right) = 0.81, \text{ as in part (a).}$$

Component	*(1) Mol %*	*(2) M. wt.*	*(3) (1) × (2)*	*(4) p_c, psia*	*(5) T_c, °R*	*(6) (1) × (4)*	*(7) (1) × (5)*
methane	79.05	16.04	12.70	673	344	531	272
ethane	10.85	30.07	3.26	709	550	77.0	59.7
propane	4.61	44.09	2.03	618	666	28.5	30.7
iso-butane	1.28	58.12	0.74	530	733	6.8	9.4
n-butane	2.04	58.12	1.19	551	766	11.2	15.6
iso-pentane	0.21	72.15	0.15	482	830	1.0	1.7
n-pentane	0.34	72.15	0.25	485	847	1.6	2.9
hexanes	0.84	86.17	0.72	434	915	3.6	7.7
heptanes +	0.78	140.0	1.09	405	1172	3.2	9.2
	100.00		22.13			664 psia	409°R

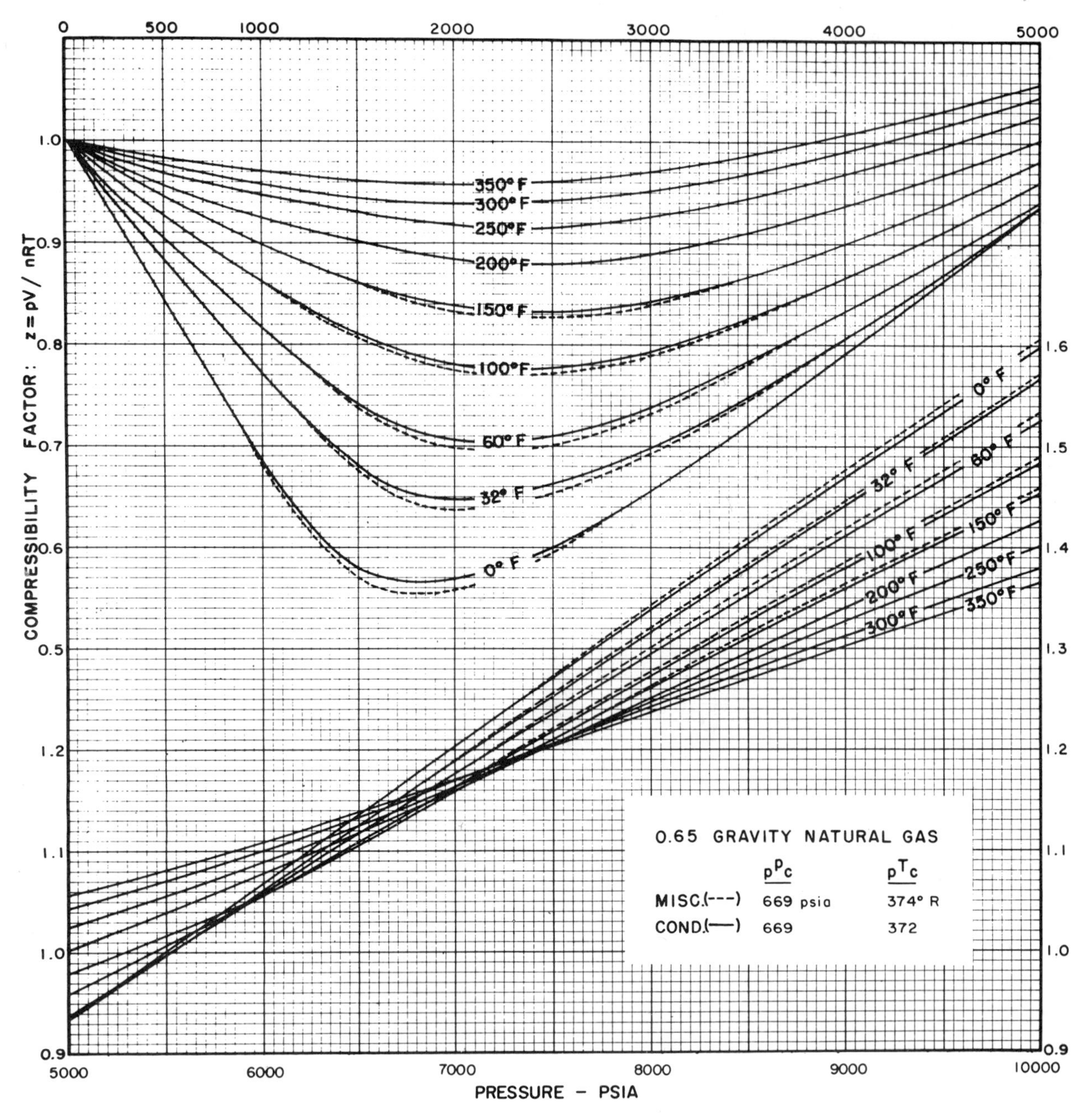

Fig. 1.8. 0.65 gravity natural gas.

3. (a) What is the density of a miscellaneous 0.90 gravity gas at 2000 psia and 150°F?

$z = 0.67$ from Figure 1.11; then, using Eq. (1.4),

$$\rho = \frac{pM}{zRT} = \frac{(2000)(29)(0.90)}{(0.67)(10.7)(610)} = 11.9 \text{ lb/ft}^3$$

(b) What is the specific volume at these conditions?

$$v = \frac{1}{\rho} = \frac{1}{11.9} = 0.084 \text{ ft}^3/\text{lb}$$

4. A cylindrical tank contains miscellaneous 0.80 gravity gas at 2500 psia and 100°F. The volume of the tank is 10 ft³.

(a) How many mols of gas are in the tank?

$$n = \frac{pV}{zRT} = \frac{(2500)(10)}{(0.67)(10.7)(560)} = 6.2$$

(b) What standard volume of gas is this?

$$V_s = 6.2 \times 379 = 2350 \text{ SCF}$$

or

$$\frac{p_s V_s}{T_s} = \frac{pV}{zT} \qquad (1.5)$$

and

$$V_s = \frac{pV}{zT} \times \frac{T_s}{p_s} = \frac{(2500)(10)(520)}{(0.67)(560)(14.7)} = 2350 \text{ SCF.}$$

(c) 1000 SCF of gas is released from the tank. This causes the temperature to fall to 60°F. What is the final tank pressure?

$$\text{Mols remaining} = 6.2 - \frac{1000}{379} = 3.6 \text{ mols}$$

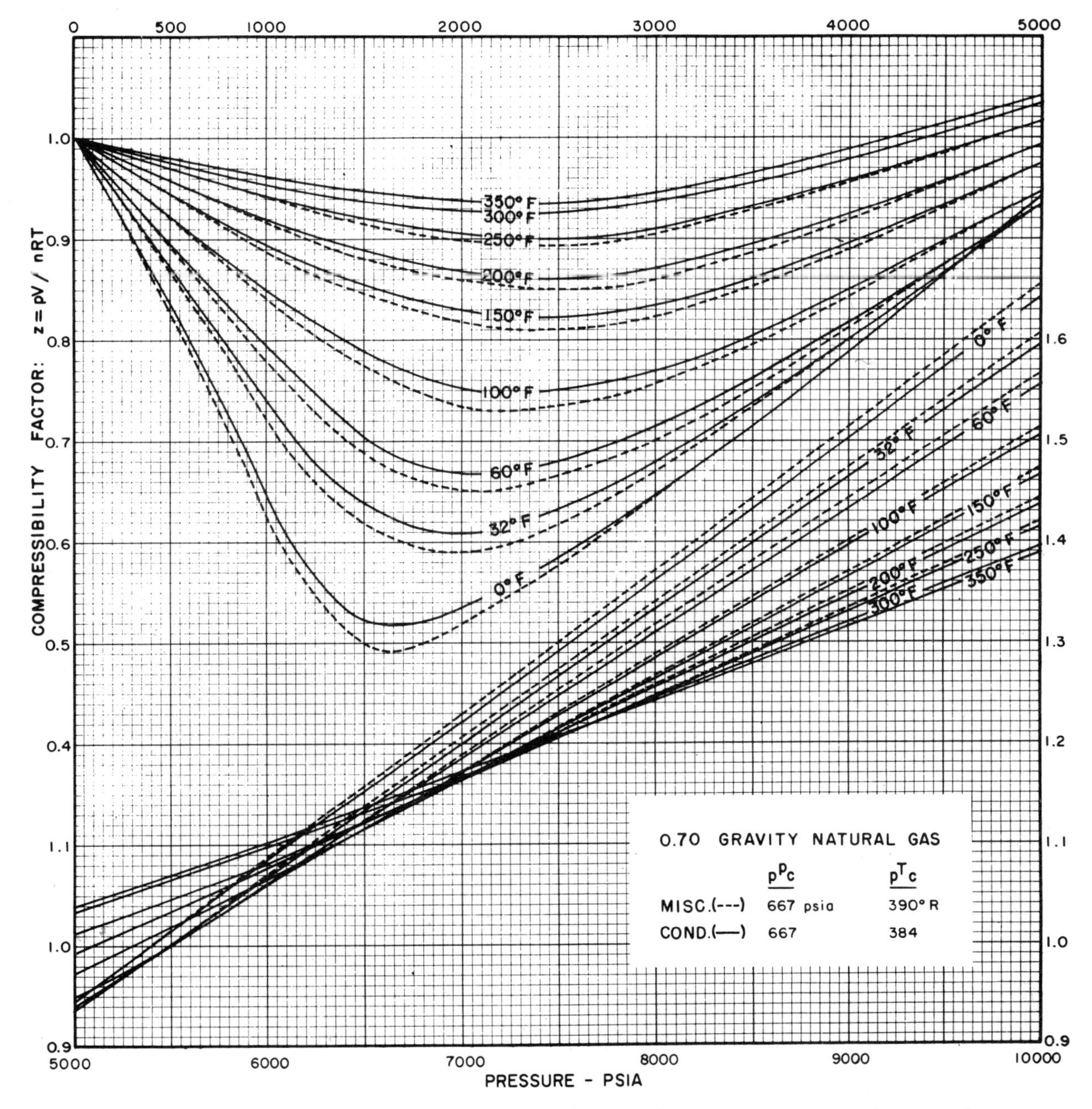

Fig. 1.9. 0.70 gravity natural gas.

$$p = \frac{znRT}{V}$$

But z = function of p, therefore the solution requires trial and error.

$$\frac{p}{z} = \frac{(3.6)(10.7)(520)}{10} = 2000$$

Assume

(1) $p = 1500$, then $z = 0.581$*

$$\frac{p}{z} = 2580 \text{ (high)}$$

(2) $p = 1300$, then $z = 0.617$*

$$\frac{p}{z} = 2100 \text{ (high)}$$

(3) $p = 1250$, then $z = 0.627$*

$$\frac{p}{z} = 1955 \cong 2000 \text{ (close enough)}$$

final $p = 1250$ psia

5. The *PVT* behavior of a reservoir oil is shown in Figure 1.16. Calculate the reservoir volume occupied by one barrel of tank oil and its originally dissolved gas at (a) 2000 psig, (b) 1000 psig. The reservoir temperature = 220°F, and gas gravity = 1.06.

---

*Values of z are obtained at 60°F at the assumed pressures from the 0.80 gravity chart.

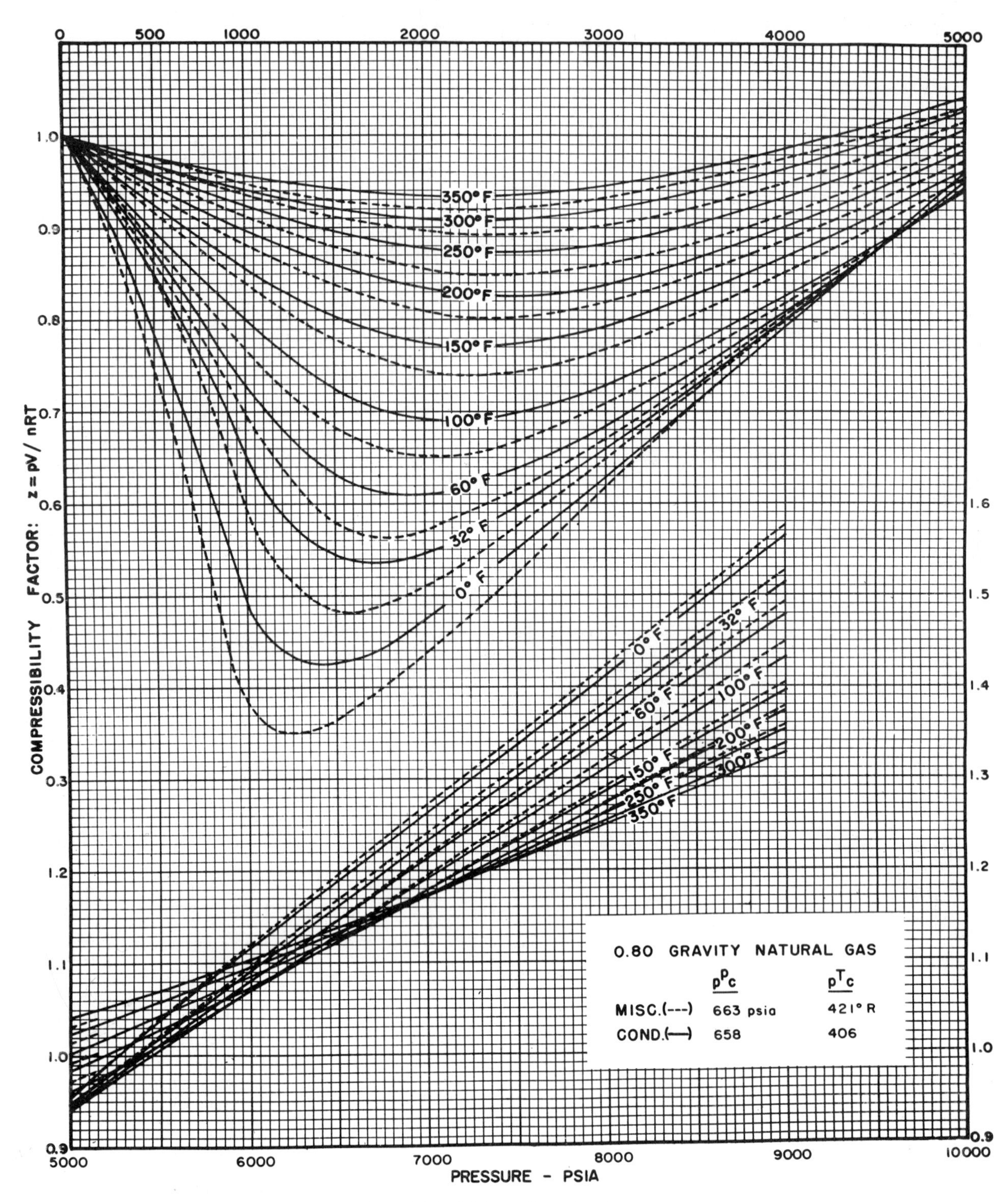

Fig. 1.10. 0.80 gravity natural gas.

Solution:

$$B_t = B_o + V_g$$

where B_t = Total formation volume factor of one barrel tank oil + originally associated gas, and

B_o, V_g = Reservoir volume of oil and gas respectively.

At any p,

$$V_g = B_g(R_{sb} - R_s)$$

where B_g = Reservoir volume factor for gas, i.e., volume reservoir space occupied per volume of standard gas.

R_s = Solution gas oil ratio (see Figure 1.3)

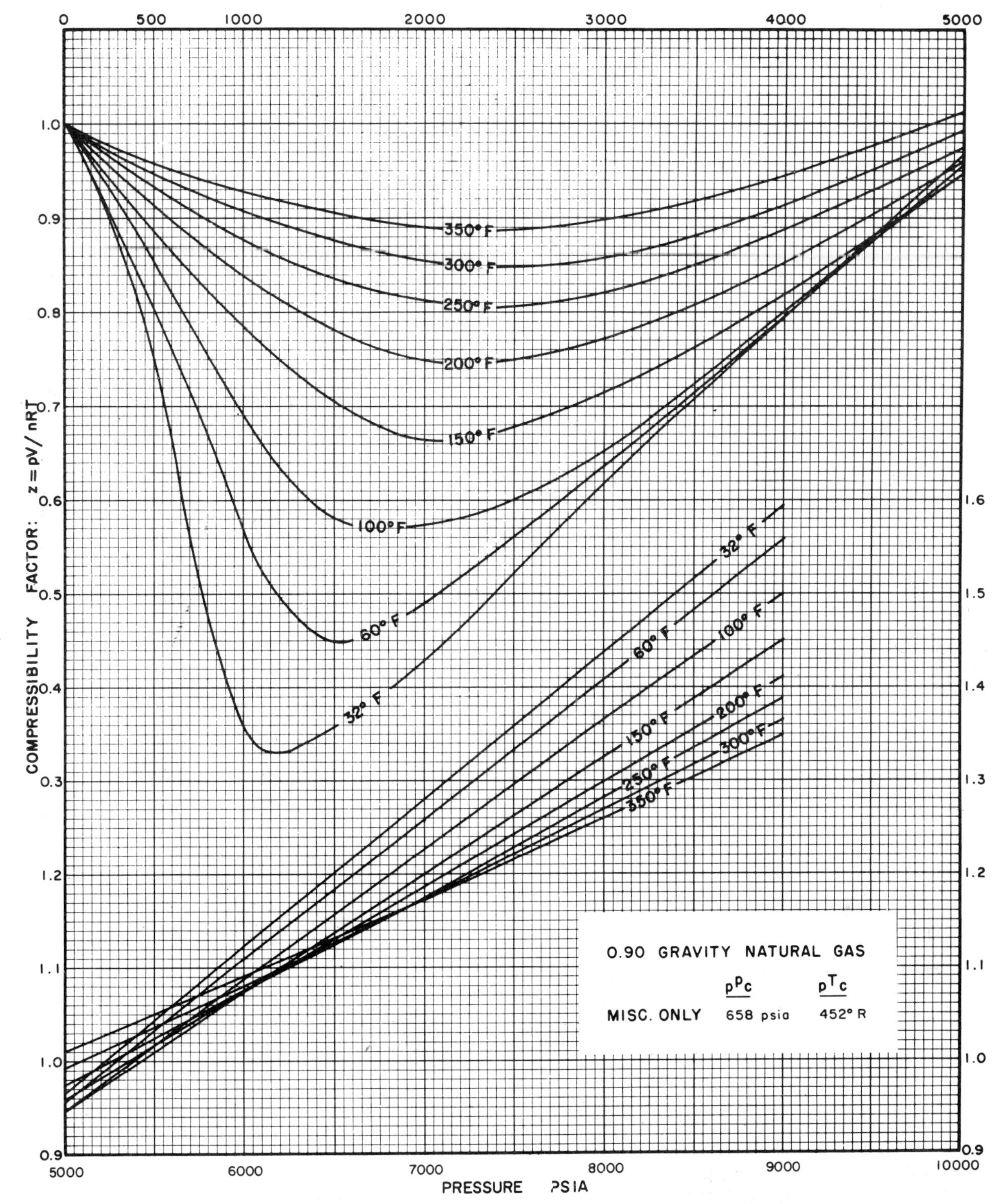

Fig. 1.11. 0.90 gravity natural gas, miscellaneous only.

From the Gas Law,

$$\frac{p_s V_s}{T_s} = \frac{p_R V_R}{z_R T_R}$$

or

$$\frac{V_R}{V_s} = \frac{p_s z_R T_R}{p_R T_s} = B_g$$

where subscript R denotes reservoir conditions. Substituting:

$$p_s = 14.7 \text{ psia}$$

$$T_s = 520°\text{R}$$

$$B_g = 0.0283 \frac{z_R T_R}{p_R}$$

(a) From Figure ~~1.6~~ 1.5,

$$_pp_c = 650 \text{ psia}$$

$$_pT_c = 502°\text{R}$$

Then, $p_r = \frac{2015}{650} = 3.10$

$$T_r = \frac{680}{502} = 1.35$$

$$z = 0.67 \text{ (Figure 1.4)}$$

From Figure 1.16,

$$B_o = 1.32$$

$$R_{sb} - R_s = 157 \text{ ft}^3\text{/bbl (gas liberated)}$$

$$B_t = 1.32 + 0.0283 \frac{(0.67)(680)}{2015} \times (157) \times \frac{1}{5.61}$$

$$= 1.32 + 0.18 = 1.50 \text{ bbl reservoir space}$$

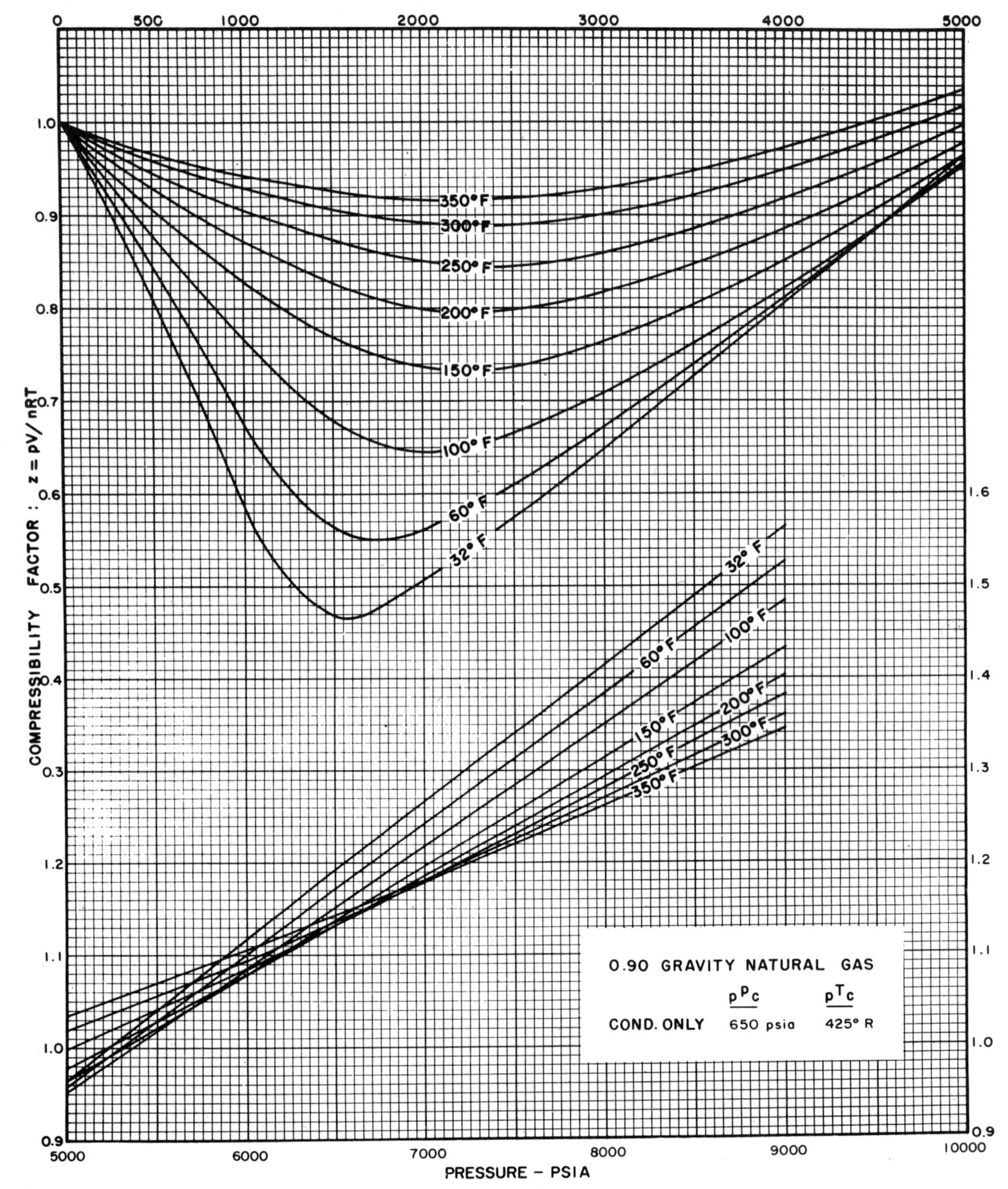

Fig. 1.12. 0.90 gravity natural gas, condensate only.

(b) $p_r = \dfrac{1015}{650} = 1.56$, and $z = 0.786$.

$$B_t = 1.23 + 0.0283\,\frac{(0.786)(680)}{1015} \times (357) \times \frac{1}{5.61}$$

$$= 1.23 + 0.95 = 2.18 \text{ bbl.}$$

PROBLEMS

1. What is the density of methane at standard conditions? *Ans.* 0.0424 lb/ft³.

2. What is the density of air at standard conditions? *Ans.* 0.0765 lb/ft³.

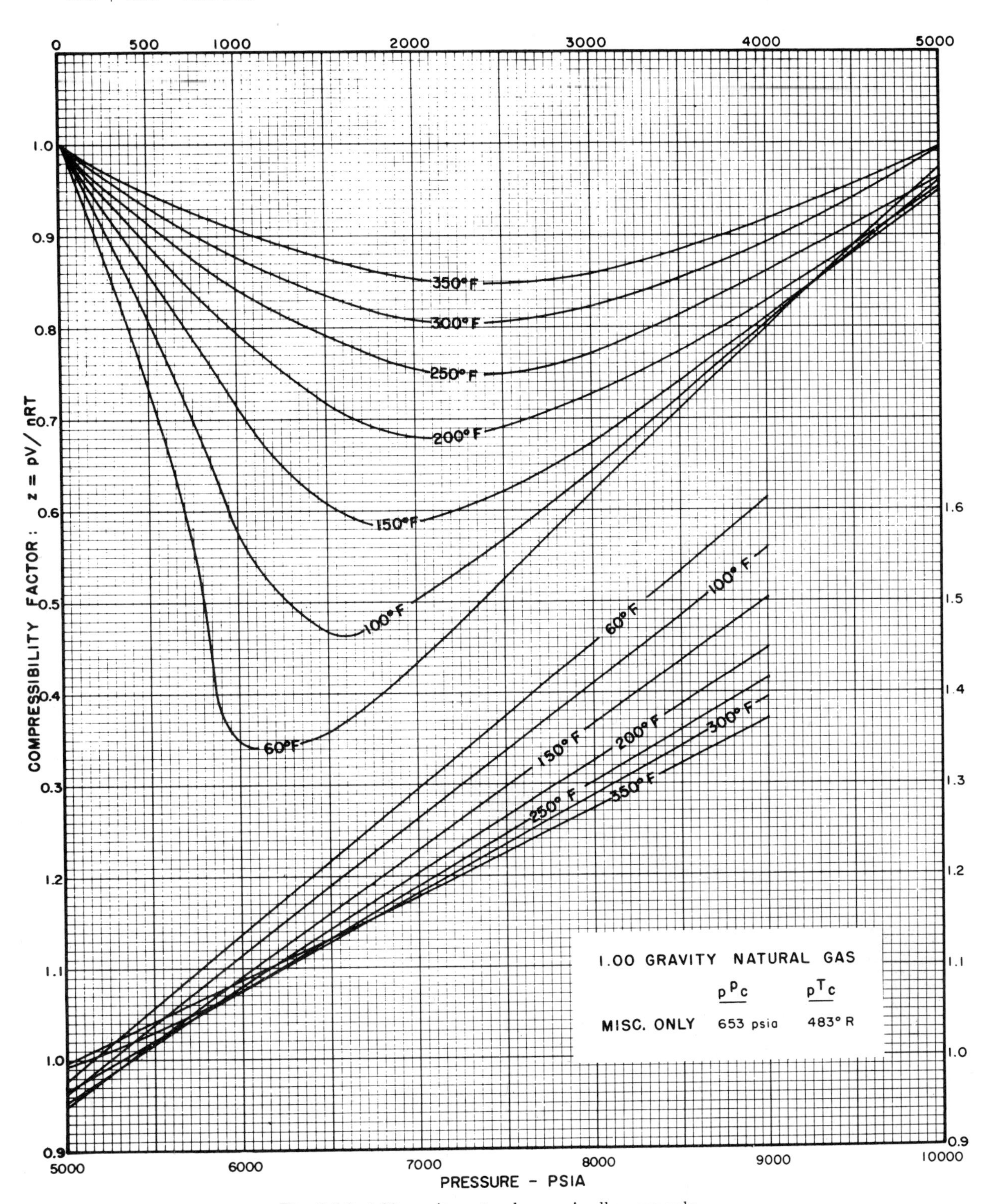

Fig. 1.13. 1.00 gravity natural gas, miscellaneous only.

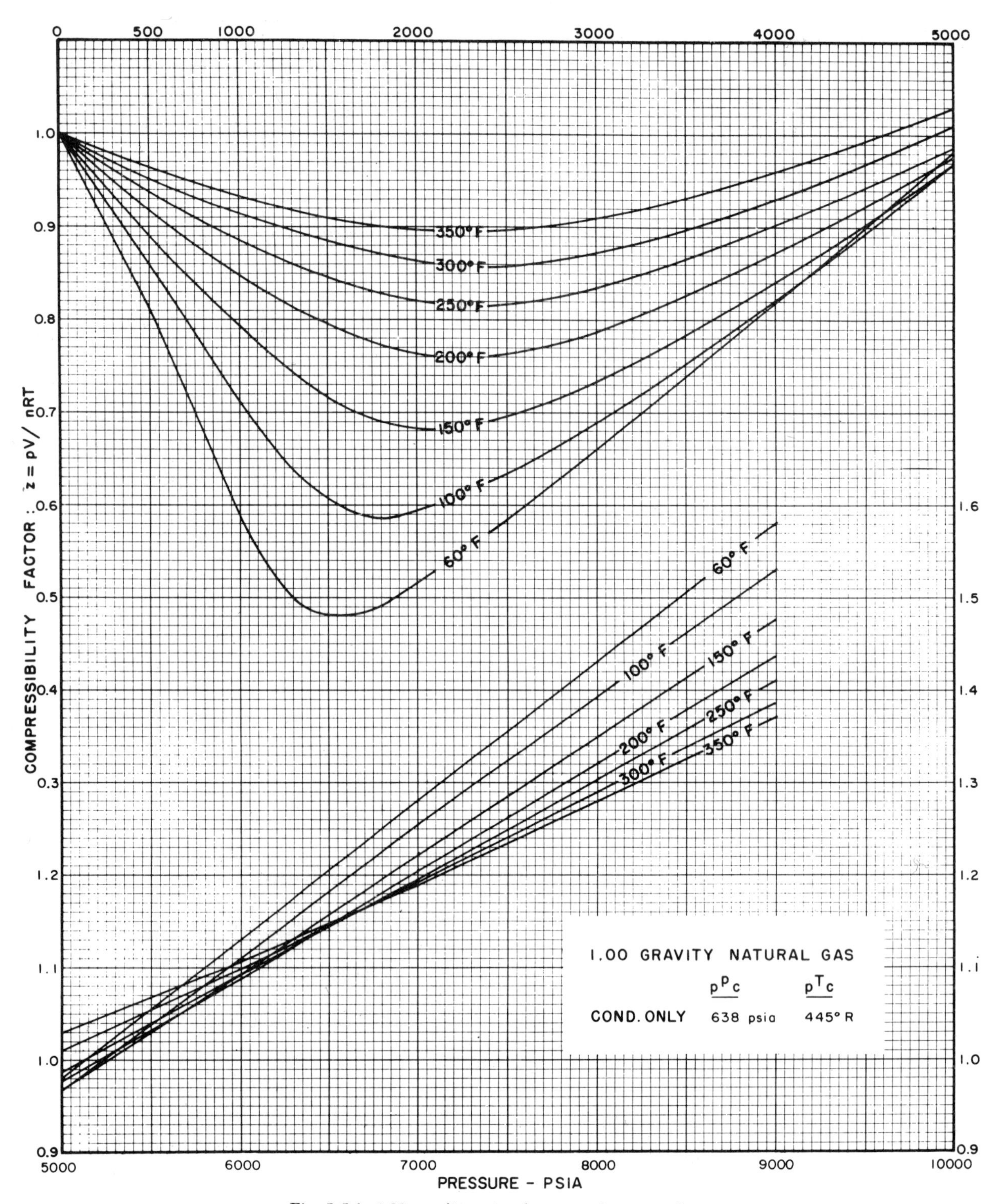

Fig. 1.14. 1.00 gravity natural gas, condensate only.

3. Given the following analysis of a natural gas:

Component	*Mol %*
C_1	83.19
C_2	8.48
C_3	4.37
iC_4	0.76
nC_4	1.68
iC_5	0.57
nC_5	0.32
C_6	0.63

Compute:

(a) Gas gravity. *Ans.* 0.70.
(b) Pseudo-critical pressure and temperature. *Ans.* $_pp_c$ = 668 psia, $_pT_c$ = 393°R.
(c) What is the ratio of actual volume to ideal volume for this gas at 4000 psig and 200°F? *Ans.* $z = 0.916$.
(d) What volume will 100 lb of this gas occupy at the conditions in (c)? *Ans.* 7.9 ft³.
(e) What is the gas density at the same conditions? *Ans.* ρ = 12.7 lb/ft³.

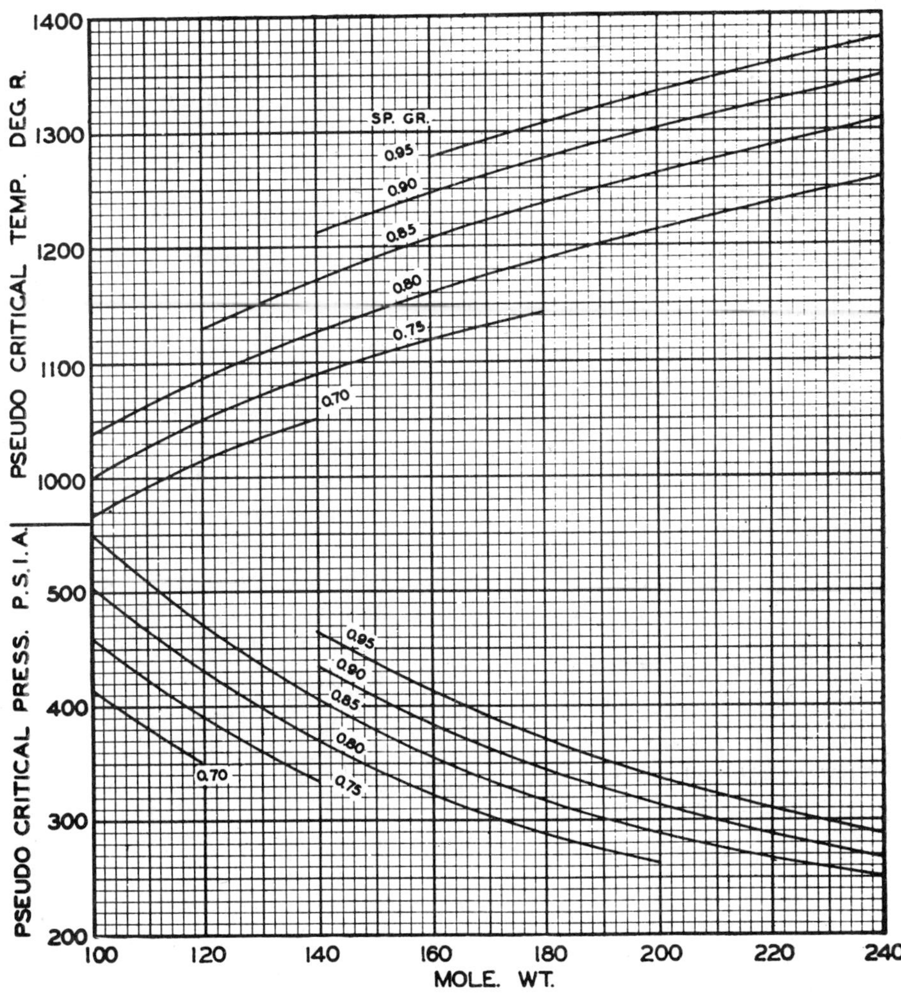

Fig. 1.15. Pseudo-critical properties for C_7^+ natural gas as functions of molecular weight and specific gravity. After Matthews, Roland, and Katz,[11] courtesy Natural Gasoline Association of America.

4. A steel tank has a volume of 20 ft³. It contains a 0.90 gravity miscellaneous gas at p = 3000 psia and T = 150°F. Plot the depletion history of the tank as gas is withdrawn isothermally. Note: Show this as a plot of p/z vs G_p, where G_p = cumulative standard (14.7 psia and 60°F) gas volume withdrawn.

5. Prove that the plot of Problem 4 is a linear function having a slope,

$$m = -\frac{14.7T}{520V}$$

where m = slope of p/z vs G_p plot

T = tank temperature, °R

V = tank volume, cu ft.

6. Methane exists at 150°F in a tank of unknown volume. Initial pressure is 2000 psia. It is noted that withdrawal of 20 standard cubic feet drops the tank pressure to 1900 psia. Calculate (a) the tank volume, and (b) the standard volume of methane it contained originally. (c) how much C_1 remains when p = 1000 psia?

7. Given the data of Figure 1.16 (p. 18). What total volume will one barrel of tank oil and its originally dissolved gas occupy at

(a) 4000 psig

(b) 1500 psig

(c) 500 psig

(d) 0 psig (d) *Ans.* $B_t = 1.09 + \dfrac{638}{5.61} = 115$ bbl

(e) Show this behavior in a graph of B_t vs p.

8. Assuming that air is a mixture of oxygen and nitrogen, what volume percentage of each is required to make the average mol. wt. of air = 29?

9. For ideal gases, prove volume per cent = mol per cent.

10. What volume is occupied by any perfect gas at 32°F and 14.7 psia? *Ans.* 359 cu ft.

11. What is the API gravity of a 1.11 specific gravity salt water?

12. Which has the higher density, a 20° or a 30° API oil?

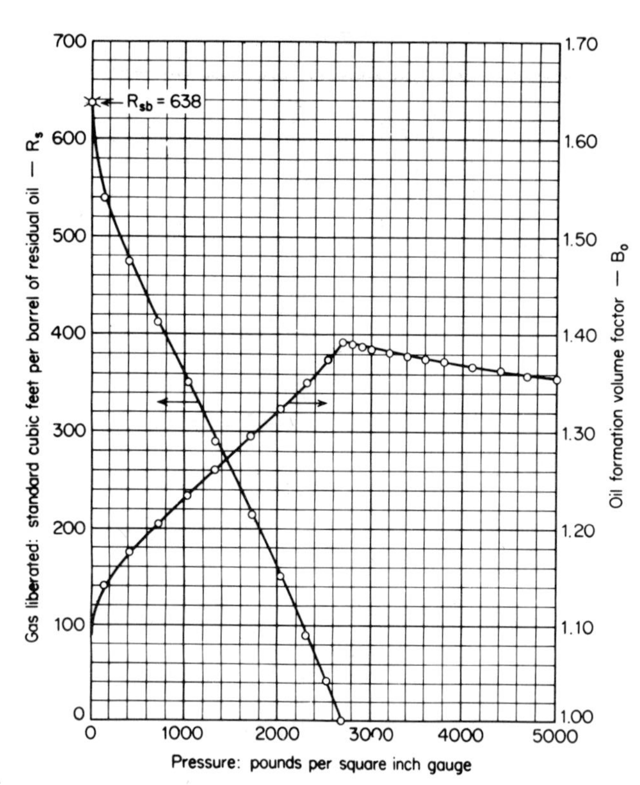

Fig. 1.16. Differential vaporization of an east Texas reservoir fluid. Courtesy Core Laboratories, Inc., Dallas, Texas.

13. Would you normally expect reservoir oil or tank oil to be more dense? Why?

14. What is the density of the reservoir oil depicted in Figure 1.16 at the bubble point? Assume a tank oil gravity (14.7 psia and 60°F) of 40° API. (Realize that the volume is B_{ob} and the total weight is the sum of the weights of 1 bbl tank oil plus the dissolved 1.06 sp. gr. gas.)

15. What is the API gravity of the same reservoir oil at 2000 psia? At atmospheric pressure?

REFERENCES

1. Fieser, L. F., and M. Fieser, *Organic Chemistry*, 2nd ed., p. 32. Boston: D. C. Heath and Company, 1950.
2. Standing, M. B., "A Pressure-Volume-Temperature Correlation for Mixtures of California Oils and Gases," *API Drilling and Production Practices*, 1947, pp. 275–287.
3. Borden, G., Jr.; and M. J. Rzasa, "Correlation of Bottom Hole Sample Data," *Trans. AIME*, Vol. 189, 1950, pp. 345–348.
4. Recommended Practice for Measuring, Sampling, and Testing Natural Gas, API RP 50A, 3rd ed. Dallas: American Petroleum Institute, 1953.
5. Brown, G. G., Katz, D. L., Oberfell, G. G., and R. C. Alden, *Natural Gasoline and the Volatile Hydrocarbons*, Section 1. Tulsa: Mid-West Printing Co., 1948.
6. Standing, M. B., *Oil Field Hydrocarbon Systems*. New York: Reinhold Publishing Corp., 1952.
7. Burcik, E. J., *Properties of Petroleum Reservoir Fluids*. New York: John Wiley & Sons, Inc., 1957.
8. Standing, M. B., and D. L. Katz, "Density of Crude Oils Saturated with Natural Gas," *Trans. AIME*, Vol. 146, 1942, p. 159.
9. Gatlin, C., "Simplified Compressibility Factor Charts for Natural Gas Calculations," *The Pipeline Engineer*, Aug. 1957, p. D-31.
10. Buthod, P., "Compressibility Factors and Specific Heats of Natural Gases," *Oil and Gas Journal*, Sept. 8, 28, 1949.
11. Mathews, T. A., Roland, C. H., and D. L. Katz, "High Pressure Gas Measurement," *Proceedings of Natural Gas Association of America*, 1942, p. 41.

Chapter 2 . . .

Concepts of Petroleum Geology and Basic Rock Properties

Petroleum is not found in underground lakes or rivers, but it exists within the void space of certain rocks. Unfortunately, these oil-bearing rocks are a definite minority, and the determination of their whereabouts is the basic problem confronting geologists and geophysicists.

The mechanism of the origin and accumulation of petroleum is not completely understood and is the subject of much controversy. However, a great deal has been learned about the habitat of oil and gas, i.e., the nature of the rocks in which they exist. In this chapter we will consider these problems by discussing the requirements for oil accumulations and the rock properties which are essential to these accumulations.

2.1 Requirements for Commercial Oil Accumulations

Certain requirements must be fulfilled for a commercial petroleum deposit to be present. These are

1. A source: material from which oil is formed.
2. Porous and permeable beds (reservoir rocks) in which the petroleum may migrate and accumulate after being formed.
3. A trap: subsurface condition restricting further movement of oil such that it may accumulate in commercial quantities.

2.11 Source of Petroleum

A complete understanding of the origin of petroleum would be of great benefit to exploration operations, but unfortunately this has not yet been attained. Many theories on the origin of petroleum have been proposed and are normally classified into two general groups:

1. Inorganic theories — these are primarily of historical interest only and will not be discussed here.
2. Organic theories.

At the present time most authorities overwhelmingly favor the organic approach. Their principal reasons are the following:[1]

1. No inorganic theory can account for the necessary quantities of carbon and hydrogen needed to form large petroleum deposits. The abundance of plant and animal life present in sediments is a sufficient source.
2. Many crude oils contain porphyrins and nearly all contain nitrogen. The presence of these materials strongly suggests organic origin as they are present in all organic matter. Also, porphyrins of vegetable origin have been found to be more plentiful than those of animal origin.
3. Petroleum rotates the plane of polarized light. This property is restricted primarily to organic materials known as optical isomers and further suggests organic matter as the source of petroleum.

The complete process of alteration whereby organic materials are transformed into petroleum is not known. The main factor which prohibits complete laboratory verification of the theory is the inability to reproduce the million or so years during which the process occurs. The role of anerobic bacteria in promoting this alteration may be considerable, as suggested by ZoBell.[2]

The evidence from studies of thousands of oil fields has led most geologists to the following general conclusions:[3]

1. Petroleum originates from organic material, primarily vegetable, which has been altered by heat, bacterial action, pressure, and other agents over long periods of time.
2. Conditions favoring petroleum formation are found only in sedimentary rocks.

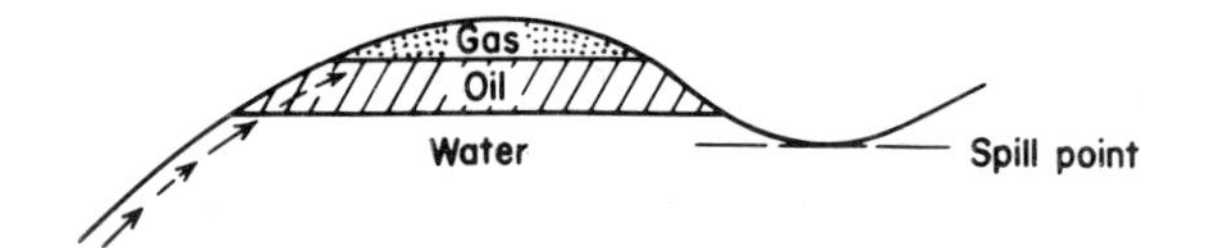

Stage 1. Gas, oil, and water above spill point. Both oil and gas continue to be trapped while water is displaced. This stage ends when oil-water interface reaches spill point.

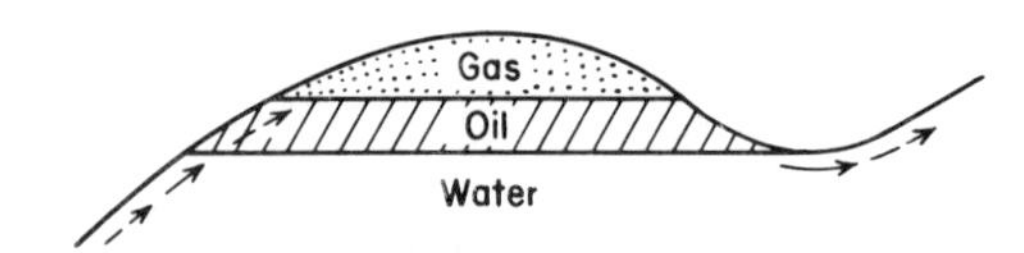

Stage 2. Stage of selective entrapment and gas flushing. Gas continues to be trapped but oil is spilled up dip. This stage ends when oil-gas interface reaches spill point.

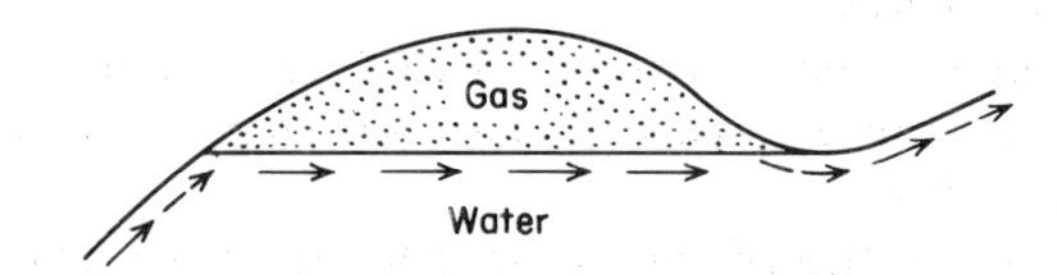

Stage 3. End stage. Trap filled with gas. Excess gas spills up dip as more gas enters trap. Oil bypasses trap and continues upward migration.

Fig. 2.1. Illustration of differential entrapment principle, showing various stages of hydrocarbon accumulation in an anticline. Solid and dashed arrows denote oil and gas movement respectively. After Gussow,[4] courtesy AAPG.

3. The principal sediments generally considered as probable source rocks are shales and limestones that were originally muds under saline water.

2.12 Porous and Permeable Beds (Reservoir Rocks)

After its formation, petroleum may migrate from the source rock into porous and permeable beds where it accumulates and continues its migration until finally trapped. The forces causing this migration are

1. Compaction of sediments as depth of burial increases.
2. Diastrophism: crustal movements causing pressure differentials and consequent subsurface fluid movements.
3. Capillary forces causing oil to be expelled from fine pores by the preferential entry of water.
4. Gravity which promotes fluid segregation according to density differences.

Considerable differences of opinion exist as to the distance which petroleum may cover in its migration from source to trap. An interesting theory which logically explains some types of accumulation is that of Gussow[4] which is illustrated by Figures 2.1 and 2.2.

It is certain, however, and universally agreed that if a petroleum deposit is to be of commercial significance, it must be found in a porous and permeable reservoir. The terms *porous* and *permeable* denote two distinct rock properties whose measurements and quantitative definitions have comprised much of the technical literature of the oil industry. Consequently, they must be defined and discussed in some detail.

Porosity

Porosity is a measure of the void space within a rock expressed as a fraction (or percentage) of the bulk volume of that rock.

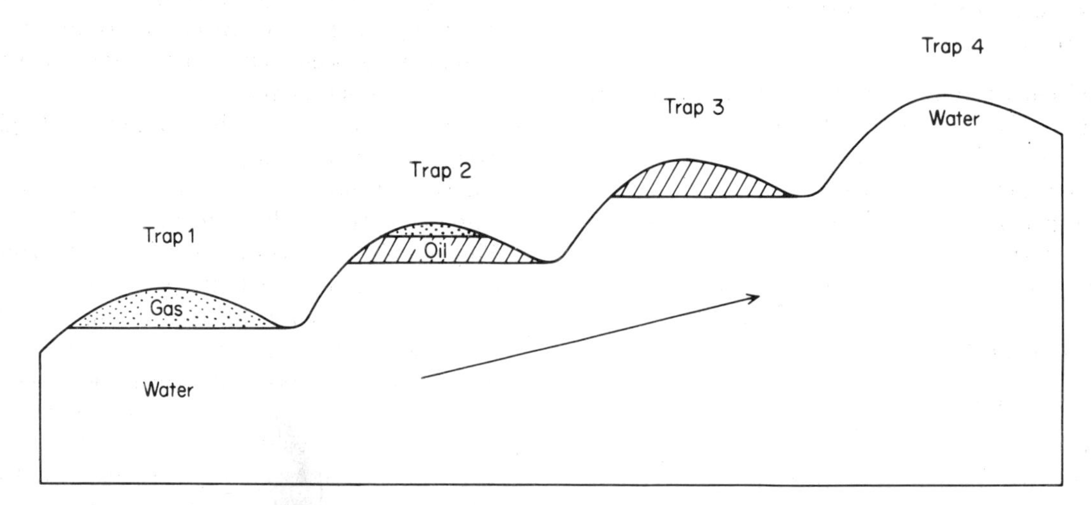

Fig. 2.2. Final condition of differential entrapment in a series of interconnected traps. After Gussow,[4] courtesy AAPG.

The general expression for porosity is

$$\phi = \frac{V_b - V_s}{V_b} = \frac{V_p}{V_b} \tag{2.1}$$

where ϕ = porosity

V_b = bulk volume of the rock

V_s = net volume occupied by solids (also called grain volume)

V_p = pore volume = the difference between bulk and solid volumes

To illustrate this principle consider Figure 2.3 which shows various arrangements of packed spheres and their computed porosities.[5] The bulk volume of each configuration is the figure formed by connecting the centers of eight adjacent spheres (the four shown in the front views and four behind them). Each bulk volume so formed contains one net sphere as the enclosed solid or grain volume. Note that the grain size (sphere diameter) is immaterial to the porosity value.

In actual rocks porosity is classified as

A. *Absolute porosity:* total porosity of a rock, regardless of whether or not the individual voids are connected, and

B. *Effective porosity:* only that porosity due to voids which are interconnected.

It is the effective porosity which is of interest to the oil industry, and all further discussion will pertain to this form. In most petroleum reservoir rocks the absolute and total effective porosity are, for practical purposes, the same. Actually, this restriction implies the property of permeability which will be discussed shortly.

Geologically, porosity has been classified in two types, according to the time of formation:

1. *Primary porosity* (intergranular): Porosity formed at the time the sediment was deposited. The voids contributing to this type are the spaces between individual grains of the sediment.
2. *Secondary porosity:* Voids formed after the sediment was deposited. The magnitude, shape, size, and interconnection of the voids bear no relation to the form of the original sedimentary particles.

Primary Porosity

The sedimentary rocks which typically exhibit primary porosity are the clastic (also called fragmental or detrital) rocks which are composed of erosional fragments from older beds. These are normally classified by grain size, although much looseness of terminology exists. The most widely used of these classifications is Wentworth's Scale.[6] Typical clastic rocks which are common reservoir rocks are sandstones, conglomerates, and oölitic limestones.

Secondary Porosity

Porosity of this type has been subdivided into three classes based on the mechanism of formation.

1. *Solution porosity:* voids formed by the solution of the more soluble portions of the rock in percolating surface and subsurface waters containing carbonic and

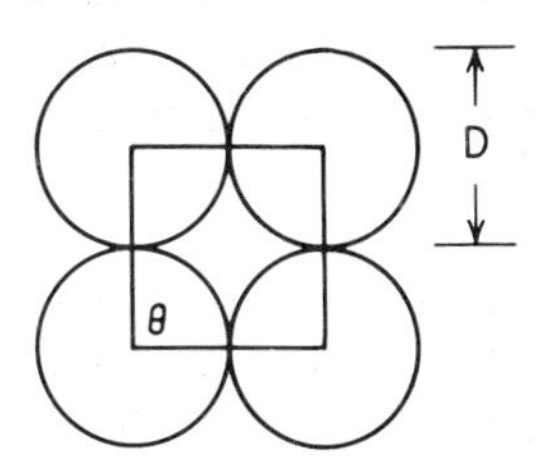

(A) Cubic packing. Top row is directly above bottom row.

$\theta = 90°$

$V_b = D^3$

$V_s = \pi D^3/6$

$\therefore \phi = \frac{D^3(1 - \pi/6)}{D^3}$

$= 0.476 = 47.6\%$

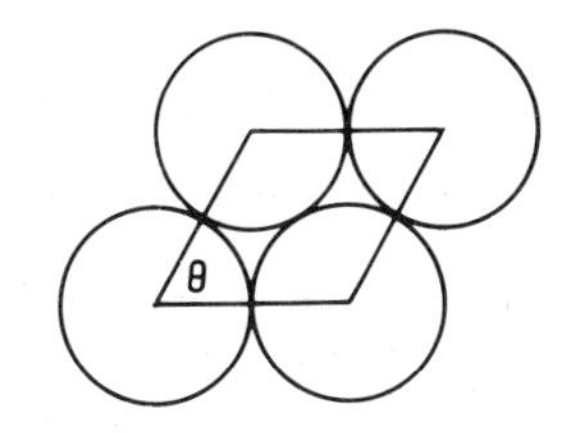

(B) Hexagonal packing. Top row has been moved one radius to the right.

$\theta = 60°$

$V_b = D^2 \cdot D \sin \theta$

$= 0.866D^3$

$V_s = \pi D^3/6$

$\phi = \frac{D^3(0.866 - \pi/6)}{0.866D^3}$

$= 0.395 = 39.5\%$

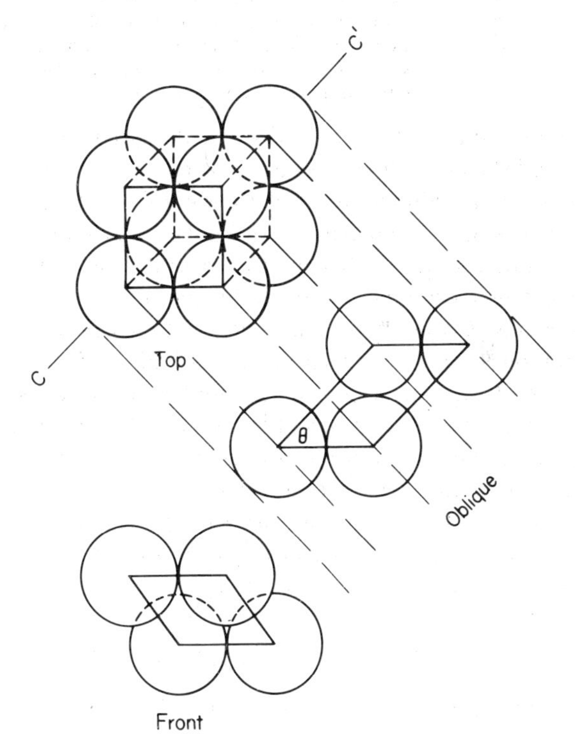

(C) Rhombohedral packing. Top row has been moved to left and forward one radius as shown in front view

$\theta = 45°$

$V_b = D^2 \cdot D \sin \theta = D^3/\sqrt{2}$

$V_s = \pi D^3/6$

$\phi = \frac{D^3(1/\sqrt{2} - \pi/6)}{D^3/\sqrt{2}}$

$= 0.259 = 25.9\%$

Fig. 2.3. Three methods of packing spheres and the resultant porosities. Cubic packing **(A)** is the most porous and **(C)** rhombohedral packing the least porous arrangement possible.

(A) (B) (C)

Fig. 2.4. Core samples of various producing formations. Courtesy Core Laboratories, Inc., Dallas, Texas. **(A)** Ellenburger dolomite, Triple "N" Field, Andrews County, Texas. $\phi = 20.2\%$, $k = 850$ md. for this sample. Excellent example of secondary porosity. **(B)** Pennsylvanian reef, Dawson County, Texas. $\phi = 5\%$, $k = 186$ md. Note the low porosity value despite presence of a few large voids. **(C)** Cardium conglomerate and sand, Pembina Field, Alberta. Conglomerate: $\phi = 8.8\%$, $k = 3.0$ md. Sand: $\phi = 18.0\%$, $k = 9.0$ md. Note that the sand is twice as porous as the conglomerate which illustrates the effect of grain size uniformity.

other organic acids. This is also called vugular porosity and the individual holes are called vugs. Unconformities in sedimentary rocks are excellent targets for zones of solution porosity. Voids of this origin may range from small vugs to cavernous openings. An extreme example is Carlsbad Cavern.

2. *Fractures, fissures, and joints:* voids of this type are common in many sedimentary rocks and are formed by structural failure of the rock under loads caused by various forms of diastrophism such as folding and faulting. This form of porosity is extremely hard to evaluate quantitatively due to its irregularity.
3. *Dolomitization:* This is the process by which limestone ($CaCO_3$) is transformed into dolomite $CaMg(CO_3)_2$. The chemical reaction explaining this change is:

$$2\,CaCo_3 + MgCl_2 \rightarrow CaMg(CO_3)_2 + CaCl_2$$

It has long been observed that dolomite is normally more porous than limestone. This is the reverse of what might normally be expected since dolomite is less soluble than calcite. The best explanation of this seems to be that of Hohlt[7] who has shown that under pressure calcite crystals tend to orient their "*C*" axes in the plane of bedding while dolomite crystals are always in a random packing. Consequently, ground waters more readily percolate through dolomite, which has more and larger intercrystalline voids, and attack more rock surface. Porosity formed by dolomitization is then due to solution effects enhanced by a previous chemical change in limestone.

It should be realized that primary and secondary porosity often occur in the same reservoir rock. Figure 2.4 shows actual examples of various porosity forms.

Typical Porosity Magnitude

A typical value of porosity for a clean, consolidated, and reasonably uniform sand is 20%. The carbonate rocks (limestone and dolomite) normally exhibit lower values with a rough average near 6 to 8%. These values are approximate and certainly will not fit all situations. The principal factors which complicate intergranular porosity magnitudes are

1. Uniformity of grain size: The presence of small particles such as clay, silt, etc. which may fit in the voids between larger grains greatly reduces the porosity. Such rocks are called dirty or shaly.
2. Degree of cementation: Cementing material deposited around grain junctions reduces porosity.
3. Packing: This effect is illustrated in the systems of spheres of Figure 2.3. Geologically young rocks are

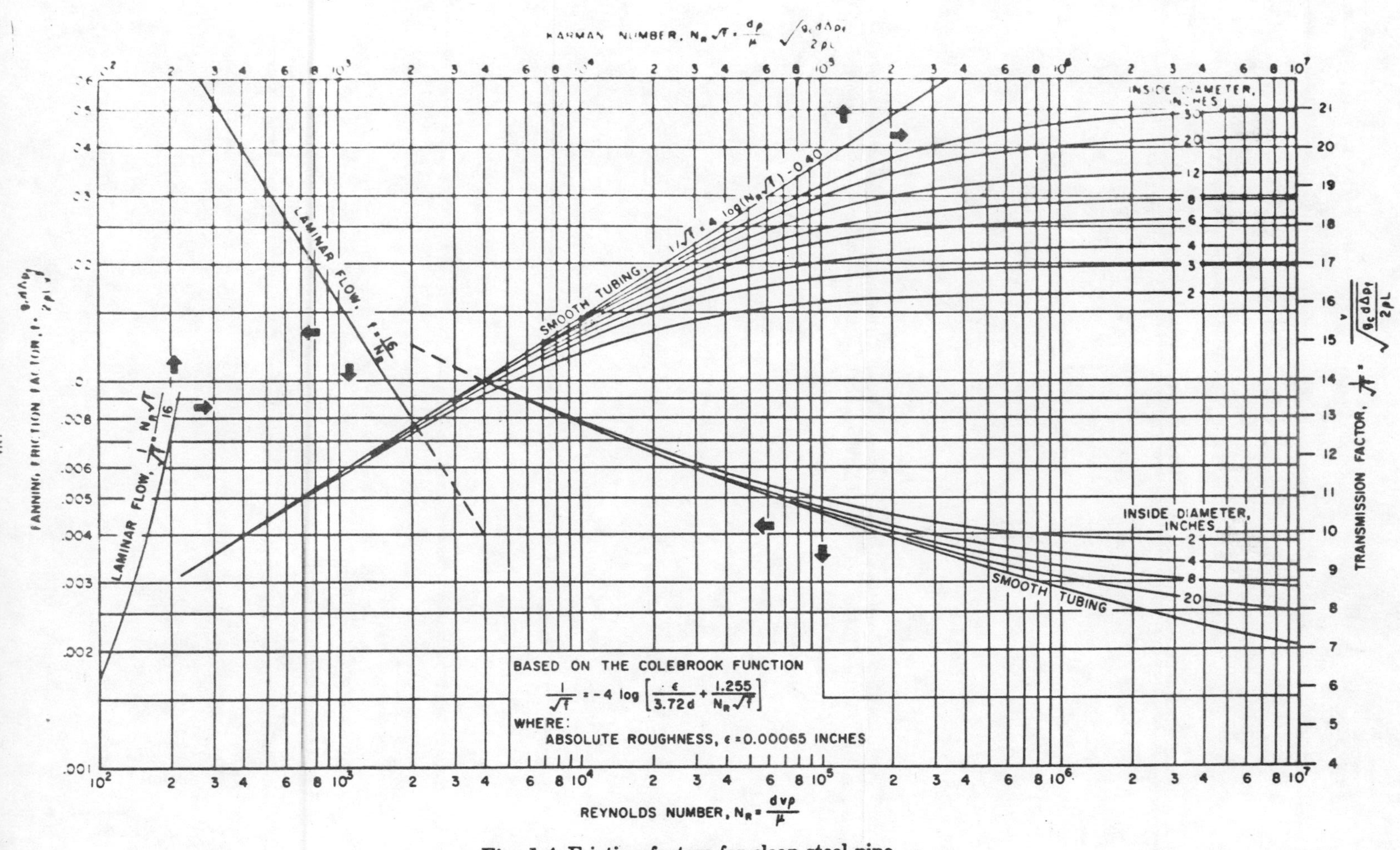

Fig. 1.4 Friction factors for clean steel pipe.

$N_{Re} = 928 \frac{\rho v d}{\mu}$

Karman #

$N_R\sqrt{f} = 928 \frac{d\rho}{\mu}\sqrt{\frac{25.8 d \Delta P_f}{\rho L}}$

Transmission Factor

$\frac{1}{\sqrt{f}} = \frac{v}{\sqrt{\frac{25.8 d \Delta P_f}{\rho L}}}$

$\Delta P = \frac{f \rho L v^2}{25.8 d}$

often packed in an inefficient manner and are as a result highly porous.

4. Particle shape.

There are a number of references available which have compiled considerable porosity data for various formations.[8,9]

Quantitative Use of Porosity Data

The subject of how porosity measurements are performed will be discussed in a later chapter on core analysis. For the present, it will be assumed that such measurements have been made and that the porosity is known. As defined previously, porosity is a measure of the void space within a rock, and as such may be used to determine the quantity of fluid which may be stored within that rock.

Consider a bulk volume of rock with a surface area of one acre and a thickness of one foot. This constitutes the basic rock volume measurement used in oil field calculations, an acre-foot. It is also standard practice to express all liquid volumes in terms of barrels. The following conversion factors are useful:

$$1 \text{ acre} = 43{,}560 \text{ ft}^2$$

$$1 \text{ acre-ft} = 43{,}560 \text{ ft}^3$$

$$1 \text{ bbl} = 42 \text{ gal} = 5.61 \text{ cu ft}$$

$$1 \text{ acre-ft} = \frac{43{,}560}{5.61} = 7758 \text{ bbl}$$

It is then obvious that the pore space within a rock is equal to $7758 \times \phi = V_p$ (bbl/acre-ft) where ϕ = porosity of the rock in question. Further reasoning as shown by Figure 2.5 results in the well-known *Volumetric Equation of Oil in Place:*

$$N = \frac{7758\, \phi\, S_o}{B_o} = \frac{7758\, \phi\, (1 - S_w)}{B_o} \tag{2.2}$$

where N = tank oil in place, bbl/acre-ft

S_o = fraction of pore space occupied by oil (the oil saturation)

S_w = the water saturation

B_o = the formation volume factor for the oil at the reservoir pressure, barrels reservoir space/barrel tank oil.

Determining the proper values of S_w in Eq. (2.2) is considerably more difficult than obtaining ϕ. For the present, let it suffice to say that some water will always exist within the reservoir rock and that its volume must be subtracted from the space available for oil. This water is commonly called the connate water and is assumed incompressible in this equation. Note also that the pore space is assumed to be occupied by either oil or water, and that no free gas is present. Consequently, the equation as given must be applied to the

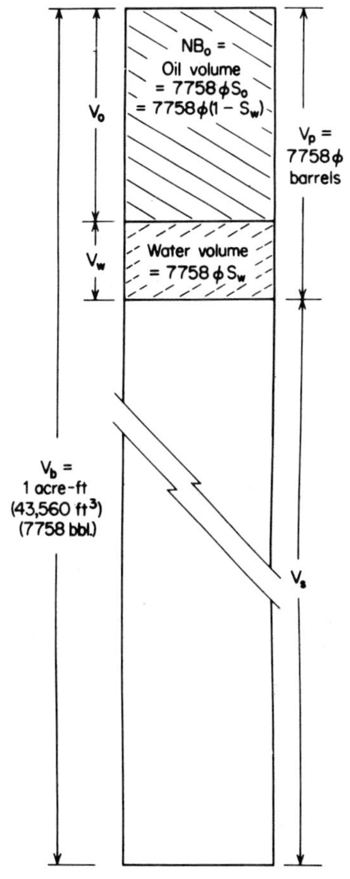

Apparent relationships

(1) $S_o = \dfrac{V_o}{V_p}$, (2) $S_w = \dfrac{V_w}{V_p}$ $\Big\}$ $S_o + S_w = 1$

(3) $V_o + V_w = V_p$

or

$NB_o + 7758\, \phi\, S_w = 7758\, \phi$

from which

(4) $N = \dfrac{7758\, \phi\, (1 - S_w) \text{ barrels/acre-ft.}}{B_o}$

Fig. 2.5. Concepts of volumetric equation.

reservoir at or above the bubble point, and is generally used to compute the initial oil in place.

A similar expression may be derived for the amount of gas stored in a particular sand. In this case it is convenient to express the gas volume in terms of SCF or in MCF (thousands of standard cubic feet). Recall the form of the Gas Law

$$\frac{p_s V_s}{T_s} = \frac{p V_p}{z T} \tag{2.3}$$

where subscript, s, denotes standard conditions. z_s = 1.0, and is not shown. Then:

$$V_s = G = V_p \times \frac{p T_s}{z T p_s} \tag{2.4}$$

where G is the standard gas volume contained in V_p at conditions p, T, z.

But: $V_p = 43{,}560\phi(1 - S_w)$ cu ft/acre-ft
$T_s = 460° + 60° = 520°\text{R}$
$p_s = 14.7$ psia

Substitution of these values in (2.4) gives:

$$G = 43{,}560\phi(1 - S_w) \times \frac{520}{14.7} \times \frac{p}{zT} \tag{2.5}$$

or

$$G = \frac{1540\phi(1 - S_w)p}{zT} \text{ MCF/Acre-ft} \tag{2.5a}$$

Permeability: Darcy's Equations

In addition to being porous, a reservoir rock must be permeable; that is, it must allow fluids to flow through its pore network at practical rates under reasonable pressure differentials. Permeability is defined as a measure of a rock's ability to transmit fluids. The quantitative definition of permeability was first given in an empirical relationship developed by the French hydrologist Henry D'Arcy who studied the flow of water through unconsolidated sands.

This law in its differential form is:

$$v = -\frac{k}{\mu}\frac{dp}{dL} \tag{2.6}$$

where v = apparent flow velocity
μ = viscosity of the flowing fluid
dp/dL = pressure gradient in the direction of flow
k = permeability of the porous media

Consider the linear system of Figure 2.6.

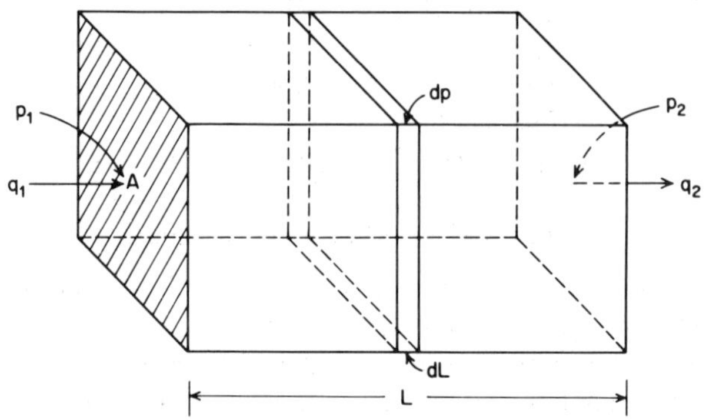

Fig. 2.6. Linear flow system illustrating notations of Darcy's equation. For incompressible fluids, $q_1 = q_2$.

The following assumptions are necessary to the development of the basic flow equations:

1. Steady state flow conditions exist.
2. The pore space of the rock is 100% saturated with the flowing fluid. Under this restriction k is the absolute permeability.
3. The viscosity of the flowing fluid is constant. This is not true since $\mu = f(p,T)$ for all real fluids. However, this effect is negligible if μ at the average pressure is used, and if conditions 4–6 hold:
4. Isothermal conditions prevail.
5. Flow is horizontal and linear.
6. Flow is laminar.

With these restrictions in mind, let

$$v = \frac{q}{A} \tag{2.7}$$

where q = volumetric rate of fluid flow
A = total cross-sectional area perpendicular to flow direction

This is a further assumption since only the pores, and not the full area, conduct fluid. Hence, v = an apparent velocity. The actual velocity, assuming a uniform medium, $= v/\phi$.

Case I: Linear Incompressible Fluid Flow

Substitution of (2.7) in (2.6) gives

$$q/A = -\frac{k}{\mu} \times \frac{dp}{dL} \tag{2.8}$$

Separation of variables and insertion of the limits depicted by Figure 2.6, gives

$$\frac{q}{A}\int_0^L dL = -\frac{k}{\mu}\int_{p_1}^{p_2} dp \tag{2.9}$$

which, when integrated, is

$$q = \frac{kA(p_1 - p_2)}{\mu L} \tag{2.10}$$

or

$$k = \frac{q\mu L}{A\Delta p}$$

Note: The negative sign used to denote a negative pressure gradient in the direction of integration has been removed by reversing the pressure limits.

Equation (2.10) is basic and the following units serve to define the darcy.

If q = 1 cc/sec
A = 1 cm²
μ = 1 centipoise
$\Delta p/L$ = 1 atmosphere/cm

then

$$k = 1 \text{ darcy}$$

A permeability of one darcy is much higher than that commonly found in reservoir rocks. Consequently, a more common unit is the millidarcy, where

$$1 \text{ darcy} = 1000 \text{ millidarcys}$$

Case II: Linear Compressible Fluid Flow

Consider the same linear system of Figure 2.6, except that the flowing fluid is now compressible; then $q \neq$ constant, but is a $f(p)$.

Assuming that Boyle's law is valid ($z = 1$):

$$p_1 q_1 = p_2 q_2 = pq = \text{constant} \tag{2.11}$$

where subscripts denote point of measurement; then

$$pq = \frac{-kA}{\mu} \times \frac{dp}{dL} p = p_2 q_2 \tag{2.12}$$

$$q_2 \int_0^L dL = -\frac{kA}{\mu} \times \frac{1}{p_2} \int_{p_1}^{p_2} p\, dp \tag{2.13}$$

from which

$$q_2 = \frac{kA}{\mu L} \times \frac{p_1^2 - p_2^2}{2} \times \frac{1}{p_2} \tag{2.14}$$

Expressing Eq. (2.14) in terms of $\bar{q}_g$, the rate of gas flow at the average pressure in the system is

$$\bar{q}_g = \frac{kA}{\mu L} \times \frac{p_1^2 - p_2^2}{2} \times \frac{1}{\bar{p}} \tag{2.15}$$

But

$$\bar{p} = \frac{p_1 + p_2}{2}, \text{ and } \frac{p_1^2 - p_2^2}{2} = \frac{(p_1 + p_2)(p_1 - p_2)}{2}$$

Therefore,

$$\bar{q}_g = \frac{kA\Delta p}{\mu L} \tag{2.16}$$

which is exactly the same as Eq. (2.10).

An expression for the standard flow rate, q_{gs}, is obtained from Charles' law:

$$\frac{p_s q_{gs}}{T_s} = \frac{p_2 q_2}{T_f} = \frac{kA}{2\mu L}(p_1^2 - p_2^2)\frac{1}{T_f} \tag{2.17}$$

where T_s = 60°F (520°R)
p_s = 1 atm
T_f = flowing temperature

Thus,

$$q_{gs} = \frac{kA(p_1^2 - p_2^2)}{2\mu L} \times \frac{T_s}{T_f} \times \frac{1}{p_s} \tag{2.18}$$

Note: For standard units, p_s = 1 atm and is often omitted from (2.18). The use of these linear equations is generally limited to laboratory testing. The equations which normally apply to field flow calculations are based on a radial system as shown in Figure 2.7.

Case III: Radial Incompressible Fluid Flow

We will start with the differential form of Eq. (2.8) with notations and sign convention as applied to Figure 2.7.

$$\frac{q}{A} = \frac{k}{\mu} \times \frac{dp}{dr} \tag{2.19}$$

But for radial flow,

$$A = 2\pi r h$$

where r = radius or distance from center, cm
h = thickness of bed, cm

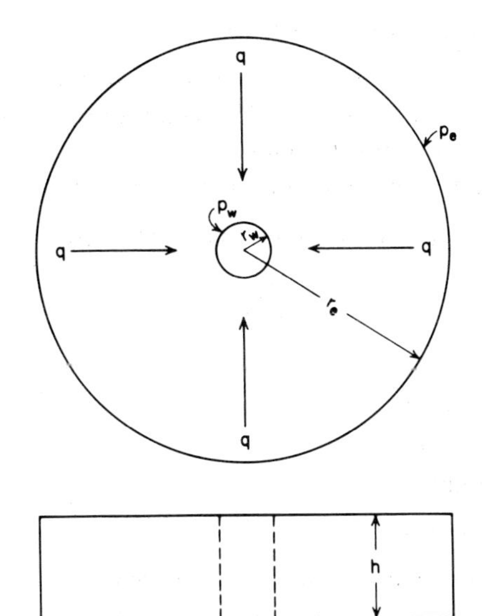

Fig. 2.7. Ideal radial flow system.

Substitution of $2\pi rh$ for A and separation of variables gives

$$q \int_{r_w}^{r_e} \frac{dr}{r} = \frac{2\pi hk}{\mu} \int_{p_w}^{p_e} dp \tag{2.20}$$

which when integrated is

$$q = \frac{2\pi hk(p_e - p_w)}{\mu \ln (r_e/r_w)} \tag{2.21}$$

Note: ln = natural logarithm

Equation (2.21) is the basic expression for the steady state radial flow of a liquid. The units are the same as previously defined; however, any consistent set of units may be used for r_e and r_w.

Case IV: Radial Compressible Fluid Flow

By proceeding in the same manner as in Case II, the radial equations for gases may be obtained. Again, Boyle's law is assumed valid; hence,

$$p_e q_{ge} = p_w q_{gw} = \bar{p}\bar{q}_g = \text{constant}$$

where subscripts refer to position at which q is specified: well, external boundary, etc.

$$q_{gw} \int_{r_w}^{r_e} \frac{dr}{r} = \frac{2\pi hk}{\mu p_w} \int_{p_w}^{p_e} p dp \tag{2.22}$$

and

$$q_{gw} = \frac{\pi kh(p_e^2 - p_w^2)}{\mu \ln (r_e/r_w)} \times \frac{1}{p_w} \tag{2.23}$$

Similarly,

$$\bar{q}_g = \frac{2\pi hk(p_e - p_w)}{\mu \ln (r_e/r_w)} \tag{2.24}$$

and

$$q_{gs} = \frac{\pi hk(p_e^2 - p_w^2)}{\mu \ln (r_e/r_w)} \times \frac{T_s}{T_f} \times \frac{1}{p_s} \tag{2.25}$$

Conversions to Practical Units

The standard units which define the darcy are useful in laboratory calculations. For computations pertaining to field problems it is more convenient to convert to practical units by use of an appropriate constant.

For example, let us convert Eq. (2.21) to

$$q = \frac{Chk(p_e - p_w)}{\mu \ln (r_e/r_w)} \tag{2.26}$$

where h = ft
k = darcys
p_e, p_w = psia
μ = cp
q = barrels/day

The following conversion factors are needed:

1 barrel = 159,000 cc
1 ft = 30.48 cm
1 atm = 14.7 psi

Then:

$$q \text{ bbl/d} \times \frac{159{,}000 \text{ cc/bbl}}{24 \times 3600 \text{ sec/d}} = \frac{2\pi\left(h \text{ ft} \times 30.48 \frac{\text{cm}}{\text{ft}}\right)k\left[(p_e - p_w)\text{psi} \times \frac{1}{14.7}\frac{\text{atm}}{\text{psi}}\right]}{\mu \ln (r_e/r_w)}$$

or,

$$q = \frac{(24)(3600)(2\pi)(30.48)}{(159{,}000)(14.7)} \times \frac{hk(p_e - p_w)}{\mu \ln r_e/r_w}$$

$$q = \frac{7.07\, hk\ (p_e - p_w)}{\mu \ln (r_e/r_w)}, \text{ where } C = 7.07. \tag{2.27}$$

Values of C for some other equations and systems of units are given in Table 2.1B.

TABLE 2.1

A. Summary of Darcy Equations for Steady State, Isothermal, Homogeneous Fluid Flow

Case	*Flow Type*	*Equation*
I	Linear, Incompressible	$q = \frac{kA\Delta p}{\mu L}$
II	Linear, Compressible	(a) $q_2 = \frac{kA(p_1^2 - p_2^2)}{2\mu L} \times \frac{1}{p_2}$
		(b) $\bar{q}_g = \frac{kA\Delta p}{\mu L}$
		(c) $q_{gs} = \frac{kA(p_1^2 - p_2^2)}{2\mu L} \times \frac{T_s}{T_f p_s}$
III	Radial, Incompressible	$q = \frac{2\pi hk(p_e - p_w)}{\mu \ln(r_e/r_w)}$
IV	Radial, Compressible	(a) $q_{gw} = \frac{\pi hk(p_e^2 - p_w^2)}{\mu \ln(r_e/r_w)} \times \frac{1}{p_w}$
		(b) $\bar{q}_g = \frac{2\pi hk(p_e - p_w)}{\mu \ln(r_e/r_w)}$
		(c) $q_{gs} = \frac{\pi hk(p_e^2 - p_w^2)}{\mu \ln(r_e/r_w)} \times \frac{T_s}{T_f p_s}$

The above equations are in standard units:

q = cc/sec	p = atm	μ = centipoise
k = darcys	L = cm	r_e, r_w = consistent
A = sq cm	h = cm	ln = base e
		T = absolute

B. Coefficients C to Convert Darcy Equations to Listed Practical Units

Case	q	A	h	p or Δp	L	C
I	bbl/day	ft²	—	psi	ft	1.127
II (a)	ft³/day*	ft²	—	psia	ft	3.17
II (b)	ft³/day*	ft²	—	psia	ft	6.33
II (c)	ft³/day*	ft²	—	psia	ft	112 (p_s, T_s included)
III	bbl/day	—	ft	psi	—	7.07 (3.07 for $\log_{10}$)
IV (a)	ft³/day*	—	ft	psia	—	19.9 (8.65 for $\log_{10}$)
IV (b)	ft³/day*	—	ft	psia	—	39.8
IV (c)	ft³/day*	—	ft	psia	—	704 (p_s, T_s included)

k = darcys
μ = cp

*For gas volumes in bbl/day divide C by 5.61.

Electrical Analogy

Darcy's law is analogous to Ohm's law defining the flow of electric current.

Ohm's law, for a linear system

$$I = \frac{EA}{RL}$$

Darcy's linear equation

$$q = \frac{kA\Delta p}{\mu L}$$

where E = voltage drop
A = cross-sectional area
R = resistivity
L = length
I = current flow

These analogs for the two cases are apparent:

$$I \sim q$$
$$A \sim A$$
$$R \sim \mu/k$$
$$L \sim L$$
$$E \sim \Delta p$$

Dimensional Analysis of Permeability

The physical concept of the darcy is enhanced by a dimensional analysis of Darcy's equation. This is accomplished by resolving all quantities into their basic dimensions of mass M, length L, and time t, and solving for the dimensions of k.

$$k = \frac{q\mu L}{A\Delta p}$$

where q = volume/time = L^3/t
$L = L$
A = area = L^2

$$\Delta p = \text{force/area} = \frac{\text{mass} \times \text{acceleration}}{\text{area}}$$

$$\Delta p = \frac{ML/t^2}{L^2} = \frac{M}{Lt^2}$$

$$^*\mu = M/Lt$$

$$\therefore \quad k = \frac{(L^3/t)(M/Lt)(L)}{(L^2)(M/Lt^2)} = L^2$$

The fact that permeability is not a dimensionless quantity, but has the units of area, gives an insight into its nature. Its magnitude is primarily dependent on pore size. Consequently a large void, such as a vug or fracture, may have far more permeability than thousands of small voids. The pore size of a clastic rock is largely dependent on its grain size. Hence a fine grained rock will have a lower permeability than a coarse grained rock of the same porosity if other factors, primarily cementation, are constant.

Typical Permeability Magnitudes

In general, rocks having a permeability of 100 md or greater are considered fairly permeable, while rocks with less than 50 md are considered tight. Values of several darcys are exhibited by some poorly cemented or unconsolidated sands. Many productive limestone and dolomite matrices have permeabilities below 1 md, however; these have associated solution cavities and fractures which contribute the bulk of the flow capacity. Current stimulation techniques of acidizing and hydraulic fracturing allow commercial production to be obtained from reservoir rocks once considered too tight to be of interest. We shall have more to say of this in later chapters. The illustrations of Figure 2.4 show values of k for those samples.

2.13 Traps

In order for petroleum to accumulate in commercial quantities, it must, in its migration process, encounter a subsurface rock condition which halts further migration and causes the accumulation to take place. These subsurface conditions are numerous in type ranging from very simple to extremely complex forms. Numerous systems of trap classification exist; the following system proposed by Sanders[10] appears as simple and useful as any.

1. Structural traps: those traps formed by deformation of the earth's crust by either faulting or folding.
2. Stratigraphic traps: those formed by changes in lithology, generally a disappearance of the containing bed or porosity zone.
3. Combination traps: traps having both structural and stratigraphic features.

Virtually all conceivable trap forms have been found in nature and are covered in detail in a number of petroleum geology texts. Only two typical examples of each type are shown here in Figure 2.9.

A feature of all traps is the impermeable cap rock which forms the top of the trap. Geologically this is often discussed as a separate feature rather than as a part of the trap. This distinction is, however, beyond the present treatment; and we will include the cap rock as merely a part of the trap.

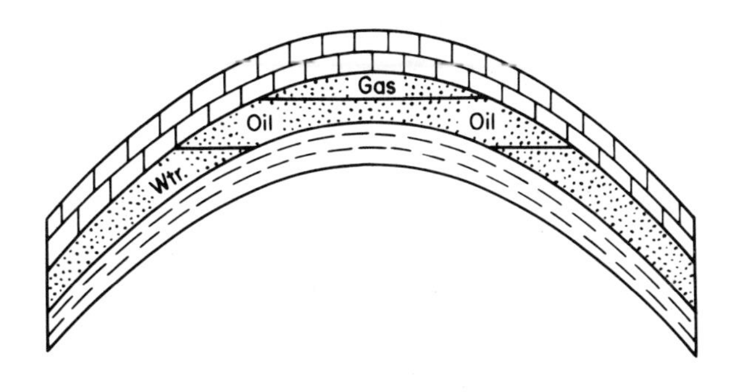

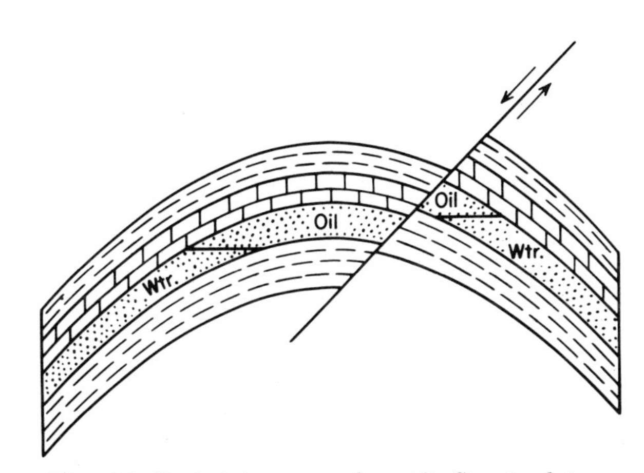

Fig. 2.9. Typical trap examples. **(A)** Structural traps. Top is simple anticline. Bottom is faulted anticline. Hydrocarbons accumulate at structurally high positions.

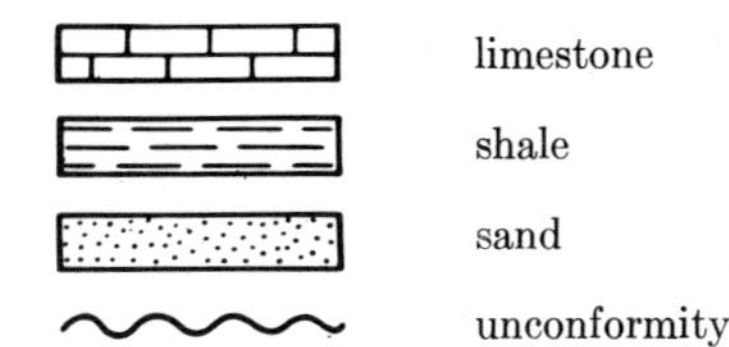

*The dimensions of dynamic viscosity are obtained by analyzing the forces exerted on a plate of area A, moving on a viscous liquid with velocity, v, caused by a force, F.

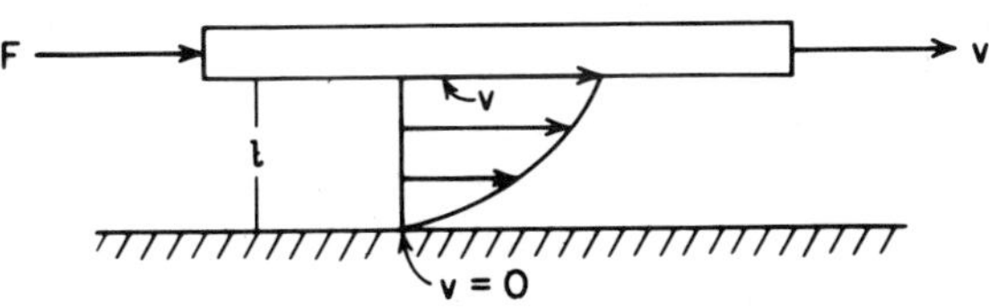

Fig. 2.8. Flow between plates as means of defining viscosity. Lower plate is stationary. Fluid velocity between plates varies from v at top to zero at bottom.

$$\mu = \frac{\text{shearing stress}}{\text{rate of shearing strain}} = \frac{F/A}{dv/dl} = \frac{\frac{ML/t^2}{L^2}}{\frac{L/t}{L}} = \frac{M}{Lt}$$

2.2 Subsurface Pressures

It is readily seen in Darcy's equations that fluid movement from a reservoir rock to a well bore can take place only if a pressure differential can be established between the reservoir and the well. This requires that the fluids

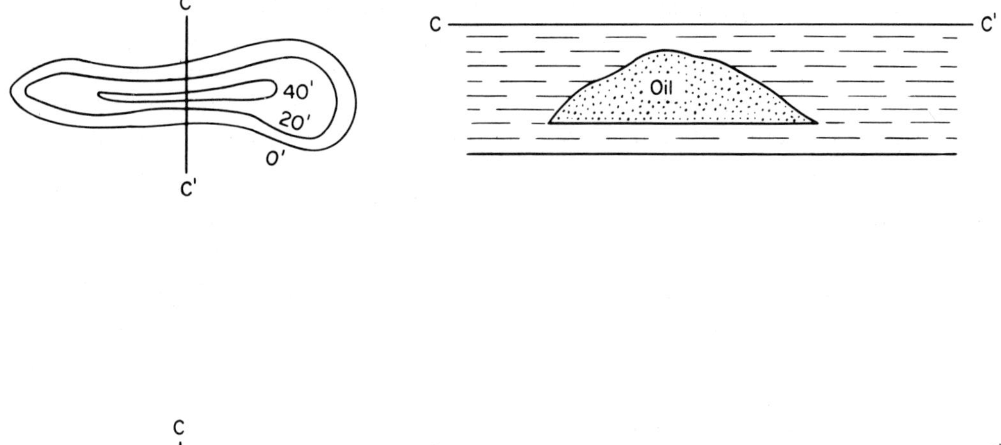

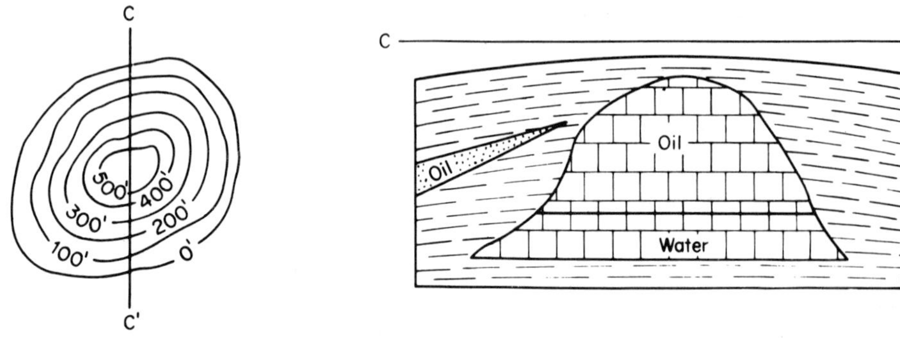

Fig. 2.9 (B). Stratigraphic traps. Top is classic shoestring sand type of Kansas and Oklahoma. These were originally offshore bars similar to those now existing off the southeast Texas coast. The sand thickness or isopac map to the left is typical. Bottom example is a reef deposit similar to those found in west Texas. Note flank sand which is also a stratigraphic form.

in the rock be confined under an elevated pressure. Fortunately, it is the general case to find such a condition prevailing. The measurement of subsurface (bottom hole, reservoir, rock, etc.) pressures and utilization of pressure data are important phases of petroleum engineering and will be further discussed in later chapters.

Source of Subsurface Pressures

In general, the elevated pressures encountered with depth are due to one or both of two causes:

1. Hydrostatic pressure imposed by the weight of fluid (predominantly water) which fills the voids of the rocks above and/or contiguous with the reservoir in question.
2. Overburden pressure due to the weight of the rocks and their fluid content existing above the reservoir.

It is more common to find subsurface pressures varying as a linear function of depth with a gradient close to the hydrostatic gradient of fresh to moderately saline water. Departures from this behavior, both higher and lower, are considered abnormal. It is, however, the abnormally high pressures which are more important as a source of serious drilling and production hazards.

Overburden pressures are most common in soft rock areas such as the Gulf Coast. Here the sediments are geologically young and are not as consolidated as the hard rocks of the Mid-Continent and West Texas areas. Consequently, due to compaction, repacking, or other sedimentary changes, it is not uncommon to find that the rocks have failed to support the weight of the overburden with the result that this total weight is supported by the enclosed fluid.

Magnitude of Subsurface Pressures

Pressure-depth relationships are commonly spoken of in terms of gradients. The hydrostatic gradient in fresh water is 0.433 psi/ft of depth which is the quotient of 62.4 lb/ft^3 $\div$ 144 in.^{2}/ft^2. Since most subsurface waters are saline, it is common to find the gradient to be above 0.433 as in the case shown in Figure 2.10.[11] Pressure data from over 100 high pressure wells in the Texas-Louisiana Gulf Coast are shown in Figure 2.11.[12] Note that the earth or overburden pressure gradient is shown as 1.0. This figure is commonly used and may be obtained by using an average water saturated rock specific gravity of 2.3. Hence the overburden gradient is

$$2.3 \times 0.433 \cong 1.0 \text{ psi/ft}$$

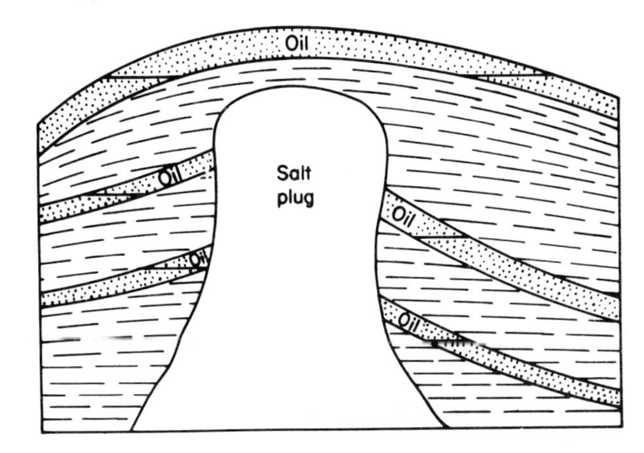

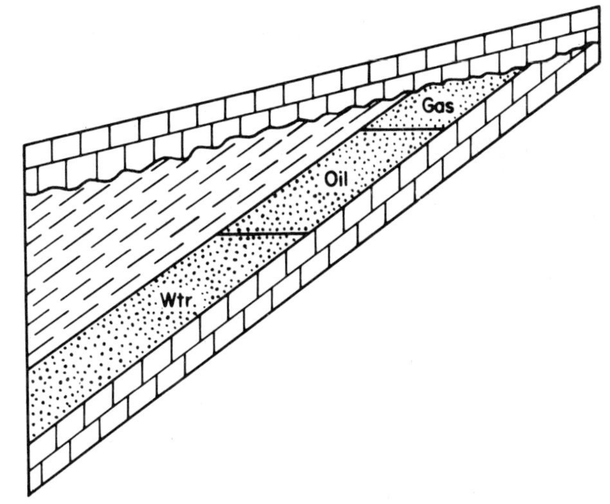

Fig. 2.9(C). Combination traps. Top is piercement salt dome type which is typical of many Gulf Coast fields. Oil is found in flank sands, upper beds, or the caprock. Lower is an unconformity trap. Hydrocarbons have been trapped by impervious bed above the unconformity. The largest field in the U. S., the East Texas Field, is of this type.

2.3 Subsurface Temperatures

Since the earth is assumed to contain a molten core, it is logical to assume that temperature should increase with depth. This temperature-depth relationship is commonly a linear function of the form:

$$T_D = T_a + \alpha D$$

where T_D = temperature of the reservoir at any depth, D

T_a = average surface temperature

α = temperature gradient, degrees/100 ft

D = depth, hundreds of ft

Nichols[13] has compiled much data on the South Central United States as shown in Figure 2.12. A normal gradient seems to be about 1.6°F/100 ft, although it should be noted that considerable variations occur in various areas. West Texas is a known abnormally low temperature area while many Gulf Coast temperatures are higher than normal.

Several devices for measuring subsurface temperatures are available and will be discussed under temperature logging. A knowledge of down hole temperatures and temperature gradients is a fundamental piece of engineering data.

2.4 Sample Problems

1. How much tank oil exists in the following oil field?
Area of field = 640 acres = 1 section
Average sand thickness = 20 ft
Average porosity, ϕ = 20%
Average connate water saturation, S_w = 30%
Formation volume factor for oil, B_o = 1.20

Solution:

$$N = \frac{7758(.20)(1-.30)}{1.20}(640)(20) = 11.6 \times 10^6 \text{ bbl}$$

2. (a) Given the following data for a cylindrical core sample, compute its porosity.
Clean, dry weight of sample = 311 gm
Wt. of sample with pores completely filled (100% saturated) with a 1.05 sp. gravity brine = 331 gm
Diameter of sample = 4.0 cm. Length of sample = 10.0 cm

Solution:

$$\phi = \frac{V_p}{V_b} \qquad V_b = \pi(2)^2(10) = 40\pi\text{cc}$$

$$V_p = \frac{331 - 311}{1.05} = 19.05\text{cc}$$

$$\phi = \frac{19.05}{40\pi} = 15.1\%$$

(b) What is the density of the rock grains?

$$\rho_s = \frac{311}{(.849)(40\pi)} = 2.9 \text{ gm/cc}$$

(c) What gross storage space exists in this rock?

$$V_p = (7758)(.151) = 1170 \text{ bbl/acre-ft}$$

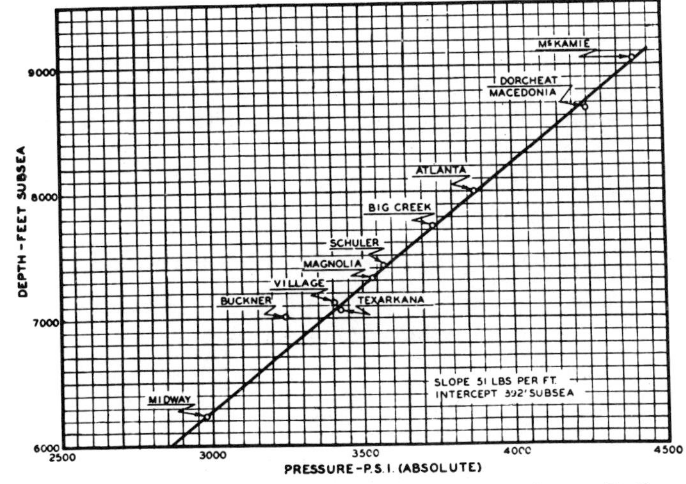

Fig. 2.10. Original pressures in Smackover pools vs. depth. Pressures are corrected to water-table level. After Bruce,[12] courtesy AIME.

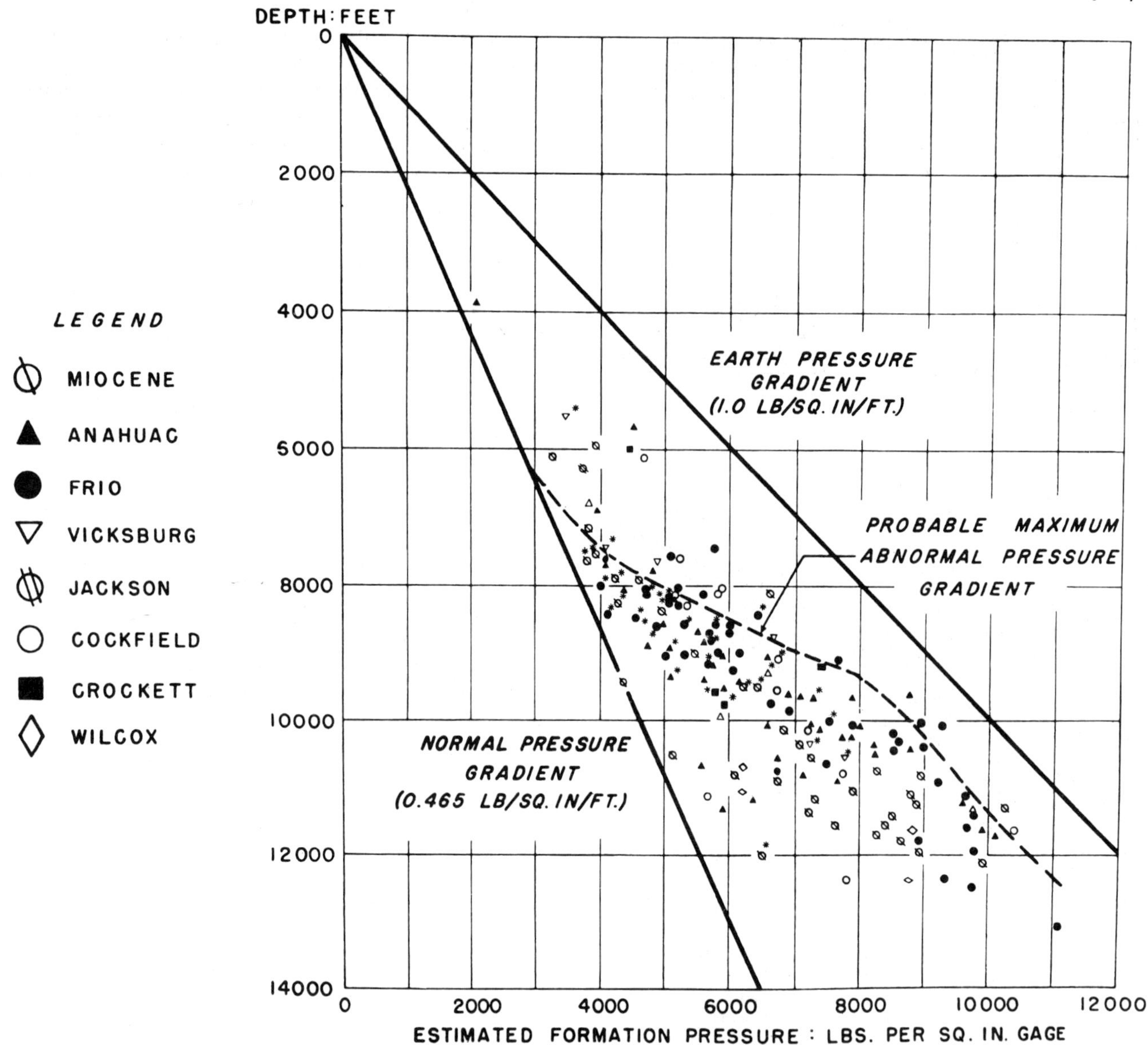

Fig. 2.11. Magnitude of some abnormal pressures encountered in the Gulf Coast area. After Cannon and Sullins,[12] courtesy API.

3. The sample of Problem 2 was subjected to a laboratory linear flow test using distilled water as the flowing fluid. It is assumed that this did not alter the rock; the following data were obtained:

$\mu_w = 1$ cp

$p_1 = 50$ psig

$p_2 =$ atm pressure, 14.7 psia

$q = 0.50$ cc/sec

What is the permeability of this sample?

Solution:

Assuming Eq. (2.10) to apply,

$$k = \frac{q\mu L}{A\Delta P} = \frac{(0.50)(1.0)(10.0)}{(4\pi)(50/14.7)} = 0.117 \text{ darcys} = 117 \text{ md}$$

4. Compute the ideal production rate of the following well which was completed in the above sand.

Spacing = 1 well per 40 acres

Sand thickness = 10 ft

Static reservoir pressure, $p_e = 2000$ psia

Bottom hole producing pressure, $p_w = 1500$ psia

Reservoir oil viscosity, $\mu_o = 3$ cp

Well diameter = 8 in.

Solution:

Using equation (2.27),

$$q_o = \frac{7.07hk(p_e - p_w)}{\mu_o \ln (r_e/r_w)}$$

$$= \frac{(7.07)(10)(0.117)(500)}{(3) \ln (700/.33)} = 180 \text{ barrels oil per day}$$

Note: The drainage radius $r_e = 700$ ft is slightly more than half the 1320 ft between 40 acre spaced wells. This is a common approximation. Actually $\ln (r_e/r_w)$ is relatively insensitive to reasonable changes in r_e.

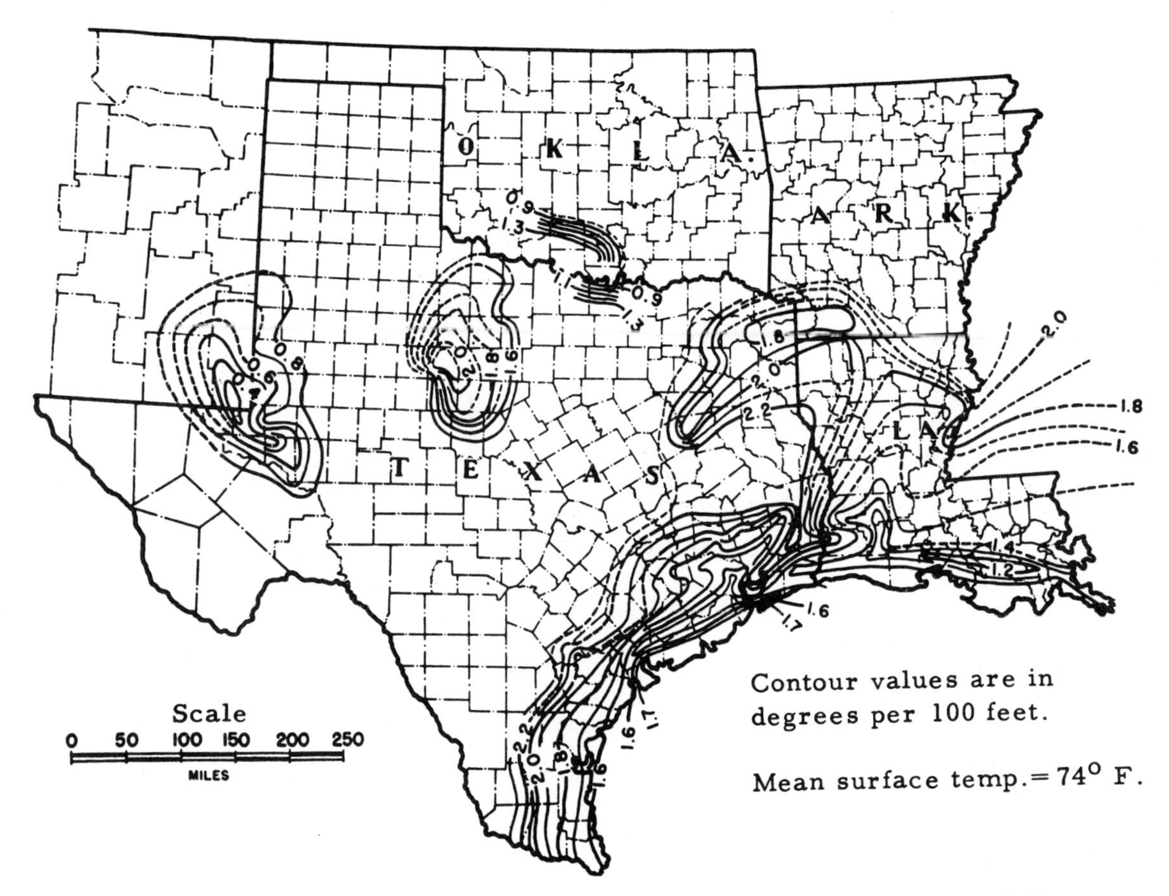

Fig. 2.12. Values of geothermal gradient for the South Central U. S. After Nichols,[13] courtesy AIME.

PROBLEMS

1. Check the constants in Table 2.1B.

2. A regular section of land is 1 mile square and contains 640 acres. The spacing of a well is commonly denoted by the acreage assigned to it (10 acres, 40 acres, etc.) and the well is normally located in the center of its tract. Show that the following are the distances between wells in the same row for the given spacings:

Spacing, acres	*Distance between wells, ft*
10	660
40	1320
160	2640

3. A cylindrical core sample was subjected to the following linear flow test conditions. Compute the sample's permeability in md.

$L = 10.0$ cm $\mu = 1.0$ cp

diam $= 2$ cm $q = 12$ cc/min

$p_1 = 44.1$ psig

$p_2 = 14.7$ psia

Ans. $k = 210$ md.

4. Show that the total pressure drop Δp_t across a linear system composed of individual rock segments of length L_n each having a permeability k_n is given by

$$\Delta p_t = \frac{q\mu}{A}\left(\frac{L_1}{k_1} + \frac{L_2}{k_2} + \frac{L_3}{k_3} + \dots\right)$$

5. A linear flow system 10 cm long contains two cores of length, L_1 and L_2. Cores 1 and 2 have permeabilities of 100 and 400 md, respectively. A flow test was conducted under a total pressure differential of 10 psi, 8 psi of which occurred across core 1. What are L_1 and L_2? *Ans.* 5 cm.

6. Plot the pressure vs. path length distribution of the flow test in Problem 5.

7. Show that the ratio of the respective pressure drops across two cores in linear series is given by

$$\frac{\Delta p_1}{\Delta p_2} = \left(\frac{L}{L_2} - 1\right)\frac{k_2}{k_1}$$

where $L = L_1 + L_2$.

8. What linear pressure gradient is required to cause a flow of 0.10 cc/sec of a 5 cp fluid through a rock having a permeability of 50 md and an area of 5 cm^2? *Ans.* $dp/dL = -2$ atm/cm.

9. (a) What is the ideal rate of oil flow into the following well?
   Spacing = 1 well per 160 acres
   Sand thickness = 30 ft
   k = 200 md
   Depth to center of sand = 5000 ft
   Pressure gradient for area = 0.45 psi/ft of depth
   Well diameter = 12 in.
   Oil viscosity = 5 cp
   Bottom hole producing pressure = 2000 psig
   (b) What would be the standard rate of gas flow into the same well?
   μ_g = 0.02 cp
   Temperature gradient in area = 1.4°F/100 ft, and T_a = 70°F. *Ans.* (a) q_o = 270 bbl/d; (b) $q_{gs} = 48 \times 10^6$ SCF/day; for r_e = 1350 ft

10. (a) Compute the initial tank oil in place for the following field.
   Area = 1000 acres
   Average porosity, ϕ = 18%
   Average connate water saturation, $\bar{S}_w$ = 25%
   Formation volume factor for oil, B_o = 1.15
   Average sand thickness = 15 ft
   (b) Compute the standard quantity of 0.70 gravity gas which could be contained in the same field.
   Reservoir depth = 10,000 ft
   Static pressure gradient for area = 0.50 psi/ft
   Temperature gradient = 1.5°F/100 ft
   Average surface temperature = 75°F
   (c) Repeat part (b) for a 1.0 gravity gas. *Ans.* (a) $N = 13.6 \times 10^6$ bbl tank oil; (b) $G_i = 22.8 \times 10^9$ SCF = 22,800 MMCF (millions of std. cu ft).

11. To what value would the well diameter of Problem 9 have to be increased in order to double the flow rate?

12. Plot the producing history of gas field of Problem 10 (b) as was done for the tank of Problem 4, Ch. 1. Compute the total gas-containing pore volume from the slope of the p/z vs G_p curve and check it against that obtained from the volumetric equation.

13. Show that the pressure gradient in an ideal radial system is given by

$$\frac{dp}{dr} = \frac{c}{r}$$

where $c = q\mu/2\pi hk$. Where is dp/dr a maximum? Minimum?

14. Show that a plot of p vs $\ln r$ is linear from $r = r_w$ to $r = r_e$.

15. Using the data of Problem 9 (a) plot the pressure distribution during steady state flow conditions. What fraction of the total pressure drop has been expended at r = 20 ft? When r = 50 ft? When r = 100 ft?

16. Two core samples of identical dimensions are connected in parallel and steady state, incompressible flow is established through the system. The permeabilities are 100 and 500 md, respectively. What fraction of the total flow passes through each core? *Ans.* 1/6 and 5/6.

17. Three cores of equal area and the following properties are connected in parallel.

Core No.	*L, cm*	*k, md*
1	2	100
2	3	200
3	4	300

(a) If the area of each core is 2 cm² and water (μ = 1 cp) is flowed through the system under a pressure drop of 2 atm, what will be the total rate of flow through all three cores? Hint: Show first that the total flow rate, q, is

$$q = q_1 + q_2 + q_3 = \frac{A\Delta p}{\mu}\left(\frac{k_1}{L_1} + \frac{k_2}{L_2} + \frac{k_3}{L_3}\right)$$

(b) What fraction of total flow passes through each core? *Ans.* 6/23, 8/23, 9/23.

REFERENCES

1. Levorsen, A. I., *Geology of Petroleum.* San Francisco: W. H. Freeman and Co., 1954, p. 478.
2. ZoBell, C. E., "The Role of Bacteria in the Formation and Transformation of Petroleum Hydrocarbons," *Science,* Vol. 102, Oct. 12, 1945, pp. 364–369.
3. DeGolyer, E., ed., *Elements of the Petroleum Industry.* New York: AIME, 1940, p. 26.
4. Gussow, W. C., "Differential Entrapment of Oil and Gas: A Fundamental Principle," *Bulletin of American Association of Petroleum Geologists,* Vol. 38, May, 1954, No. 5, pp. 816–853.
5. Graton, L. C., and H. J. Fraser, "Systematic Packing of Spheres — With Particular Relation to Porosity and Permeability," *The Journal of Geology,* Vol. 43, No. 8, Part 1, Nov.-Dec., 1935, pp. 785–909.
6. Wentworth, C. K., "A Scale of Grade and Class Terms for Clastic Sediments," *Journal of Geology,* Vol. 30, July-Aug., 1922, pp. 377–392.
7. Hohlt, R. B., "The Nature and Origin of Limestone Porosity," *Colorado School of Mines Quarterly,* Vol. 43, 1948, No. 4.
8. Rall, C. G., Hamontre, H. C., and D. B. Taliaferro, "Determination of Porosity by a Bureau of Mines Method: A List of Porosities of Oil Sands," *Bureau of Mines Report of Investigations 5025,* U. S. Department of the Interior, Feb. 1954.
9. Bulnes, A. C., and R. U. Fitting, Jr., "Introductory Discussion of Reservoir Performance of Limestone Formations," *Trans. AIME,* Vol. 160, 1945, p. 179.
10. Sanders, C. W., "Stratigraphic Type Oil Fields and Proposed New Classification of Reservoir Traps," *Bulletin of American Association of Petroleum Geologists,* 27, 1943, pp. 539–550.
11. Bruce, W. A., "A Study of the Smackover Limestone Formation and the Reservoir Behavior of Its Oil and Condensate Pools," *Trans. AIME,* Vol. 155, 1944, p. 91.

12. Cannon, G. E., and R. S. Sullins, "Problems Encountered in Drilling Abnormal Pressure Formations," *API Drilling and Production Practices*, 1946, p. 30.

13. Earl A. Nichols, "Geothermal Gradients in Mid-Continent and Gulf Coast Oil Fields," *Trans. AIME*, Vol. 170, 1947, pp. 44–50.

SUPPLEMENTARY READINGS

1. Levorsen, A. I., op. cit., a complete text on Petroleum Geology which covers most of the items of this chapter in detail. Chapters 3–7, 11, 12 are particularly pertinent at this point.

2. Cox, B. B., "Transformation of Organic Material into Petroleum under Geological Conditions — the Geologic Fence," *Bull. American Assoc. of Petroleum Geologists*, Vol. 30, May 1946, pp. 645–659. A fine discussion of this subject with complete bibliography.

3. Pirson, S. J., *Elements of Oil Reservoir Engineering*, 2nd ed. New York: McGraw-Hill Book Co., Inc., 1958. A standard text for Petroleum Engineers. Chapters 1 and 2 treat many of the subjects of this chapter. Excellent diagrams of traps.

4. Muskat, M., *Physical Principles of Oil Production*, 1st ed. New York: McGraw-Hill Book Co., Inc., 1949. A complete and highly mathematical treatment of what is commonly called reservoir engineering. Chapter 1 covers rock properties.

5. Fancher, G. H., "The Porosity and Permeability of Clastic Sediments and Rocks," in *Subsurface Geologic Methods*, 2nd ed., ed. by L. W. Leroy. Golden: Colorado School of Mines, 1951, pp. 685–713. Discussion of porosity and permeability parallel to this chapter. Treatment of permeability is extended to basic concepts of multiphase flow.

6. *Recommended Practice for Determining Permeability of Porous Media, API RP. 27*, 3rd ed. Dallas, Texas: American Petroleum Institute, 1952. Recommended laboratory procedure for permeability determinations, sample calculations, basic theory. 51 references cited.

Chapter 3 . . .

Petroleum Exploration Methods and General Leasing Practices

As could be expected, the field of oil exploration has attracted the talents of many individuals who propose to find oil with the aid of forked sticks, clothes hangers, and numerous other "witching" devices. The most successful methods are, however, techniques somewhat more scientific in nature and are the ones which will be briefly discussed in this chapter under the following general headings:

1. Direct Indications
2. Geological Methods
3. Geophysical Methods

The relative success of various technical and nontechnical exploration methods are compiled annually by the American Association of Petroleum Geologists.[1] The success percentage for technical methods has been relatively constant for the last 15 years at about $12\frac{1}{2}\%$, or approximately one success per eight wildcats. This constancy indicates improvements in technique, since each field discovered leaves one less to find. The success of nontechnical methods averages 4.4% or about one-third as successful as technical methods. The average success for all exploration methods is about one producing well per nine wildcats.

3.1 Direct Indications

Nearly all of the great oil provinces of the world exhibit some surface evidence of the presence of petroleum. Typical of these indications are natural seepages of oil, outcrops of oil-bearing rocks, and various forms of gas seepages such as mud volcanoes. Certainly, the visible presence of hydrocarbons suggests that an area deserves attention; it does not, however, necessarily prove that oil exists in commercial quantities. In the United States these types of occurrences have been quite thoroughly explored and no longer exist as an exploration tool of any importance. DeGolyer[2] has pointed out, however, that there is one surface indication which is still of primary importance in any area, namely, an oil well.

3.2 Geological Exploration Methods

A petroleum geologist's main job is to select promising sites for the drilling of exploratory wells based on his prediction of an area's subsurface stratigraphy and structure. In order to make these predictions he normally prepares maps, both surface and subsurface, on which known points are used to extrapolate the probable conditions at unknown points. Surface features such as elevations, dips and strikes of outcrops, and lithological changes may be mapped as clues to subsurface features. Aerial photographs also prove valuable in locating subsurface structures in many areas. The value of surface mapping is generally limited to shallow beds, as the deeper structures are not often reflected by surface features. Consequently, surface work is generally restricted to new areas. Aerial photographs are often used to locate the most promising of these which are then mapped in detail by field geologists.

The current depths to which exploratory wells are being drilled are such that the petroleum geologist must prepare maps from subsurface data in attempting to predict the conditions at these depths. Subsurface maps are numerous in variety and type; the following are, however, typical, basic forms.

1. Structural contour maps: maps composed of lines connecting points of equal elevation above or below a datum (normally sea level).
2. Isopachous maps — maps composed of lines connecting points of equal bed thickness.
3. Cross sections — a form of subsurface presentation which depicts the position and thickness of various strata.

In addition to being useful as an exploration tool, subsurface maps are a necessary part of any reservoir engineering study; and petroleum engineers, as well as geologists, must be completely familiar with their construction and interpretation.

The data for subsurface maps are obtained from a number of sources, such as

1. Well logs: representations of some rock property or properties versus depth. Some of these are listed below and will be discussed in later chapters.
(a) Sample logs
(b) Drilling time logs
(c) Electric logs
(d) Radioactivity logs
(e) Caliper logs
2. Core drilling: shallow, small hole drilling for information purposes only. The formations encountered are cored, i.e., obtained as small cylindrical samples which are readily and accurately identified.
3. Strat tests: deep exploratory holes drilled primarily for information.

The construction of subsurface maps requires great interpretive skill; when prepared by a competent worker, they are an extremely valuable tool.

3.3 Geophysical Exploration

Exploration methods falling in this category are those employing a physical measurement of subsurface conditions made from a surface location. The methods to be discussed briefly are

1. Gravitational
2. Magnetic
3. Seismic

3.31 Gravitational Methods

This type of geophysical prospecting is based on Newton's hypothesis that every particle in the universe attracts every other particle in the manner defined by the equation

$$F = \gamma \frac{m_1 m_2}{r^2}$$

where F = attractive force
$m_{1,\,2}$ = masses of particles in question
r = distance between particles
γ = gravitational constant (6.67×10^{-8} in cgs units)

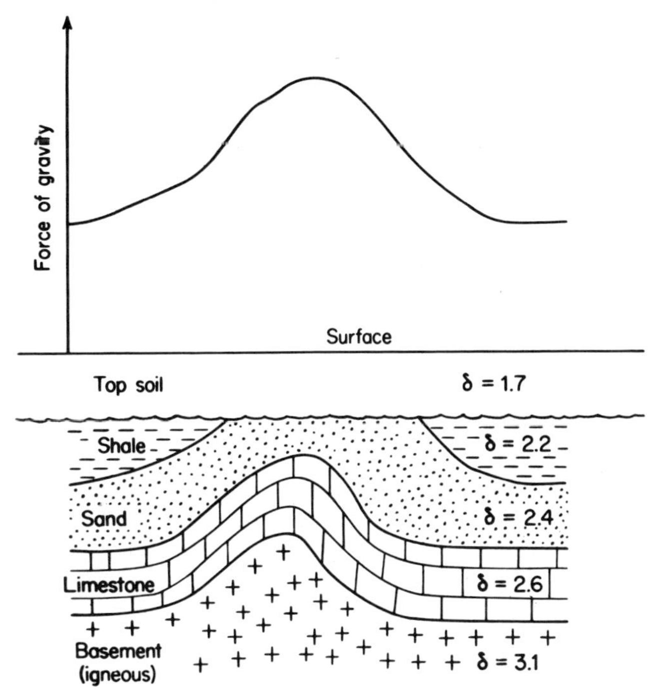

Fig. 3.1. Schematic illustration of gravity effect over a buried anticline. δ = specific gravity of various rocks. From Stommel,[3] courtesy Colorado School of Mines.

Consider Figure 3.1, which portrays a buried anticline. The denser rocks are closer to the surface and cause a higher gravitational force in their vicinity. Such an area, when mapped, shows a *gravity high* indicating a possible structure. The instruments used to record these small variations must be extremely sensitive and yet rugged and portable enough for field use. In recent years, the gravimeter has become the standard instrument for this use. It is essentially a weighing scale consisting of a spring supported mass. Variations in gravitational pull cause a slight vertical motion of the spring, which is greatly magnified by an intricate linkage; gravitational variations may be measured to approximately one part in 10 million with this instrument.

Other instruments used in gravity work are the torsion balance and the pendulum. These instruments, however, have been practically replaced by the gravimeter.

3.32 Magnetic Prospecting

This method seeks to map anomalies in the earth's magnetic field and to correlate these with underground structure. Sedimentary rocks are essentially nonmagnetic; consequently, any magnetic irregularities

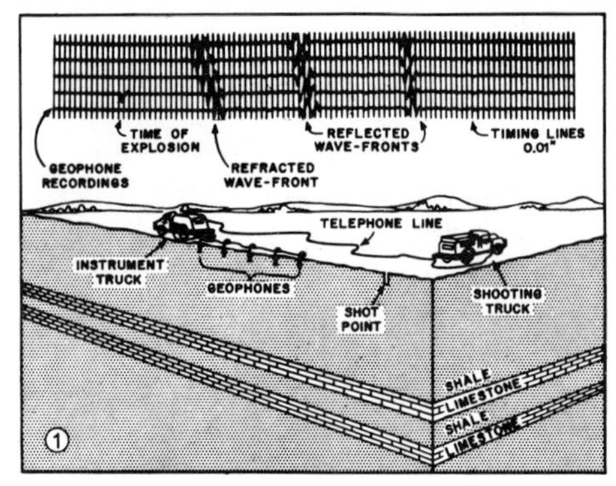

General setup of equipment and description of seismogram.

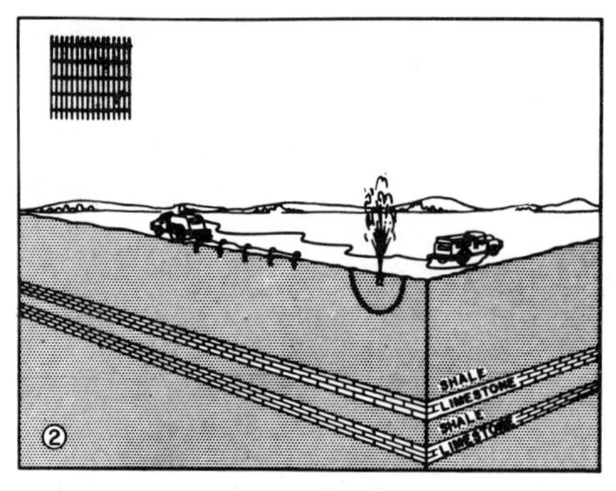

Notice the recording of explosion time and expanding wave front.

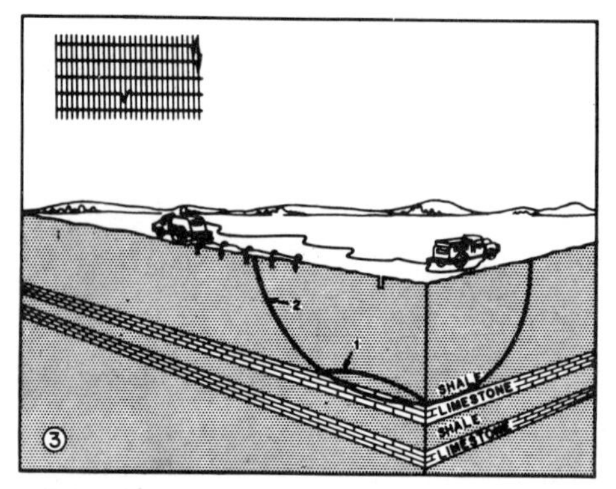

(1) Reflected wave front returning to surface.
(2) Refracted wave front has passed geophones.

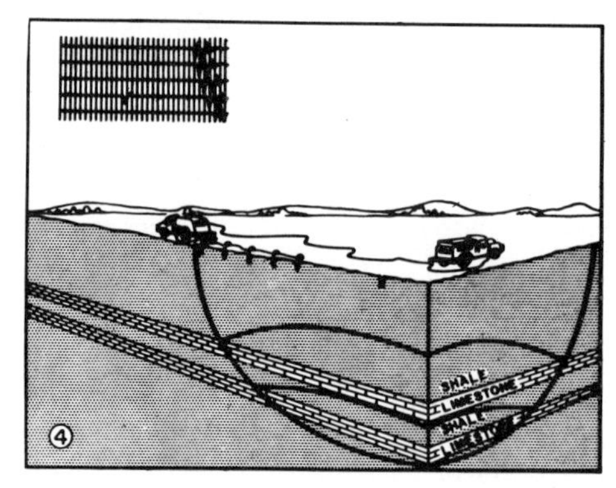

Wave front reflected from upper horizon returning to surface.

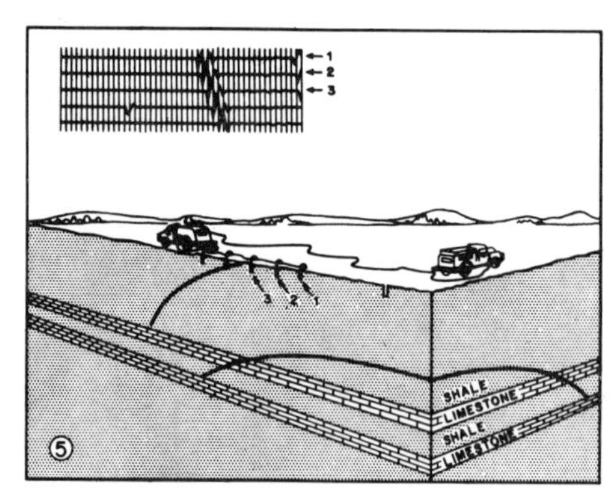

Wave front from upper horizon reaches geophones.

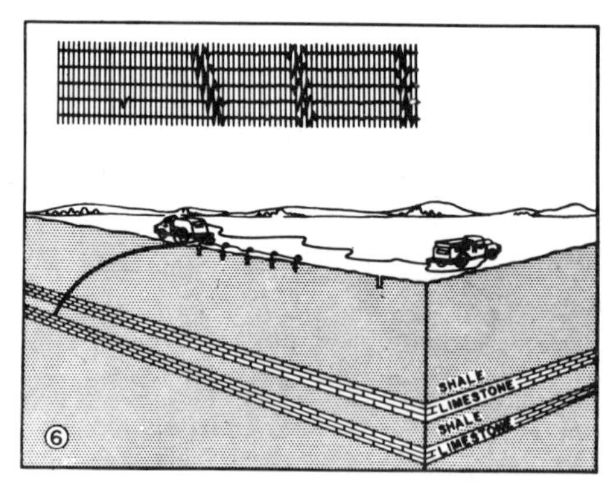

Reflected energy from lower horizon has been recorded.

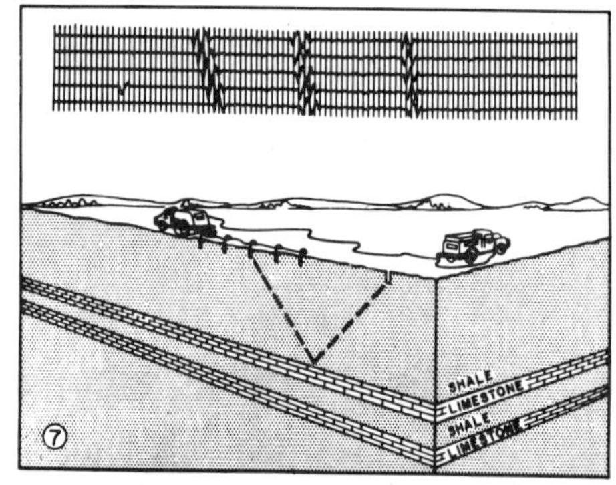

Small portion of wave front reaching geophone took an approximate "V" shaped path.

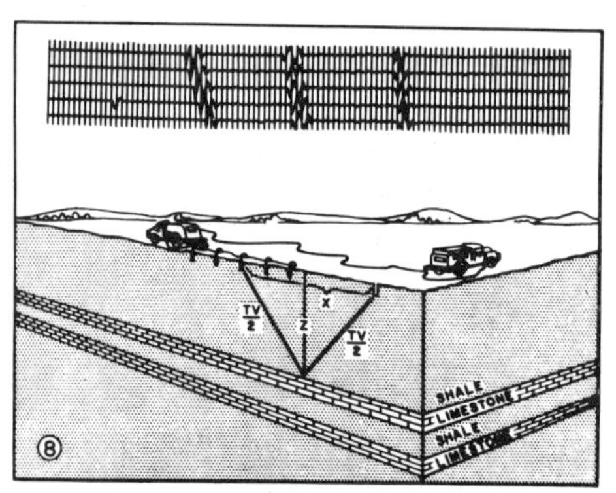

Depth calculation can be made from this equation

$$Z = \frac{1}{2}\sqrt{T^2V^2 - X^2}$$

where Z equals depth, T equals time, V equals velocity of travel, X equals distance between shot point and geophone.

Fig. 3.2. Sequence of events following seismic explosion illustrating reflection method. Courtesy Seismograph Service Corporation.

found are normally attributed to depth variations of basement rocks. The vertical component of the magnetic field is of primary importance in this method.

The instruments used in magnetic prospecting vary from a simple dip needle (essentially a vertical compass) to elaborate airborne magnetometers, by which large areas may be quickly and economically mapped. Numerous oil fields have been located by this method, a notable example being the Hobbs, New Mexico, pool. Present usage of this method is largely restricted to reconnaissance work.

3.33 Seismic Methods

The seismograph is the most successful and widely applied geophysical tool in exploration history. This method is based on the difference in propagation velocity of artificially induced elastic waves through various subsurface strata. These waves are generated by explosion of dynamite in shallow shot holes. A pattern of recording detectors arranged at known distances from the shot points record the arrival time of the refracted and reflected longitudinal waves induced by the explosion. If the wave paths, which follow the basic laws of optics, and wave velocities are known, their travel distances may be computed from the arrival times recorded by the detectors. This computed distance allows the depth and variations in depth of various strata to be computed. Variations in depth from a common surface elevation indicate structure. Figure 3.2 illustrates these principles.

Two basic seismic techniques exist, namely the refraction and the reflection methods. The refraction method had spectacular success in the 1924-30 period when it located many Gulf Coast salt dome fields. In this technique, the wave travelling along a boundary between rocks of different elastic properties is utilized. The reflection method uses the waves reflected from such boundaries. Currently, the reflection method is of principal importance; refraction shooting is used primarily as a reconnaissance tool to select areas and obtain interpretative data for the more detailed reflection method.

The main difference in instrumentation of the two methods lies in the distances from shot point to recorders. In refraction shooting, spacings of two to eight miles are used, while the reflection technique spacing is generally less than the depth to the first reflecting bed, commonly less than one mile.

The interpretation of seismic data requires the use of specialized, highly skilled personnel and involves the analytical application of mathematics, physics, and geology.

3.34 Geochemical Methods

Geochemical prospecting is based on the assumption that the hydrocarbons contained in an oil pool tend to migrate upwards, due to their low density; it is also assumed that eventually some molecules will reach the surface. Analyses of soil samples taken above known oil fields have, in many cases, shown a comparatively high percentage of hydrocarbons present. By the same line of thinking, higher than average chloride contents might be expected around the edge of the pool left by waters which have migrated upwards and evaporated. These variations are small and require extremely precise analysis techniques. Although these phenomena have been observed over known fields, attempts to find new fields in this manner have been generally unsuccessful; the method has little current application. It is of interest, however, because of its direct approach: i.e., it is searching for oil, not structure.

3.4 Leasing and Scouting Activities

Petroleum is legally a mineral. In the United States the landowner owns any minerals found beneath his land, providing, of course, that a previous owner has not sold or retained these mineral rights. Normally, the landowner does not desire to drill and develop his own property. Consequently, he leases his land to another party, commonly an oil company, who agrees to do this for him under the terms imposed by their particular agreement or lease.

3.41 The Oil and Gas Lease

According to Paine[4] an "oil and gas lease records a delegation by the owner of the minerals, called a lessor, to a lessee of the right to develop and produce from the property, subject to a rental payment of either money and/or a portion of the output." Oil operators usually attempt to use standard lease agreements, the terms of which have been specifically defined by court actions. However, it is not always possible to do this, and special lease agreements of almost infinite variety are in existence.

A necessary part of any lease is a complete legal description of the land to which it pertains. The United States General Land Office system is used in all areas of the country except the thirteen original states plus Tennessee, Kentucky, and Texas. In these states local methods are used. Departures from the standard method also exist in Louisiana, because of the difficulty of surveying the swamps and bayous. In the General Land Office system, the township is the largest unit of area and includes 36 sections, numbered as in Figure 3.3. The position of a township is defined by reference to a meridian (north-south line) and an east-west line called a base. Figure 3.4 is a map of Oklahoma which illustrates this system. For example, Township 25 North, Range 9 East is located in central Osage County.

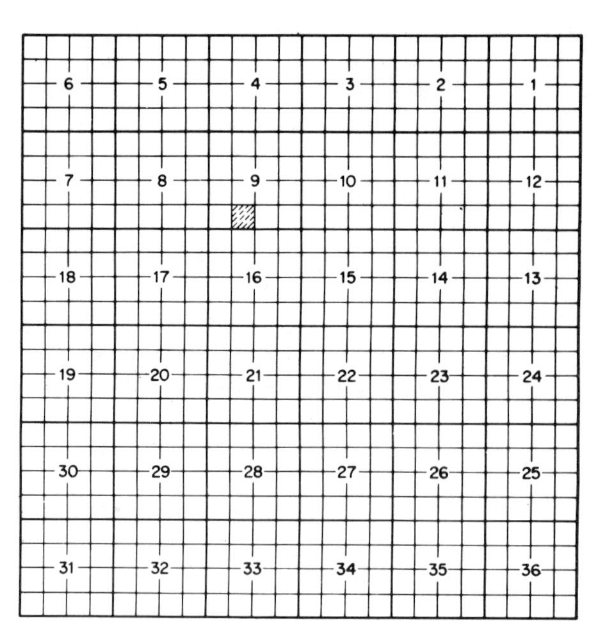

Fig. 3.3. Numbering system for sections within a township. The smallest squares cover 40 acres.

A 40 acre tract in section 9 of that township is shown by the hachured area in Figure 3.3. This could be described as the Southeast quarter of the Southwest quarter of section 9, or in an abbreviated form as the SE/4, SW/4, Sec. 9, T25N, R9E, Osage County, Oklahoma.

An oil and gas lease contains a number of clauses stating the duties and obligations of both lessor and lessee. Some of these are:

1. Term of the lease: The time period for which it is valid. This may require that a well be started within a certain period, or it may stipulate a rental payment in lieu of drilling. A "thereafter" clause is normally present, allowing the lessee to produce any oil and gas he finds, as long as it is profitable.
2. Development provisions: This clause is primarily for the lessor's protection and normally states the obligation of the lessee to drill offset wells to adjacent producers, thereby preventing drainage from under the lease in question.
3. Assignment clause: This sets forth the terms under which the lessee may trade or sell the lease. It is to the lessee's advantage to have a free hand in these matters; in some leases, however, all assignments must meet the approval of the lessor.

Other provisions are found in oil and gas leases: the surrender clause, allowing the lessee to give up the lease and be free of further obligation; clauses stating the amount of the royalty payments, tax obligations, distribution of any unusual handling costs, and numerous other liabilities and obligations of lessor and lessee. For an interesting coverage of these topics the reader is referred to Paine's book.[4]

3.42 Types of Oil and Gas Ownership

There are two general categories of ownership of oil and gas: operating interests, and royalty interests.

The operating interest is that owned by the lessee who is responsible for the proper operation of the property. This interest bears the cost of producing the oil and maintaining the lease.

The royalty interests normally pay no part of the operating costs and receive the full revenue from their share of the oil. The original royalty form is the landowner's royalty which is the lessor's consideration. This is commonly one-eighth of the production. Another frequent royalty form is the overriding royalty. Quite often, an overriding royalty is given to a lease broker as his compensation for acquiring the lease. The principal difference between these two types of royalties is that the overriding royalty is affixed to a particular lease and expires with that lease, while the landowner's royalty will be perpetuated as a mineral interest and may reappear in subsequent leases.

It is obvious that a royalty interest is worth more than the same fraction of the operating interest. Royalty interests must, however, pay their proportionate share of taxes, and in some cases bear their share of transportation charges.

Both forms of ownership are often subdivided by numerous trades and sales. For example, a person may buy one undivided royalty acre under a section. He then owns 1/640th of 1/8th or 1/5120th of the oil produced from that section. While this latter fraction may not seem to be very large, its value under the Long Beach Field, which has produced over 500,000 barrels per acre, is considerable.

3.43 Scouting Activities, Trading and Promotion

As stated earlier, the right of assignment is a necessary part of an oil and gas lease, for without it one of the most lively segments of the industry would be missing. It behooves any oil operator to keep track of his competitor's operations. To achieve this many companies employ scouts whose job it is to report on all activities in their areas.

Consider a case in which a company starts to drill a wildcat in a location where considerable unleased acreage is available. Immediately scouts, landmen, and promoters of all kinds descend on the area. The operator drilling the well may classify it as a tight hole which means that no information on the well is to be given out. He may also fence in the location and allow no one to enter. Naturally these precautions serve to heighten interest in the well.

But despite precautions, the desired information nearly always leaks out, although it may be extremely difficult to sift fact from fancy. Scouts may still park

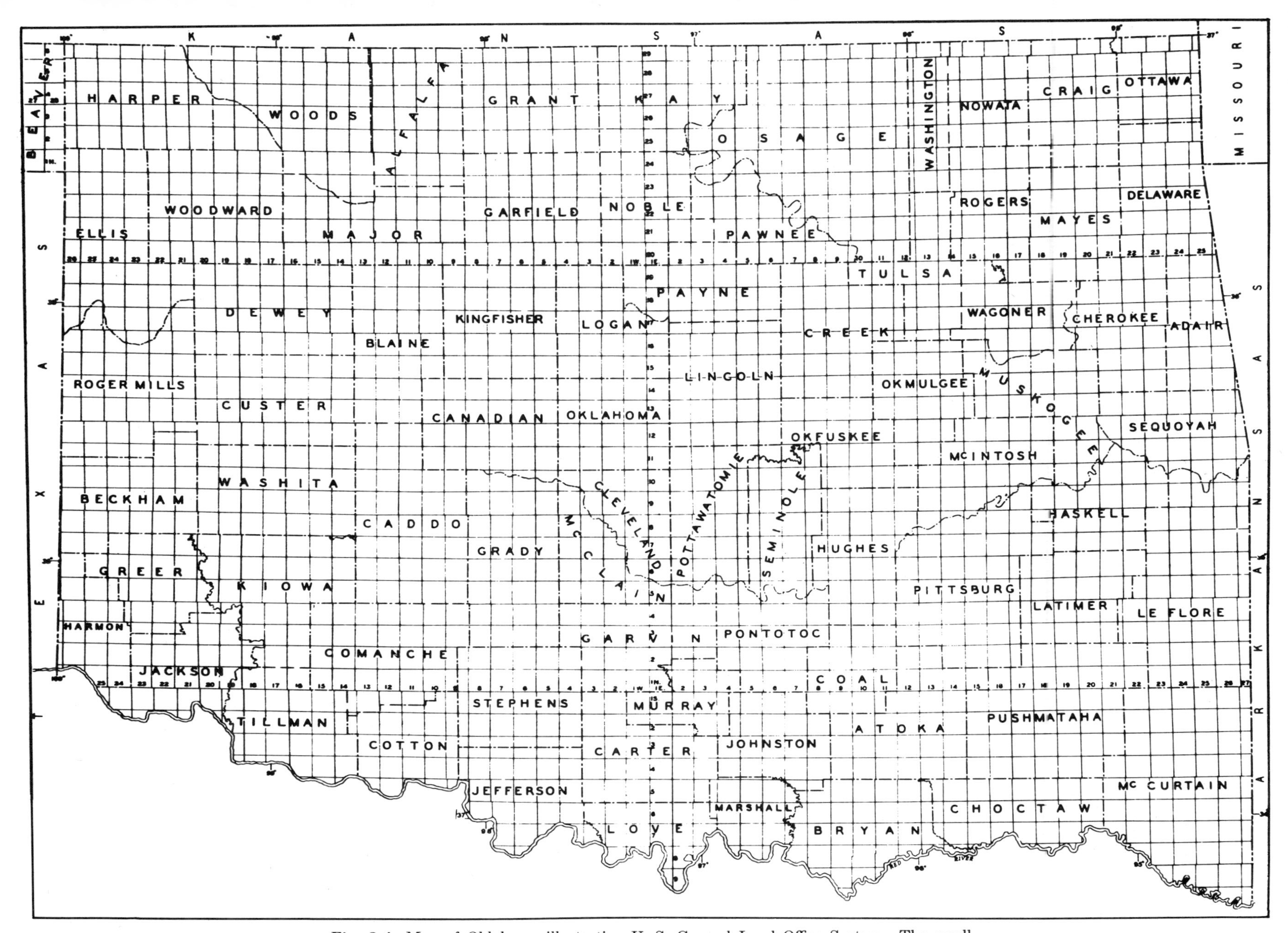

Fig. 3.4. Map of Oklahoma illustrating U. S. General Land Office System. The small squares are townships of 36 square miles each.

in the nearest public road and watch testing operations through field glasses. Crew members are often approached and persuaded to divulge information. In extreme cases, telephones have been tapped.

Many small companies operate on farmouts from major companies; this arrangement generally involves the subleasing of part of a large acreage block to the small company, providing they will drill an exploratory well. This is one way for the large company to meet a drilling requirement, and if production is found they will still hold substantial acreage in the area.

Many times, operators share the expense of drilling an exploratory well by making "dry hole" contributions to the company doing the drilling. This contribution is commonly so much per foot, and is paid only if the well is dry. In contrast to this is "bottom hole" money, which is paid whether the well is dry or not.

For those who wish to pursue these topics further, a number of advanced readings are listed at the end of this chapter.

REFERENCES

1. Lahee, F. H., "Exploratory Drilling in 1955," *Bulletin of American Association of Petroleum Geologists*, Vol. 40, No. 6, June, 1956, pp. 1057–1075.
2. DeGolyer, E., "Direct Indications of the Occurrence of Oil and Gas," in *Elements of the Petroleum Industry*, ed. E. DeGolyer. New York: AIME, 1940, pp. 21–25.
3. Stommel, H. E., "Subsurface Methods as Applied to Geophysics," in *Subsurface Geologic Methods*, 2nd ed., ed. L. W. LeRoy. Golden: Colorado School of Mines, 1951. Ch. 14, pp. 1038–1119.
4. Paine, P., *Oil Property Valuation*. New York: John Wiley & Sons, Inc., 1942, p. 20.

SUPPLEMENTARY READINGS

1. LeRoy, L. W., "Graphic Representations," in *Subsurface Geologic Methods*, 2nd ed., ed. L. W. LeRoy. Golden: Colorado School of Mines, 1951, pp. 856–893. Shows numerous examples and methods of subsurface representations.
2. Dobrin, M. B., *Introduction to Geophysical Prospecting*. New York: McGraw-Hill Book Co., Inc., 1952. Excellent survey of geophysical prospecting designed for readers other than geophysicists.
3. Paine, P., op. cit. Ch. II, pp. 13–43. Covers types of property, ownership, and leases.
4. Sullivan, R. E., *Handbook of Oil and Gas Law*. Englewood Cliffs, N. J.: Prentice-Hall, Inc., 1956. A complete treatment of the legal aspects of the petroleum industry with emphasis on production.
5. DeGolyer, E., ed., *Elements of the Petroleum Industry*. New York: AIME, 1940; pp. 90–115. Covers leasing, trading and promotion, and royalties.
6. Campbell, John M., *Oil Property Evaluation*. Englewood Cliffs, N. J.: Prentice-Hall, Inc., 1959.

Chapter 4 . . .

Cable Tool Drilling

4.1 Introduction

The first oil well in the United States was drilled with cable tools in 1859 to a depth of 65 feet. This was the historic Drake well located near Titusville, Pennsylvania; it is credited with having started the American petroleum industry. The cable tool (also called churn or percussion) drilling method, however, did not originate in this country, but is believed to have been employed first by the early Chinese in the drilling of brine wells.[1]

In this method, drilling is accomplished by the pounding action of a steel bit which is alternately raised by a steel cable and allowed to fall, delivering sharp, successive blows to the bottom of the hole. This principle is the same as that employed in drilling through concrete with an air hammer, or in driving a nail through a board.

The original percussion drilling apparatus consisted of a spring pole anchored into the ground at an angle, with the bit suspended from the free end by a rope. To impart the necessary reciprocating action to the bit, the Chinese employed a number of men who alternately jumped on and off the spring pole beam from a ramp. Many early brine wells in the United States were drilled in the same manner, except that the spring pole was equipped with stirrups where two or three men stood and literally kicked the well down.

As more and deeper wells began to be drilled, efforts were made to improve the drilling equipment. Steam engines began to be used; walking beams replaced the spring pole; steel cables replaced manila ropes; and other improvements followed. Although the modern cable tool rig is a far cry from the ancient Chinese model, the changes have been in materials and equipment, for the basic operating principle is unchanged.

4.2 Equipment and Basic Technique

To discuss the equipment and technique of this method we will refer to Figure 4.1 which shows the American

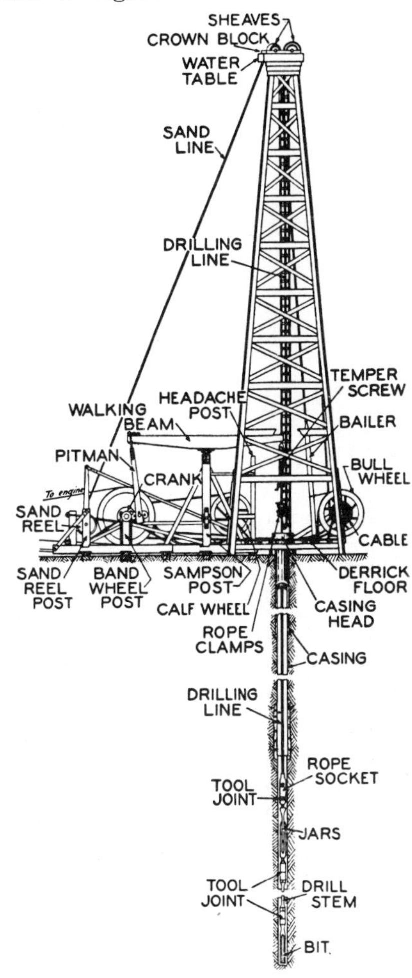

Fig. 4.1. American standard cable tool rig. After Brantly,[2] courtesy AIME.

standard cable tool rig. Although rigs of this type have been almost completely replaced by lighter, more portable models, the standard rig is still the yardstick of percussion drilling.

4.21 The Drill String

The drill string of a cable tool rig is composed of the bit, drill stem, jars, and a rope socket enabling their attachment to the drilling line or cable. The main parts of the drill string are

1. The drill bit: A heavy steel bar, generally four to eight feet long, having the lower or drilling end dressed to varying degrees of sharpness depending on the formation to be drilled. Sharper bits are used in hard rock drilling while soft rock bits are quite blunt. Cable tool bits are made from high carbon and molybdenum-silicon alloy steels in a number of patterns by various manufacturers. Naturally these bits require frequent sharpening or dressing which is either performed at the well by the driller and the tool dresser or in the nearest blacksmith shop. Two typical cable tool bits are shown in Figure 4.2.

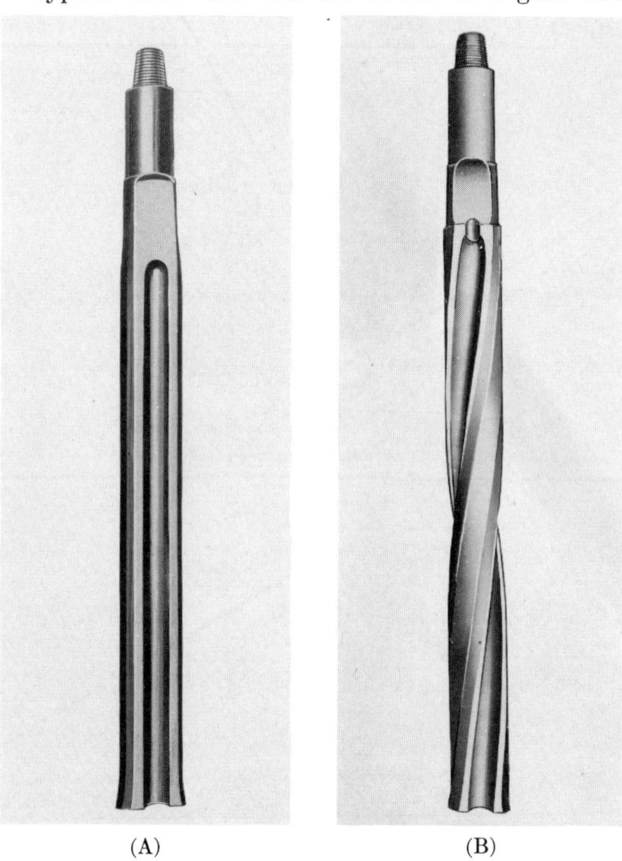

Fig. 4.2. Typical cable tool bits. Courtesy Spang and Company. **(A)** Straight regular pattern. **(B)** Twisted Mother Hubbard pattern.

2. The drill stem: A cylindrical steel bar generally 10 to 20 feet long which is screwed directly above the bit. Its diameter depends on the hole size and the amount of weight desired. The purpose of this member is to furnish additional weight for the downward drilling blow.
3. Jars: Heavy steel links which telescope within each other much like two links in a chain. Their function is to produce a sharp upward blow on the tools, causing them to be jerked loose from soft, sticky formations, and allowing a clean, sharp drilling blow. Long stroke jars having two to six feet of telescope action are often used in fishing jobs (retrieving of tools, etc. lost or stuck in the hole). Drilling jars normally have strokes of less than one foot and are often omitted in hard rock drilling. This piece of equipment is responsible for the affectionate title *Jarheads* which rotary drillers often apply to their cable tool counterparts.
4. Tool joints: Connectors for the bit, drill stem, etc. These consist of tapered, coarse threaded connectors machined on the ends of the tools. The thread design allows easy makeup, and the necessary tightness is obtained from the metal-to-metal fit at the flat shoulders of the joint. Proper tightness of these joints is essential to prevent the severe drilling vibrations from unscrewing the tools.

The drilling line is also part of the drill string, but for convenience it is discussed in the next section.

4.22 Rig Lines

The standard rig has three lines or cables used for various purposes. These are the drilling line, the sand line, and the calf or casing line. Before discussing the individual functions of these, let us look at wire lines in general.

A steel cable is composed of a number of wire strands wound helically around a hemp or independent wire rope center with a uniform length of lay. A lay is the length of rope required for individual strands to make one revolution about the center and is further specified as right, left, or Lang lay, as shown in Figure 4.3.

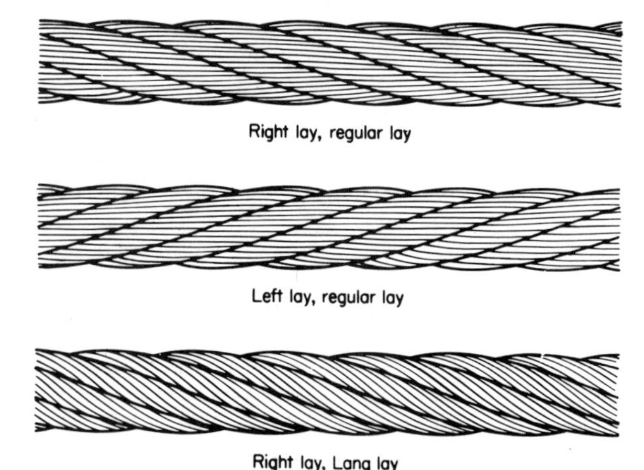

Fig. 4.3. Illustration of right and left, regular and Lang lay. Courtesy API.[3]

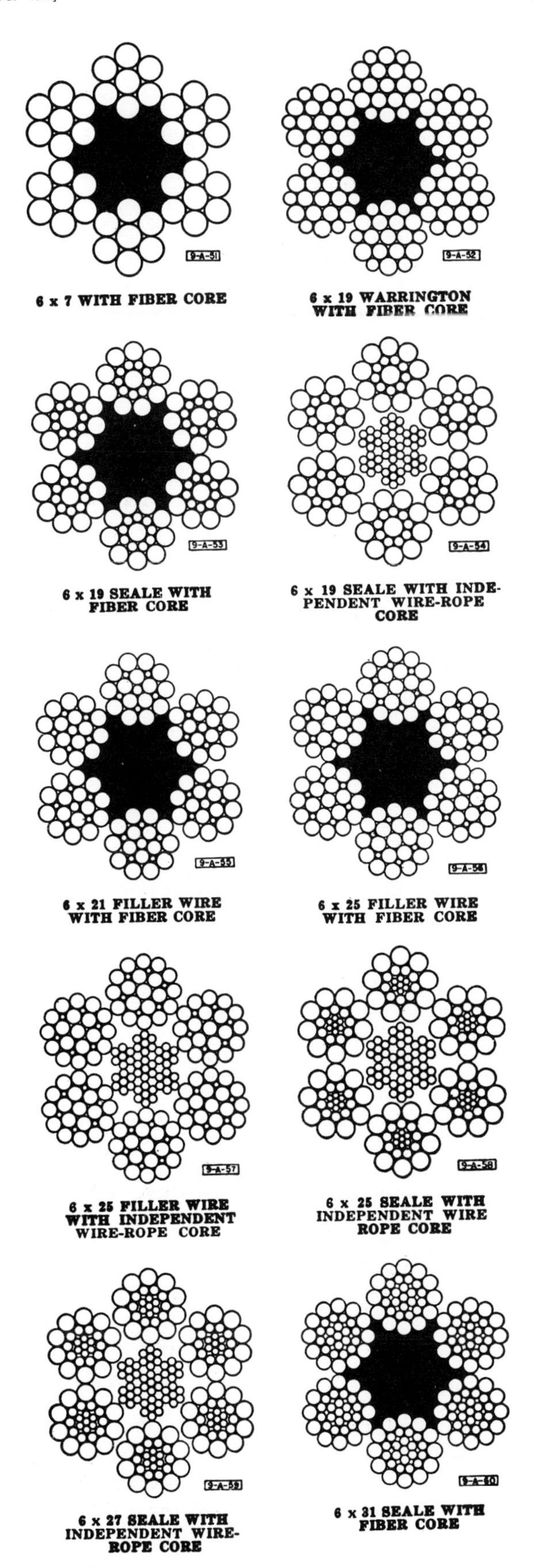

Fig. 4.4. Typical strand constructions for oil field wire ropes. Courtesy API.[3]

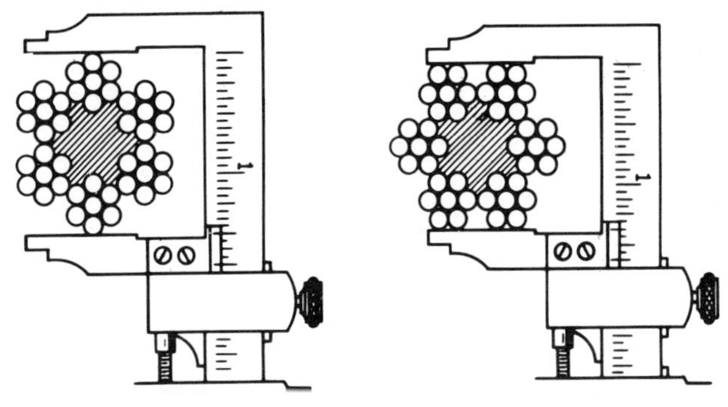

Fig. 4.5. Measurement of wire line diameter. Left is correct method. Courtesy API.[3]

Typical strand constructions used in oil field service are shown in Figure 4.4. Wire lines are specified by the nominal diameter as shown in Figure 4.5.

The net steel area of several strand constructions may be computed from the data of Table 4.1. Table 4.2 lists the wire line types commonly used for oil field purposes.

Detailed data and recommended practices for all types of wire ropes may be found in the API Standards.[3,4]

TABLE 4.1

FACTORS FOR COMPUTATION OF NET STEEL AREAS OF VARIOUS WIRE ROPES*

	C	
Strand construction	*Fiber core*	*Independent wire rope center*
6×7	0.380	0
6×19	0.405	0.455
6×31	0.400	0.460
6×37	0.400	0.460

*Net metallic area of any cable $= Cd^2$, where d = nominal rope diameter, in. Table reproduced by courtesy of the Bethlehem Steel Co.

Drilling Lines

The cable tool drilling line is spooled on the bull wheel, passed over a sheave at the top of the derrick, and suspended from the walking beam by the temper screw where the driller pays it out as drilling progresses. This line is subjected to extremely severe service. It must not only support the tools plus its own weight, but is also subjected to the jarring, fluctuating loads of drilling as well as severe abrasion against the casing and hole wall. Consequently, its life is comparatively short despite the extreme care given it.

The most common type of cable used for this service is a 6×19 fiber core plow or improved plow steel cable. Sizes of $\frac{3}{4}$ to 1 in. are normally used although larger tapered strings may be found on deep wells. Tapered strings consist of sections of different line sizes with the largest at the top and the smallest at the bottom, such as a $1\frac{1}{8} \times 1 \times \frac{7}{8} \times \frac{3}{4}$ in. line. This practice allows lower beam loads and puts the heavy line at the top where the tensile load is maximum.

TABLE 4.2

TYPICAL SIZES AND CONSTRUCTIONS OF WIRE ROPE FOR OIL-FIELD SERVICE*

1	2	2
Service and Well Depth	*Wire-Rope Diameter, in.*	*Wire-Rope Description*
Rod and Tubing Pull Lines		
Shallow	$\frac{1}{2}$ to $\frac{3}{4}$ incl.	6 x 31 or 18 x 7[1] or 6 x 25 FW,
Intermediate	$\frac{3}{4}$, $\frac{7}{8}$	PS or IPS, PF or NPF,
Deep	$\frac{7}{8}$ to $1\frac{1}{8}$ incl.	LL[2], FC or IWRC
Rod Hanger Lines	$\frac{1}{4}$	6 x 19, IPS, PF, RL, FC
Sand Lines		
Shallow	$\frac{1}{4}$ to $\frac{1}{2}$ incl.	6 x 7
Intermediate	$\frac{1}{2}$, $\frac{9}{16}$	PS or IPS, Bright or Galv.,[3]
Deep	$\frac{9}{16}$, $\frac{5}{8}$	PF or NPF, RL, FC
Drilling Lines — Cable Tool (Drilling and Cleanout)		
Shallow	$\frac{5}{8}$, $\frac{3}{4}$	6 x 21 FW or 6 x 19 Seale,
Intermediate	$\frac{3}{4}$, $\frac{7}{8}$	PS or IPS, NPF,
Deep	$\frac{7}{8}$, 1	RL or LL, FC
Casing Lines — Cable Tool		
Shallow	$\frac{3}{4}$, $\frac{7}{8}$	6 x 25 FW,
Intermediate	$\frac{7}{8}$, 1	IPS, PF or NPF,
Deep	1, $1\frac{1}{8}$	RL, FC or IWRC
Drilling Lines — Coring and Slim-Hole Rotary Rigs		
Shallow	$\frac{7}{8}$, 1	6 x 19 Seale or 6 x 25 FW,
Intermediate	1, $1\frac{1}{8}$	PS or IPS, PF or NPF, RL, IWRC or FC
Drilling Lines — Large Rotary Rigs		
Shallow	1, $1\frac{1}{8}$	6 x 19 Seale or 6 x 21 FW,
Intermediate	$1\frac{1}{8}$, $1\frac{1}{4}$	PS or IPS, PF or NPF,
Deep	$1\frac{1}{4}$ to $1\frac{1}{2}$ incl.	RL, IWRC or FC
Winch Lines — Heavy Duty	$\frac{5}{8}$ to $\frac{7}{8}$ incl.	6 x 31 Seale or 6 × 21 FW, IPS, PF, RL, IWRC or FC
Horsehead Pumping-Unit Lines		
Shallow	$\frac{1}{2}$ to $1\frac{1}{8}$ incl.[4]	6 x 37 or 18 x 7[1] or 8 x 19 IPS, PF, IWRC
Intermediate	$\frac{5}{8}$ to $1\frac{1}{8}$ incl.[5]	6 x 19 IPS, PF, IWRC

Abbreviations:

FW — Filler-wire construction
PS — Plow steel
IPS — Improved plow steel
PF — Preformed
NPF — Non-preformed
RL — Right lay
LL — Left lay
FC — Fiber core
IWRC — Independent wire rope core

[1] 18 x 7 wire rope is non-rotating and is furnished in right lay and fiber core construction only.

[2] Single line pulling of rods and tubing requires left lay construction. Either left lay or right lay may be used for multiple line pulling. See footnote 1.

[3] Bright wire sand lines are regularly furnished; galvanized finish is sometimes required.

[4] Applies to pumping units having one piece of wire rope looped over an ear on the horsehead and both ends fastened to a polished-rod equalizer.

[5] Applies to pumping units having two vertical lines (parallel) with sockets at both ends of each line.

*From the *Tool Pusher's Manual*, Courtesy American Oil Well Drilling Contractors Association

Sand Lines

The sand line is spooled on the sand reel, passed over a sheave at the top of the derrick, and is normally attached to a bailer which stands vertically to one side of the derrick floor while drilling is underway. Periodically, the drill string must be removed from the hole and the bailer run to remove accumulated fluid and cuttings. The load and service requirements on the sand line are very light compared to the drilling line, allowing smaller and less expensive lines to be used for this purpose. Normally, $\frac{7}{16}$ to $\frac{5}{8}$ in., 6 × 7 construction, plow steel cables are used as sand lines.

Calf Lines

The calf line is spooled on the calf wheel and is used to run casing into the well. Consequently, it may be subjected to the greatest loads of any rig line. This loading is not, however, as severe as the drilling service and may be more accurately predicted. Line loads may also be controlled within desired limits by varying the number of lines strung between the block and derrick.

Computations of this type will be illustrated in a later chapter. Calf lines are normally $\frac{3}{4}$ to 1 in. diameter, 6×25 construction, improved plow steel cables.

The upkeep and replacement of these rig lines represents a large expense item and every effort is made to insure their long life.

4.23 Surface Equipment

Some of the surface equipment of a standard rig was mentioned in connection with items previously discussed. The bull and calf wheels and the sand reel, which house or spool their respective lines, have been mentioned. These wheels are merely reels, properly constructed to handle the loads and line lengths for which they are rated. The walking beam is supported by the sampson post, and imparts the reciprocating motion to the drilling line. It is connected to the large band wheel by the crank and pitman. The band wheel is in turn rotated by a belt drive from the prime mover.

Prime Movers

The first prime movers used in cable drilling were steam engines; currently, though, the internal combustion engine is the most common. Electric motors have been used but never to any great extent. The single cylinder, heavy flywheel steam engine, while satisfactory for this purpose, is bulky, requires large quantities of reasonably high quality water, and has a low overall efficiency. On account of these disadvantages, it has been largely replaced by multi-cylinder, internal combustion engines. The ability of light multi-cylinder engines to follow the load (race on downstroke, slow on pickup stroke) has been cited as an aid to faster drilling. Any of the above types are satisfactory for this service and the choice is generally governed by the operating costs in the area of use.

Bailers and Sand Pumps

As mentioned before, it is necessary to periodically remove the accumulated cuttings from the well. To accomplish this, a pipe equipped with a bail at the top and a valve at the bottom is run into the hole on the sand line. The bottom valve, generally a disc or flapper type, is opened and closed by a protruding stem which strikes the bottom of the hole as the bailer is alternately raised and lowered. This motion causes the cuttings to wash into and be trapped within the bailer, where they remain until dumped at the surface. Sometimes, if the cuttings are coarse, it is necessary to use a sand pump to recover them. This device is similar to a bailer except that it incorporates a piston which creates sufficient vacuum to suck the cuttings inside. The inspection of these cuttings gives valuable geological and operating information.

Derricks

The derrick is the structure which affords the vertical clearance necessary for conducting such drilling operations as withdrawing the tools and running casing. It must be strong enough to withstand the loads imposed upon it. These loads vary with depth and the size of hole and casing. The ratings of derricks and computation of loads will be discussed under rotary drilling equipment.

4.24 Portable Cable Tool Rigs

As previously mentioned, the American standard rig of Figure 4.1 is no longer in general use. Contractors, in an attempt to reduce operating costs, have switched to portable, unitized rigs such as shown in Figure 4.6. On this particular portable rig the three reels have been retained and the travelling block is tied out of the way when not in use. The prime mover and drawworks

Fig. 4.6. Portable cable tool rig drilling well in Kentucky. Courtesy Bucyrus-Erie Company.

(drums plus controls) are close coupled, and the entire rig is mounted either on a skid or trailer for maximum portability. Standard derricks have been replaced by telescoping double pole masts which may be laid down on top of the rigs for hauling. Masts of this type which can handle loads of 170,000 pounds are available.

Portable rigs capable of drilling to 7500 feet or performing workover jobs to 10,000 feet are in use. The principal advantages of portable equipment, as the name implies, stem from savings in transportation costs and rig-up time. Many improvements in materials, prime movers, mode of power transmission, clutches, brakes, and other controls have made it possible for portable equipment to do the job in a generally more efficient and economic manner.

4.3 Drilling Technique

It would appear that for a given formation and depth, there should be an optimum stroke length, speed, and tool weight at which the maximum drilling rate is obtained. These conditions must also fall within the safe and economic operating limits of the rig equipment. Normally, the choice of these rests with the driller who makes such adjustments according to his experience. The fact that some drillers make more footage than others implies that the optimum conditions are not always held to.

In the past, most analyses of these problems have assumed that the best operating speed (strokes per minute) is that which is in resonance with the natural frequency of the drill string. At these speeds the amplitude of the downhole stroke, and hence tool velocity, is theoretically a maximum. Obviously, maximum velocity implies the hardest blow on the formation. We shall first look at this approach.

4.31 Harmonic Vibration Approach

Sprengling and Stephenson[5] have presented the following formula for the natural vibration frequency of a cable tool drill string:

$$F = 93.9\sqrt{\frac{eA}{LW_s}} \tag{4.1}$$

where F = natural vibration frequency, cycles/minute
e = modulus of elasticity of the drilling line, psi
A = cross-sectional area of steel in line, in.^{2}
$W_s = 3\,W_t + W_l$
W_t = tool weight, lb, (including bit, drill stem, etc.)
W_l = weight of drilling line
L = length of cable (or depth of hole), ft

This is the same equation as that presented by Mills[6] which defines W in a different manner.

$$F = 54.3\sqrt{\frac{eA}{LW_s'}} \tag{4.2}$$

where W_s' = tool weight + $\frac{1}{3}$ line weight, lb

The derivation of these equations may be found in any engineering mechanics text,[7] and are actually forms of

$$f = \sqrt{\frac{kg}{W}} \tag{4.3}$$

where f = natural frequency of a spring supported weight under free vibration
k = spring constant
g = gravitational constant
W = weight supported by the spring

By this theory, the drilling speed should be the natural frequency of the system or an even multiple or fraction of it, i.e., $\frac{1}{4}$, $\frac{1}{2}$, 1, 2, 4, etc. Practical considerations generally limit the actual drilling speed to 20–40 spm and a value within this range satisfying Eq. 4.1 is chosen.

As pointed out by Griffith,[8] the greatest amplitude does not necessarily create the most economic drilling conditions. It is quite possible that shortened bit and cable life may more than offset any gains in rate of penetration. Further, the indeterminate damping effects of well fluids and wall friction greatly reduce the bit travel. In the similar problem of selecting sucker rod pumping speeds, most operators ignore resonance, which appears to exert a negligible effect in most cases.

The harmonic vibration analysis leaves much indeterminate, for instance proper tool weight and stroke length. It is not necessarily correct to assume that harmonic speeds give the most economic drilling performance.

4.32 Bonham's Analysis

In an attempt to solve the cable tool problem Bonham[9] has analyzed the drilling cycle using entirely different assumptions. The following development is taken largely from his paper as it originally appeared.

This analysis begins at the instant the tools have struck the formation at the end of the drop stroke, and just before the pickup begins. At this instant, the tension in the drilling line must be sufficient to lift and accelerate the tools upward. The following nomenclature will be used:

A = net steel area in cable, in.^{2}
d = cable diameter, in.
D = bit or hole diameter, in.
S = cable stretch, ft
e = modulus of elasticity of drilling line, 15×10^6 psi
g = 32.2 ft/sec^2
G = net striking energy of tools at impact, ft-lb

h = height of free fall, ft
L = cable length, approximately the hole depth, ft
N = drilling speed, strokes per minute
S_w = static stretch of cable due to tool weight, W, ft
S_l = portion of drilling stroke through which tools fall under restraint, ft
S_y = portion of drilling stroke through which tools fall without restraint, ft
t = time, sec
T = cable tension, lb, at any point and time
V = tool velocity, ft/sec
w = cable weight, lb/ft
W = net weight of tools (gross less buoyant effect of drilling fluid)
K = T/W, ratio of cable tension to tool weight at any time
$\tan\theta$ = spring constant for drilling line, lb/ft

Figure 4.7 portrays conditions at the instant of impact; the load T is measured at the rope socket just above the tools. The line 1–3–4 is a load-deflection curve for the cable during the S_l portion of the drilling stroke, i.e., the restrained fall portion.

Optimum Tool Weight

From Figure 4.7:

$$(4.4)\qquad S_l = \frac{KW}{\tan\theta} = \frac{KWL}{eA}$$

where $\tan\theta = \frac{eA}{L}$ *

Similarly,

$$(4.5)\qquad S_w = \frac{W}{\tan\theta} = \frac{WL}{eA}$$

Consider the area under the load-deflection curve as representing the energy stored in the drilling cable by its stretching, or as the work done on the cable by the load W.
Then:

$$(4.6)\qquad \text{Area 1–2–3} = (W/2)\times S_w = \frac{W^2L}{2eA} = \text{Area 1–6–3}$$

Also

$$(4.7)\qquad \text{Area 2–3–4–5} = \frac{W+KW}{2}(S_l - S_w)$$

Substitution of (4.4) and (4.5) in (4.7) gives

$$(4.8)\qquad \text{Area 2–3–4–5} = \frac{W+KW}{2}\times\frac{WL\,(K-1)}{eA} = 0, \text{ when } K = 1.$$

*Recall that

$$e = \frac{\text{Stress}}{\text{Strain}} = \frac{KW/A}{S_l/L} = \frac{KWL}{AS_l} \text{ or } S_l = \frac{KWL}{eA}$$

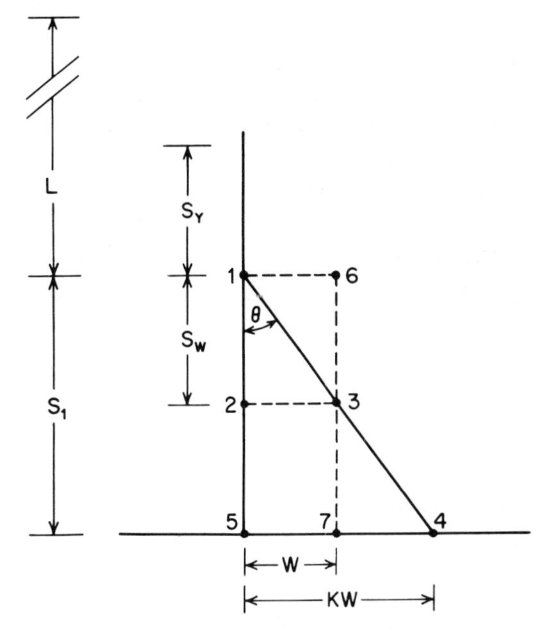

Fig. 4.7. Load-deflection diagram for drilling line at impact with pickup impending. After Bonham,[9] courtesy *Petroleum Engineer.*

When $K > 1$,

$$(4.9)\qquad \text{Area 1–5–4} = \frac{KW}{2}\times S_l = \frac{K^2W^2L}{2eA}$$

The net area, or net energy available for useful work is

$$(4.10)\qquad G = W(S_l + S_y) - \frac{K^2W^2L}{2eA}$$

Or,

$$S_l + S_y = \frac{G}{W} + \frac{K^2WL}{2eA} = \text{stroke of tools necessary to produce net energy input of } G \text{ ft-lb to formation}$$

Equation (4.10) contains seven variables; the following restrictions, however, may be made:

A, e = constant for a given cable
L = slowly varying quantity and may be assumed constant to study conditions at any particular depth
K, G = f (drilling speed, tool weight, etc.). These will be discussed later, but will be considered constants for the time being.

Then we take the derivative of $(S_l + S_y)$ with respect to W

$$(4.11)\qquad \frac{d(S_l + S_y)}{dW} = \frac{-G}{W^2} + \frac{K^2L}{2eA}$$

Equating Eq. (4.11) to zero and solving for W, we get

$$(4.12)\qquad W = \pm\sqrt{\frac{2GeA}{K^2L}}$$

Since $d^2(S_1 + S_y)/dW^2 = 2G/W^3$ which is positive

and $\neq 0$, the function of Eq. (4.11) has a minimum value where $W = W_0 = \sqrt{2GeA/K^2L}$ ·
Equation (4.12) permits the calculation of tool weight for the shortest possible stroke for any energy input, G, at any depth, L. Bonham logically assumes that the minimum value of $S_1 + S_y$ for any condition will result in minimum cable wear and allow the maximum strokes per minute giving the greatest economy and footage. This is the reverse of the harmonic approach which implies maximum cable stretch.

Minimum Stroke Length

The minimum value of $S_l + S_y$ is

$$(S_l + S_y)_{\min} = \frac{G}{W_0} + \frac{K^2W_0L}{2eA} \tag{4.13}$$

Substitution of Eq. (4.4) in Eq. (4.13) for $W = W_0$ gives

$$S_y = \frac{G}{W_0} + \frac{W_0L}{eA}\frac{(K^2 - 2K)}{(2)} \tag{4.14}$$

For the special case when $K = 2$

$$S_y = \frac{G}{W_0} \tag{4.15}$$

Rewriting Eq. (4.10) for the condition $S_1 + S_y = (S_1 + S_y)_{\min}$ and $K = 2$,

$$(S_l + S_y)_{\min} = \frac{2G}{W_0} \tag{4.16}$$

Therefore, for this case,

$$S_l = S_y = \frac{G}{W_0} \tag{4.17}$$

Assumptions and Restrictions

The following restrictions and assumptions will be made:

1. $K = 2$. This means that the cable tension at the start of the pickup stroke is sufficient to give the tools an upward acceleration of g.
2. The upper limit for G is fixed at 1325 D. This is arbitrarily chosen as the limit imposed by bit life between sharpenings, and must be evaluated for different formations from field experience.

Horsepower Input to Formation

(4.18) Work per unit time $= GN = 1325\ DN$

The input horsepower to the formation may therefore be expressed as:

$$\text{HP} = \frac{GN}{33{,}000} = \frac{1325\ DN}{33{,}000} \tag{4.19}$$

Drilling Speed

If it is assumed that the drilling motion is simple harmonic, a simple relationship between stroke length and drilling speed may be obtained for the case where $K = 2$. Consider Figure 4.8 which schematically shows the tools of weight W being accelerated upward by the cable tension T. It is obvious that

$$T = W + \frac{W}{g} \times a = W(1 + a/g) = KW$$

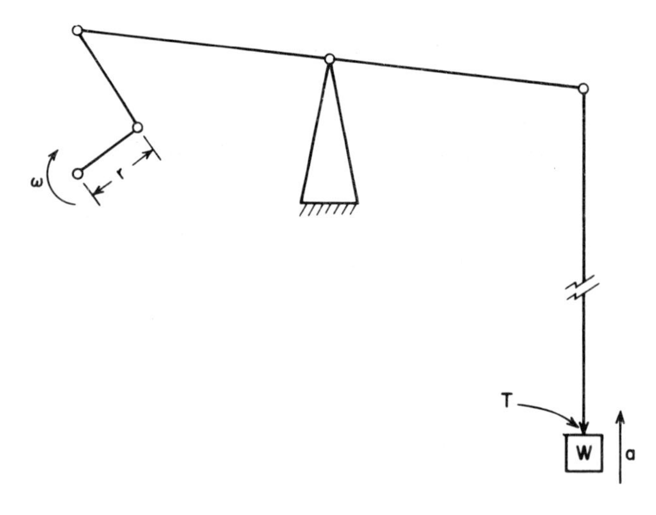

Fig. 4.8. Schematic diagram illustrating ideal situation of simple harmonic motion.

The maximum value of acceleration for simple harmonic motion is

$$a_{\max} = r\omega^2$$

where ω = angular velocity of the crank, radians/second

r = radius of crank (one-half surface stroke length), ft

Converting this expression to our units and nomenclature:

$$K = 1 + \left[\frac{\left(\frac{2\pi N}{60}\right)^2 \times \frac{S_l}{2}}{32.2}\right] = 1 + \frac{S_lN^2}{5870}$$

where $S_l = S_y$ = nominal surface stroke.

Hence, for $K = 2$,

$$N = \frac{76.6}{\sqrt{S_l}} \tag{4.20}$$

Equation (4.20) agrees closely with Bonham's equation:

$$N = \frac{77.3}{\sqrt{S_l}} \tag{4.21}$$

which he derived in a different manner.

Thus far, expressions for optimum tool weight, stroke length, and drilling speed have been presented for the conditions noted. These equations are recommended for outlining drilling programs below 1000 feet. At shallower depths, the indicated tool weights and drilling speeds are too high for practical application. The restriction of $G_{\max} = 1325\ D$ is an approximation

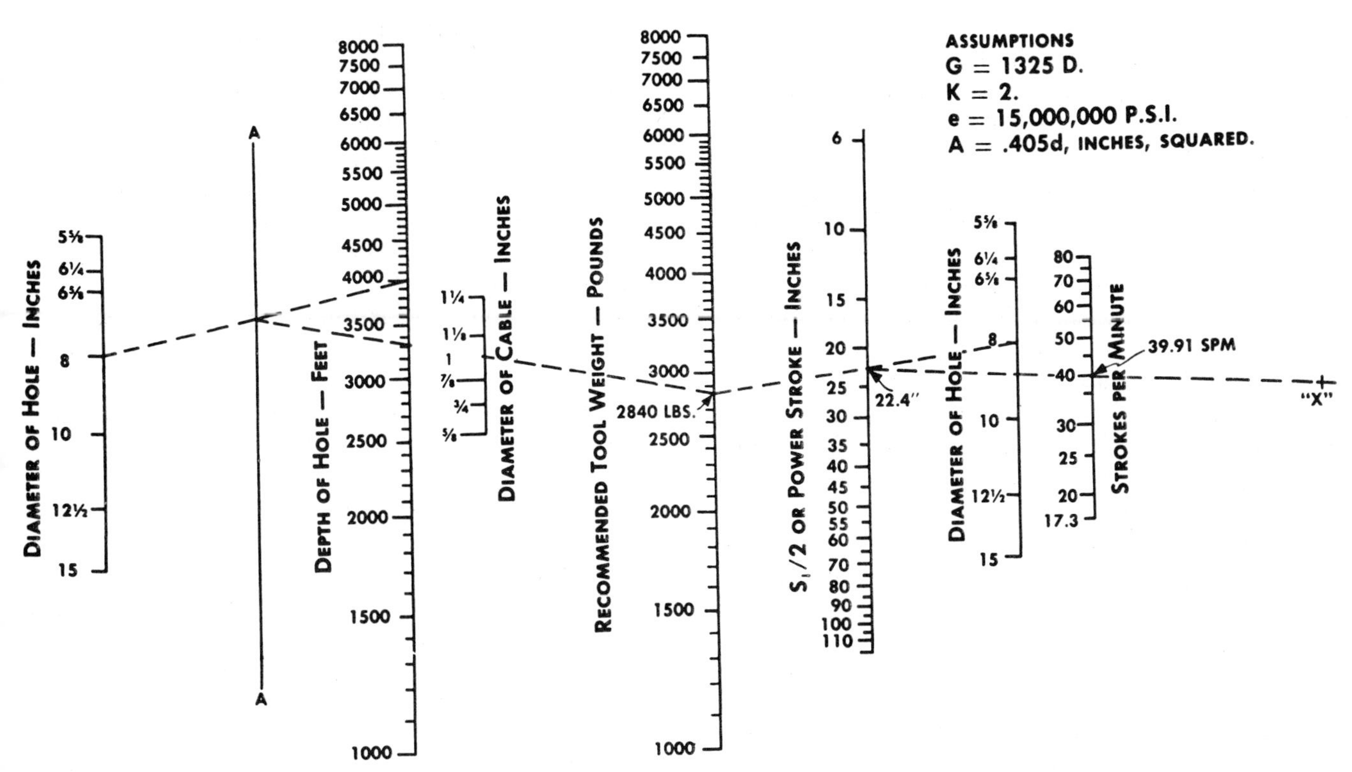

How to Use Alignment Chart to solve cable tool drilling problems: (1) Connect hole diameter with depth (2) from intersection with A-A', connect with diameter of cable and extend to scale showing recommended tool weight (1) connect this point with the hole diameter and read $\frac{S_1}{2}$ or power stroke (4) connect point on power stroke curve with point "X" to determine strokes per minute. Example: What is the correct weight of tools for drilling an 8-in. diameter hole at 4000 ft using 1-in. cable. From the alignment chart, the answer is 2840 lb. Using this weight of tools, what is the correct length of power stroke? Answer: 22.4 in. With this length of power stroke, the above weight of tools, etc., what is the proper drilling speed in strokes per minute? Answer: 40 spm.

Fig. 4.9. Alignment chart for solution of cable tool drilling formulae. After Bonham,[9] courtesy *Petroleum Engineer*.

based on Bonham's experience, but needs further evaluation for various formation types. Solutions of Eqs. (4.12), (4.17), and (4.21) are readily obtained from the nomograph of Figure 4.9. Note that the power stroke, $\frac{S_l}{2}$, is obtained and is half the nominal surface stroke. This nomenclature comes from the assumption that the acceleration of the tools takes place in the first half of the pickup stroke with the last half merely keeping the cable out of the tool's way.

Certainly the equations presented here cannot solve all the problems of cable tool drilling and were not presented as such by their originator. Situations undoubtedly exist where the analysis will not hold. In soft formations where the bit tends to drill faster than the cuttings can be mixed, adjustments in tool weight, hitch, and drilling speed may be required. Hitch is the term describing the point in the drop stroke at which the tools strike bottom, i.e., a tight hitch allows the bit to strike just at the end of the drop stroke. This analysis does furnish, however, an analytical solution to problems previously unsolved except by rig floor trial and error. Further field verification of the theory is needed; such experiments could possibly aid the economic application of cable tools.

4.4 Relative Merits of Cable Tool Drilling

Advantages

The principal factor governing the choice of drilling method is cost. In many shallow areas cable tool holes may be drilled more cheaply than rotary holes. This economy is due primarily to the following factors:

1. The lower initial equipment cost, hence lower depreciation.
2. The lower daily operating expense: lower fuel and maintenance costs, fewer personnel, low water requirement.
3. Lower transportation costs.
4. Lower rig-up time and expense.
5. Drilling rates comparable to those of rotary in hard, shallow areas.

Small truck-mounted rotaries, such as are commonly associated with core and shot-hole drilling, have been able to compete with and partially replace cable tools in many shallow areas of Eastern Oklahoma and Kansas, and North Texas.

More precise sample (cuttings) data are obtained from a cable tool hole. By running the bailer or sand pump every few feet a very accurate record of formation type vs depth may be obtained. Further, the presence or oil- or gas-bearing strata are readily detected by the relatively unrestricted entry of these fluids into the bore hole. The importance of these particular advantages has been reduced in the last few years by the development of better logging and sampling techniques for rotary holes. However, since the cable tool method requires no circulating fluid and consequently exerts little back pressure on encountered formations, these advantages definitely do exist. Another closely related advantage is the minimum contamination of the pay section which again is due to the absence of a drilling fluid in the hole. Consequently, there is little permeability loss around the bore hole caused by the filtration loss of a mud. This is one reason why many wells are drilled to the pay section with a rotary and *tailed-in* with cable tools.

Disadvantages

The principal disadvantages of churn drilling are the limitations on drilling rate and depth. Although a cable tool well has been drilled to a depth of 11,145 feet,[10] the time and cost involved make it generally unattractive in most areas at the present time. For example, cable tool drilling rates commonly vary from 5 to 150 feet per day depending on rock type, depth, and other factors. In contrast to this are areas where rotary drillers sometimes drill 2000 feet in eight hours.

Other disadvantages stemming largely from the absence of a drilling fluid are

1. Lack of automatic control over high pressures which may be encountered, and consequent greater danger of blow outs with accompanying hazards to personnel and equipment.
2. Lack of control over unconsolidated and caving formations; multiple casing strings may be required.

Frequent failures of the drilling line and the resultant fishing jobs also reduce the applicability of cable tools to deeper wells.

Competent cable tool drillers and tool dressers are difficult to obtain in many areas, particularly those in which rotary drilling is dominant. The use of less skilled personnel can prove costly.

It should be pointed out that many of the advantages and disadvantages of this method are due to either the absence or presence of a drilling fluid. We shall have more to say of this necessary evil in later chapters.

4.5 Current Applications of Cable Tools

At the present time cable tools account for the drilling of slightly more than ten per cent of all wells in this country.[11] These tools are utilized mainly in the shallow areas of Pennsylvania, West Virginia, Ohio, Michigan, Eastern Kansas, and Northeastern Oklahoma. On a footage basis, however, cable tool usage is considerably less significant.

There are, however, a number of auxiliary purposes for which cable tool equipment is used. Some of these are:

1. Drilling-in operations: Drilling the desired footage of the pay section only. This is particularly useful in mud-sensitive zones, thin pay sections with bottom water, low pressure and lost circulation zones.
2. Remedial and cleanout work: Post-completion work on wells which can be economically performed with cable tool equipment. Low per day costs often give cable tools the advantage in this work.
3. Preparatory work for rotary drilling: In some areas it has proved economical to set the short string of surface pipe with cable tools, enabling rotary equipment to commence drilling without this delay.
4. Water well drilling: Shallow water wells used to supply rotary rigs are generally drilled with cable tools.

Many other auxiliary jobs of cable tool equipment have been taken over by truck-mounted, well-servicing units.

PROBLEMS

1. Show that Eq. 1.4 may be expressed as

$$F = \frac{2.31 \times 10^5 d}{W_s L}$$

for a 6 × 19, fiber core, drilling line, where d = line diameter, in. Note: Use $A = 0.405\ d^2$.

2. A 1 in., 6 × 19, FC drilling line is used at 2000 ft depth with a tool weight of 10,000 lb; calculate the proper drilling speed (strokes/min) based on the harmonic vibration approach. Note: line weight = 1.6 lb/ft. *Ans.* $N = 28$.

3. At what depth should $N = 40$? Use W_s of Problem 2. *Ans.* $L = 1000$ ft.

4. Using Bonham's method, outline a drilling program for a 6000 ft cable tool well. This is to include tool weight, surface stroke length, and drilling speed. Assume a 1 in. line. Work this in 500 ft increments commencing at 1000 ft. (Use Figure 4.9.) Show the results as graphs of W, S_l, and N vs L. Hole sizes are to be:

Depth interval, ft	*Hole size, in.*
0 — 2000	15
2 — 4000	10
4 — 6000	8

5. Compute the input horsepower to the formation for the above conditions. Show this as a function of depth.

6. Using the same tool weights and stroke lengths as in Problem 4, compute the corresponding drilling speeds by the harmonic vibration method. Compare with those of Problem 4.

7. Note in Eq. 4.12 that tool weight W_0 is proportional to $\sqrt{A/L}$, all other factors being constant. Using this observation, show that for a given W_0 the relationship between depth L and drilling line diameter d is given by

$$\frac{L_2}{L_1} = \left(\frac{d_2}{d_1}\right)^2$$

8. For a 1 in., 6 × 19, fiber core cable ($C = 0.405$) show that, according to Bonham's assumptions, the tool weight may be expressed as

$$W_0 = 6.34 \times 10^4\sqrt{\frac{D}{L}}$$

9. Assume that the total work which must be applied to drill a given footage varies as the rock volume removed and is therefore proportional to D^2. From experience it is also known that the allowable blow energy varies linearly with D as indicated by the restraint, $G = 1325D$. Using these plus the further restriction that the blow frequency N must be the same for both cases, show that penetration rate R_p varies inversely as hole size; that is, that

$$\frac{R_{p1}}{R_{p2}} = \frac{D_2}{D_1}$$

Hint: $R_p = \dfrac{\text{working rate}}{\text{work required per foot}} = \dfrac{1325\,DN}{kD^2}$

10. Show that Bonham's analysis also implies that the total number of blows required to drill a given footage is directly proportional to hole diameter.

11. Show that input horsepower to the formation is given by

$$\text{HP} = \frac{W_0\sqrt{S_l}}{430}$$

REFERENCES

1. Uren, L. C., *Petroleum Production Engineering, Oil Field Development*, 4th ed. New York: McGraw-Hill Book Co., Inc., 1956, p. 81.
2. Brantly, J. E., "Oil-Well Drilling Machinery and Practices," in *Elements of the Petroleum Industry*, 1st ed., ed. E. DeGolyer. New York: AIME, 1940, p. 122.
3. *API Std. 9A*, 13th ed. New York: American Petroleum Institute, 1953.
4. *API R.P. 9B*, 2nd ed. New York: American Petroleum Institute, 1954.
5. Sprengling, K., and E. A. Stephenson, "Cable Tool Drilling," *API Drilling and Production Practices*, 1940, pp. 64–72.
6. Mills, K. N., "Mechanics of Cable Tool Drilling," *World Oil*, Sept. 1952, p. 123.
7. Singer, F. L., *Engineering Mechanics*. New York: Harper & Brothers, Publishers, 1943, p. 444.
8. W. H. Griffith, discussion of Sprengling and Stephenson paper, loc. cit., p. 72.
9. Bonham, C. F., "Engineering Analysis of Cable Tool Drilling," *The Petroleum Engineer*, Dec. 1955, pp. B–93 to B–100.
10. Inghram, E. C., "The World's Deepest Cable Tool Well," *Drilling*, Aug. 1955, p. 145.
11. *The Oil and Gas Journal*, Jan. 30, 1956, p. 167.

SUPPLEMENTARY READINGS

1. Uren, L. C., *Petroleum Production Engineering Oil Field Development*, 4th ed. New York: McGraw-Hill Book Co., Inc., 1956. On pp. 78–87, 136–175 cable tool drilling is discussed. Numerous illustrations of equipment and discussion of operations.
2. Fox, J. A., "Problems in Deep Cable Tool Drilling—Appalachian District," *The Oil Weekly*, Apr. 22, 1940, p. 23. Problems are discussed and a possible solution is offered. Results cited here are listed as evidence supporting Bonham's analysis.
3. Pennington, J. V., "Faster Drilling Awaits Upping of Energy at Hole Bottom," *The Oil and Gas Journal*, Nov. 16, 1953, p. 245. Results of research on high frequency percussion drilling and possible field application.
4. *API Std. 2*, 11th ed., 1955, and *API Rec. Pract. 3*. Dallas: American Petroleum Institute, Div. of Prod. These standards list many specifications of cable tool equipment plus recommendations for tool care and handling.

Chapter 5 . . .

Rotary Drilling: General Method and Equipment

5.1 Introduction and Basic Operations

The rotary drilling method is comparatively new, having first been practiced by Leschot, a French civil engineer, in 1863. United States patents on rotary equipment were issued as early as 1866 but, as was the case with cable tools, the early application was for water well drilling. It was not until approximately 1900 that two water well drilling contractors, M. C. and C. E. Baker, moved their rotary equipment from South Dakota to Corsicana, Texas where it found use in the soft rock drilling of that area.[1] In Texas in 1901 Captain Lucas drilled the Spindletop discovery well with rotary tools. This spectacular discovery is credited with initiating both the Southwest's oil industry and the widespread use of the rotary method. The inherent advantages of this method in the soft rock areas of Texas and California insured its acceptance, and it was in general use by the early 1920's. It is interesting to note that in the 1914-18 period, cable tools drilled 90% of all U.S. wells. At the present time these figures are approximately reversed.

In the rotary method, the hole is drilled by a rotating bit to which a downward force is applied. The bit is fastened to, and rotated by, a drill string, composed of high quality drill pipe and drill collars, with new sections or joints being added as drilling progresses. The cuttings are lifted from the hole by the drilling fluid which is continuously circulated down the inside of the drill string through water courses or nozzles in the bit, and upward in the annular space between the drill pipe and the bore hole. At the surface, the returning fluid (mud) is diverted through a series of tanks or pits which afford a sufficient quiescent period to allow cutting separation and any necessary treating. In the last of these pits the mud is picked up by the pump suction and repeats the cycle. Figure 5.1 shows the basic components of a rotary drilling rig.

Making a connection, the process of adding a new joint of pipe to the drill string is shown in Figure 5.2. Periodically the pipe must be removed from the hole in order to replace the bit. This operation is illustrated in Figure 5.3. Here the pipe is pulled in *stands* of four joints each. Only two or three joints per stand will be pulled when using shorter derricks or masts.

The truck mounted rig shown in Figure 5.4 serves to illustrate the high degree of portability which has been attained by equipment manufacturers. This particular rig is designed for *slim hole drilling* to depths of 10,000 feet. "Slim hole drilling" refers to operations in which the hole size is smaller than usual.

5.2 Basic Rig Components

Rotary drilling equipment is complex and any detailed discussion would of necessity involve intricate mechanical design problems. Since petroleum engineers, for whom this book is designed, are normally neither required nor qualified to solve these problems, this chapter will stress the jobs to be performed rather than the equipment itself. To do this we will consider only the basic rig components in the following order:

1. Derricks, masts, and substructures
2. Drawworks
3. Mud pumps
4. Prime movers

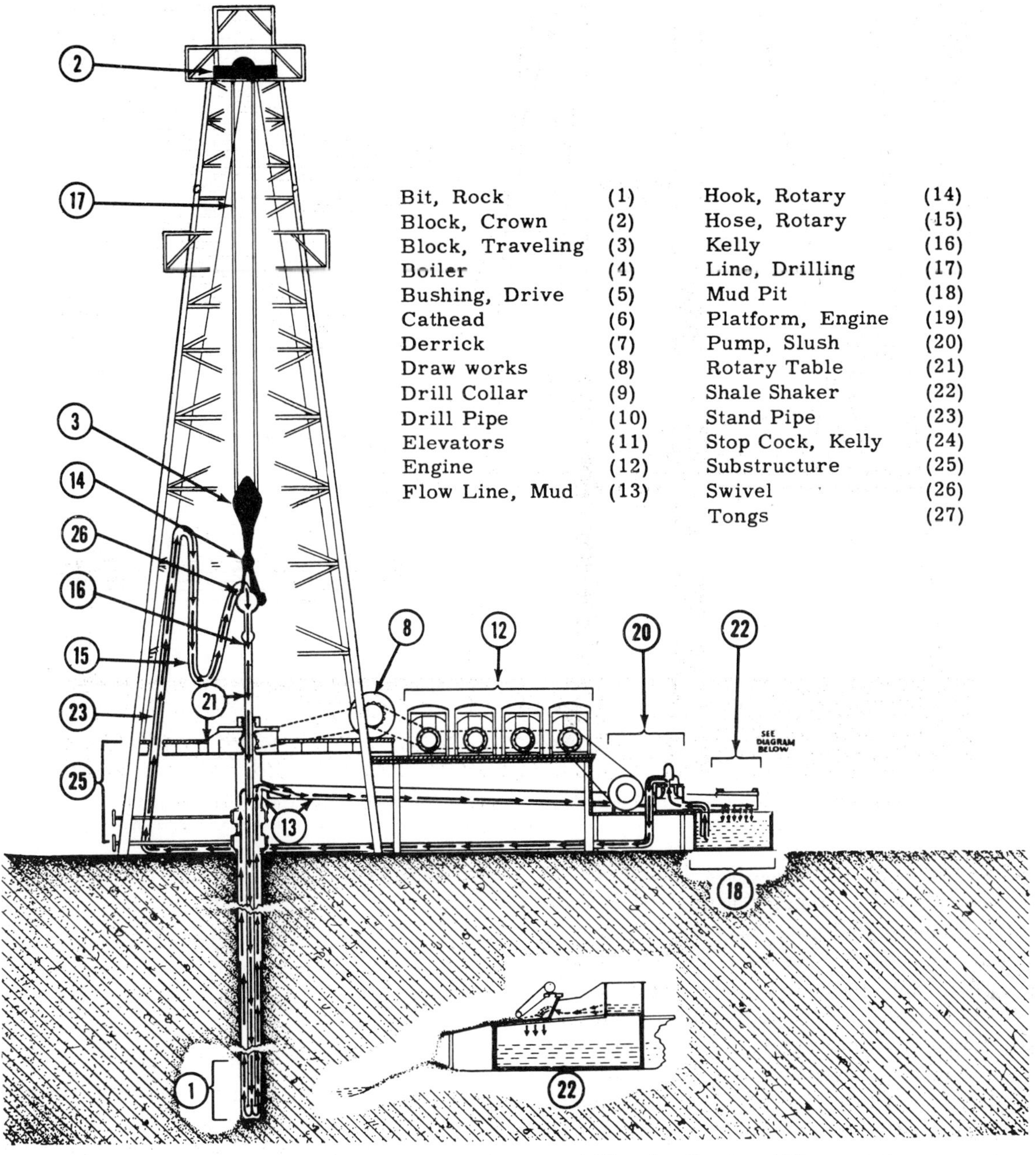

Fig. 5.1. Basic components of a rotary drilling rig. Courtesy API.

5. The drill string
6. Bits
7. Drilling line
8. Miscellaneous and auxiliary equipment

5.21 Derricks, Masts, and Substructures

The function of a derrick is to provide the vertical clearance necessary to the raising and lowering of the drill string into and out of the hole during the drilling operations. It must be of sufficient height and strength to perform these duties in a safe and expedient manner. Derricks are of two general types, standard and portable.

Standard Derricks

A standard derrick cannot be raised to a working position as a unit, i.e., it is of bolted construction and must be assembled part by part. Likewise, it must be disassembled if it is to be transported. Exceptions to this are those relatively short moves in which the entire rig is *skidded* to a nearby location. Detailed specifications for the nine API standard derricks may be found in the *API Std. 4A.*[2]

Derricks are rated according to their ability to withstand two types of loading:

(1) Compressive loads
(2) Wind loads

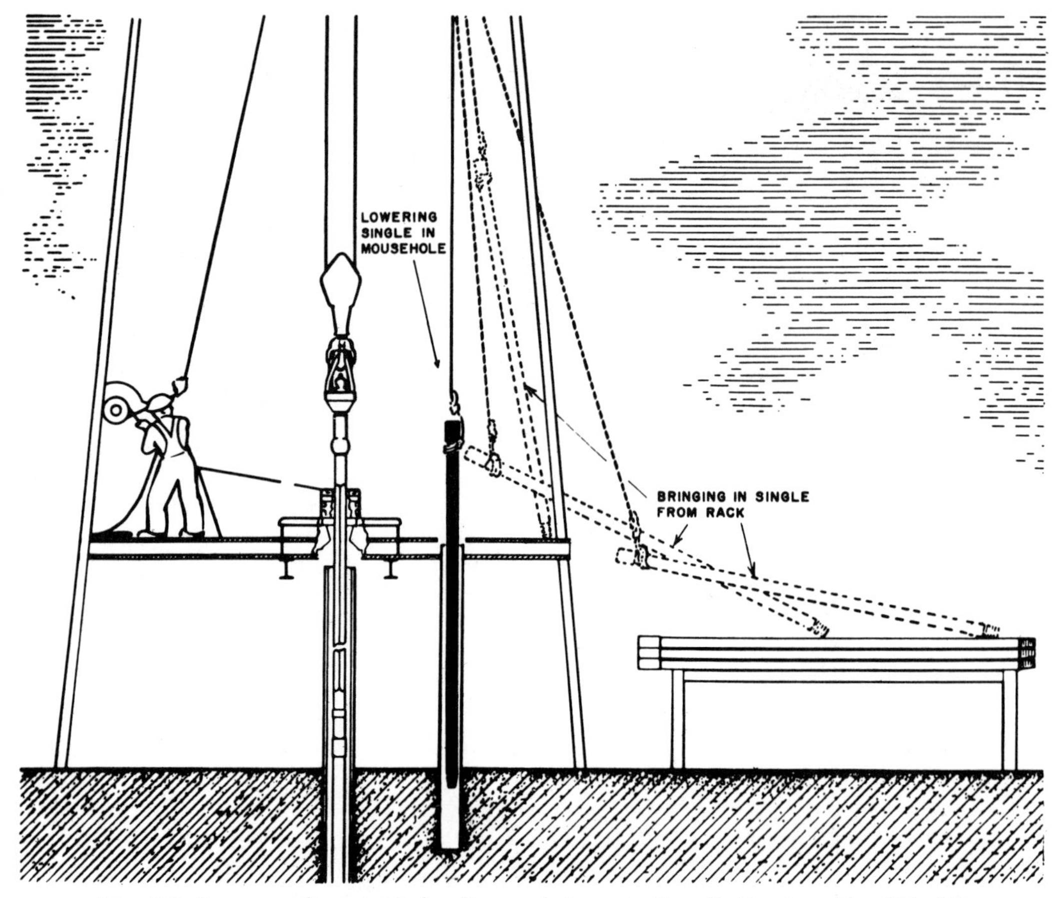

Fig. 5.2. Sequence of events during "mouse hole connections." Courtesy The Ohio Oil Company.

The allowable compressive load of a derrick is computed as the sum of the strengths of the four legs. In these calculations, each leg is treated as a separate column and its strength computed at the weakest section. This load, (excluding the weight of the derrick) with its safety factor of 2, is the API safe load capacity. Capacities for reinforced derricks are computed in a similar manner by the formulae of *API Std. 4B*. Derricks with load capacities from approximately 86,000 to 1,400,000 lb, depending on steel grade and leg size, are available.

Allowable wind loads for API derricks are specified in two ways, with or without pipe setback. For the drill pipe to stand vertically stable during a trip, the tops of the stands must lean outward against the fingers at the pipe racking platform. This results in an overturning moment applied to the derrick at that point. If the wind is blowing perpendicular to the setback, which is essentially a pipe wall, a further overturning moment is applied. This is the worst possible condition, i.e., wind and setback loads acting in the same direction. The minimum allowable wind loads for 136 ft and shorter derricks is 11.76 lbs/ft² (54 mph) with pipe setback. Minimum wind loads for taller derricks are 22.50 lbs/ft² (75 mph) with setback and 52.90 lbs/ft² (115 mph) without setback. Wind loads are calculated by the formula:

$$p = 0.004\,V^2 \tag{5.1}$$

where p = wind load, lb/ft²
V = wind velocity, mph

Calculation of Derrick Loads

The block and tackle arrangement for a rotary rig is shown schematically in Figure 5.5. If we assume the system as frictionless, the following relationships are apparent:

$$F_d = \frac{n+2}{n} W \tag{5.2}$$

where F_d = total compressive load on the derrick
n = number of lines through the travelling block (those supporting W)
W = hook load

Hence it is seen that the derrick load is always greater than the hook load by the factor $(n + 2)/n$ due to the two additional lines (drawworks and anchor) exerting downward pull. Further it may be noted that during hoisting:

$$v_L = n\,v_h \tag{5.3}$$

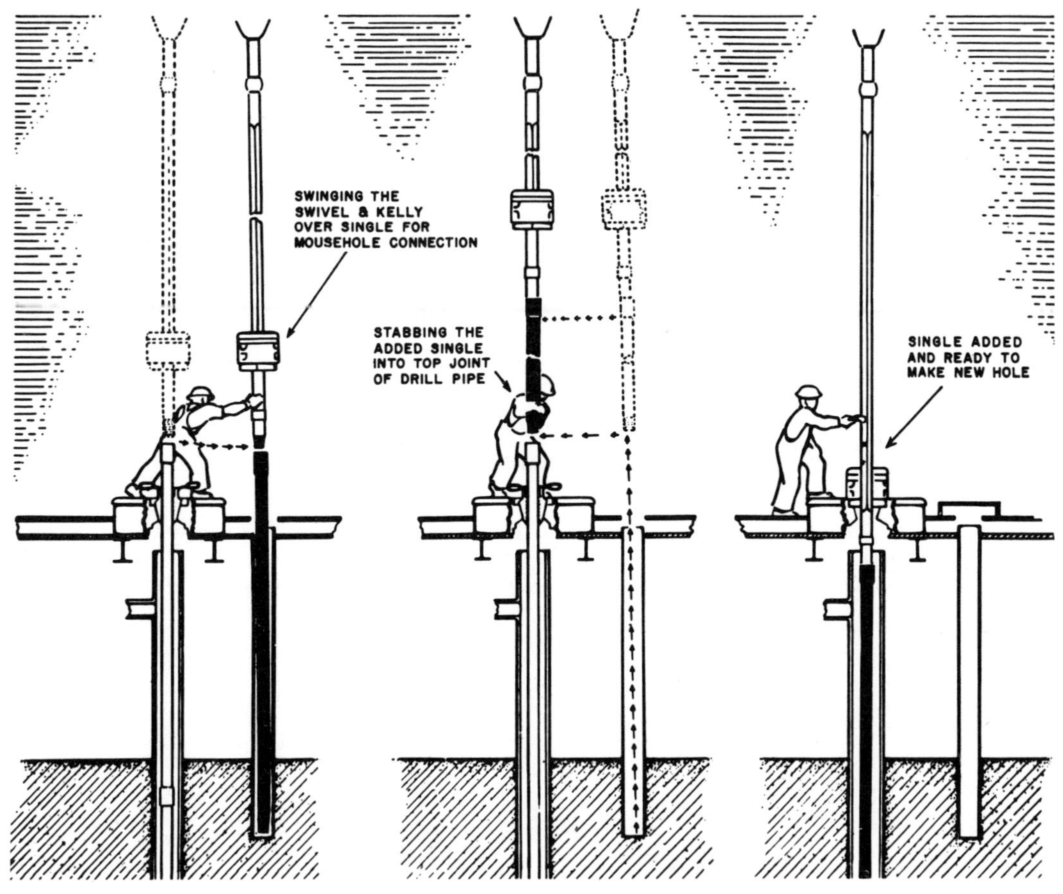

Fig. 5.2 (cont.)

where v_L = velocity of line being spooled (or unspooled) at the drawworks during hoisting

v_h = hook velocity

Portable Derricks (Masts)

A portable derrick or mast is one capable of being erected as a unit. The telescoping type in Figure 5.4 is one example. Specifications for portable masts are given in *API Standard 4D*.[3] These specifications are not as precise as those for standard derricks, and more variations of design are in evidence. The computations of wind, setback, and compressive loads are the same as for standard derricks. Minimum wind loads are specified as 11.76 lb/ft^2 (54 mph). Detailed instructions as to allowable hook loads, operational features such as raising and lowering procedures, guying instructions, and recommendations for care and lubrication are furnished for each mast by the individual manufacturers.

Choice of Derrick — Type and Size

Portable derricks have virtually replaced the standard types in many areas on account of the savings from reduced erection and tear down time, and reduced transportation costs. Since the development of portable well servicing equipment, derricks are no longer left standing over completed wells such as may be observed in the older areas. Consequently, standard derricks are currently used for special applications such as

(1) Situations where laydown room is not available: town lots, offshore platforms, etc.
(2) Deep wells requiring floor capacity for racking pipe
(3) Deep, hard areas where the trip time saving by using a tall derrick may more than offset moving costs
(4) Any other application where portability and quick rig up time are not the primary considerations.

The size derrick for a specific application will normally depend on

(1) Maximum compressive load anticipated (Normally the casing load), and
(2) maximum wind velocities expected in area of use.
(3) Derrick height will normally be governed by trip frequency and the relative saving from pulling *fourbles* versus triples versus doubles, etc.

Substructures

The substructure is, as the name implies, the support on which the derrick rests. This must be of sufficient strength to support the anticipated loads with adequate safety factors. Its height must be sufficient to house and afford access to the blowout preventers. A typical

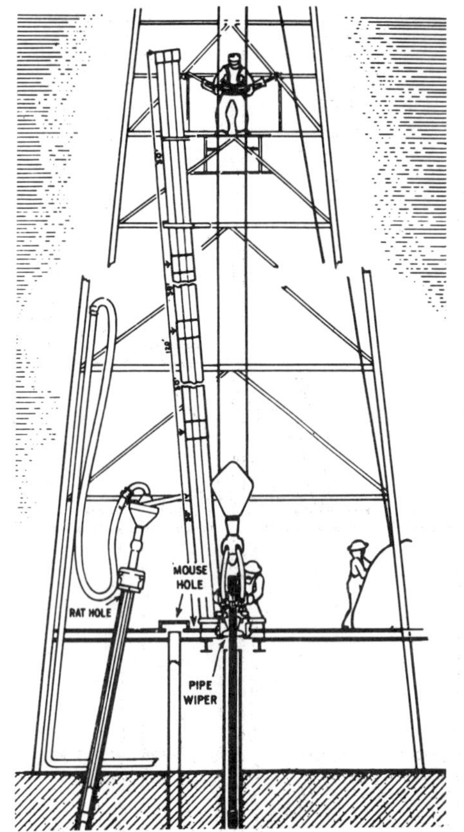

Fig. 5.3. Making a "trip." Courtesy The Ohio Oil Company.

substructure is shown under the rig in Figure 5.1. Although the API has adopted three substructure types as set forth in *API Standard 4A*,[2] numerous individual designs may be found. The choice of type is often governed by the soil condition in the area of application.

No standards have been adopted for use with portable masts; consequently, the types used here are even more numerous. For shallow, low pressure areas such as parts of Kansas and Oklahoma where blowout preventers are not often required, the substructure may be a simple I-beam skid.

5.22 Drawworks (Hoists)

The drawworks or hoist is the key piece of equipment on a rotary rig. When a contractor is asked what kind of a rig he has, the answer is invariably given as the manufacturer of the drawworks, even though the rest of the rig may have been obtained from other companies. The eminence of this rig component may be explained by observing the functions of the drawworks, which are:

(1) It is the control center from which the driller operates the rig. It contains the clutches, chains, sprockets, engine throttles, and other controls which enable the rig power to be diverted to the particular operation at hand.
(2) It houses the drum which spools the drilling line during hoisting operations and allows feed-off during drilling.

The drawworks location on a rotary rig is illustrated in Figure 5.1. The prime mover type, whether steam, internal combustion, or electric, and the type of power transmission will alter the individual drawworks design. For example, a direct current electric motor driven hoist may require only two forward speeds or gear ratios due to the large speed-torque range of the prime mover. A comparable internal combustion engine driven drawworks may require eight forward speeds. A typical operating curve for a six speed drawworks with torque converter is shown in the graph of Figure 5.6 (page 58).

Power Rating

Drawworks are commonly designated by a horsepower and depth rating. Any depth rating must also specify the size drill pipe to which the rating pertains. The drawwork horsepower output required in hoisting is

$$\text{HP} = \frac{W\,v_h}{33{,}000} \times \frac{1}{e} \tag{5.5}$$

where W = Hook load, lb
v_h = Hoisting velocity of traveling block, ft/min
33,000 = ft-lb/min per horsepower
e = Hook to drawworks efficiency

Hook to drawworks efficiencies are commonly between 80 and 90%, depending on the number of lines in use. Other variables governing this factor are condition of sheave bearings, and time since the last grease job. A deduction of 2% per working line is commonly applied. (6 lines = 88%, 8 lines = 84%, etc.)

5.23 Mud Pumps

The function of the mud or slush pumps is to circulate the drilling fluid at the desired pressure and volume. The pump normally used for this service is the reciprocating piston, double-acting, duplex type shown in Figure 5.7. Piston and valve action are illustrated in Figure 5.8. The term "double-acting" denotes that each side of the piston does work, while "duplex" refers to the number of pistons (two). Triplex (three pistons) types have been used but to no great extent.

The superiority of the piston type pump for drilling service is largely due to the following features:

(1) Ability to handle fluids containing high percentages of solids, many of which are abrasive.
(2) Valve clearance will allow passage of large solid particles (typically lost circulation materials) without damage.
(3) Ease and simplicity of operation and maintenance. Liners, pistons, and valves may be replaced in the field by the rig crew.
(4) Wide range of volume and pressure available by using different liner and piston sizes.

Fig. 5.3 (cont.)

Fig. 5.4. Truck mounted rotary rig in road position. The engine mounted at the rear furnishes power for drilling and also serves as the motor for the vehicle. Courtesy Cardwell Manufacturing Company.

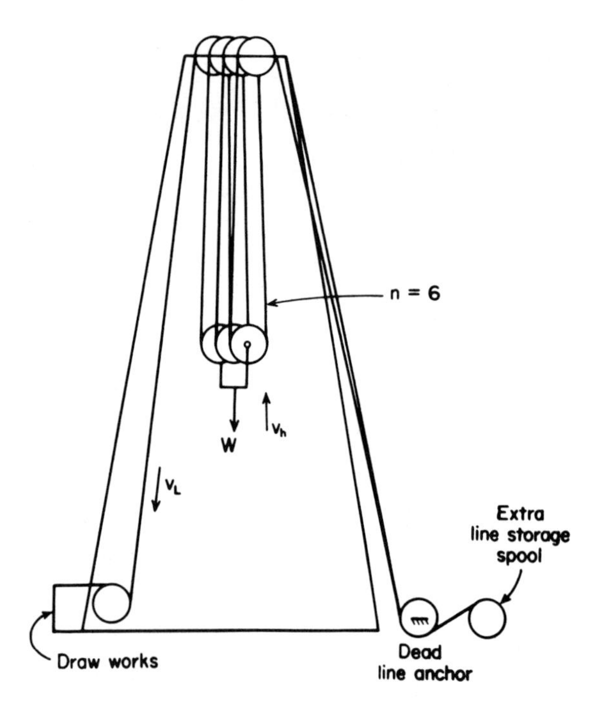

Fig. 5.5. Schematic block and tackle arrangement for rotary rig.

The main disadvantage of these pumps is that the discharge flow is pulsating, which causes periodic impact loads on discharge lines. This effect is minimized by air filled surge chambers located on the discharge line.

Mud pumps are commonly denoted by bore and stroke. An 8 × 18-in. power pump has a piston diameter (liner size) of 8 in. and a stroke length of 18 in. Direct acting steam pumps require an extra dimension for their description; hence, a 16 × 8 × 18-in. pump is a steam pump having a 16 in. steam piston opposing the 8 in. mud piston with an 18 in. stroke. These pumps differ from those in Figure 5.7 in that the power end of the steam type consists of another set of pistons against which high pressure steam exerts sufficient force to power the pump. By making the steam pistons larger than those at the mud end, one can hold the operating steam pressure much lower than the mud discharge pressure (commonly down at about one-half). The relative merits of steam pumps versus power pumps are about the same as those of steam engines versus internal combustion engines, as is discussed in the next section.

Pump Ratings and Capacities

Pumps are commonly rated by hydraulic horsepower as defined by

$$\text{HP} = \frac{D^2SNp}{107{,}000} \quad (5.6)$$

where D = liner diameter, in.

S = stroke length, in.

N = revolutions per minute of crank $\left(\frac{\text{piston strokes/min}}{4}\right)$

p = discharge pressure, psig

This expression is derived assuming the following to be true:

(1) The piston rod area can be neglected.

(2) Suction pressure is atmospheric.

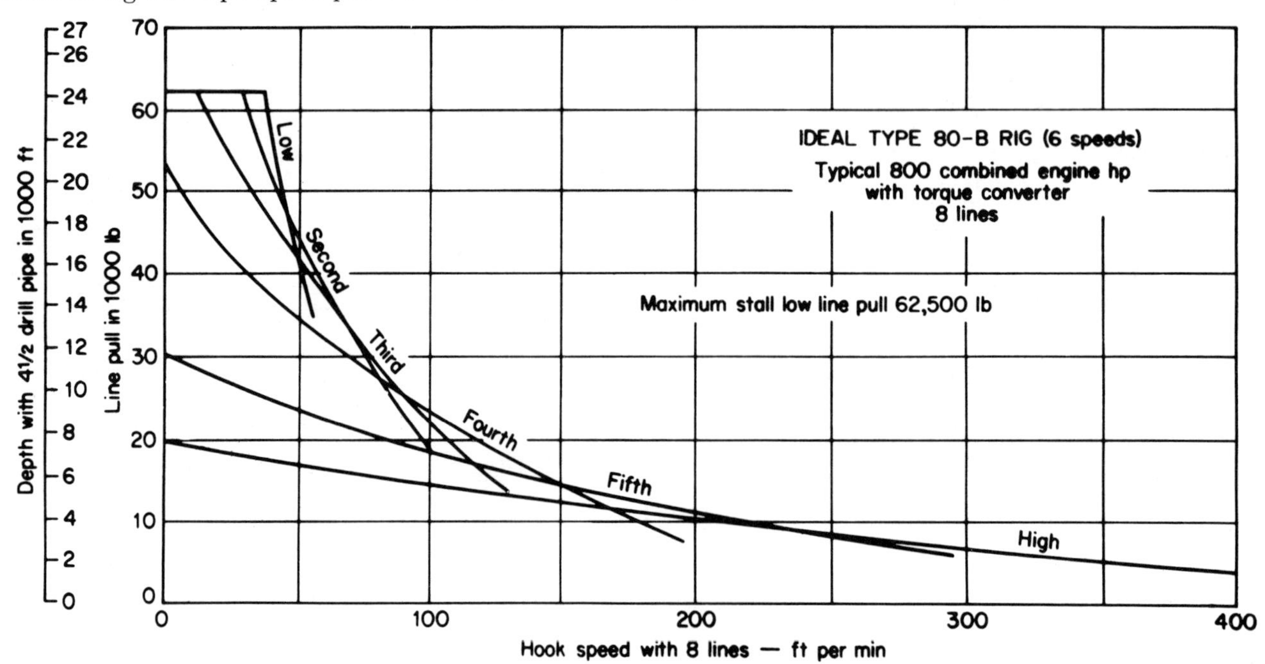

Fig. 5.6. Operating curve for typical 6500–10,000 ft rotary rig. Courtesy National Supply Company.

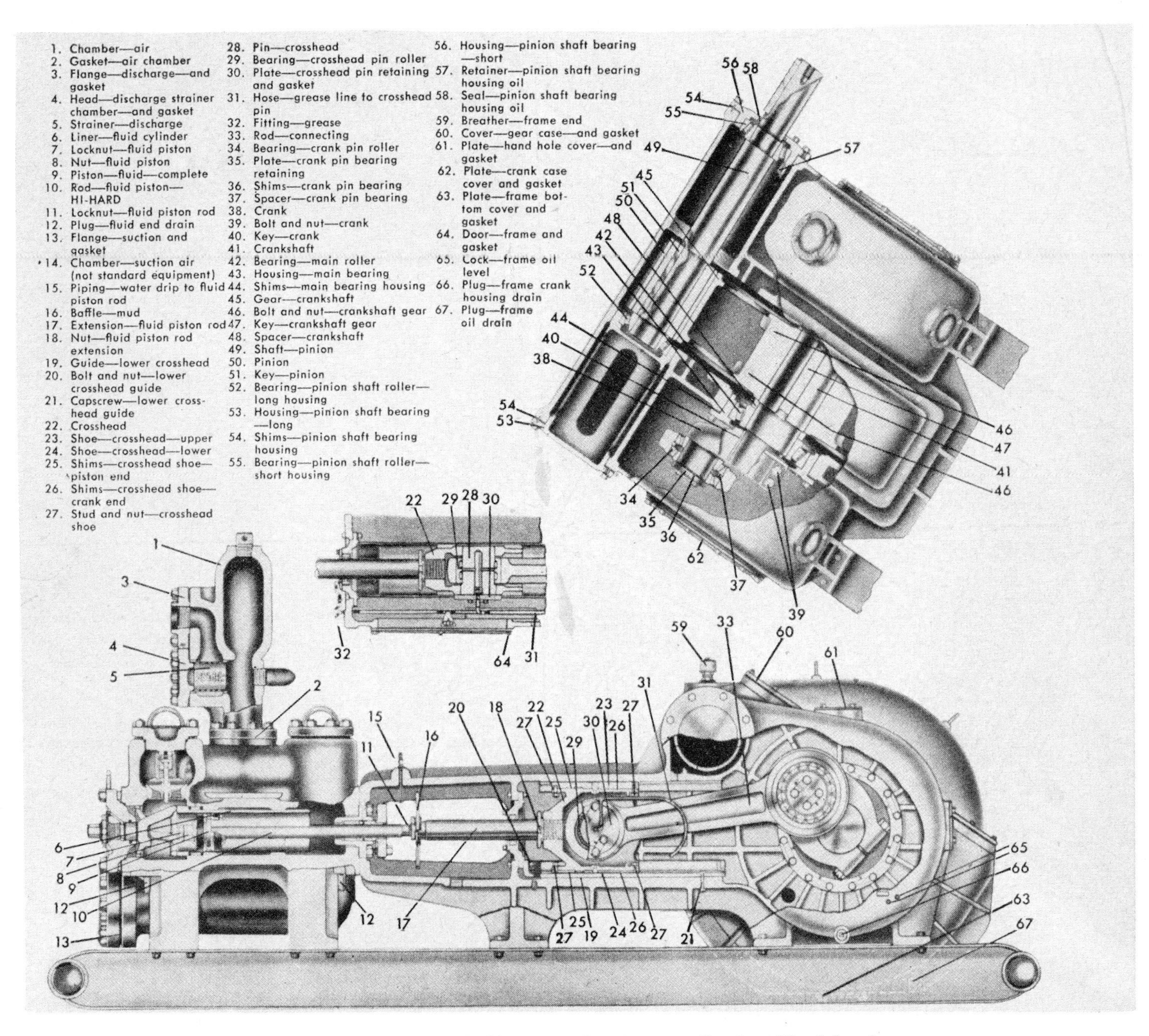

Fig. 5.7. Cross section of mud (slush) pump and part names. Courtesy Oilwell Supply.

$$\text{Work per piston stroke} = p \times \frac{\pi D^2}{4} \times \frac{S}{12} \text{ ft-lb}$$

Since both pistons make a stroke in each direction for each revolution of the crank, i.e., there are 4 individual piston strokes per crank revolution:

$$\text{Work per complete stroke} = p \times \frac{\pi D^2}{4} \times \frac{S}{12} \times 4N$$

For a mechanical efficiency of 85%,

$$\text{(5.6a)} \quad \text{HP} = \frac{p \times \dfrac{\pi D^2}{4} \times \dfrac{S}{12} \times 4N}{33{,}000 \times 0.85} = \text{Eq. (5.6)}$$

A simple expression for output horsepower may be developed by noting that

$$\text{Volume per minute } q = \frac{\pi D^2}{4} \times S \times 4N \times \frac{1}{231} \text{ gal/min}$$

where 231 in.^{3} = 1 gal.

Substitution of

$$231\, q = \frac{\pi D^2}{4} \times S \times 4N$$

in Eq. (5.6) gives

$$\text{(5.7)} \qquad \text{HP} = \frac{qp}{1714}, \text{ for } e = 100\%$$

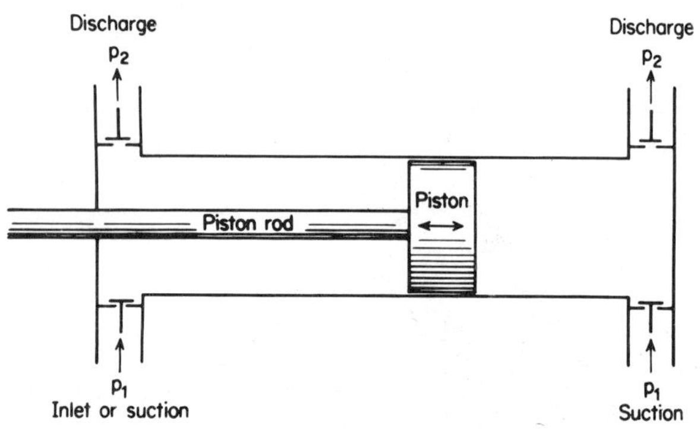

Fig. 5.8. Schematic valve operation for double acting slush pump.

The displacement of a duplex mud pump may be computed from:

$$(5.8)\quad q = \left[2\left(\frac{\pi D^2}{4}\right)S + 2\left(\frac{\pi D^2}{4} - \frac{\pi d^2}{4}\right)S\right]N \times \frac{1}{231} \times e$$

$$= .00679\ SN(2D^2 - d^2)e$$

where q = gal/min discharge

d = piston rod diameter, in.

e = volumetric efficiency (commonly taken as 90% for power or 85% for steam pumps)

Proper pump selection and utilization is extremely important to the over-all efficiency of a drilling operation. The circulation requirements in the area of use should be carefully analyzed before the final selection is made. Typical hydraulic calculations which emphasize this point will be discussed in later chapters.

5.24 Prime Movers

Before discussing the prime movers in common use, let us look more closely at the drilling services which must be performed. The bulk of rig power is consumed by two operations:

(1) Circulation of the drilling fluid

(2) Hoisting

Fortunately these requirements do not occur at the same time, and the same engines perform both jobs. The power consumption of the circulation system and rotary table is essentially constant over reasonable time periods. Such is not the case with hoisting, as is illustrated in Figure 5.9. Here it may be seen that rotary rig prime movers must be capable of handling highly variable loads at rapid acceleration over a wide torque and speed range.

Steam Engines

The first rotary drilling prime movers were steam engines. Although these are no longer dominant in the drilling industry, they are still commonly used in many areas. The common engine for this service is a twin-cylinder, double-acting type, which in basic principle operates much like a power mud pump in reverse. High pressure steam is furnished by a boiler plant located near the rig. The desirable steam pressure is governed by the power requirement (depth, circulation rate, etc.) and the piston size against which the steam works.

Electric Motors

Both direct and alternating current motors have been used for drilling purposes. The speed-torque range and ease of control of the d-c type make it an ideal prime mover. Its principal disadvantage is, of

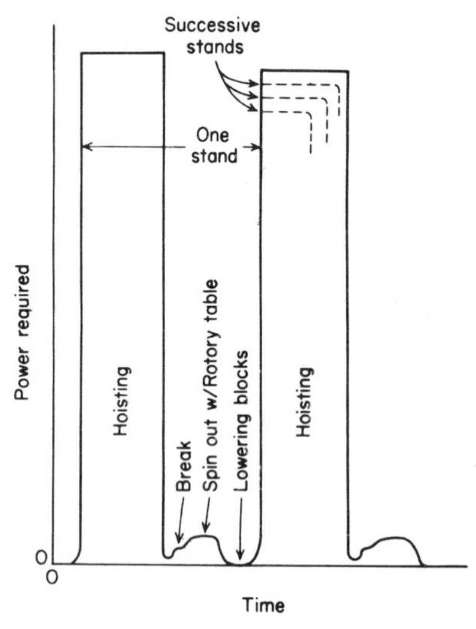

Fig. 5.9. Load-time diagram illustrating variation during hoisting cycle. After Crake,[4] courtesy *Petroleum Engineer.*

course, the inavailability of such power. Recent improvements in the design of diesel-electric power units have overcome many disadvantages of previous motor-generator systems; and these units are finding considerable application, particularly in offshore drilling. The operating characteristics of alternating current motors do not fit drilling requirements sufficiently well to warrant any particular use. They have been used for special applications such as in city drilling where power availability and noise restrictions have made it practical.

Internal Combustion Engines

Engines of this type are the most commonly used in the drilling industry. This has not been the result of any superiority of performance, as they are inferior in this respect to both steam and D.C. electric. Torque-speed characteristics may be improved by the use of torque converters. The internal combustion engines used are automotive type (multicylinder, light flywheel) diesel and gas engines capable of rapid accelerations.

Relative Merits of Prime Mover Types

Table 5.1 is an attempt to consolidate the general advantages and disadvantages of various prime mover

TABLE 5.1

ADVANTAGES AND DISADVANTAGES OF ROTARY DRILLING PRIME MOVER TYPES AND PRINCIPAL AREAS OF USE

Prime mover type	*Advantages*	*Disadvantages*	*Principal areas of use*
Steam	1. Excellent operating characteristics; torque at stall point, flexibility. 2. Simplicity, ease, and economy of maintenance. 3. Lower initial cost than other types.	1. Lack of portability boiler plant, piping, etc. 2. Water requirements. 3. Overall efficiency is low, hence higher fuel requirements.	Gulf Coast, California
D. C. electric (diesel generated)	1. Excellent operating characteristics. 2. Ease of control.	1. Power must be generated — reduction in overall efficiency. 2. Less portable than internal combustion. 3. Requires highly skilled maintenance; crews require special training. 4. Personnel hazard. 5. Higher initial cost.	Offshore, Gulf Coast. Use in other areas is increasing, as a result of design improvements in engine-generator units.
Internal Combustion (spark plug and diesel)	1. Portability 2. Fuel availability and costs. 3. Simplicity of maintenance compared to electric. 4. Torque converter transmission gives operating characteristics approaching those of steam — also reduces necessary gear ratios through drawworks.	1. Poor operating characteristics; limited speed-torque range. Improved by torque converters at expense of efficiency.	Inland drilling

types. The importance of these will vary with the intended area of use, and final selection is normally based on an economic appraisal.

5.25 The Drill String

The rotary drill string includes the components shown in Figure 5.10. The drill string is an extremely expensive rig component and must be replaced periodically; consequently, every care should be exercised to insure its long life. Most drill string failures are due to material fatigue which has been aggravated by corrosion and improper care and handling.

Kelly Joint

The kelly joint is the topmost joint in the drill string. It is commonly square (see Figure 5.1) but may be hexagonal or even octagonal. The kelly passes through snugly fitting, properly shaped bushings in the rotary table allowing the table's rotation to be transmitted to the entire drill string. This is its primary function.

Drill Pipe and Tool Joints

The drill pipe furnishes the necessary length for the drill string and serves as a conduit for the drilling fluid. Drill pipe sections (or joints) are hollow, seamless tubes manufactured from high grade steel. The tool joints or connectors for the drill string are a separate component and are attached to the pipe after its manufacture. The API has listed specifications for a number of tool joint types in its *Standard 7*;[5] however, individual manufacturers have many design variations. Figure 5.11 shows a few common types. Drill pipe, as well as all other oil field tubular goods, is specified by its outside diameter, weight per foot, steel grade, and range (length).[6] The most common joint length is range 2, or 27 to 30 ft. Physical properties of the API drill pipe steel grades D and E are given in Table 5.2.

TABLE 5.2

PHYSICAL PROPERTIES OF API DRILL PIPE[6]

Physical property	*Grade D*	*Grade E*
Tensile yield strength (minimum*) psi	55,000	75,000
Ultimate yield strength (minimum*) psi	95,000	100,000
Percentage elongation in 2 in. (minimum)		
Strip specimens	18	18
Full section specimens	20	20

*Minimum strengths are 80% of the average strengths of all specimens tested. Yield strength is the tensile stress required to produce a total elongation of 0.5% of the original specimen length.

Drill Collars

Drill collars are heavy walled, large O.D. steel tubes whose function is to furnish the compressive load on the

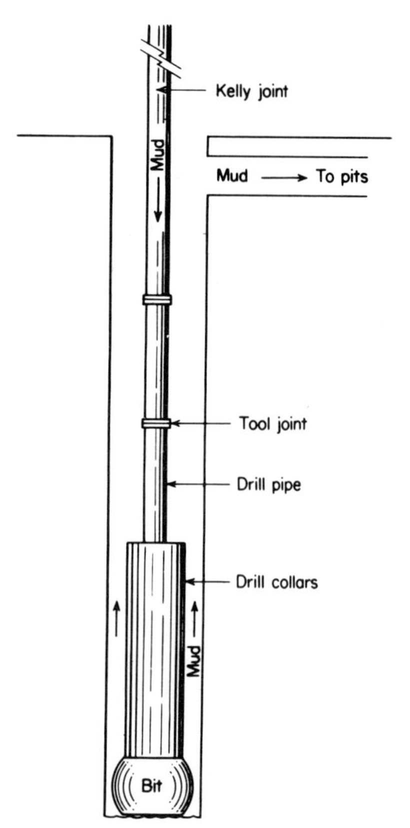

Fig. 5.10. Schematic diagram of drill string components and bit. Bit load is furnished by heavy-walled, large-diameter drill collars.

bit, allowing the lighter drill pipe to remain in tension. Using these, drillers have been able to increase penetration rates and drill straighter holes with fewer drill string failures. We shall have more to say on this topic in a later chapter.

5.26 Rotary Bits

Numerous individual rotary bit designs are available from a number of manufacturers. All are designed to give optimum performance in various formation types according to the ideas and experience of each company. Needless to say, there is no universal agreement on this subject; advocates of each design may be found. The severity of drilling requires that precise control over steel quality and heat treatment be exercised.[7]

For the purpose of our discussion we will classify bits into three general types as illustrated in Figure 5.12. These are:

(1) Drag type
(2) Rolling cutter (roller bits)
(3) Diamond

Drag Bits

Drag bits have no moving parts and drill by the shovelling action of their blades on the encountered formation. Their water courses are placed such that the drilling fluid is directed on the blades, keeping them clean. Bits of this type were once widely used for drilling soft, sticky formations, but in recent years have been largely replaced by rolling cutter types. The blades are manufactured from various alloy steels and are normally hard-faced with tungsten carbide.

Rolling Cutter (Rock) Bits

The first successful rolling cutter bit was designed by Howard R. Hughes in 1909. This bit, with subsequent improvements, allowed the rotary method to compete with cable tools in hard formations which are undrillable with drag bits. It may be noted in Figure 5.12 that the soft formation, three cone bits have relatively long, widely spaced teeth with interruptions in the pattern. Tooth length, spacing, and pattern are balanced to obtain the fastest penetration rate with a minimum of balling between teeth. Note also that the cones are offset, i.e., the cone axes do not intersect at a common point. This offset imparts a drag bit action to the teeth as the bit is rotated. Bits designed for harder formations have successively shorter, more closely spaced teeth (allowing higher loading) and diminishing cone offset which reaches zero on the hardest formation types (2-C). The bit illustrated in Figure 5.13 has tungsten carbide inserts or buttons instead of teeth. It was designed for drilling the hardest of formations such as quartzite or chert and has had excellent success in many areas.

Jet Bits

Jet bits are rolling cutter bits which have been equipped with fluid nozzles. Each nozzle directs a high velocity fluid jet directly on the hole bottom which rapidly removes the cuttings. This allows each bit tooth to strike new formation rather than expend some of its energy in regrinding previously loosened chips. The pressure losses through these nozzles are considerable and require extra pump capacity. Figure 5.14 shows a jet bit.

Diamond Bits

Diamond bits drill by a scraping, drag-bit action of the stones which protrude from a steel matrix. Their use is justified in many areas where their long life and the consequent reduction in trip time affords sufficient advantage to offset the higher bit cost. The actual cost of a diamond bit is the initial cost less a salvage which is paid according to the weight of undamaged diamonds remaining after the bit's use. This commonly ranges from 25 to 75 percent of the initial cost. Diamond bits are normally used in hard formations. Diamond core heads are widely used and will be discussed under coring.

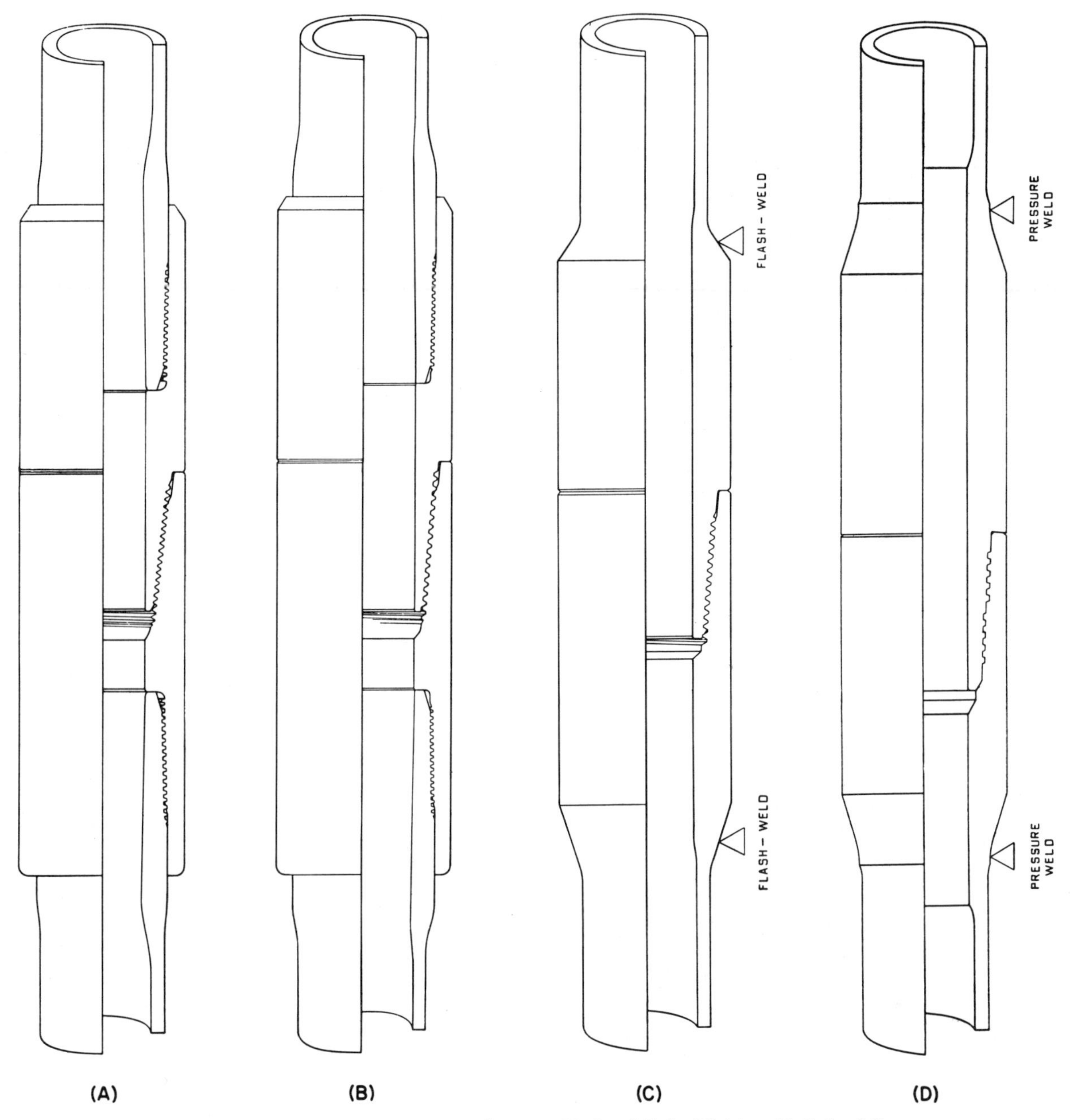

Fig. 5.11. Typical tool joint designs. Courtesy National Tube Division, U. S. Steel Corp. **(A)** Internal upset drill pipe with full-hole shrink-grip tool joint. **(B)** External upset drill pipe with internal-flush shrink-grip tool joint. **(C)** External upset drill pipe with flash-weld unitized tool joint. **(D)** External-internal upset drill pipe with hydril-pressure welded tool joint.

5.27 Drilling Line

The rotary drilling line affords a means of handling the loads suspended from the hook during all drilling operations. Normally, the maximum load occurs when running casing, although fishing operations frequently require line pulls in excess of the drill string weight. The wire line most commonly used for this service is the 6 × 19, Seale construction, fiber core, plow steel cable (Figure 4.4). Where dictated by high load requirements, premium grade lines with an independent wire rope center are used.

Care and Maintenance

Much has been written concerning the proper care and maintenance of drilling lines. This is a large expense item, and every effort is made to prolong its life. Wire line service may be increased by periodically placing a new portion of line in the system. This tends to distribute the same degree of wear over the entire line. In this procedure excess line is moved through the system at such a rate that it is evenly worn and reaches the drawworks drum for cut-off just as it is worn out. The length of line purchased is then a compromise

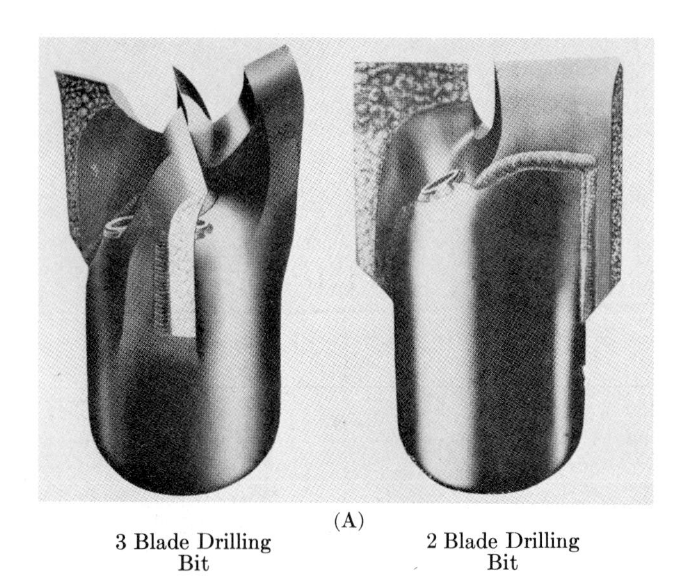

(A)

3 Blade Drilling Bit   2 Blade Drilling Bit

Fig. 5.12. Rotary drilling bits. Courtesy Christensen Diamond Products, Hughes Tool Company, and Reed Roller Bit Company. **(A)** Drag bit (Reed).

between service life and handling costs (transportation, etc.). The ton-mile method of line service evaluation has been adopted for determination of cut-off frequency. This method is based on the assumption that a line will safely perform so much work (tons × miles). There are several ways of evaluating ton-mile service, as is pointed out in the *Tool Pusher's Manual:*[8]

(1) API formulae, in its *Recommended Practices (RP) 9B*[9]
(2) Slide rules furnished by various manufacturers
(3) Tables such as are presented in the *Tool Pusher's Manual.*

5.28 Miscellaneous Rig Equipment

Although space does not permit any detailed discussion of all equipment items, there are a few other components which should be mentioned.

(B)(1)

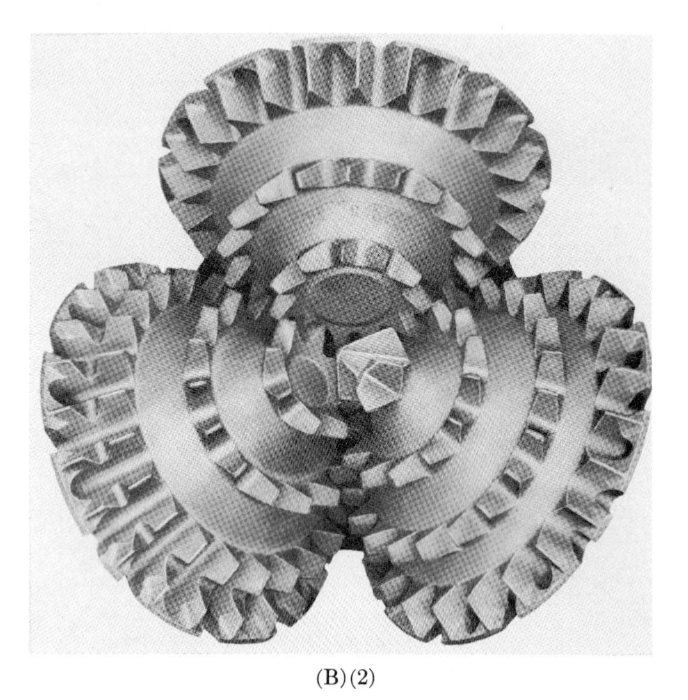

(B)(2)

(B)(3)

Fig. 5.12(B). Rolling cutter bits. **(1)** Cross roller type (Reed). **(2)** Three cone hard rock bit (Hughes W7R). **(3)** Two cone rock bit (Hughes LW-3).

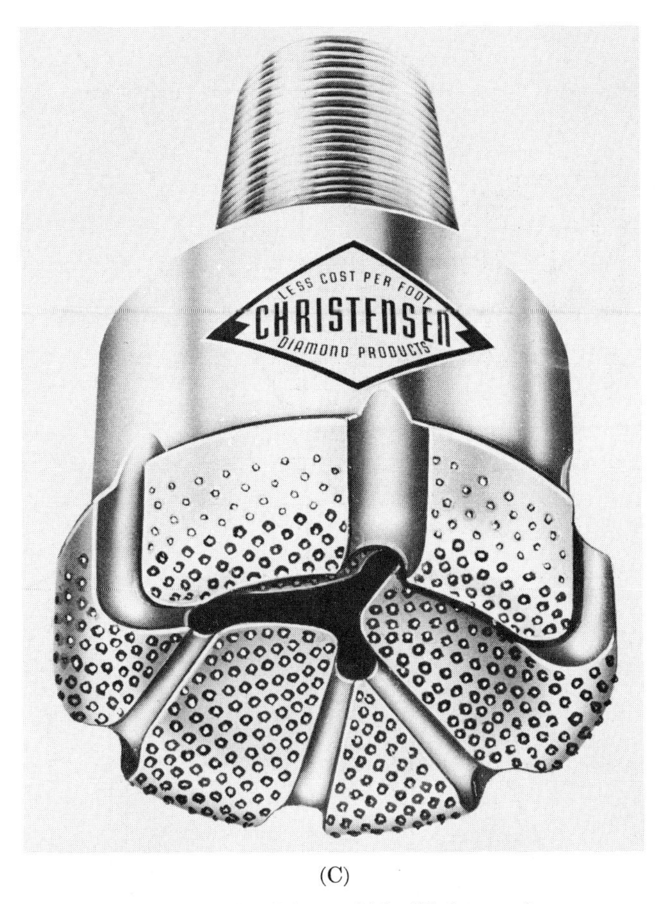

(C)

Fig. 5.12(C). Diamond bit (Christensen).

Rotary Tables

The rotary table has two primary functions:

(1) It transmits the rotation to the drill string by turning the kelly joint

(2) It suspends the pipe weight during connections and trips.

Normally, the table is chain-driven through the drawworks, although a smaller, separate engine is sometimes used as an independent unit.

Travelling Block, Hook, and Swivel

The travelling block is merely the travelling pulley (sheave) assembly which connects the drilling line to the hook and swivel. This may be combined with the hook as a unit, or they may be separate parts. The swivel must suspend the drill string and allow rotation at the same time. Further, it affords drilling fluid passage from the mud hose into the drill string. These items and their relative position are shown in Figure 5.1.

Blowout Preventers

It is not always possible to predict the exact magnitude of pressures which will be encountered in the drilling of a well. Consequently, it is not uncommon to encounter pressures greater than those imposed by the drilling fluid, with the result that formation fluids flow into the bore hole and eventually to the surface. This effect is called a *blowout,* and is one of the most feared and expensive accidents which can occur in well drilling. The main function of blowout preventers is to furnish a means of closing off the annular space between the drill pipe and casing. Many design variations

Fig. 5.13. Tungsten carbide button bit for hardest drilling. Courtesy Hughes Tool Company.

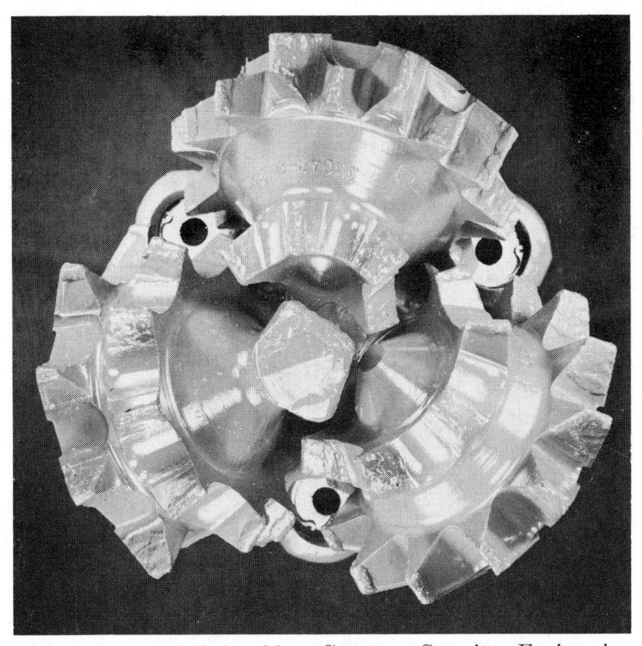

Fig. 5.14. Typical jet bit. Courtesy Security Engineering Division, Dresser Industries.

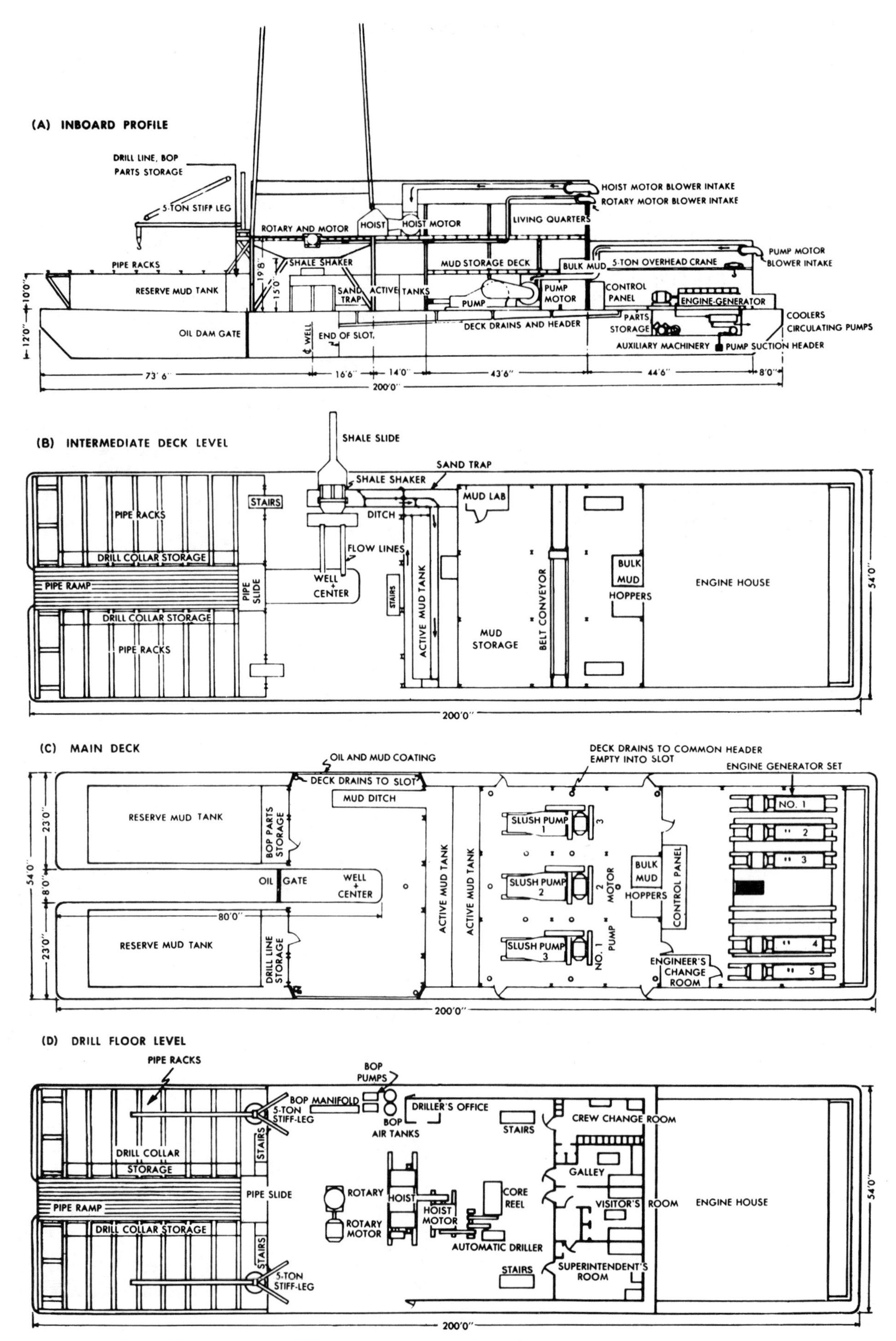

Fig. 5.15. Inland water drilling barge with heavy duty diesel-electric rig. After Rowan,[10] courtesy *Petroleum Engineer*.

are available; however, all are essentially valves designed for closing the annular passage between two concentric pipes. Most blowout preventers are either hydraulically or pneumatically operated, with manual operation available as a safety precaution. The pressure rating of blowout preventers is dictated by the hazards of the area and depth in which they are used.

Many rig items have not even been mentioned in this chapter and those which have, are not covered in any detail. We have mentioned, however, the main rig components and the general method involved. Those readers who desire detailed equipment data are directed to the references at the end of this chapter.

5.30 Inland Water and Offshore Drilling Equipment

The primary difference between land and water drilling installations occurs in the supporting structure for the rig and its auxiliary equipment. Other rig components are similar to those used in land drilling with added emphasis on operational ease, safety precautions, and standby equipment. We will discuss water drilling equipment in two general categories:
(1) Inland water equipment
(2) Offshore (open water) equipment.

5.31 Inland Water Equipment

Inland water drilling includes swamp, dredged marsh, bay, and lagoon locations where the waters are relatively shallow and protected from appreciable wave action. The first structures used in this service were piling-supported platforms. These were unsatisfactory because of cost, construction time, and the attendant difficulties of moving the rig between wells. Consequently, the submersible drilling barge was developed and has been used extensively since the mid 1930's. A typical barge of modern design is shown in Figure 5.15.

The submersible barge is a platform capable of supporting the drilling equipment which may be towed from location to location with a minimum of lost time. In moving to a new location, the barge is towed to the site and spotted with the location stake directly under the rotary opening. Piles are then driven around the well site which properly space the drilling slot of the barge. The barge is then sunk, levelness being attained by controlling the amount of water in each barge compartment. After the well is completed, the procedure is reversed; then the entire assembly is towed to the next location.[1c]

Inland barges are designed for drilling in 10 to 20 ft of water. Drilling in deeper water is accomplished by construction of a shell or gravel fill for the barge to rest on. Moving costs are reduced by holding the barge's draft and width to a minimum, thus allowing it to be towed through shallow and narrow waterways.

5.32 Offshore Equipment

Large Self-Contained Platforms

The first offshore (unprotected water) drilling structure was built in 1937 and was located $1\frac{1}{2}$ miles from shore in 15 feet of water. This was a large self-contained platform supported by numerous creosote treated timber piling. The Creole Field was discovered in 1938 by a well drilled from this platform; this field eventually contained 19 wells drilled from ten surface locations.[11] The self-contained platform, with design improvements, is still in use. As the name implies, these platforms are sufficiently large to house the entire rig, crew quarters, and supplies. The cost of such an installation is enormous, but may be quite economical on a per well basis, since numerous wells may be directionally drilled from its deck. This practice, however,

Fig. 5.16. Self-contained offshore drilling platform. Courtesy *Oil and Gas Journal*.

has the very dangerous disadvantage that a blowout and/or fire of any well jeopardizes the entire installation. A more recent self-contained platform is shown in Figure 5.16.

While these structures lend themselves to development drilling (drilling in known fields), they are poorly adapted to exploratory work. It would be extremely costly to build such a structure only to find no oil accessible from its surface. Consequently, smaller, less expensive platforms with floating tenders evolved to offset this and other disadvantages of the self-contained platforms.

Platform Drilling Tender Installations

In this operation the relatively small platform carries the substructure, drawworks, rotary table, engines, and other small items. The barge or tender houses the rest

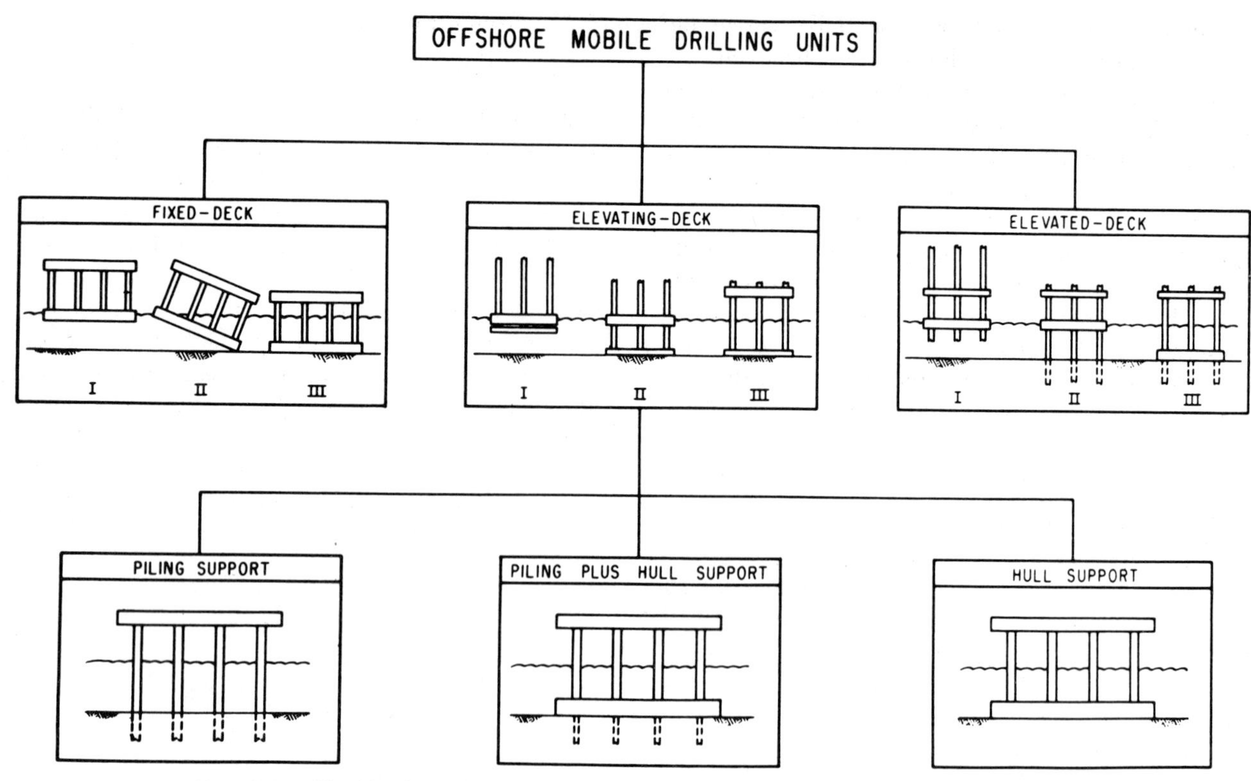

Fig. 5.17. Classification of offshore mobile drilling units. After Howe,[12] courtesy API.

which includes the mud pits, supplies (dry mud, chemicals, etc.), pipe racks, cementing equipment, fuel and water, and crew quarters. The principal advantages of these installations are greater mobility and lower installation costs. A disadvantage is the higher percentage of lost time due to high wind and waves; during periods of rough weather, the tenders must pull away to prevent collision with the platform. Large self-contained platform operations may proceed in any weather short of a hurricane.

Completely Mobile Offshore Units

The next development in offshore equipment was the completely mobile offshore unit. This is based on an extension of the inland barge principle, in that the entire rig is floated to the location as a unit. Figure 5.17 shows the classifications of these units as given by Howe.[12] Just as in land drilling, the economic advantages of complete mobility are considerable; units of this type are rapidly replacing former installation types.

The offshore drilling topic could be expanded over the entire length of this book if we were to attempt anything like a complete coverage of the subject. No mention has been made of such problems as corrosion, wave forces, and meteorological statistics, all of which are extremely important to offshore equipment design. The offshore oil potential is great; undoubtedly, an increasingly greater proportion of U. S. production will come from these fields in future years. The drilling and production problems of these operations offer a challenge to the entire industry.

PROBLEMS

1. According to the API formula, what wind velocity is required to exert a wind loading of 0.2 psi? *Ans.* 85 mph

2. A drilling rig has eight lines strung through the travelling block. A hook load of 240,000 lb is being hoisted at a velocity of 50 ft/min. Calculate:
(a) the velocity of the line being spooled at the drawworks (this is called the fast line)
(b) the line pull at the drawworks assuming frictional losses of 2% per working line
(c) the output horsepower of the drawworks. *Ans.* (a) 400 ft/min (b) 36,000 lb (c) 430 HP

3. For the data of Problem 2 calculate the derrick load, assuming
(a) a frictionless system
(b) an actual system (no friction loss occurs at the dead line, since it is static) considering the fast line load from above. *Ans.* (a) 300,000 lb (b) 306,000 lb

4. A 6 × 19 Seale, improved plow steel, FC cable has a breaking strength of 41.8 tons. Develop the following figures as taken from the *Tool Pusher's Manual.* (2% per line for friction)

No. of lines	*Safety factor*	*Hook load*	*Fast line load*	*Dead line load*	*Derrick load*
6	4	109,600	20,900	18,300	148,800
8	4	140,600	20,900	17,550	179,050
10	4	169,300	20,900	16,950	207,150

5. Assuming the drawworks operating curves of Figure 5.6, what is the highest gear which would handle the hoisting conditions of Problem 2? Which gear could handle the same load at 100 ft/min hoisting speed?

6. At what depth can an 800 HP rig hoist $4\frac{1}{2}$ in., 16.6 lb/ft drill pipe at 100 ft/min? Assume 8 lines and 55% engines to hook efficiency. *Ans.* 8,800 ft

7. Plot the allowable volume vs discharge pressure for a 600 HP duplex mud pump at a volumetric efficiency of 90%.

8. A $6 \times 14''$ power pump is operating at 40 strokes per minute. (In common nomenclature a stroke is synonymous with a crank revolution.) The piston rod diameter is 2 in. Calculate:

(a) output volume at 90% volumetric efficiency. *Ans.* 233 gpm

(b) input horsepower to pump at 85% mechanical efficiency and a discharge pressure of 500 psig. *Ans.* 89 HP

9. An 8 in. hole is being drilled with $4\frac{1}{2}$ inch O. D. drill pipe. If an upward velocity of 150 ft/min is required in the annulus, what is the required pump output in gal/min? If circulating pressure is 1000 psi, what is the output HP requirement at the pump?

REFERENCES

1. McCaslin, L. S., Jr., "Southwest U. S. A. — Birthplace of Rotary Drilling," *Oil and Gas Journal*, Anniversary Number, May 1951, p. 226.
2. *Specifications for Steel Derricks, API Std. 4A*, 14th ed. New York: American Petroleum Institute, Jan. 1952.
3. *Specifications for Portable Masts, API Std. 4D*, 3rd ed. New York: American Petroleum Institute, Mar. 1955.
4. Crake, W. S., "Application of Internal Combustion Engine Power to Rotary Drilling Rigs," *The Petroleum Engineer*, Dec. 1947, p. 70.
5. *Specifications for Rotary Drilling Equipment, API Std. 7*, 11th ed. New York: American Petroleum Institute, Apr. 1953.
6. *Specifications for Casing, Tubing, and Drill Pipe, API Std. 5A*, 22nd ed. New York: American Petroleum Institute, Mar. 1958.
7. Bentson, H. G., "Rock-Bit Design, Selection, and Evaluation," *API Drilling and Production Practices*, 1956, p. 288.
8. *Tool Pushers Manual.* Dallas, Texas: American Association of Oil Well Drilling Contractors, 1955.
9. *Recommended Practice on Application, Care, and Use of Wire Rope for Oil Field Service, API RP 9B*, 2nd ed. New York: American Petroleum Institute, Jan. 1954.
10. Rowan, C. L., "Submersible Barges," in *Fundamentals of Rotary Drilling*. Dallas, Texas: The Petroleum Engineer Publishing Co., 1955, p. 55.
11. McGee, D. A., "A Report on Exploration Progress in the Gulf of Mexico," *API Drilling and Production Practices*, 1949, p. 38.
12. Howe, R. J., "Some Factors in the Engineering Design of Offshore Mobile Drilling Units," *API Drilling and Production Practices*, 1955, p. 209.

SUPPLEMENTARY READINGS

1. Brantly, J. E., *Rotary Drilling Handbook*, 5th ed. New York: Palmer Publications, 1952, pp. 10–235.
2. *Tool Pusher's Manual.* Dallas, Texas: American Association of Oil Well Drilling Contractors, 1955.
3. *Fundamentals of Rotary Drilling.* Dallas, Texas: *The Petroleum Engineer*, 1955. This includes the following articles on equipment:
   Moon, J., "Masts, Derricks, and Substructures," p. 14.
   Kirberger, R. E., "Draw Works," p. 23
   Boggs, J. H., Fitch, E. C., and A. T. Woods, "Pumps," p. 26
   Bohneberg, H. A., "Rotary Table," p. 36
   Freeman, O. B., "Power Plants," p. 38
   Graham, C. R., "Hook, Swivel, and Kelly," p. 42
   Jones, M. R., "Blowout Preventers," p. 45
   Rowan, C. L., "Submersible Barges," p. 55
   Malott, R. A., "Ships and Power Slips," p. 60
   Schall, F. W., "Drill Pipe and Drill Collars," p. 61
4. Uren, L. C., *Petroleum Production Engineering. Oil Field Development*, 4th ed. New York: McGraw-Hill Book Co., Inc., 1956, pp. 176–251.
5. Payne, J. M., "Mobile Units for Off Shore Drilling," *API Drilling and Production Practices*, 1954, p. 257.

Chapter 6 . . .

The Composition, Functions, and General Nature of Rotary Drilling Fluids

6.1 Introduction

Early spring-pole drillers, including the ancient Chinese, added water to the borehole as an aid to *rock softening* and cutting removal. Since the drilling fluid is a comparatively unimportant feature of the cable tool method, little thought was given this subject until the advent of rotary drilling. As pointed out previously, the continuous circulation of the drilling fluid (mud) was a principal reason for the early success of rotary tools in areas considered undrillable by the percussion method. Consequently, the advancement of drilling fluid technology has contributed greatly to the success of rotary drilling.

The first rotary drilling fluid was water. When quicksand threatened the progress of the famous Spindletop well, a muddy fluid was mixed by driving cattle through a shallow water pit. When used as a drilling fluid, this clay-water mixture lined the borehole, prevented caving of the quicksand, and allowed drilling to continue.[1] From this crude beginning has evolved the drilling fluid technology of today.

The extent to which drilling mud properties must be controlled varies with geologic conditions. In soft rock areas successful completion of a well may require very precise control of mud properties, and the mud used is often an expensive and complicated chemical mixture.

In hard rock areas plain water may be a satisfactory and even superior drilling fluid. In addition to liquid muds, both air and gas are used as drilling fluids with spectacular results in many areas. Therefore, we may surmise that the selection of mud type is governed by the specific requirements of the geologic area in question, and depends on the drilling fluid's ability to perform the functions necessary in that area.

6.2 Testing of Drilling Fluids

The purpose of testing drilling muds is to determine their ability to perform certain necessary functions. The industry standard for mud testing is the *API Recommended Practices No. 29*, from which some of this section is taken. Also, several excellent manuals and books are available which cover testing procedures in detail.[2–5]

6.21 Mud Density

The density of drilling muds is normally measured with a mud balance or scale such as shown in Figure 6.1. These instruments are rugged, easily calibrated, and lend themselves to field use. Hydrometers are also used, although not so commonly. Calibration of either device is performed by using fresh water having a known density of 8.33 lb/gal (or 62.4 lb/ft^3). The general practice is to report density in pounds per gallon, hence a 10 pound mud is one having a density of 10 lb/gal. It has been suggested that density be reported as a pressure gradient; this, however, is not yet common.

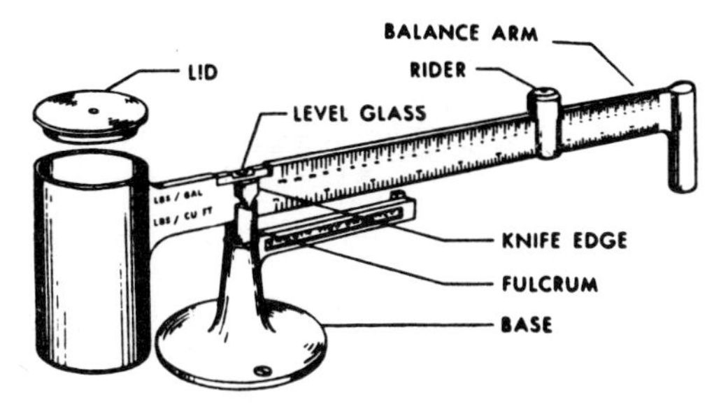

Fig. 6.1. Diagram of typical mud balance.

6.22 Mud Viscosity

Mud viscosity is difficult to measure; several measuring devices are in common use. Further complication generally ensues because each of the methods yields a different value. Before discussing the modes of measurement, let us define what is meant by drilling mud viscosity.

Most liquid drilling muds are either colloids and/or emulsions which behave as plastic or non-Newtonian fluids. The flow characteristics of these differ from those of Newtonian fluids (such as water, light oils, etc.) in that their viscosity is not constant, but varies with the rate of shear. (Recall Figure 2.8, Chapter 2, for the definition of viscosity.) This general behavior is shown in Figure 6.2. Note that for plastic fluids a certain value of stress (true yield point) must be exceeded in order to initiate movement. This is followed by a transition zone of decreasing slope in which the flow pattern changes from plug to viscous flow. The viscosity of a true (Newtonian) liquid is constant and equal to the slope of the line depicting its stress-strain behavior. Therefore, if the viscosity of a plastic fluid is measured in a conventional manner, i.e., the ratio of shearing stress to rate of shearing strain, the value obtained will depend on the rate of shear at which the measurements were taken.

Marsh Funnel

Various field instruments used for viscosity measurements are shown in Figure 6.3. The oldest of these is the Marsh funnel. In this method the funnel is filled to the upper mark (1500cc) with freshly collected, well agitated mud. The operator then notes the time, removes his finger from the discharge and measures the time for one quart (946 cc) to flow out. This time in seconds is recorded as the Marsh funnel viscosity of the mud. Marsh funnels are manufactured to precise dimensional standards and may be calibrated with water which has a funnel viscosity of 26 ± 0.5 sec. Funnel viscosities are of little quantitative use but have general comparative value. From experience, it is known that for certain mud types, certain funnel viscosities are desirable. Oil base mud viscosities are sometimes reported as the 500 cc discharge time, although the tendency is to standardize on the quart basis.

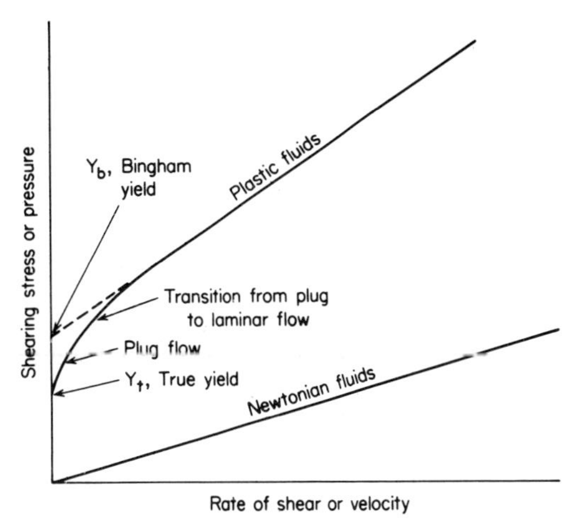

Fig. 6.2. Flow behavior of plastic and Newtonian fluids.

Stormer Viscosimeter

A more quantitative measurement of viscosity is obtained with the Stormer viscosimeter. This device consists of a spindle which is rotated in a test cup by a set of gears driven by a falling weight. The revolutions of the spindle are recorded by a revolution counter. By trial and error, weights are added to the line until a stabilized rotating speed of 600 rpm is obtained. The weight or driving force in grams is then used with a calibration chart to obtain the mud viscosity. This value is the apparent viscosity of the mud measured at a rate of shear corresponding to 600 rpm of the Stormer instrument.

Multispeed Rotational Viscosimeters

Realization that neither funnel nor Stormer viscosity measurements adequately defined the flow properties of plastic fluids led to the development of multispeed measuring devices.[6-10] One of these is the Fann V-G (viscosity-gel) meter. In principle, the Fann meter is like the Stormer, in that the basic measurement is the torque necessary to revolve an inner rotor in a stationary, mud-filled test cup. The spindle is driven by a two speed synchronous motor. Gear changes plus the two motor speeds allow six operating spindle rpm's. Torque readings are obtained directly from a dial on the instrument. This multiplicity of measurements allows a more complete definition of the plastic flow curve (shearing stress versus rate of shearing strain). The instrument constants have been adjusted so that the slope of the linear portion of the flow curve may be obtained as the difference between the 600 and 300 rpm torque readings.[8] This slope is defined as plastic viscosity (or rigidity) and is given in centipoise.

Apparent viscosity at 600 rpm is obtained as one half the 600 rpm torque reading. These and other relationships are illustrated by the following equations and Figure 6.4.

(A)

(C)

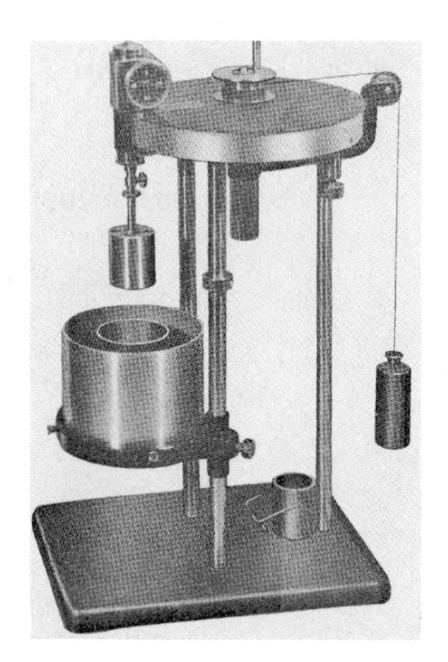

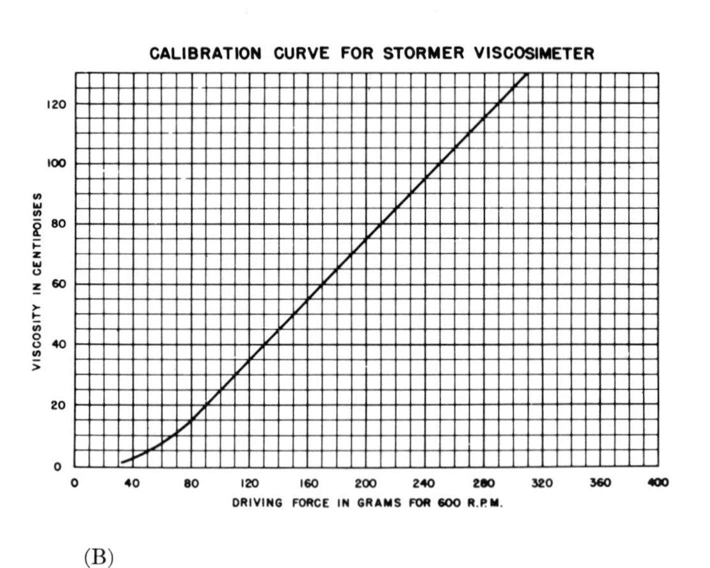

(B)

Fig. 6.3. Viscosity measuring devices. **(A)** Marsh funnel and mud measuring cup. **(B)** Stormer viscosimeter and calibration chart. **(C)** Fann V-G meter.

(6.1) $$\mu_p = \Phi_{600} - \Phi_{300}$$

(6.2) $$\mu_{aF} = \tfrac{1}{2}\Phi_{600}$$

(6.3) $$Y_b = \Phi_{300} - \mu_p$$

where μ_p = plastic viscosity, cp

μ_{aF} = apparent viscosity, cp

Y_b = Bingham yield point, lb/100 ft^2

Φ = torque readings from instrument dial at 600 and 300 rpm.

From these relationships:

(6.4) $$Y_b = 2(\mu_{aF} - \mu_p)$$

or

(6.4a) $$\mu_{aF} = \mu_p + \tfrac{1}{2}Y_b$$

True yield point (Figure 6.2) is normally defined by equation (6.5) for plastic or Bingham fluids.[6,11]

(6.5) $$Y_t = \tfrac{3}{4}Y_b$$

In summary, the Marsh funnel viscosity is useful as a comparative value and is recorded in seconds. Stormer viscosity is an apparent viscosity expressed in centipoise, but has limited value since it is valid at only one rate of shear. The Fann viscosity is the plastic viscosity and represents the rate of change of shearing stress with respect to shearing strain over the linear portion of the consistency curve. The latter, along with yield point, is useful in hydraulic calculations, as is shown in the next chapter.

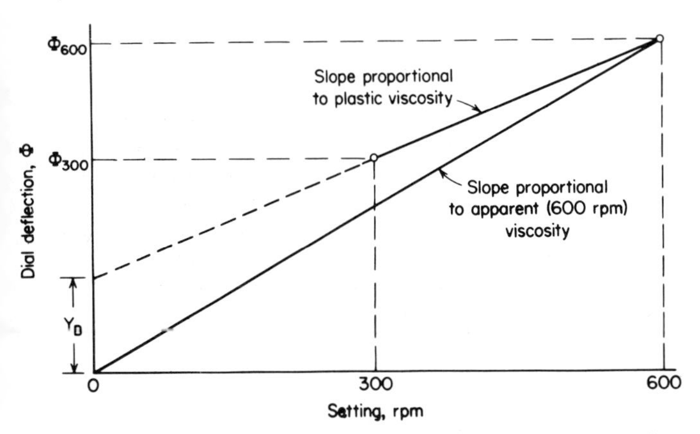

Fig. 6.4. Measurement of plastic flow properties with the Fann V-G meter. By adjustment of instrument constants, μ_p and Y_B are readily obtained from 600 and 300 rpm dial readings.

Plastic viscosity is due to friction between solid particles in the mud and the viscosity of the dispersed phase (base liquid). Normally, it is not appreciably changed by chemical treatment and is almost entirely dependent on solid content. Exceptions to this rule are surfactant muds in which the clay solids exist as dispersed flocs (see Figure 6.17). This state of dispersion lowers plastic viscosity by reducing inter-solid friction.[12] Ordinary thinners (viscosity reducers) lower apparent viscosity by reducing yield point, and have little or no effect on plastic viscosity.[3] (Note Eq. 6.4a). No means of computing either apparent or plastic viscosity from Marsh funnel time is available. The mud temperature at which these measurements are made should be recorded.

6.23 Gel Strength

The gel strength of a mud is a measure of the shearing stress necessary to initiate a finite rate of shear. These measurements are normally taken and reported as initial gel strength (zero quiescent time) and final gel strength (ten minutes quiescent time). Fundamentally, for Bingham fluids initial gel strength and true yield value should be the same; however, such is not the observed case.[7] This discrepancy is probably due to

(1) the failure of drilling muds to behave as Bingham fluids at low rates of shear

(2) the impossibility of measuring initial shear at exactly zero quiescent time.

Initial gel strength has qualitative usefulness but should not be confused with true yield point. A correlation between initial gel strength and Bingham yield is shown in Figure 6.5. Gel strengths are measured with either the Stormer or Fann instruments as the shear stress necessary to cause spindle movement at a very low shearing rate.

6.24 Filtration Test

The filtration, water loss, or wall building test is conducted with a filter press as indicated in Figure 6.6.

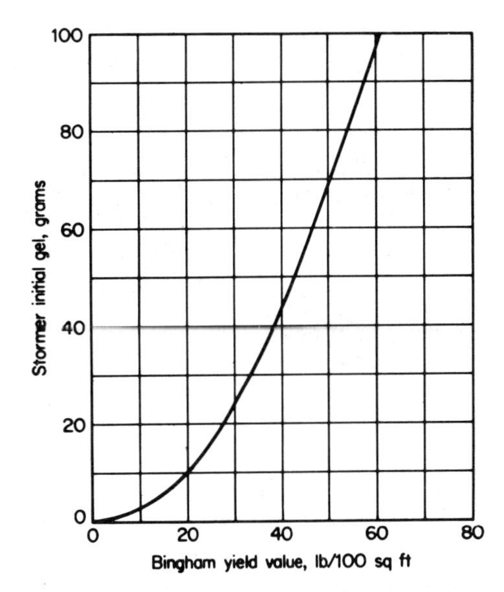

Fig. 6.5. Correlation of stormer initial gel strength with Bingham yield value. After Melrose and Lilienthal[7], courtesy AIME.

It consists of determining the rate at which filtrate (continuous liquid phase of the mud) is forced from the sample under a specified pressure (100 psig). Both the filtrate volume (cc's) and the mud cake thickness (32nd of an inch) are reported. The standard time of the test is 30 minutes, hence the water loss is reported as cc's per 30 minutes.

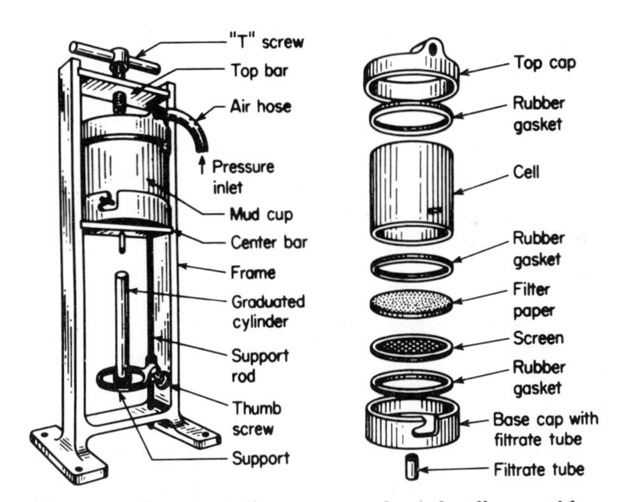

Fig. 6.6. Standard filter press and mud cell assembly.

In field testing, it is common practice to double the $7\frac{1}{2}$ minute filtration loss and report this as the 30 minute figure. This procedure is based on the observation that

$$V_2 = V_1 \sqrt{t_2/t_1} \tag{6.6}$$

where V_2 = filter loss at t_2, cc

V_1 = filter loss at t_1, cc

t_1, t_2 = filtration times, min

This procedure does not account for the initial spurt (high filter loss) period which occurs before the mud solids bridge on the filter paper. For some muds this may be a considerable volume and should be handled by the procedure indicated in Figure 6.7.

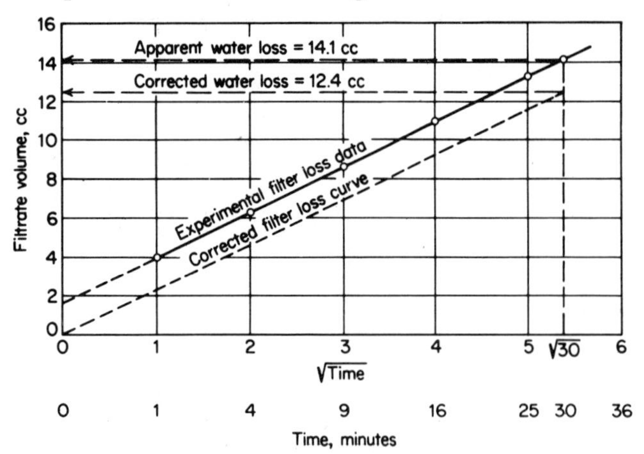

Fig. 6.7. Method of correcting observed water loss for *initial spurt period*. Experimental data is plotted (solid line). A corrected line parallel to the actual and through the origin is used to determine the corrected or API water loss. Note that correction of 1.7 cc is made.

Filtration loss may be corrected for temperature changes by

$$V_2 = V_1 \sqrt{\mu_1/\mu_2} \tag{6.7}$$

where V_2 = corrected water loss at T_2
V_1 = measured water loss at T_1
μ_1 = viscosity of liquid phase at T_1
μ_2 = viscosity of liquid phase at T_2
T_1, T_2 = temperatures in question

Stated simply, the filter loss of a mud is proportional to the square root of the filtrate viscosity. This is true only if the colloidal properties of the mud are unaltered, and hence filtrate viscosity is the only variable. The viscosity of water at various temperatures is given in Figure 6.8.

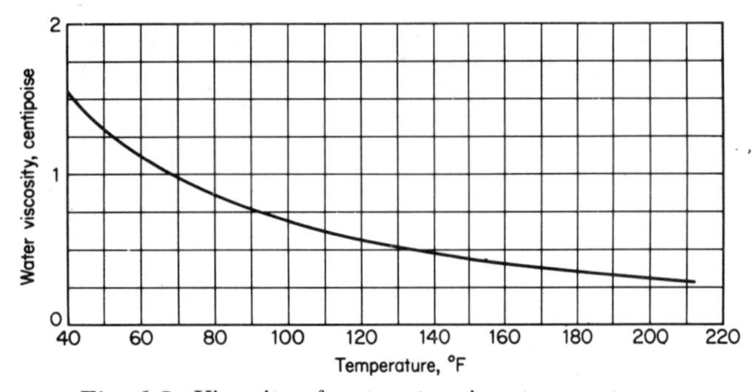

Fig. 6.8. Viscosity of water at various temperatures.

Later, we will have a closer look at filtration theory with regard to formation contamination.

6.25 Other Mud Tests

Other tests performed on drilling muds are given in Table 6.1.

TABLE 6.1
MISCELLANEOUS MUD TESTS*

Test	*Apparatus*	*Purpose*
pH (acidity or alkalinity)	pH papers or pH meters	Guide to chemical treatment. Certain treating materials require certain pH range for proper functioning.
Filtrate analysis	Standard chemicals such as acid, indicators, etc.	Determination of contaminant and selection of chemical treatment necessary.
Sand content	Screens, measuring tube, centrifuge	Determination of sand content of mud, to prevent abrasion of pump and drill pipe.
Oil, water, solids determination	Distillation kit	Determination of these contents as guide to control of desired properties.

*These tests and others are completely discussed in references 2-5.

6.3 Basic Functions of the Drilling Fluid

The general functions of a drilling fluid are:

1. To cool and lubricate the bit and drill string
2. To remove and transport cuttings from the bottom of the hole to the surface
3. To suspend cuttings during times when circulation is stopped
4. To control encountered subsurface pressures
5. To wall the hole with an impermeable filter cake

In performing these duties a drilling fluid should not:

6. Require excessive pump pressure at the desired circulation rate
7. Hamper evaluation of the productive possibilities of encountered formations
8. Adversely affect the productive potential of possible pay sections
9. Allow suspension and continual circulation of excess and/or abrasive solids such as cuttings, encountered clays, and fine sand.
10. Corrode the drill string
11. Reduce rate of penetration

As we shall see, it is probably impossible for any one drilling fluid to satisfy all those requirements optimally. Consequently, the choice of mud type for a specific instance is governed by those functions which are the most critical to the well in question, and some sacrifice of other desirable properties must be made. The relative ability of muds to perform most of these duties is, in general, determined by the standard tests just discussed.

6.31 Cooling and Lubrication of Bit and Drill String

The heat generated by friction at the bit and other contact points between the drill string and hole wall is absorbed by the drilling fluid. The ability of a mud to absorb heat depends on its specific heat and the circulation volume. Normal circulation requirements as dictated by other functions are adequate for this purpose and ordinarily it is given no special consideration, even in air or gas drilling.

Improper lubrication of the bit and drill string will cause excessive torque and reduced bit bearing life. A relatively recent development to improve drill string lubrication are the extreme pressure (E.P.) lubricants developed by Gulf Oil Corp.[13] These products coat metal surfaces such as bit bearings with a high strength lubricating film, thereby reducing metal-to-metal friction and reducing wear. Very striking results have been observed in both laboratory and field tests, and these products represent a considerable contribution to reduction of drilling costs. The addition of oil, such as in oil emulsion muds, also improves down hole lubrication and reduces bit wear.

Although there is no standardized test procedure for measuring the lubricity of drilling muds, the Gulf researchers have adapted the Timken Test Machine for this purpose with highly satisfactory results. The measurement obtained is the wear experienced by metal-to-metal surfaces at various imposed normal forces.

6.32 Cutting Removal and Transportation

The rapidity with which cuttings or chips are removed from below the bit has a considerable effect on drilling efficiency and hence on the rate of penetration. It is the mud velocity, however, rather than any specific property, which chiefly governs this factor. In this section we will be concerned only with the upward transportation of cuttings in the drill string-borehole (or casing) annulus.

Numerous investigators have studied the ability of mud to lift cuttings.[14-17] As in all fluid flow problems, different equations are applied depending on whether the flow is laminar (viscous) or turbulent. Figure 6.9 shows ideally the velocity profiles in an annulus. In laminar or viscous flow, there is a much larger velocity variation across the annulus than in turbulent flow. In laminar flow, a cutting fortunate enough to ride the center (point of maximum velocity) will reach the surface quickly, while those close to the walls will move upward quite slowly: Furthermore, flattish particles will tend to overturn, thereby presenting a minimum area to the direction of mud flow, and will alternately tumble and rise, as is shown in Figure 6.10. This greatly decreases their net upward velocity. The tendency to overturn is caused by unequal fluid velocity at the particle edges as depicted in Figure 6.11. Most bit cuttings are flattish and will behave in this manner. Exceptions are sand grains, which are more nearly spherical.

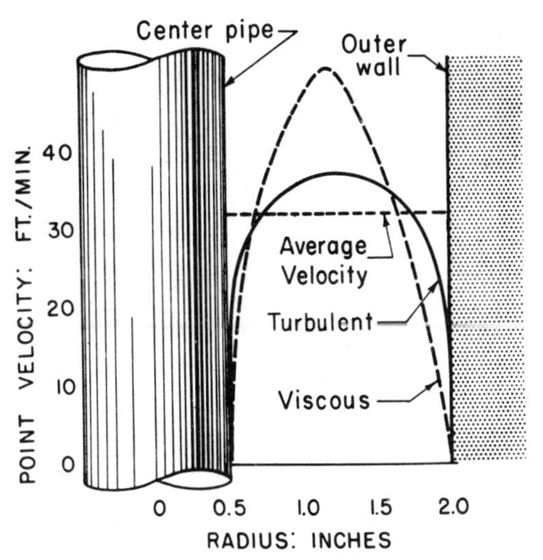

Fig. 6.9. Comparison of viscous and turbulent velocity distributions in annulus. After Williams and Bruce,[17] courtesy AIME.

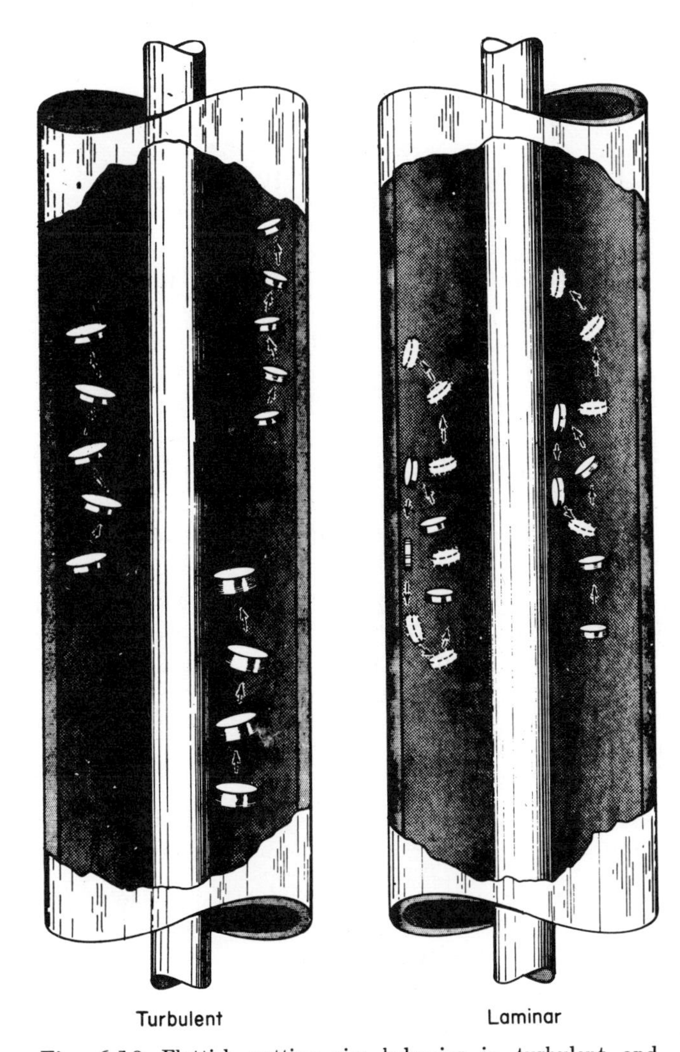

Fig. 6.10. Flattish cutting rise behavior in turbulent and laminar flow. After Williams and Bruce,[17] courtesy AIME.

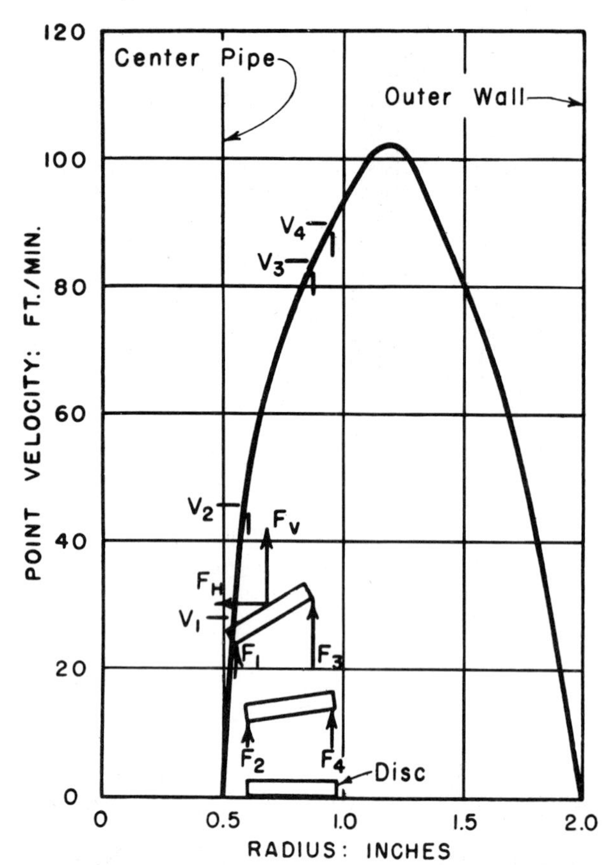

Fig. 6.11. Overturning torque effect on flattish cuttings in laminar flow. After Williams and Bruce,[17] courtesy AIME.

In turbulent flow, the velocity profile is much flatter and the overturning will not occur (see Figure 6.9). Under this condition the particles will keep their maximum area normal to the mud flow and will rise steadily.

Williams and Bruce[17] conclude that low viscosity, low gel strength muds are the most effective cutting lifters at a given flow velocity. This follows from the general Reynold's number definition of viscous or turbulent flow, since the velocity at which turbulent flow occurs is lower for low viscosity muds. This conclusion is discussed analytically in Chapter 7.

Due to the alternate rise and fall of flat cuttings in laminar flow, no equation will be entirely satisfactory. Pigott[14] has applied Stokes' law:

$$(6.8)\quad v_{ls} = \frac{148\, d_c^2(\rho_s - \rho_m)}{\mu} \quad \text{for spherical particles}$$

and

$$(6.9)\quad v_{ls} = \frac{57.5\, d_c^2(\rho_s - \rho_m)}{\mu} \quad \text{for flat cuttings}$$

where v_{ls} = maximum (or terminal) slip velocity of cutting in laminar flow, ft/min

d_c = cutting diameter, in.

ρ_m, ρ_s = mud and cutting density, respectively, lb/gal

μ = mud viscosity, cp (normally used as the Stormer 600 rpm value)

148 and 57.5 = dimensional constants which also include drag coefficients (flat cuttings fall about 40% as fast as spheres)

Turbulent slip velocities are obtained from alterations of Rittinger's equation. Williams and Bruce[17] have used

$$(6.10)\quad v_c = 170\sqrt{\frac{d_c(\rho_s - \rho_m)}{\rho_m}} \quad \text{for spheres}$$

$$(6.11)\quad v_c = 133\sqrt{\frac{t_c}{d_c}}\sqrt{\frac{d_c(\rho_s - \rho_m)}{\rho_m}} \quad \text{for flat cuttings}$$

where v_c = uncorrected cutting slip velocity in turbulent flow, ft/min

t_c/d_c = thickness to diameter ratio of cutting

The values of v_c must be corrected for wall effect, or the fact that the cutting itself detracts from the annular area. This is accomplished by the empirical expression

$$(6.12)\quad v_{ts} = \frac{v_c}{1 + d_c/d_a}$$

where v_{ts} = turbulent slip velocity of cutting, ft/min

d_a = hydraulically equivalent diameter of annulus (hole diameter — pipe diameter), in.

Example 6.1

(a) What is the slip velocity of a 0.50 in. diameter sphere under laminar flow conditions in a 30 cp, 10 lb/gal mud? Assume ρ_s = 21.7 lb/gal.

Solution:

Using Eq. (6.8):

$$v_{ls} = \frac{148(0.5)^2(21.7 - 10)}{30} = 14.4 \text{ ft/min}$$

(b) What would the maximum falling rate be for the system in turbulent flow? Ignore wall effect.

Solution:

Using Eq. (6.10):

$$v_c = 170\left[\frac{(0.5)(21.7 - 10)}{10}\right]^{\frac{1}{2}} = 130 \text{ ft/min}$$

(c) What is the turbulent slip velocity corrected for wall effect? Hole size = 8 in.; drill pipe = $4\frac{1}{2}$ inches.

Solution:

Using Eq. (6.12):

$$v_{ts} = \frac{130}{1 + \frac{0.5}{3.5}} = 104 \text{ ft/min}$$

In actual practice little specific attention is given to calculation of cutting slip velocity and rule of thumb annular velocities are applied from experience. It is generally considered that 80 to 150 ft/min annular mud velocity is sufficient to clean the hole under the various conditions. The upper value might be applied in fast drilling areas and/or where hole enlargement exists. The lower could apply to slow drilling, medium hard rock areas providing hole enlargement may be ignored.

6.33 Cutting Suspension

The ability of a mud to suspend cuttings during periods of non-circulation is primarily dependent on its gel strength. Gel strength is an exhibition of thixotropy, or the ability to thicken with quiescence time. A mud behaving like household gelatin would have excellent cutting suspension properties; however, its hydraulic properties would hardly be suitable. Further, the drilling mud thixotropy must be reversible, i.e., it must revert to liquid upon agitation.

Cutting suspension is desirable in the borehole as a preventive to the drill pipe's sticking during shutdowns. Cuttings settling above the bit or at the top of the drill collars can cause the pipe to become stuck and result in an expensive fishing job. It is also desirable for cuttings, sand, or other excess entrained solids to drop out in the mud pits prior to re-entering the pump suction. It is indeed difficult to prepare a practical mud which will gel in the hole, yet not in the pits. Consequently, most treatment of muds is aimed at reducing gel strength rather than increasing it. Again, the importance of gel strength for cutting suspension is largely dependent on the specific case at hand. In hard rock areas where penetration rate is low and hence a relatively small quantity of cuttings is in the hole, gel strength is of minor importance. This is proved by the widespread use of plain water as a drilling fluid. In a soft rock, high penetration rate area, the quantity of cuttings in the hole at any instant is quite large, and gel strength may be given more consideration. It should be mentioned that shale shakers (see diagram in Figure 5.1, page 53) or screens are generally used to remove large cuttings.

Other mud properties affecting cutting suspension are viscosity and density. The viscosity effect on slip velocity in laminar flow is obvious from Stokes' equation. Density affects static suspension since cutting fall is the result of the density contrast between solid and liquid.

6.34 Control of Encountered Subsurface Pressures

We have stated earlier that an important advantage of the rotary method of drilling is automatic control it affords over encountered subsurface pressures. This pressure control is of course due to the pressure exerted by the drilling fluid which may be expressed as

$$p_m = \frac{\rho_m}{8.33} \times 0.433\ D = 0.052\rho_m D \tag{6.13}$$

where p_m = static pressure exerted by mud column at depth D, psig

ρ_m = mud density, lb/gallon

D = depth, ft

It should be noted that p_m is the static pressure and is less than the circulating pressure by an amount equal to the annular hydraulic losses. A mud's borehole density may be greater than its measured ρ_m because of the cutting content acquired while in the hole. Normally this is negligible.

Mud density is then the primary property governing the pressure control function. In many instances water plus encountered clays will be sufficiently dense to control pressures. In other cases large amounts of weighting materials must be used.

6.35 Wall Building Properties

An important mud function is to wall the borehole with a relatively impermeable filter cake. This cake is composed of mud solids which bridge over the minute pores in the rock, with additional solids being deposited as more mud filtrate is forced through. It is often desirable to have a mud which will very quickly produce a highly impermeable wall cake, thereby reducing the filtration loss and cake thickness.

If the pores, fractures, vugs, etc. are too large for the mud solids to bridge, whole mud is accepted by the formation. This is the separate but related problem of lost circulation. In following sections (and chapters) we shall discuss further the possible consequences of poor filtration control. The filtration or wall building test is diagnostic of this function.

The selection of mud properties for a specific job is a compromise between functional and operating considerations. Any benefit due to the improvement of some property must be weighed against the cost of obtaining it. As in all engineering problems, the final solution is dependent on economic considerations. The drilling fluid's effect on formation evaluation, productive potential, and rate of penetration will be covered in later chapters. Pumping requirements are analyzed in the next chapter.

Prevention or retardation of drill string corrosion is an important mud function in many areas. Oil, or chemical inhibitors such as sodium chromate or sodium dichromate are often used for this purpose.

6.4 Composition and Nature of Common Drilling Muds

Water or oil are satisfactory drilling fluids in some instances. In general, however, the functions to be performed require mud properties which cannot be obtained from ordinary liquids. Consequently, a typical mud consists of

(1) a continuous phase (liquid base)
(2) a dispersed gel-forming phase such as colloidal solids and/or emulsified liquids which furnish the desired viscosity, thixotropy, and wall cake.
(3) other inert dispersed solids such as weighting materials, sand, and cuttings
(4) various chemicals necessary to control properties within desired limits.

TABLE 6.2
COMMON DRILLING MUD ADDITIVES

Use	*Trade names*	*Chemical or mineral name*
Weighting materials	Magcobar, Baroid, Milbar, Controlbar, Maccowate	Barite or Barium Sulfate
Clays	Magcogel, Aquagel, Controlgel, Wyo-Jel	Wyoming bentonite
	High Yield, Baroco, Green Band, Controlclay, Macco Kernco	High-yield drilling mud clay
	Salt Gel, Zeogel, Salt Water, Brinegel, Salt Clay	Attapulgite for salt water muds
Lubricant	Bit Lube, Mud Lube	E. P. Lubricants
Thinners	Q-X Quebracho, Tannex, Tanco, MCL Quebracho, Maccotan	Quebracho-organic dispersant mixture
	TannAthin, Carbonox, Controltan, Maccolig	Lignite, mineral lignin
	Magcophos, Barafos, Oilphos, Maccofos	Sodium tetraphosphate
	Kembreak, Lignox, Kembreak, Controlcal, Q-Broxin, Spersene	Modified ligno-sulfonate (wood by-product)
	Alkatan, Hydrotan	Reacted caustic-tannin (dry)
	Emulsite, Hydrocarb, Ligco, Polytone	Reacted caustic-lignite (dry)
	Man-Kem, Man-Tan	Mangrove bark
Fluid Loss Control	My-Lo-Jel, Impermex, Milstarch, Controloid, Macco Starch	Pregelatinized drilling-mud starch
	Driscose	Sodium carboxy-methyl cellulose
Lost Circulation Materials	Mud Fiber, Fibertex, Milfiber, Strata Fiber	Blend of cane and wood fibers
	Fiber-Seal	Blended long-fiber product
	Cell-O-Seal, Jel Flake, Milflake, Sealflakes	Shredded cellophane flakes
	Leather-Floc, Leath-O	Short-fiber, leather product
	Magco-Mica Micatex, Mica	Graded mica
	Magco-Fiber, Silvacel, Controlfiber	Shredded tree bark or wood fiber
	Tuf-Plug	Ground walnut shells
	FormAplug	Time-setting clay cement
	Cord-Seal, Rubber Seal	Shredded rubber tires
	Bridge bag	Graded aggregate
	Magco-Wool, Controlwool	Fibrous mineral wool
	Pana-Seal, Controlite, Strata-Seal	Perlite: a light heat-expanded mineral

TABLE 6.2 (Continued)
COMMON DRILLING MUD ADDITIVES

Use	*Trade names*	*Chemical or mineral name*
Oil Base* and Emulsion† Muds	Jel-Oil, Ken Oil	Oil-base mud
	Control Emulsion	Non-soap o/w and w/o emulsifier
	No-Blok	Dry concentrate for w/o emulsion muds
	Jel-Oil "E"	Oil and emulsifier for o/w emulsion muds
	Speedy-Drill, Seeco-Mul, L.S.D.	Soap-type emulsifier
	Salt Kem	Emulsifier for salt-water muds
For anhydrite or gypsum contamination	Anhydrox	Barium carbonate
For treating cement contamination	Smentox	Sodium bicarbonate
Starch Preservative	My-Lo-Jel Preservative, Impermex Preservative	Formaldehyde
Higher alcohol anti-foaming agent	Magconol, Anti-Foam	Ethyl hexanol
Anti-foaming agents	Controlfoam, Magco-Defoamer	Corning silicone

*Black Magic is an oil-base mud sold by Oil Base, Inc. This firm handles a complete line of products for making and treating oil-base muds.

†The reacted and straight lignite products listed under thinners are also used in the preparation of o/w emulsion muds.

Drilling mud chemistry is quite complicated and, unfortunately, is incompletely understood. This is true of the entire field of surface chemistry and should not be construed as a failing unique to this industry. Most drilling muds are colloids and/or emulsions and it is the gel-forming components which receive the most attention. Certain formations (salt, gypsum, anhydrite, etc.) contaminate many muds, and it is necessary to use various chemicals to restore desired mud properties, and/or inhibit further effect of the formations. From laboratory experimentation and field experience, the mud engineer is able to diagnose these problems and prepare muds suitable for almost any application. One of the complicating factors in this work is the extreme temperature and pressure variation between the surface and the wellbore. A mud exhibiting excellent properties at the surface may be unsuitable at 350°F and 10,000 psi. Normally, the pressure effect is not critical; many mud treating materials, however, have limited temperature (depth) ranges of application because of thermal instability.

In this section we will discuss the compositional nature of some common mud types and those properties

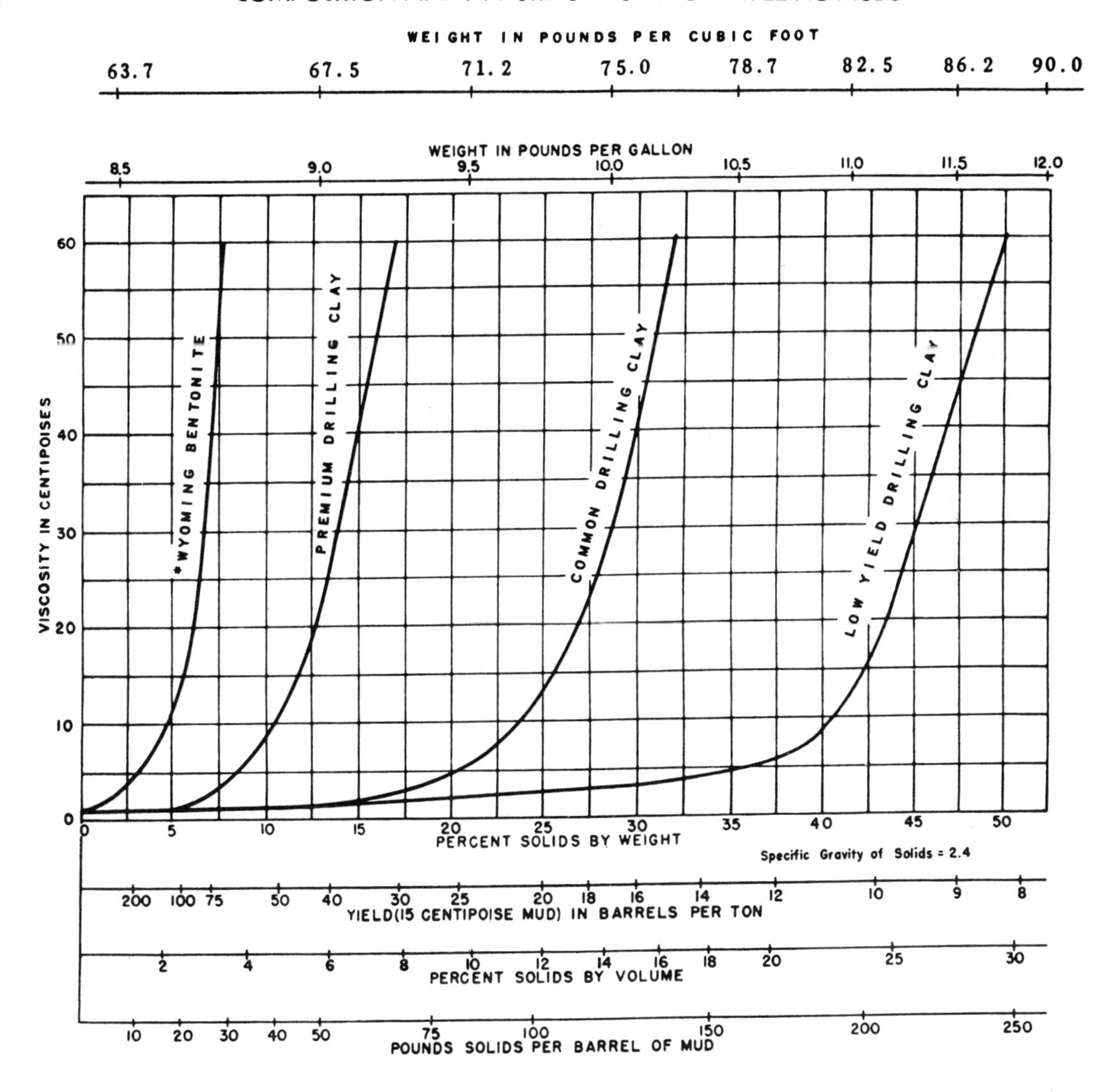

Fig. 6.12. Yield curves for typical clays.[4]

which make their use advantageous. Common chemical treatments will be mentioned for each of these muds and reference will be made to the mud additives listed in Table 6.2. The complexity of drilling mud chemistry is such that it cannot be covered in the allotted space, and the interested reader is referred to Rogers' text[5] as a starting point.

Seemingly, no drilling mud classification scheme is entirely satisfactory and some overlapping always occurs. For our purposes we will include all muds in the following categories:

1. Fresh Water Muds
   a. Simple clay-water mixtures
   b. Chemically treated clay-water mixtures
   c. Calcium treated muds
2. Salt Water Muds
3. Emulsion Muds
   a. Oil-in-water emulsions
   b. Water-in-oil (inverted) emulsions
4. Oil Base Muds
5. Surfactant Muds

6.41 Fresh Water Muds

The basic ingredients of these muds are fresh water and suspended clays. Desirable properties for specific conditions are obtained by adding various materials to this basic mixture.

Clay-Fresh Water Suspensions

Certain clay minerals, when ground to colloidal size and contacted with water, readily hydrate (adsorb water) to form stable colloids. The relative ability of various clays to form viscous colloidal mixtures is shown in Figure 6.12. Sodium bentonite has the highest fresh water yield of all clays. In mud work, yield is defined as the barrels of 15 centipoise mud obtained per ton of clay; for bentonite, this yield is approximately 100 barrels. Bentonite is chiefly composed of the clay mineral montmorillonite. Wyoming is the principal source of sodium bentonite; that particular bentonite in which Na^+ is the dominant ion. This grade adsorbs more water, yielding higher viscosity at lower clay content, than do other clay minerals. In other words, it forms a better colloid in fresh water.

In some areas, the natural clays encountered have sufficient quality to form satisfactory fresh water muds and no commercial clays need to be added. It is often necessary, however, to supplement natural clays with sufficient bentonite to obtain the desired properties. Such additions increase viscosity and gel strength, and reduce filtration loss. In general, simple water-clay systems are suitable for shallow or upper hole drilling in areas where contaminating beds are no problem. Lost circulation and/or weighting materials may be added as needed.

When contaminating formations such as salt (NaCl), gypsum ($CaSO_4 \cdot nH_2O$), and/or anhydrite ($CaSO_4$) are encountered, bentonite treated muds are unsatisfactory. These contaminants are a source of calcium ions which cause flocculation of the clay particles. This flocculation causes increases (sometimes drastic) in water loss, viscosity, and gel strength; and chemical additives must be used to restore the desired properties. Where the source of calcium contamination is small and of predictable magnitude, such as a cement plug, it may be treated out of the system directly by using sodium bicarbonate or barium carbonate. These remove the soluble calcium as precipitated calcium carbonate. Various polyphosphates, caustic soda and quebracho, and lignin or humic compounds are commonly used thinners or viscosity reducers. Organic colloids such as pregelatinized starch or carboxymethylcellulose (CMC) are used to reduce water loss. Muds which have been treated with the caustic-quebracho mixture turn red and are commonly called *red muds*. If considerable salt is encountered, its concentration will eventually increase to the point where the mud must be classed as a salt water mud. The dividing salinity between fresh and salt water types is rather vague, but is generally considered as 10,000 parts per million (1% salt by weight).

Calcium Treated Fresh Water Muds

If considerable calcium ion (Ca^{++}) contamination is anticipated, it is a common and economical practice to pretreat the system with calcium. Such muds have been called lime base, limed, lime treated, low lime, gypsum, and calcium treated muds, depending on the calcium source and the manner and degree of treatment. Therefore the difference between calcium contaminated and calcium treated muds is that in the latter the calcium is not treated out, but is allowed to remain in the mud for special benefits. These muds in general have the ability to tolerate other flocculating salts (up to 50,000 ppm NaCl), they contain high solid percentages at low viscosities, and have low and relatively easily maintained filter loss and gel strengths. These properties have made their use exceedingly widespread, particularly in deep drilling areas.

Figure 6.13 shows the effect of lime, $Ca(OH)_2$, on the viscosity of a 6% bentonite mud. Note that small additions greatly increase viscosity, but that further additions reverse this trend until finally, little effect is noted. This is the lime hump, the magnitude of which depends on natural clay content, bentonite content, mud density, and previous chemical treatment. The low viscosity exhibited by lime muds is due to the combined effect of base-exchanging the clay from sodium to calcium base and the resultant partial loss of adsorbed water, as is shown in Figure 6.14. The clay particle's effective size (clay + adsorbed water) reduction allows greater freedom of movement and hence reduces the viscosity.

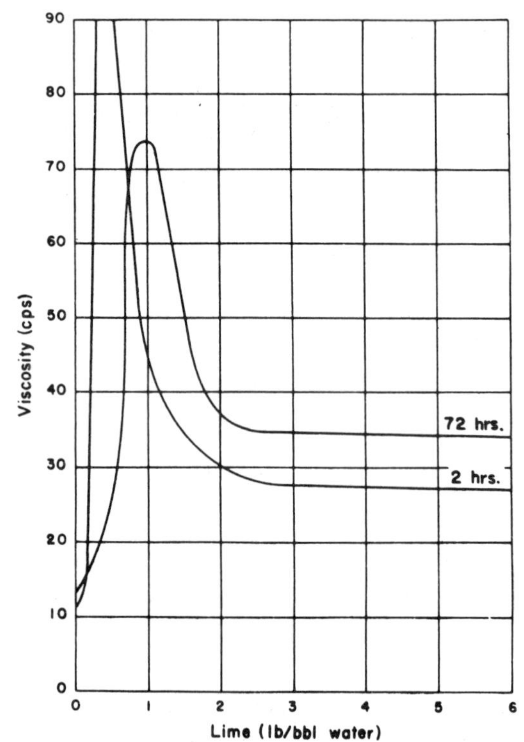

Fig. 6.13. Effect of lime on the viscosity of a 6% bentonite — fresh water mud.[3]

Various thinners, organic colloids, and other chemicals may be used to further control mud properties. Caustic soda (NaOH) and quebracho mixtures are powerful and commonly used thinning solutions. Other compounds such as lignite and modified lignosulfonate are also satisfactory. Polyphosphates are not used in

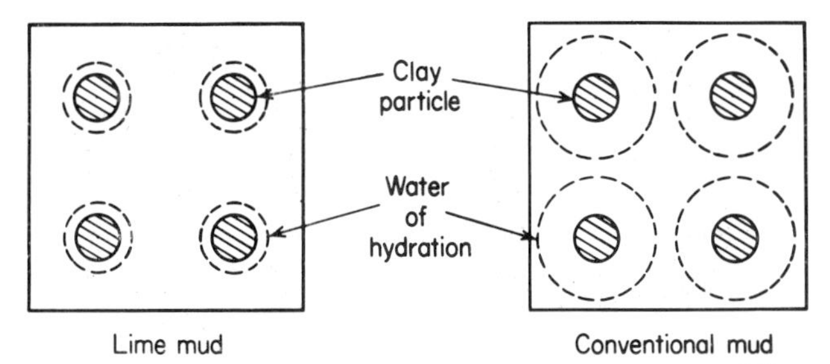

Fig. 6.14. Relative thickness of adsorbed water layer of lime treated and conventional fresh water muds.[3]

calcium treated muds. Filter loss may be controlled with either starch and/or CMC, depending on the pH, salinity, and water loss desired. In muds having pH > 12 and/or salinity > 250,000 ppm, starch may be used without a preservative (bactericide). At values below these a preservative must be added to prevent its fermentation. CMC is effective in all muds except those of high salinity. The choice between the two is economic, with starch being generally preferable if very low water loss is desired. Figure 6.15 illustrates the relative effectiveness of these additives in calcium treated muds.

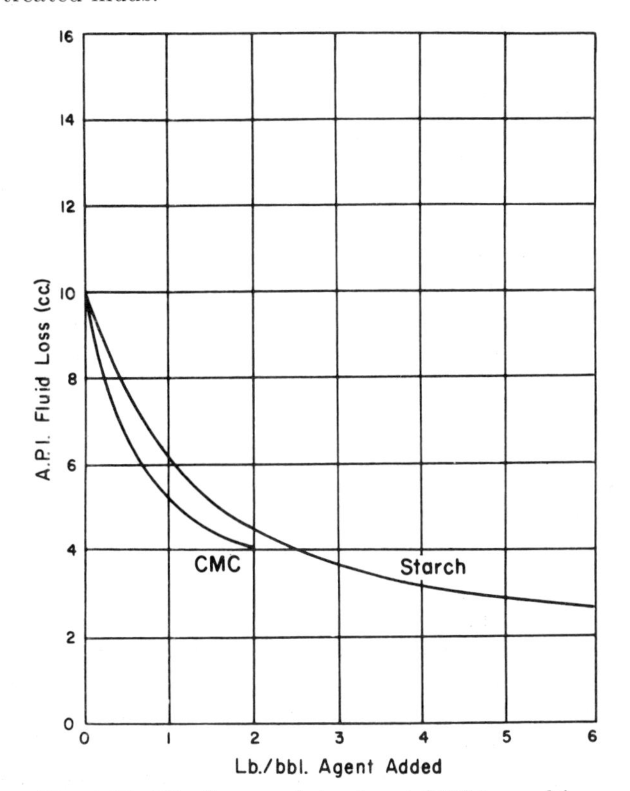

Fig. 6.15. Effectiveness of starch and CMC in a calcium treated mud.[3]

Lime treated muds are differentiated by the degree and type of treatment they have received. Several classifications are used, but we will consider only two sub-types:

(1) High lime-treated muds
(2) Low lime-treated muds

High Lime-Treated Muds

Muds in this category generally contain 2 to 4 lb/bbl caustic soda, 2 to 4 lb/bbl of organic thinner, and 2 to 20 lb/bbl of slaked lime. The pH is generally above 12 and the lime content is sufficient to saturate the system. Such muds have excellent properties, such as

(1) Low viscosity and gel strength
(2) Easily maintained low filter loss
(3) High resistance to contamination
(4) May be weighted up to 20 lb/gal, i.e., they may contain high solid concentrations at low viscosities.

Their principal disadvantage is a tendency to solidify at high temperatures.[18] Normally this tendency can be predicted from pilot tests;[19] this factor, however, has led to the use of the so-called low lime muds.

Low Lime-Treated Muds

Low lime muds contain less caustic soda, lime, and clay than high pH, high lime muds. Modified lignosulfonate is commonly used as the thinner. These muds have a reduced tolerance for contaminants and exhibit higher viscosities. Closer attention must be given such systems and more frequent treatment is necessary. Their pH is normally from 11.5 to 12.5.

Other classifications of calcium treated muds are sometimes used, such as gypsum-treated or *gyp-muds*. These designations stem from the Ca^{++} source, which is either encountered gypsum and anhydrite or pre-added plaster (commercial calcium sulphate). These are often used when drilling gypsum and anhydrite beds; treatment is similar to the normal lime muds. With the development of new modified lignosulfonate thinners, gyp muds are now widely used in the Gulf Coast and other areas, replacing lime treated muds to a great extent.[43] The system is maintained at relatively low pH (around 9.5), with gel strengths and flow properties being easily maintained. The higher calcium content in the filtrate is more inhibitive to clay swelling, making it desirable for drilling bentonitic shale sections. These muds are also superior to lime muds in high temperature stability.

6.42 Salt Water Muds

Sodium bentonite does not form a satisfactory colloid in salt water. Salt concentrations neutralize the electric charge on dispersed bentonite particles. The positive charges surrounding the solid nucleus are driven closer to the particle which neutralizes the negative charges, allowing flocculation (formation of particle aggregates which are larger than colloidal size). The clay mineral, attapulgite, will, however, hydrate and form a stable suspension in salt water. Such clays are commonly called salt-clays and are used in saline water in about the same manner as bentonite in fresh water. The yield curve for bentonite (Figure 6.14) is used for attapulgite in salt water. Salt clays do not, however, furnish the wall building properties of bentonite, but exhibit thick mud cakes and high water loss. Consequently, starch (or CMC in moderate salinity muds) is normally required for filtration control.

The principal use of salt water muds is in those areas where considerable salt sections must be drilled or in localities where abundant salt water is available for use, such as in offshore, swamp, and seaside locations.[20] Pre-saturation of the system will eliminate down hole solution of salt, thereby keeping the hole more nearly to gauge. Salt water muds, properly treated with organic colloids and attapulgite clays, may exhibit water loss as low as one cc and may be weighted as high as 19 lb/gal. Foaming tendencies may be reduced by adding surface active agents to the system. In general then, the difference between fresh and salt water muds is the type of clay used as the gel-forming phase.

6.43 Emulsion Muds

Oil-in-Water Emulsions

The most common emulsion muds are oil-in-water types, in which the oil is the dispersed phase and exists as small individual droplets. The base mud may be any type of fresh or salt water mud. The stability of such emulsions depends on the presence of emulsifying agents in the mixture. Soaps, lignin compounds, organic colloids such as starch or CMC, and other colloidal solids are common emulsifying agents. While the precise mechanism of emulsion formation is not completely understood, it is generally agreed that the emulsifying agent, whether molecular or colloidal, orients itself around the dispersed fluid droplet as indicated in Figure 6.16. This orientation furnishes like charges to the dispersed droplets causing their mutual repulsion and hence, a state of stable dispersion. Considerable agitation and mixing is required to obtain complete stability and several cycles through the mud system are normally required.

Virtually all types of oil have been used in these muds; however, refined diesel oil having the following properties is considered the most satisfactory:

(1) Uncracked, for stability
(2) High flash point, to minimize fire hazard
(3) High aniline number (> 155), to reduce deterioration of rubber parts in the circulation system
(4) Low pour point, for use in varying temperature.

A further advantage of refined oil is that its odor and fluorescence are more readily distinguishable from that of crude oil. This is of considerable value to the geologist who must inspect the cuttings for oil bearing possibilities.

Normally, the quantity of oil added to the base mud averages near 10%. This amount is dependent on the

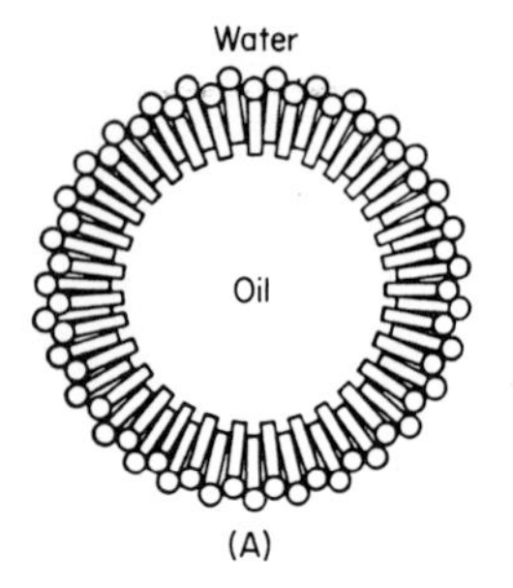

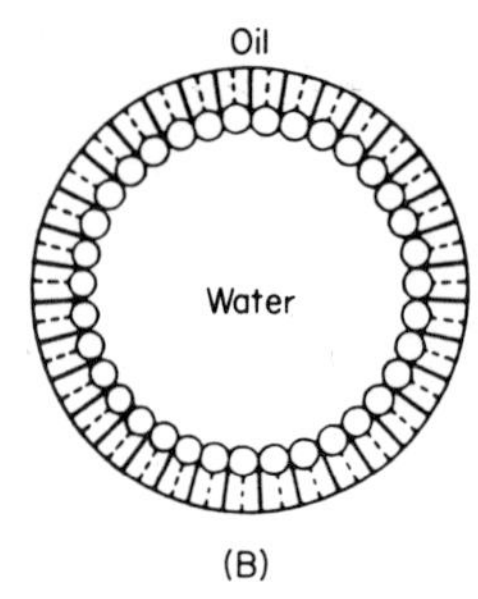

Fig. 6.16. Molecular (or colloidal) orientation of emulsifying agent around dispersed phase droplet in emulsion muds.[3] **(A)** Oil in water. **(B)** Water in oil.

desired properties and is normally determined by pilot testing in a specific case. The addition of oil does not make a good mud from a poor one, but will improve a satisfactory mud. The principal benefits of properly prepared oil-in-water emulsion muds are[21,22]

(1) Increased rate of penetration
(2) Better lubrication of bit and drill string, hence less torque, longer bit life
(3) Retardation of drill string corrosion
(4) Improved mud properties: reduced viscosity and pump pressure, reduced water loss, thin, slick filter cake
(5) Less injury to pay sections (this is actually reflection of (4))
(6) Less *hole trouble*, such as bit balling, heaving shales, hole enlargement.

The disadvantages of these muds are the additional cost of the oil and the previously mentioned problems in cutting analysis. Neither disadvantage is particularly serious and emulsion muds are widely used. The maintenance of emulsion muds is the same as that of the base mud in so far as chemical treatment is concerned. Periodic checks will indicate the necessity of further emulsifying agent or oil additions. The water loss test is a good indicator of emulsion stability; if no free oil appears in the filtrate, the globules are still dispersed and are blocked by the filter cake.

Inverted or Water-in-Oil Emulsion Muds

Muds of this type are the reverse of the above in that water is the dispersed phase. These are not general purpose muds but are used in special applications which justify their additional cost and closer supervision. A typical mud of this type contains from 30 to 60% water (either fresh or saline) emulsified in either diesel or crude oil. The emulsifying agent is generally added and mixed with the oil prior to the addition of water. Clay solids and weighting materials may be added if desired.

The desired mud properties may be obtained by changing the water to oil ratio as indicated by pilot tests. Mud viscosity is normally high; the gel strengths, however, are low. Filtration rates are quite low, usually zero cc by the API test; and any filtrate obtained should be oil. The temperature stability of these muds is good, and they may be used to approximately 200°F. Application at temperatures above this requires special attention.[23,24]

The principal applications of invert emulsion muds are the same as those of oil base muds which are discussed in the next section. These muds do have, however, certain advantages over conventional oil base muds, namely, greater ease of wellsite handling and mixing, and lower cost. Further, any water and most crude oils may be used as materials. The principal disadvantages are possible instability at temperatures above 200°F, possible reversion to an oil-in-water type mud if excess solids or weighting materials accumulate, and the requirement of closer supervision.

6.44 Oil Base Muds

These muds are in general composed of high flash diesel oil (the continuous phase), oxidized asphalt, organic acids, an alkali, various stabilizing agents, and from 2 to 5% water. The asphalt is the colloidal fraction, and provides the wall building property. The combination of the organic acid and the alkali forms an unstable soap which governs the viscosity and gel strength of the mixture. Organic acid additions thin the mud, while alkali additions cause it to thicken. A wide viscosity range is available by these treatments. The water content is contained as a tightly emulsified dispersed phase and cannot be removed without destroying the mud.

Oil base muds are expensive and are used as a special purpose drilling fluid. They are quite insensitive to common contaminants such as salt, gypsum, and anhydrite, since these compounds are insoluble in oil. Their principal uses are

(1) Drilling and coring of possible productive sections, particularly zones whose permeability, and hence productive capacity, is impaired by water, or zones where the interstitial water content of the pay section is to be determined from core analysis.
(2) Drilling of bentonitic (heaving) shales which continually hydrate, swell, and slough into the hole when contacted with water.
(3) High temperature drilling, where possible solidification or other problems make other muds undesirable.
(4) Perforating fluid. Normally, a few barrels *spotted* opposite the zone to be perforated will prevent contamination of the section after it is perforated.
(5) Numerous miscellaneous uses, such as freeing of stuck pipe, corrosion prevention, and remedial work on producing wells.

Plain crude oil is also used as a drilling fluid; however, its lack of viscosity and gel strength, high filtration loss, and volatility (fire hazard) make its applicability quite limited. Oil base muds and their variant, invert emulsion muds, overcome these difficulties in that they have low fluid losses, and may be weighted in excess of 18 lb/gal if desired. Salvaging of this mud for subsequent drilling greatly reduces its per well cost and may make the use of it economically attractive in the light of other benefits.

The primary advantages of all muds having oil as the continuous phase are minimum contamination of potentially productive zones, and relative insensitivity to the common contaminants.

6.45 Surfactant Muds

Muds of this type are variations of previously discussed water base systems and are prepared by adding various nonionic surfactants to these mixtures.[12] The electrolyte present (sodium chloride, calcium sulfate, etc.) governs the choice of surfactants in a specific instance. These surface active agents change the colloidal state of the clays from that of complete dispersion (Figure 6.17(A)) to one of controlled flocculation (Figure 6.17(C)).

The additives used in surfactant muds are marketed under the trade names of DMS (drilling mud surfactant) and DME (drilling mud emulsifier). DMS is used to obtain the desired clay particle flocculation. DME is used in emulsion muds only.

The principal advantages of these systems are

(1) lower plastic viscosity
(2) reduced tendency for solid buildup
(3) high thermal and chemical stability, in excess of 400°F
(4) elimination of ordinary thinners; hence pH may be carried at natural level (normally 7 to 9)
(5) minimum swelling effect on clay bearing zones by filtrate; hence less damage to potential pay zones; also, less trouble in drilling bentonitic (heaving) shales
(6) greater effectiveness of filtration additives

Although these muds were designed for use in deep, high temperature wells, their inherent advantages may extend their use to other areas. Detailed treating and precautionary measures for these systems appear in various mud manuals.[2,3]

6.5 Drilling Hazards Dependent on Mud Control

Certain hazards of drilling may be either avoided or overcome by proper control of mud properties. Some of these were discussed briefly in conjunction with the advantages of certain mud types. Let us, however, re-emphasize certain of these and discuss some which have not been mentioned, namely:

1. Salt section hole enlargement
2. Heaving shale problems

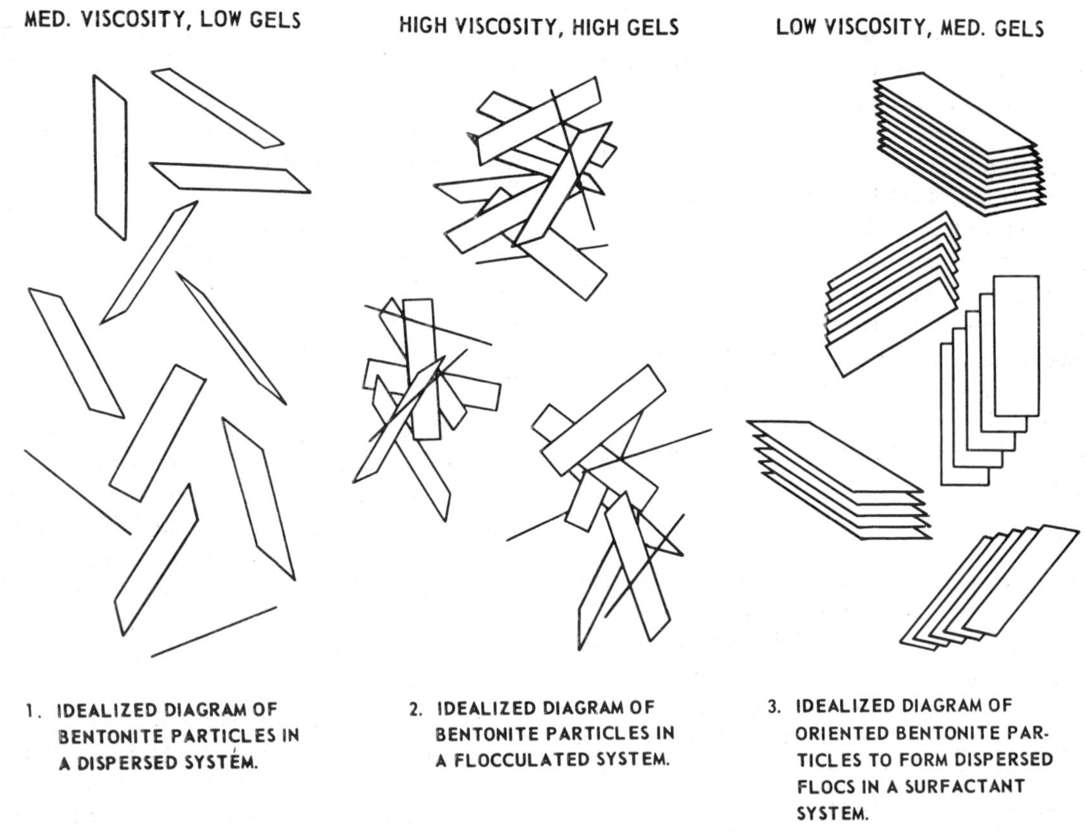

Fig. 6.17. Idealized dispersion states of bentonite particles in dispersed, flocculated, and controlled flocculated systems.[3]

3. Blowouts
4. Lost circulation

6.51 Salt Section Hole Enlargement

In many areas considerable thicknesses of rock salt must be penetrated. Solution and erosion of these beds can cause excessive hole enlargement which in turn may be a source of future trouble and expense.

(1) In case of drill string failure, the enlarged hole makes *fishing operations* (attempts to retrieve the drill string) exceedingly difficult.
(2) Larger mud volumes are required to fill the system, hence treating costs are higher.
(3) Large cement volumes are required for casing operations if fill-up through the section is to be attained.

The principal means of avoiding these problems is to prepare a salt saturated mud system prior to drilling the salt, thus avoiding the dissolving effect.

6.52 Heaving Shale Problems

Some areas are characterized by shale sections containing bentonite or other hydratable clays which continually adsorb water, swell, and slough into the hole. Such beds are referred to as heaving shales and constitute a severe drilling hazard when encountered. Pipe sticking, excessive solid buildup in the mud, and hole bridging are typical resultant problems. Various treatments are sometimes successful, such as

(1) Changing mud system to inhibitive (high calcium content) type such as lime, gyp, etc., which reduces tendency of the mud to hydrate water sensitive clays
(2) Increasing circulation rate for more rapid removal of particles
(3) Increasing mud density for greater wall support
(4) Decreasing water loss of mud
(5) Changing to oil emulsion mud
(6) Changing to oil base mud

Depending on the severity of the occurrence, any of the above may be satisfactory. The last resort is changing to oil base or water-in-oil emulsion mud.

6.53 Blowouts

A blowout occurs when encountered formation pressures exceed the mud column pressure which allows the formation fluids to blow out of the hole. This is the most spectacular, expensive, and highly feared hazard of drilling. Needless to say, proper mud density is the principal factor in avoiding this problem; however, borehole pressure reductions below mud column pressures are in many instances caused by too rapid withdrawal of the drill string. This is known as pipe pulling suction (swabbing) and has become recognized as a large factor promoting blowouts. This is particularly true in areas where a very delicate overbalance of formation pressure is necessary. The magnitude of the pulling suction depends on speed of pipe withdrawal, hole-pipe clearance, and mud viscosity and gel strength. This is a further argument for keeping mud viscosity at a minimum.

6.54 Lost Circulation

Lost circulation is defined as the loss of substantial quantities of whole mud to an encountered formation. This is evidenced by the complete or partial loss of returns (returning annular mud flow). The annular mud level may drop out of sight and stabilize at a pressure in equilibrium with formation pressure. Lost circulation occurs when formation permeability is sufficiently great to accept whole mud; the voids are too large to be plugged by the solids (clay, cuttings, etc.) in the mud. A further obvious requirement is that the mud column pressure must exceed the formation pressure. Some of the undesirable effects of lost circulation are the following:

(1) Mud costs prohibit continuance of drilling without returns. Per well mud bills in excess of $100,000 have been reported for wells in which this problem was severe.
(2) The drop in annular mud level may cause a blowout.
(3) No information on the formation being drilled is available since no cuttings are obtained.
(4) The possibility of sticking the drill pipe with a resulting fishing job is increased.
(5) Loss of drilling time and consequent cost increase is incurred.
(6) If the lost circulation zone is a potential pay zone, considerable productivity impairment may result.

The types of formations to which circulation may be lost have been classified in three groups:[4]

(1) Coarsely permeable rocks, such as gravels, reef, and irregular limestones.
(2) Faulted, jointed, and fissured formations such as:
    (a) those with naturally occurring fractures
    (b) those in which the fractures are induced or caused by mud column pressures,
(3) Cavernous and open fissured formations. As an extreme example of this type, consider the effect of drilling into the Carlsbad Caverns.

Bugbee[25] has shown that permeabilities in excess of 300 darcys are necessary for the loss of mud solids to permeable sands. This conclusion was based on a relationship between pore size and permeability (recall from Chapter 2 that dimensionally, permeability is an area). The only clastic rocks which can exhibit such permeabilities are very coarse gravels. These beds are rarely encountered in drilling and they are probably a minor contributor to lost circulation in general. Vugular limestones having finger size (and larger) holes, however, are a common source of trouble in many areas (West Texas, Mid-Continent).

Faulted, jointed, and fissured zones may occur in any rock type and are probably the most common source of lost circulation. In these zones, the fractures are held closed by the forces existing in the earth at that depth. However, when the mud column pressure exceeds the stress holding a fracture closed, it opens, accepts the mud, and loss of circulation occurs. Further, rocks which are not naturally fractured may be *broken down* (fractured) by excessive bore hole pressures. High mud density contributes to this; numerous writers have shown, however, that pressure surges caused by rapid pipe running are a major cause of induced fractures.[26-28] This is the exact opposite of the pulling suction which was mentioned as a blowout cause.

The hazards created by cavernous or open fissured formations need no explanation, as it is obvious that they may constitute the most severe case of lost circulation. Natural caverns and open fissures several feet thick have been drilled into. In Illinois, old mine tunnels are occasionally encountered which furnish a considerable problem. Severe cases in this category cannot be successfully plugged with bridging materials and cement. Consequently, drilling without returns (blind drilling) is continued for some distance below the cavernous section. Casing is then set through the zone and cemented; after this, normal drilling is resumed.

Methods of Combating Lost Circulation

The alteration of mud properties, primarily density, may help or even cure some lost circulation problems. This approach includes the use of air, gas, and aerated mud in those cases where they are applicable. Fluid flows from the formation (including blowouts) may eliminate this course of action. Sometimes a waiting period allowing the formation to adjust to the new pressure conditions may alleviate the situation.[29] The *spotting* of plugs containing mud, cement, and lost circulation materials opposite and into the problem zone is a common cure.[30] *Spotting* involves pumping the material down the drill pipe with a sufficient (calculated) volume of displacing fluid behind it to position the plug at the desired level in the hole. This requires a knowledge of the zone's location, as it is not necessarily at the bottom of the hole. The location of the loss zone may be determined with various downhole surveying devices.[31]

Lost circulation materials are commonly circulated in the mud system both as a cure and a continuous preventive. These materials are undesirable from a mud property and pumping equipment standpoint, but are tolerated as a necessary evil. While nearly everything has been used for this purpose, the more common materials are listed in the following classifications:

(1) Fibrous materials: hay, sawdust, bark, cottonseed hulls, cotton bolls, cork.
(2) Lamellated (flat, platy) materials: mica, cellophane.
(3) Granular bridging materials: nut shells, perlite, pozzolanic materials, ground plastic.

Fibrous and lamellated materials are most effective in coarsely permeable rocks where the voids are relatively small. Larger openings require use of a granular material having sufficient strength to form a bridge across the void. These effects are well illustrated by the work of Howard and Scott.[32] In their experiments, mud containing various concentrations and types of lost circulation materials was circulated through different sizes of slots or simulated fractures. The slots were considered sealed when the plugging materials withstood 1000 psi differential across the fracture. The main conclusions of this study were:

(1) Granular bridging materials are the most effective lost circulation agent in fractured rocks.
(2) Concentrations of 20 lb/bbl gave maximum results and little increase in plugging will result from higher concentrations.

The materials evaluated and the maximum size fractures sealed by each are shown in Figure 6.18.

Fundamental Concepts of Bridging

Fractures cause a major portion of severe lost circulation problems and since, in general, their sealing requires a granular material having sufficient strength to bridge the opening, let us consider the mechanics of bridging. An early worker on this subject was Coberly,[33] who was primarily concerned with determining the proper liner slot size necessary to prevent the entry of unconsolidated sand into the bore hole of producing oil wells. For the time being we will skip Coberly's problem and look only at the results of his work as pertaining to lost circulation. His problem, then, was to determine the size slot necessary to exclude certain sized sand (a granular, bridging material). This is the exact reverse of studying what granular particle sizes will bridge a given slot, and the results should be equivalent.

From numerous bridging tests both with steel balls and actual oil sands, Coberly reached the following conclusions which have long been verified by practice:

(1) Spherical grains form stable bridges on slots up to twice their own diameter.
(2) Angularity does not greatly affect the bridging powers of such particles.
(3) In mixtures, the bridging influence of the larger grains is great.
(4) For oil sands (or other mixtures having large variations of particle size) the grains will bridge a slot diameter approximately equal to twice the screen size which will pass 90% (retain 10%) of the grains.

Applying these principles to lost circulation additives, one would expect the action of bridging materials to follow the same laws. Application of these principles also requires that the granular materials have sufficient strength to stand the pressure differential across the bridge. This lack of strength explains the failure of many materials to seal large slots.

Properly designed bridging materials then should seal fractures wider than themselves. The bridge is initiated when two of the larger particles start into the fracture at the same time and then lodge against each other.

MATERIAL	TYPE	DESCRIPTION	CONCENTRATION Lb/bbl	LARGEST FRACTURE SEALED Inches (0 .04 .08 .12 .16 .20)
Nut shell	Granular	50%-3/16+10 Mesh 50%-10+100 Mesh	20	
Plastic	"	"	20	
Limestone	"	"	40	
Sulphur	"	"	120	
Nut shell	"	50%-10+16 Mesh 50%-30+100 Mesh	20	
Expanded Perlite	"	50%-3/16+10 Mesh 50%-10+100 Mesh	60	
Cellophane	Lamellated	3/4 inch Flakes	8	
Sawdust	Fibrous	1/4 inch Particles	10	
Prairie hay	"	1/2 inch Fibers	10	
Bark	"	3/8 inch Fibers	10	
Cotton seed hulls	Granular	Fine	10	
Prairie hay	Fibrous	3/8 inch Particles	12	
Cellophane	Lamellated	1/2 inch Flakes	8	
Shredded wood	Fibrous	1/4 inch Fibers	8	
Sawdust	"	1/16 inch Particles	20	

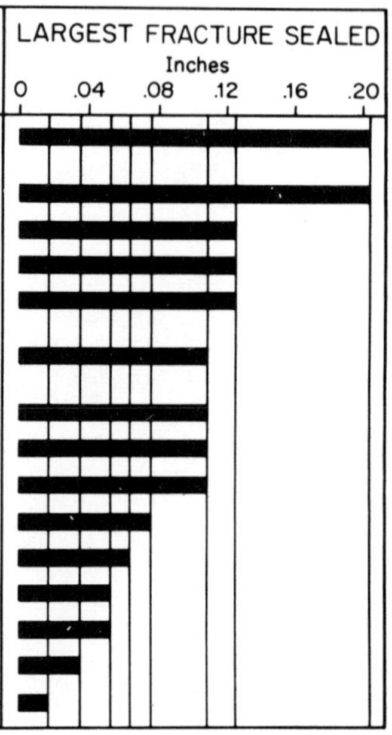

Fig. 6.18. Summary of lost circulation tests. After Howard and Scott,[32] courtesy AIME.

Other smaller particles will then bridge in the openings between the larger, previously bridged particles. This process will continue until the openings become quite small and the problem becomes one of filtration. Visualization of this process implies that the particle size distribution should be considered, so that the proper sizes to fit into successively smaller interparticle voids will be present. Many have mentioned an optimum particle size distribution for lost circulation materials, but no definition is apparent.

A similar problem exists in the concrete mixing industry in the selection of aggregate (gravel) for mortar. Here it is desired to choose an aggregate whose packing gives maximum density (minimum voids or porosity) so that a maximum strength concrete will be obtained. Theoretical work by Furnas [34] substantiated by experiments[35] has shown that the best continuous grading of aggregate sizes is that in which a fixed ratio exists between amounts of successive sizes. In other words, each screen should retain more than the next smaller one by a constant factor (1.05, 1.10, etc.). The grading curve (particle size versus cumulative percent) is then a geometric series.

It would appear that a granular bridging material capable of forming the most dense (hence least porous and permeable) bridge should be the most successful. This approach to lost circulation material sizing needs to be examined more closely.

Other Considerations

Mixtures of material types may be advantageous in many instances. For example, cotton seed hulls, cellophane, etc. can be an aid in sealing the smaller voids after a bridge is formed.

Granular materials chosen should meet requirements other than strength and size.[13] Low specific gravities (< 1.5) will minimize settling in ordinary muds. Pump abrasion is a further consideration and is largely avoided by use of materials of low hardness (2 to 4 on Moh's scale). Both of these factors eliminate the use of sand and gravel which would otherwise serve well.

6.6 Drilling Mud Calculations

The most common mud engineering calculations are those concerned with the changes of mud volume and density caused by the addition of various solids or liquids to the system. The first step is to compute the system volume, which is the sum of the mud in the hole and surface pits. While the surface volume is readily obtained from the pit size, the downhole volume is difficult to determine. Boreholes are not always cut *to gauge* (the same size as the bit) and unless a caliper log is available, which is unusual at the time of drilling, the true hole size must be estimated. In hard rock areas little error may result from assuming bit size to exist; in salt or sloughing sections, however, this will be a gross error. With experience in the area, the mud engineer is able to make reasonable approximations. Those lacking this experience may compute hole volume (borehole less drill string volume) and apply any correction factor deemed applicable.

Consider then the volume and density change of a mud (or water) resulting from the addition of solids. Two basic assumptions must be made:

(1) The volumes of each material are additive. This may immediately raise a question concerning bentonite and water mixtures since it is known that bentonite swells when wet. This expansion is due, however, to the adsorption of water; hence the clay volume increase is at the expense of water volume, and the total volume (clay plus water) is, for practical purposes, unchanged.

(2) The weights of each material are additive.

Writing expressions for these assumptions:

$$V_s + V_{m1} = V_{m2} \tag{6.14}$$

$$\rho_s V_s + \rho_{m1} V_{m1} = \rho_{m2} V_{m2} \tag{6.15}$$

where V_s = volume of solid
V_{m1} = volume of initial mud (or any liquid)
V_{m2} = final volume of mixture
ρ_s = density of solid
ρ_{m1} = density of initial mud
ρ_{m2} = density of final mud

Solving for V_s:

$$V_s = \frac{V_{m2}(\rho_{m2} - \rho_{m1})}{\rho_s - \rho_{m1}} \tag{6.16}$$

As to units, the densities may be in any consistent set; lb/gal or gm/cc are commonly used in field or laboratory problems, respectively. V_s will then be in the same units chosen for V_{m2}; bbl, cc, etc. Equation (6.16) is not particularly useful as it stands, since the net volume of a powdered solid is not readily measurable. However, the corresponding weight to add is

$$\rho_s V_s = \frac{\rho_s V_{m2}(\rho_{m2} - \rho_{m1})}{\rho_s - \rho_{m1}} \tag{6.17}$$

Example 6.2

A 9.5 lb/gal mud contains clay (sp. gr. = 2.5) and fresh water. Compute (a) the volume % and (b) the weight % clay in this mud.

Solution:

(a) Altering Eq. (6.16):

$$\text{Volume \% solids} = \frac{V_s}{V_{m_2}} \times 100 = \frac{\rho_{m_2} - \rho_{m_1}}{\rho_s - \rho_{m_1}} \times 100$$

$$= \frac{9.5 - 8.33}{(2.5)(8.33) - 8.33} \times 100 = 9.4\%$$

(b) Weight % solids $= \dfrac{\rho_s V_s}{\rho_{m_2} V_{m_2}} \times 100 = \dfrac{\rho_s(\rho_{m_2} - \rho_{m_1})}{\rho_{m_2}(\rho_s - \rho_{m_1})} \times 100$

$$= \frac{20.8\,(9.5 - 8.33)}{9.5\,(20.8 - 8.33)} \times 100 = 20.6\%$$

Example 6.3

For laboratory purposes, it is desired to mix one liter of bentonite-fresh water mud having a viscosity of 30 cp. (a) What will be the resulting mud density? (b) How much of each material should be used?

Solution:

From Figure 6.12, solid content = 3.0% by volume. Again, using altered forms of Eq. (6.16):

(a) $.03 = \dfrac{\rho_{m_2} - \rho_{m_1}}{\rho_s - \rho_{m_1}} = \dfrac{\rho_{m_2} - 1.0}{2.5 - 1.0}$

from which

$$\rho_{m_2} = 1.045 \text{ gm/cc} = 8.7 \text{ lb/gal}$$

(b) $V_s = \dfrac{1000(1.045 - 1.0)}{2.5 - 1.0} = 30 \text{ cc} = 2.5 \times 30 = 75 \text{ gm}$

Also:

$$V_{m_1} = V_{m_2} - V_s = 1000 - 30 = 970 \text{ cc water}$$

For certain types of problems it is convenient to express Eq. (6.16) in a different form. Suppose that the quantity of solids (V_s) necessary to increase (or decrease) the density of an initial mud is desired. Then:

$$V_s = \frac{(V_{m1} + V_s)(\rho_{m2} - \rho_{m1})}{\rho_s - \rho_{m1}} \tag{6.16a}$$

where $V_{m1} + V_s = V_{m2}$ (volumes additive)

Solving for V_s gives

$$V_s = \frac{V_{m1}(\rho_{m2} - \rho_{m1})}{\rho_s - \rho_{m2}} \tag{6.18}$$

Example 6.4

(a) How much weighting material ($BaSO_4$, the mineral barite, sp.gr. = 4.3) should be added to the mud of Example 6.3 to increase its density to 10 lb/gal? (b) What will the resulting volume be?

Solution:

(a) $V_s = \dfrac{1000\,(10 - 8.7)}{35.8 - 10} = 50.4 \text{ cc or } 4.3 \times 50.4 = 217 \text{ gm}$

(b) $V_{m_2} = 1000 + 50.4 = 1050 \text{ cc}$

Since barite is so universally used as a weighting material, it is useful to express Eq. (6.18) in field units. Barite is sold in 100 lb bags or sacks. Such a sack contains 100/(4.3) (62.4) = 0.373 cu ft, or 0.373/5.61 = 0.0665 barrels of net material. Therefore 1 barrel (net) of barite = 1/0.0665 ≅ 15 sacks.

Let S_B = sacks of barite necessary to increase the density of 100 bbl of mud from ρ_{m1} to ρ_{m2}. Substituting these special conditions into Eq. (6.18):

$$\frac{S_B}{15} = \frac{100\,(\rho_{m2} - \rho_{m1})}{35.8 - \rho_{m2}} \quad \text{or}$$

$$S_B = \frac{1500\,(\rho_{m2} - \rho_{m1})}{35.8 - \rho_{m2}} \tag{6.18a}$$

Example 6.5

(a) How many sacks of barite are necessary to increase the density of 1000 bbl of mud from 10 to 14 lb/gal?
(b) What will be the final mud volume?

Solution:

(a) Using Eq. (6.18a)

$$S_B = \frac{1500\,(14 - 10)}{35.8 - 14} = 275 \frac{\text{sacks}}{100 \text{ bbl}} = 2750 \frac{\text{sacks}}{1000 \text{ bbl}}$$

(b) $V_{m_2} = 1000 + \dfrac{2750}{15} = 1180 \text{ bbl}$

Example 6.6

(a) How much fresh water must be added to 1000 bbl of 12 lb/gal mud to reduce its density to 10 lb/gal?
(b) What will the resulting volume be?

Solution:

(a) $(1000 \text{ bbl})(42 \text{ gal/bbl})(12 \text{ lb/gal}) + (V_w)(42)(8.33) = (1000 + V_w)(42)(10)$

from which $V_w = 1200$ bbl

(b) $V_{m_2} = 1000 + 1200 = 2200 \text{ bbl}$

Equations (6.19) and (6.20) are other commonly used forms of Eq. (6.16).

$$V_w = \frac{V_{m1}(\rho_{m1} - \rho_{m2})}{\rho_{m2} - 8.33} \tag{6.19}$$

$$S_c = \frac{875\,(\rho_{m2} - \rho_{m1})}{20.8 - \rho_{m2}} \tag{6.20}$$

where V_w = barrels of water necessary to reduce density of V_{m1} barrels initial mud from ρ_{m1} to ρ_{m2}.

S_c = sacks (100 lb) of clay (sp. gr. = 2.5) necessary to change density of 100 bbl initial mud from ρ_{m1} to ρ_{m2}.

In working with laboratory size samples, it is convenient to measure quantities in grams or cubic centimeters. For field use, it is necessary to express these results in pounds per barrel. It is then useful to realize:

$$1 \text{ lb/bbl} \times \frac{454 \text{ gm/lb}}{3785 \text{ cc/gal} \times 42 \text{ gal/bbl}} = \text{gm/cc}$$

or

$$\text{gm/350 cc} = \text{lb/bbl} \tag{6.21}$$

For laboratory or pilot testing purposes, it is convenient to work with a 350 cc quantity so that treating agent additions in gm/per 350 cc of mud will be equivalent to field additions in lb/bbl.

Example 6.7

A mud engineer finds from pilot tests that 2.0 gm of CMC is required to obtain the desired water loss reduction for a one liter mud sample. How much CMC should be added to the actual 1000 barrel system?

Solution:

$$\text{CMC needed} = \frac{350}{1000} \times 2.0 \times 1000 = 700 \text{ lb}$$

In making recommendations for mud treating, it is necessary to know the time required for the entire mud system to make a complete cycle. This is called the cycle time and is computed from a knowledge of pumping rate and system volume. Recalling Eq. (5.8):

$$q = 0.00679\, SN\, (2D^2 - d^2)e \tag{5.8}$$

where q = pump discharge rate, gal/min
S = stroke length, in.
N = complete strokes per minute
D = piston (liner) diameter, in.
d = piston rod diameter, in.
e = pump volumetric efficiency, commonly used as 90% for power pumps and 85% for steam.

Cycle time is then expressed as

$$t_c = \frac{6180\, V_m}{SN\, (2D^2 - d^2)e} \tag{6.22}$$

where t_c = cycle time, min
V_m = system volume, bbl

Example 6.8

What is the cycle time for the following conditions?

V_m = 1000 bbl

Pump liners = $7\frac{1}{2}$ in. diam

Stroke length = 16 in.

Piston rod diameter = $2\frac{1}{4}$ in.

N = 40 strokes per minute

Power pump is used

Solution:

$$t_c = \frac{(6180)(1000)}{(16)(40)[(2)(7.5)^2 - (2.25)^2](0.90)} = 100 \text{ min}$$

Treating materials would then be added at a rate allowing their uniform distribution in the system. In example (6.8), if 20 sacks of material were needed, they could be added at the approximate rate of one sack per five minutes.

6.7 Field Maintenance of Mud Systems

Field mud systems must be checked frequently to insure that desired properties are maintained. Normally a member of the drilling crew (usually the derrick man) performs routine tests at specified intervals during his tour (or shift) and records the results on the tour report. This furnishes the mud engineer with all pertinent information as to what occurred during his absence. If precise mud control is necessary, the engineer may be required to be on continuous duty at the wellsite.

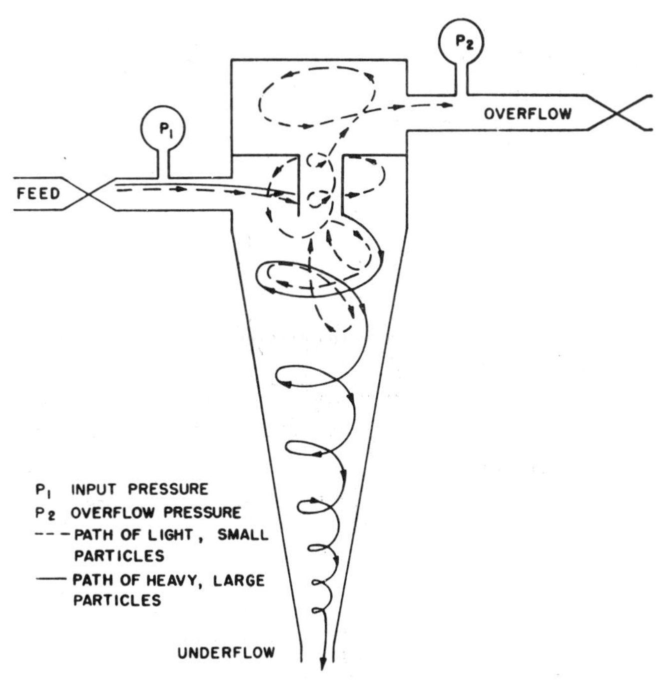

Fig. 6.19. Flow pattern in mud cyclone. After Scott and Lummus,[37] courtesy *Oil and Gas Journal.*

A considerable problem in mud systems is the gradual accumulation of drilled solids which pass through the shale shaker, remain suspended in the pits, and are continuously recirculated. This continuous increase in solid content impairs flow properties (increases viscosity, gel strength, and density), retards rate of penetration, causes undue pump wear, and reduces bit life. Periodically, then, part of the system must be *jetted* into the reserve pit (discarded) with extra water being added to make up the required volume. Additional weighting materials must then be added to re-establish the desired density. The weighting material in the rejected mud represents a financial loss. To prevent this waste, mud centrifuges (called cyclone separators in mud work) have been developed which will reject only the low density clays (sp.gr. = 2.5 – 2.7) and retain the high density barite.[36,37] One device of this type is shown in Figure 6.19; Figure 6.20 is a diagram of a typical cyclone installation. Properly designed installations will reject up to 75% of the clay while retaining 80 to 90% of the barite. Under proper conditions, coarse lost circulation materials may also be retained.

Cyclone separators contain no moving parts; the centrifugal force is furnished by the tangential entry and consequent swirling motion of the mud. The fluid volume and solid contents of overflow and underflow

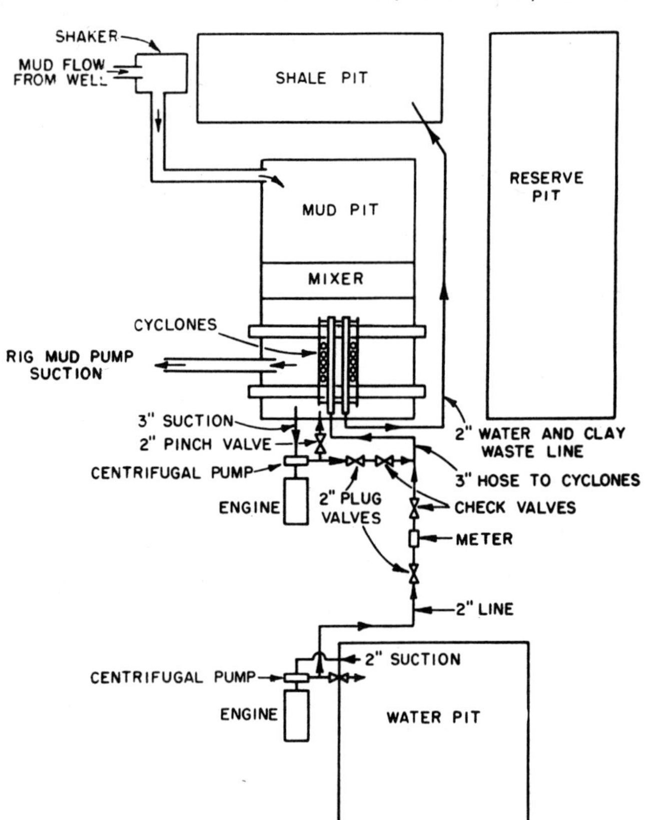

Fig. 6.20. Cyclone equipment layout. After Scott and Lummus,[37] courtesy *Oil and Gas Journal.*

may be controlled by altering the pressure drop across the cyclone. The overflow is rejected and flows to the reserve pit, while the underflow is normally discharged back into the mud pit at atmospheric pressure. Use of these or similar devices has greatly reduced mud costs, particularly in Gulf Coast wells where mud forming clays and shales plus the necessity for precise density control make their use extremely attractive.

6.8 Air, Natural Gas, and Aerated Mud as Drilling Fluids

Spectacular results have been obtained in many instances by using air, natural gas, or highly aerated muds as drilling fluids. The principal benefit from such practices is the economy resulting from a large increase in penetration rate. Other benefits such as less pay zone contamination and elimination of lost circulation are also realized in many cases. Consideration of the use of these fluids requires an analysis of their ability to perform the functions most necessary to the specific area in question. For example, cutting lifting, and bit and drill string cooling and lubrication are adequately handled by gases; however, cutting suspension, walling of the hole, and pressure control are not. Consequently, these techniques are applicable only to those wells where the latter functions are not particularly critical.

6.81 Air and Gas as Drilling Fluids

The most severe restriction on air and gas as drilling fluids is their inability to control encountered subsurface pressures. When a permeable zone is drilled, its fluid content readily enters the borehole and interferes with normal circulation. Water entry is the most common problem and its removal may require prohibitive air circulation rates. Small quantities of water from low permeability formations, while posing no removal problem, do cause cutting balling and general hole stickiness which may result in stuck drill pipe. Many authorities consider this more of a problem than large water volumes. Much work is being done to develop materials and techniques for effectively sealing encountered water formations. Such methods will have to be fast, safe, and must be performed with a minimum of special equipment in order to be economically feasible. Reed reports that water flows of over 50 bbl/hr have been handled in the Texas panhandle by the use of foaming agents.[38] This technique holds considerable promise; the cost of the foaming agent, however, must be balanced by the increased penetration rate.

One of the most successful applications of air and gas drilling is in the San Juan Basin of New Mexico. This is a gas producing area in which the pay section consists of approximately 800 ft of alternating shale and sand encountered at depths from 4000 to 5000 ft. Water sensitive clay content, low matrix permeability, and a natural fracture system all contribute to drilling and completion problems. The superiority of gas drilling over other techniques in this area is illustrated in Table 6.3. Another major benefit is the greater well productivity brought about by decreased permeability damage. A further economic incentive is that drilling gas is normally furnished by adjacent producing wells; this eliminates the need for compressors.

TABLE 6.3

COMPARISON OF DRILLING METHODS FOR THE MESA VERDE SECTION, SAN JUAN BASIN[39]

Drilling method	*Time required*	*Bits used*	*Remarks*
Rotary, conventional mud	10-20 days	8-10	Severe formation damage
Cable tool	5-8 weeks	—	Continuous fishing jobs and waste of gas
Rotary, gas drilling	$4\frac{1}{2}$ days	3-4	Higher well productivity

It should be realized that such results are not always obtainable, and that all cases are not so favorable. In summary, air and gas are satisfactory drilling fluids in areas where sloughing formations, high pressures, and water production are not insurmountable problems.

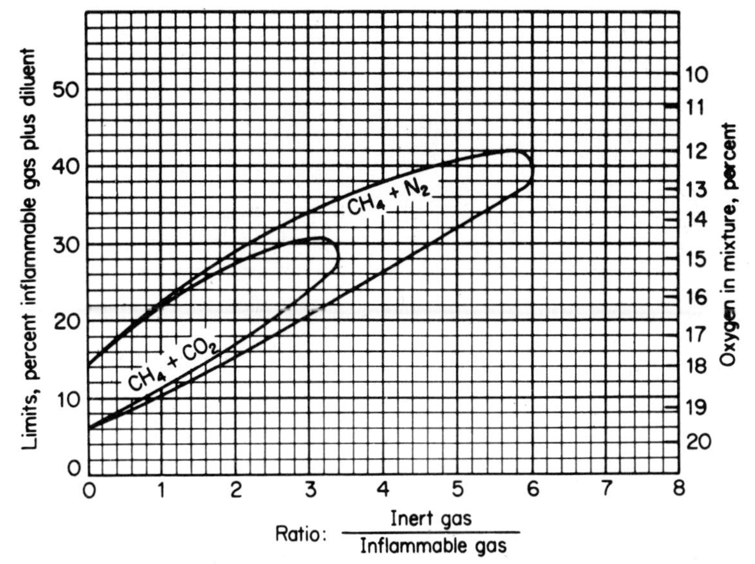

Fig. 6.21. Limits of inflammability of methane when mixed with various proportions of air, nitrogen, and carbon dioxide. After Nicolson,[39] courtesy *Petroleum Engineer.*

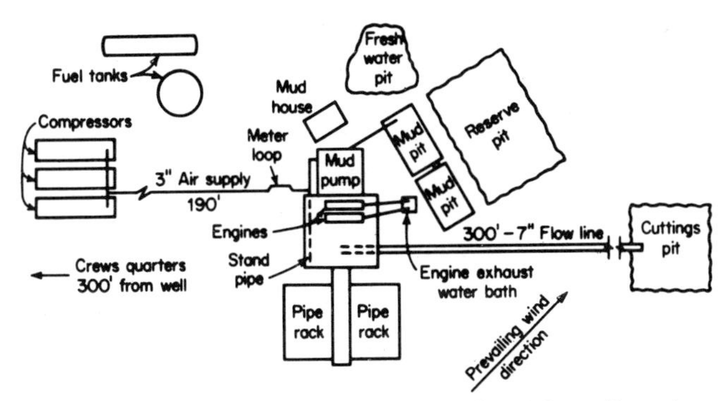

Fig. 6.22. Air-drilling location layout. After Adams,[40] courtesy API.

Hazards of Air and Gas Drilling

The danger of explosions and/or fires is inherent in air and gas drilling, and extra safety precautions are necessary.[40,41] Certain critical air-gas mixtures are explosive and should be avoided (Figure 6.21). The exhaust line should be located several hundred feet downwind (prevailing wind direction) from the rig. Other precautions, such as spark-proof ignition systems on rig engines, fire extinguishers, fans, and spinning ropes rather than chains, should be taken. A mud pump and a supply of mud should be on hand in case it becomes desirable to kill the well. Figure 6.22 is a typical location layout illustrating these safety precautions.

6.82 Aerated Mud Drilling Fluids

Some of the advantages of air drilling may be combined with conventional mud drilling by using an aerated mud. The principal advantages of this technique over straight air drilling are the following:[42]

(1) low pressure and/or low permeability water zones may be drilled without danger of pipe sticking.

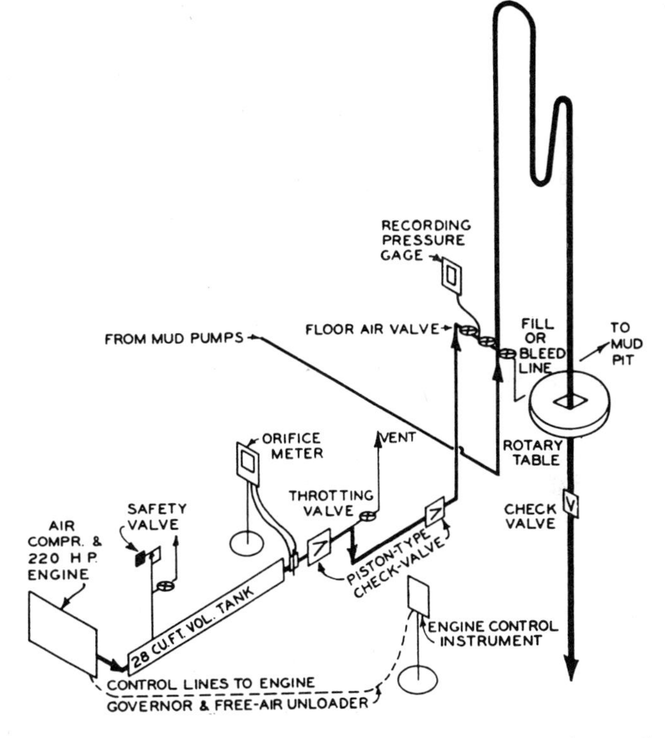

Fig. 6.23. Aerated mud system. After Bobo, Ormsby, and Houch,[42] courtesy *Oil and Gas Journal.*

(2) greater safety: quick weight buildup is quickly obtained by shutting off the injected air. Explosions and/or fire possibilities are minimized by the water in the mixture.

A typical layout for this method is shown in Figure 6.23. Air is injected into the system between the mud pumps and the standpipe. The air-mud ratio may be controlled by increasing or reducing inputs of each fluid. Such a system is highly versatile and offers considerable promise in some areas. All air, all mud, or any mixture in between may be obtained.

Low solids, low viscosity muds are the best for this purpose, as they are most easily de-aerated in the pits. Foaming and emulsification of muds are undesirable, since the high compressibility of such mixtures makes mud pump operation very inefficient. Drill string corrosion may be controlled by using saturated lime water in the drilling mud.

A further discussion of these techniques will be presented later, when we consider the factors affecting rate of penetration. Air and gas and their variant, aerated mud, are being widely applied as special drilling fluids.

PROBLEMS

1. Derive Eq. (6.4) from Eqs. (6.1), (6.2), and (6.3).

2. The following are Fann V-G meter measurements for a particular mud: $\Phi_{600} = 20$, $\Phi_{300} = 12$.
Compute: (a) plastic viscosity, (b) apparent viscosity, (c) Bingham yield value, (d) true yield value. *Ans.* (d) 3 lb/100 ft².

3. The corrected API (30 minute) filter loss of a water base mud was determined as 10 cc at a temperature of 86°F. What would it have been at 140°F? At 212°F? *Ans.* 14.6 cc; 18.8 cc.

4. A mud exhibited a water loss of 5.0 cc in 10 minutes. Assuming no initial spurt occurred, what would be reported as the API filter loss? *Ans.* 8.7 cc.

5. What would the 30 minute water loss of a mud be at 70°F if its corrected 15 minute water loss at 100°F was 8.0 cc? *Ans.* 9.5 cc.

6. How much fresh water and clay must be mixed to prepare 1 liter of 1.05 sp.gr. mud? *Ans.* 83.2 gm. clay, 967 cc water.

7. A fresh water-clay mud has a density of 9.0 lb/gal.
Compute: (a) the weight percent clay in the mud (b) the volume percent clay in the mud. *Ans.* (a) 12.5% (b) 5.4%.

8. (a) How many sacks of barite should be added to 500 bbl of the mud in Problem 7 to increase its density to 12 lb/gal?
(b) How many gm per liter is this?
(c) What is the resulting volume and weight percent solids in the 12 lb mud? *Ans.* (a) 945 sacks (b) 540 gm/liter (c) 16% by volume; 41.7% by weight.

9. What will be the final density of an oil emulsion mud composed of 20% (by volume) diesel oil and a 10 lb/gal water base mud? Sp.gr. of oil = 0.75. *Ans.* 9.25 lb/gal.

10. An 8 × 16 in. duplex mud pump is operating at 50 strokes per minute. The mud system contains 1200 bbl of 10 lb/gal mud. It is desired to add sufficient water to lighten this mud to 9.5 lb/gal. Piston rod diameter = 2 in.
(a) What volume of the original mud should be discarded to have the same final system volume?
(b) At what uniform rate should the water be added in order to complete the operation in two cycles of the system? (Assume 90% pump volumetric efficiency.) *Ans.* (a) 359 bbl (b) 1.33 bbl/min.

11. Show that for a given t_c/d_c, ρ_s, and ρ_m, a plot of turbulent slip velocity vs cutting diameter may be represented as a linear plot on log-log graph paper.

12. Prepare a set of curves showing V_c vs d_c for $\rho_m = 8$, 12, and 16 lb/gal. Assume $\rho_s = 21.7$ lb/gal (sp.gr. = 2.6) and $t_c/d_c = 0.20$.

13. Assuming a desired upward cutting velocity of 50 ft/min, what should the annular mud velocity be if:

$$d_c = 0.50 \text{ in. (max)} \qquad \rho_m = 12 \text{ lb/gal}$$
$$d_c/t_c = 0.20 \qquad \rho_s = 21.7 \text{ lb/gal}$$

14. A pilot test on a 600 cc mud sample indicated that 1.5 gm of starch gave the desired water loss. How much starch should be added to the 1800 bbl mud system?

REFERENCES

1. "The History of Drilling Mud," Sec. 100 in *Drilling Mud Data Book,* Baroid Sales Division, National Lead Co.
2. "Field Testing," Sec. 900 in *Drilling Mud Data Book,* Baroid Sales Division, National Lead Co.
3. *Training Course for Mud Engineers,* Magnet Cove Barium Corp.
4. *Principles of Drilling Mud Control,* Petroleum Extension Service, University of Texas.
5. Rogers, W. F., *Composition and Properties of Oil Well Drilling Fluids.* Houston: Gulf Publishing Co., 1948.
6. Beck, R. W., Nuss, W. F., and T. H. Dunn, "The Flow Properties of Drilling Muds," *API Drilling and Production Practices,* 1947, p. 9.
7. Melrose, J. C., and W. B. Lilienthal, "Plastic Flow Properties of Drilling Fluids — Measurement and Application," *Trans. AIME,* Vol. 192 (1951), p. 159.
8. Savins, J. G., and W. F. Roper, "A Direct-indicating Viscosimeter for Drilling Fluids," *API Drilling and Production Practices,* 1954, p. 7.
9. Cardwell, W. T., Jr., "Drilling-Mud Viscosimetry," *API Drilling and Production Practices,* 1941, p. 104.
10. Garrison, A. D., and K. C. ten Brink, "A Study of Some Phases of Chemical Control in Clay Suspensions," *Trans. AIME,* Vol. 136 (1940), 2nd reprinted ed., p. 175.
11. Bingham, E. C., *Fluidity and Plasticity.* New York: McGraw-Hill Book Co., 1922.
12. Burdyn, R. F., and L. D. Wiener, "That New Drilling Fluid for Hot Holes," *Oil and Gas Journal,* Sept. 10, 1956, p. 104.
13. Rosenberg, M., and R. J. Tailleur, "Increased Drill Bit Life Through Use of Extreme Pressure Lubricant Drilling Fluids," AIME Paper 1152–G, Presented Houston, Oct. 1958.

14. Pigott, R. J. S., "Mud Flow in Drilling," *API Drilling and Production Practices,* 1941, p. 91.

15. Hall, H. N., Thompson, H. and F. Nuss, "Ability of Drilling Mud to Lift Bit Cuttings," *Trans. AIME,* Vol. 189, (1950), p. 35.

16. MacDonald, G. C., "Transporting Rotary Bit Cuttings," *World Oil,* Apr. 1949, p. 114.

17. Williams, C. E., Jr., and G. H. Bruce, "Carrying Capacity of Drilling Muds," *Trans. AIME,* Vol. 192, (1951), p. 111.

18. Gray, G. R., Neznayko, M., and P. W. Gilkeson, "Some Factors Affecting the Solidification of Lime-treated Muds at High Temperatures," *API Drilling and Production Practices,* 1952, p. 73.

19. Watkins, T. E., and M. D. Nelson, "Measuring and Interpreting High-Temperature Shear Strengths of Drilling Fluids," *Trans. AIME,* Vol. 198, (1953), p. 213.

20. O'Brien, T. B., "Formulation and Use of Sea-water Muds," *API Drilling and Production Practices,* 1955, p. 86.

21. Perkins, H. W., "A Report on Oil-Emulsion Drilling Fluids," *API Drilling and Production Practices,* 1951, p. 349.

22. Lummus, J. L., Barrett, H. M., and H. Allen, "The Effects of Use of Oil in Drilling Muds," *API Drilling and Production Practices,* 1953, p. 135.

23. Lummus, J. L., "Multipurpose Water-in-Oil Emulsion Mud," *Oil and Gas Journal,* Dec. 13, 1954, p. 106.

24. Nelson, M. D., Crittendon, B. C., and G. A. Trimble, "Development and Application of a Water-in-Oil Emulsion Drilling Mud," *API Drilling and Production Practices,* 1955, p. 235.

25. Bugbee, J. M., "Lost Circulation — A Major Problem in Exploration and Development," *API Drilling and Production Practices,* 1953, p. 14.

26. Goins, W. C., Jr., Weichert, J. P., Burba, J. L., Jr., Dawson, D. D., Jr., and A. J. Teplitz, "Down-the-hole Pressure Surges and Their Effect on Loss of Circulation," *API Drilling and Production Practices,* 1951, p. 125.

27. Cardwell, W. T., Jr., "Pressure Changes in Drilling Wells Caused by Pipe Movement," *API Drilling and Production Practices,* 1953, p. 97.

28. Ormsby, G. S., "Calculation and Control of Mud Pressure in Drilling and Completion Operations," *API Drilling and Production Practices,* 1954, p. 44.

29. Goins, W. C., Jr., "How to Combat Lost Circulation," *Oil and Gas Journal,* June 9, 1952.

30. Messenger, J. U., and J. S. McNiel, Jr., "Lost Circulation Corrective: Time Setting Clay Cement," *Trans. AIME,* Vol. 195, (1952), p. 59.

31. Shumate, H. J., "Lost Circulation — Its Causes and What to Do About It," *Oil and Gas Journal,* Oct. 18, 1951.

32. Howard, G. C., and P. P. Scott, Jr., "An Analysis and the Control of Lost Circulation," *Trans. AIME,* Vol. 192, (1951), p. 171.

33. Coberly, C. J., "Selection of Screen Openings for Unconsolidated Sands," *API Drilling and Production Practices,* 1937, p. 189.

34. Furnas, C. C., "Mathematical Relations for Beds of Broken Solids of Maximum Density," *Industrial and Engineering Chemistry,* Vol. 23, Sept. 1931, p. 1052.

35. Anderegg, F. O., "The Application of Mathematical Formulae to Mortars," *Industrial and Engineering Chemistry,* Sept. 1931, p. 1058.

36. Bobo, R. A., and R. G. Hoch, "Mechanical Treatment of Weighted Drilling Muds," *Trans. AIME,* Vol. 201, (1954), p. 93.

37. Scott, P. P., Jr., and J. L. Lummus, "Cyclones Save Barite, Reject Clay Solids," *Oil and Gas Journal,* Oct. 8 and Oct. 15, 1956.

38. Reed, R. M., "Air Drilling with Foam Combats Water Influx," *The Petroleum Engineer,* May 1958, p. B–57.

39. Nicolson, K. M., "Air and Gas Drilling," *Fundamentals of Rotary Drilling.* Dallas, Texas: *The Petroleum Engineer,* 1954, p. 86.

40. Adams, J. H., "Air and Gas Drilling in the McAlester Basin Area," API Paper no. 851–31–N. Presented Tulsa, Apr. 1957.

41. Fuller, L. S., "Hazards of Drilling with Natural Gas," *World Oil,* Jan. 1954.

42. Bobo, R. A., Ormsby, G. S., and R. S. Hoch, "Phillips Tests Air-Mud Drilling," *Oil and Gas Journal,* Jan. 24, 1955.

43. Weiss, W. J., Graves, R. H., and W. L. Hall, "A Fundamental Approach to Well Bore Stabilization," *The Petroleum Engineer,* Apr. 1958.

Chapter 7 . . .

Rotary Drilling Hydraulics

7.1 Introduction

Proper utilization of mud pump horsepower is of considerable importance to rotary drilling operations. Analytical appraisal of the rig's circulating system requires an understanding of the components which consume power, so that the available energy may be used as advantageously as possible. The standard hydraulics approach to such analyses is hindered by numerous factors, among which are:

(1) Mud flow property peculiarities, as discussed in Chapter 6

(2) Irregularities of the circulating system.

Drilling mud leaves the pump discharge, passes through the surface lines, standpipe, and mud hose, and finally enters the drill string at the top of the kelly joint. Here it begins the long downward travel through the drill pipe and drill collars, is expelled through the water courses or nozzles of the bit, and returns up the annulus. The annular area is relatively small around the drill collars and becomes larger in the portion containing drill pipe. Since the mud enters the drill string and leaves the annulus at essentially the same elevation, the only pressure required is that necessary to overcome the frictional losses in the system. Hence the discharge pressure at the pump is defined by:

$$\Delta p_t = \Delta p_s + \Delta p_p + \Delta p_c + \Delta p_b + \Delta p_{ac} + \Delta p_{ap} \tag{7.1}$$

where Δp_t = pump discharge pressure

Δp_s = pressure loss in surface piping, standpipe, and mud hose

Δp_p = pressure loss inside drill pipe

Δp_c = pressure loss inside drill collars

Δp_b = pressure loss across bit water courses or nozzles

Δp_{ac} = pressure loss in annulus around drill collars

Δp_{ap} = pressure loss in annulus around drill pipe

The solution to Equation (7.1) is rather tedious, in that separate calculations for each section are required. However, with a little practice and understanding the task is not particularly formidable.

Before discussing plastic fluid flow calculations let us first review the fundamental equations of Newtonian fluid flow.

7.2 Newtonian Fluid Flow Calculations

Fluid flow through pipes is considered as either laminar or turbulent. In laminar (viscous) flow the fluid moves in parallel layers or laminae which are at all times parallel to the direction of flow. In turbulent flow, secondary irregularities and eddys are imposed on the main or average flow pattern. Calculation of pressure drop for pipe flow requires a knowledge of which flow pattern pertains to the specific case, since different equations apply for each situation. Definition of the existing flow pattern is given by a dimensionless quantity known as the Reynolds number:

$$R_e = \frac{928\ \rho \bar{v} d}{\mu} \tag{7.2}$$

where R_e = Reynold's number

$\bar{v}$ = average velocity of flow, ft/sec, = $q/2.45d^2$

ρ = fluid density, lb/gal

d = pipe inside diameter, in.

μ = fluid viscosity, cp

q = circulating volume, gal/min

It is commonly considered that if:

$R_e < 2000$, flow is laminar

$R_e > 4000$, flow is turbulent

$2000 < R_e < 4000$, flow is in transition, and is neither laminar nor turbulent.

The onset of turbulence is accelerated by any irregularity or entrance condition which will distort the flow pattern.

The pressure drop in laminar flow is given by the Hagan-Poiseuille law; this, in practical units, is

$$\Delta p = \frac{\mu L \bar{v}}{1500\, d^2} \tag{7.3}$$

where Δp = laminar flow pressure drop, lb/in.^{2}
L = length of pipe, ft

For turbulent flow, Fanning's equation applies:

$$\Delta p = \frac{f \rho L \bar{v}^2}{25.8\, d} \tag{7.4}$$

where Δp = turbulent flow pressure drop, lb/in.^{2}
f = Fanning friction factor

The friction factor f is a function of R_e and pipe roughness, and has been evaluated experimentally for numerous materials (see Figure 7.1). Care must be exercised, because in some texts $4f$ is plotted, instead of the f used here. This may be readily checked by noting that for all pipes, $f \cong 0.013$ at $R_e = 2000$.

In summary, one can calculate pressure drop for Newtonian fluid flow systems in the following manner:
(1) Calculate R_e from Eq. 7.2.
(2) If $R_e < 2000$, use Eq. (7.3) to calculate the pressure drop.
(3) If $R_e > 2000$, use Eq. (7.4). In this case the friction factor f is obtained from Figure 7.1 or its equivalent.

Detailed treatments of Newtonian fluid flow calculations may be found in numerous hydraulics texts and handbooks.

7.3 Plastic Fluid Flow Calculations

As was pointed out in Chapter 6, the flow behavior of plastic drilling fluids is complicated by the variation of apparent viscosity with rate of shear or flow. Consequently, the Newtonian fluid equations must be altered for application to typical drilling mud systems. An early worker on this subject was Pigott,[1] whose data have been widely used for the construction of hydraulic tables and curves. Let us first consider the development of Beck, Nuss, and Dunn[2] which incorporates the classic work of Bingham[3] with the commonly applied Newtonian fluid flow equations just presented.

Laminar Flow Region

The typical pressure-velocity behavior of a plastic fluid flowing through a pipe was shown in Figure 6.2. A definite pressure (Y_t) is required to initiate flow. True laminar flow is represented by the linear portion of the curve, the equation of which is:

$$144 \Delta p = \frac{4}{3} Y_t + m \bar{v} \tag{7.5}$$

where $144\Delta p$ = pressure drop, lb/ft^2
$\frac{4}{3} Y_t = Y_b$, the Bingham yield value, lb/ft^2 (commonly called yield point)
m = slope of linear portion which is proportional to the plastic viscosity, μ_p. *Note:* $m = \mu L/1500\ d^2$, from Eq. (7.3).

For practical values of $\bar{v}$, the behavior of Bingham fluids may be expressed as:

$$\Delta p = \frac{L Y_b}{300 d} + \frac{\mu_p \bar{v} L}{1500 d^2} = \frac{L}{300 d}\left(Y_b + \frac{\mu_p \bar{v}}{5d}\right) \tag{7.6}$$

where Y_b = yield point, lb/100 ft^2
μ_p = plastic viscosity, cp

Equation (7.6) may be used in cases where laminar flow exists. Determination of flow characteristic (laminar or turbulent) is made by comparing the actual velocity with a calculated critical velocity.

Critical Velocity Calculation

If Eqs. (7.3) and (7.6) are equated, an equivalent Newtonian viscosity in terms of d, $\bar{v}$, μ_p, and Y_b is obtained:

$$\mu = \frac{5 d Y_b}{\bar{v}} + \mu_p \tag{7.7}$$

Substituting Eq. (7.7) for μ in the Reynold's number Eq. (7.2), equating the resulting equation to 2000, and solving for $\bar{v}$ gives:

$$v_c = \frac{1.08 \mu_p + 1.08\sqrt{\mu_p^2 + 9.3 \rho d^2 Y_b}}{\rho d} \tag{7.8}$$

where v_c = critical velocity, ft/sec, above which turbulent flow exists and below which the flow is laminar.

Equation (7.8) assumes that turbulence occurs at $R_e = 2000$. This is reasonable, since drill pipe rotation and tool joint irregularities promote turbulence at this lower value. Therefore, if:

$\bar{v} < v_c$, flow is laminar
$\bar{v} > v_c$, flow is turbulent

Turbulent Flow Calculations

Fanning's equation may be used for turbulent flow calculations providing the Reynold's number expression is altered by substitution of

$$\mu_t = \frac{\mu_p}{3.2} = \text{turbulent viscosity of plastic fluids} \tag{7.9}$$

This is the experimental finding of Beck, Nuss, and Dunn, i.e., that the viscosity of plastic fluids in turbulent flow is constant, as defined by Eq. (7.9). This premise has been criticised by Havenaar,[4] who proposed a different expression based on the volume percent

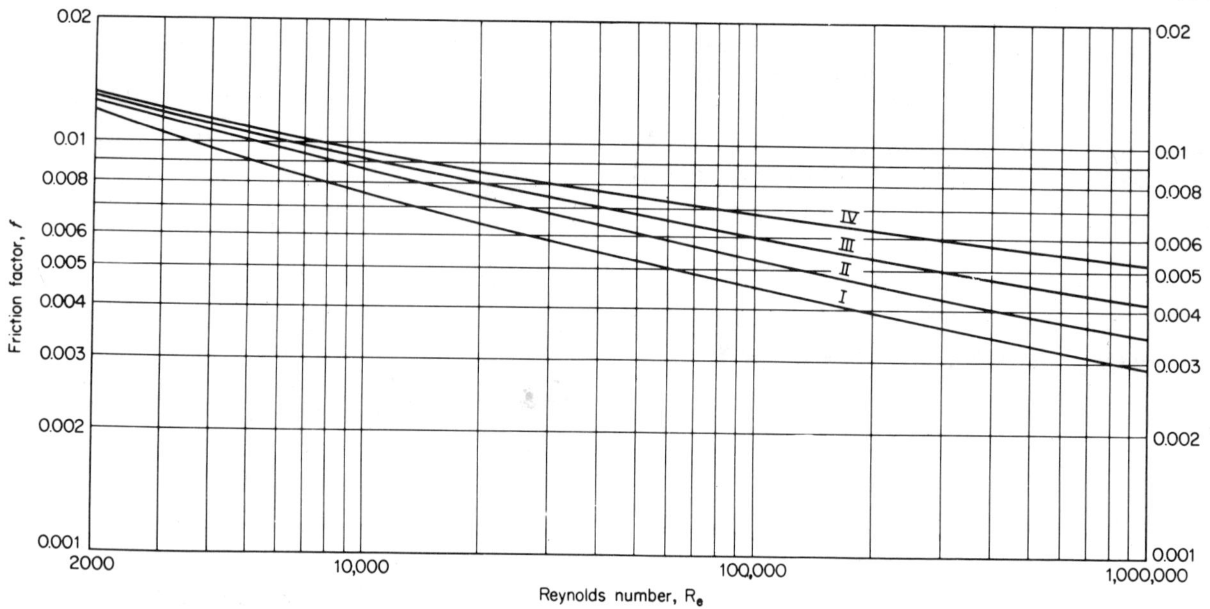

Fig. 7.1. Friction factor vs. Reynolds number for mud flow calculations. After Ormsby,[13] courtesy API.

solids in the mud. Fortunately, reasonable errors in viscosity determination have little effect on turbulent flow calculations, and Eq. (7.9) will be used for our present purpose.

Substitution of μ_t for μ in the general Reynold's number expression (Eq. 7.2) gives:

$$R_e = \frac{928\rho\bar{v}d}{\mu_t} = \frac{2970\rho\bar{v}d}{\mu_p} \tag{7.10}$$

Figure 7.1 allows determination of the friction factor f for drill pipe and open hole combinations. This f may then be used in Eq. (7.4) for calculation of pressure drop.

In summary, one can calculate pressure drop for plastic fluids as follows:

(1) Calculate v_c from Eq. (7.8)
(2) If $\bar{v} < v_c$, flow is laminar, and Eq. (7.6) applies
(3) If $\bar{v} > v_c$, flow is turbulent, requiring:
  (a) Calculation of R_e from Eq. (7.10)
  (b) Determination of f from Figure 7.1 at the calculated R_e for the conduit in question
  (c) Calculation of pressure drop from Eq. (7.4a).

Example 7.1

Mud is flowing through $4\frac{1}{2}$ inch OD, internal flush drill pipe. Calculate the frictional pressure drop per 1000 ft of pipe.

Mud properties

$\rho_m = 10$ lb/gal Pipe i.d. = 3.640 in.

$Y_b = 10$ lb/100 ft² Circulating rate $q = 400$ gal/min

$\mu_p = 30$ cp

Solution:

(1) $$v_c = \frac{(1.08)(30) + (1.08)\sqrt{(30)^2 + (9.3)(10)(3.64)^2(10)}}{(10)(3.64)}$$

$$= 4.3 \text{ ft/sec}$$

(2) $$\bar{v} = \frac{Q \text{ ft}^3/\text{sec}}{A \text{ ft}^2} = \frac{q \text{ gal/min} \times \frac{1}{7.48}\frac{\text{ft}^3}{\text{gal}} \times \frac{1}{60}\frac{\text{min}}{\text{sec}}}{(\pi/4)(d/12)^2}$$

$$= \frac{q}{2.45d^2}$$

$$= \frac{400}{2.45(3.64)^2} = 12.3 \text{ ft/sec}$$

(3) $\bar{v} > v_c$, and flow is turbulent.

(a) $$R_e = \frac{(2970)(10)(12.3)(3.64)}{30} = 44{,}300$$

(b) $f = 0.0062$ from Curve II, Figure 7.1

(c) $$\Delta p_p = \frac{(0.0062)(10)(1000)(12.3)^2}{(25.8)(3.64)}$$

$$= 100 \text{ psi/1000 ft}$$

Hydraulically Equivalent Annulus Diameter

For annular flow it is necessary to use a hypothetical circular diameter, d_a, which is the hydraulic equivalent of the actual annular system. The hydraulic radius concept is satisfactory for this purpose and is defined as:

$$r_h = \frac{\text{cross-sectional area of flow stream}}{\text{wetted perimeter of conduit}} = \text{hydraulic radius}$$

For an annulus,

$$r_h = \frac{\pi(r_1^2 - r_2^2)}{2\pi(r_1 + r_2)} = \frac{r_1 - r_2}{2}$$

where r_1 and r_2 are the large and small radii, respectively. For a circular pipe,

$$r_h = \frac{\pi r^2}{2\pi r} = \frac{r}{2}$$

The frictional loss in an annulus is equal to the loss in a circular pipe having the same hydraulic radius; hence, in general terms,

$$(7.11) \qquad r_a = r_1 - r_2 \qquad \text{or} \qquad d_a = d_1 - d_2$$

where r_a and d_a are the hydraulically equivalent radius and diameter of a circular pipe which may be used in the previous pressure drop equations.

Example 7.2

Mud is flowing through the annulus between an 8-in. hole and $4\frac{1}{2}$-inch drill pipe. What value of d should be used in Eqs. 7.2, 7.3, 7.4, 7.6, 7.7, 7.8? *Ans.* $(8 - 4\frac{1}{2}) = 3\frac{1}{2}$ in.

7.4 Pressure Drop Across Bit Nozzles and Watercourses

Figure 7.2 illustrates the flow of an incompressible fluid through a converging tube (nozzle, orifice, etc.). Assuming steady state, adiabatic, and frictionless conditions:

$$(1) \qquad \frac{P_1}{w} + \frac{\bar{v}_1^2}{2g} = \frac{P_2}{w} + \frac{\bar{v}_2^2}{2g}$$

where P_1, P_2 = pressure, lb/ft²
w = density, lb/ft³
$\bar{v}_1$, $\bar{v}_2$ = velocities at points 1 and 2, ft/sec

or

$$(1a) \qquad \frac{\Delta P}{w} = \frac{\bar{v}_2^2 - \bar{v}_1^2}{2g}$$

Practically, $\bar{v}_2^2 - \bar{v}_1^2 \cong \bar{v}_2^2$, hence

$$(2) \qquad \bar{v}_2^2 = 2g\frac{\Delta P}{w}$$

The ideal rate of flow, $Q_i = A_2\bar{v}_2$. The actual flow rate Q is:

$$(3) \qquad Q = CQ_i$$

where C is the flow or nozzle coefficient for a particular design. With these substitutions, Eq. (2) becomes

$$(7.12) \qquad \Delta P = \frac{wQ^2}{2gC^2A_2^2}$$

Altering Eq. 7.12 to practical units for mud flow, we have:

$$(7.13) \qquad \Delta p = \frac{q^2\rho}{7430C^2d^4}$$

where d = nozzle or watercourse diameter, in.

Eckel and Bielstein[5] have shown that C may be as high as 0.98 for properly designed jet bit nozzles; however, 0.95 is commonly used for field purposes. For ordinary watercourses, which are merely flat drilled holes, $C = 0.80$.

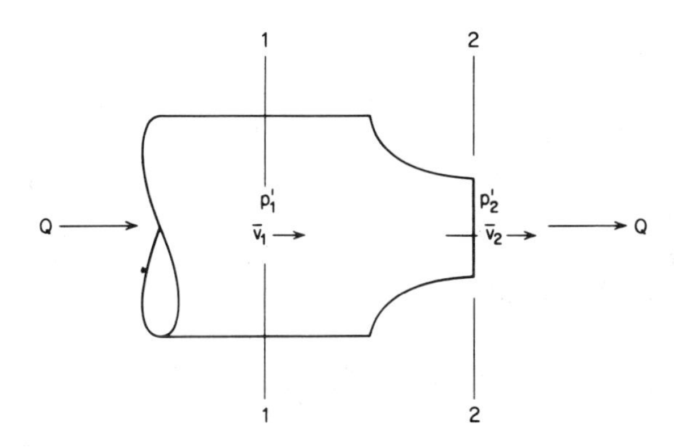

Fig. 7.2. Schematic sketch of incompressible fluid flowing through a converging tube or nozzle.

Multiple Nozzles

Normally a jet rock bit has the same number of nozzles as cones. The calculation of pressure drop across a multiple nozzle bit may be simplified by substituting the sum of the nozzle areas for A in Eq. 7.12. This fact follows from Eq. 7.12.

For a single nozzle:

$$\Delta P = \frac{wQ^2}{2gC^2A^2}$$

For several nozzles, each of area A_1:

$$\Delta P_m = \frac{wQ_1^2}{2gC^2A_1^2}$$

However, $Q_1 = \frac{Q}{n}$, where n = number of nozzles.

Therefore:

$$\frac{\Delta P_m}{\Delta P} = \frac{Q_1^2}{Q^2}\frac{A^2}{A_1^2} = \frac{Q_1^2A^2}{n^2Q_1^2A_1^2}$$

It is desired to choose an A such that

$$\frac{A^2}{n^2A_1^2} = 1$$

$$\therefore A^2 = n^2A_1^2$$

or

$$(7.14) \qquad A = nA_1, \quad \text{Q.E.D.}$$

Similarly, for use in Eq. 7.13,

$$(7.15) \qquad d_e = \sqrt{nd^2}$$

Where the multiple nozzles vary in size,

$$(7.15a) \qquad d_e = \sqrt{ad_1^2 + bd_2^2 + \text{etc.}}$$

where a = number of nozzles having diameter d_1
b = number of nozzles having diameter d_2, etc.
d_e = hydraulically equivalent single nozzle diameter, in.

Example 7.3

A 10 lb/gal mud is being circulated at the rate of 500 gal/min through a tri-cone bit having three $\frac{3}{8}$-in. diameter jets. What is the pressure drop across the bit?

Solution:

$d = \sqrt{3(\frac{3}{8})^2} = 0.65$ in. (equivalent single nozzle diameter)

By Eq. (7.13),

$$\Delta p = \frac{(500)^2(10)}{(7430)(0.95)^2(0.65)^4} = 2100 \text{ psi}$$

7.5 Pressure Drop Calculations for a Typical System

To illustrate the pressure loss calculations for each system component, let us work a complete example. As stated in Eq. (7.1), the total pressure drop in the system is the sum of the losses in each section. Consider then the following operating conditions:

Operating Data

Depth = 6000 ft (5500 ft drill pipe, 500 ft drill collars)

Drill pipe = $4\frac{1}{2}$ in. internal flush, 16.6 lb/ft (i.d. = 3.826 in.)

Drill collars = $6\frac{3}{4}$ in. (i.d. = 2.813 in.)

Mud density, ρ_m = 10 lb/gal

$\mu_p = 30$ cp

$Y_b = 10$ lb/100 ft²

Bit = $7\frac{7}{8}$ in., 3 cone, jet rock bit

Nozzle velocity = at least 25 ft/sec per inch of bit diameter (this value is obtained by a commonly applied rule of thumb).

What hydraulic (pump output) horsepower will be required for these conditions?

Calculation Steps

1. Circulation rate: This is obtained from the desired annular velocity necessary for proper hole cleaning (cutting removal). Assume that this is a fast drilling, soft rock area and that 180 ft/min (3 ft/sec) upward velocity based on a gauge hole is required.
2. The flow rate q is normally desired in gal/min:

$$q = \text{Annulus area} \times \text{velocity}$$
$$= 2.45\,(d_h^2 - d_p^2)\bar{v}$$
$$= 2.45\,(62 - 20.2)(3)$$
$$= 308 \text{ gal/min}$$

3. Nozzle size: 3 nozzles (one for each cone) will be used, hence $\frac{1}{3}\,q$ will flow through each. For $\bar{v}$ = 250 ft/sec through each,

$$d = \sqrt{\frac{q/3}{2.45\bar{v}}} = \sqrt{\frac{103}{(2.45)(250)}} = 0.41$$

The nearest stock nozzle is $\frac{13}{32}$ in.; this one is then chosen (see Figure 7.9). This nozzle allows an actual velocity of:

$$\bar{v} = \frac{103}{(2.45)(13/32)^2} = 225 \text{ ft/sec}$$

4. Surface equipment losses: The surface equipment consists of the standpipe, swivel, kelly joint, and the piping between the pump and standpipe. Since this part of the system causes only a small fraction of the total losses, it will be satisfactory to choose the one of the precomputed cases shown in Figure 7.3 which most closely approximates the actual case. Assume Case 2 fits our example. Then,

$$\Delta p_s \cong 30 \text{ psi}, \quad \text{from Figure 7.3.}$$

5. Pressure drop inside drill pipe: The critical velocity is calculated from Eq. 7.8:

$$v_c = \frac{1.08\mu_p + 1.08\sqrt{\mu_p^2 + 9.3\rho_m d^2 Y_b}}{\rho_m d}$$
$$= \frac{(1.08)(30) + 1.08\sqrt{(30)^2 + (9.3)(10)(3.826)^2(10)}}{(10)(3.826)}$$
$$= 4.2 \text{ ft/sec}$$

The actual velocity inside drill pipe is:

$$\bar{v} = \frac{q}{245d^2} = \frac{308}{(2.45)(3.826)^2} = 8.58 \text{ ft/sec}$$

Since 8.58 > 4.2, flow is turbulent and Eq. 7.4a applies:

$$R_e = \frac{2970\rho\bar{v}d}{\mu_p}, \qquad \text{(Eq. 7.10)}$$
$$= \frac{(2970)(10)(8.58)(3.826)}{30}$$
$$= 32{,}500$$

From Figure 7.1, curve II

$$f = 0.0066$$

Applying equation 7.4a:

$$\Delta p_p = \frac{f\rho L\bar{v}^2}{25.8\,d} = \frac{(0.0066)(10)(5500)(8.58)^2}{(25.8)(3.826)}$$
$$= 270 \text{ psi*}$$

6. Pressure drop inside drill collars:

$$\bar{v} = \frac{308}{(2.45)(2.813)^2} = 15.9 \text{ ft/sec.}$$

$\therefore$ flow is turbulent, by inspection.

$$R_e = \frac{(2970)(10)(15.9)(2.813)}{30} = 44{,}300$$

---

*It should be realized that the drill pipe may have two inside diameters; one for the pipe body and one for the tool joint. In such cases separate pressure drop calculations must be made for the total length of each i.d. This is commonly accomplished by computing an average i.d. for the pipe-tool joint combination. This complication was avoided in our examples by specifying the internally flush joint. The possible effect of tool joint design is, however, evident from the Hughes charts shown later.

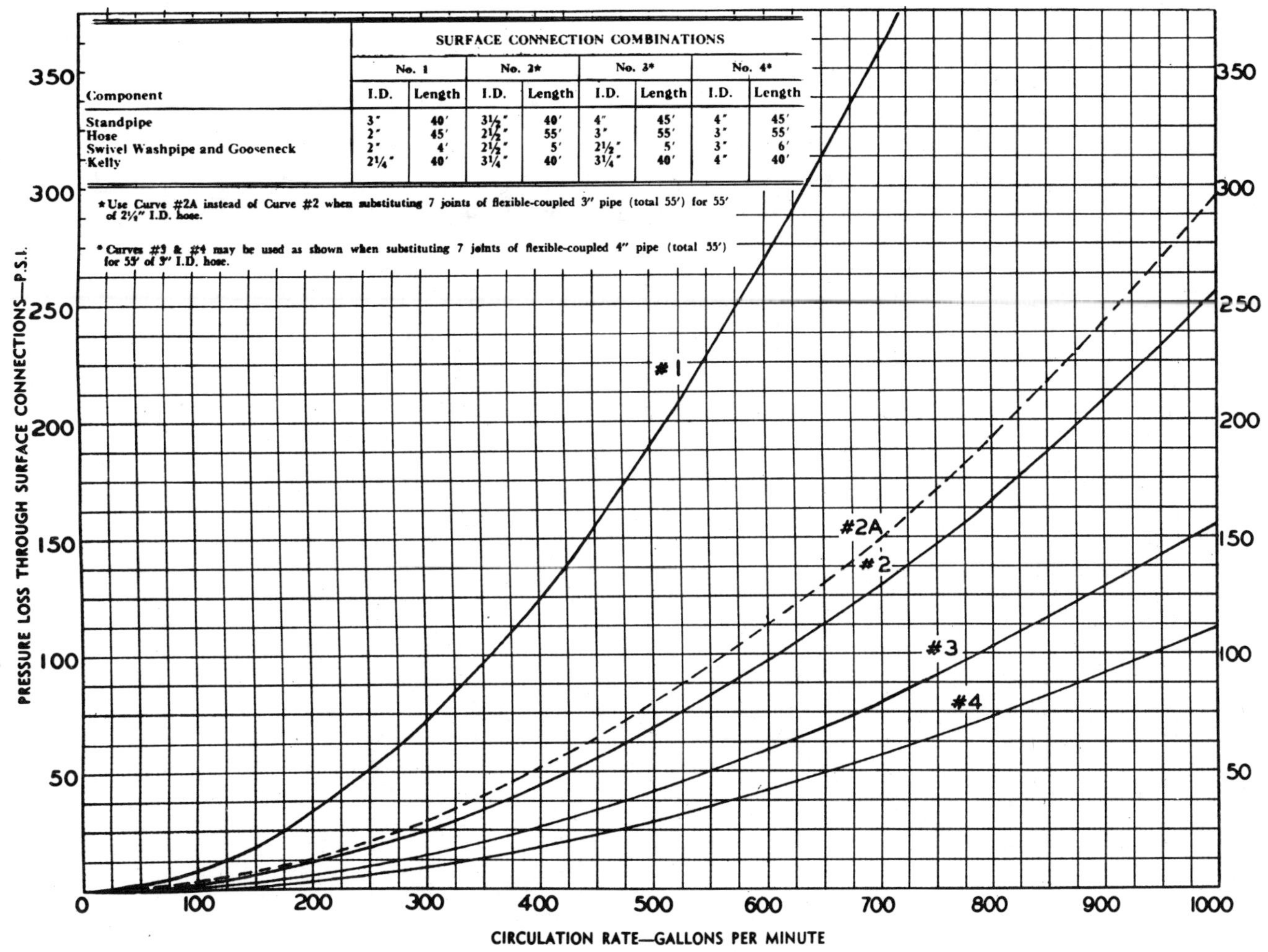

Component	SURFACE CONNECTION COMBINATIONS							
	No. 1		No. 2★		No. 3*		No. 4*	
	I.D.	Length	I.D.	Length	I.D.	Length	I.D.	Length
Standpipe	3″	40′	3½″	40′	4″	45′	4″	45′
Hose	2″	45′	2½″	55′	3″	55′	3″	55′
Swivel Washpipe and Gooseneck	2″	4′	2½″	5′	2½″	5′	3″	6′
Kelly	2¼″	40′	3¼″	40′	3¼″	40′	4″	40′

★Use Curve #2A instead of Curve #2 when substituting 7 joints of flexible-coupled 3″ pipe (total 55′) for 55′ of 2½″ I.D. hose.

*Curves #3 & #4 may be used as shown when substituting 7 joints of flexible-coupled 4″ pipe (total 55′) for 55′ of 3″ I.D. hose.

Fig. 7.3. Pressure losses in surface connections. Courtesy Hughes Tool Company.

Note: From part (2),

$$R_e = 32{,}500 \times \frac{3.826}{2.813} = 44{,}300$$

$$f = 0.0062$$

$$\Delta p_c = \frac{(0.0062)(10)(500)(15.9)^2}{(25.8)(2.813)} = 108 \text{ psi}$$

7. Pressure drop across bit: This is a jet bit with a nozzle coefficient of 0.95; therefore, Eq. 7.16, corrected for multiple nozzles, is used.

$$d = \sqrt{(3)\left(\frac{13}{32}\right)^2} = 0.704 \text{ in.} \qquad \text{(Eq. 7.15)}$$

where d = equivalent single nozzle diameter for use in Eq. 7.13.

$$\Delta p_b = \frac{q^2\rho_m}{7430c^2d^4} = \frac{(308)^2(10)}{(7430)(0.95)^2(0.704)^4}$$

$$= 580 \text{ psi}$$

Note: Step (a) may be avoided where the nozzles are all the same size by simply applying Eq. 7.13 to the q through one nozzle:

$$\Delta p_b = \frac{(308/3)^2(10)}{(7430)(0.95)^2(13/32)^4} = 580 \text{ psi}$$

8. Annular loss around drill collars: The critical velocity for this section is

$$v_c = \frac{(1.08)(30) + 1.08\sqrt{(30)^2 + (9.3)(10)(1.125)^2(10)}}{10(1.125)}$$

$$= 7.25 \text{ ft/sec}$$

Note: The hydraulically equivalent diameter of the annulus equals

$$(7.875 - 6.750) = 1.125 \text{ inches}$$

The actual velocity,

$$\bar{v} = \frac{308}{(2.45)[(7.875)^2 - (6.75)^2]} = 7.6 \text{ ft/sec}$$

$\therefore$ flow is turbulent

and $$R_e = \frac{(2970)(10)(7.6)(1.125)}{30} = 8450$$

$f = 0.0098$, curve IV, Figure 7.1

Therefore,

$$\Delta p_{ac} = \frac{(0.0098)(10)(500)(7.6)^2}{(25.8)(1.125)} = 97 \text{ psi}$$

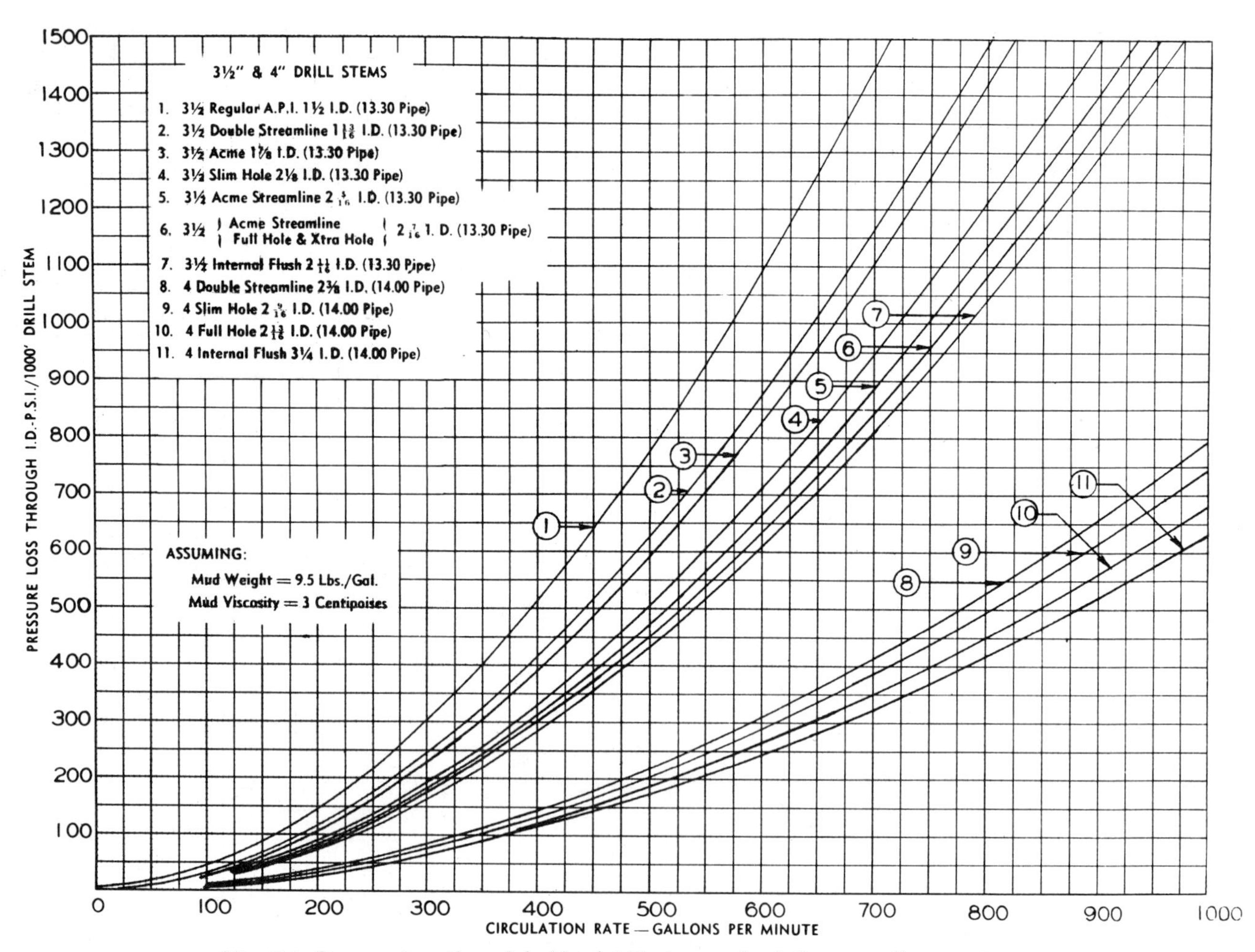

Fig. 7.4. Pressure drop through inside of drill pipe vs. circulation rate. Courtesy Hughes Tool Company.

9. Annular loss around drill pipe.

$$v_c = \frac{(1.08)(30)1.08\sqrt{(30)^2 + (9.3)(10)(3.375)^2(10)}}{(10)(3.375)}$$

$$= 4.38 \text{ ft/sec}$$

∴ flow is laminar, since circulation volume was calculated from annular velocity of 3 ft/sec. Using Eq. 7.6,

$$\Delta p_{ap} = \frac{LY_b}{300d} + \frac{\mu_p \bar{v} L}{1500d^2}$$

$$= \frac{(5500)(10)}{(300)(3.375)} \frac{(30)(3)(5500)}{(1500)(3.375)^2} = 83 \text{ psi}$$

Note: The annular pressure loss, like internal pressure drop, is composed of two parts: the loss around the pipe body and the loss around the tool joints. In most cases this is a negligible correction and the drill pipe section of the annulus may be treated as though the pipe diameter were uniform. In close diameter holes, however, the restrictions around the tool joints may be appreciable and require individual consideration. In such cases a common approximation is to consider each 1000 ft of annulus to consist of 950 ft of drill pipe and 50 ft of tool joint. Separate calculations, based on the actual annular clearances, are then performed for the proper length of each section. We will not apply this for our problems, but note it as a possible refinement.

10. The total pressure drop in the system is then:

$$\Delta p_t = 27 + 270 + 108 + 580 + 97 + 83 \cong 1170 \text{ psi}$$

The horsepower output at the pump is calculated from Equation 5.7:

$$\text{HP} = \frac{qp}{1714} = \frac{(308)(1170)}{1714} = 210$$

The input power from engine to pump (for 90% pump volumetric efficiency and 85% mechanical efficiency) is:

$$\text{HP} = \frac{210}{0.90 \times 0.85} = 275$$

As a general rule it is not necessary to make such a complete and involved calculation of circulating pressure requirement. The Hughes Tool Company has prepared sets of curves[6] and tables[7] some of which are reproduced as Figures 7.3 through 7.11. These were computed from altered forms of Fanning's equation for mud having a viscosity of 3 cp and a density of 9.5 lb/gal.

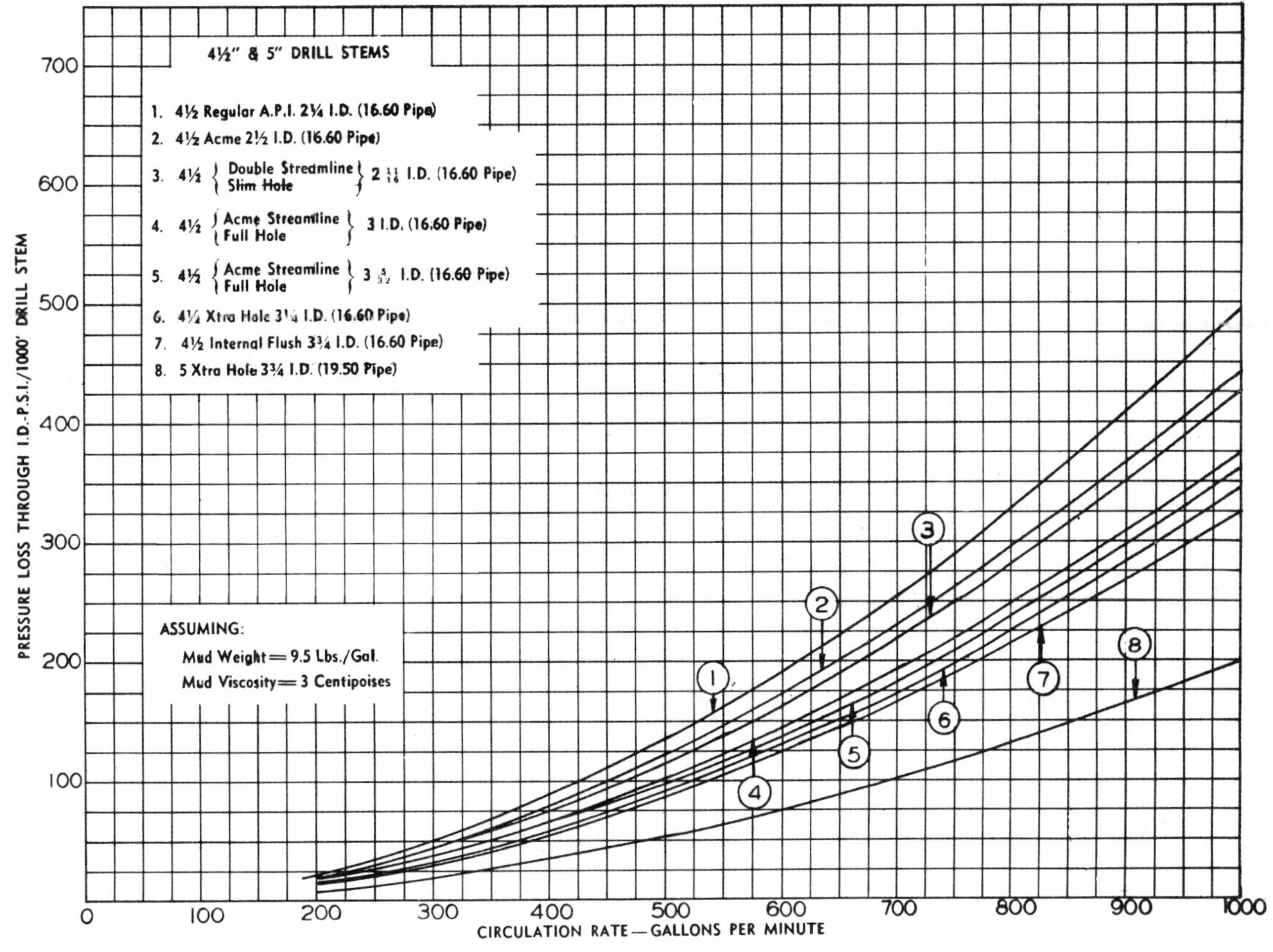

Fig. 7.5. Pressure drop through inside of drill pipe vs. circulation rate. Courtesy Hughes Tool Company.

The friction factor f has been accounted for in terms of flow rate and pipe size.

Pressure drops read from these charts may be corrected for densities other than 9.5 lb/gal by:

$$\Delta p = \Delta p_u \times \frac{\rho_m}{9.5} \tag{7.16}$$

where Δp = corrected value for mud of density ρ_m, lb/gal

Δp_u = uncorrected value from charts

Although viscosity has a relatively small effect on pressure drop in turbulent flow, it is not entirely satisfactory to assume that the value of 3 cp will be sufficiently accurate for all cases. Further, the charts (Figures 7.12 and 7.13) for annular flow are based on a turbulent condition, even though laminar flow often prevails in this section. The error due to the latter assumption is generally negligible, since annular losses are a relatively small part of total pressure loss. In some cases it may be worthwhile, however, to correct other chart values for viscosity:

$$\Delta p = \Delta p_u \left(\frac{\mu_t}{3}\right)^{0.14} \tag{7.17}$$

where μ_t = turbulent viscosity of mud in use, cp.

Consequently it will be more expedient to use the charts for most hydraulic calculations, although resort to the longer procedure may be useful in some instances. For illustration, let us rework Example 7.3 using the Hughes charts and the necessary correction factors.

Example 7.4

Data are those of the previous problem in Section 7.5. Calculate circulating pressure required.

Circulation rate q = 308 gal/min, 3 13/32 in. nozzles.
Pressure loss calculations

	System component	*Figure no.*	Δp_u, *psi*	Δp, *psi*
(1)	Surface connections	7.3 — #2	27	33 *
(2)	Inside drill pipe	7.5 — 7	176	218 *
(3)	Inside drill collars	7.7 — 2¾″	75	93 *
(4)	Bit nozzles	7.9	550	580 **
(5)	Outside drill collars	7.10	—	125
(6)	Outside drill pipe	7.10	—	77
	Circulating pressure			= 1126 psi

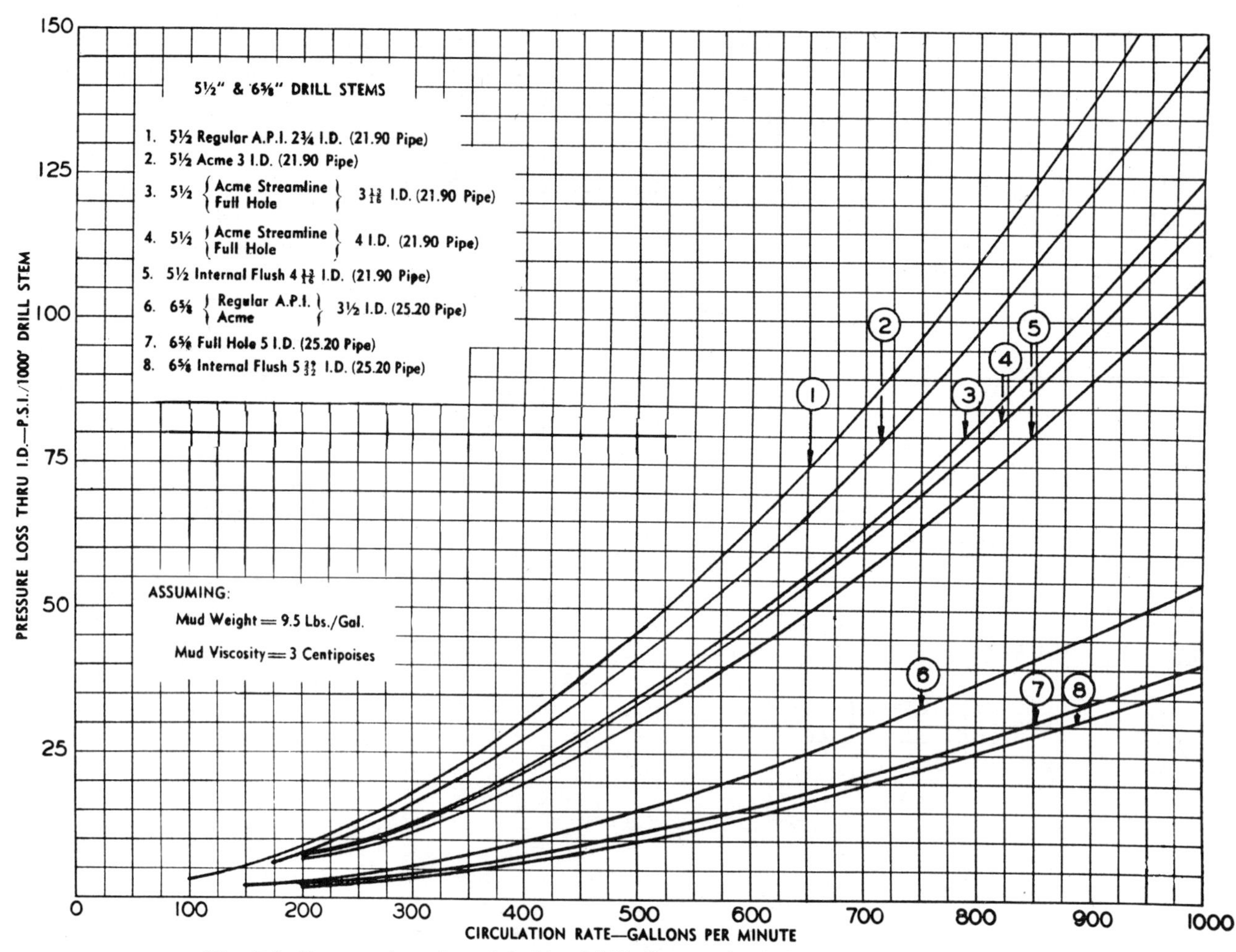

Fig. 7.6. Pressure drop through inside of drill pipe vs. circulation rate. Courtesy Hughes Tool Company.

$$^*\Delta p = \left(\frac{\mu_t}{3}\right)^{0.14} \times \frac{\rho_m}{9.5} \times \Delta p_u$$
$$= \left(\frac{30}{(3.2)(3)}\right)^{0.14} \left(\frac{10}{9.5}\right) \Delta p_u$$
$$= 1.24\ \Delta p_u$$

$$^{**}\Delta p = \frac{\rho_m}{9.5} \times \Delta p_u = 1.05 \times 550 = 580$$

Note that this result checks with that of Example 7.3 within 4%; this is sufficiently accurate for practical purposes.

There are, however, certain mud engineering problems which are best solved by the more rigorous method presented previously. This is illustrated by Example 7.5.

Example 7.5

(a) Given the following data, compute the bottom hole circulating pressure.

Hole size $= 9\frac{7}{8}''$

Depth $= 10{,}000$ ft

Drill pipe $= 4\frac{1}{2}''$ o.d.

Drill Collars = ~~600~~ 500 ft of 8 in. o.d.

Circulating rate = 400 gal/min

Mud = 12 lb/gal

Yield value (Bingham) $= \frac{60 \text{ lb}}{100 \text{ ft}^2}$

Plastic viscosity = 40 cp

Calculation of Δp_{ac}:

$$v_c = \frac{(1.08)(40) + 1.08\sqrt{(40)^2 + (9.3)(12)(1.875)^2(60)}}{(12)(1.875)}$$
$$= 9.5 \text{ ft/sec}$$

$$\bar{v} = \frac{400}{(2.45)[(9.875)^2 - (8)^2]} = 4.9 \text{ ft/sec}$$

∴ laminar flow exists.

$$\Delta p_{ac} = \frac{(500)(60)}{(300)(1.875)} + \frac{(40)(4.9)(500)}{(1500)(1.875)^2}$$
$$= 72 \text{ psi}$$

Since the flow around the drill collars is laminar, the flow in the drill pipe annulus must also be laminar, since there the velocity is less.

$$\bar{v} = \frac{400}{(2.45)[(9.875)^2 - (4.5)^2]} = 2.11 \text{ ft/sec}$$

$$\Delta p_{ap} = \frac{9500}{(300)(5.375)}\left[60 + \frac{(40)(2.11)}{(5)(5.375)}\right] = 370 \text{ psi}$$

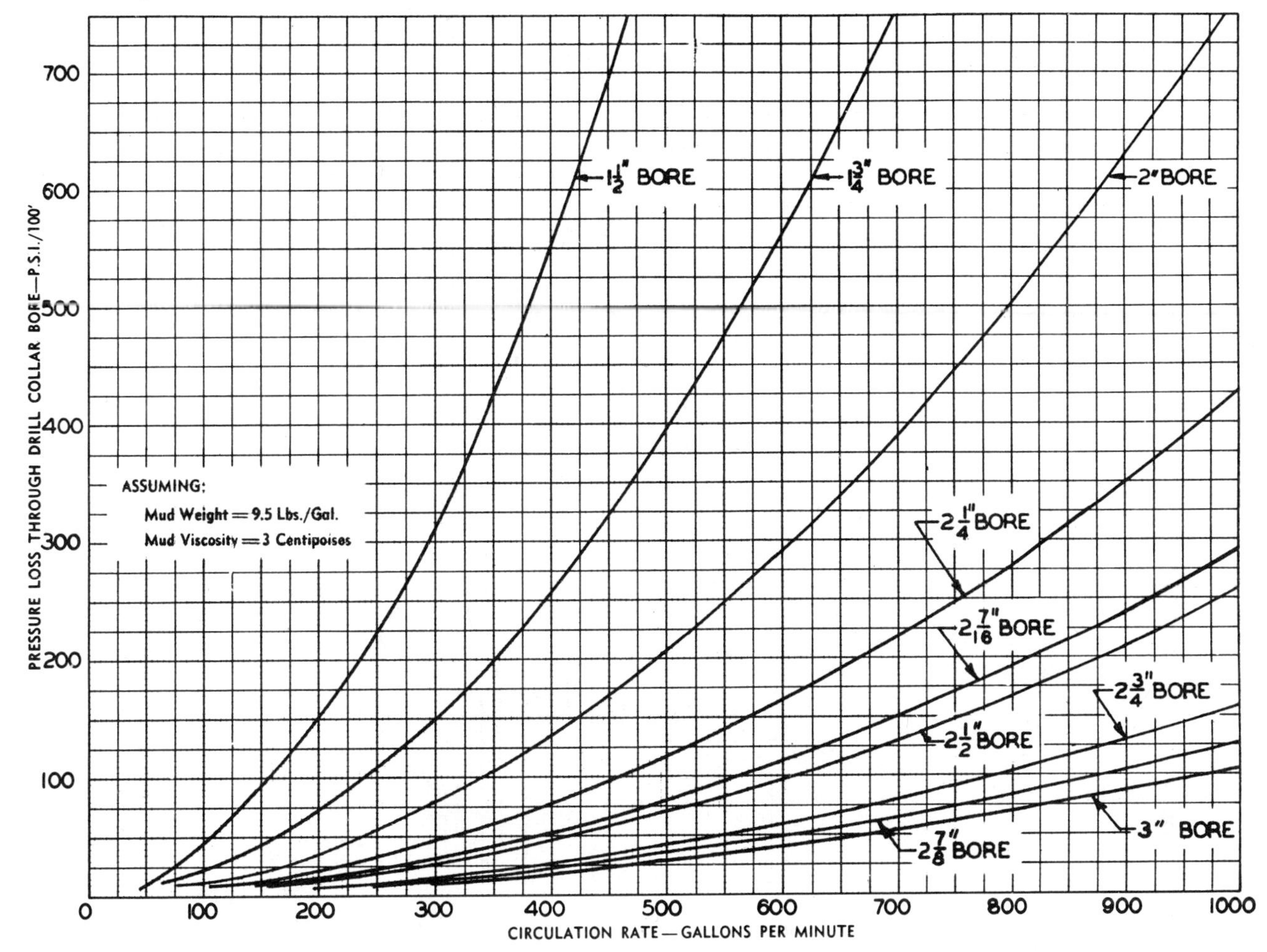

Fig. 7.7. Pressure drop through bore of drill collars vs. circulation rate. Courtesy Hughes Tool Company.

The bottom hole circulating pressure is the sum of the hydrostatic pressure plus the frictional losses:

$$\begin{aligned} \text{BHP} &= 370 + 72 + 10{,}000 \times 10/8.33 \times 0.433 \\ &= 442 + 5200 \cong 5640 \text{ psi} \end{aligned}$$

(b) By chemical treatment and removing excess clay solids, the mud properties were changed to the following values:

$\mu_p = 30$ cp, $Y_b = 5$ lb/100 ft², and $\rho_m = 12$ lb/gal.

What reduction in bottom hole circulating pressure would these changes allow?

Drill collar section

$$v_c = \frac{(1.08)(30) + 1.08\sqrt{(30)^2 + (9.3)(12)(1.875)^2(5)}}{(12)(1.875)}$$

$$= 4.0 \text{ ft/sec} \qquad \therefore \text{ flow is still laminar}$$

$$\Delta p_{ac} = \frac{500}{(300)(1.875)}\left[5 + \frac{(30)(4.9)}{(5)(1.875)}\right]$$

$$= 18 \text{ psi}$$

Drill pipe section

$$\Delta p_{ap} = \frac{9500}{(300)(5.375)}\left[5 + \frac{(30)(2.11)}{(5)(5.375)}\right]$$

$$= 43 \text{ psi}$$

Reduction in BHCP $= 442 - 61 \cong 380$ psi

Such a reduction could mean the difference between having or avoiding a lost circulation problem. It should be noted that circulation rate and density were not changed; the entire reduction was due to improved flow properties. Other calculations which further illustrate the possibilities of flow property alteration have been presented by Mesaros.[8]

Temperature Corrections on Flow Properties

The effect of elevated temperature on flow properties of drilling fluids is not generally considered in routine hydraulics calculations. The main reason for this seeming oversight is that the behavior of flow properties with temperature is apparently unique for each mud, and requires individual evaluation. Srini-Vasan[9] has compiled considerable data on water base muds which show that, in general, a plot of the log of the ratio of either plastic or apparent viscosity to the viscosity of water at the same temperature is a linear function of temperature, that is,

$$\log \frac{\mu}{\mu_w} = A + bT \qquad (7.18)$$

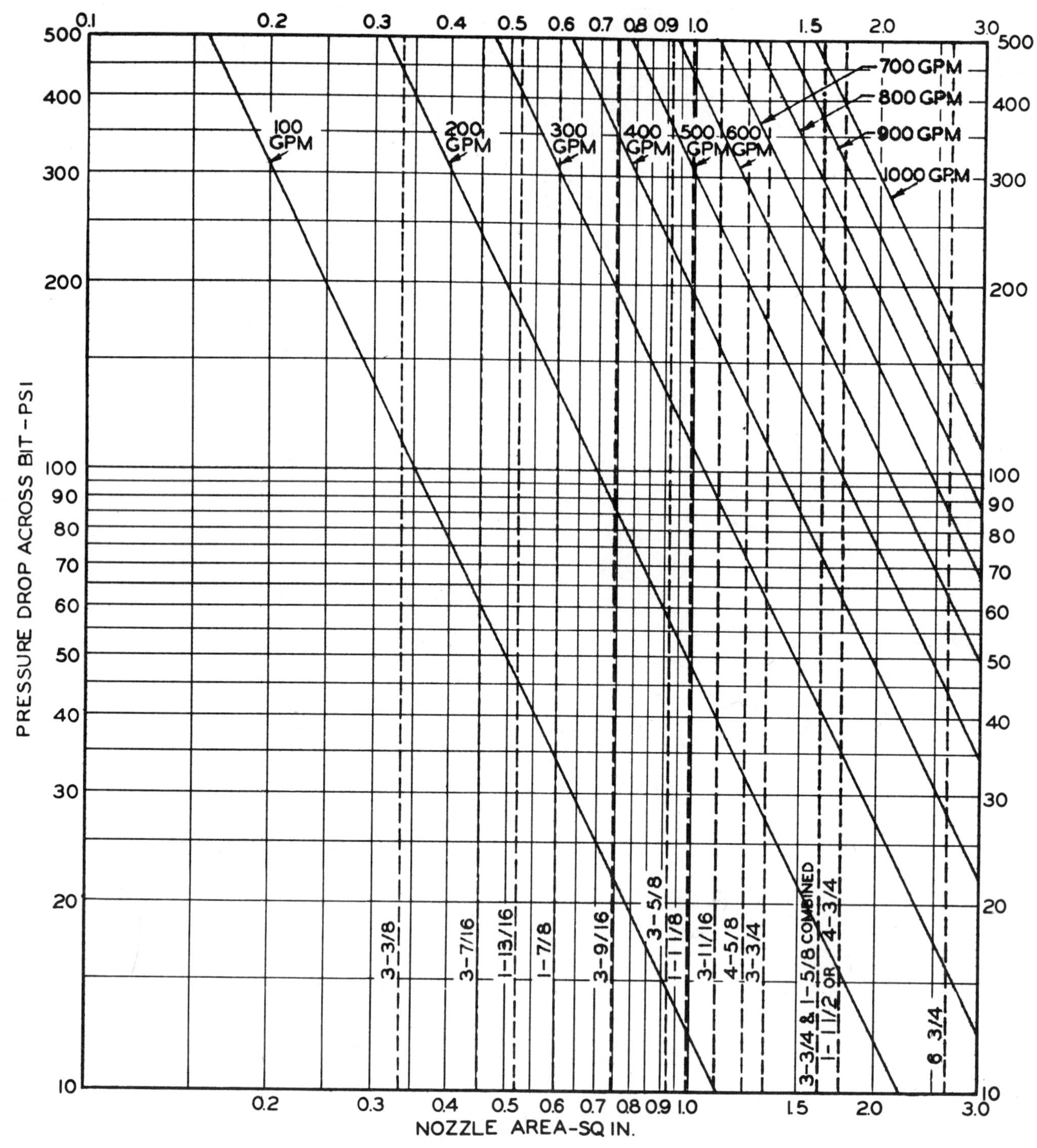

Fig. 7.8. Pressure drop across conventional water course bits. Assuming $\rho_m = 9.5$ lb/gal and $C = 0.80$. Courtesy Hughes Tool Company.

where μ = either plastic or apparent mud viscosity
μ_w = water viscosity
T = temperature, °F

Use of this equation requires knowledge of viscosity at two temperatures, so that linear plots can be established. Water viscosity data were given in Figure 6.8. Yield point at the desired temperature may then be calculated from Eq. 6.4. The average circulating temperature of the mud stream may be estimated from Figure 7.12. Note that the bottom hole circulating temperature is considerably less than the static temperature. Average circulating temperature may be taken as the arithmetic mean of bottom hole and either mud suction or mud discharge temperatures. Hence for a 12,000 ft well, the

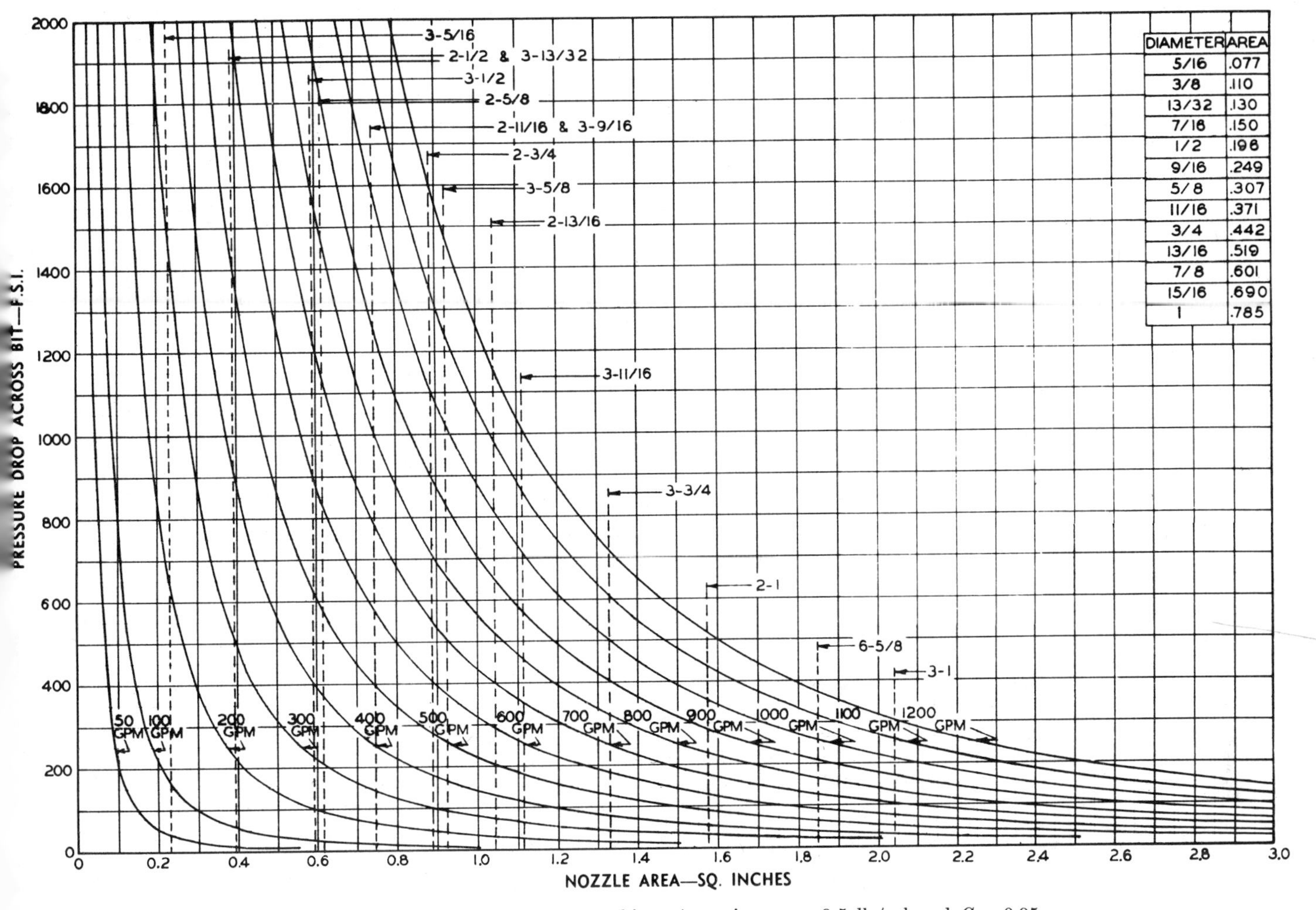

DIAMETER	AREA
5/16	.077
3/8	.110
13/32	.130
7/16	.150
1/2	.196
9/16	.249
5/8	.307
11/16	.371
3/4	.442
13/16	.519
7/8	.601
15/16	.690
1	.785

Fig. 7.9. Pressure drop across jet type bits. Assuming ρ_m = 9.5 lb/gal and C = 0.95. Courtesy Hughes Tool Company.

average temperature in the annulus is approximately 148°F. This chart is based on Gulf Coast wells, but may be altered to fit other geologic areas.

Due to the small effect of viscosity on total system pressure loss, temperature corrections on flow properties are largely academic. There may be, however, certain instances where such refinements are applicable. This discussion, however, pertains only to temperature ranges which do not cause appreciable chemical alteration of the mud, and certainly does not include cases of solidification or extreme gelation.

7.6 Hydraulics and Rate of Penetration

It has been established that the penetration rate in many formations is proportional to the hydraulic horsepower expended at the bit. Realization of this factor and the subsequent widespread use of jet bits has reversed the trend to small drill pipe sizes which existed a few years ago. Pressure drops inside the drill string are a large part of total system losses, and it is obvious that an increase of inside diameter will greatly improve hydraulic efficiency. Consequently, tool joints which have little or no internal restriction are most commonly used. The use of larger drill pipe and drill collars also entails closer hole-pipe clearances; thus the desired annular velocity can be obtained at lower circulation rates. In addition, annular velocity requirements have been reappraised and lower figures are now being applied.

The specific application of hydraulic calculations to improvement of penetration rate will be demonstrated in the next chapter.

7.7 Pressure Surges Caused by Pipe Movement

It has long been observed in the field that circulation is frequently lost after making a *trip*, particularly if previous trouble has been encountered. Originally, this loss was believed to be due to the drill string and bit knocking mud cake off the walls of the hole and re-exposing the thief zone(s). Similarly, blowouts often occurred during the pipe removal part of a trip due to formation gas (trip gas) entering the hole and lightening the mud column. Experimental[10,11] and analytical[12,13]

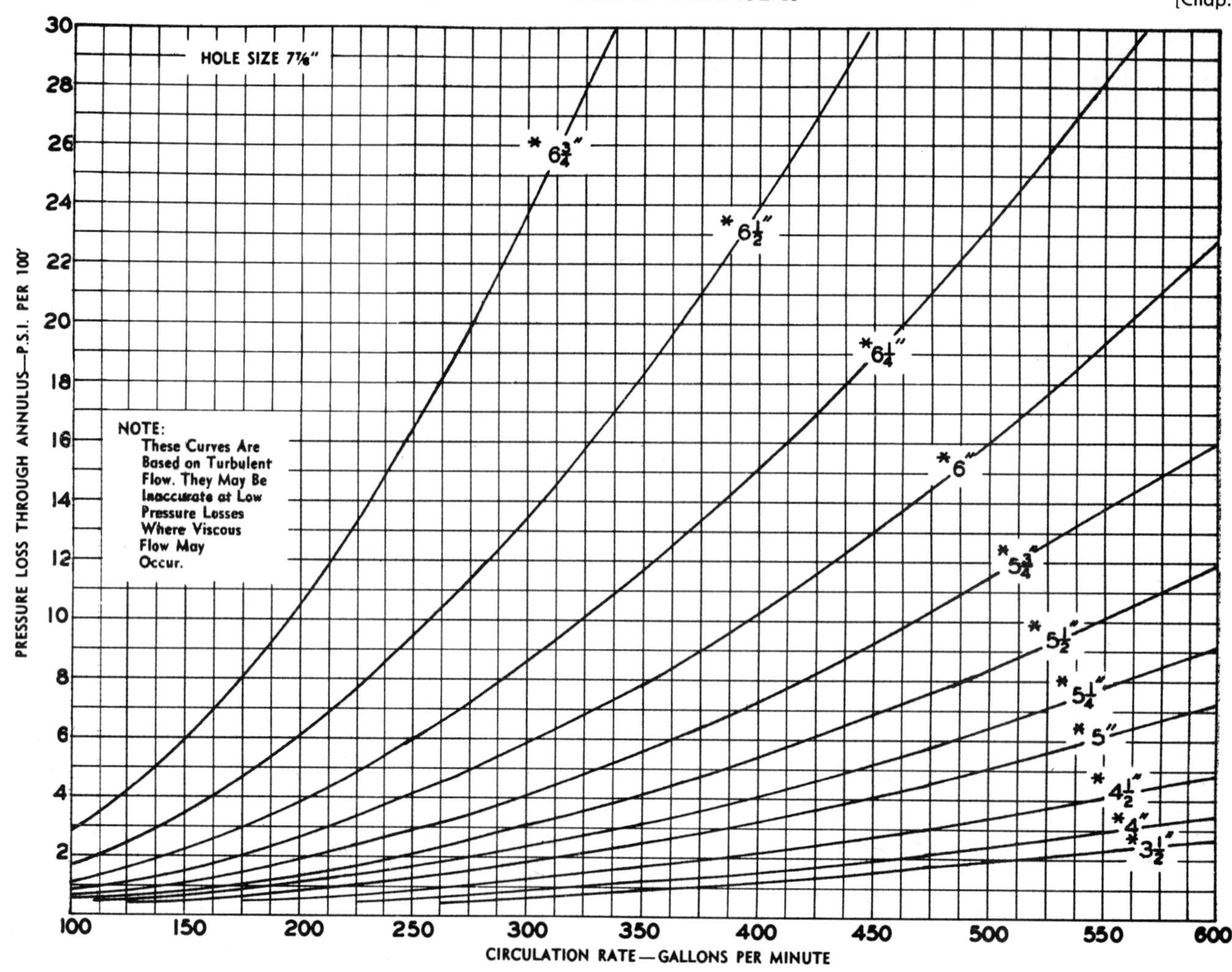

Fig. 7.10. Annular pressure drop for turbulent flow of 3 cp, 9.5 lb/gal mud *(pipe sizes). Courtesy Hughes Tool Company.

work, however, has shown these problems to be largely due to down hole pressure variations caused by the piston-cylinder action of the pipe and borehole.

Consider Figure 7.13, which shows the drill pipe being lowered in the mud-filled borehole. The velocity profile in the annulus is characterized by the counterflow pattern shown: some mud adjacent to the pipe will flow downward due to viscous drag, while the displaced volume will flow upward. The maximum pressure surge may be calculated by assuming that the drill pipe is sealed or plugged and that its entire volume (total cross-sectional area) is displaced. While this is a somewhat extreme assumption, it is comparable to running a back pressure valve (float) such as is common practice in drill stem testing or running casing. Let us, then, analyze this situation using the hydraulic principles previously presented.

The following assumptions will be made:

(1) the volume of mud flowing upward = the total volume of pipe (calculated from outside diameter) being lowered.

(2) the mud's upward velocity = velocity calculated using the total annular area (no counterflow) + 1/2 the downward pipe velocity. This is the assumption made by Ormsby[13] which was found to yield sufficiently accurate results. In equation form:

$$v_a = \frac{d_{op}^2 v_p}{d_h^2 - d_{op}^2} + \frac{v_p}{2} = v_p\left(\frac{1}{2} + \frac{d_{op}^2}{d_h^2 - d_{op}^2}\right)$$

where v_a = upward flow velocity, ft/sec
v_p = pipe velocity, ft/sec
d_{op} = pipe outside diameter, in.
d_h = hole diameter, in.

Knowing v_a, calculations of running pressure (or pulling suction) may be made in the same manner as any other mud flow problem. Example 7.6 illustrates this procedure.

Example 7.6

(a) Given the following data, calculate the actual pressure imposed on a formation at 10,000 ft. Assume pipe is closed at lower end.

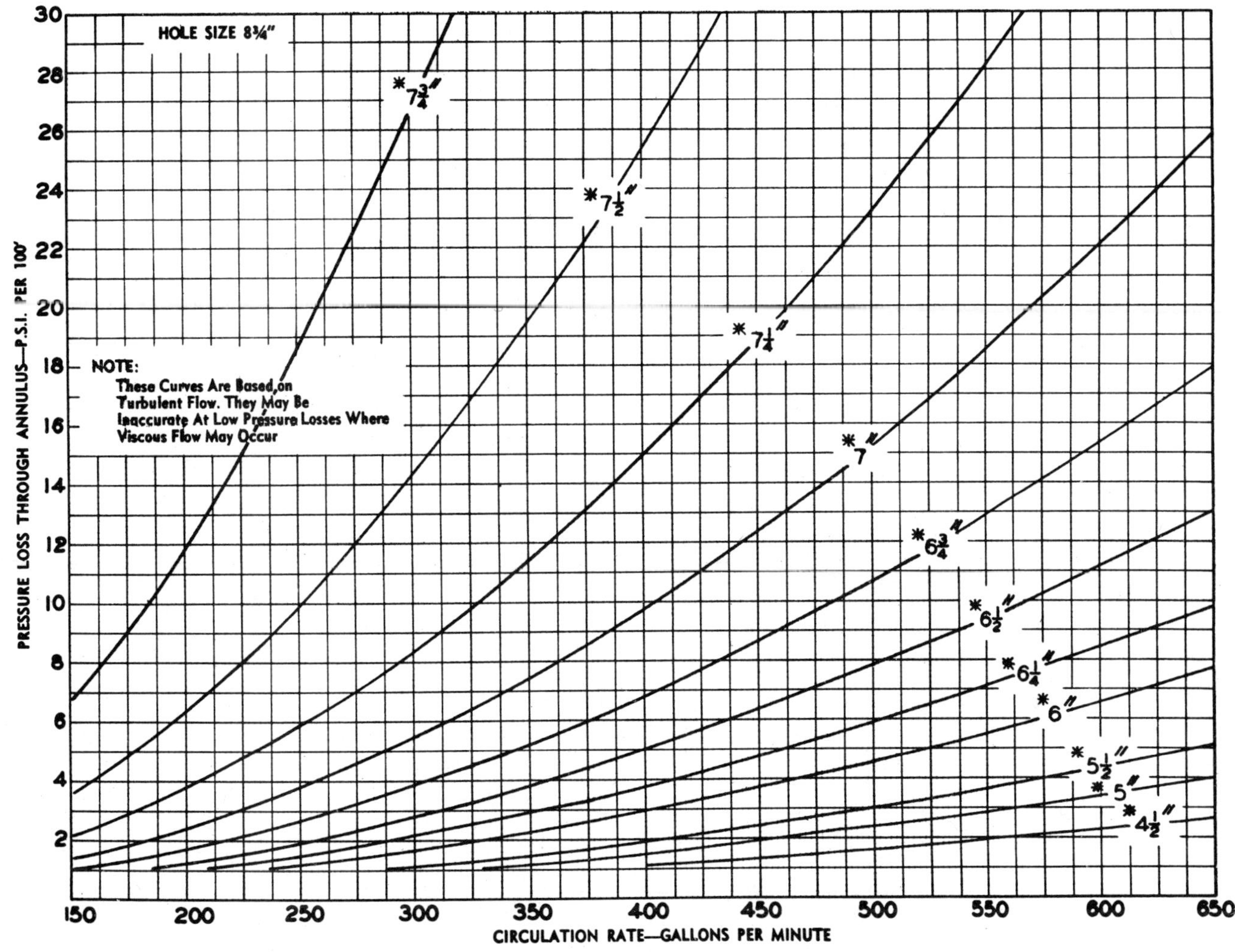

Fig. 7.11. Annular pressure drop for turbulent flow of 3 cp, 9.5 lb/gal mud *(pipe sizes). Courtesy Hughes Tool Company.

Length pipe in hole = 9500 ft
μ_p = 35 cp
Y_b = 15 lb/100 ft²
ρ_m = 10 lb/gal
Hole size = 9 in.
Pipe size = 4½ in. o.d. (ignore drill collars)
The pipe is being lowered at 400 ft/min.

Solution:

$$v_c = \frac{(1.08)(35) + 1.08\sqrt{(35)^2 + (9.3)(10)(4.5)^2(15)}}{(10)(4.5)}$$

$$= 5.0 \text{ ft/sec}$$

From Eq. (7.16):

$$v_a = \frac{400}{60}\left(\frac{1}{2} + \frac{(4.5)^2}{(9)^2 - (4.5)^2}\right) = 5.6 \text{ ft/sec}$$

∴ flow is turbulent.

Δp_{ap} for turbulent condition:

$$R_e = \frac{(2970)(10)(5.6)(4.5)}{35} = 21,400$$

$$f = 0.0084 \text{ from Figure 7.1, curve IV}$$

$$\Delta p_{ap} = \frac{(0.0084)(10)(9500)(5.6)^2}{(25.8)(4.5)} = 215 \text{ psi due to friction}$$

The total instantaneous bottom hole pressure is:

$$p = 215 + (0.052)(10{,}000)(10) = 5415 \text{ psi}$$

(b) Rework (a) including the effect of 500 ft of 7½ inch o.d. drill collars on the bottom of the drill string.

Solution:

$$v_{ac} = \frac{400}{60}\left[\frac{1}{2} + \frac{(7.5)^2}{(9)^2 - (7.5)^2}\right] = 18.5 \text{ ft/sec}$$

$$R_e = \frac{(2970)(10)(18.5)(1.5)}{35} = 23{,}400$$

$$f = 0.0083$$

$$\Delta p_{ac} = \frac{(0.0083)(10)(500)(18.5)^2}{(25.8)(1.5)} = 370 \text{ psi}$$

From part (a),

$$\Delta p_{ap} = \frac{9000}{9500} \times 215 = 203$$

$$\text{Pressure surge} = 203 + 370 \cong 570 \text{ psi}$$

$$\text{Total } p = 570 + 5200 = 5770 \text{ psi}$$

(c) What would the instantaneous bottom hole pressure be if the pipe were being removed at the same rate as in parts (a) and (b)?

(1) $p = 5200 - 215 = 4985$ psi.

(2) $p = 5200 - 570 = 4630$ psi.

In other words, the running pressure and pulling suction are equal but of opposite sign.

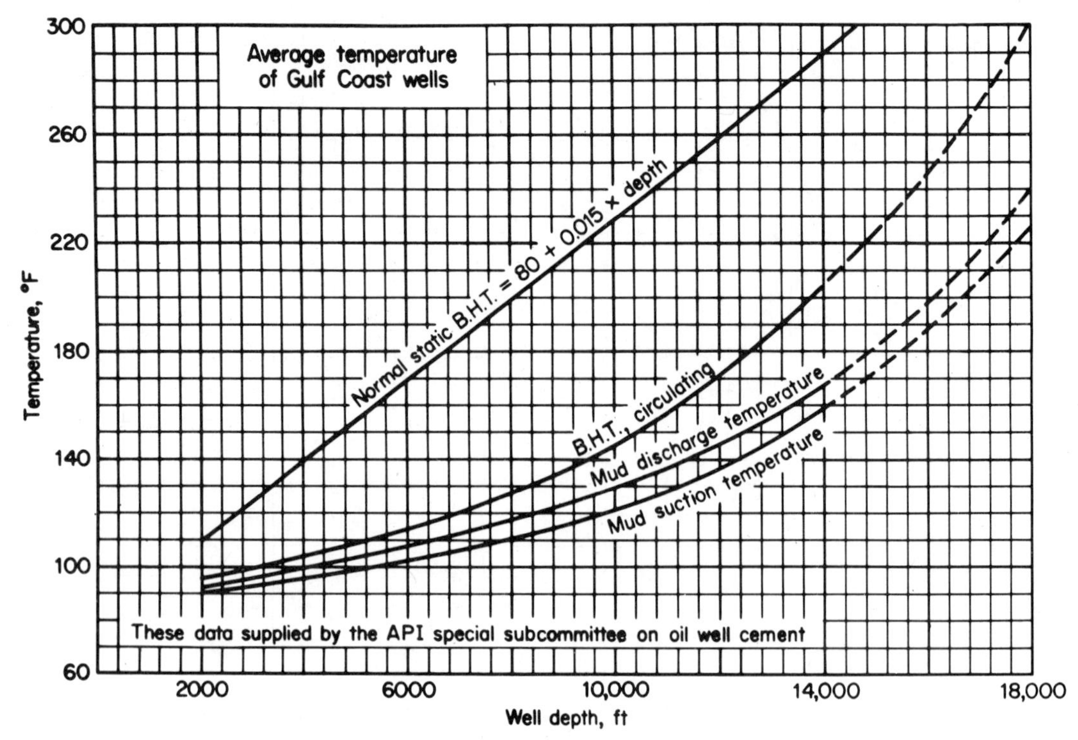

Fig. 7.12. Circulating and static wellbore temperatures for Gulf Coast wells of various depths. Courtesy Halliburton Oil Well Cementing Company.

These problems serve to illustrate the potential hazards of lowering or removing the drill string too fast. The importance of such calculations lies not in the precise magnitude of the numerical answers but in the realization of the problem which is involved. Slowing of pipe movement means longer trip time, but in many areas such action is well worth this expense. It should be noted that ignoring the presence of drill collars caused an error of approximately 100%. Closer hole-collar clearance will greatly increase this error, since turbulent pressure loss varies as $\bar{v}^2$. Note also that Δp varies as the length of pipe in the hole. Consequently the string may be pulled (or run) faster as length decreases. However, once the drill collars are in the hole, that section exerts a large and constant effect.

The use of yield point in the previous calculation may also introduce errors, particularly for muds which exhibit large gel strengths with quiescent time. A more realistic value would be the actual gel strength at the time involved, if it were known.

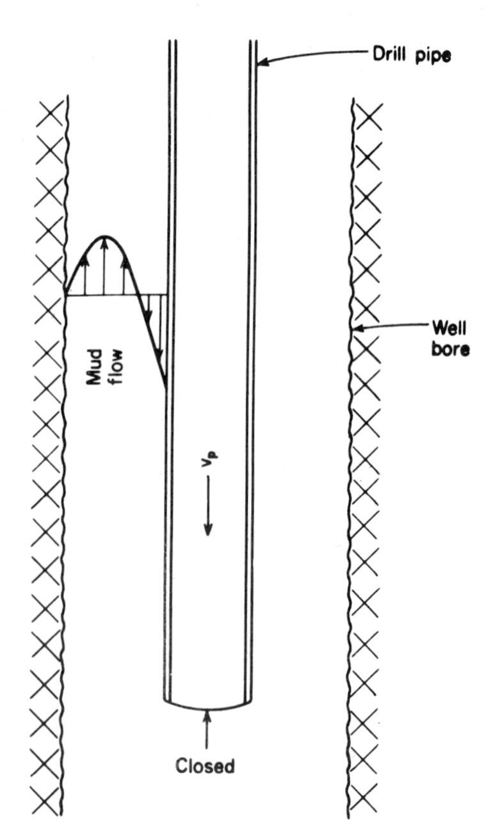

Fig. 7.13. Worst pipe running condition (closed end). Arrows indicate counter flow pattern in annulus.

7.8 Air, Gas and Aerated Mud Drilling

Air Requirements

The circulation volume required in air or gas drilling is governed by the annular velocity necessary to lift the cuttings. These velocities are such that turbulent flow always exists; thus cutting slip velocity can be defined by Rittinger's equation:

$$v_{ts} = \sqrt{\frac{4(\rho_s - \rho_f)gd_c}{3\rho_f C_d}} \tag{7.20}$$

where v_{ts} = turbulent slip (maximum free fall) velocity of a solid particle, ft/sec

ρ_s = particle density, lb/ft³

ρ_f = fluid density, lb/ft³

g = 32.2 ft/sec²

d_c = particle diameter, ft

C_d = drag coefficient for the particular particle. This is essentially a function of particle shape at high Reynolds numbers.

For our purposes, $\rho_s - \rho_f = \rho_s$. For spheres, $C_d = 0.50$, hence 7.20 may be altered to:[14]

$$v_{ts} = 2.67\left(\frac{d_c\rho_s}{\rho_f}\right)^{1/2}, \quad \text{for spheres} \tag{7.21}$$

where d_c = particle diameter, in.

Gray[15] has evaluated C_d as 0.85 for sand grains and 1.40 for flat particles such as limestone and shale cuttings. The density of a gas ρ_f may be expressed in terms of the gas law:

$$\rho_f = \frac{pM}{ZRT}$$

For air, $M = 29$, $Z \cong 1.0$ in the range of pressures and temperatures commonly encountered in air or gas drilling, and, $R = 10.7$ for p = psia and T = °Rankine. Substituting these values in Eq. 7.21 we obtain:

$$v_{ts} = 1.62\left(\frac{d_c\rho_s T}{p}\right)^{1/2} \tag{7.22}$$

Example 7.9

(a) What air velocity will just float a $\frac{3}{8}$-in. diameter, 2.6 specific gravity, spherical particle at 30 psig and 120°F?

Solution:

$$v_{ts} = 1.62\left[\frac{(0.375)(162)(580)}{44.7}\right]^{1/2} = 45 \text{ ft/sec}$$

$$= 2700 \text{ ft/min}$$

(b) What would the velocity be for a shale cutting of the same density?

Solution:

Using $C_d = 1.40$,

$$v_{ts} = 45\left(\frac{0.50}{1.40}\right)^{1/2} = 27 \text{ ft/sec} = 1620 \text{ ft/min}$$

In practice, a general rule of thumb figure for annular velocity is from 2000 to 4000 ft/min, with an average of 3000 ft/min. This is adequate to lift $\frac{3}{8}$- to $\frac{1}{2}$-in. spherical cuttings at low pressures and temperatures. Although most air drilled cuttings are quite small (actually dust particle size) when they reach the surface, this does not mean they were that small when they left bottom. The milling action of the drill string in the dry hole and impacts with other cuttings probably are largely responsible for most of this pulverizing, along with re-grinding at the bit, which works on exceptionally large particles.

Various methods for computing air (or gas) volume requirements are in use.[16–18] The method developed by R. R. Angel includes drilling rate as a parameter and will be presented here. The volume required is based on a circulation rate having a lifting power equivalent to 3000 ft/min of standard air. Table 7.1 lists the necessary factors which were obtained from computer solutions of the basic equation. The standard air or gas volume required in a given instance is:

$$Q_a = Q_0 + ND \tag{7.23}$$

where Q_a = required standard air volume, SCF/min

Q_0 = uncorrected value from Table 7.1, SCF/min

N = drilling rate factor from Table 7.1

D = hole depth, thousands of feet

Example 7.10

Calculate the circulation rate required to air drill a $7\frac{7}{8}$ in. hole with $4\frac{1}{2}$ in. drill pipe at a rate of 90 ft/hr at 12,000 ft depth.

$$Q_a = Q_0 + ND = 670 + (98.3)(12) = 1850 \text{ SCF/min}$$

Horsepower Requirements

The horsepower equation for the adiabatic compression of gases is commonly used for compressor calculations.

$$\text{HP} = \frac{4.36kp_1Q}{k-1}\left[\left(\frac{p_2}{p_1}\right)^{\frac{k-1}{k}} - 1\right] \tag{7.24}$$

where HP = horsepower output from compressor

k = specific heat ratio for gas being compressed = 1.4 for air and ≅ 1.3 for low molecular weight natural gas.

p_1 = suction or intake pressure of compressor, psia. See Figure 7.14 for pressure at various altitudes.

Q = intake air volume at intake conditions, p_1 and T_1, MCF/min

p_2 = discharge pressure, psia

The discharge pressure at the compressors is, of course, the standpipe or circulating pressure required. Calculation of this value for a flowing compressible column is rather complicated and is probably not worthwhile in air drilling, due to the uncertainty as to the quantity of water which may be encountered. Figure 7.15 includes three curves which illustrate this point. Curve 1 is the approximate theoretical pressure for circulating dry air using $4\frac{1}{2}$ inch drill pipe in $7\frac{7}{8}$- to 9-in. hole, as taken from a paper by J. O. Scott.[19] Curves 2 and 3 are equipment requirements based on experience

TABLE 7.1

DATA FOR CALCULATING APPROXIMATE CIRCULATION RATES REQUIRED TO PRODUCE A MINIMUM ANNULAR VELOCITY WHICH IS EQUIVALENT IN LIFTING POWER TO A STANDARD AIR VELOCITY OF 3,000 FT/MIN*

		AIR					*GAS, SPECIFIC GRAVITY* .60				
			Value of N					*Value of N*			
Hole Size	*Pipe OD*	Q_0	*Drilling Rate (ft/hr)*				Q_0	*Drilling Rate (ft/hr)*			
(in.)	*(in.)*	*scf/min*	0	30	60	90	*scf/min*	0	30	60	90
17½	6⅝	4,209	82.2	131	177	221	5,434	66.3	128	186	240
	5½	4,428	79.8	126	171	213	5,716	61.8	119	174	226
	4½	4,588	78.0	123	166	207	5,924	58.0	113	165	215
15	6⅝	2,905	71.7	112	151	188	3,751	64.2	118	167	214
	5½	3,124	68.7	107	143	178	4,033	58.6	108	154	197
	4½	3,285	66.0	103	137	171	4,241	54.0	100	144	185
12¼	6⅝	1,700	62.3	97.8	130	160	2,194	63.0	112	155	194
	5½	1,918	58.0	89.5	119	146	2,477	56.3	97.7	137	172
	4½	2,079	55.3	83.6	111	136	2,684	50.8	88.2	124	157
11	6⅝	1,237	60.6	94.5	124	151	1,597	64.5	112	152	188
	5½	1,456	54.8	83.8	110	135	1,880	55.5	95.4	131	163
	4½	1,616	50.6	76.9	101	124	2,087	50.0	84.4	116	146
9⅞	5½	1,079	53.0	80.3	104	126	1,393	56.4	94.7	128	157
	5	1,163	50.3	75.5	98.7	120	1,502	52.3	87.7	119	147
	4½	1,240	47.8	71.7	93.3	114	1,600	48.8	81.6	111	138
9	5	898	49.1	73.0	94.4	113	1,160	53.0	87.1	116	141
	4½	975	46.1	68.5	88.5	107	1,258	49.0	80.3	108	132
	3½	1,103	41.5	61.0	79.0	95.5	1,424	42.0	68.9	93.1	115
8¾	5	827	49.0	72.7	93.2	112	1,068	53.5	87.0	115	140
	4½	903	46.0	67.8	87.3	105	1,166	49.1	80.0	107	130
	3½	1,032	40.8	60.0	77.3	93.7	1,332	41.8	68.3	92	114
7⅞	4½	670	44.7	65.0	82.7	98.3	865	50.1	78.8	104	125
	3½	798	39.2	56.7	72.5	86.9	1,031	41.6	66.3	87.8	107
7⅜	3½	676	38.5	55.0	69.8	83.2	873	41.6	65.3	85.5	104
6¾	3½	535	37.3	52.8	66.1	78.0	690	41.5	63.8	82.3	99.0
6¼	3½	430	37.0	51.5	63.6	74.7	555	42.0	63.1	80.0	94.7
	2⅞	494	32.8	46.0	57.3	67.7	638	37.0	55.1	71.4	85.4
4¾	2⅞	229	31.6	41.3	49.5	56.5	296	37.0	51.3	62.6	72.2
	2⅜	271	27.8	37.2	44.8	51.6	350	32.3	45.6	56.3	65.5

*After Angel.[8]

in the Permian Basin as reported by Smith and Rollins[20] for dry and wet conditions, respectively. The spread between curves 2 and 3 is an uncertainty range dependent on the amount of liquids encountered. Compressor ratings of 100 to 125 psi are generally adequate for shallow work such as surface or shot hole drilling.

The number of compression stages which must be used is governed by the compression ratio, p_2/p_1. If this is less than 5 or 6, one stage will generally be sufficient; two stages will be used if $6 < p_2/p_1 < 36$. In general, the compression ratio per stage is taken as the square root of the total compression ratio.[21] More complete information on compressor sizing and selection may be found in manufacturers' literature and engineering handbooks. Compressor selection for air drilling purposes has been discussed by Morris and Ramey.[22]

Example 7.11

(a) Estimate the compressor output horsepower for the data of Example 7.10. Assume suction conditions of 80°F and an operating altitude of 3000 ft above sea level.

Solution:

The air volume of 1850 ft³/min at 60°F and 14.7 psia must be corrected to the actual suction conditions. From Figure 7.14, the atmospheric pressure at 3000 ft = 13.1 psia.

$$Q = 1850 \times \frac{14.7}{13.1} \times \frac{460 + 80}{520}$$

$$= 2150 \text{ ft}^3/\text{min} = 2.15 \text{ MCF/min}$$

From Figure 7.15(2), p_2 = 300 psig. Assuming n = 1.4 for air, and applying Eq. 7.24:

$$\text{HP} = \frac{(4.36)(1.4)(13.1)(2.15)}{(1.4 - 1)}\left[\left(\frac{300}{13.1}\right)^{.4/1.4} - 1\right]$$

$$= 625$$

(b) How many stages will be required for these conditions and what will be the compression ratio per stage?

Solution:

$$R = \frac{p_2}{p_1} = \frac{300}{13.1} = 22.9 \qquad \therefore \text{ two stages will be required,}$$

and

$$R \text{ per stage} = \sqrt{22.9} = 4.8$$

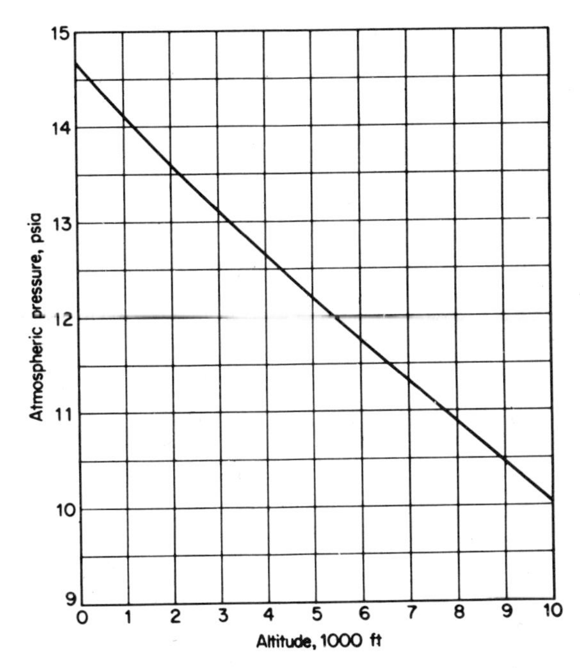

Fig. 7.14. Atmospheric pressure vs. elevation above sea level.

Since horsepower requirement is greatly influenced by the compression ratio, it could be advantageous, in some cases, to operate at higher circulating pressures, hold back pressure on the annulus, and recycle previously circulated air.[15] Practical application of this technique will require an air-dust separator capable of removing all abrasive particles from the air prior to its entering the compressors. A further advantage of such a method, if practical, is the prevention of natural gas waste which attends the current practice of discharging to the atmosphere.

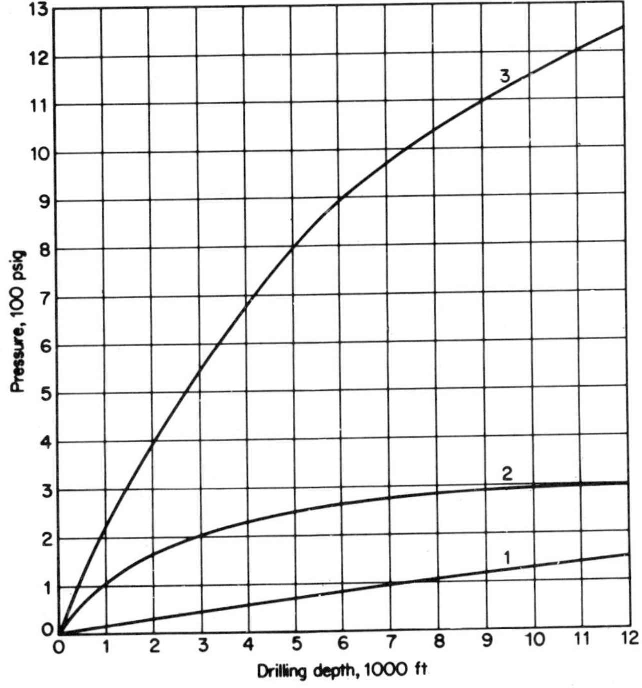

Fig. 7.15. Approximate circulating pressure requirements for air drilling in $7\frac{7}{8}$- to 9-in. holes.

(1) Theoretical dry air curve[19]
(2) Equipment ratings from experience — dry conditions[20]
(3) Equipment ratings for wet conditions[20]

Aerated Mud Drilling

The air quantity required for aerated mud drilling is primarily dependent on the desired density reduction of the mixture. Figure 7.16 allows rapid and reasonably accurate estimates to be made for normal conditions.[23] The dotted line is a sample solution for the amount of air needed to lighten mud density from 8.5 to 6 lb/gal at 5800 ft depth and an average temperature of 150°F. The circulating pressure required at the mud pumps and hence at the air compressors may be calculated by the method of Poettmann and Carpenter[24] if necessary. Empirical data are often available which give a satisfactory value in lieu of such calculations.

PROBLEMS

1. Compute the pressure drop per 1000 ft of $4\frac{1}{2}$-in. internally flush drill pipe if water (μ = 1 cp) is being used as drilling fluid. The circulation rate is 500 gal/min. Inside diam. = 3.83 in.

2. Repeat Problem 1 for a 9 lb/gal mud having: $\mu_p = 30$ cp, $Y_b = 20$ lb/(ft²·100)

3. At what velocity in Problem 2 does flow become turbulent?

4. Develop Eq. 7.8 from Eqs. 7.7 and 7.2.

5. (a) What is the hydraulic radius of a 4-in. square pipe? (b) What is the hydraulically equivalent circular diameter of such a pipe? *Ans.* (a) 1 in. (b) 4 in.

6. Develop Eq. 7.13 from Eq. 7.12.

7. What single nozzle size should be used in Eq. 7.13 if the actual bit has
(a) three $\frac{9}{16}$-in. jets
(b) two $\frac{11}{16}$-in. jets
(c) two $\frac{11}{16}$-in. and one $\frac{1}{2}$-in. jets?

8. Given the following data, compute the required circulating pressure and pump output horsepower by both methods of Section 7.5, and compare the results. Make a correction using Eq. 7.20, where applicable.

Depth = 10,000 ft (9400 ft pipe, 600 ft drill collars)

Drill pipe = $4\frac{1}{2}$-in. internally flush tool joints

Drill collars = 6-in., with $2\frac{1}{2}$-in. bore

Bit = $8\frac{3}{4}$-in., jet type, three cone, rock bit

Nozzle velocity = at least 25 ft/in. of bit diameter

Mud: $\rho_m = 12$ lb/gal

$\mu_p = 40$ cp

$Y_b = 30$ lb/100 ft²

Necessary annular velocity = 120 ft/min, based on bit size.

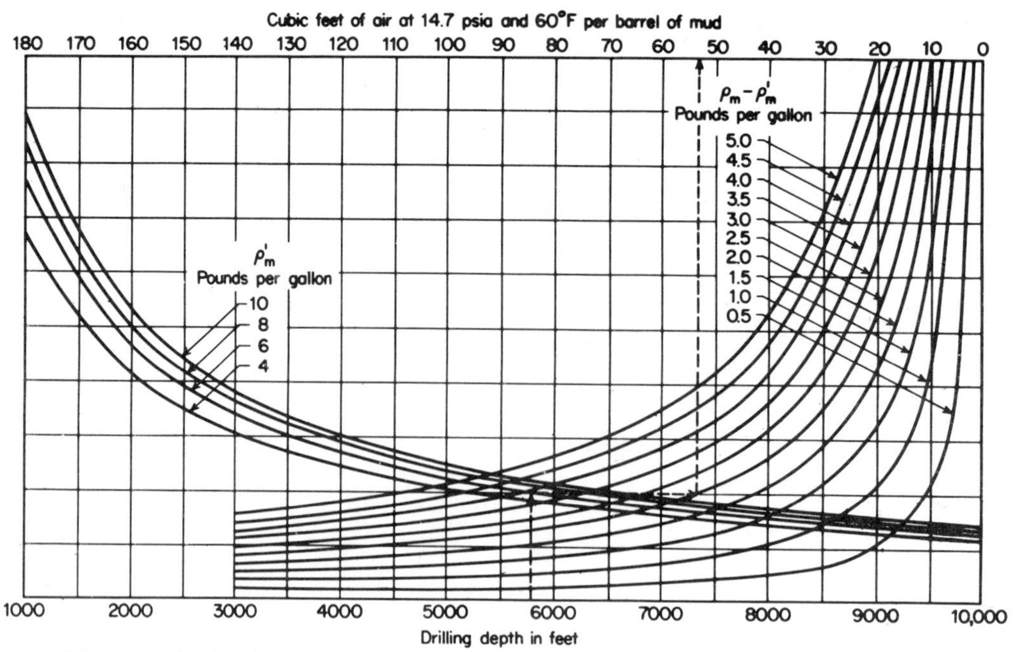

Fig. 7.16. Estimation of air requirements for aerated mud drilling. After Poettmann and Bergman,[23] courtesy *World Oil*.

Average temperature = 150°F
ρ_m' = Desired aerated density, lb/gal
ρ_m = Original mud density, lb/gal

9. Find the bottom hole circulating pressure in Problem 8, by each method of Sect. 7.5. Which do you feel is the more accurate method for this particular calculation? Why?

10. What reduction in bottom hole circulating pressure would result from lowering Y_b to 5 lb/100 ft^2, in Problem 9?

11. The plastic viscosity of a drilling mud was determined as 20 cp at 70°F and 16 cp at 100°F. Using a plot as suggested by Eq. 7.21, determine the average plastic viscosity in a 12,000 ft well having the circulating temperatures denoted by Figure 7.12. See Figure 6.8 for water viscosity data.

12. Calculate the pressure surge imposed by lowering 10,000 ft of $4\frac{1}{2}$-in. drill pipe with 500 ft of 6-in. drill collars into an 8-in. hole at

(a) 1 ft/sec
(b) 2 ft/sec
(c) 3 ft/sec

Assume the bit is plugged. The mud properties are those of Problem 8.

13. Assuming a well depth of 12,000 ft, plot the actual bottom hole pressure vs the pipe running and hoisting speeds, using data of Problem 12.

14. Develop Eq. 7.23 from Eq. 7.22.

15. Calculate the terminal slip velocity of a $\frac{1}{4}$-in. shale cutting of 2.6 sp.gr. in air at 50 psia and 100°F. $C_d = 1.40$.

16. What is the air requirement for drilling a 9-in. hole with $4\frac{1}{2}$-in. drill pipe at a rate of 1 ft/min? What is the air requirement for natural gas, considered at the same rate? Depth = 8000 ft.

17. Assuming a drilling rate of 60 ft/hr and $4\frac{1}{2}$-in. drill pipe, make a plot of air requirement vs hole size for a depth of 5000 ft.

18. Prepare a similar plot of air requirement vs depth for $4\frac{1}{2}$-in. drill pipe in $7\frac{7}{8}$-in. hole at drilling rates of 30, 60, and 90 ft/hr.

19. Prepare a plot of air requirement for $4\frac{1}{2}$-in. drill pipe in $7\frac{7}{8}$-in. hole at 10,000 ft vs drilling rate.

20. What conclusions may be drawn from the curves of 17, 18, and 19?

21. (a) Estimate the compressor capacity required to air drill at 10,000 ft under the dry and wet conditions depicted as curves 2 and 3 of Figure 7.15. Assume $4\frac{1}{2}$-in. drill pipe in $7\frac{7}{8}$-in. hole and 60 ft/hr drilling rate. Assume 2000 ft altitude and 70°F suction conditions.

(b) Estimate the necessary number of compressor stages in both cases and the compression ratio per stage.

22. (a) What injected air volume is required to lighten a 9 lb/gal mud to 5 lb/gal? Assume 150° average temperature and a mud circulation rate of 300 gal/min at 8000 ft.

(b) What compressor HP is required if the circulating pressure is 700 psig? Suction conditions are same as in Problem 21(a).

23. (a) Rework Problem 8 using the Hughes Charts for $3\frac{1}{2}$- and $5\frac{1}{2}$-in. drill pipe with the same tool joints. Inside diameters are:

$3\frac{1}{2}$-in., 13.3 lb/ft = 2.76 in.

$5\frac{1}{2}$-in., 21.9 lb/ft = 4.78 in.

(b) Plot the resulting HP vs drill pipe size from (a).

24. Inspection of Figure 7.1 shows that over a considerable range, f may be expressed as a logarithmically linear function of R_e. For example, Curve II is well approximated as $f = 0.026\, R_e^{-0.14}$. Show that equation (7.4) may be expressed as:

$$\Delta p = K \frac{\mu_t^{0.14} \rho^{0.86} q^{1.86} L}{d^{4.86}}$$

where $K = 0.72 \times 10^{-4}$. (This equation was used to compute Figures 7.4, 7.5, 7.6, 7.7 for $\rho = 9.5$ lb/gal, $\mu_t = 3$ cp, and $L = 1000$ ft.)

25. Develop a similar expression for Curve III, Figure 7.1.

26. (a) Using the information from Problem 24, develop Eq. 7.17.

(b) Similarly, show what assumption is involved in Eq. 7.16.

REFERENCES

1. Pigott, R. J. S., "Mud Flow in Drilling," *API Drilling and Production Practices*, 1941, p. 91.
2. Beck, R. W., Nuss, W. F., and T. H. Dunn, "The Flow Properties of Drilling Muds," *API Drilling and Production Practices*, 1947, p. 9.
3. Bingham, E. C., *Fluidity and Plasticity*, New York: McGraw-Hill Book Co., 1922.
4. Havenaar, I., "The Pumpability of Clay-Water Drilling Fluids," *Trans. AIME*, Vol. 201, (1054), p. 287.
5. Eckel, J. R., and W. J. Bielstein, "Nozzle Design and Its Effect on Drilling Rate and Pump Operation," *API Drilling and Production Practices*, 1951, p. 28.
6. *Hydraulics in Rotary Drilling*, Bulletin No. 1-A, Hughes Tool Company, Houston, Texas, 1954.
7. Hydraulics for Jet Bits, Bulletin No. 2-B, Hughes Tool Company, 1956.
8. Mesaros, J., "The Application of Flow Properties to Drilling Mud Problems," API Paper No. 851-31-D, presented at Tulsa, Apr. 1957.
9. Srini-Vasan, S., and C. Gatlin, "The Effect of Temperature on the Flow Properties of Clay-Water Drilling Muds," *Journal of Petroleum Technology*, Dec. 1958, p. 59.
10. Cannon, G. E., "Changes in Hydrostatic Pressure Due to Withdrawing Drill Pipe from the Hole," *API Drilling and Production Practices*, 1934, p. 42.
11. Goins, W. C., Jr., Weichert, J. P., Burba, J. L., Jr., Dawson, D. D., Jr., and A. J. Teplitz, "Down the Hole Pressure Surges and Their Effect on Loss of Circulation," *API Drilling and Production Practices*, 1951, p. 125.
12. Cardwell, W. T., Jr., "Pressure Changes Caused by Pipe Movement," *API Drilling and Production Practices*, 1953, p. 97.
13. Ormsby, G. S., "Calculation and Control of Mud Pressures in Drilling and Completion Operations," *API Drilling and Production Practices*, 1954, p. 44.
14. Nicolson, K. M., "Air Drilling in California," *API Drilling and Production Practices*, 1953, p. 300.
15. Gray, K. E., "The Cutting Carrying Capacity of Air at Pressures above Atmospheric," AIME Tech. Paper 874-G, Presented Dallas, Oct., 1957.
16. Nicolson, K. M., "Air and Gas Drilling," in *Drilling Fundamentals*, Dallas: The Petroleum Engineer Publishing Company, 1954, p. 86.
17. Martin, D. J., "Additional Calculations to Determine Volumetric Requirements of Air or Gas as a Circulating Fluid in Rotary Drilling," Engr. Bull. No. 23-A, Hughes Tool Company, Houston, Texas, June 15, 1953.
18. Angel, R. R., "Volume Requirements for Air or Gas Drilling," *Trans. AIME*, Vol. 210, (1957), p. 325.
19. Scott, J. O., "How Much Air to Put Down the Hole," *Oil and Gas Journal*, Dec. 16, 1957, p. 104.
20. Smith, F. W., and H. M. Rollins, "Air Drilling Practices in the Permian Basin," *The Petroleum Engineer*, Dec. 1956, p. B-48.
21. *N.G.S.M.A. Data Book*, 6th ed. Tulsa: Natural Gasoline Supply Men's Association, 1951, pp. 23-37.
22. Morris, J. T., and Robert T. Ramey, "Compressors for Air Drilling," *World Oil*, Nov. 1955, p. 119.
23. Poettmann, F. H., and W. E. Bergman, "Density of Drilling Muds Reduced by Air Injection," *World Oil*, Aug. 1955, p. 97.
24. Poettmann, F. H., and P. G. Carpenter, "The Multiphase Flow of Gas, Oil, and Water through Vertical Flow Strings, with Application to the Design of Gas Lift Installations," *API Drilling and Production Practices*, 1952, p. 257.

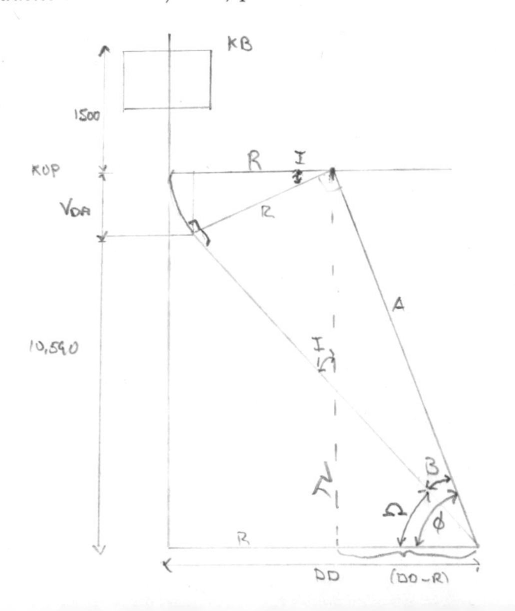

$D° = °/100\,ft$

$c = (a^2 + b^2)^{1/2}$

$\frac{D°}{360} = \frac{100}{2\pi R}$; $R =$

$\tan \phi = \frac{TV}{DD - R}$; $\phi =$

$A = \frac{TV}{\sin \phi}$

$TD = MD + A \cos B$

$\sin B = \frac{R}{A}$

$B =$

$\Omega = \phi - B$

$I = 180 - (\Omega + 90)$

$MD = \frac{I}{100\, D°} + (KB - TV)$

$V_{OA} = R \sin I$

Chapter 8 . . .

Factors Affecting Penetration Rate

8.1 Introduction

Contract drilling prices have remained essentially constant over the last ten or fifteen years despite the fact that the greatest inflationary period in United States history has occurred during the same time. This unique price stability has been due, largely, to the highly competitive and resourceful nature of the drilling industry in general, and to the rig floor, desk, and laboratory thinking and experimentation which has resulted in improved techniques and equipment for making hole faster.

The factors which affect rate of penetration are exceedingly numerous and certainly are not completely understood at this time. Undoubtedly, influential variables exist which are as yet unrecognized. A rigorous analysis of drilling rate is complicated by the difficulty of completely isolating the variable under study. For example, interpretation of field data may involve uncertainties due to the possibility of undetected changes in rock properties. Studies of drilling fluid effects are always plagued by the difficulty of preparing two muds having all properties identical except the one under observation. These and other complexities will become more apparent in later sections.

While it is generally desirable to increase penetration rate, such gains must not be made at the expense of over-compensating, detrimental effects. The fastest *on-bottom* drilling rate does not necessarily result in the lowest cost per foot of drilled hole; other factors such as accelerated bit wear, equipment failure, etc., may raise cost. These restrictions should be kept in mind during the following discussion, which deals largely with *on-bottom* drilling rate.

Some of the more recognizable variables which affect penetration rate are the following:

1. Personnel efficiency
   a. Competence
      (1) experience
      (2) special training
   b. Psychological factors
      (1) company-employee relations
      (2) pride in job
      (3) chance for advancement
2. Rig efficiency
   a. State of repair, preventive maintenance
   b. Proper size
   c. Ease of operation, degree of automaticity, and power equipment
3. Formation characteristics
   a. Compressive strength
   b. Hardness and/or abrasiveness
   c. State of underground stress (overburden pressure, etc.)
   d. Elasticity — brittle or plastic
   e. Stickiness or *balling* tendency
   f. Permeability
   g. Fluid content and interstitial pressure
   h. Porosity
   i. Temperature
4. Mechanical factors
   a. Weight on bit
   b. Rotating speed
   c. Bit type
5. Mud properties
   a. Density
   b. Solid content
   c. Flow properties
   d. Fluid loss

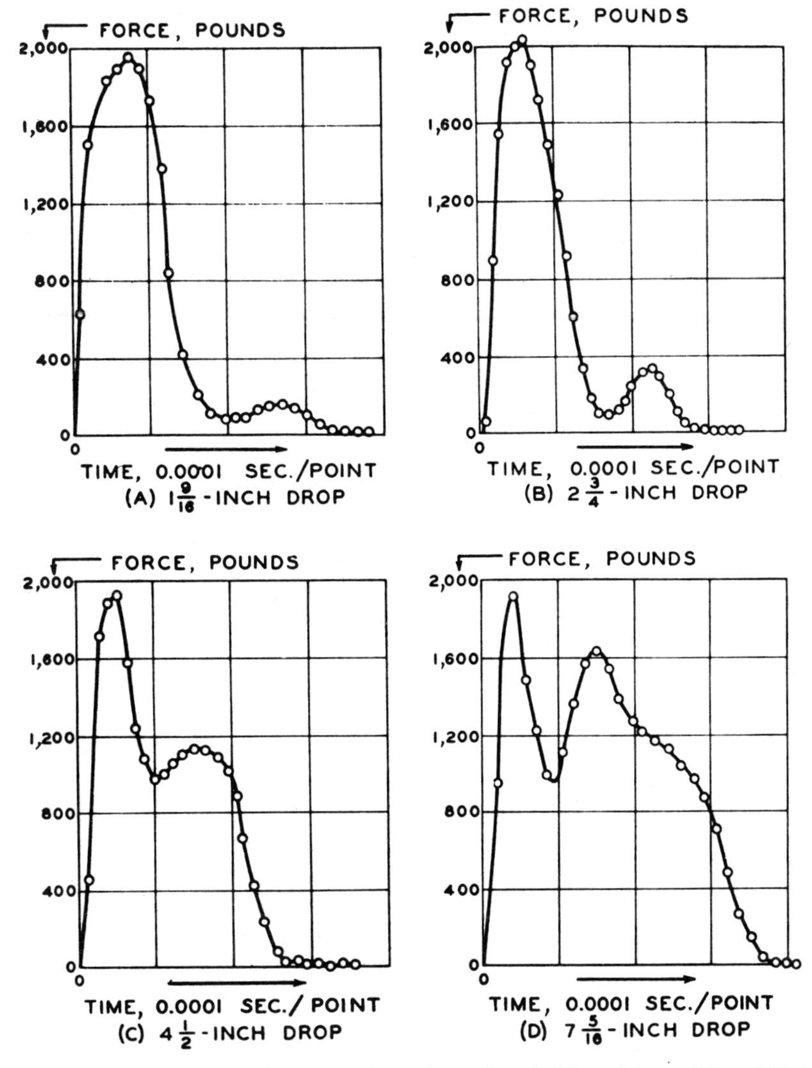

Fig. 8.1. Tracings of typical force waveforms for various heights of drop with an 0.03-in. bit and weight of 15 lb for single blows on Indiana limestone. After Pennington,[1] courtesy API.

e. Oil content
f. Surface tension—wettability
6. Hydraulic factors—essentially bottom hole cleaning

8.2 Fundamentals of Rock Failure

For drilling purposes, rocks may be classified into three general types, namely:

(1) Soft rocks: soft clays and shales, unconsolidated to moderately cemented sands
(2) Medium rocks: some shales, porous limestones and dolomites, consolidated sands, gypsum
(3) Hard rocks: dense limestones and dolomites, highly cemented sands, quartzite, and chert.

As mentioned in our earlier discussion of bits, soft rocks may be drilled by the scraping-cutting action of drag-type bits, or by the combined grinding-scraping action of offset cone-angle, rolling cutter bits. The harder formations are drilled mostly by the crushing penetration of the bit teeth. Since it is in the latter type of formation that penetration rates are lowest, let us further consider the failure mechanism of these rocks.

Experiments on the failure of elastic rocks conducted by Battelle Memorial Institute for Drilling Research, Inc. are of considerable fundamental interest.[1–3] In part of these studies a drop tester, consisting of a weighted bit on a rod, was dropped from various heights, striking the rock specimen below. Strain gauges close to the bit allowed the force waveforms to be recorded by an oscilloscope camera. Four such patterns for single blows at various drop heights are shown in Figure 8.1. Note that the first force peak is 2000 lb for all cases, with the magnitude of the second peak increasing with drop height. Figure 8.2 shows the first energy peak with an expanded time scale. The curve between O and A corresponds to the bit crushing small surface irregularities on the rock. Between A and B, the constant steep slope denotes that elastic deformation is being undergone by the rock. If the impact force is not above point B, the rock does not fail and only a

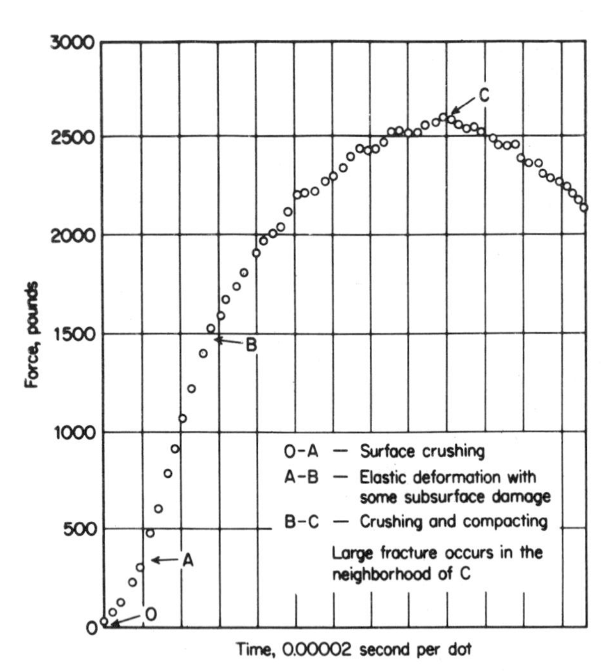

Fig. 8.2. Expanded first force pulse of a force waveform for a single blow with an 0.03-in. bit on Indiana limestone. After Pennington,[1] courtesy API.

surface mark is left. Between points *B* and *C*, the narrow wedge of rock beneath the bit is crushed, and the blow energy is transmitted to the rock around the wedge, and chips are formed. Beyond point *C*, the energy level drops until the bit once again contacts solid rock. If sufficient energy is left, another peak occurs, etc. It was found that the amount of rock removed was linearly related to the blow energy. Other experiments with static rather than impact loading gave similar results, except that a lower energy level was required to fracture a given amount of rock. The results in Figures 8.1 and 8.2 are exactly analogous to cable tool drilling (or any percussive device) in brittle rocks.

The sequence of pictures in Figure 8.3 shows the drilling action of a rolling cutter bit in sandstone at atmospheric conditions.[4] Cuttings are removed by air circulation. In the first picture, a force is applied to the center tooth as the cone rolls. The rock does not fail until the fourth picture, when the applied load finally exceeds the rock's strength. The failure is a small explosion, with chips flying in all directions. This continues as the tooth sinks deeper and is complete, in the eleventh picture. It may thus be surmised that the rotary drilling mechanism is not too basically different from the straight percussive (cable tool) method.

8.3 Rock Characteristics

The rock properties which govern drilling rate are not completely understood. Furthermore, correlation is lacking between strength and elastic properties as measured at laboratory conditions, and those which exist at the depths of interest to the oil industry. Considerable data and an extensive bibliography on rock properties has been published by Wuerker.[5] Other investigations concerned with the effect of pressure on rock drillability are also available.[6,7]

In general, penetration rate varies inversely with the compressive strength of the rock being drilled. The related property of hardness or abrasiveness enters the picture because of its effect on bit life. The hardest and also the strongest sedimentary rocks are chert (a form of quartz, SiO_2) and various quartzites which offer severe drilling problems when encountered.

The elastic properties of various formations are greatly influenced by the state of stress at which they exist. The behavior of most shales is typical of this effect, because they become increasingly difficult to drill at greater depths. For explanation of this behavior, consider Figure 8.4 which shows a thin impermeable element of formation directly below the bit. If the hole is filled with liquid, the upper surface of the element is subjected to a pressure which is dependent on mud density and depth. Practically speaking, this pressure tends to prevent removal of the element much as though the rock's strength were increased by the applied pressure. Therefore it may be expected that the effect of such superimposed stresses will be more pronounced in weak (soft, relatively compressible) rocks than in stronger, more competent beds [see Figure 8.5(A)]. Note that drilling rate reaches an essentially minimum value at some confining pressure, with little change occurring at subsequent higher pressures. This is particularly evident for the soft shale shown in Figure 8.5(C). Also note that the effect of confining pressure on drilling rate is greater at higher bit weights, as is shown in Figure 8.6.

Payne and Chippendale[8] have reported test results which lend further insight into the mechanism of rock failure at high vs. atmospheric pressures. In these experiments, crushing loads were applied to rocks with a round (5 mm diameter) tungsten carbide penetrator, the size and type of indentation being the object of study. Typical results are shown in Figure 8.7. The letters *A* and *H* denote imposed hydrostatic pressures of atmospheric and 5000 psi; the numbers refer to the applied load in pounds. In the atmospheric tests visible cracks appeared, and the chips around the indentation were separate, easily removable flakes. Under 5000 psi, however, the failure was more plastic, as evidenced by the extruded material which piled up around the depression and could not easily be removed. Hence, the sandstone was brittle at low pressure and plastic at high pressure. Not all rocks behaved in this manner. Quartzite, granite, and dolomite experienced the same type of failure at the higher pressure; however, the

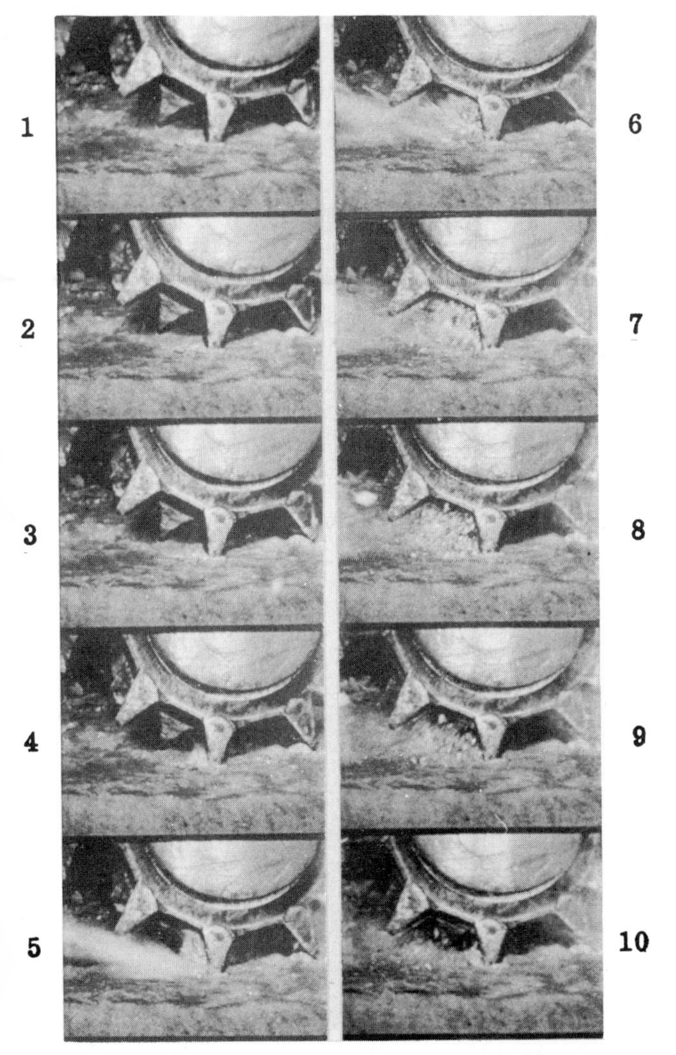

Fig. 8.3. Elastic rock failure as caused by rolling cutter bit. Conditions are atmospheric pressure with cuttings being blown away by air. Starting in picture 1, a tooth starts to apply pressure to the rock as the cone rolls forward. The rock does not break immediately. Then in picture 4, the pressure of the tooth exceeds rock strength and failure starts. The failure is practically an explosion of rock. Chips continue to fly as the tooth sinks deeper. Finally, by picture 10 particle discharge diminishes and then ceases. Filmed by Hughes Tool Company with a $9\frac{3}{4}$-in. O.S.C. bit cutting Berea sandstone. After Murray and MacKay,[4] courtesy Hughes Tool Company.

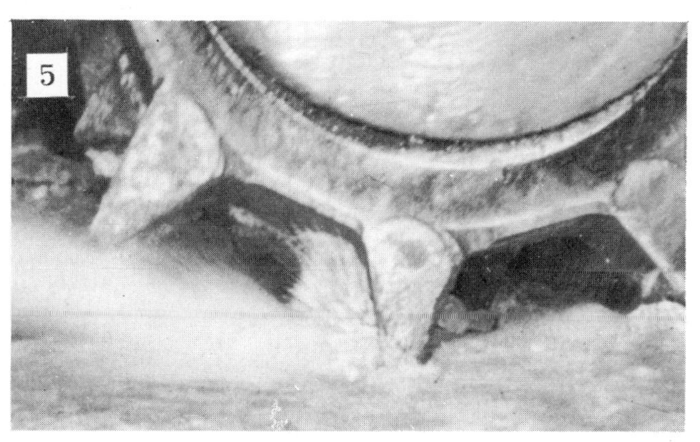

Fig. 8.3 (cont.). Failure is explosive when bit pressure finally exceeds strength of rock.

loads necessary to produce the fracture were 50 to 100% higher than at atmospheric conditions.

The balling tendency of a formation is primarily dependent on its mineral composition. Hydratable clays, bentonite in particular, form a sticky, pasty mixture with water which becomes imbedded between bit teeth and surrounds the cones and the entire bit. This reduces tooth penetration and, consequently, drilling rate.

Permeability affects the drillability of rocks through its effect on the relief of imposed pressures. Consider again the rock element shown in Figure 8.4. If this rock were sufficiently permeable to the drilling fluid, no appreciable pressure differential would exist across a thin element; hence the pressure effect would be minimized. This is the conclusion of Bredthauer,[9] and Murray and Cunningham,[7] and is in agreement with their experimental evidence, i.e., that permeable rocks which allowed pressure equalization ahead of the bit showed no appreciable change in drillability with pressure. From these data, it may be concluded that any other factors facilitating more rapid pressure relief will also decrease the effect of pressure on drilling rate. A rock completely saturated with incompressible fluids (water, oil) should be less sensitive to borehole pressure effects than one containing a gas, since in the former a small quantity of mud filtrate is sufficient to equalize pressure for a substantial distance.

A porous zone drills faster than a dense section of the same rock. This fact has long been used by drillers and geologists for detecting the presence of such zones. A major portion of this effect is probably due to the lower compressive strength of porous zones.

The effect of temperature (in the range of our interest) on rock properties is not generally considered. However, rock failure becomes more plastic as temperature

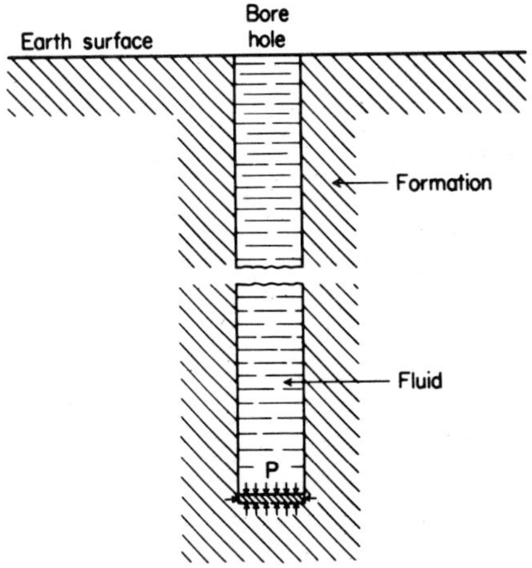

Fig. 8.4. Element of formation beneath rock bit. After Murray and Cunningham,[7] courtesy AIME.

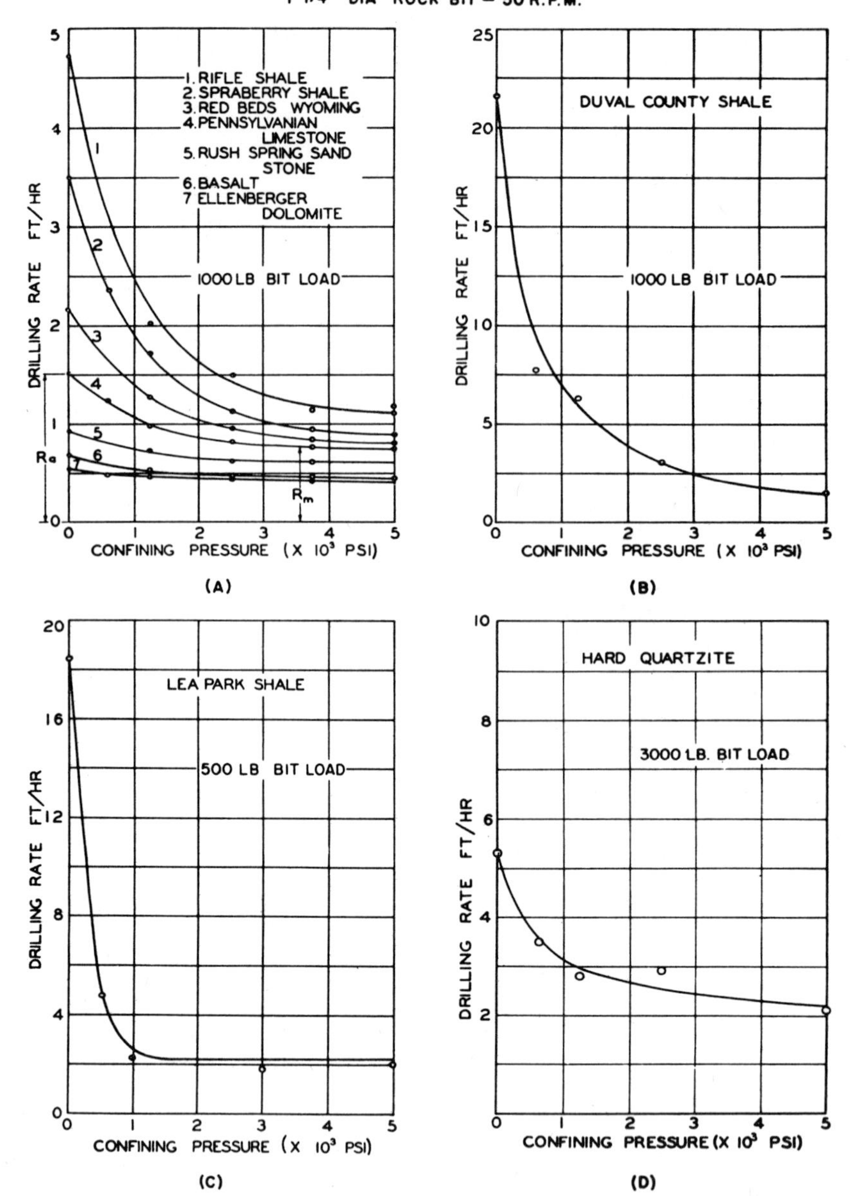

Fig. 8.5. Penetration rate vs. confining pressure for various rock types. After Murray and Cunningham,[7] courtesy AIME.

increases; such effects may be of interest in the future, as drilling goes deeper.

No mention has been made of such properties as grain size, Poisson's ratio, modulus of elasticity, modulus of resilience, or other more involved measurements of strength and elastic behavior.[5] These properties are, in general, unknown for sedimentary rocks at the conditions of interest in this field. A knowledge of these factors could lead to a more thorough knowledge of the mechanics of rock failure.

It might seem to the reader that too much space has been allotted to rock properties, which are beyond human control. It should be realized, however, that it is these rock properties which completely govern drilling practices in different geologic areas. Hole size, bit type, drilling fluid selection, and general operating procedures are all dictated by the nature of the rocks to be drilled. Consequently, rock characteristics are of prime importance, with operational procedures being secondary.

8.4 Mechanical Factors

In all areas, penetration rate is governed by the weight on the bit and/or the rotary speed which may be applied. Normally, these limits are imposed by either crooked hole, equipment, and/or hydraulic considerations. Let us postpone our discussion of crooked hole problems until Chapter 9, and deal with the latter factors in this chapter. Other variables will be considered constant or of no consequence, except as noted.

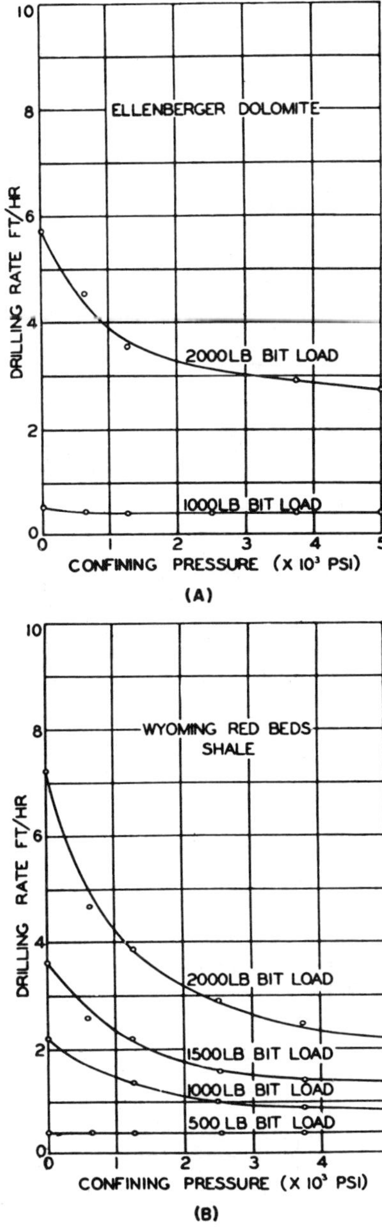

Fig. 8.6. Drilling rate vs. confining pressure, variable bit load. After Murray and Cunningham,[7] courtesy AIME.

Weight on Bit

The general effect of this variable on rate of penetration is shown in Figure 8.8.[10] Note that there are two distinct drilling regions:

(1) drilling at bit loads below the compressive strength of the rock — out to approximately 30,000 lb, for the pink quartzite.

(2) drilling at bit loads above this critical weight. Obviously bit weights in region (2) are desirable and are applied if possible. This principle is well illustrated by Figure 8.9, which shows the increase in penetration rate and footage when sufficient bit weight to overcome the rock's compressive strength was applied.

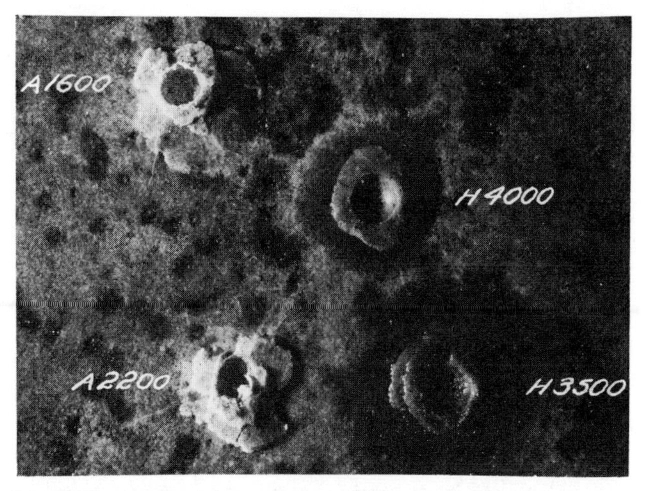

Fig. 8.7. Rock failure under different imposed pressures. Letters A, H denote atmospheric and 5000 psi pressure, respectively. Numbers indicate applied load, lbs. After Payne and Chippendale,[8] courtesy Hughes Tool Company.

From Figure 8.8 it is evident that for each rock type the points above the critical load lie on approximately straight lines:[11]

$$R_p = a + bW \tag{8.1}$$

where R_p = instantaneous or *on bottom* penetration rate, ft/hr

W = weight on bit, lb

a, b = intercept and slope respectively, which are dependent on rock properties, bit size and type, drilling fluid properties, etc.

Other data, both field[12–14] and laboratory [15,16] tend to substantiate the linearity of R_p versus W over reasonable ranges of bit loading. This relation will hold true only if other factors are constant. In particular, adequate cleaning and cutting removal must be maintained. This does not necessarily mean constant circulation rate or constant nozzle velocity, since at higher values of W more and probably larger cuttings are being generated. Consequently, circulation inadequacies will affect this relationship, unless the flow rate is high enough to perform its function equally well at all W's of interest. This will be further emphasized in a later section devoted to hydraulic factors.

Rotational Speed

The effect of this variable on penetration rate is not too well established. In general, on bottom drilling rate increases with increased rotary speed. Figures 8.10 and 8.11 are typical of the rotary speed effect and suggest that the following relationship exists:

$$R_p = f(N)^n \tag{8.2}$$

where f = some function

N = rotary speed, rpm

$n < 1$

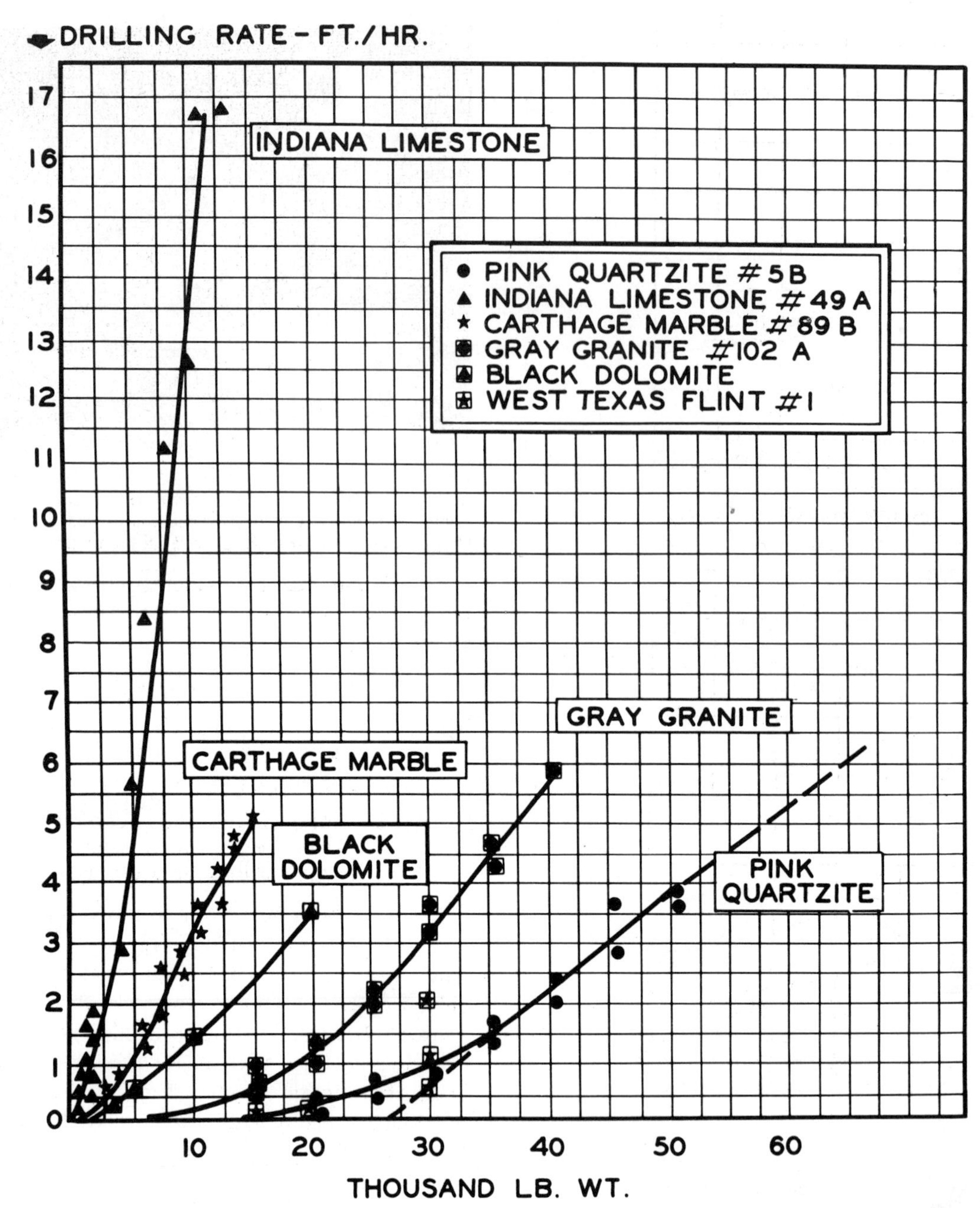

Fig. 8.8. The effect of bit loading on penetration rate for various rocks. Laboratory data — atmospheric pressure. Courtesy Hughes Tool Company.[10]

Again, the true relationship may be obscured by the hole cleaning variable. A satisfactory nozzle velocity at 100 rpm may not be adequate at 200 rpm, since at the latter speed a cutting has only half as much time to clear bottom prior to the next tooth impact.

The Battelle Institute study[2,3] set forth a formula for a rate of advance predicting linearity between R_p and N. It was assumed that each tooth impact removes essentially the same amount of rock. Fundamentally, this would seem logical; however, available data do not confirm this analysis for rotary (roller bit) drilling. Turbodrill data to be cited later confirms Eq. (8.2) for rotating speeds up to 750 rpm.[17]

Combined Effect of W and N

An empirical combination of Eq. 8.1 and 8.2 indicates:

$$R_p = e + fWN^{1/2} \tag{8.3}$$

where e, f = constants for a given set of conditions.

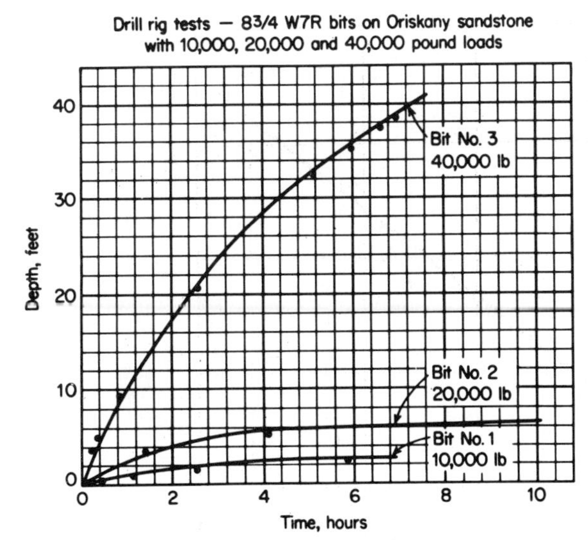

Fig. 8.9. Increased penetration facilitated by proper bit weight application. Laboratory data — atmospheric pressure. Courtesy Hughes Tool Company.[10]

Figure 8.12 is a replot of Figure 8.10, except that W and N have been combined. Other laboratory[15] and turbodrill field data[17] exhibit essentially the same value of n.

The compressive strengths of many rocks encountered in oil well drilling are low enough that little error is involved by assuming R_p to be directly proportional to W, that is, that $a = 0$ in Eq. 8.1. It is also generally considered that the effect of bit size may be eliminated by substituting the applied weight per inch of diameter[7] for W in the equations, with the following result:

$$R_p = f\frac{W}{d}(N)^n = fwN^{1/2} \tag{8.4}$$

where d = bit diameter, in.

w = lb/in. of bit diameter

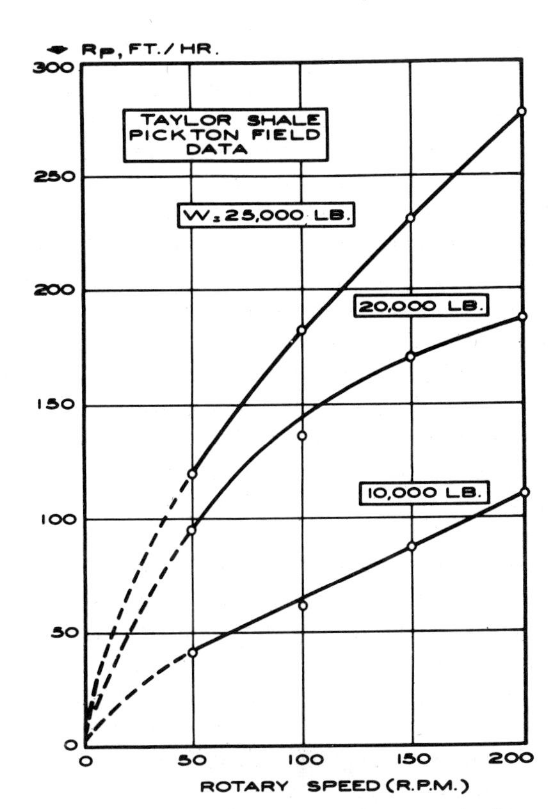

Fig. 8.10. Effect of rotary speed on penetration rate at various bit weights. Replotted from data of Bielstein and Cannon,[13] courtesy API.

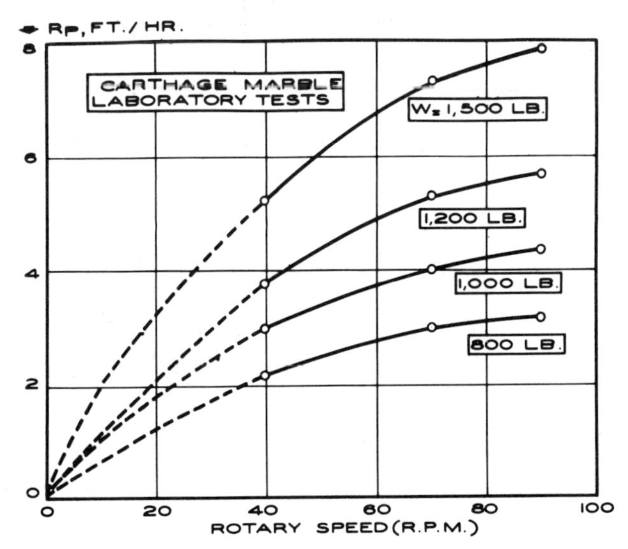

Fig. 8.11. Effect of rotary speed on penetration rate at various bit weights. Replotted from data of Wardroup and Cannon,[16] courtesy API.

Equation (8.4) may be considerably in error for extremely hard rock drilling. Analysis of West Texas field tests showed the effect of bit weight to be expressible as a power function:[18]

$$R_p = f'W^mN^{1/2} \tag{8.5}$$

where $m \simeq 1.2$.

Eqs. 8.3, 8.4, and 8.5 are purely empirical, and may not fit all situations. Certainly, these equations are not valid for W's too small to cause rock failure; however, such values are of little practical importance.

Practical Limitations on W and N

These factors must, of course, be held within the operational limits imposed by economic drill string and bit life. Over-all penetration rate is greatly affected by increased trip frequency, caused by accelerated bit wear due to operating at excessive W and/or N values. The practical limitation on bit weight is primarily governed by the allowable tooth stress of the bit itself, and the current maximum which has been applied to some bit types is about 10,000 lb/in. Recall that crooked hole considerations are ignored at this time. Rotational speed also affects bit performance, primarily through bearing and tooth wear, and through the severity of vibrational loads.

Most drill string failures are attributed to material fatigue which has been aggravated by corrosion and/or

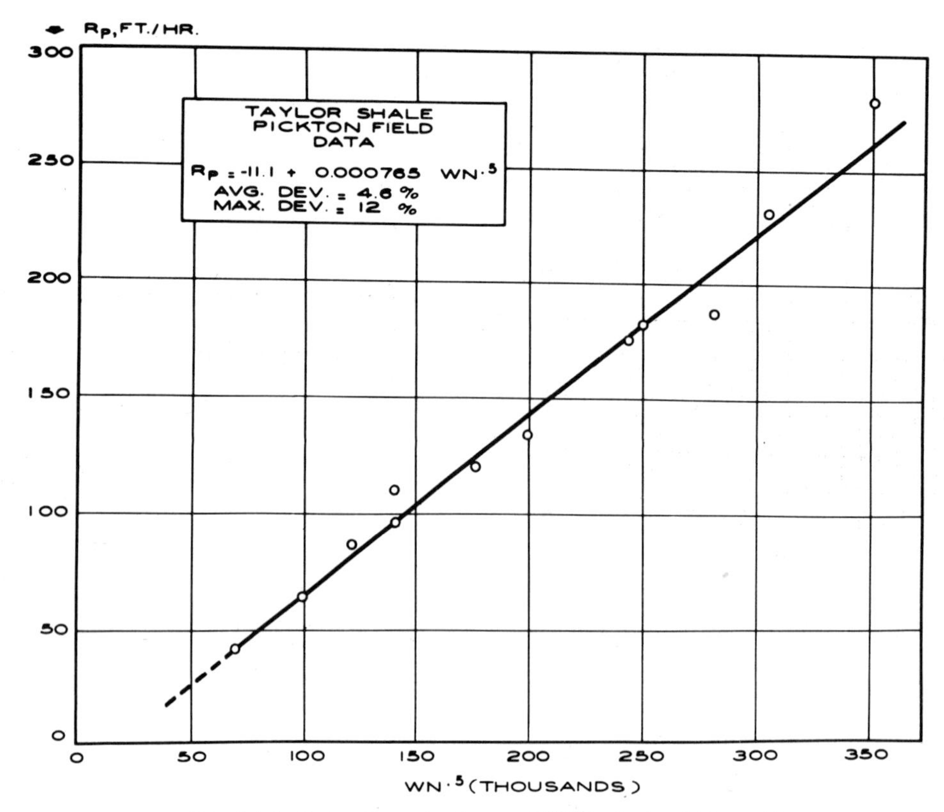

Fig. 8.12. Rate of penetration vs. $WN^{1/2}$. Data of Fig. 8.10.

improper care and handling.[19,20] Since fatigue failures are related to both the number of stress reversals and the magnitude of the stress, it should be expected that both table speed and weight on bit will affect the frequency of such failures. Of the two, table speed seems to be the more critical, provided that excessive drill collar-hole clearance is avoided. The drill collars and the adjacent sections of drill pipe receive the worst punishment from vibrational loads; it is in these sections that most failures occur.

Theoretically, every drill string has a natural frequency at which vibrational effects are maximum. It is generally agreed that table speeds in resonance with this will result in severe vibrations and accelerated failure frequency. Main[21] has presented Eq. 8.6 for computing this critical speed range, based on certain simplifying assumptions. This formula has not, however, been widely applied.

$$N_c = \frac{258{,}000}{L} \tag{8.6}$$

where N_c = critical rotational speed at which vibratory effects are maximum, rpm

L = length of drill string, ft

Note: Higher order harmonic vibrations will also occur at $4N_c$, $9N_c$, etc.

Crane[22] has suggested a drilling practice based on a constant weight-times-speed factor. This criterion is in qualitative agreement with field practices: as weight is increased, the rotating speed is decreased. But the question arises, what is the limiting value which should be applied: $WN = ?$

A satisfactory solution in one area will not necessarily apply to another where conditions are different. It is, however, possible to estimate this factor from standard operating practices. For example, Figure 8.13 shows common operating weight-speed ranges for various bit sizes. Constant WN curves added to these approximate the operating trend with a reasonable degree of accuracy. The WN values shown were chosen arbitrarily in an attempt to approximate the indicated maxima. Speer[23] has also derived a similar solution for an *optimum* WN value; however, his curve falls above that suggested in Figure 8.13, and is not quite symmetrical. This approach will be shown later. If one assumes that these or similar data represent the current limiting conditions, then a constant WN factor appears to be a reasonable solution to the problem of allowable table speed–bit weight combinations. Other evidence indicates, however, that the constant weight-speed product approach may be used for determining an *optimum* N at a *given* W; however, it may be unsatisfactory for determining the proper W for a given N.[24]

Bit size may be eliminated as a factor from the suggested WN values of Figure 8.13 by again expressing bit loading as lb/in. of diameter:

$$wN = K \tag{8.7}$$

where $K \cong 5 \times 10^5$ lb-rpm/in., from Figure 8.13.

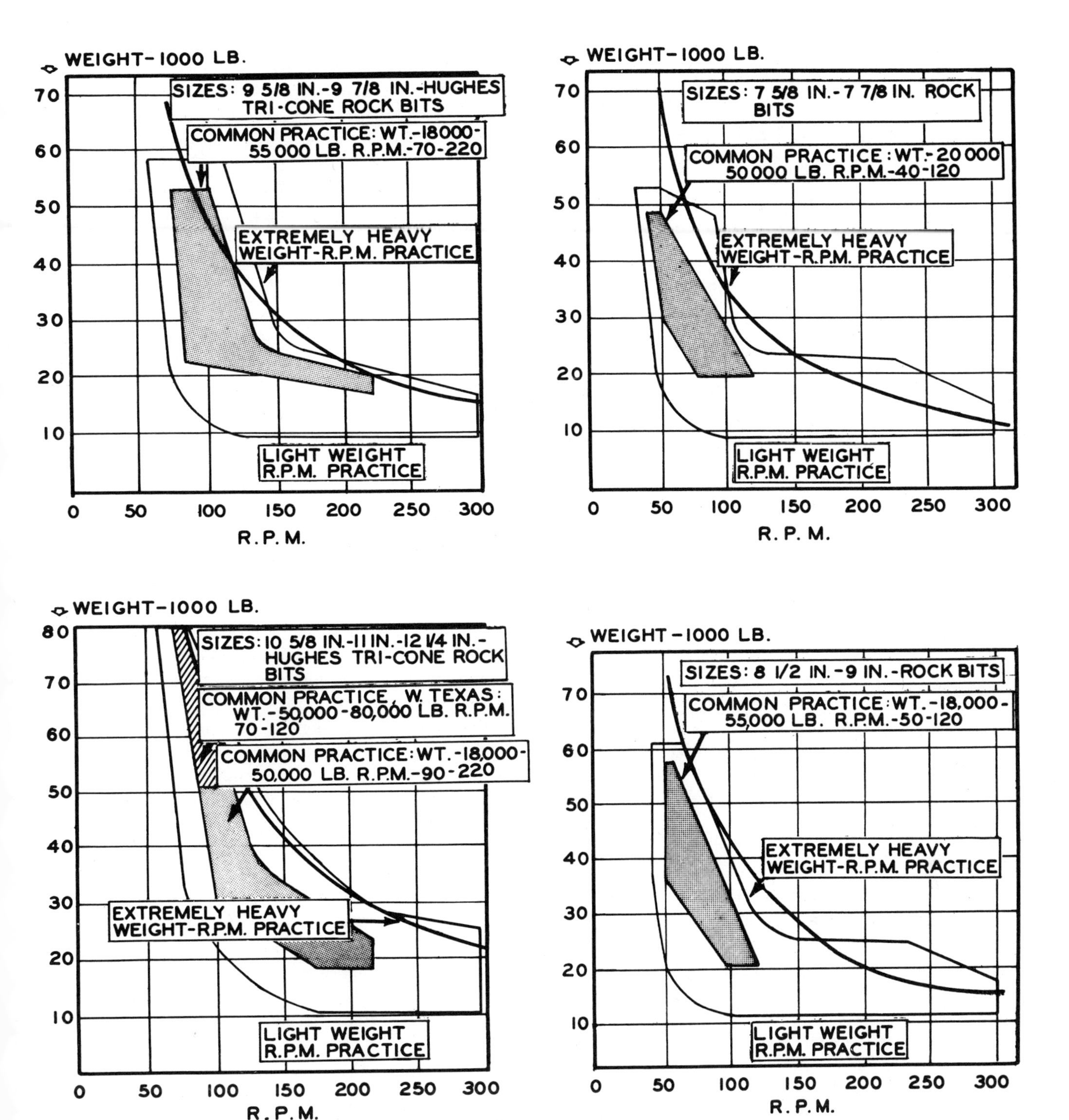

Fig. 8.13. Drilling practice curves showing applied range of weight and rotary speed combinations. Courtesy Hughes Tool Company. Curves of constant weight-speed product have been added as follows:

Bit size (in.)	WN
$7\frac{5}{8}$ to $7\frac{7}{8}$	3.6×10^6 lb-rpm
$8\frac{1}{2}$ to 9	4.0×10^6
$9\frac{5}{8}$ to $9\frac{7}{8}$	4.8×10^6
$10\frac{5}{8}$ to $12\frac{1}{4}$	6.4×10^6

Acceptable values of K will, in general, vary with bit size, since the smaller bits will not tolerate unit loadings as high as will the larger sizes. For the size range shown, however, there appears to be little difference in field practices.

Using Eq. (8.7) as a means of relating w and N, we may now alter Eq. (8.4) to

$$R_p = aw^{1/2} = \beta N^{-1/2} \tag{8.8}$$

where $a = fK^{1/2}$

$\beta = fK$

This relationship shows that penetration rate increases proportionally with $w^{1/2}$ and $N^{-1/2}$ if the product wN is restricted to a constant value. It must be realized that this entire treatment is empirical, and like most empirical equations, has a limited scope of applicability. The drilling practice curves indicate a range of applicability of $50 < N < 300$.

The uniformity with which the force W is applied to the bit is normally dependent on the driller's handling of the brake. The drilling line should be payed out smoothly so that the bit weight is constant. The ideal operation, if achieved, requires the full attention of the driller or another crew member. The use of an automatically controlled feed to insure constant bit loading has increased drilling rates by over 10% in hard rock areas.[16]

Bit Type

The precise nature of the drilling action of roller bits is quite involved and, is not, apparently, completely understood. As was mentioned in Chapter 5, soft formation roller bits have longer, widely spaced teeth with considerable cone offset. These impart a *drag bit cutting action* along with *nearly static percussive* effects. Thus the drilling action is a complex combination of two mechanisms. Hard rock roller bits have zero cone offset; thus they approach true roll, and hence drill by an essentially percussive process. Tooth stresses prohibit utilization of any large degree of shovelling or cutting action in hard rocks.

Numerous factors affect bit performance, such as tooth geometry (tooth type, wedge angle, spacing, row to row patterns, etc.), cone offset, number of cones, bearing designs, metallurgical control, etc. An informative paper concerning bit design factors is that by H. G. Bentson,[25] to which the reader is referred for detailed study.

Drag bits are little used for oil well drilling at this time. This decline in usage is not due to these bits' unsatisfactory cutting action; the drag bit is an efficient drilling tool, and is still widely used in the mining industry for small hole work. It seems logical that any rock which behaves plastically could best be drilled by this type of bit. For example, one can pound on modelling clay with a hammer and merely deform it; however, it is readily cut with a knife. Since at high confining pressures even the harder rocks may behave plastically, it again would seem that percussive bit action would be a poor choice, and that cutting or shovelling should be more efficient. The real problem has been to design a drag bit or an offset cone angle roller bit capable of standing the stresses and abrasive wear necessary for employing cutting action in hard, high strength rocks.

Diamond bits drill in much the same manner as drag bits, each stone acting as an individual tooth. Large stone, widely spaced patterns are used on relatively soft rocks; the hardest drilling requires smaller stones and more closely spaced designs. Usage of diamond bits is, of course, largely restricted to hard rock areas and/or deep wells where reduction in trip frequency is a prime aim.

8.5 The Effect of Drilling Fluid Properties on Penetration Rate

In Chapter 6 it was intimated that a drilling fluid property desirable for one function might be undesirable for other reasons, and that the mud selected was generally a compromise based on the considerations most critical to the area in question. This will become more apparent as we consider the effect of mud properties on the rate of penetration.

The effect of mud density has already been covered since it governs the pressure imposed on the hole bottom. This is probably the main factor contributing to the success of air drilling in such areas as depicted by Figure 8.14.[26] The effect of mud density is probably not completely expressed in terms of the pressure exerted by the mud column. Rather, it should be considered in terms of the pressure differential between the interstitial fluid pressure in the rock and the imposed pressure.[27] All rocks are porous, although some are so slightly porous that we consider them non-porous. Therefore, reconsider Figure 8.4 for the case where no mud column pressure is exerted. In this case, internal pressures

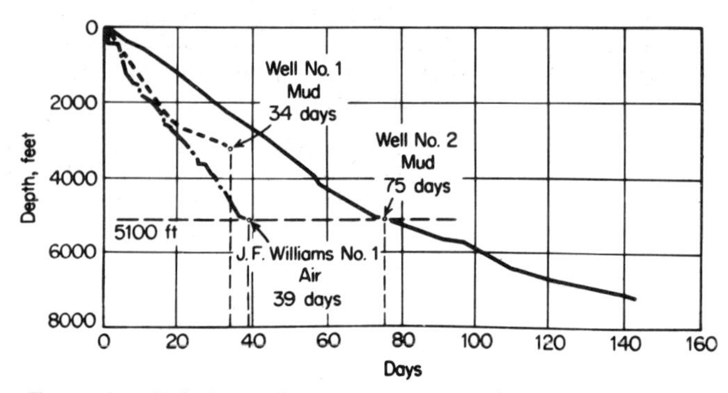

Fig. 8.14. Relative drilling times of air vs. mud drilled wells in Atoka County, Oklahoma. After Adams,[26] courtesy API.

within the rock will help remove the small element below the bit, and thus increase drillability.

Rapid removal of cuttings from below the bit allows each successive tooth impact to attack new rock; this, of course, leads to faster drilling. Research has shown that highly turbulent flow conditions at the hole bottom are desirable for this purpose.[28] Since density appears in the Reynolds number expression, it follows that, if all other factors are constant, an increase in density will decrease the bottom clearance time of cuttings. This is a minor effect as far as density is concerned, and does not warrant specific attention.

Solid Content

Many properties of a mud vary with its solid content and it is thus difficult to isolate data which relate specifically to this factor. It has long been observed in West Texas that very small amounts of clay solids in the drilling fluid greatly reduce the drilling rates obtainable by use of water. Such decreases are completely beyond those expected from the relatively small density increase, and it is apparent that other factors must be involved. More recently, it has been found that low-solids muds in which small quantities of CMC are wholly or partially substituted for larger quantities of bentonite allow increased drilling rates,[29] despite the fact that all other mud properties remain the same. It is then logical to assume that the solid content may itself be a separate factor. Perhaps the solid particles cushion the bit tooth-rock contact such that a clean, sharp impact is not obtained. At any rate, it is the opinion of many experts that the percentage of solids in muds exerts a separate and distinct effect on drilling rate.

Flow Properties

The same pumps are normally used for drilling the entire hole. This means that the available hydraulic horsepower must be essentially constant. An increase in yield point and viscosity increases system frictional losses thereby reducing the pressure drop (and velocity) which can be applied across the bit. Chip clearance time is thus increased and penetration rate decreased. Viscosity influences bottom hole turbulence in the same way; its effect on the Reynolds number is much more drastic than that of density, due to its wider range of values. For example, viscosity at bottom hole conditions commonly ranges from 0.50 cp for water to 50 cp for colloidal muds, a ratio of 100. Density variation for this same viscosity range might be from 8 to 16 lb/gal, or a ratio of 2. Therefore viscosity exerts a considerable effect on the attainment of turbulent flow.

Results from the laboratory experiments of Eckel[30] are shown in Figure 8.15. Here the relative drilling rate (based on the rate obtained using water as drilling fluid) is shown as a function of Stormer viscosity. Other variables including per cent solids and filtration rate were also involved, so that viscosity is not the only factor influencing results. For normal muds, viscosity and solid content are so interrelated that isolation of either factor is difficult. From the standpoint of lubrication between bit teeth and the rock, high viscosity causes high film pressures, and hence reduces the chances of having rock-to-tooth friction. Surface tension also enters these considerations.

The bit teeth must shear and displace drilling fluid before contacting the rock. The energy expended for this purpose is dependent on viscosity; very high viscosity muds may provide an effective viscous cushion which softens bit teeth impact. Yield point effects may be considered in regard to their contribution to apparent or equivalent Newtonian viscosity.

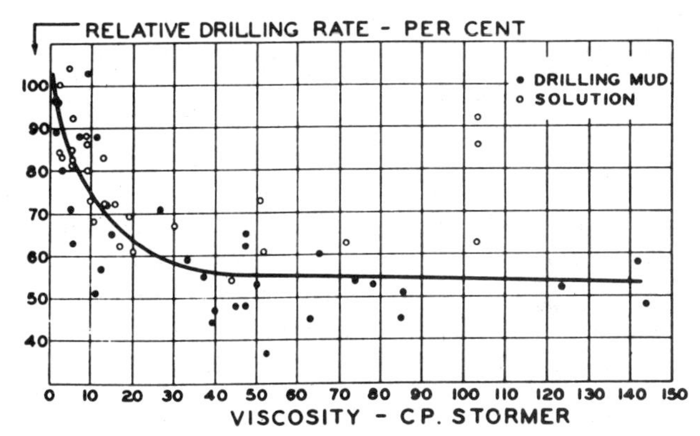

Fig. 8.15. Laboratory data showing effect of viscosity on drilling rate. After Eckel,[30] courtesy API.

Filtration Loss

The effect of filtration on the drillability of permeable rocks has been explained by Murray and Cunningham[7] in conjunction with their studies on hydrostatic pressure effects. They found that penetration rate was not affected by imposed borehole pressures if such pressures were equalized ahead of the bit. Water, for example, may readily enter a permeable rock ahead of the bit so that no pressure differential exists across the thin element being drilled. Low water loss muds, however, almost instantaneously deposit a tough, low permeability filter cake on the hole bottom, allowing a definite pressure differential to exist. This has two detrimental effects: the dynamic filtration pressure effect already discussed, and the requirement that the bit tooth penetrate the filter cake prior to contacting the rock. Also, the loosened cuttings are trapped in the pasty filter cake mixture and require longer bottom clearance times.

Figure 8.16 shows the observed effect of water loss on the penetration rate in shale. As stated by Cunningham and Goins,[31] it is difficult to explain this effect in almost impermeable rocks; however, even the extremely low

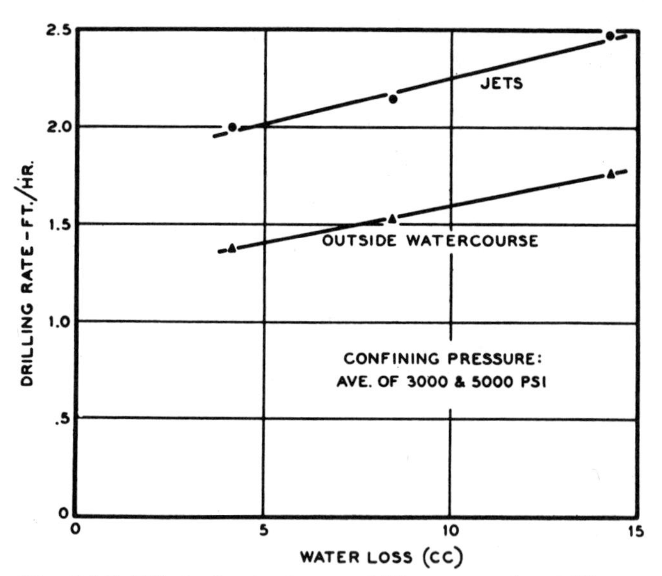

Fig. 8.16. Effect of water loss on drilling rates when drilling a Wilcox shale with a $1\frac{1}{2}$-in. bit in a lime base mud. Laboratory data: bit weight — 750 lb; speed — 50 rpm. Each point is average of 24 tests. After Cunningham and Goins,[31] courtesy *Petroleum Engineer*.

permeability of shales may allow some pressure equalization effect. Also, the increased starch content used to reduce water loss could conceivably cause the observed behavior. Thus the low fluid loss which is desirable from the formation damage standpoint is undesirable from the standpoint of penetration rate.

Oil Content

It has long been observed in the field that the addition of oil to water base muds almost always improves penetration rate in virtually all types of rocks.[32,33] The largest increases appear to occur in the soft rock areas, with smaller increases being noted in hard rock drilling. Generally, these increases have been attributed to better lubrication (increased bit life), better bore-hole conditions (less enlargement, minimum heaving shale troubles, less hole-pipe friction and drag), and less bit balling by hydratable clays and shales. At present, it is considered that prevention of balling is the most important factor, in soft rock areas.

Increased bit life means that less non-productive rig time is expended in making trips; thus the over-all penetration rate is increased. It has also been observed that increases in *on bottom* drilling rate are also obtained, despite the conflicting facts that oil additions to water base muds generally decrease water loss and increase apparent viscosity. Therefore, other factors exist

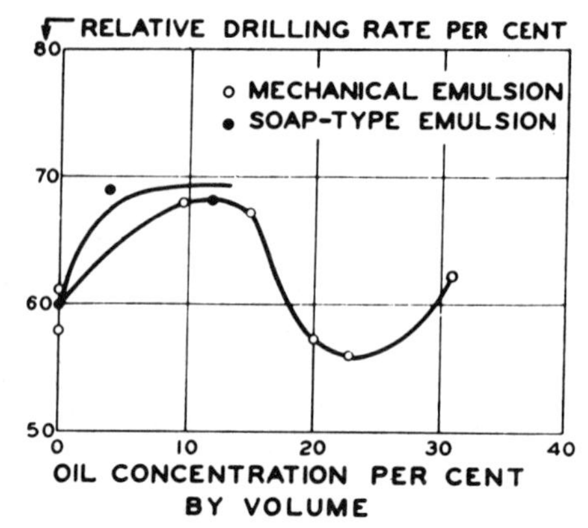

Fig. 8.18. Relative drilling rate vs. oil content — West Texas field data. After Eckel,[30] courtesy API.

which more than compensate for the latter detrimental effects. Density and solid content decreases may be compensating factors in some instances.

Laboratory test results showing the effect of oil percentage on the drillability of two Gulf Coast shales are shown in Figure 8.17. Note the optimum oil concentration for shale *B* which is not evident for shale *A*. This optimum percentage (shale *B*) compares very closely with the West Texas field data of Figure 8.18. Figure 8.19 shows the variation of drilling rate in Miocene shale (*B*) as a function of both oil content and jet velocity. This suggests that the optimum oil percentage may vary slightly with hydraulic conditions. In these tests much of the increased drilling rate was due to the decrease in balling shown in Figure 8.20. The presence of oil is thought to reduce balling by virtue of its preferential wetting of steel resulting in less clay-steel adhesion. A similar effect on the powdered, paste forming, shale particles may also contribute to reduced balling. Additional tests performed on powdered shale showed that the interstitial water content of the shale was insufficient to form a paste and that additional water was required for balling conditions. It should be noted that balling is evidently not the only factor involved, since penetration rate decreased above 15% oil content despite the

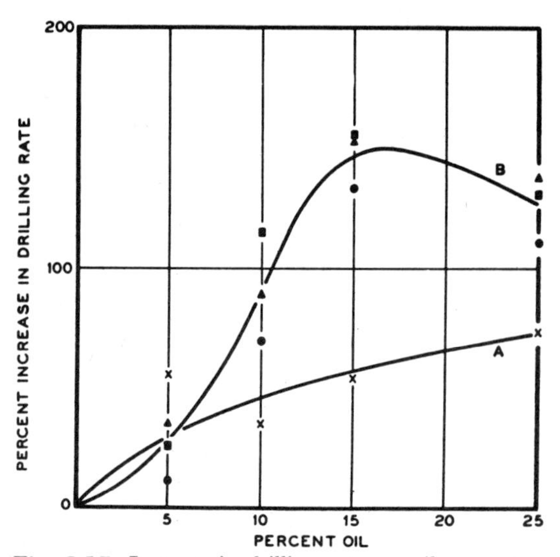

Fig. 8.17. Increase in drilling rate vs. oil concentration. Curve (A) is for a Vicksburg shale; curve (B) is for a Miocene shale (laboratory data). After Cunningham and Goins,[31] courtesy *Petroleum Engineer*.

continued decrease of balling. Part of the increased drilling rate may be due to deeper tooth penetration for a given bit load when an oil film is present to reduce the rock-steel frictional coefficient.

Just why penetration rate does decrease at oil contents above an optimum value in some cases, or why it doesn't in others, is unknown at this time. Other factors apparently become dominant under proper conditions, and cause irregular drilling rate behavior. Available evidence shows that oil concentrations of 15 to 20% are optimum in many cases.

Surface Tension

Russian authors have reported that certain electrolytes and surface active agents act as rock hardness reducers with 30 to 60% increase in drilling rate being obtained by their proper use. Presumably, these materials act to produce more complete wetting of the rock by the liquid. This suggests that microscopic cracks, which tend to heal themselves after the bit load is released, may be held open by the wetting film of liquid; this would facilitate chipping by the next tooth impact.[28] Friction between bit tooth and rock would be reduced, also. Tests conducted by the Battelle group using sodium hydroxide solutions confirmed the results claimed by the Russian scientists.

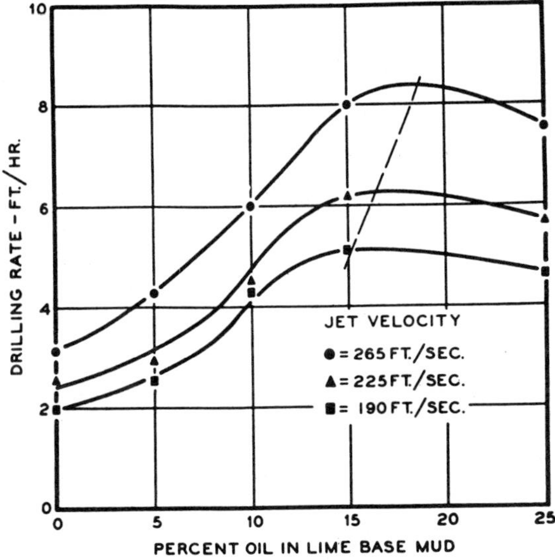

Fig. 8.19. Drilling rate vs. percent oil in a Miocene shale from Louisiana. Laboratory data: bit was $1\frac{1}{4}$-in., 2-cone jet, with 500 lb weight and 50 rpm. Hydrostatic pressure was 5000 psi. Each point is average of 16 tests. After Cunningham and Goins,[31] courtesy *Petroleum Engineer*.

The use of emulsifying agents in emulsion muds appears to affect drilling rate, in some cases. Lignites, salts of higher fatty acids, and other detergents or surfactants are common chemical emulsifiers. Hard rock drilling rates using fresh water have sometimes shown marked increases when surfactants were added

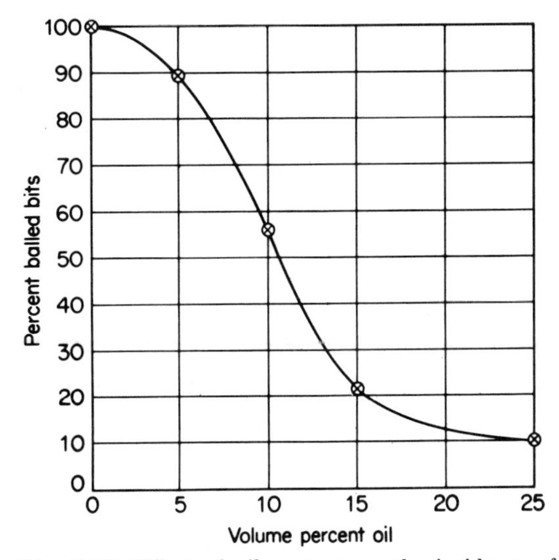

Fig. 8.20. Effect of oil content on the incidence of balled bits in laboratory drilling of Miocene shale. After Cunningham and Goins,[31] courtesy *Petroleum Engineer*.

to the water.[16] These increases were due to both increased bit life and higher on-bottom penetration rates. Both the type of detergent and the character of the rocks govern the effect obtained. It then appears that alterations in wettability and surface tensions play some part in rock drillability. This subject warrants further investigation and may offer some economic promise.

8.6 Hydraulic Factors

In this section we will be primarily concerned with the rapidity of cutting or chip removal from below the bit. Instantaneous removal of these particles is of course impossible; however, proper application of available hydraulic energy can minimize regrinding and increase penetration rate. This is the principle which has been stated earlier as pertaining to the widespread acceptance of jet bits. The jets themselves do not drill the hole but merely expedite cutting removal.

The effect of hydraulic factors on the drilling rate of drag-type bits has been rather extensively studied.[34–36] The principal conclusions from these studies were the following:

(1) Rate of penetration is directly proportional to nozzle velocity (at constant circulation rate)

(2) Rate of penetration is directly proportional to circulation volume (at constant nozzle velocity).

The direct proportionalities noted were due, essentially, to increases in the allowable weight on the bit resulting from better hole cleaning at the higher volumes and velocities. It should be noted that these results applied to soft rocks and drag-type bits. This type of drilling is carried on at relatively low bit loads, with penetration rate being almost entirely dependent on the drilling fluid's ability to remove large cutting volumes and prevent bit balling.

The effects of hydraulic factors on rock bit drilling rates have also been investigated.[13,16,23,31,37-40] In soft and medium rocks, increases in penetration rate have generally resulted from increased hydraulic horsepower. Some question exists as to whether these increases correlate best with the impact of the jet stream as indicated by drag-bit experience, or with the hydraulic horsepower expended across the bit. *Impact force* depends on flow rate times velocity:

$$F = Mv = \frac{\rho_m q v}{60g} \tag{8.9}$$

where F = continuous force on bottom exerted by jet streams, lb
M = mass rate of flow, lb-sec/ft
ρ_m = mud density, lb/gal
q = mud flow rate, gal/min
v = nozzle velocity, ft/sec
g = 32.2 ft/sec^2.

Field data on penetration rate in soft and medium strength rocks are largely manifestations of increased weight on bit made possible by improved bit and bottom hole cleaning. Figure 8.21 shows the typical increases in bit weight facilitated by increased nozzle velocity, as experienced in California drilling. The curvature of the lines indicates balling, which becomes more severe as bit weight is increased.

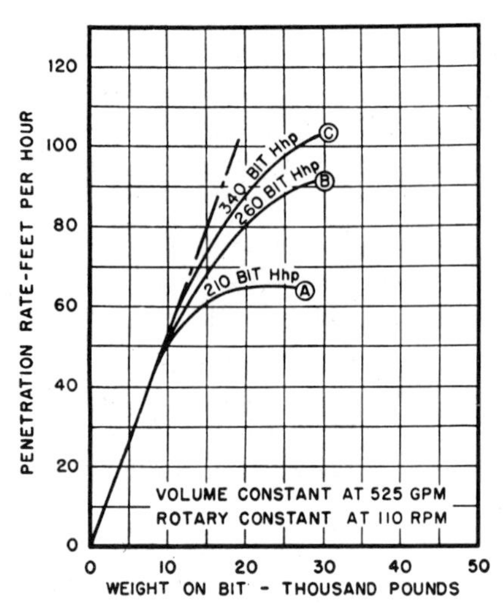

Fig. 8.21. Combined hydraulic and bit weight effect on penetration rate in California drilling. After Thompson,[38] courtesy API.

In hard rock drilling, the maximum bit weights applied are dependent on equipment considerations, rather than on bit balling. Consequently, it appears that increased nozzle velocity aids drilling rate by minimizing regrinding, with this effect disappearing as soon as no appreciable regrinding occurs. Nozzle velocities above this latter maximum effective value will result in no further increase in penetration rate. From results in West Texas, Bromell[39] has concluded that bit horsepower is an unimportant variable below penetration rates of 14 ft/hr, as long as adequate bottom hole cleaning is provided. Total pump horsepower may be reduced with no reduction in penetration rate resulting, if the circulation volume is cut down and the nozzle size is decreased, as shown in Table 8.1. The reductions are, of course, limited by the minimum annular velocity which may be tolerated.

TABLE 8.1

ILLUSTRATION OF CONSTANT PENETRATION RATE AT ESSENTIALLY CONSTANT BIT HORSEPOWER AND REDUCED CIRCULATION VOLUMES*

Surface pressure	*q, gal/min*	*Nozzle area, in.*^{2}	*Nozzle velocity, ft/sec*	*Required engine horse-power*	*Bit horse-power*	*Drilling rate, ft/hr*
1150	410	0.620	215	440	95	41.9
1125	345	0.488	230	340	90	41.3
1025	295	0.359	265	260	100	41.7

*After Bromell[39], courtesy API.

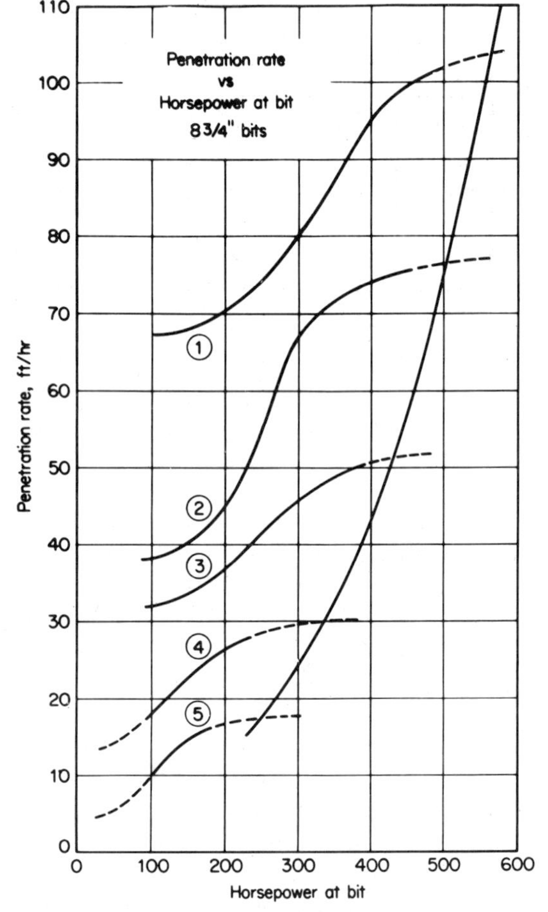

Fig. 8.22. *S* curve relationship between bit horsepower and penetration rate — Texas Gulf Coast area. After Keating,[40] courtesy API.

Considerable attention has been focused on the determination of minimum nozzle velocities which must be applied to make the use of jet bits advantageous over conventional types. It has been consistently demonstrated that penetration rate undergoes a sharp increase as jet velocity is increased through some critical range. Maximum benefit is obtained above the critical range, while lower velocities may prove less effective than normal circulation through conventional watercourse bits. Such relationships may be seen in Figure 8.22 which shows an S curve relationship between horsepower and drilling rate.[40] At low horsepowers, little cleaning is done, and little penetration benefit is obtained by increasing jet velocity. However, at some intermediate value, sharp rate increases are achieved from relatively small velocity increases. This effect diminishes as maximum cleaning is obtained, as indicated by the second slope reduction at the top of the S.

Minimum nozzle velocities of 200 to 250 ft/sec appear to be necessary in order to obtain substantial drilling rate increases in most rocks. Maximum nozzle velocities are not so well defined and have probably not been attained for soft rock drilling. The practical limit is imposed by available horsepower and nozzle erosion. Maximum values proposed for any case must depend on the bottom cleaning required; this is a function of bit type, the nature of the rocks being drilled, the rotary speed, and the weight on the bit.

It is apparently not possible to develop any general mathematical relationship between drilling rate and bit hydraulics which will fit all situations. The best operation will, in most cases, be a program which utilizes maximum bit horsepower at all times. In general, the available pump horsepower is constant; therefore, since the drill string losses increase with depth, the available bit horsepower decreases with depth:

(1) $$HP_B = HP_t - HP_x$$

where HP_B = bit horsepower

HP_t = total pump horsepower

HP_x = system horsepower exclusive of bit

Since

(2) $$HP_t = \frac{q \Delta p_t}{1714}$$

(3) $$q \Delta p_t = \text{constant}$$

The circulating rate, q, is governed by hole size and minimum allowable annular velocity, and is often treated as a constant in a given case. Thus for these operational restrictions, the total pressure drop, Δp_t, must also be constant. Example 8.1 illustrates this approach.

Example 8.1

Constant HP, Minimum q
Hole size = $8\frac{3}{4}$ in.
Drill pipe = $4\frac{1}{2}$ in., 16.6 lb, IF.
Drill collars = 500 ft long, $6\frac{3}{4}$ in. o.d., $2\frac{3}{4}$ in. bore
Bit = 3 cone, jet type, rolling cutter rock bit
Pump capacity = 400 HP (output)
Required annular velocity = 150 ft/min (from experience)

Calculate (1) Operating conditions for utilization of 400 HP, (2) Nozzle size at 6,000 ft depth, (3) Nozzle size at 11,500 ft depth.

Solution:

In lieu of more specific data we will assume the mud properties of the Hughes charts in Chapter 7.

(1) $$q = 2.45(d_h^2 - d_p^2)\bar{v}_A$$
$$= 2.45[(8.75)^2 - (4.5)^2](150/60)$$
$$= 350 \text{ gal/min}$$

Since

$$HP = \frac{q \Delta p_t}{1714}$$

$$\Delta p_t = \frac{(1714)(400)}{350} = 1960 \text{ psi}$$

These are then the imposed operating limits for circulating volume and pressure.

(2) Calculating hydraulic losses,

Δp_s = 20 psi (Figure 7.3, #3)
Δp_p = 220 psi (Figure 7.5, #7)
Δp_c = 100 psi (Figure 7.7, $2\frac{3}{4}$ in.)
Δp_{ac} = 26 psi (Figure 7.11, $6\frac{3}{4}$ in.)
Δp_{ap} = 55 psi (Figure 7.11, $4\frac{1}{2}$ in.)
Δp_x = 421 psi
Δp_b = 1960 − 421 = 1540 psi

From Figure 7.9,

$A_N = 0.260$ in.^{2} = total area of 3 nozzles;

hence

$$d_N = \left(\frac{4A_N}{3\pi}\right)^{1/2} = 0.332 \text{ in.}$$

The nearest larger stock nozzle size would normally be selected.

(3) At 11,500 ft,

Δp_s = 20 psi (same as above)
Δp_p = 440 psi (twice)
Δp_c = 100 psi (same as above)
Δp_{ac} = 26 psi (same as above)
Δp_{ap} = 110 psi (twice)
Δp_x = 696 psi
Δp_b = 1960 − 696 = 1264

Hence,

$A_N = 0.290$ in.^{2}
$d_N = 0.351$

Note that nozzle size is relatively insensitive to depth changes under these conditions.

We now consider a second approach in which total pressure loss, Δp_t, is held constant at some maximum value as dictated by pump and/or surface connection ratings.[34,41,42] The flow rate q will be treated as a restrained variable free to vary above some minimum value, i.e., $q \geq q_{\min}$; where $q_{\min}$ depends on the necessary annular velocity and hole-pipe clearance.

Assuming turbulent flow and that annular losses are relatively small, we may write:

$$\Delta p_x = K_1 q^{1.86} \tag{4}$$

where K_1 is dependent on pipe length, size, tool joint, and mud properties, but is constant for a given set of operating conditions. The power expenditure is

$$\text{HP}_x = \frac{q\Delta p_x}{1714}$$

$\therefore$

$$\text{HP}_x = \frac{K_1 q^{2.86}}{1714} \tag{5}$$

Also,

$$\text{HP}_t = \frac{q\Delta p_t}{1714} \tag{6}$$

Substituting Eqs. (5) and (6) into Eq. (1), we arrive at an expression for bit horsepower:

$$\text{HP}_B = \frac{q\Delta p_t}{1714} - \frac{K_1 q^{2.86}}{1714} \tag{7}$$

Differentiating (7) with respect to q, noting that $\Delta p_t =$ constant, and solving for maximum Δp_B which is the desired operating condition:

$$\Delta p_x = 0.35 \Delta p_t \tag{8.10}$$

or

$$\Delta p_b = 0.65 \Delta p_t \tag{8.10a}$$

This shows that bit horsepower is maximum when approximately 65% of the total pressure drop occurs across the bit, pump pressure being constant. It is generally desirable to operate with maximum bit HP, if feasible. We must now determine some rational expression for the corresponding value of q. This is readily obtained from Eq. (8.10), since Δp_t is known from pump rating, etc.

$$\Delta p_x = K_1 q^{1.86} = 0.35 \Delta p_t$$

from which

$$q_0 = 0.57\left(\frac{\Delta p_t}{K_1}\right)^{0.54} = \text{optimum } q \tag{8.11}$$

This expression allows calculation of the corresponding nozzle size, since from Chapter 7,

$$\Delta p_b = \frac{\rho q_0^2}{7430 C^2 d_e^4} \tag{7.13}$$

where ρ = mud density, lb/gal

C = nozzle coefficient = 0.95 for jet bits

d_e = hydraulically equivalent single nozzle diameter, in.

Substituting $\Delta p_b = 0.65\ \Delta p_t$, $C = 0.95$, and solving for d_e

$$d_e = 0.123\ q_0^{1/2}\left(\frac{\rho}{\Delta p_t}\right)^{1/4} \tag{8.12}$$

Example 8.2 Constant Pressure Operation

Let us rework Example 8.1 assuming the same pipe sizes, etc., and a depth of 6000 ft. This time we will let $\Delta p_t = 2000$ psig and q vary, according to our recent discussion. Immediately we see that $\Delta p_x = 700$, and $\Delta p_b = 1300$ psig.

Solution:

$$q_0 = 0.57\left(\frac{2000}{K_1}\right)^{0.54}$$

To determine K_1 we must consider both the drill pipe and drill collar section.

It can be shown (see Problem 6b) that

$$\Delta p_p + \Delta p_c = \Delta p_p\left[1 + \frac{L_c}{L_p}\left(\frac{d_p}{d_c}\right)^{4.86}\right]$$

where L_c, L_p = length of drill collars and pipe, respectively, ft

d_c, d_p = internal diameters of same, in.

Since

$$\Delta p_p = \frac{0.58 q^{1.86} L}{1000 d^{4.86}}$$ = turbulent pressure loss inside pipe, psi, according to Hughes charts (see Problem 6)

It can be shown (see Problem 6b)

$$K_1 = \frac{0.58 L_p}{1000 d_p^{4.86}}\left[1 + \frac{L_c}{L_p}\left(\frac{d_p}{d_c}\right)^{4.86}\right]$$

Substitution of values yields

$$K_1 = \frac{(0.58)(5.5)}{(3.83)^{4.86}}\left[1 + \frac{0.5}{5.5}\left(\frac{3.83}{2.75}\right)^{4.86}\right]$$
$$= 0.00682$$

Solving for q_0:

$$q_0 = 0.57\left(\frac{2000}{0.00682}\right)^{0.54} = 507 \text{ gal/min}$$

The corresponding nozzle size is

$$d_e = 0.123(507)^{1/2}\left(\frac{9.5}{2000}\right)^{1/4} = 0.728 \text{ in.}$$

Assuming 3 equal size nozzles, the actual size of each should be:

$$d = \frac{0.728}{\sqrt{3}} = 0.42 \text{ in.}$$

It is interesting to compare this solution to that of Example 8.1.

Ex. 8.1 $\Delta p_b = 1540$ psi, $\text{HP}_b = 314$

Ex. 8.2 $\Delta p_b = 1300$ psi, $\text{HP}_b = 384$

We note that the bit pressure loss has dropped 240 psi, but due to the increased flow rate (and total input HP), the bit power has been increased by 70 HP. Actually, Example 8.1 is the more efficient, in terms of per cent of input power utilized at the bit; however, bit HP is less.

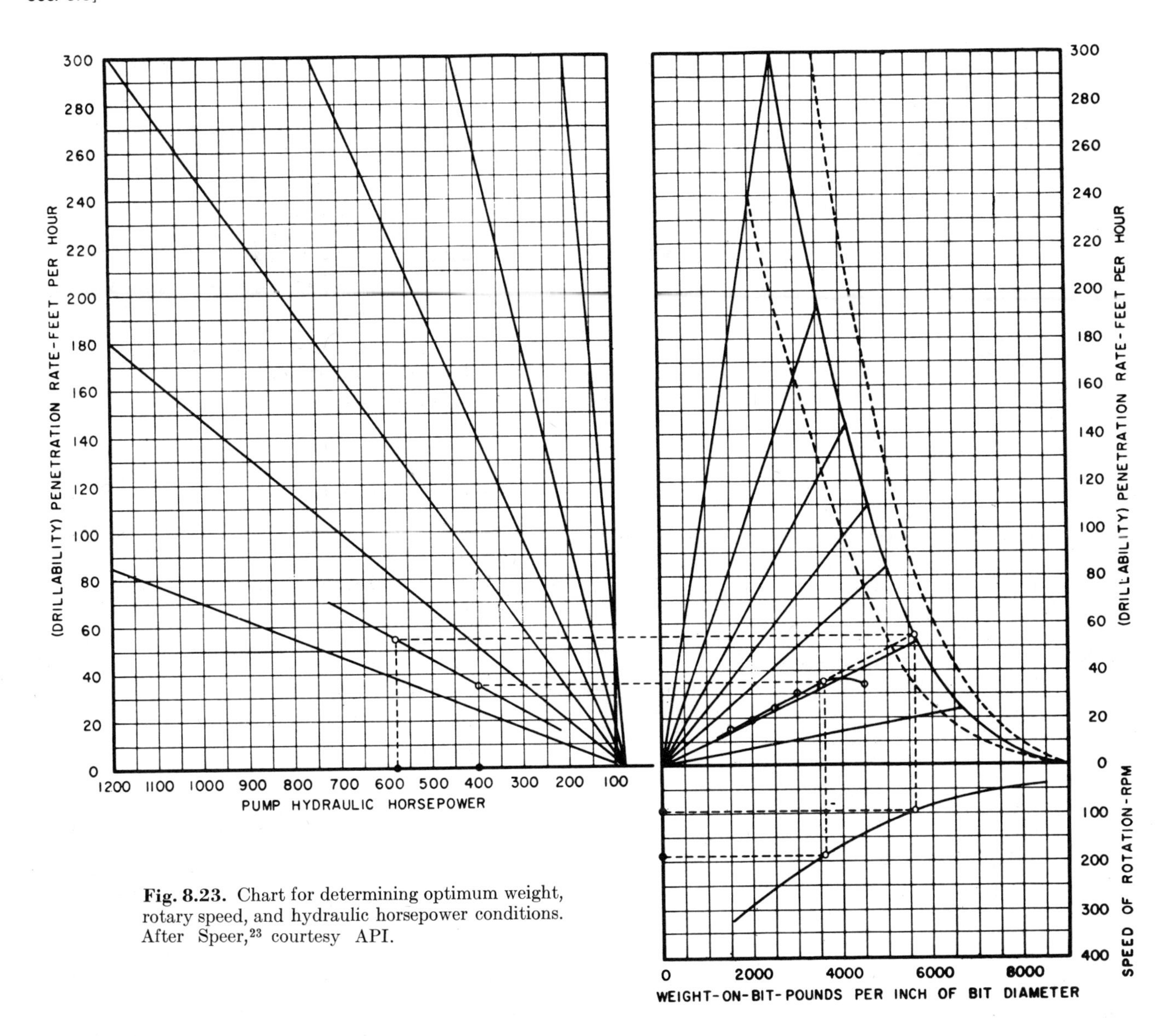

Fig. 8.23. Chart for determining optimum weight, rotary speed, and hydraulic horsepower conditions. After Speer,[23] courtesy API.

Depending on the situation, rig equipment, etc., either approach may be desirable. In some cases constant pressure operation (variable q and HP) may be permissible to some particular depth, at which point the optimum rate may become less than the minimum value permissible for hole cleaning. If this occurs, the operation necessarily changes to that for constant HP case. The article by Colebrook[42] contains a number of useful curves illustrating the effect of certain variables on constant pressure operation.

Neither of these approaches may be desirable in hard rock drilling where maximum benefit from bottom cleaning may be obtained at a lower hydraulic level than that predicted by either method. This is well illustrated by Bromell's data in Table 8.1.

An over-all approach to selection of optimum drilling conditions has been developed by John Speer[23] who has presented a comprehensive analysis of a large number of data. This method is illustrated by Example 8.3 which is solved graphically with Figure 8.23.

Example 8.3 (After Speer[23])

A certain rig is available to drill a 1,000-ft homogeneous shale section. It is desired to determine: (a) optimum operating conditions for the available rig; and (b) the equipment required to drill the shale section most economically. The given parameters are as follows:

1. 400 HP pump
2. $4\frac{1}{2}$-in. drill pipe
3. $9\frac{7}{8}$-in. hole
4. three $\frac{1}{2}$-in. bit nozzles

When the shale section is drilled, short-duration tests of weight vs. penetration rate are made (1) to establish a drillability index for the formation and (2) to determine at what weight the bit begins to ball up with the available pump (a tentative index must first be established to select an appropriate rotary speed). These test values are plotted on the weight vs. penetration rate chart (upper righthand section) and an

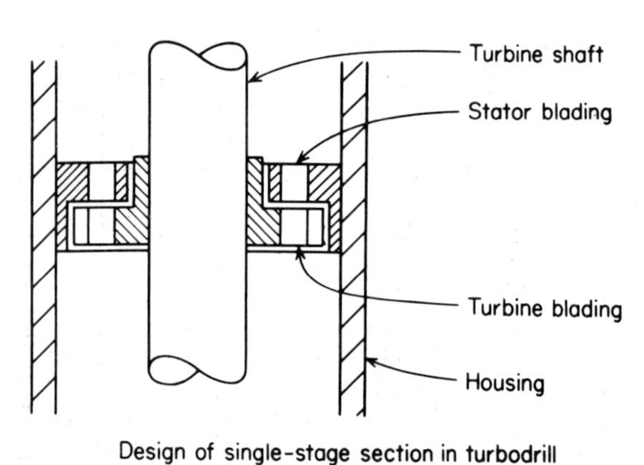

Design of single-stage section in turbodrill

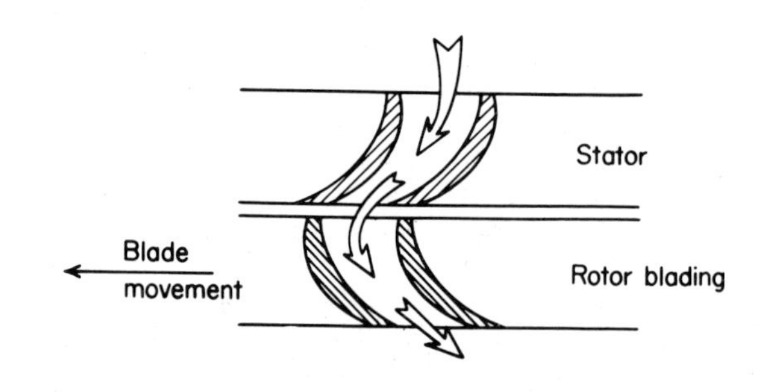

Typical cross section of turbine blading showing the flow of fluid through one stage of the unit

Fig. 8.24. Elementary diagram of turbine components. After LeVelle,[44] courtesy *Petroleum Engineer*.

appropriate curve is drawn through them. The low weight section of this curve should be linear until the bit begins to ball up. From the point on this curve where the bit begins to ball up, a horizontal line is drawn left to intersect with 400 HP. The single point thus obtained is sufficient; a linear relationship can now be drawn between hydraulic horsepower and penetration rate for this shale section. It will be noticed that this index is in a sense dimensionless, inasmuch as it includes hole size, type of bit, drillpipe size, etc. As a next step, a vertical line is drawn from the weight where the bit begins to ball up downward, to intersect with the *optimum* W vs N curve. The intersection shows that this formation would be most economically drilled with a weight of 3600 lb/in. of bit diameter (35,000 lb) and 190 rpm rotary speed, with the 400 HP pump.

Now return to the weight vs. penetration rate index, and extend the linear section of the curve to intersect the optimum weight on bit curve. From this point, a horizontal line is drawn left to intersect the hydraulic horsepower index curve already established for this formation. This intersection gives the hydraulic horsepower required to maintain sufficient bottom-hole cleaning to permit use of optimum weight. A vertical line is also drawn downward to determine optimum speed of rotation for the new weight. These manipulations show that minimum footage cost could be achieved in this formation with a 580 HP pump, used in conjunction with a weight of 5,600 lb/in. of bit diameter (55,000 lb) and 95 rpm. By working backwards from the hydraulic horsepower index curve, the best weight and rotary speed practices for any hydraulic horsepower can also be determined in the same way.

Inspection of Figure 8.23 shows this method to be based on the linear relationship of hydraulic horsepower and penetration rate. This premise has been questioned by Moore[43] who, among others, feels that hydraulic effects are best expressed by impact force (Eq. 8.9).

The reader should not infer that any of the various empirical expressions presented in this chapter are infallible. They are presented as means of estimating the effect of certain factors; the refinement of various parameters must be attained under individual well conditions.

8.7 Other Drilling Methods

While considerable effort has been exerted to increase conventional drilling rates, the possibility of developing other drilling methods has not been ignored. These range from mere variations of the conventional rotary technique to radical departures from basic methods. The future application of any new technique will depend on the standard American question: can it do the job more efficiently?

8.71 The Turbodrill

The turbodrill is a rotary drilling device employing the same basic mechanism as the conventional method. The principal difference is that the bit's rotation is provided by a down hole multistage turbine which is powered by the drilling fluid; thus rotation of the drill string is eliminated. The turbine section contains matching sets of stators and rotors, each pair being referred to as one stage. Current models utilize 100 or more stages. Drilling mud is deflected by the stator and strikes the rotor blades, causing them to rotate (see Figure 8.24).[44] The design principles governing nozzle and blade angles are the same as in other types of turbines. Variations in these factors allow the design of turbines with different operating characteristics as dictated by different applications. Figure 8.25 is a cutaway view of the French (Neyrpic) model which has had considerable success in European tests.[45]

Turbodrills are not new as far as the idea of their application to drilling is concerned. The first patent relating to turbodrilling was issued to C. G. Cross in 1873, almost thirty years before the use of rotary tools in oil well drilling. Subsequent development in this country was sporadic; although various turbodrill designs were developed and field tested, they are generally unsuccessful.[46,47] The main failing of American designs

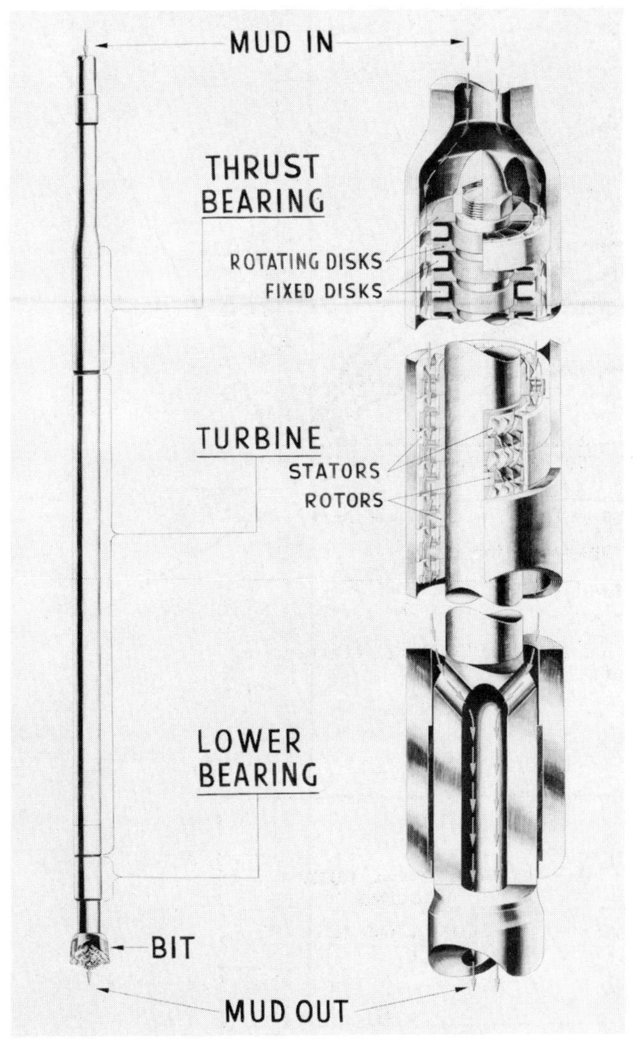

Fig. 8.25. Cutaway view of the French (Neyrpic) turbodrill. After O'Connor,[45] courtesy *The Monitor*.

was excessive bearing failure brought about mainly by the action of abrasive mud solids. The Russians are credited with defeating this problem; they developed rubber thrust bearings which can tolerate abrasive solids and are, in fact, lubricated by the drilling fluid. Over 80% of all Russian drilling is performed with turbines which, according to their claims, are far superior to conventional equipment.[48,49] The Russian comparisons are not indicative of relative over-all performance, however, due to the superiority of U.S. drill pipe, drill collars, and bits.

Typical operating characteristics for the turbodrill are shown in Figure 8.26. Maximum power and efficiency occur between 500 and 700 rpm, which are normal operating speeds. Speed is controlled by the weight applied to the bit (this governs torque). Surface tachometers are being developed which will enable closer control of turbine performance. The best drilling fluid for turbine use appears to be water, both from turbine operation and penetration rate standpoints. This is in line with our previous discussion. Mud may, however, be used if necessary.

Penetration Rate Comparisons

Figure 8.27 graphically portrays comparative tests between the conventional rotary and the Neyrpic drills which were conducted in France in 1956. It should be noted that the bit weight applied in the conventional tests was less than half that on the turbodrill and is considerably below standard practices in this country. Analysis of these data by Scott[17] showed that drilling rate varied linearly with bit weight and the square root of rotating speed, as predicted by Eq. 8.4. Accordingly, the drilling rates of the two methods should be approximately equal when

$$W_t\sqrt{N_t} = W_c\sqrt{N_c}$$

where W = weight on bit
N = rotating speed
t,c = turbine and conventional methods, respectively

The relative rates of the two methods may be approximated by:

$$\frac{R_{pt}}{R_{pc}} = \frac{W_t}{W_c}\sqrt{\frac{N_t}{N_c}}$$

where $R_{pt,pc}$ = turbine and conventional on bottom drilling rates, respectively.

The comparison as indicated by the French tests is not a fair picture, as there is no apparent reason why the rotary drill could not carry the same weight on bit as the turbodrill; in fact, from a standpoint of bit life, it is probable that higher loading could be tolerated. This implies that equal crooked hole problems are encountered with both techniques, which appears to be a sound assumption. Therefore, insofar as penetration rate is concerned, turbodrill benefits in this country must come primarily from increased rotational speeds.

Hydraulic Considerations

Hydraulic losses in turbodrilling include the same losses as are incurred in ordinary rotary drilling, plus the additional drop across the turbine blades. The magnitude of this additional component depends on the individual design (blade angle, number of stages, etc.) and the operating conditions. This additional pressure drop results in thrust, T:

$$T = \frac{\pi}{4}d^2\Delta p \tag{8.14}$$

where T = thrust, lb
d = average turbine blade diameter, in.
Δp = pressure drop across turbine, psi

This thrust is consumed as part of the weight on bit. The hydraulic efficiency is improved by the use of

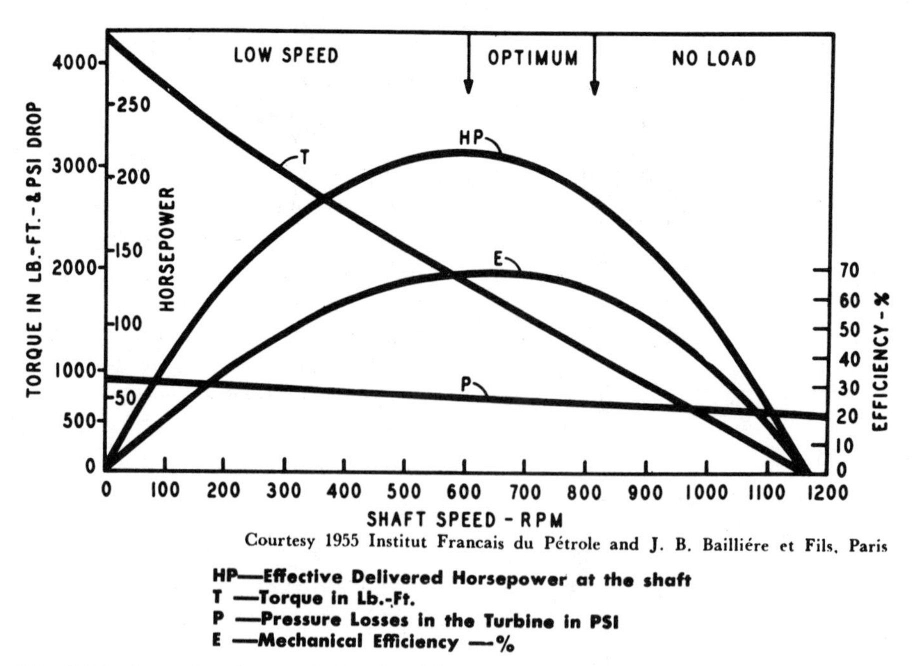

Fig. 8.26. Operating characteristics of a 10-in. turbodrill. After Thacker and Postlewaite,[47] courtesy *World Oil.*

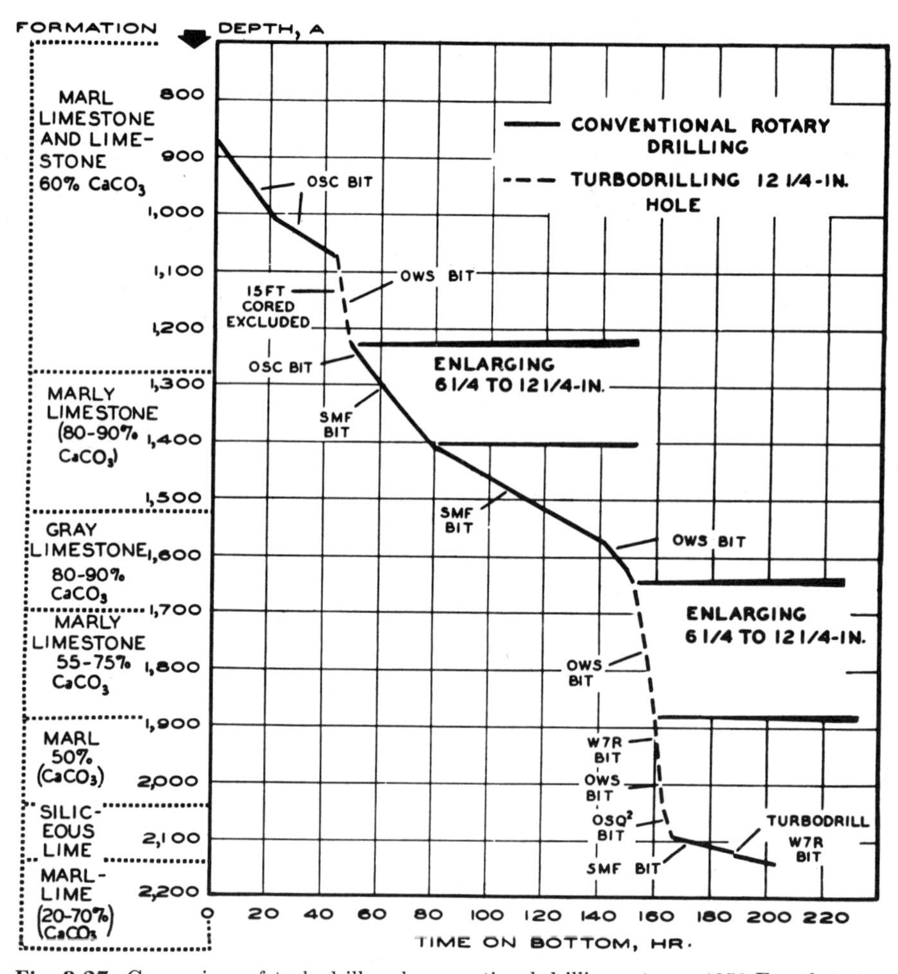

Fig. 8.27. Comparison of turbodrill and conventional drilling rates — 1956 French tests. After Scott,[17] courtesy *Oil and Gas Journal.*

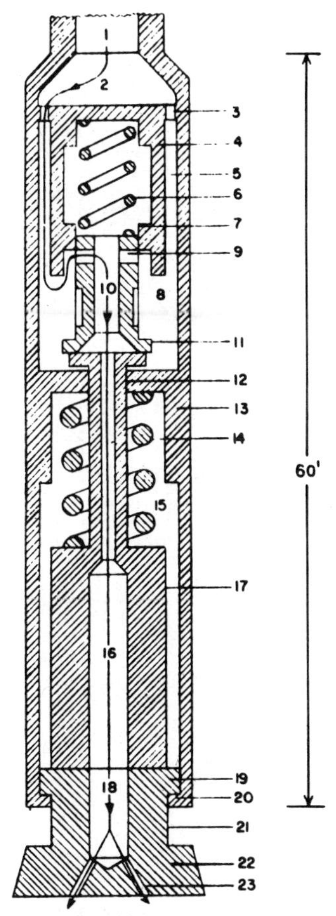

A — Hanging position, tool is open and is being lowered to bottom. Fluid is bypassed through passages 9, 10, 16, and 18 so that circulation can be established while reaming to bottom.

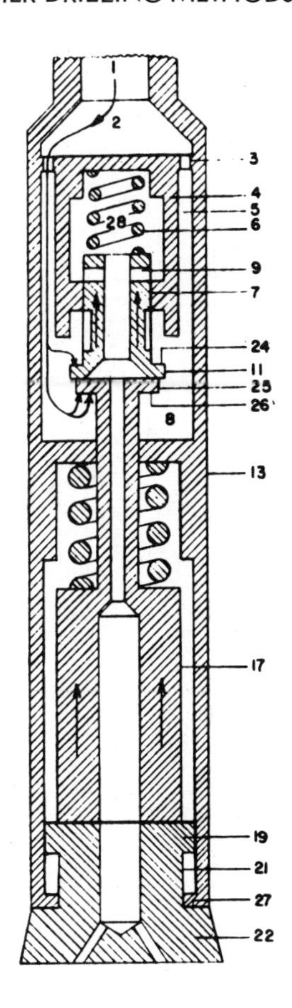

B — Starting position. Tool is set on bottom to collapse anvil 21 which has a hexagonal slip joint. Fluid is now trapped in chamber 8, and clamps valve and hammer lips together.

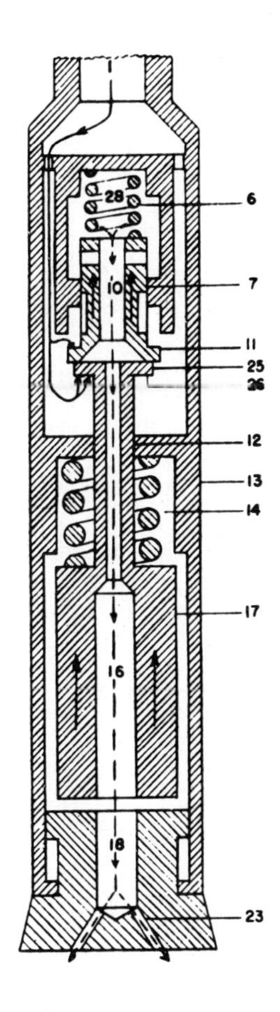

C — Hammer and valve moving up under fluid pressure due to net diameter of valve piston 7 being greater than diameter of hammer stem 12. Resultant pressure acts upward on net area.

Fig. 8.28. Operating cycle of Gulf's hammer drill. Courtesy *Petroleum Engineer*.[50]

thinner walled, larger inside diameter drill pipe made possible by its lighter, nonrotating duty.

Future Use of the Turbodrill

The highly developed status of United States rotary drilling equipment and techniques makes it a considerably more formidable opponent to the turbodrill than the European or Russian rotary techniques. Certain advantages are inherent in turbodrilling, however:

(1) Higher penetration rates due to increased rotating speed
(2) Elimination of drill string rotation
  (a) lighter and cheaper drill pipe and tool joints are permissible
  (b) quieter operation — no rotary table noise
  (c) less fishing jobs caused by drill string failure.

The main disadvantages are:

(1) Turbodrill cost, both initial and maintenance
(2) Greater pump capacity and/or pressure required
(3) Greater care required to remove abrasive solids from the mud stream
(4) The pressure drop across the lower part of the turbine must be kept low, thus bits with high jet velocities cannot be used.

Considerable doubt exists as to the ability of current bit designs to withstand both the high speeds and weights which must be applied to compete with United States conventional rotary practices. Many operators feel that the application of turbodrilling depends primarily on the development of bits which will withstand such punishment. It appears, at this time, that future U.S. usage of the turbodrill will be restricted to hard rock areas, or to certain special applications, with the extent of such use depending on whether the indicated drilling rate increases and other advantages can offset the disadvantages.

8.72 Combination Rotary and Percussion Methods

It is recognized that large increases in bit energy may be obtained if high frequency percussive blows are

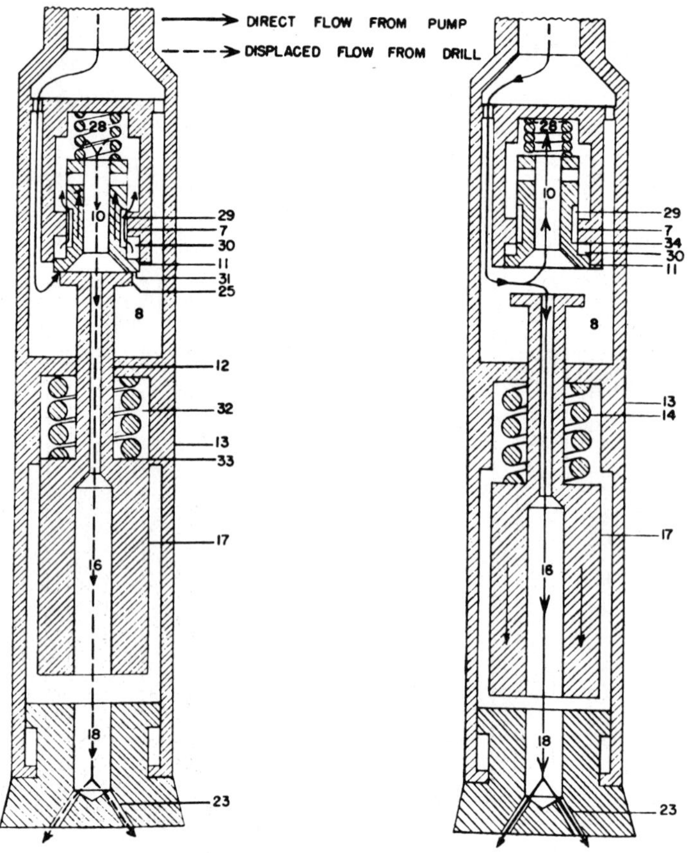

D — **Hammer reversing.** Pressure above valve lip is relieved through communication of port 29 with low pressure chamber 28. Area 31 is now at higher pressure than area 24 and valve accelerates, hammer drops.

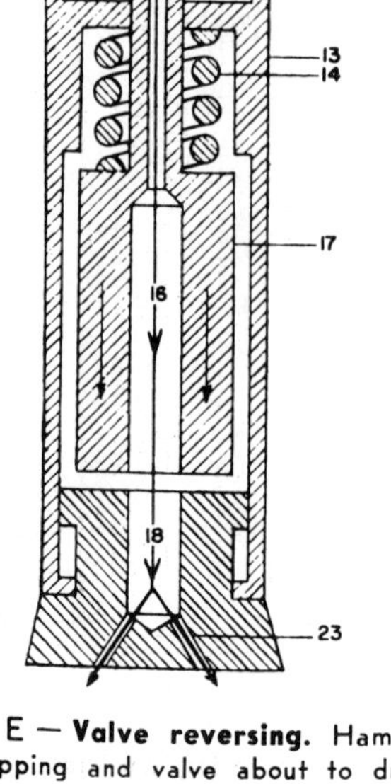

E — **Valve reversing.** Hammer dropping and valve about to drop. A conventional tungsten carbine tri-cone bit is attached to the anvil in full-scale laboratory and field tests.

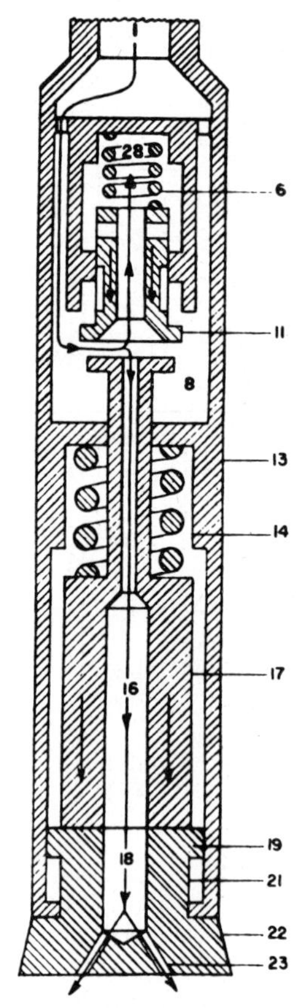

F — **Hammer striking anvil.** Hammer is driven down by gravity, hammer spring 14 and pressure drop of fluid through small passage above chamber 16. Fluid passes through chamber 18 into bit.

Fig. 8.28 (cont.).

combined with bit rotation. Such a technique has the basic features of both rotary and cable tool drilling.

A recent development employing this drilling mechanism is the *hammer drill* designed and tested by the Gulf Oil Corporation.[50,51] The bit is rotated under applied weight, percussive blows being furnished by a *mud engine* powered by the drilling fluid. The percussion frequency is on the order of 600 strokes/min. The mechanical operation of the engine is shown in the series of Figure 8.28. Considerable laboratory and field test data indicate that large increases in penetration rate may be obtained. Comparative laboratory results from drilling in granite are shown in Figure 8.29.

Although still in the experimental and development stage, this device holds considerable promise. Advantages of increased drilling rate, longer bit wear (attributed to lower rpm and lighter bit weights), and less hole deviation have been claimed by the tool's developers. Field trials have indicated that the mud engine can operate for 60 to 100 hours without maintenance. At this time it appears that the mud-powered hammer drill shows considerable promise for hard rock drilling. Further field tests are required for industry's acceptance.

The Battelle Memorial Institute scientists have also experimented with a high frequency percussive drilling device in which vibrations (several hundred reversals per second) are imparted to the bit by an electro-

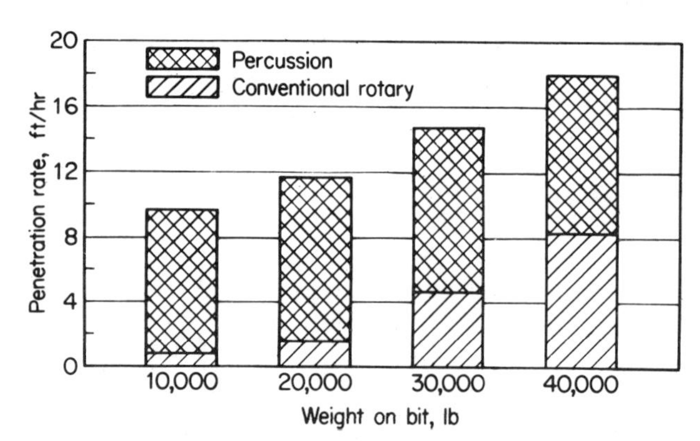

Fig. 8.29. Comparative penetration rates of conventional vs. the hammer drill. Laboratory data — atmospheric pressure. Courtesy *Petroleum Engineer*.[50]

magnetic mechanism. This is powered by alternating current transmitted down specially designed cables, which fit inside the drill pipe and remain stationary while the pipe turns.[2]

Very little data on the penetration rates obtained with this vibratory drill are available. Equipment failures and other operating troubles have prevented its thorough testing. This particular drill must be classed as highly experimental until further data are available.

8.73 The Pellet Impact Drill

An extremely interesting and radical departure from conventional drilling methods is the pellet impact drilling technique developed by the Drilling Methods Section of the Carter Oil Company Research Laboratory in Tulsa, Oklahoma.[52] This method of drilling utilizes the high velocity, random impact of steel pellets to cause rock failure. The process may be visualized from Figure 8.30. The high velocity jet stream from the primary nozzle draws drilling fluid and pellets into the secondary nozzle and discharges them against the rock. The pellets are then lifted off bottom by the drilling mud; they then re-enter the aspirator section, and are recycled. Additional pellets are suspended as a cloud above the primary nozzle due to the high ascending velocity alongside the enlarged secondary nozzle section.

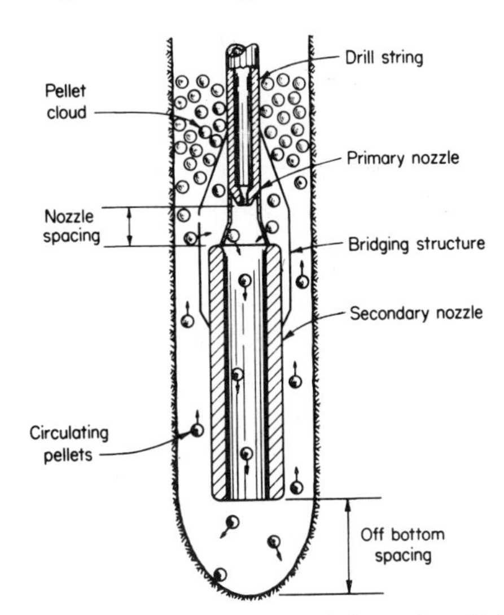

Fig. 8.30. Schematic operation of pellet impact drill. After Eckel,[52] courtesy AIME.

The apparent advantages of this method are elimination of drill pipe rotation and the attendant benefits; in particular, there is no necessity for frequent drill pipe removal, since the pellet supply can be replenished from the surface. This latter factor could result in the elimination of heavy hoisting equipment.

The experimental difficulties encountered and the necessary modifications made by the Carter engineers and scientists required an estimated 37 man-years of research prior to publication of the results cited here. Some of these problems were:

(1) Determination of the optimum off bottom spacing for the secondary nozzle. This was found to be from 2.8 to 3.4 nozzle diameters.

(2) Nozzle diameters

(a) The primary nozzle had to be large enough to permit the desired flow rate at the available pressure drop.

(b) The secondary nozzle had to be large enough to prevent jamming of the pellets: the ratio of pellet diameter to nozzle diameter had to be less than 0.5. The drilling rate as related to nozzle area ratio and input fluid horsepower is shown in Figure 8.31.

(3) Drilling fluid: water was superior to all those tested, including air.

(4) Pellet charge or total quantity of pellets in the system. The drilling rate was insensitive to this, from saturation charge (maximum drilling rate) to $2\frac{1}{2}$ times the saturation charge.

These are only a few of the problems encountered, but they serve to illustrate the complexity of the experiments.

Some drilling results are shown in Table 8.2. Certainly these penetration rates are lower than those obtainable by conventional methods. The pellet impact drill is not considered economic at the present time; however, it is of extreme fundamental importance, as it demonstrates a completely new concept of oil well drilling. As stated by the authors of the reference cited, the purpose of the paper was to make the findings of this work a matter of record, in the hope that the general knowledge of the process might be a guide for other developments.

TABLE 8.2

Results of 9 in. Pellet Impact Drilling Tests

Rock	*Hole diameter, in. Min.*	*Max.*	*Avg. depth drilled per test, ft*	*Average drilling rate, ft/hr*
Okla. marble (soft rock)	$9\frac{3}{4}$	$11\frac{1}{2}$	2	7.5
Virginia limestone (medium rock)	$9\frac{3}{4}$	10	1.5	4
Pink quartzite (hard rock)	$9\frac{1}{2}$	$9\frac{7}{8}$	1.25	0.5

8.74 Use of Retractable Bits

Since trips consume a large amount of rig time, considerable thought has been given to designing retractable bits which can be changed without withdrawing the drill string. Field tests of this process have indicated economic feasibility, and this technique should receive increased attention in future years.[53] Casing of the proper size is used as drill pipe and is never removed from the hole. Worn bits are replaced

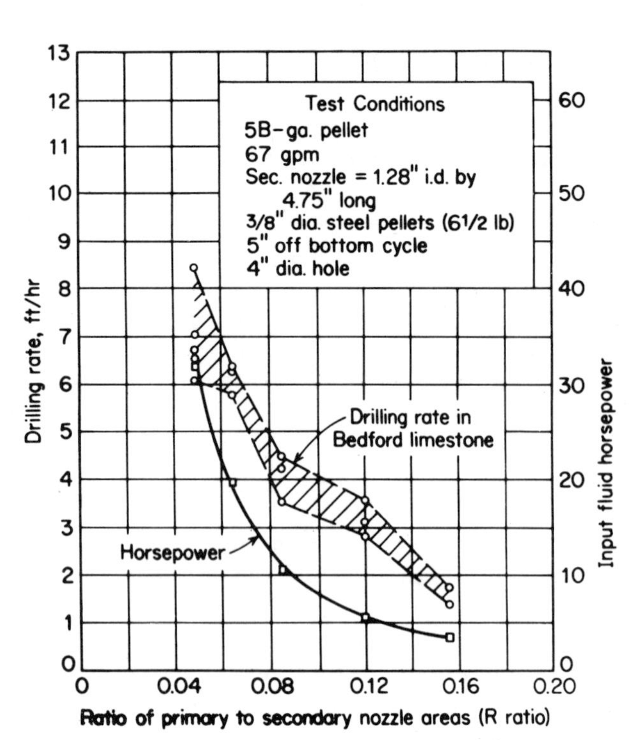

Fig. 8.31. Drilling rate and horsepower as functions of nozzle area ratios. After Eckel,[52] courtesy AIME.

by wire line operations. Future improvements in bit design are indicated, and savings of 25% of drilling costs may be possible by this method.

8.75 Simultaneous Drilling

This modification involves the simultaneous drilling of two directional wells with a single rig and drilling crew. This technique has reportedly been successfully applied by the Russians in some fields.[54] The main saving is again due to reduction of trip time.

The schematic rig layout shown in Figure 8.32 requires an enlarged floor area, two rotary tables, and a special crown block which allows the block location to be readily transferred from hole to hole. As pipe is withdrawn from hole 1 it is lowered into hole 2; hence one hole stands idle half the time. This is a principal disadvantage, as it may cause hole trouble in some areas.

Theriot[54] has estimated that 60% of trip time, 80% of cementing time, and 90% of waiting on cement time may be eliminated in this manner by proper timing of these operations. These and other savings may amount to approximately 34% of total rig time for an offshore, directionally drilled well. Application of this method

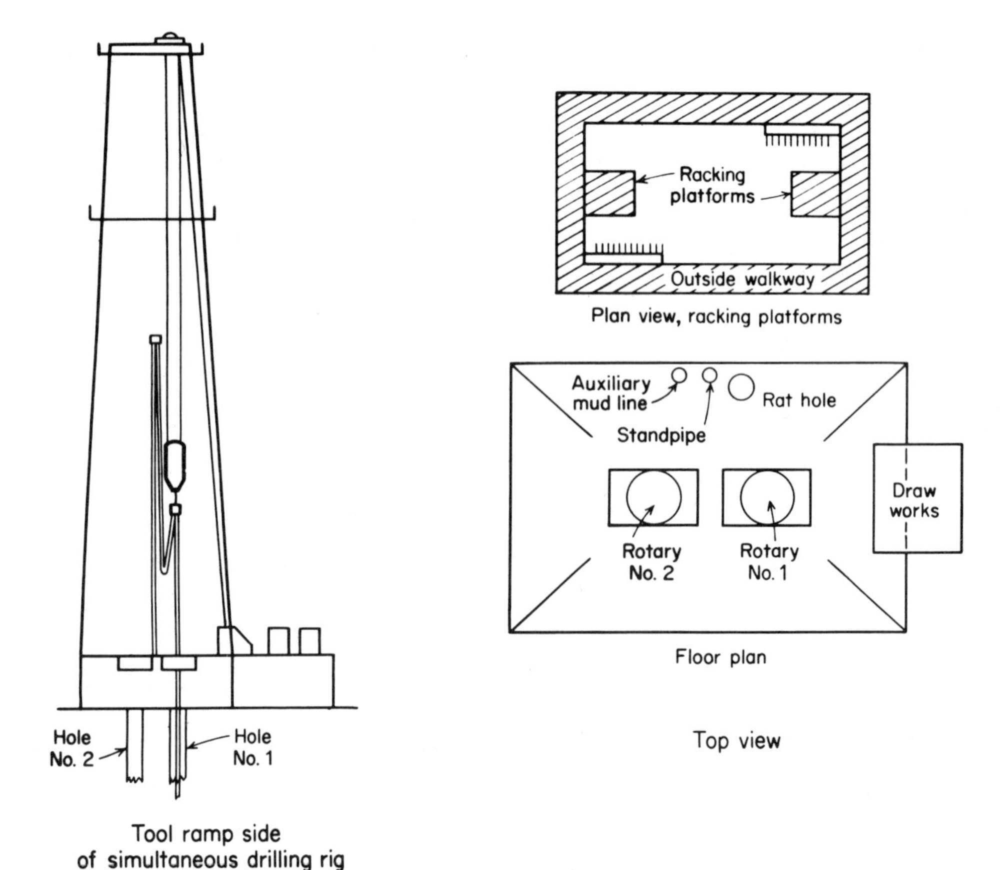

Fig. 8.32. Schematic rig layout for simultaneous drilling operations. After Theriot,[54] courtesy AIME.

will, however, depend on the elimination of operational problems and the development of proper equipment. Certainly this is an interesting possibility for certain areas, particularly offshore operations.

Suggested improvements in the basic methods of drilling have utilized either or both of two approaches:
(1) Higher energy input at the bit
(2) Minimization of nonproductive rig time
Economical improvement of either factor offers considerable benefit to the entire industry. None of the methods cited in this section are intended to replace completely the standard method. Some methods, however, are expected to supplement it in certain areas. Continued improvements in the conventional technique make it increasingly difficult for new methods to compete favorably. The success of any drilling method or technique hinges on its ability to minimize total drilling costs as defined by Eq. 8.15:

$$(8.15)\quad C_F = \frac{C_R\left(t_{mt} + \dfrac{D}{\bar{R}_p}\right) + C_B + C_A + C_C + C_M}{D}$$

where C_F = average direct drilling and completion cost/ft
C_R = rig operating cost, \$/hr
t_{mt} = maintenance and trip time, hr
D = depth interval
C_B = bit costs
C_A = auxiliary equipment costs, such as turbodrill, etc.
C_M = mud costs
C_C = completion costs, including casing, cementing, etc.
$\bar{R}_p$ = average *on bottom* penetration rate, ft/hr

Clearly, C_F does not depend on R_P only, and it is quite possible to have increases in $\bar{R}_p$ resulting in overcompensating increases in t_{mt}, C_B, C_A, C_C, and/or C_M. The definition of a minimum C_F for a given set of conditions is an extremely elusive figure and it is improbable that such has ever been precisely calculated. Attempting to reduce C_F is, of course, the primary job of those engaged in the drilling industry, regardless of their capacity.

PROBLEMS

1. A medium strength formation is observed to drill at 40 ft/hr at $W = 24{,}000$ lb and $N = 150$ rpm. Bit size is 8 in. Assuming adequate bottom hole cleaning and that Eq. 8.4 applies:
(a) What penetration rate could be attained at $W = 36{,}000$ lb? ($N = 150$)
(b) At $N = 225$, $R_p = ?$ ($W = 24{,}000$)

2. Assume that $WN = 3.6 \times 10^6$ rpm-lb is the operating limit for the conditions of Problem 1. Plot the rate of penetration vs W and N for the range $12{,}000 \leq W \leq 60{,}000$ and $60 \leq N \leq 300$. Show these plots on both cartesian and log-log paper. Assume adequate bottom hole cleaning and that Eq. 8.8 applies.

3. At what rate would you expect the same formation to drill using a 10 in. bit at:
(a) $W = 24{,}000$ lb
(b) $W = 30{,}000$ lb
(c) $W = 40{,}000$ lb

4. Field tests indicate that under common conditions a dense dolomite section drills according to:

$$R_p = 3.8 \times 10^{-5} w^{1.2} N^{1/2} \quad \text{ft/hr}$$

where w = bit load in lb/in. of diam.
If operating restrictions are imposed such that $wN = 5 \times 10^5$,
(a) What R_p can be obtained if $N = 50$ rpm?
(b) Suppose that crooked hole restrictions cause bit loading to be fixed at $w = 5000$ lb/in. What reduction in R_p will this cause, based on part (a)? *Ans.* (b) 16.9 – 10.3/16.9 = 39% reduction.

5. A rig has 700 HP available for drilling an 8¾ in. hole from 6000 to 12,000 ft. The drill string is composed of:
Surface connections: case 3, Figure 7.3
Drill pipe: 4½ in., IF
Drill collars: 6¾ in. O.D., 2¾ in. bore
Assume mud conditions of Hughes charts in Chapter 7, and constant maximum pump HP operation.
(a) Calculate bit horsepower, equivalent single nozzle size, and actual size for 3 nozzles at 10,000 ft for required annular velocities of (1) 100 ft/min (2) 150 ft/min (3) 200 ft/min. Show these data graphically.
(b) For an annular velocity of 150 ft/min, calculate the equivalent single nozzle size at 6000, 8000, 10,000, and 12,000 ft and show these results graphically.
(c) Discuss any conclusions you might draw from (a) and (b).

6. In Problem 24 of the previous chapter the method for developing Figure 7.4 through 7.7 was given:

$$(1)\qquad \Delta p = \frac{K\mu_t^{0.14}\rho^{0.86}q^{1.86}L}{d^{4.86}}$$

By substituting $K = 0.72 \times 10^{-4}$, $\mu_t = 3$ cp, $\rho = 9.5$ lb/gal we obtained:

$$(2)\qquad \Delta p = \frac{0.58q^{1.86}L}{1000d^{4.86}}$$

which is the equation used in constructing Figures 7.4–7.7, where proper average values of d were used for the various tool joints. This shows turbulent flow pressure loss to be directly proportional to $q^{1.86}$. Assuming turbulent flow throughout the entire system, show that:
(a) $HP_x \cong K_1 q^{2.86}$, where K_1 is an appropriate constant for given pipe sizes, etc.
(b) The sum of drill pipe and drill collar pressure loss may be expressed as

$$\Delta p_p + \Delta p_c = \Delta p_p\left[1 + \frac{L_c}{L_p}\left(\frac{d_p}{d_c}\right)^{4.86}\right]$$

where L_c = length of drill collars of diameter d_c
L_p = length of drill pipe of diameter d_p

7. (a) Using the chart developed by Speer (Figure 8.23) determine proper drilling conditions from following data:

w, *lb/in.*	R_p, *ft/hr*
2000	40
3000	60
4000	75
5000	75

The Pump HP = 400.

(b) What hydraulic HP will allow utilization of optimum w and N? What would be the resulting R_p? *Ans.* (a) $W \simeq 4000$, $N \simeq 170$, (b) 500 HP, $R_p \simeq 95$ ft/hr.

8. Suppose that in a hard rock area application of Figure 8.23, balling did not occur, that is, R_p continued to increase nearly proportionally with w until the maximum allowable value was reached. What might then be done concerning pump horsepower? (See Bromell's data in Table 8.1.)

9. Which of the operations in Examples 8.1 and 8.2 results in the higher impact force on bottom? (See Eq. 8.9.) Will this always be the case? Explain.

10. Using the same pipe and hole sizes as in Example 8.1 and the mud properties of the Hughes Charts, calculate the optimum flow rate q_0 at various depths to 15,000 ft. Keep the same drill collar length and vary pipe length only. Show these data as a plot of q_0 vs depth. Discuss the results.

11. Repeat Problem 10 for a pressure limit of 3000 psig. (See Reference 42 for similar plots.)

REFERENCES

1. Pennington, J. V., "Some Results of DRI Investigations — Rock Failure in Percussion," *API Drilling and Production Practices*, 1953, p. 329.
2. *1954 Report.* Drilling Research Inc., Battelle Memorial Institute.
3. Simon, R., Cooper, D. E., and M. L. Stoneman, "The Fundamentals of Rock Drilling," API Paper 826-27-H, Presented Columbus, Ohio, Apr. 1956.
4. Murray, A. S., and S. P. MacKay, "Water Still Poses Tough Problem in Drilling with Air," *Oil and Gas Journal*, June 10, 1957, p. 105.
5. Wuerker, R. G., "Annoted Tables of Strength and Elastic Properties of Rocks," Petroleum Branch, AIME, Dec. 1956.
6. Cunningham, R. A., "The Effect of Hydrostatic Stress on the Drilling Rates of Rock Formations," unpublished M.S. thesis. Houston, Texas: Rice Institute, 1955.
7. Murray, A. S., and R. A. Cunningham, "Effect of Mud Column Pressure on Drilling Rates," *Trans. AIME*, Vol. 204, (1955), p. 196.
8. Payne, L. L., and W. Chippendale, "Hard Rock Drilling," *The Drilling Contractor*, June, 1953.
9. Bredthauer, R. O., "Strength Characteristics of Rock Samples under Hydrostatic Pressure," unpublished M.S. thesis. Houston, Texas: Rice Institute, 1955.
10. Catalog No. 21, Hughes Tool Company, Houston, Texas, 1955-56.
11. Gatlin, C., "How Rotary Speed and Bit Weight Affect Rotary Drilling Rate," *Oil and Gas Journal*, May 20, 1957, p. 193.
12. Brantly, J. E., and E. H. Clayton, "A Preliminary Evaluation of Factors Controlling Rate of Penetration in Rotary Drilling," *API Drilling and Production Practices*, 1939, p. 8.
13. Bielstein, W. J., and G. E. Cannon, "Factors Affecting the Rate of Penetration of Rock Bits," *API Drilling and Production Practices*, 1950, p. 61.
14. Speer, J. W., "Drilling Time Reduced 31 Percent," *Oil and Gas Journal*, Oct. 11, 1954, p. 130.
15. Eckel, J. R., "Effect of Mud Properties on Drilling Rate," *API Drilling and Production Practices*, 1954, p. 119.
16. Wardroup, W. R., and G. E. Cannon, "Some Factors Contributing to Increased Drilling Rate," *Oil and Gas Journal*, Apr. 30, 1956.
17. Scott, J. O., "What Those French Turbodrill Tests Show," *Oil and Gas Journal*, Feb. 11, 1957, p. 121.
18. Woods, H. B., and E. M. Galle, "Effect of Weight on Penetration Rate," *The Petroleum Engineer*, Jan. 1958, p. B-42.
19. Grant, R. S., and H. G. Texter, "Causes and Prevention of Drill Pipe and Tool Joint Troubles," *API Drilling and Production Practices*, 1941, p. 9.
20. McGhee, E., "How to Get Your Money's Worth From Your Drill String," *Oil and Gas Journal*, Oct. 9, 1956, p. 133.
21. Main, W. C., "Discussion of Texter and Grant Paper (Reference 19), ibid.
22. Crane, F. S., "Drilling Based on Constant Weight-Speed Factor," *World Oil*, Mar. 1956, p. 142.
23. Speer, J. W., "How to Get the Most Hole for Your Money," *Oil and Gas Journal*, Mar. 31, 1958 and Apr. 7, 1958, pp. 90 and 148, respectively.
24. Woods, H. B., personal communication.
25. Bentson, H. G., "Rock-Bit Design, Selection, and Evaluation," *API Drilling and Production Practices*, 1956, p. 288.
26. Adams, J. H., "Air and Gas Drilling in the McAlester Basin Area," API Paper No. 851-31-N, Presented Tulsa, Apr. 1, 1957.
27. Cunningham, R. A., and J. G. Eenink, "Laboratory Study of Effect of Overburden, Formation and Mud Column Pressures on Drilling Rate," AIME T.P. 1094-G, presented Houston, Texas, Oct. 1958.
28. "Effects of Drilling Fluid on Penetration of Rock Bits," Excerpts from Battelle Memorial Institute Report to the American Association of Oil Well Drilling Contractors, in *The Petroleum Engineer*, Jan. 1956, p. B-85.
29. Mallory, H. E., "How Low Solid Muds Can Cut Drilling Costs," *The Petroleum Engineer*, Apr. 1957, p. B-21.

30. Eckel, J. R., "Effect of Mud Properties on Drilling Rate," *API Drilling and Production Practices,* 1954, p. 119.

31. Cunningham, R. A., and W. C. Goins, Jr., "How Mud Properties Affect Drilling Rate," *The Petroleum Engineer,* May 1957, p. B–119.

32. Perkins, H. W., "A Report on Oil-Emulsion Drilling Fluids," *API Drilling and Production Practices,* 1951, p. 349.

33. Lummus, J. L., Barrett, H. M., and H. Allen, "The Effects of Use of Oil in Drilling Muds," *API Drilling and Production Practices,* 1953, p. 135.

34. Nolley, J. P., Cannon, G. E., and D. Ragland, "The Relation of Nozzle Fluid Velocity to Rate of Penetration with Drag-Type Rotary Bits," *API Drilling and Production Practices,* 1948, p. 22.

35. Eckel, J. R., and J. P. Nolley, "An Analysis of Hydraulic Factors Affecting the Rate of Penetration," *API Drilling and Production Practices,* 1949, p. 9.

36. Eckel, J. T., and W. J. Bielstein, "Nozzle Design and its Effect on Drilling Rate and Pump Operation," *API Drilling and Production Practices,* 1951, p. 28.

37. Hellums, E. C., "The Effect of Pump Horsepower on the Rate of Penetration," *API Drilling and Production Practices,* 1952, p. 83.

38. Thompson, G. D., "A Practical Application of Fluid Hydraulics to Drilling in California," *API Drilling and Production Practices,* 1953, p. 123.

39. Bromell, R. J., "Bit Hydraulics for Hard Rock Drilling," API Paper No. 906-1-J, Presented Fort Worth, Mar. 1956.

40. Keating, T. W., Clift, W. D., and J. Cutrer, "Report on Hydraulics of Rotary Drilling," API Paper No. 926-1-F, Presented San Antonio, Texas, Mar., 1956.

41. Bobo, R. A., and R. S. Hoch, "Keys to Successful Competitive Drilling," Part 5C, *World Oil,* Nov. 1957, p. 112.

42. Colebrook, R. W., "Downhole Power Speeds Drilling," *Oil and Gas Journal,* Nov. 17, 1958, p. 172.

43. Moore, P. L., "5 Factors That Affect Drilling Rate," *Oil and Gas Journal,* Oct. 6, 1958, p. 141.

44. LeVelle, J. A., "An Engineer's Look at Turbine Drilling," *The Petroleum Engineer,* Oct. 1956, p. B–39.

45. O'Connor, J. B., "Turbodrills for American Drilling," *Monitor,* Nov. 1956, p. 2.

46. Parsons, C., "Drilling by the Turbine Method," *API Drilling and Production Practices,* 1950, p. 38.

47. Thacher, J. H., and W. R. Postlewaite, "Turbodrill Development Past and Present," *World Oil,* Dec. 1956, p. 131.

48. Rosu, G. G., "Turbine Drilling Systems and Its Application in USSR," *World Petroleum,* Mar. 1955 and April 1955, p. 84 and p. 38, respectively.

49. Trebin, F. A. "Turbine Drilling in USSR," *Petroleum Time,* Oct. 1955, p. 1088.

50. "New Percussion Drill Shows Great Promise," *The Petroleum Engineer,* July, 1957, p. B–32.

51. Topanelian, E., Jr., "The Application of Low Frequency Percussion to Hard Rock Drilling," *Journal of Petroleum Technology,* July, 1958 , p. 55.

52. Eckel, J. E., Deily, F. H., and L. W. Ledgerwood, Jr., "Development and Testing of Jet Pump Pellet Impact Bits," *Trans. AIME,* Vol. 207, (1956), p. 1.

53. Camp, J. M., Ortloff, J. E., and R. H. Blood, "Wireline Retractable Rock Bits," *World Oil,* Oct. 1957, p. 190.

54. Theriot, W. A., "Simultaneous Drilling," *Journal of Petroleum Technology,* Apr. 1958, p. 13.

SUPPLEMENTARY READINGS

Those desiring to study more thoroughly rock penetration by various drilling devices will find the following references quite helpful. These will also serve to introduce the petroleum engineer to the considerable mining engineering literature on these subjects.

1. Hartman, H. L., "Fundamental Studies of Percussion Drilling," *Mining Engineering,* Jan. 1959, p. 68.

2. Goodrich, R. H., "High Pressure Rotary Drilling Machines," Missouri School of Mines Symposium, Bull. 94, 1956, p. 25.

3. Fairhurst, C., "The Design of Rotary Drilling Bits," *Mine and Quarry Engineering,* June 1954, p. 271.

4. Lacabanne, W. D., and E. P. Pfleider, "Research in Rotary Percussion Drilling," *Mining Engineering,* July 1957, p. 766.

Chapter 9 . . .

Rotary Drilling Techniques

Rotary drilling includes many separate techniques or practices which have been developed through field experience and/or analytical appraisals. In this chapter we will discuss three such operations, namely:

(1) Control of hole deviation in essentially vertical wells: vertical drilling

(2) Control of hole deviation in wells which are intentionally aimed at horizontally displaced bottom hole targets: directional drilling

(3) Retrieving undesirable objects from the hole; in particular, portions of the drill string and/or bit: fishing

As we shall see, the first two items are variations of the same thing. Fishing is essentially a separate topic.

9.1 Vertical Drilling

In oil well drilling there is no such thing as a truly vertical hole; however, wells which aim at a target directly below their surface location are considered to be vertical wells. That is to say, their deviation from vertical is held to small angles.* The compass direction of deviation is of secondary importance (and generally is not even measured); principal consideration is given to the angle between the hole and the vertical.

Hole crookedness was considered a serious disadvantage to the early use of rotary tools. Contrary to much popular belief, cable tool holes may also be crooked.[1,2] Geologists, who rely on depth measurements for subsurface mapping, found contour mapping almost impossible in some instances. The plan view of the 14 wells shown in Figure 9.1 illustrates such problems. Depths obtained from drill pipe measurements in these wells would hardly allow accurate subsurface mapping. Wells have also been known to run into each other during drilling. In the early days at Seminole, Oklahoma, two wells 660 ft apart at the surface ran together at 1900 ft. Two California wells 2000 ft apart at the surface ran together at 6115 ft.[3] These and other similar occurrences spurred the development of down hole surveying instruments which could measure hole deviation from the vertical. One of the earliest of these was the acid bottle, in which hydrofluoric acid etched a horizontal line on the inside of a partially filled glass bottle. Other instruments were designed around the plumb-bob or pendulum principle. One of these later types and its operation is shown in Figure 9.2. These instruments measure only the vertical deviation and not its compass direction. Similar instruments which incorporate compass readings are also available and will be mentioned under directional drilling.

The extreme crookedness of early wells caused the industry to become quite straight hole conscious. As soon as reliable surveying instruments became available, severe restrictions were imposed on hole deviation by the producing companies. Drilling contracts commonly specified 3 to 5° as the maximum acceptable deviations in vertical holes. As a result, lighter bit weights had to be used, and penetration rate was consequently reduced. Drilling personnel then began to look for other means of minimizing crooked hole problems. The use of longer sections of drill collars, which furnished all of the bit weight, helped to a great extent. Various types of stabilizers were used with indifferent success, probably because no one knew for sure where to place them. Other approaches such as bit alterations and numerous changes in operating techniques were tried. Despite the empirical knowledge gained from such experimenting,

*In common field language, such holes are called *straight*, which is something of a misnomer.

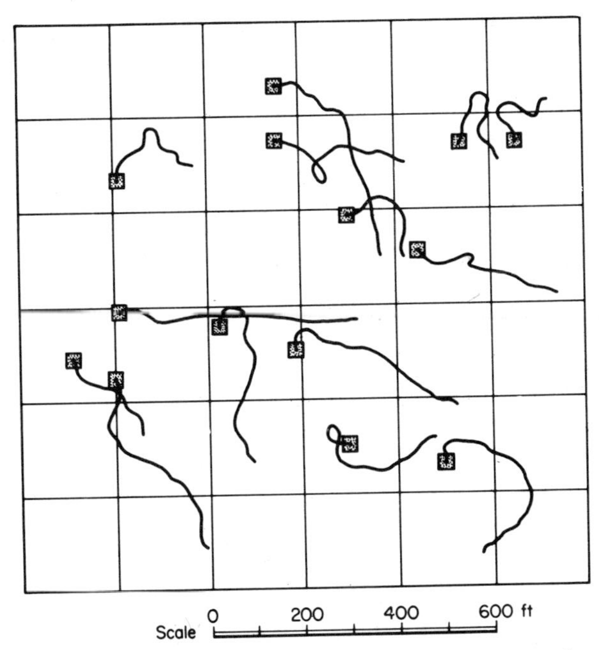

Fig. 9.1. Plan view of 14 *vertical* wells drilled to 6000 ft. After Suman,[3] courtesy AIME.

no approach was widely successful and no complete agreement existed as to the basic causes of hole deviation.

It was not until 1950 when Arthur Lubinski published his analytical treatment of drill string buckling[4] that a sound basis for solving hole deviation problems became available. This work, however, as well as a subsequent one,[5] was based upon a simplifying assumption of perfect hole verticality. This assumption was quickly removed[6–8] and has resulted in charts and tables[9] for universal use. Field experience has proved the theoretical findings and standard operating procedures have been based on them.

Fundamental Principles

The mathematical treatment of hole deviation is quite long and involved and will not be presented here. There are, however, certain basic concepts which must be understood. Consider Figure 9.3, which shows the lower portion of the drill string in a *straight** but inclined hole whose angle of inclination with respect to the vertical is α. It is assumed that the drill string lies on the low side of the hole and contacts the wall at the point of tangency T. The force with which the bit acts on the formation (frictional and rotational effects are ignored) is F_B, applied at an angle ϕ with the vertical. The force F_B may be resolved into two components, namely the longitudinal force F_1 in the direction of the axis of the hole, and the lateral force F_2 perpendicular to the axis of the hole. F_2 may either act on the low side of the hole [Figure 9.3(A)], or be nil [Figure 9.3(B)], or act on the high side of the hole [Figure 9.3(C)]. Whenever F_2 acts on the low side of the hole [Figure 9.3(A)], the hole deviation with respect to vertical will decrease. Conversely, whenever F_2 acts on the high side of the hole [Figure 9.3(C)], the hole deviation will increase. Finally, if F_2 is nil [Figure 9.3(B)], i.e., if ϕ and α are equal, then a stable condition occurs, and drilling will proceed in the prolongation of the axis of the hole, which means that hole deviation α will be maintained. (Actually, as will be explained further in this section, the above statements are valid for isotropic formations only.) In the event that the hole deviation is decreasing [Figure 9.3(A)], the lateral force F_2 will become smaller and smaller until it is nil. Thereafter, a stable condition is reached for a smaller value of α. Similarly, if the hole deviation is increasing, a stable condition will be reached for some larger value of α.

*The word *straight* is used here in its geometrical meaning and not its common oil-field meaning, *nearly vertical*.

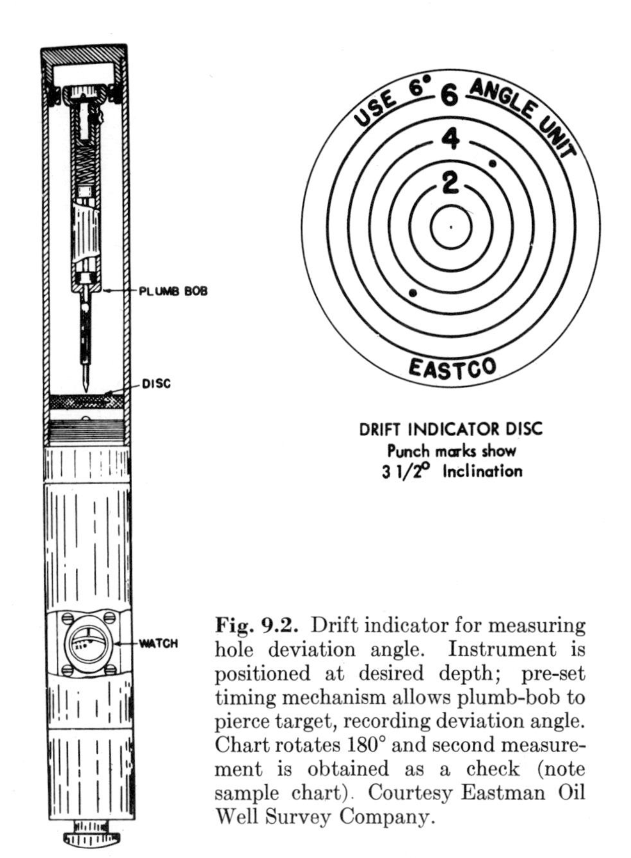

Fig. 9.2. Drift indicator for measuring hole deviation angle. Instrument is positioned at desired depth; pre-set timing mechanism allows plumb-bob to pierce target, recording deviation angle. Chart rotates 180° and second measurement is obtained as a check (note sample chart). Courtesy Eastman Oil Well Survey Company.

Let F_p denote the buoyant weight of the section of drill collars below the point of tangency. F_p is applied at the center of gravity of that section. In the case of Figure 9.3(A), an increase of F_p results in an increase of F_2. In the case of Figure 9.3(C), an increase of F_p results in a decrease of F_2, which may even become negative, i.e., the case of Figure 9.3(C) may become the case of Figure 9.3(A). From the above, it is clear that an increase of F_p results in a smaller equilibrium angle α.

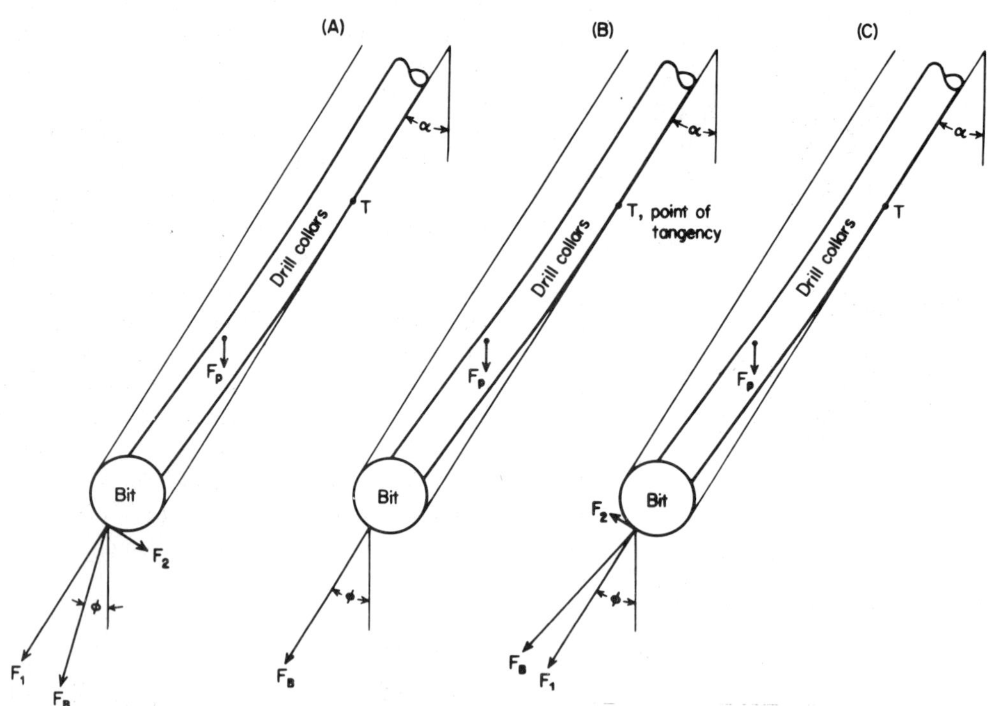

Fig. 9.3. Idealized sketch of forces affecting hole deviation angle.

Therefore, F_p has a beneficial effect which is often called the *pendulum effect.*

In isotropic formations, the value of the equilibrium angle α is dependent on three variables:

(1) Weight on bit
(2) Drill collar size
(3) Hole size

Let us qualitatively consider the separate effects of these factors.

Weight on Bit

An increase in the weight on bit increases the bending of the unsupported portion of the collar string above the bit, which moves T closer to the bit and decreases the weight F_p. Therefore, it is apparent that increased weight on bit results in increased hole deviation.

Drill Collar and Hole Size

These two factors are interrelated through their mutual effect on clearance, which is the difference between hole and drill collar diameters. First, consider the effect of drill collar size with a constant clearance.

For the same weight on bit, large drill collars, being stiffer, are less subject to bending. In other words, for large collars, the point T is located higher. Therefore, the length of the portion of the string below T is greater. Both the fact that this length is greater and the fact that the weight per unit length is greater result in a greater F_p, thereby reducing the equilibrium angle.

The effect of hole size will be considered with drill collar size held constant. This is equivalent to considering hole-drill collar clearance. A larger clearance requires a larger lateral deflection before the hole wall is contacted. Hence, the point T moves up the hole as bit size is increased. This results in a greater force F_p, which should reduce the equilibrium angle, α. However, another factor acting in the opposite direction must be considered. A large deflection results in a greater angle between the axis of the bit and the vertical. This, in turn, results in a greater angle ϕ between the force F_B and the vertical; this has a tendency to increase the equilibrium angle, α. For all except small clearances, the second factor dominates and hole deviation is

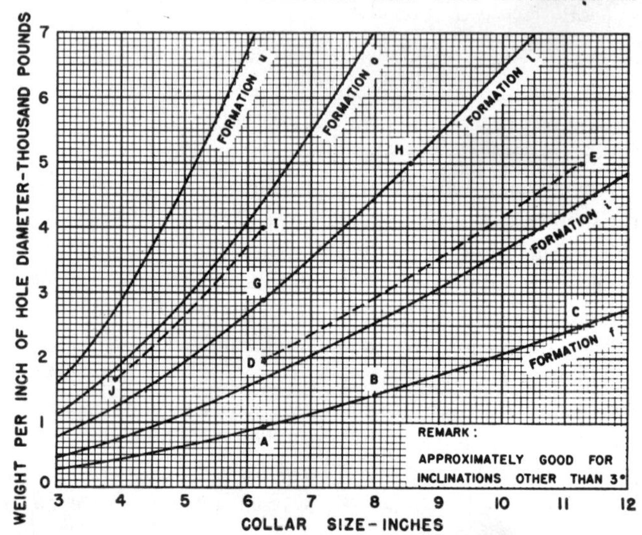

Fig. 9.4. Effect of collar size on allowable bit weight for 3° hole deviation and 1-in. clearance. After Woods and Lubinski,[7] courtesy API.

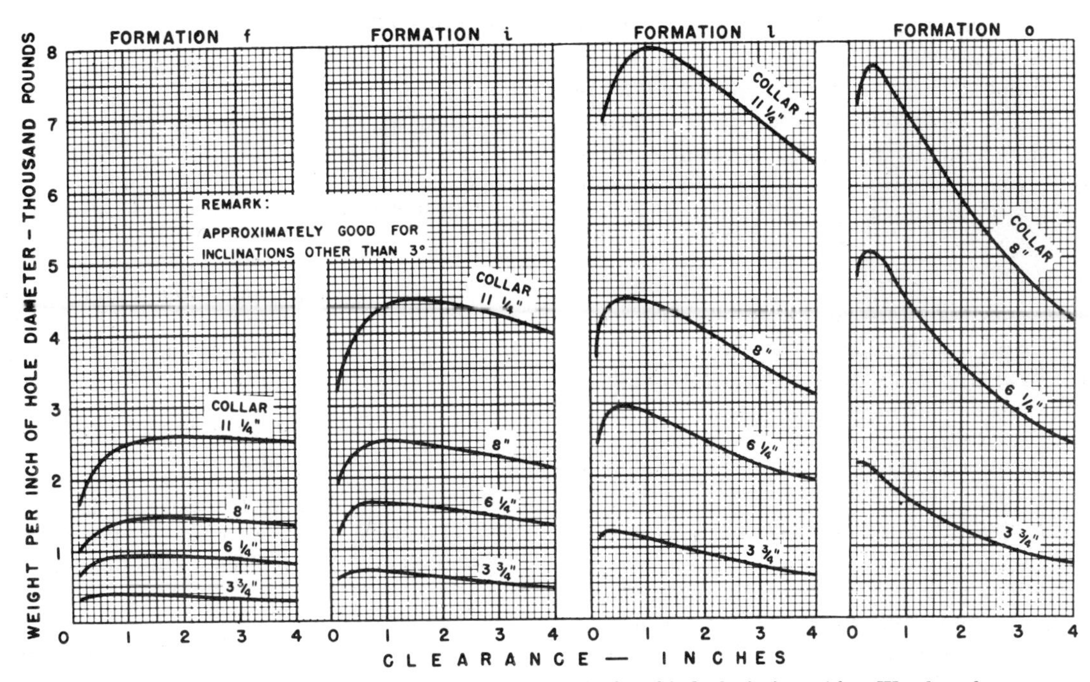

Fig. 9.5. Effect of clearance on allowable bit weight for 3° hole deviation. After Woods and Lubinski,[7] courtesy API.

increased. The effect of collar size and hole clearance is demonstrated in Figures 9.4 and 9.5.

Thus far we have considered only isotropic formations, that is, formations having identical properties in all directions. In other words, we have not considered the possibility of formation characteristics influencing the magnitude of hole deviation. In general, the bit tends to drill updip, which implies that formations are drilled more easily perpendicular to the bedding planes than parallel to them. As a result, the direction of drilling is no longer that of the force F_B with which the bit acts on the formation. In an anisotropic formation, the direction of drilling under equilibrium conditions is more inclined with respect to the vertical than the direction of F_B. Figure 9.6 serves to clarify this concept. Therefore, some force F_2, directed as in Figure 9.3(A), exists for equilibrium conditions.

It has been explained above that, for isotropic formations, the equilibrium angle α is dependent on the weight on bit, hole size, and drill collar size. For anisotropic formations, α depends also on additional factors. These were resolved by Lubinski and Woods[6] in terms of dip and anisotropic index. The greater the dip and/or the anisotropic index, the more *crooked* is the formation. The letters *f*, *i*, *l*, *o*, and *u* in Figures 9.4, 9.5, and 9.7 denote the following variations of formation crookedness, which may be due to various combinations of dips and anisotropic indices.

The foregoing discussions have been somewhat simplified. Those desiring a more rigorous and complete treatment should study the original references.

Crookedness of formation	*Designation of formation*	*Weight (approx. lb) to maintain 3° with 6¼-in. collars and 1-in. clearance*	*Weight (approx. lb) per in. of hole diameter*
Very Severe	*f*	6,500	900
Severe	*i*	12,000	1,600
Moderate	*l*	20,000	3,000
Mild	*o*	30,000	4,000
Very Mild	*u*	50,000	7,000

Problems in Hole Deviation

Figure 9.7 allows the rapid solution to practical hole deviation problems in terms of the five variables previously mentioned. This chart is based on a mud density of 10 lb/gal and a drill collar with inside-to-outside diameter ratio of 0.375. Corrections for other mud weights may be ignored; the same holds true for other diameter ratios, except for the very smallest collars (errors may be appreciable for 4 in. OD drill collars with a large bore). Figures 9.8(A) and (B) show or indicate the solution to the following example problems. These have been taken directly from the original Woods and Lubinski paper.[7]

Problems similar to the following may be solved by Figure 9.7. In a given formation 10° hole inclination is maintained by carrying 4,000 lb with 5-in. drill collars in a 9-in. hole. Formation dip is 45°. What weight may be carried with 11-in. drill collars in a 12-in. hole, if formation dip does not change and the same angle of 10° is maintained?

This and any other problem solvable by Figure 9.7 must contain the following elements:

1. Established data: These are the numerical values of drill collar OD, weight on bit, hole size, and formation dip that resulted in a given hole inclination. They are obtained from past drilling experience in the formation under consideration, and establish its degree of crookedness.
2. Problem data: These concern the hole to be drilled. They are the numerical values for all but one of the quantities listed under established data. At least one of the quantities is different in the problem data than in the established data.
3. Unknown: The numerical value of the quantity not given in the problem data.

Problems Involving Neither Change in Hole Inclination Nor Formation Dip

If there is neither change in formation dip nor in hole inclination between the established data and the problem data, the problem may be solved without the actual knowledge of dip and the two righthand sections of the chart are not needed.

For clarity, the data of this and the subsequent problems will be tabulated as shown below.

Example 9.1

	Established data	*Problem data*
Collar OD, in.	5	11
Weight, lb	4,000	?
Hole size, in.	9	12
Hole inclination, °	10	10
Formation dip, °	Same in both data	

Solution:

Using the established data, locate points and draw lines as follows [see Figure 9.8(A)]: Line M_1N_1 (5-in. collars); point A_1 (4,000 lb); point B_1 (9-in. hole); point C_1 (10° hole inclination). Complete rectangle $A_1B_1C_1D_1$.

Using problem data, proceed as follows: M_2N_2 (11-in. collars); B_2 (12-in. hole); C_2 (10° hole inclination); D_2 (point located on the same curve as D_1). Complete rectangle $A_2B_2C_2D_2$.

Read unknown weight at A_2: 22,000 lb.

Example 9.2

	Established data	*Problem data*
Collar OD, in.	5	11
Weight, lb	4,000	22,000
Hole size, in.	9	?
Hole inclination, °	10	10
Formation dip, °	Same in both data	

Solution:

A construction of two rectangles similar to that in Example 9.1 may be used. Read unknown hole size at B_2: 12 in.

Example 9.3

	Established data	*Problem data*
Collar OD, in.	5	?
Weight, lb	4,000	22,000
Hole size, in.	9	12
Hole inclination, °	10	10
Formation dip, °	Same in both data	

Solution:

This problem cannot be directly solved. It is necessary to try a few collar sizes and proceed for each of them as in Example 9.1, until a collar size is found that satisfies the other problem data.

Problems Involving a Change in Either Hole Inclination or Formation Dip

When the problem involves a change in either hole inclination or formation dip, we must use formation dip in both established data and problem data. Consider the following:

Example 9.4

	Established data	*Problem data*
Collar OD, in.	7	11
Weight, lb	13,500	?
Hole size, in.	9	12
Hole inclination, °	5	10
Formation dip, °	45	30

The solution is indicated in Figure 9.8(B). The first part of the problem is the same as before: viz., from the established

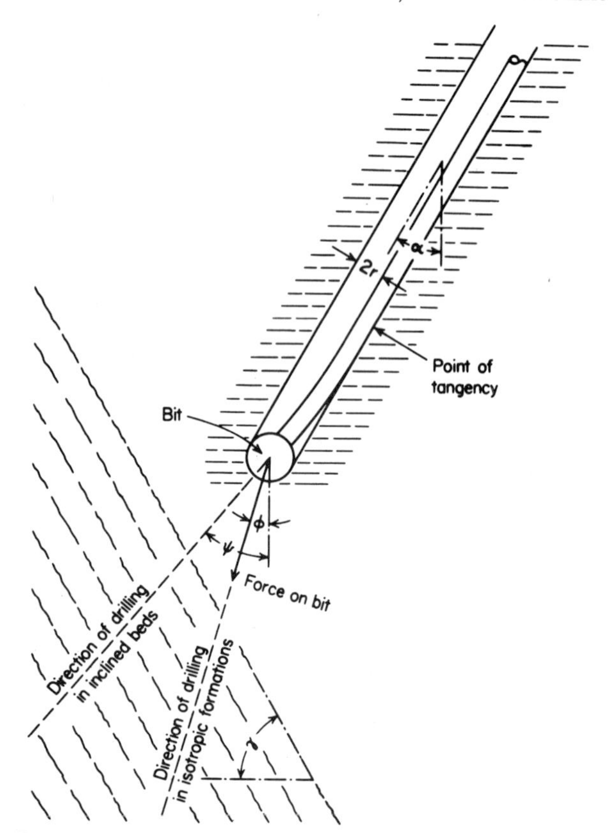

Fig. 9.6. The effect of formation characteristics on hole deviation. γ is the dip angle. After Lubinski and Woods,[6] courtesy API.

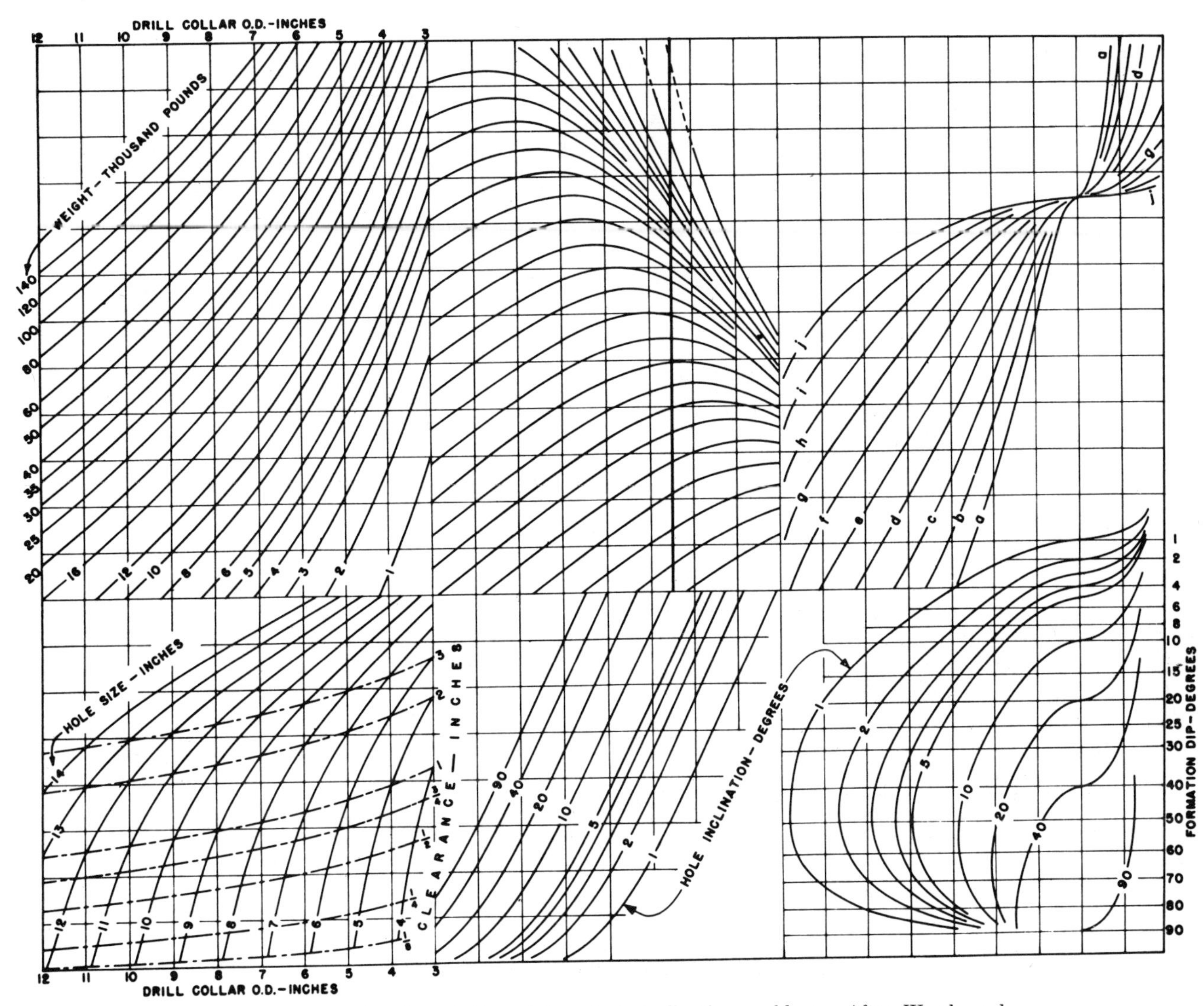

Fig. 9.7. Chart for the solution of bore hole inclination problems. After Woods and Lubinski,[7] courtesy API.

data, exclusive of formation dip, construct the rectangle $A_1B_1C_1D_1$ as in Example 9.1. In Example 9.1, for which hole inclination and formation dip were constant, D_2 was located on the same curve as D_1. On the other hand, if either hole inclination or formation dip change, D_2 is not necessarily located on the same curve as D_1. Therefore, we must establish the curve on which D_2 is located. Using established data for formation dip and hole inclination, locate points and draw lines as follows [see Figure 9.8(B)]: E_1 (intersection of the curve on which D_1 is located with the vertical reference line); F_1 (45° dip and 5° hole inclination); G_1 (intersection of lines through E_1 and F_1).

Using problem data, proceed as follows: F_2 (30° dip and 10° hole inclination); G_2 (located on the same curve as G_1); E_2 (located on the reference line); M_2N_2 (11-in. drill collars); B_2 (12-in. hole); C_2 (10° hole inclination); D_2 (point located on the same curve as E_2). Complete rectangle $A_2B_2C_2D_2$. Read unknown weight at A_2: 90,000 lb.

Example 9.5

	Established data	*Problem data*
Collar OD, in.	7	11
Weight, lb	13,500	90,000
Hole size, in.	9	?
Hole inclination, °	5	10
Formation dip, °	45	30

Solution:

This problem may be solved by a construction similar to the one used in Example 9.4. Read unknown hole size at B_2: 12-in.

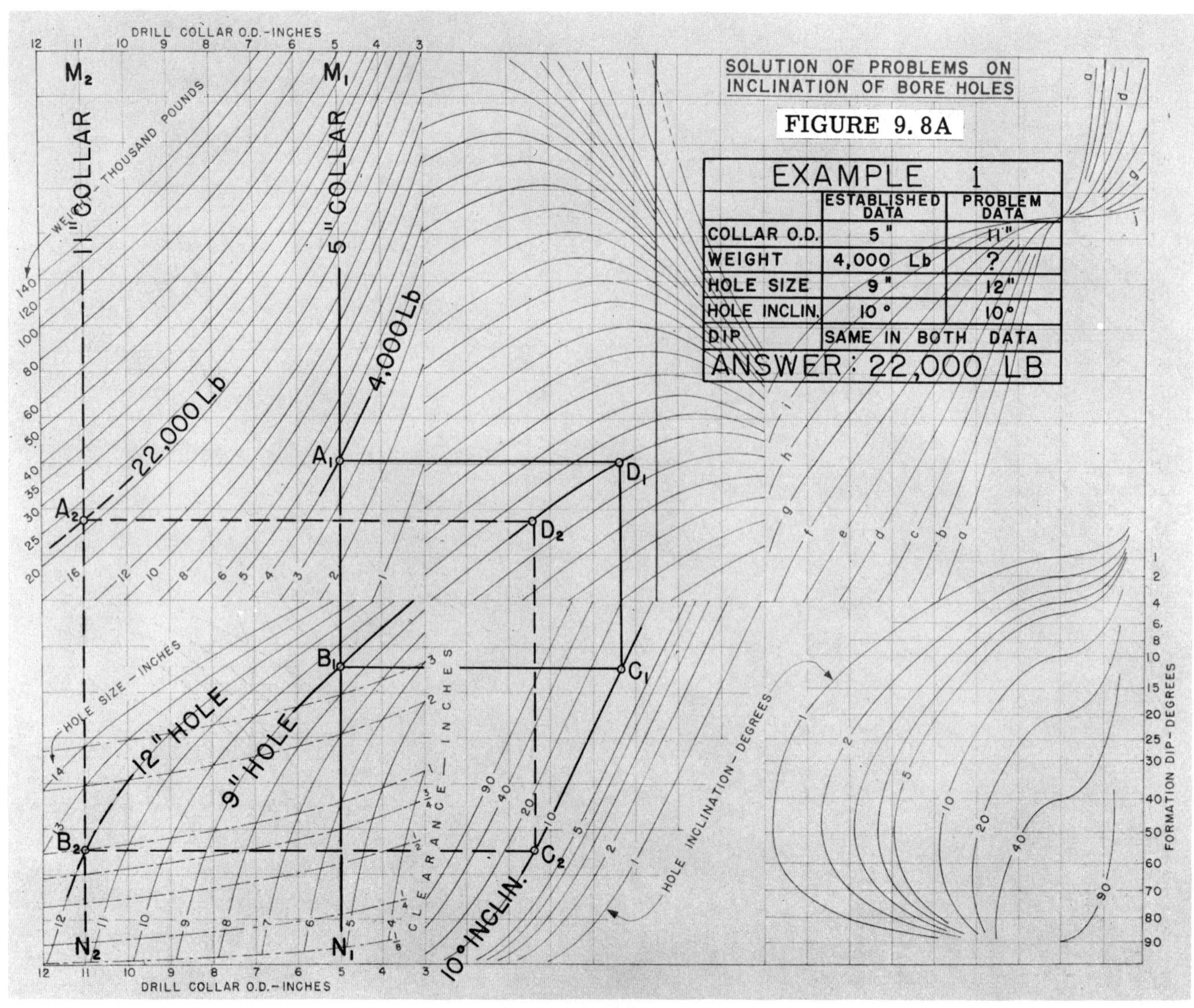

Fig. 9.8(A). Sample solution for example problems.

Example 9.6

	Established data	*Problem data*
Collar OD, in.	7	?
Weight, lb	13,500	90,000
Hole size, in.	9	12
Hole inclination, °	5	10
Formation dip, °	45	30

Solution:

This problem cannot be directly solved. It is necessary to try a few collar sizes and proceed for each of them as in Example 9.4.

Example 9.7

	Established data	*Problem data*
Collar OD, in.	7	11
Weight, lb	13,500	90,000
Hole size, in.	9	12
Hole inclination, °	5	?
Formation dip, °	45	30

Solution:

This problem cannot be directly solved. It is necessary to try a few hole inclinations and proceed for each of them as in Example 9.4.

Example 9.8

	Established data	*Problem data*
Collar OD, in.	7	11
Weight, lb	13,500	90,000
Hole size, in.	9	12
Hole inclination, °	5	10
Formation dip, °	45	?

Solution:

Using established data, construct rectangle $A_1B_1C_1D_1$ as in Example 9.1 or 9.4. Similarly, using problem data, construct rectangle $A_2B_2C_2D_2$, then locate points and draw lines as follows: Points E_1 and E_2 (located on the reference line and on the same curves as D_1 and D_2, respectively); F_1 (45° and 5° hole inclination): G_1 (intersection of lines through E_1 and F_1);

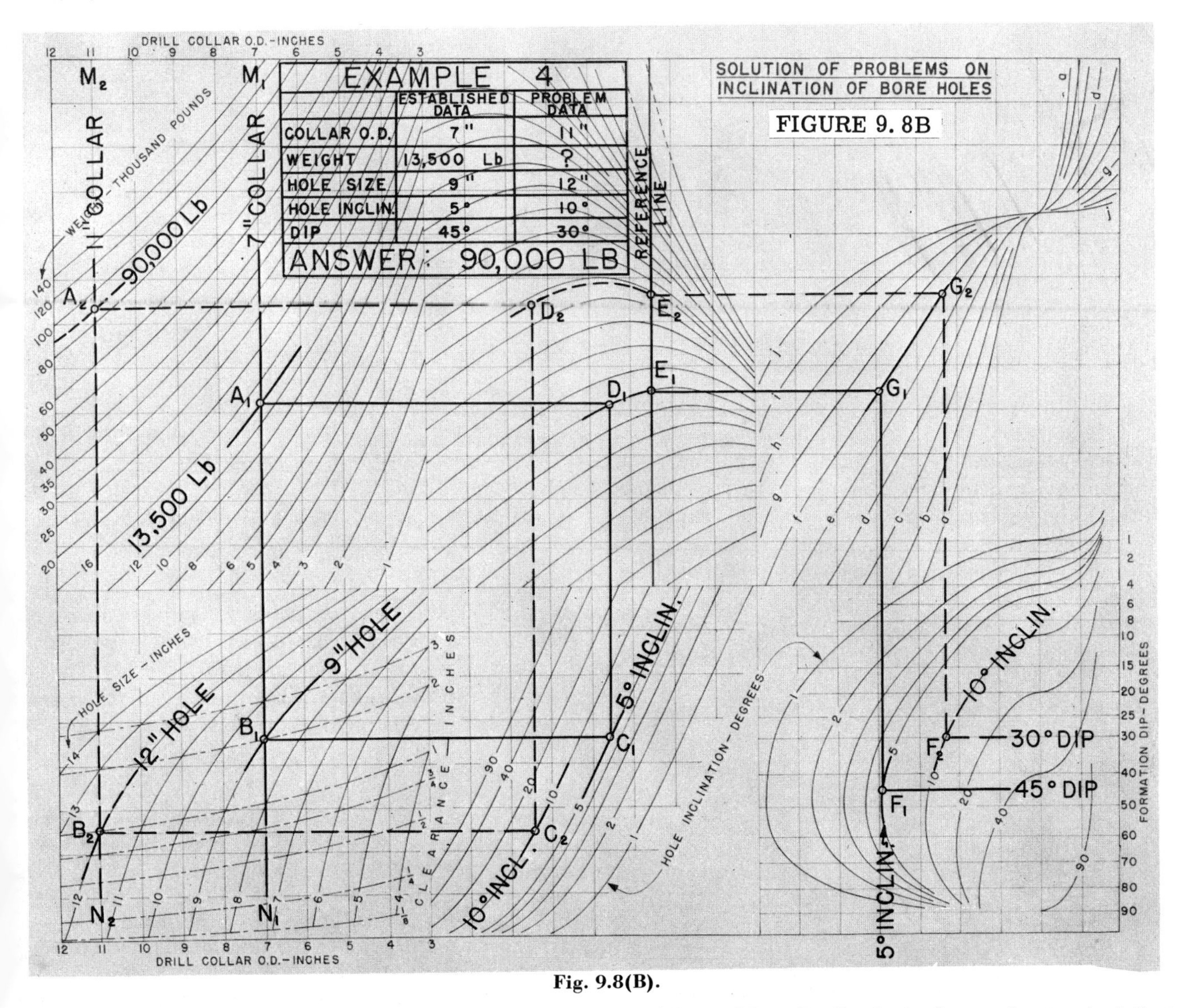

Fig. 9.8(B).

G_2 (located on the same curve as G_1 and a line through E_2); F_2 (10° hole inclination).

Read unknown dip at F_2: 30°.

The individual effect of changing one of the variables may be studied by solving a series of problems similar to the examples just shown. Figures 9.4 and 9.5 were prepared in this manner. Numerous papers have pointed out other specific applications of these principles to actual field problems.[10–12]

The Use of Stabilizers

Stabilizers are used as a means of controlling the location of the contact point between the hole and drill collars. The potential benefit of such action may be visualized from Figure 9.9. Part (A) shows the normal position of the drill string in an inclined hole, as previously described. If a stabilizer is used as indicated in parts (B) and (C), the effective contact point is moved up the hole; this increases the hole-straightening force imposed by the pendulum weight. Whether the optimum stabilizer location is as shown in 9.9 (B) or (C) depends on drill collar and hole size, hole inclination, and the weight on bit.[8] Qualitatively, it may be surmised that for a given hole deviation, the use of a single, properly located stabilizer will allow a higher weight on bit.

Multiple Stabilizers

The effect of using several closely spaced stabilizers near the bit has also been studied. In practice, this effect has also been attained by welding continuous steel strips outside the drill collars. This procedure may be analyzed as though the diameter were decreased by the same amount. Figure 9.5 shows that very small clearances reduce the allowed bit loading for a given equilibrium deviation angle. Therefore continuous stabilization (very low clearance) will result in a large equilibrium hole inclination; the rate of angle buildup is, however, quite slow. Consequently, this technique may be used to:

(1) prevent sudden changes in deviation (dog-legs)
(2) drill relatively short sections of extremely crooked hole formations; this means that although the

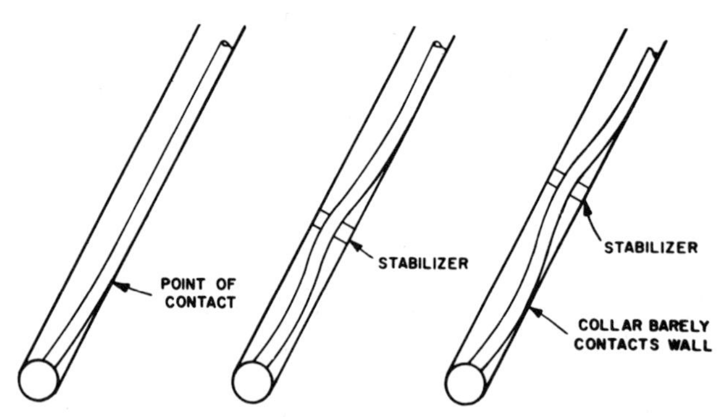

Fig. 9.9. The use of a stabilizer in reducing deviation angle. After Woods and Lubinski,[8] courtesy API.

equilibrium angle is very large, the rate of buildup is so slow that the section may be drilled before it becomes excessive. In fact, this technique may result in a smaller deviation than other methods, with the further advantage of higher allowable bit weight.

(3) drill when building up angle to some desired or maximum value.

These statements pertain to stabilizers at the extreme lower end of the string, with the lowest being just above the bit. Multiple stabilizers have little or no effect if the lowermost is at the ideal position for a single stabilizer.

Hole Deviation Problems Using Stabilizers

The charts of Figures 9.10 to 9.17 may be used to determine proper stabilizer location and the resulting bit weight increase for commonly encountered sizes. Note that the additional bit weight is expressed as a percentage increase; such an expression requires a knowledge of the allowable bit load without the stabilizer. This is obtained either directly from previous drilling experience in the field, or by use of Figure 9.7,

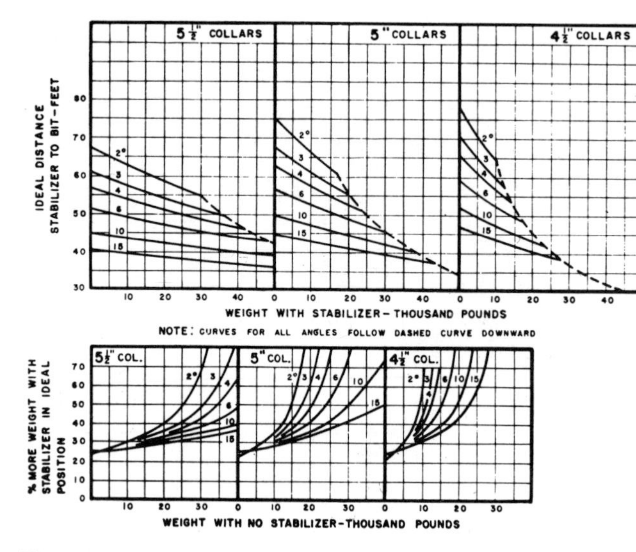

Fig. 9.10. Use of stabilizer. After Woods and Lubinski,[8] courtesy API.
Hole = $6\frac{1}{8}$ in.

if such experience was obtained with some other combination of collar size, hole size, etc. Similar charts for other sizes may be constructed from dimensionless charts presented in the original work.[8] Use of Figures 9.10 to 9.17 is illustrated by the following examples.

Example 9.9

The following data refer to an extremely severe crooked hole formation. Hole size is $8\frac{3}{4}$ in. The maximum allowable hole deviation of 3° is reached with a weight on bit of only 3,700 lb. Collar size = 6 in., W = 3,700 lb, and α = 3°.

(a) What is the allowable bit weight, with stabilizer?

(b) Determine the position of an ideally located stabilizer for these conditions.

(c) Repeat the calculation for 7-in. and 8-in. drill collars with and without ideally located stabilizers.

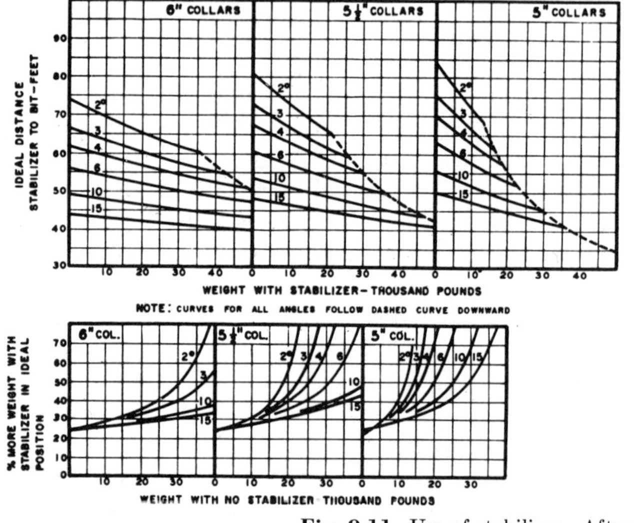

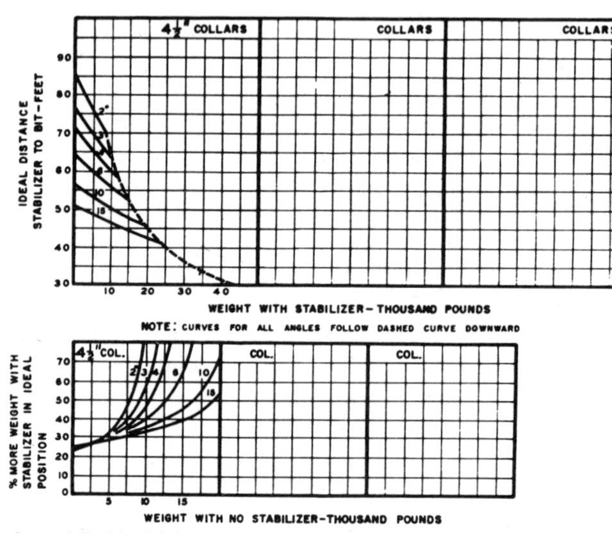

Fig. 9.11. Use of stabilizer. After Woods and Lubinski,[8] courtesy API.
Hole = $6\frac{3}{4}$ in.

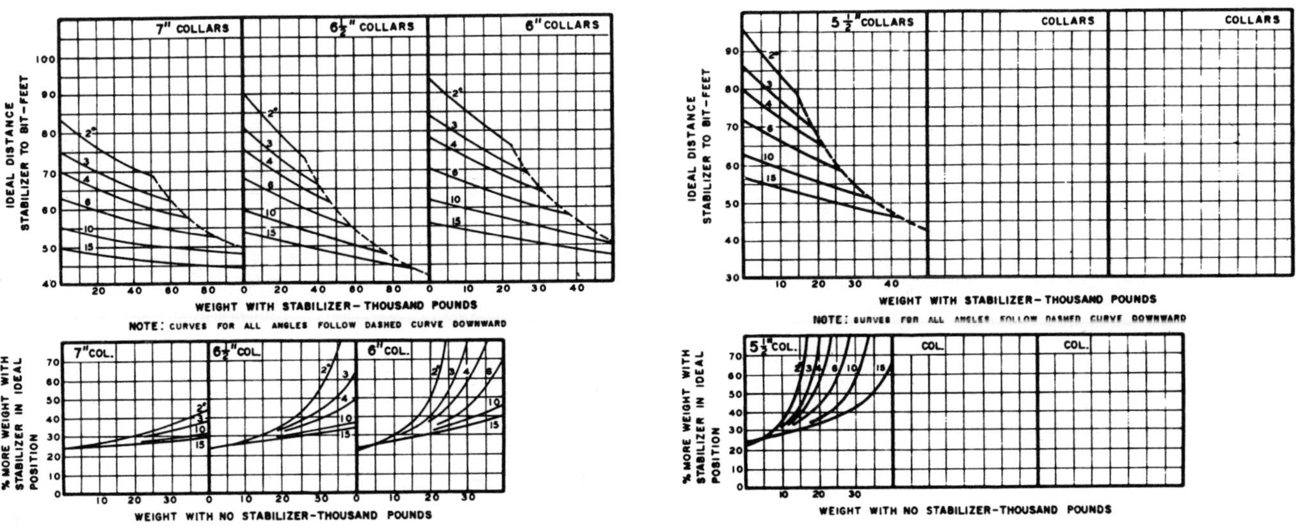

Fig. 9.12. Use of stabilizer. After Woods and Lubinski,[8] courtesy API. Hole = $7\frac{7}{8}$ in.

Solution:

		Weight for 3° deviation, lb	
Collar size, in.	*Ideal stabilizer location from bit, ft*	*Without stabilizer*	*With stabilizer*
6	90	3,700	4,600†
7	87	5,000*	6,300†
8	76	5,900*	7,400†

Example 9.10

Make the same comparison as in Example 9.9 for a more moderate crooked hole formation. The hole deviation of 3° is obtained with 19,000 lb on bit using 6-in. collars with no stabilizer.

Solution:

		Weight for 3° deviation, lb	
Collar size, in.	*Ideal stabilizer location from bit, ft*	*Without stabilizer*	*With stabilizer*
6	67	19,000	27,500†
7	74	30,000*	41,800†
8	68	42,500*	56,500†

From these and other similar examples, Woods and Lubinski make the following general conclusions concerning the use of stabilizers:

(1) In very severe crooked hole formations, allowable bit weights for maintaining a specific deviation are increased approximately 25% for both packed holes and conventional clearance.
Note: Packed holes are those cases where the diametral clearance between collars and hole is 1 in. or less. Conventional clearance is 2 in. or more.

(2) In very mild crooked hole formations (where heavy weights may be carried) the percentage improvement varies considerably with clearance.
(a) For most packed holes: 30 to 40% weight increase
(b) For conventional clearances: 40 to 80%

(3) About the same weight may be carried by running either
(a) oversized collars and no stabilizer
(b) a stabilizer and the largest collar size which can be washed over. ("Washed over" refers to a fishing operation utilizing a sleeve-type retrieving tool which fits around the drill collars, and hence requires a certain clearance between the collars and the hole.)

In actual drilling, it would be very impractical to keep moving the stabilizer as the ideal position changes. For this reason, a further theoretical study was conducted to determine ranges within which the stabilizer could be placed without losing the major portion of the benefit which would be obtained if the stabilizer were at the ideal position. Combining the results of this study with another one pertaining to the effect of hole stabilizer clearance (Figures 9.10 to 9.17 are for zero clearance), it was shown that the following recommendation could be made: place the stabilizer between the ideal position, according to Figures 9.10 to 9.17, and a position 10% closer to the bit. When, however, there is very light weight on bit in nonpacked holes, the stabilizer should be placed no more than 5% closer to the bit than the ideal position.

Example 9.11

Correct the ideal stabilizer locations in Examples 9.9 and 9.10 to practical positions.

Solution:

(a) Example 9.9 dealt with a light weight condition; therefore, for the 6-in. and 7-in. collars the stabilizer should be

*obtained from Figure 9.7.
†obtained from Figure 9.13.

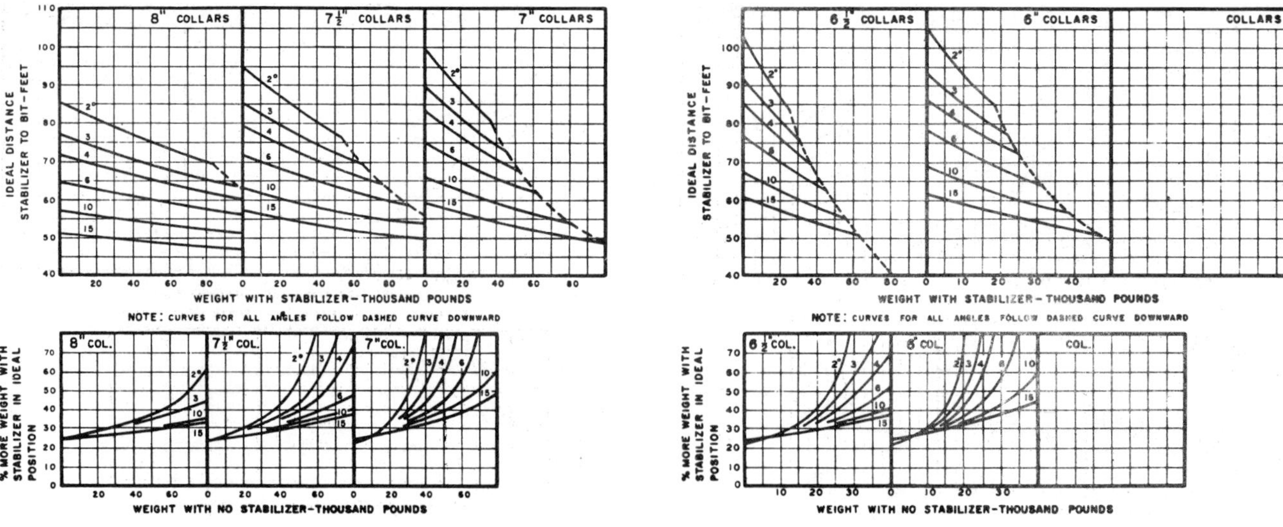

Fig. 9.13. Use of stabilizer. After Woods and Lubinski,[8] courtesy API. Hole = $8\frac{3}{4}$ in.

placed between the locations indicated in Example 9.9 and locations which are 5% closer to the bit. For 8-in. collars, a packed hole condition permits a less stringent stabilizer location, and the above value of 5% becomes 10%.

(b) Example 9.10 included moderate weights on bit; locations between the ideal one and 10% below are permissible.

	Example 9.9		*Example 9.10*	
Collar	*Stabilizer location, ft*		*Stabilizer location, ft*	
size, in.	*Ideal*	*Practical*	*Ideal*	*Practical*
6	90	85 to 90	67	60 to 67
7	87	83 to 87	74	67 to 74
8	76	69 to 76	68	61 to 68

The Hazards of Dog-Legs

Thus far we have considered only the magnitude of the deviation angle and have not been concerned with the abruptness of its change. This latter factor is, however, more important from the standpoint of operational problems than is the angle itself. An abrupt change in hole deviation (vertical and/or horizontal direction angles) results in a troublesome situation which is commonly referred to as a *dog-leg*.

Consideration of this problem introduces the difficulty of defining a crooked hole. The old practice of specifying a maximum allowable deviation from the vertical is inadequate, since a severe dog-leg can occur while staying well within the contract limit. Further, a dog-leg can result from a change in horizontal angle (compass direction) with little or no change in vertical angle. Since it is uncommon to measure horizontal direction in so-called vertical drilling, the detection of such a hole condition is generally postponed until trouble occurs. When drilling below a dog-leg, the drill pipe presses against the shoulder with an increasingly large force which results in accelerated tool joint and pipe wear. This often results in the situation known as key-seating, in which the drill pipe wears a

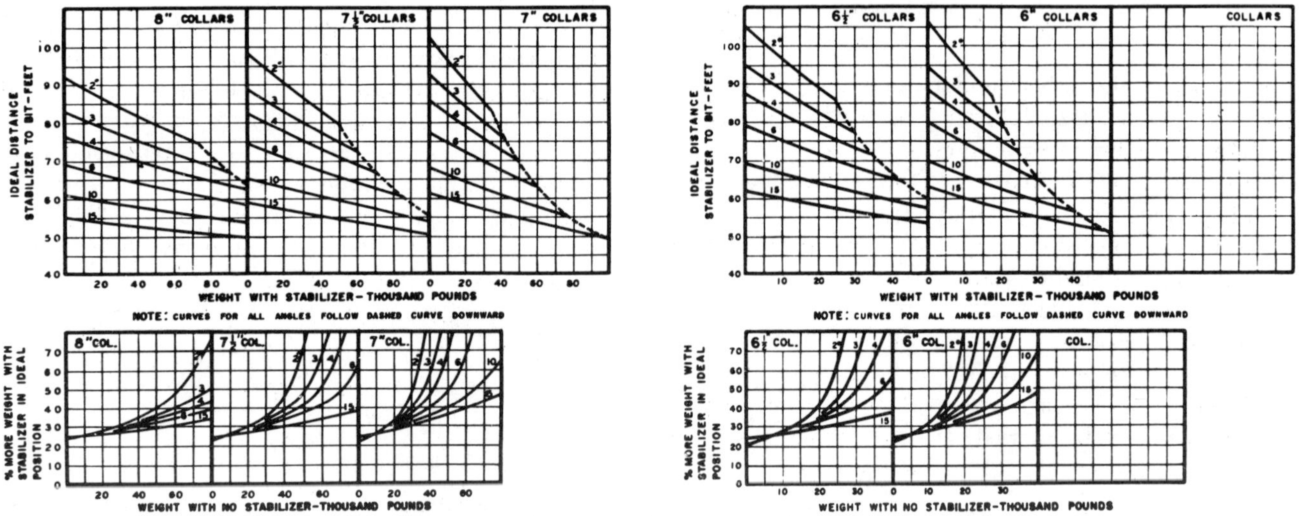

Fig. 9.14. Use of stabilizer. After Woods and Lubinski,[8] courtesy API. Hole = 9 in.

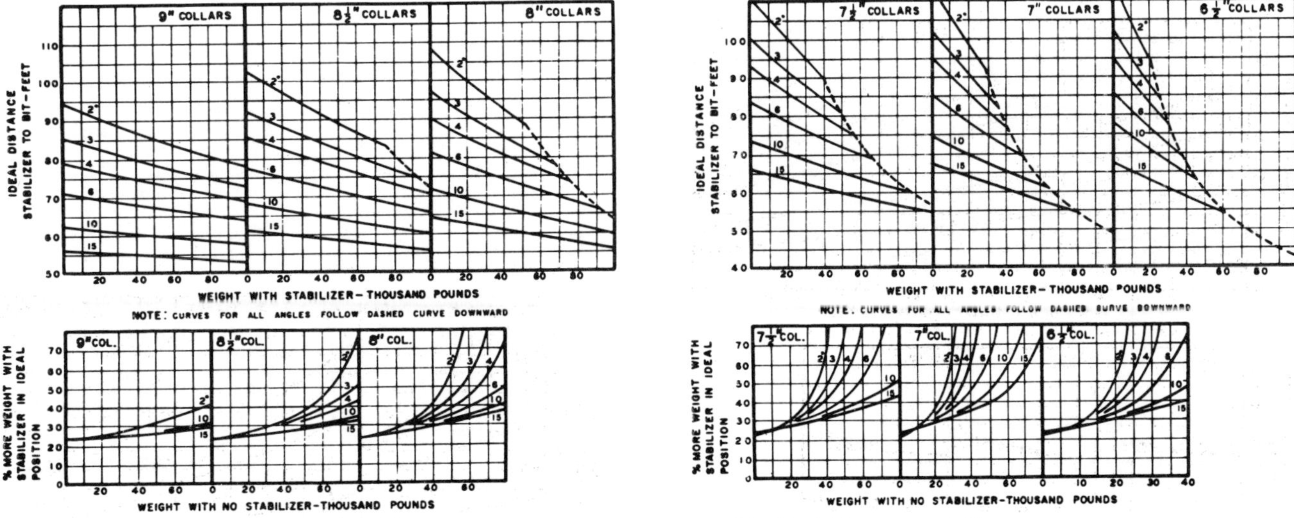

Fig. 9.15. Use of stabilizer. After Woods and Lubinski,[8] courtesy API. Hole = $9\frac{7}{8}$ in.

groove in the high side of the hole. On coming out of the hole, the drill collars may jam in this groove or seat and become stuck. Figure 9.18 shows a dog-leg with the resulting key-seat.

Recognition of these and other problems has caused the industry to strive for a more rigorous definition of a crooked hole.[5,13] For example, a 4° hole with the deviation constant in one direction is straight but not vertical. A 2° hole which spirals is vertical but crooked, etc. The practical definition is that the hole is crooked when conformance with the specified angle retards penetration rate. The best definition is probably one which combines vertical and horizontal deviations in a rate of change or abruptness restriction. Conformance to this restriction will require both vertical deviation and horizontal direction to be measured. Devices for making both measurements are shown in the directional drilling section.

A suggested method for defining dog-leg severity has been presented.[14] This technique requires the following measurements:

(1) Horizontal directions at two survey stations.
(2) Vertical deviations at the same two stations.
(3) The distance between stations.

The dog-leg severity in degrees per 100 ft is obtained from this information applied to Figure 9.20, as illustrated by Example 9.12. Figure 9.19 may be used for other problems.

Example 9.12 (After Lubinski[14])

Given the following data, compute the average hole curvature (dog-leg severity) between stations *1* and *2*.

	Stations	
	1	*2*
Depth, ft	3,666	3,696
Vertical deviation	3.5°	4.5°
Horizontal direction	N 11° E	N 23° E

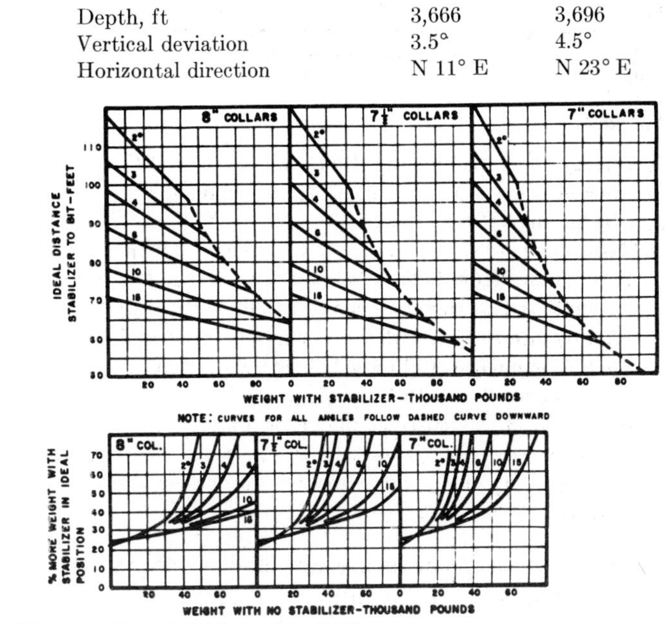

Fig. 9.16. Use of stabilizer. After Woods and Lubinski,[8] courtesy API. Hole = $10\frac{5}{8}$ in.

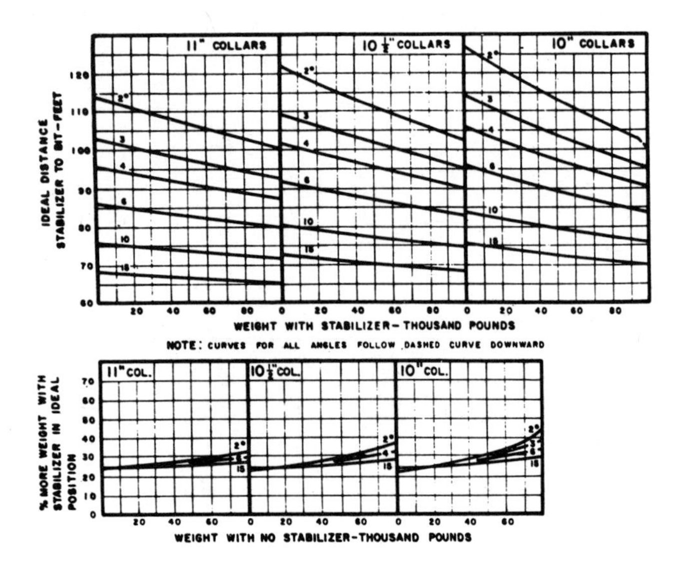

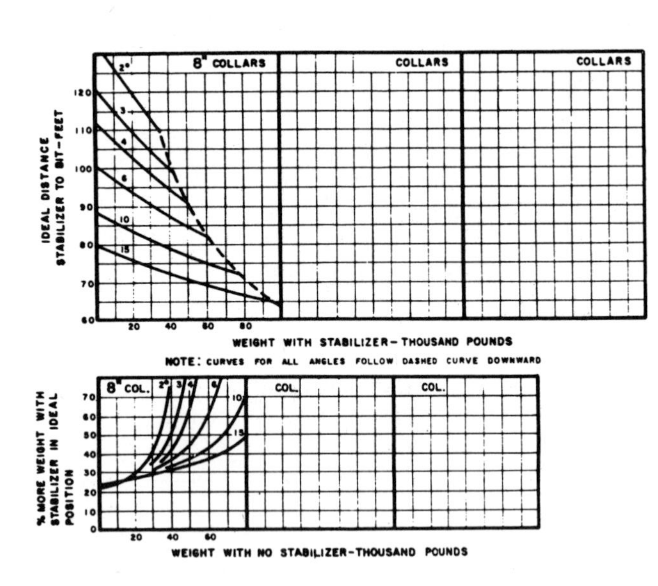

Calculations:

1. Change in horizontal direction: $23° - 11° = 12°$
2. Average vertical deviation: $1/2(3.5 + 4.5) = 4°$
3. Change in vertical angle: $4.5 - 3.5 = 1°$
4. Distance between survey pts.: $3{,}696 - 3{,}666 = 30$ ft
5. Proceed on Figure 9.20 as follows: The change of horizontal angle is 12°, therefore construct the vertical line MN. Find its intersection with the 4° curve at A (4° is the average hole inclination). Proceed along the horizontal line AP to its intersection with the 1° curve at B (1° is the change in inclination). Construct the vertical line BQ. The horizontal line from R (the distance between stations is 30 ft) intersects BQ at C; this indicates the following answer: Dog-leg severity = 4.2°/100 ft

There appears to be no precise definition of just what degree of severity constitutes an operational hazard. Future experience coupled with the above or a similar approach should lead to such a definition.

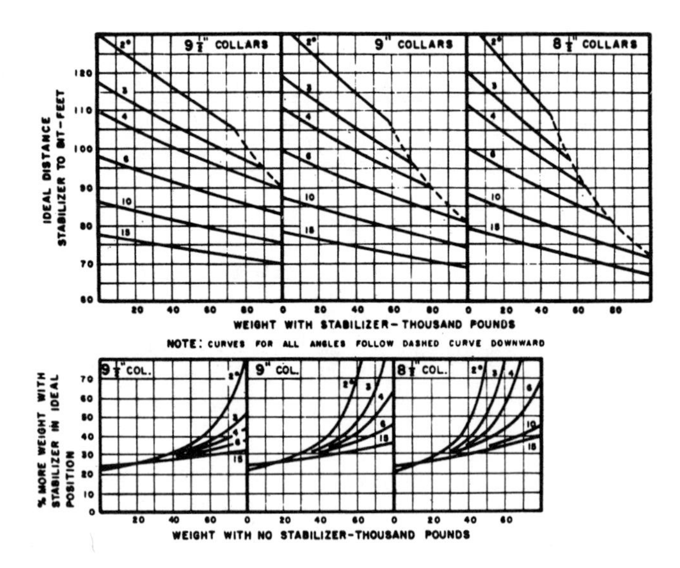

Fig. 9.17. Use of stabilizer. After Woods and Lubinski,[8] courtesy API.
Hole = $12\frac{1}{4}$ in.

Practical Significance of Hole Deviation Studies

It would be difficult to evaluate the total savings realized by the oil industry resulting from studies of crooked hole problems. The works of Lubinski, and Woods and Lubinski stand as milestones in the solution

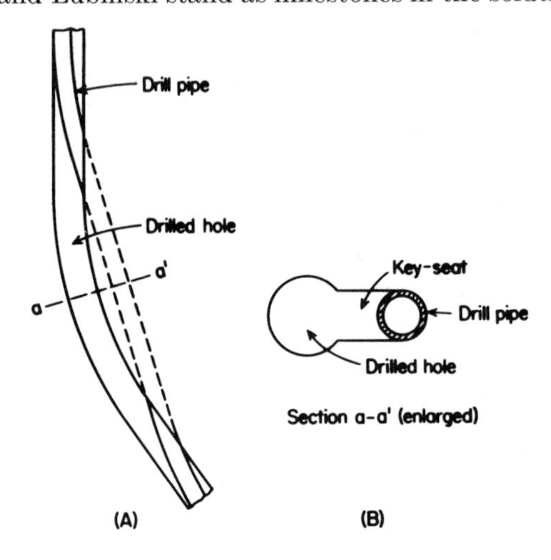

Fig. 9.18. Dog-leg resulting in the formation of a key-seat. After Stearns, courtesy *Oil and Gas Journal.*

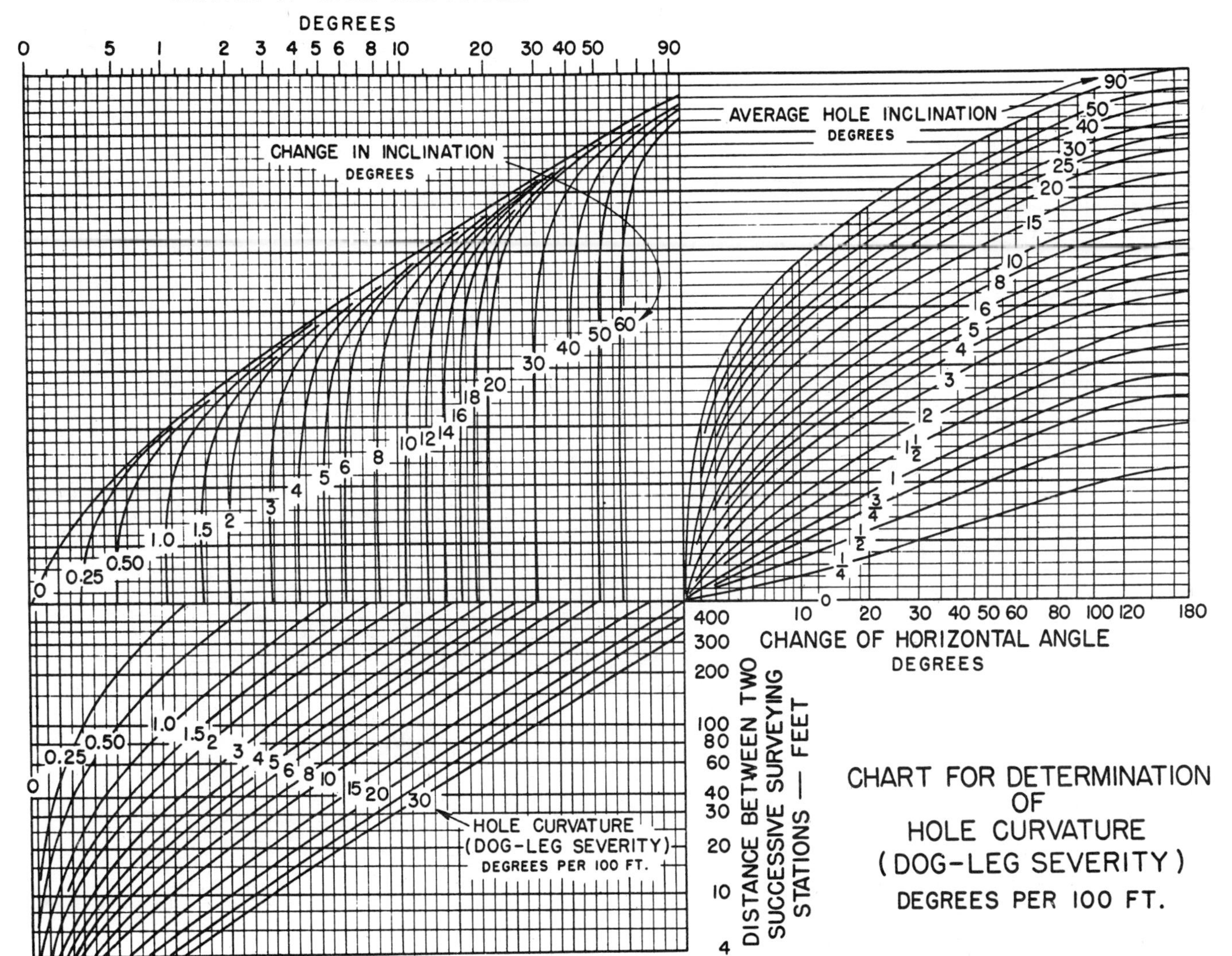

Fig. 9.19. After Lubinski,[14] courtesy API.

of problems as old as rotary drilling. The basic understanding of the problem and the factors affecting it have led to many beneficial changes in drilling practices, and others will undoubtedly follow.

The charts presented in this chapter allow the quantitative solution of hole deviation problems which could previously be attacked only by trial and error methods. This ability to predict hole deviation quantitatively has also shown the practical limits of equipment variations, such as in the use of drill collars and stabilizers. Once these limits were understood, it became apparent that still greater benefits could be obtained by relaxing hole deviation requirements. It has also been noted that the rate of change of hole deviation may be more important than the deviation itself. Further, Geological Departments are beginning to realize that the hole does not have to bottom below the rig as long as its actual location is known. Thus more of the so-called vertical wells are taking on the aspects of directional holes when such practices result in decreased drilling costs.

An excellent example of the liberalized deviation angle trend is a 15,000 ft Oklahoma wildcat.[15] This well was located in a steeply dipping, crooked-hole area which had always been troublesome to drill. The drilling contract allowed a deviation of 1°/1000 ft, i.e., at 10,000 ft a deviation of 10° was acceptable. Also, the rate of deviation change was restricted to $1\frac{1}{2}$°/100 ft. Analysis of the subsurface geology coupled with the expected updip drift of the bit resulted in the surface location being offset to account for the anticipated lateral displacement. The hole was drilled under close supervision and was completed 962 ft away (horizontally) from the surface location. Penetration rate was relatively high and a considerable saving resulted. This and other similar cases demonstrate the mutual

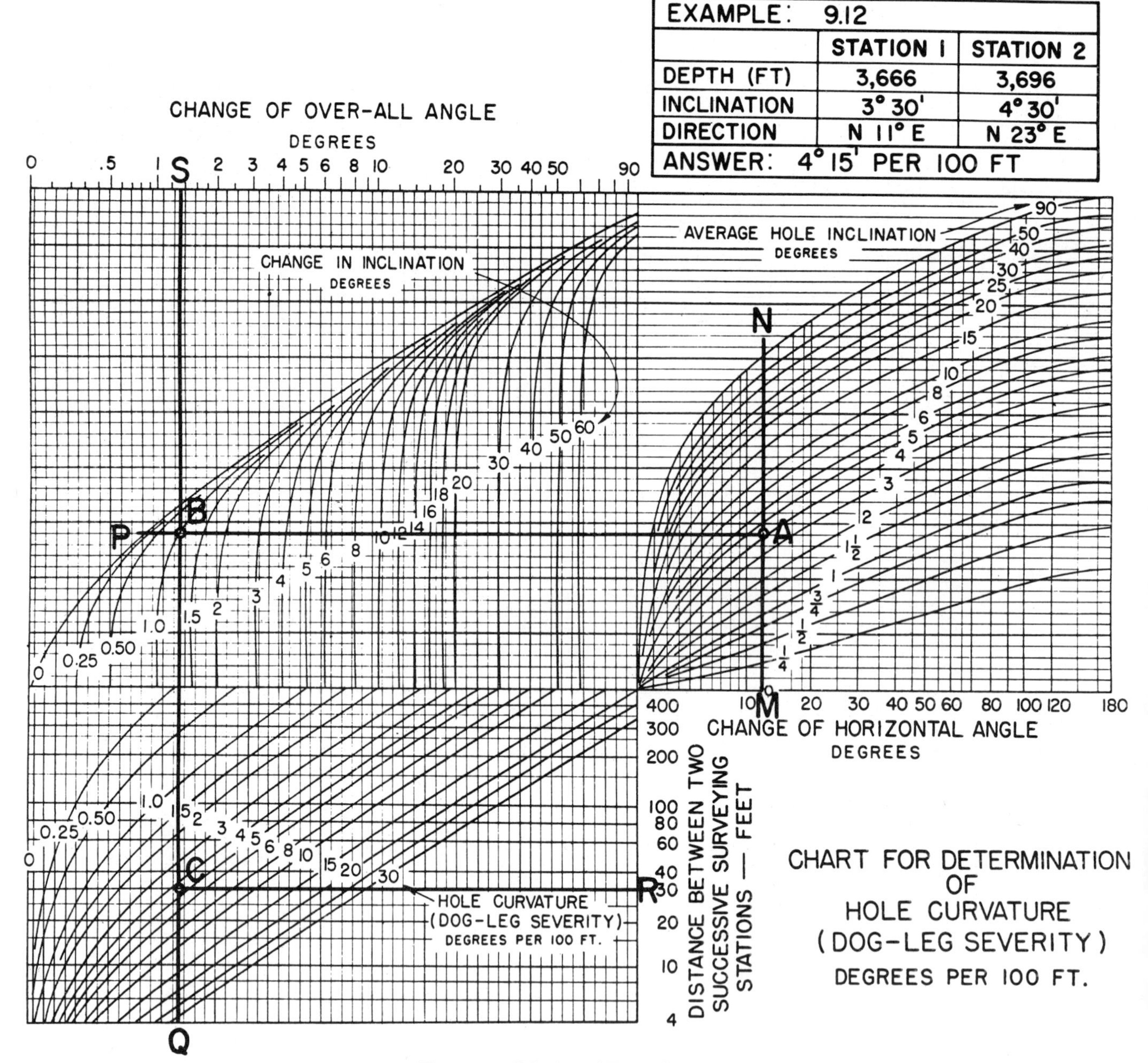

Fig. 9.20. Solution of Example 9.12.

advantages which may be derived from closer producer-drilling contractor cooperation in developing modern contract specifications.

9.2 Directional Drilling[16–22]

The general purpose of directional drilling is to place the bottom of a well under an inaccessible surface location. In such drilling, both the vertical and horizontal deviation are carefully controlled within precomputed limits. Some specific applications of directional drilling are shown in Figure 9.21.

Deflection Tools

A number of specialized tools are used in directional work for initiating and maintaining the desired hole direction. These are commonly called primary deflection tools. It is desirable to use deflection tools as infrequently as possible, because of the cost and rig time involved. Generally, the first deflection tool is used to obtain the initial deflection in the desired direction. Further hole progress is then controlled as much as possible by variations in bit weight, the use of stabilizers and reamers, drill string variations, and special bits. The Woods and Lubinski charts are useful in this

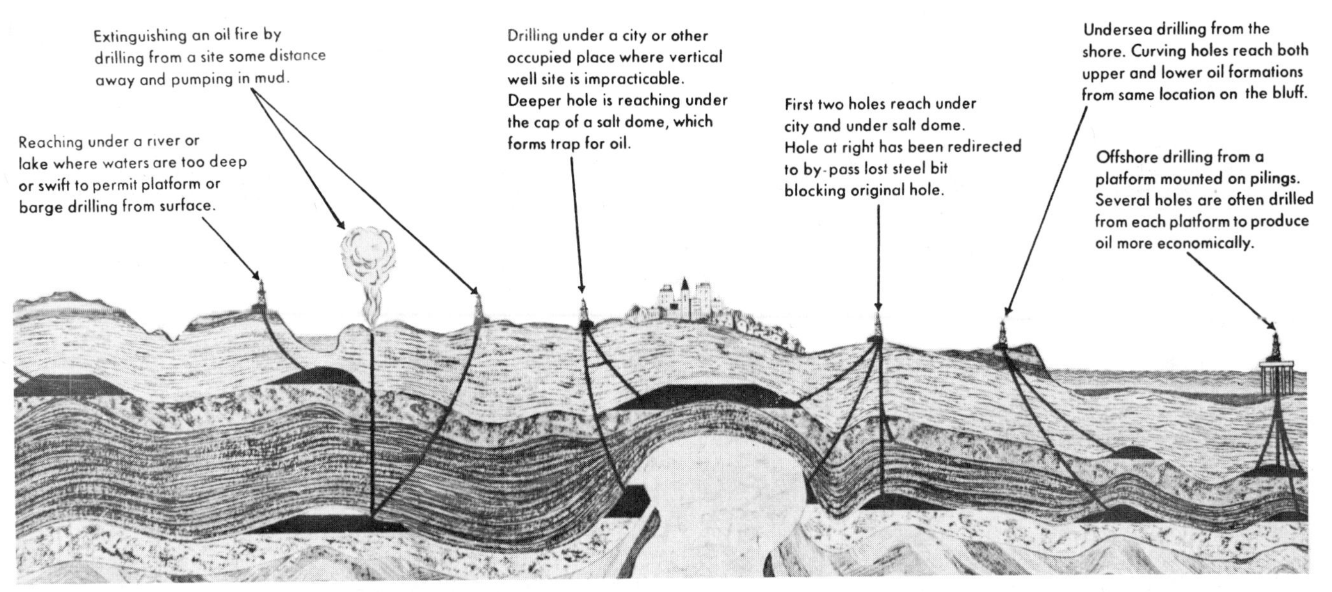

Fig. 9.21. Specific applications of directional drilling. Courtesy Eastman Oil Well Survey Company.

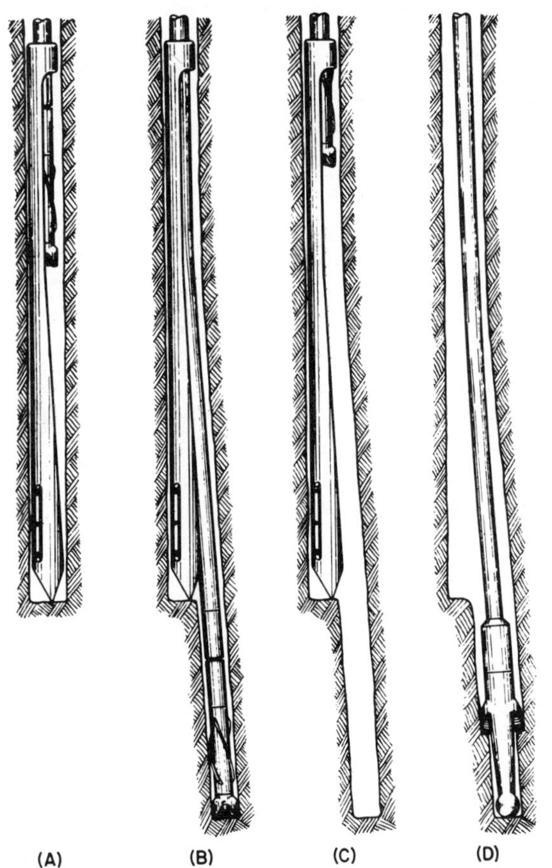

A–On bottom in oriented position before pin is sheared. B–Drilling assembly in rat hole. C–Whipstock in pick-up position. D–Reaming rat hole to full gauge with hole opener.

Fig. 9.22. Removable whipstock operation. Courtesy Eastman Oil Well Survey Company.

respect; it should be realized, however, that these pertain to the vertical angle only and not to direction.

Whipstocks are of two general types, removable and fixed. The removable one, as the name implies, is withdrawn from the hole with the drill pipe, while the fixed type stays in the hole as a permanent installation. At present, the latter type is never used in open holes. It is, however, used in sidetracking an old hole when the desired initial deflection point is inside the casing. The removable type has a collar which rigidly attaches it to the drill string with a shear pin. After the tool is run and oriented, drill string weight is slacked off; this

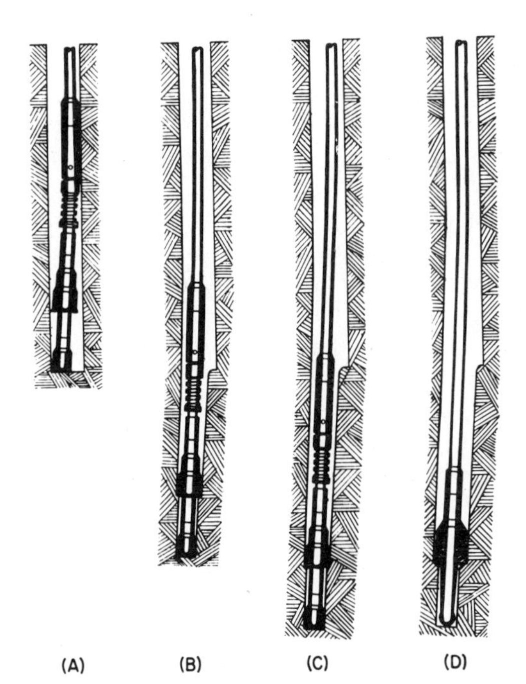

A–On bottom, in oriented position. B–Starting new hole. C–Completion of knuckle joint run. D–Enlarging hole to full gauge with hole opener.

Fig. 9.23. Operation of knuckle joint. Courtesy Eastman Oil Well Survey Company.

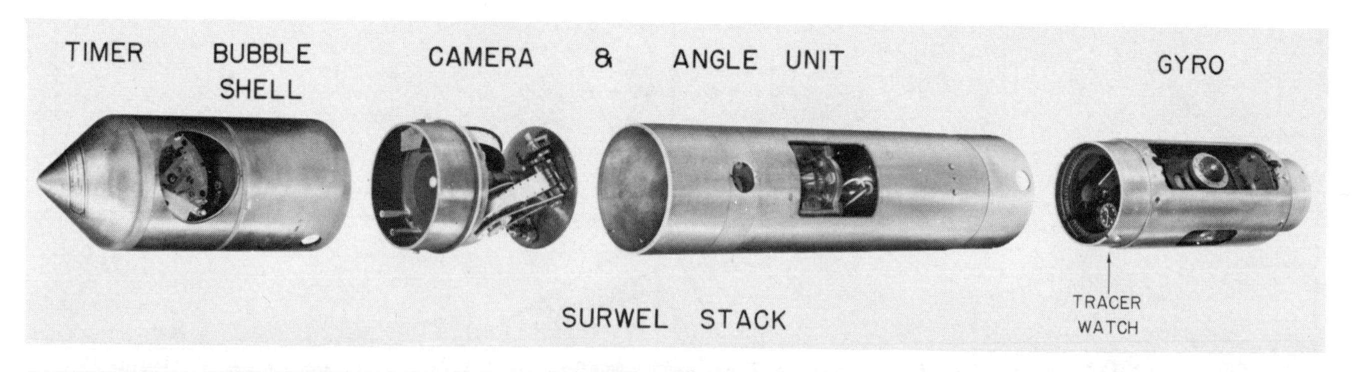

Fig. 9.24. Gyroscopic multiple shot surveying instrument (Surwel). Film shows direction by arrow, vertical deviation by cross, and time (depth) at which each picture was taken. Courtesy Sperry-Sun Well Surveying Company.

drives the chisel-shaped nose into the formation. When sufficient weight is applied, the shear pin fails, allowing the bit to rotate and drill off the face, as shown in Figure 9.22. The face is a concave groove with the desired deflection. After a few feet (10 to 20) have been drilled, a survey instrument is run to determine the angle and direction of the hole. If this is satisfactory, drilling is continued. If not, remedial measures must be taken.

The knuckle joint is essentially an extension of the drill string incorporating a universal joint at the junction. Thus rotation at different angles is possible, as illustrated in Figure 9.23. The principal disadvantage of this tool is that the angular change may be quite abrupt, and cause a dog-leg. Deviation changes of 5 to 7° in 15 or 20 ft may occur.

The most widely used of these tools is the removable whipstock. Those interested in a more detailed coverage of deflection tools are referred to Brantly's Handbook.[16]

Surveying Instruments

Directional drilling requires the measurement of both vertical deviation and horizontal direction. This is accomplished with various devices which combine the plumb-bob or pendulum reading with a simultaneously recorded compass reading. The compass used is either magnetic or gyroscopic. The instruments used are complicated mechanisms and different designs are available from various companies. They are often classified as single shot or multiple shot, depending on the number of readings obtainable from a single run.

Instruments utilizing a magnetic compass require shielding from the magnetic disturbance caused by the drill string. This is commonly accomplished by using a special nonmagnetic drill collar made of K-monel metal, which is run just above the bit to house the instruments. This metal is permanently nonmagnetic and has physical properties equal to the best drill collar steel. It is, however, quite expensive. Another way to eliminate magnetic disturbances is to use core or trigger type bits which allow the surveying instrument to protrude below the bit in the open hole.

The magnetic type single- and multi-shot instruments may be run into the hole in several ways, depending on the situation. Some of the more common methods are:

(1) Free drop or *go-devil* operation: the instruments are housed in special shock absorber barrels such as those used for the inclinometers previously mentioned. Single-shot types are retrieved with an overshot on a wire line, or by pulling the pipe. The latter procedure is used only if a trip is to be made anyway. Multi-shot surveys of this type always require pulling of the pipe, so that the instrument remains in the K-monel collar. A time vs. depth log (using a synchronised surface watch) is kept, so that the depth of each picture may be determined.

(2) Wire line operation: the single shot instrument is lowered into the open hole on a wire line (commonly a steel measuring line), positioned on bottom, held stationary until the recording is made, and then retrieved. Multishot instruments are run in the open hole on an electric cable. Again, an accurate depth vs. time log is kept.

(3) Drill pipe or tubing operation: the multi-shot instrument may be run in drill pipe or tubing. This operation is much like that described in (1). This procedure allows the magnetic instrument to function inside the casing because of the protective shielding of the K-monel collar.

A unique surveying device is the gyroscopic instrument (Surwel) shown in Figure 9.24. This utilizes the ability of a gyroscope to maintain the same directional orientation over a considerable time period. Rotation at 10,000 to 15,000 rpm is induced by power from batteries contained in the instrument. The timing device is set to take pictures at the desired intervals with a record of time, vertical angle, and direction being recorded. A separate log of depth vs. time is again kept at the surface with a synchronized watch. Surveys are taken both going in and coming out of the hole. This is the equivalent of two separate sets of measurements which serve to check each other. This device may be run on drill pipe, tubing, or other standard rig equipment, and is the only instrument which can be run inside the casing on a wire line.

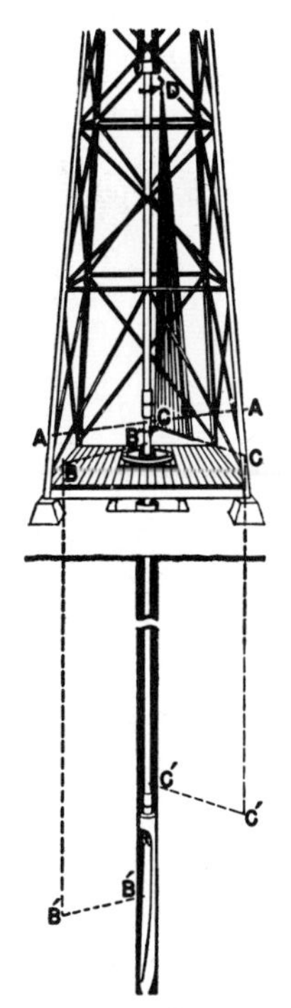

Diagrammatic representation of drill pipe alignment procedure

Fig. 9.25. Drill pipe alignment method for orientation of deflection tools. Courtesy Eastman Oil Well Survey Company.

Orientation Methods

An interesting feature of directional drilling is the orientation of primary deflection tools. It is desirable, of course, to start the initial deflection in the proper direction. This requires that the bottom hole position of the whipstock or knuckle joint be known, before it is set. Two basic methods of orienting deflection tools are in use:

(1) Drill pipe alignment method: this consists of keeping accurate track of the drill pipe's rotation as it is run into the hole. The following procedure refers to Figure 9.25. The deflection tool is affixed to the drill pipe and faced toward the desired direction (*B-B*). A sighting bar is fastened to the drill pipe and aimed at some convenient point such as the derrick leg (point *C*). The next stand of pipe is then attached to the string. The derrick man puts a second clamp near the top of this section and aligns it with the lower clamp's sighting bar by means of a cross-hair telescope (*D*). After the section is aligned, the telescope, lower clamp, and sighting bar are removed and the stand is lowered into the hole. The removed clamp is sent up on the elevators to the derrick man. The sighting bar is now inserted in the new floor clamp and the process is repeated until all the pipe is in the hole. This procedure keeps a constant angle between the tool face and the sighting bar. When the tool reaches bottom, the desired orientation is obtained by rotating the pipe to its original position.

This method is slow and is subject to error in deep wells due to unavoidable torsional stresses in the pipe which do not allow it to hang perfectly free. Consequently, this technique is little used at this time, except in shallow wells.

(2) Bottom hole orientation methods: there are several instruments and techniques used which allow the rapid and accurate bottom hole orientation of deflection tools. Only one rather typical procedure will be discussed as illustrated by Figure 9.26. A special K-monel substitute (or sub) is inserted in the string just above the deflection tool. This sub contains permanent magnets as shown, the positions of which are known with respect to the tool face. A single shot instrument containing two magnetic compasses is run with one compass opposite the magnets in the sub. The second compass is sufficiently removed so that it records magnetic north. The photographic record superimposes these readings so that the position

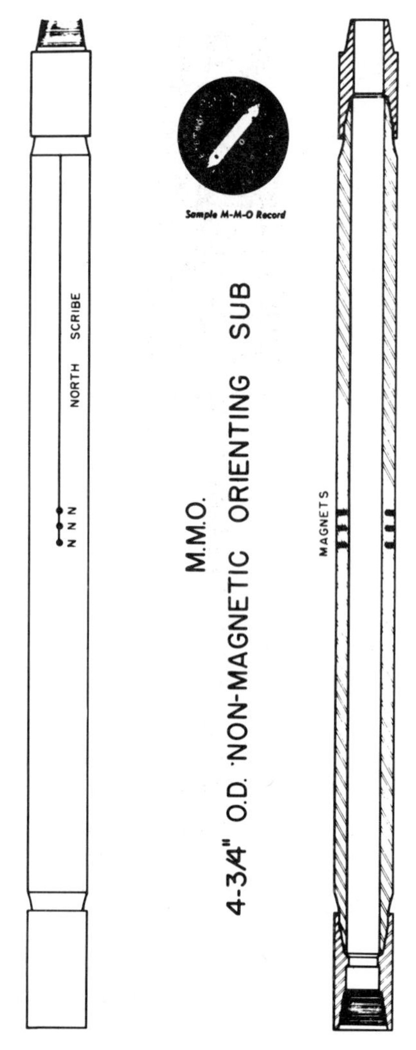

Fig. 9.26. Bottom hole orientation substitute and photographic record. The arrow indicates that the north pole of the sub magnets is pointed at S. 40° E. Courtesy Sperry-Sun Well Surveying Company.

of the north pole of the substitute magnets is clearly shown. The pipe may then be rotated as desired. A second shot may be made to insure proper alignment.

Planning a Directional Hole

One of three deflection patterns is commonly chosen in the planning of a directional well.[17] These are shown as Types 1, 2, and 3 in Figure 9.27.

Type 1. The desired deflection is obtained in the surface zone and maintained to total depth. This pattern is particularly applicable to areas of moderate depth where the desired deflections are small, and where an intermediate casing string is not required. The presence of the surface casing prevents the danger of keyseating. Casing wear is minimized by using rubber protectors on the drill pipe within the casing.

Type 2. This pattern is normally applied to deep wells which require an intermediate casing string. It permits more uniform spacing in multi-pay areas, as well as easier geological interpretation. The directional part of the hole is drilled in the more shallow, hence usually softer, formations where penetration rates are faster and trips less costly and frequent.

Type 3. This is normally applied in multi-pay fields where separate wells are drilled for each pay.

The tolerated deviation of the actual well from its proposed course is often represented by a cylinder, as shown in Figure 9.28. This is known as cylinder drilling, with the radius of the cylinder being the maximum allowable course variation. Lateral or vertical departures approaching or exceeding this limit are corrected by deflection tools when changes in operating techniques prove inadequate. In the example shown, two whipstocks were used, the first at 4993 ft measured depth (M.D.) which was the kick-off point (K.O.P., depth at which directional drilling began), and the last at 6147 ft when the hole exhibited a left turn tendency as shown in the plan view.

The calculation of well survey results requires the three dimensional location of the bottom hole position at each survey point. This is accomplished with these equations which follow from Figure 9.29:

$$Z = MD \cos \alpha \tag{9.1}$$

$$H = MD \sin \alpha \tag{9.2}$$

$$Y = H \cos \beta \tag{9.3}$$

$$X = H \sin \beta \tag{9.4}$$

where Z = true vertical depth between survey points 1 and 2.

MD = measured depth (length of drill string)

H = horizontal displacement of hole

Y = latitude, distance north or south of the east-west axis

X = departure, distance east or west of the north-south axis

α = vertical deviation angle

β = horizontal angle (compass direction)

Note: Normally the surface location of the well is the center of the directional coordinate system.

These equations assume that the hole interval is straight rather than an arc. This is satisfactory, since the distance between survey points is relatively small. The calculations are made for successive increments with their sum being the hole's position at any depth. This entire procedure is illustrated by the hypothetical case of Example 9.13.

Directional drilling involves certain risks not inherent in vertical drilling. Use of this technique is based strictly on the economic consideration dictated by the relative cost of obtaining a surface location directly above the objective.

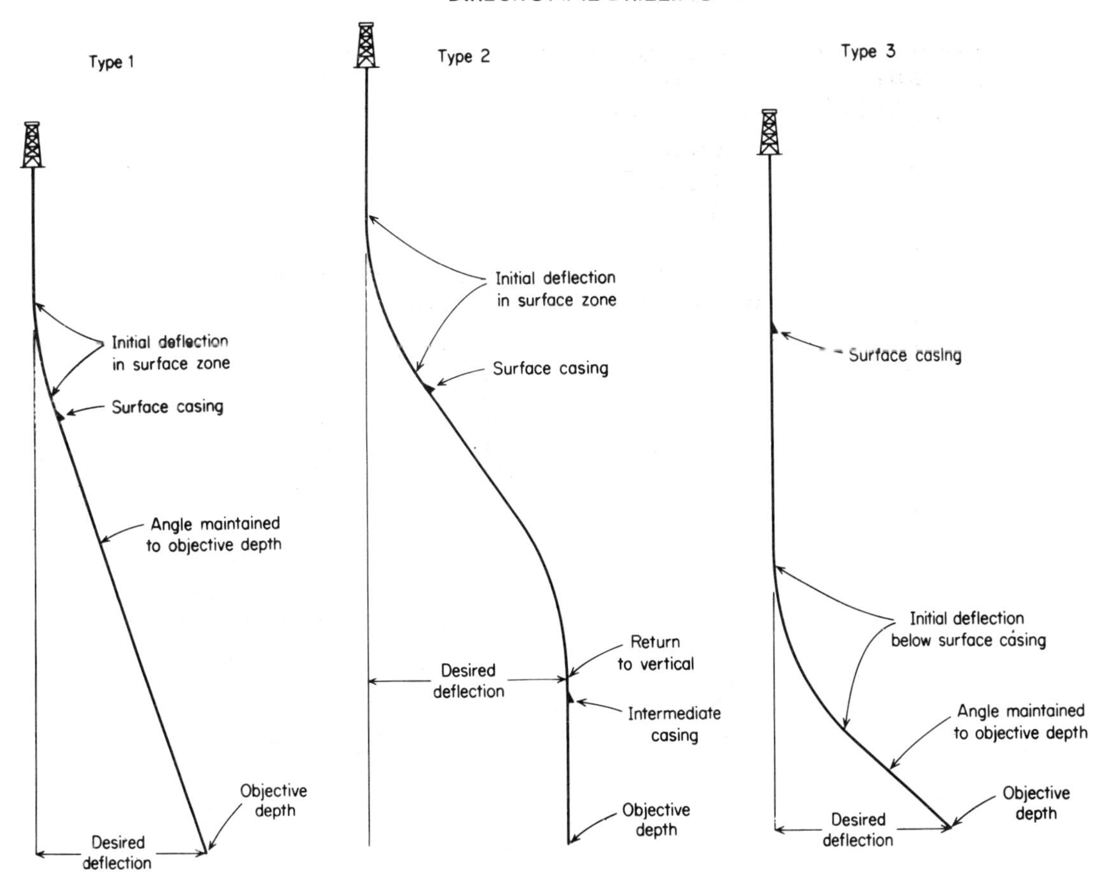

Fig. 9.27. Typical deflection patterns for directional wells. After Cook,[17] courtesy AIME.

Example 9.13

Observed and Calculated Data from Directional Well Survey*

					Vertical depth							Course				Total			
			α,					β,				y, latitude		x, departure		y, latitude		x, departure	
Test no.	Depth	Course length	Deviation angle	Cos α	Course	Total	Sin α	Course deviation	Magnetic Bearing β	Cos β	Sin β	N	S	E	W	N	S	E	W
1	100	100	1°15′	.9998	99.98	99.98	.0218	2.18	N 8°00′W	.9903	.1392	2.16	..	....	0.30	2.16	..	....	0.30
2	200	100	1°45′	.9995	99.95	199.93	.0305	3.05	N30°30′W	.8616	.5075	2.63	..	....	1.55	4.79	..	....	1.85
3	300	100	1°45′	.9995	99.95	299.88	.0305	3.05	N50°30′W	.6361	.7716	1.94	..	....	2.35	6.73	..	....	4.20
4	400	100	2°45′	.9988	99.88	399.76	.0480	4.80	N20°00′W	.9397	.3420	4.51	..	....	1.64	11.24	..	....	5.84
5	500	100	4°30′	.9969	99.69	499.45	.0785	7.85	N15°00′E	.9659	.2588	7.58	..	2.03	..	18.82	..	....	3.81
6	600	100	7°30′	.9914	99.14	598.59	.1305	13.05	N 9°30′E	.9863	.1650	12.87	..	2.15	..	31.69	..	....	1.66
7	700	100	9°45′	.9855	98.55	697.14	.1693	16.93	N19°00′E	.9455	.3256	16.01	..	5.51	..	47.70	..	3.85	..
8	800	100	13°15′	.9734	97.34	794.48	.2292	22.92	N47°00′E	.6820	.7313	15.63	..	16.76	..	63.33	..	20.61	..
9	900	100	19°15′	.9441	94.41	888.89	.3297	32.97	N53°00′E	.6018	.7986	19.84	..	26.33	..	83.17	..	46.94	..
10	1000	100	22°00′	.9272	92.72	981.61	.3746	37.46	N75°00′E	.2588	.9659	9.69	..	36.18	..	92.86	..	83.12	..
11	1100	100	24°30′	.9100	91.00	1072.61	.4147	41.47	N84°00′E	.1045	.9945	4.33	..	41.24	..	97.19	..	124.36	..
12	1200	100	26°00′	.8988	89.88	1162.49	.4384	43.84	S86°00′E	.0698	.9976	....	3.06	43.73	..	94.13	..	168.09	..

*After Stearns.[8]

9.3 Fishing Operations

The term fishing applies to all operations concerned with the retrieving of equipment or other objects from the hole. Portions of the drill string, bit, drill string accessories, and inadvertently dropped hand tools are typical items which may require fishing. The most common fishing job is that of recovering a portion of the drill string left in the hole due to either its failing or becoming stuck.

The best fishing technique is elimination of the cause; this eliminates the problem. Periodic equipment inspections such as internal corrosion surveys and magnetic flux testing minimize drill string failures by detecting faulty joints before they fail.[23] The alertness of the drilling crews in detecting crooked joints, in cleaning and lubricating tool joint threads properly, and in exercising good housekeeping and safety precautions is a further deterrent to this problem. But despite even the most rigorous precautions, fishing jobs do occur frequently. Most of them are relatively simple, and the only lost time is that required to run the fishing tool and retrieve the fish. A few such jobs,

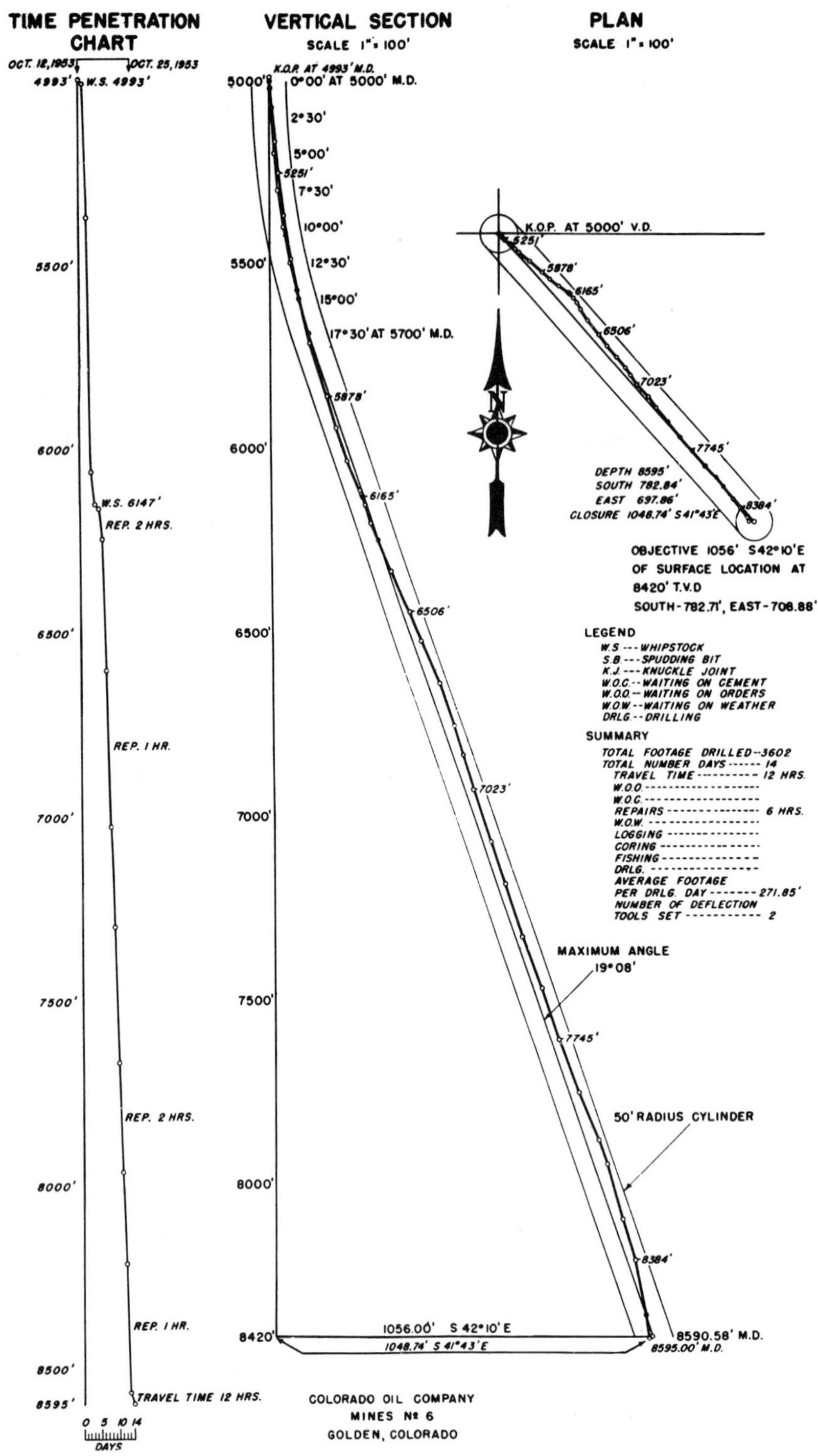

Fig. 9.28. Typical directional-drilling completion report showing cylinder of permissible variation. Courtesy Eastman Oil Well Survey Company.

however, become extremely costly and time consuming, and even result in loss of the hole.

Stuck Pipe

Many fishing jobs start with the drill pipe becoming stuck during a trip. Some of the causes of stuck pipe are

(1) foreign objects or *junk* in the hole
(2) key-seating
(3) sloughing formations (heaving shales, etc.)
(4) bit and drill collar balling
(5) pressure differential sticking
(6) cuttings settling above the bit or drill collars

All of these phenomena are quite obvious except (5)

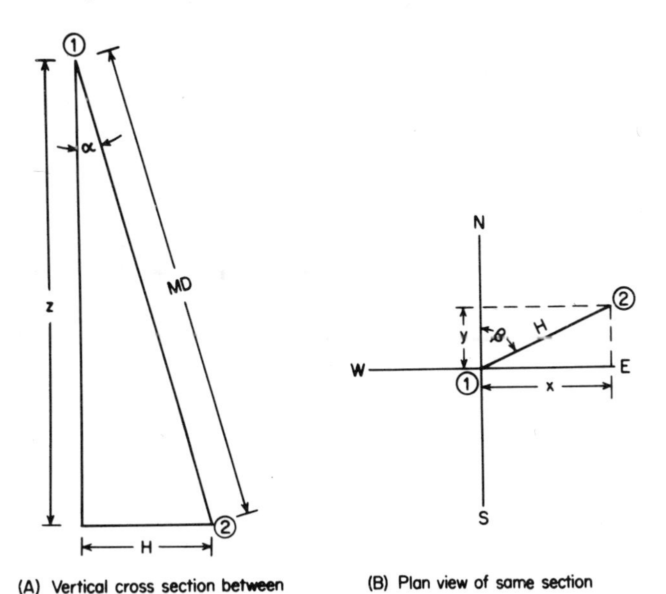

Fig. 9.29. Three-dimensional picture of bottom hole position.

which requires further clarification. Experimental work[25] verified by field observations has shown that the pressure differential between the mud column and a permeable formation exerts a considerable force against the drill pipe and literally *glues it* to the wall. This is shown in Figure 9.30. The force necessary to free the pipe was shown to vary directly with the pressure differential and also to increase with time, because of the mud cake buildup.

If the driller is unable to free the stuck pipe, other remedial measures must be applied. These are facilitated by a knowledge of the depth at which the pipe is stuck. This depth is also referred to as the free point. It may be calculated from relatively simple measurements taken on the rig floor.

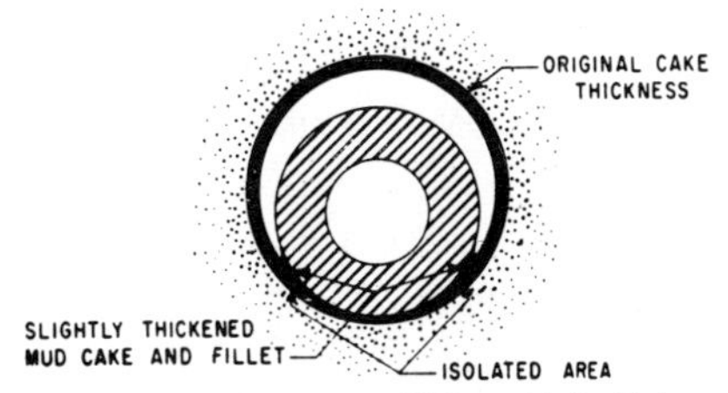

Fig. 9.30. Pressure differential sticking of drill pipe. Stuck pipe results when it becomes motionless against permeable bed, due to hole-formation Δp acting on isolated area.

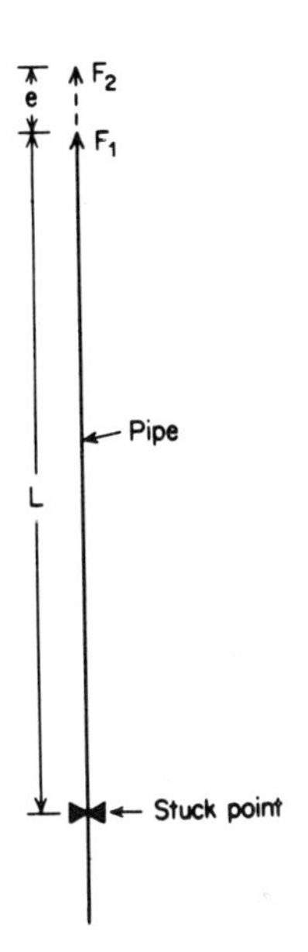

Fig. 9.31. Procedure for calculating free point of stuck pipe.

Consider Figure 9.31, which shows the pipe stuck at some depth, L. The following procedure is carried out:

(1) An upward force F_1 is applied to the pipe. This must be greater than the total pipe weight, to insure that the entire string is in tension.

(2) A reference point is marked on the pipe at the surface — normally this is done at the top of the rotary table.

(3) A larger upward force F_2 is applied causing the free portion of the drill string to stretch by an amount e. The stretch is measured as the movement of the reference point in (2). F_2 is of course limited by the yield strength (elastic limit) of the pipe.

These measurements facilitate calculation of L, as is shown by the following derivation.

Recall the definition of Young's modulus for steel:

$$E = \frac{\sigma}{\delta} \tag{9.5}$$

where $E = 30 \times 10^6$ psi

σ = tensile stress, psi

δ = unit stretch or strain, length per length

Also,

$$e = \delta \times 12L \quad \text{or} \quad \delta = \frac{e}{12L} \tag{1}$$

where e = inches of stretch measured

L = depth to stuck point, ft

The stress causing the differential stretch e is

$$\sigma = \frac{F_2 - F_1}{A} = \frac{\Delta F}{A} \tag{2}$$

where F = force, lb

A = cross sectional area of steel in the drill pipe, in.^{2}

Substituting (1) and (2) in Eq. (9.5) and solving for L, we obtain

$$L = \frac{EAe}{12\Delta F} \tag{9.6}$$

To save looking up or computing the area A it is more convenient to use the weight per foot of the pipe, which is always known.

(3) $w = (A)(12)(0.283)$ or $A = w/(12)(0.283)$

where w = weight of drill pipe, lb/ft

0.283 = density of steel, lb/in.^{3}

Substitution of (3) for A in Eq. (9.6) gives:

$$L = \frac{Ewe}{40.8\Delta F} = 735 \times 10^3 \frac{we}{\Delta F} \tag{9.7}$$

Eq. (9.7) is commonly known as the stuck pipe formula.[24] Example (9.14) illustrates its application.

Example 9.14

A string of $3\frac{1}{2}$-in., 13.3 lb/ft, grade E drill pipe is stuck in a 10,000 ft hole. The driller obtains the following data as described previously:

F_1 = 140,000 lb, which is greater than the weight of string.

F_2 = 200,000 lb, which is less than the yield strength of the pipe

e = 4 ft

Where is the pipe stuck?

Solution:

Applying Eq. (9.7),

$$L = 735 \times 10^3 \frac{(13.3)(4)(12)}{60,000} = 7,800 \text{ ft}$$

It should be noted that Eq. 9.7 was derived by neglecting hole-pipe friction above the free point. Depending on the particular case, small or considerable errors may result from this simplification. The calculation is simple, and provides an expedient rig floor solution which is sufficiently accurate for many purposes.

A very precise definition of the free point may be obtained with electromagnetic devices which are available from various service companies. These consist, essentially, of two electromagnets connected with a telescopic joint. These are run into the hole on an electric cable and lowered to some starting depth. The electric current is then turned on, causing the two magnets to attach themselves to the inside wall of the pipe. A pull is exerted at the surface, causing the pipe above the stuck point to stretch. If the magnets are above the stuck point, the distance between them is elongated. This elongation is measured by a sensitive electronic strain gauge between the magnets and transmitted to the surface for measurement. If the magnets are below the stuck point, no stretch occurs between them. A few rapid settings quickly and accurately define the free point.

A knowledge of the free point is useful in problems other than those arising from stuck drill pipe. The recovery of casing from dry and/or abandoned wells is another common application. The use of a measuring instrument is recommended whenever accurate free point definition is necessary.

Once the point at which the drill pipe is frozen is known, various procedures are followed. A few barrels of oil circulated to a point opposite the stuck point will often lubricate the area sufficiently to work the pipe free. The technique of placing a quantity of fluid at some desired position in the hole is called *spotting.* The success of this approach depends largely on the cause of the trouble.

The hydration of heaving shales is prevented by filling that section of the hole with oil. It has also been shown that the preferential wetting of the pipe by oil is a great aid in minimizing pressure differential sticking. Proper wetting agents added to the oil improve its steel wetting ability and further aid in reducing pressure differential sticking.[25] In addition, the lubricating action of the oil is of general value in reducing friction over the entire section covered by the spot.

If attempts to free the pipe fail, the string is unscrewed as near the stuck point as possible, so that fishing operations can begin. This is generally accomplished by bump-shooting the pipe at the first tool joint above the free point. Left hand torque is applied to the string at the surface, and a small powder charge is electrically exploded in the desired tool joint. Some types of charges do not require the application of torque. The explosion produces much the same effect as a hammer blow and causes the joint to unscrew at the proper point. The freed portion is then removed and the fishing begins in earnest.

Fishing Tools

The choice of fishing tools and equipment is dictated by the size and shape of the fish, the anticipated severity of the problem, and the experience of the fisherman. In past years, the tool pushers and drillers handled these problems almost exclusively, their main tool being their own ingenuity. Much of this work is currently handled by service companies which offer complete lines of fishing tools and experienced operators (fishermen) on a rental basis. Many fishing jobs are turned over to these experts, who aid and advise the tool pushers until the job is completed.

Fishing tools are quite numerous, and only a qualitative description of a few basic types will be attempted here:

(1) Overshots: these are cylindrical bowl shaped tools which telescope over the fish. An internal slip arrangement grasps the outside of the fish with the grip tightening as pull is increased. Their operation is quite analogous to that of the straw thimble novelties used by children for finger pulling. Hole-pipe clearance must be sufficient to accommodate the wall thickness of the overshot. Small drill collar-hole clearances as dictated by crooked-hole considerations are a fishing hazard, since packed holes will not allow an overshot to be used. Modern overshots have a releasing mechanism to be used if the fish cannot be pulled.

(2) Spears: these devices pass inside the fish, grasping its inner wall with expanding slips, which can be set or released by rotating the pipe. Spears are generally used when the hole clearance will not tolerate an overshot. Their principal advantage over the tapered tap is their releasing mechanism.

(3) Tapered taps: these are among the oldest fishing tools used in drilling. The tap is run into the fish and rotated until sufficient threads are cut for a firm hold.

(4) Washover pipe: this is a section of pipe having sufficient inside diameter to telescope outside the fish. It is used when annular cleaning around the fish is required.

(5) Inside and outside cutters: these are pipe cutters which can be actuated by surface manipulation of the drill pipe. Inside cutters cut the fish from the inside out, while outside cutters do the reverse. They are used when a considerable length of drill string must be retrieved in sections. The common sequence of events is washing over, cutting, and retrieving of the cut off portion. The cutting is performed by a set of knives rotated by the drill pipe which operate much like the pipe cutters used by plumbers.

Small items of junk such as bit teeth, cones, or other small pieces of steel may be retrieved with a permanent magnet run on the drill pipe. Various *junk baskets* run just above the bit are also used for this purpose.

Accessory Equipment

Jarring devices (jars) which provide a hammer type impact at the desired downhole location are extremely effective in loosening stuck pipe. They are commonly run in conjunction with overshots, spears, etc., to aid in loosening the fish once it is caught. These devices utilize the energy of compressed fluids which drive a free moving piston or hammer against the top of the jar. This compression is obtained by proper surface movement of the drill pipe. With the nitrogen type of jar, no shock loads are transmitted to the surface equipment. Other types of mechanical and torque jars are also in use.

Safety joints are often run with the fishing string for use in case the fishing tool becomes stuck. These are merely specially designed tool joints which will unscrew either with less torque or in the opposite direction to the regular tool joints. These joints are placed in the string at the most desirable point for subsequent fishing operations.

The mechanical operation of fishing tools is quite varied and complicated. Detailed operating instructions are best obtained from the manufacturers or supplying companies. Also, other more complete references on this general topic are available.[26–29]

Successful fishing operations require proper planning and the right combination of tools. Drilling practices are always influenced by the possibility that anything run into the hole may have to be fished out. For drilling in remote areas, the wellsite equipment should include fishing equipment for the hole and pipe sizes in use. In more established areas, nearby tool rental and service establishments may be relied on for quick delivery of fishing tools as they are needed.

PROBLEMS

Hole deviation problems will be presented in the same manner as the examples.

1. (a)

	Established data	*Problem data*
Collar OD, in.	$6\frac{3}{4}$	8
Weight, lb	40,000	?
Hole size, in.	9	9
Hole inclination, °	8	8
Formation dip, °	Same	

(b) What degree of crookedness would be assigned to this formation? *Ans.* (a) 65,000 lb; (b) mild.

2. (a)

	Established data	*Problem data*
Collar OD, in.	6	8
Weight, lb	20,000	40,000
Hole size, in.	9	?
Hole inclination, °	5	5
Formation dip, °	Same	

(b) What degree of crookedness does this formation exhibit? *Ans.* (a) $8\frac{3}{4}$-in. hole.

3. Work Example 9.3 in the manner indicated.

4.

	Established data	*Problem data*
Collar OD, in.	8	9
Weight, lb	15,000	?
Hole size, in.	10	12
Hole inclination, °	5	10
Formation dip, °	30	15

Ans. W = 96,000 lb.

5. Work Example 9.6.

6. Work Example 9.7.

7. A West Texas formation was drilled as follows:
Hole size = $7\frac{7}{8}$ in.
Drill collars = $6\frac{5}{8}$ in. OD
Bit weight = 30,000 lb
Equilibrium α = $12\frac{1}{2}°$
Formation dip = 15°
Stabilizers = none

(a) Using the charts of this chapter, show that the following conditions would also result in $12\frac{1}{2}°$ hole deviation:

Collars, in.	*Hole, in.*	*Stabilizer*	*Bit Weight* *lb/in.*	*lb*
$6\frac{5}{8}$	$7\frac{7}{8}$	None	3,810	30,000
$6\frac{5}{8}$	$7\frac{7}{8}$	One	4,950	39,000
8	$8\frac{3}{4}$	None	5,250	46,000
8	$8\frac{3}{4}$	One	6,500	57,000
$11\frac{1}{4}$	$12\frac{1}{4}$	None	9,300	114,000
$11\frac{1}{4}$	$12\frac{1}{4}$	One	11,300	138,000

(b) Which of the above conditions would result in the highest on bottom penetration rate?

8. What hole inclination would result if 9,300 lb/in. were carried on the $7\frac{7}{8}$ in. bit with $6\frac{5}{8}$ in. collars and no stabilizer? *Ans.* $\alpha = 14\frac{1}{4}°$. (Note that this is a relatively small increase.)

9. Suppose that the drilling conditions of Problem 8 existed, but that the dip was 45° instead of 15°. What value of α would now exist if 9,300 lb/in. were carried on the $7\frac{7}{8}$ in. bit with $6\frac{5}{8}$ in. collars? *Ans.* $\alpha = 29\frac{1}{2}°$. (Note that this is a much larger increase.)

10. What conclusions might be drawn from the results of Problems 8 and 9? (Recall that hole crookedness depends on both dip and anisotropic index.)

11. Investigate further the effect of bit weight reduction on α for the following conditions: $7\frac{7}{8}$ in. hole, $6\frac{5}{8}$ in. collars, and dips of 15° and 45°, respectively; other conditions are those given in Problem 7. Show the results as curves of W vs α for each dip over a range of $0 \leq W \leq 60{,}000$ lb. Discuss the curves.

Note: Problems 7 to 11 are from Reference 12, in which the answers and detailed discussion may be found.

12. Compute the average hole curvature or dog-leg severity for the following case:

	Station 1	*Station 2*
Depth	3,666 ft	3,696 ft
Vertical Deviation	3°	3°
Horizontal Direction	North	S 40° E

Ans. 19.5°/100 ft.

13. Sketch a plan view and a vertical cross section of the directional well data in Example 9.13.

14. 10,000 ft of $3\frac{1}{2}$ in., 13.3 lb drill pipe is stuck in the hole. The following measurements were obtained for the purpose of calculating the free point:

$$F_1 = 133{,}000 \text{ lb}$$

$$F_2 = 200{,}000 \text{ lb}$$

$$e = 4 \text{ ft}$$

Where is the pipe stuck? *Ans.* At approximately 7000 ft.

REFERENCES

1. Smith, L. E., "Crookedness of Deep Holes Determined by Acid Bottle Method," *National Petroleum News*, July 21, 1926, p. 50–A.
2. Miller, B. R., "Deep Rotary Drilling Applied to the Appalachians," *API Drilling and Production Practices*, 1946, p. 18.
3. Suman, J. R., "Drilling, Testing, and Completion," in *Elements of the Petroleum Industry*. New York: AIME, 1940, p. 160.
4. Lubinski, A., "A Study of the Buckling of Rotary Drilling Strings," *API Drilling and Production Practices*, 1950, p. 178.
5. MacDonald, G. C., and Arthur Lubinski, "Straight-hole Drilling in the Crooked-hole Country," *API Drilling and Production Practices*, 1951, p. 80.
6. Lubinski, A., and H. B. Woods, "Factors Affecting the Angle of Inclination and Dog-Legging in Rotary Bore Holes," *API Drilling and Production Practices*, 1953, p. 222.
7. Woods, H. B., and A. Lubinski, "Practical Charts for Solving Problems in Hole Deviation," *API Drilling and Production Practices*, 1954, p. 56.
8. Woods, H. B., and Arthur Lubinski, "Use of Stabilizers in Controlling Hole Deviation," *API Drilling and Production Practices*, 1955, p. 165.
9. Rollins, H. M., "Studies of Straight Hole Drilling Practices," *Drilling*, April, 1956.
10. Bromell, R. J., "Lick Those Crooked-Hole Problems," *Oil and Gas Journal*, Nov. 8, 1954, p. 149.
11. Speer, J. W., and G. H. Holliday, "Crooked Holes—And How to Help Them Go Straight," *Oil and Gas Journal*, Aug. 1, 8, and 29, 1955, pp. 80, 98, 106, respectively.
12. Lubinski, A., and K. A. Blenkarn, "Usefulness of Dip Information in Drilling Crooked Formations," *The Drilling Contractor*, Apr. 1956, p. 53.
13. McCloy, R. B., "The Need for a Definition of a Crooked Hole," *Oil and Gas Journal*, Mar. 31, 1952.
14. Lubinski, A., "Chart for Determination of Hole Curvature (Dog-Leg Severity)," Presented 1956, API Mid-Continent District Study Committee on Straight Hole Drilling. Published as follows:
"How to Spot Dog-Legs Easily," *Oil and Gas Journal*, Feb. 4, 1957, p. 129.
"How to Determine Hole Curvature," *The Petroleum Engineer*, Feb. 1957.
"How Severe is that Dog-Leg?" *World Oil*, Feb. 1, 1957.
"Dog-Leg Severity can be Determined by Use of a Chart," *Drilling*, Feb. 1957.
15. Norris, F., "Drilling a Crooked Hole Intentionally," *Oil and Gas Journal*, June 4, 1956, p. 92.
16. Brantly, J. E., *Rotary Drilling Handbook*, 5th ed. New York: Palmer Publications, 1952, pp. 352–382.

17. Cook, W. H., "Offshore Directional Drilling Practices Today and Tomorrow," AIME Paper No. 783-G, presented New Orleans, Feb. 1957.

18. Roberts, D. L., "Directional Drilling in the Los Angeles Basin," *API Drilling and Production Practices*, 1949, p. 60.

19. Graser, F. A., "The Fundamental Mechanics of Directional Drilling," *API Drilling and Production Practices*, 1949, p. 71.

20. Weaver, D. K., "Practical Aspects of Directional Drilling," *API Drilling and Production Practices*, 1946, p. 9.

21. Stearns, G., *Engineering Fundamentals in Modern Drilling*. Tulsa: *Oil and Gas Journal*, 1953, pp. 69–81.

22. Parks, G. B., "Directional Drilling in Offshore Operations," *World Oil*, June, 1949.

23. Main, W. C., "Detection of Incipient Drill-pipe Failures," *API Drilling and Production Practices*, 1949, p. 89.

24. Hayward, J. T., "Methods of Determining How Much of a Frozen or Cemented Column of Pipe is Free," *API Drilling and Production Practices*, 1935, p. 16.

25. Helmick, W. E., and A. J. Longley, "Pressure-Differential Sticking of Drill Pipe," *Oil and Gas Journal*, June 17, 1957, p. 132.

26. Medders, W. L., "Fishing Tools and Techniques," *World Oil*, a series of six articles commencing Feb. 1, 1957, p. 89.

27. Moore, E. E., "Fishing and Freeing Stuck Drill Pipe," *The Petroleum Engineer*, Apr. 1956, p. B–54.

28. Briggs, F., "Cutting and Fishing," in *Drilling Handbook*. Tulsa: *Oil and Gas Journal*, 1949, p. 26.

29. Brantly, J. E., Rotary Drilling Handbook, 5th ed. New York: Palmer Publications, 1952, pp. 383–400.

Chapter 10 · · ·

Coring and Core Analysis

Analyses of rock samples yield data basic to the evaluation of the productive potential of a hydrocarbon reservoir. Bit cuttings are, of course, rock samples; their small size, however, precludes their furnishing more than qualitative information. The desire to obtain and examine larger, unbroken pieces of reservoir rock led to the development of coring techniques, by which relatively large reservoir rock samples are obtained, either from the bottom during drilling, or from the side of the bore hole wall after drilling. The development of coring and core analysis techniques has played a large part in the elevation of petroleum engineering to its current status. All phases of the profession rely to some extent on a knowledge of rock properties and the factors which affect them.

10.1 General Coring Methods and Equipment

Two basic rotary coring methods are applied: coring at the time of drilling (bottom coring) and coring after drilling (side-wall coring). All bottom coring methods utilize some type of open center bit which cuts a doughnut shaped hole, leaving a cylindrical plug or *core* in the center. As drilling progresses, this central plug rises inside a hollow tube or core barrel above the bit where it is captured and subsequently raised to the surface. Further classification of bottom coring is commonly based on a more specific description of the equipment used:

1. Conventional Coring
   a. conventional core head (other than diamond)
   b. diamond core head
2. Wireline Retrievable Coring

Conventional Coring

Conventional coring equipment requires that the entire drill string be pulled to retrieve the core. This is a disadvantage; however, the corresponding advantage is that large cores, 3 to 5 in. in diameter and 30 to 55 ft long, may be obtained. The $3\frac{1}{2}$-in. diameter core is probably the most common.

Two typical conventional coreheads are shown in Figure 10.1; the fish tail (soft formations) and rolling cutter (hard formation) types. Their usage parallels that of the regular bits from which they have been adapted. Figure 10.2 shows three different conventional diamond core bits as a sample of the numerous types available from different companies.

The core barrels used with diamond heads are generally longer than those used with conventional heads, commonly 55 ft as compared with 30 ft. A typical diamond core barrel is shown in Figure 10.3. Note that it is composed of an outer barrel which acts as a drill collar, and a freely rotating inner barrel which houses the core. The drilling fluid passes outside the inner barrel and is discharged through watercourses at the bit face. Mud trapped above the core is expelled through the check valve atop the inner barrel.

Diamond bits cost much more, but will drill more total footage and when worn out may be returned to their supplier for salvage. This trade-in value is based on the weight of undamaged stones remaining, and may amount to 50% or more of the original cost. The core recovery with diamond heads is generally higher than with conventional heads, particularly in hard rock areas. Rate of penetration may, however, be lower in soft formations than that of the rolling cutter heads.

Wireline coring denotes the method whereby the core (and inner barrel) may be retrieved without *pulling* the drill string. This is accomplished with an overshot run down the drill pipe on a wire line. The diamond heads used in this technique have much smaller openings than those shown in Figure 10.2. The core barrels used are

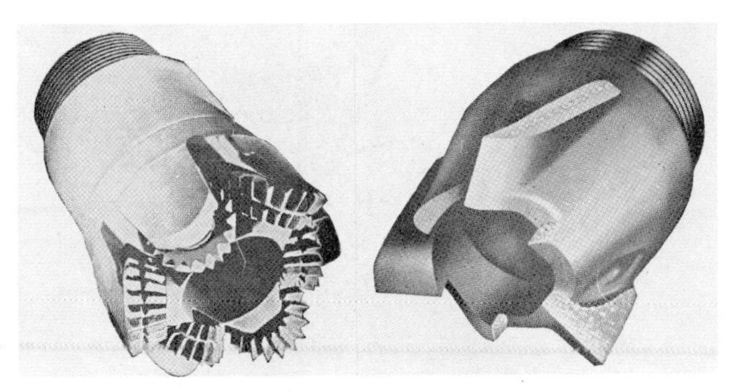

Hard Formation Cutter Head — Soft Formation Cutter Head

Fig. 10.1. Conventional core bits. Courtesy Hughes Tool Company.

somewhat variable, but are basically similar to conventional types. The cores obtained by this method are small, commonly $1\frac{1}{8}$ to $1\frac{3}{4}$ in. in diameter and 10 to 20 ft long. The main advantage of this method is the saving in trip time, as mentioned before. The durability

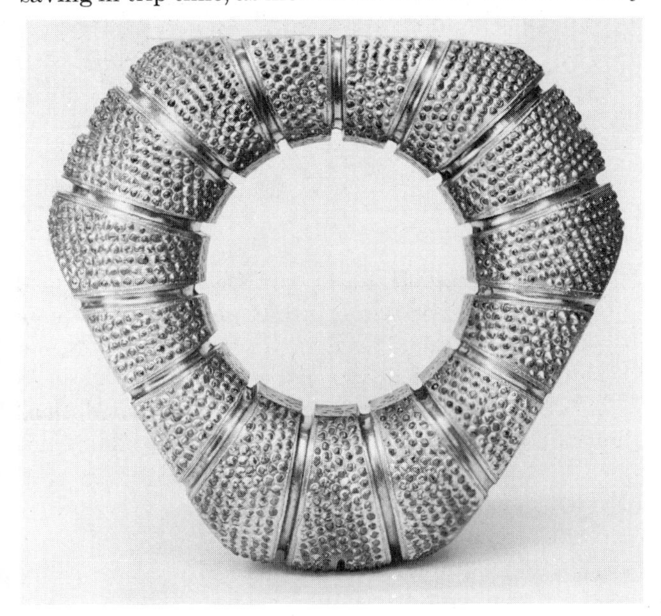

Fig. 10.2. Typical diamond core heads. Courtesy Christensen Diamond Products; and Drilling and Service, Inc.

of the diamond bit coupled with the wireline feature allows a thick section to be cored with no time lost in making trips. This is particularly beneficial in deep wells.

Sidewall Coring

It is often desirable to obtain core samples from a particular zone or zones already drilled. This is commonly accomplished by the use of a device such as that shown in Figure 10.4. A hollow *bullet* which imbeds itself in the formation wall is fired from an electric control panel at the surface. A flexible steel cable retrieves the bullet and its contained core. Samples of this type are normally $\frac{3}{4}$ or $1\frac{3}{16}$ in. in diameter and $\frac{3}{4}$ to 1 in. long. Sidewall coring is widely applied in soft rock areas where hole conditions are not conducive to drill stem testing. The zones to be sampled are normally selected from electric logs.

10.2 Operational Procedures

While this section is not offered as an operating manual, there are certain general operational considerations which warrant discussion. The following are recommendations which apply to conventional coring.[1,2]

1. Every precaution should be taken to insure that the hole is junk free; small pieces of steel (bit teeth, tong dies, etc.) will quickly ruin a core bit, whether diamond or conventional. Running a subtype junk basket with the last two or three bits will usually provide safe conditions.
2. Just as in normal drilling, sufficient drill collars should be run to furnish the bit weight. Stabilizers have been successfully applied in some cases to prevent drill string wobbling. The core barrel itself should be inspected for straightness. A crooked inner barrel will cause eccentric action on bottom.
3. Core heads should be run into the hole at a safe speed to avoid plugging or damage from hitting a bridge or dog-leg.
4. During coring the weight should be fed off smoothly and uniformly — not in bunches. This requires the full attention of a crew member at the brake, unless an automatic feed control is available.
5. Coring should begin at light bit weight and low rotary speed; these may be increased as soon as cutting action is established. Normally the applied bit weights and table speeds should be held within the limits given in Tables 10.1 and 10.2, unless specific experience in the area dictates otherwise. Circulating volumes for conventional core bits approach those of regular bits of the same size. Diamond bits require less fluid volume, and may actually be pumped and bounced off bottom by excessive circulating rates. Also, severe erosion of the water courses and bit matrix may occur. Table 10.2 includes recommended circulating rates for diamond coring.
6. Pump pressure should be closely watched during diamond coring as an indication of whether drilling fluid is passing over the face of the bit. With the bit on bottom, pressure should be higher than when the bit is off-bottom. This is essential to bit cleaning and performance. A sudden pump pressure increase not alleviated by raising the bit off bottom, may mean that the core barrel is plugged by *trash* in the mud; if this happens, it should be pulled for inspection.

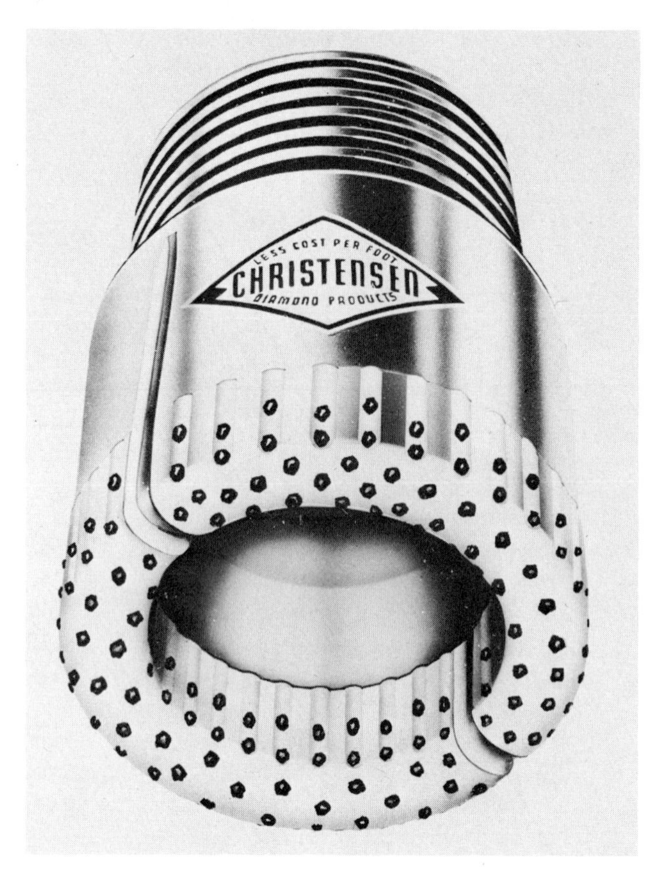

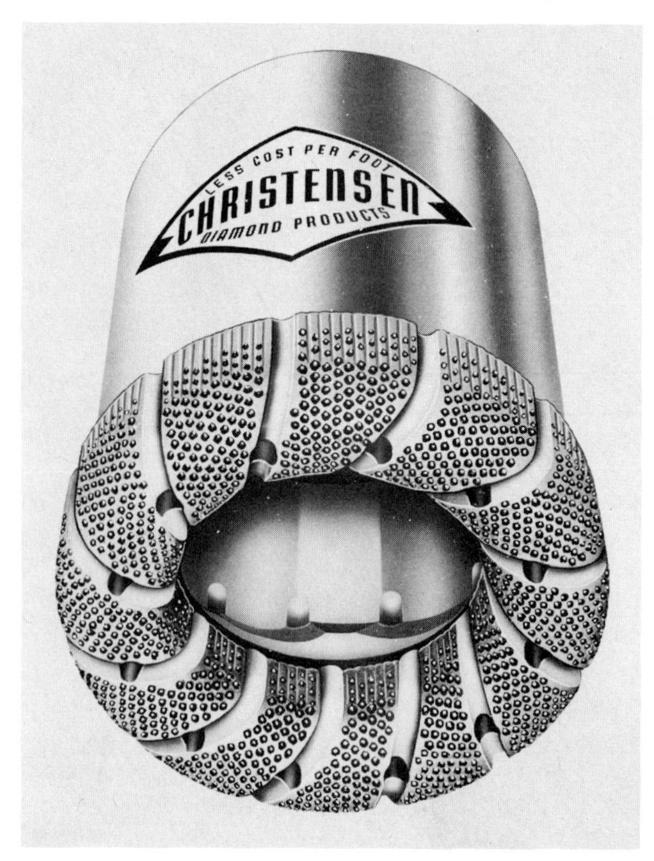

Fig. 10.2 (cont.).

7. A sudden decrease in penetration rate which cannot be attributed to a formation change may mean that the inner barrel is jammed or plugged and the assembly should be pulled for inspection. Also, the barrel may be full, unless it is certain that no errors have been made in measurement.

8. The drill string should be retrieved slowly to avoid excessive pulling suction which may suck the core out of the barrel. This is particularly important in a reduced diameter or *rat-hole* section.

These have been general operating recommendations that may be altered to fit particular cases. Experience

TABLE 10.1

Recommended Practices for Conventional (Non-Diamond) Core Bit Use
(Courtesy Hughes Tool Company)

Size range, in.	*Maximum weight recommended, lb*	*Maximum rpm*
$5\frac{1}{4} - 6$	10,000	45
$6\frac{1}{8} - 7\frac{1}{2}$	12,000	50
$7\frac{5}{8} - 8\frac{3}{4}$	15,000	60
$9\frac{5}{8} - 11\frac{7}{8}$	18,000	60

Formation	*Cutter head*	*Pump Speed*	*Rotary speed*	*Weight required on bit*	*Remarks*
Soft shale, sand gumbo	Soft formation	Fast as necessary to prevent balling up	25-35	Light weight	Too much weight will "ball up" bit
Soft with hard streaks — chalk	Soft formation	Fast as necessary to prevent balling up	25-35	Light to medium	Use light weight in soft formation
Broken formation, sulphur bearing rock	Hard formation	Medium	25-35	Light weight	Heavy weight and high speed breaks up core
Anhydrite, chert, hard lime, granite, quartzite	Hard formation	Full	30-40	Heavy weight	On granite, and such rocks, more hole per cutter head can be obtained with slow rotation

TABLE 10.2
RECOMMENDED DIAMOND CORING PRACTICES
(COURTESY CHRISTENSEN DIAMOND PRODUCTS)

	Drilling weight			*Rotary speed*			*Circulation rate*	
Bit size range OD, in.	*Recommended starting weight, lb*	*Recommended minimum drlg. wt., lb*	*Recommended drlg. wt., lb*	*Recommended starting rpm*	*Recommended minimum rpm*	*Recommended drilling rpm*	*Recommended minimum gal/min*	*Recommended maximum gal/min*
4 − 5	4,000	4,000	8,000	50	30	125	100	250
5 − 6	4,000	4,000	10,000	50	30	125	125	300
6 − 7	4,000	4,000	12,000	50	30	100	125	300
7 − 8	8,000	8,000	15,000	40	20	100	150	350
8 − 9	8,000	8,000	20,000	40	20	80	150	400
9 − 10	8,000	8,000	25,000	40	20	80	200	450
10 and larger	8,000	8,000	30,000	40	20	60	200	450

in an area is always the best guide to successful coring. Complete core recovery is the rule rather than the exception in hard, non-fractured formations, but it becomes successively more difficult to achieve in fractured and/or less consolidated formations. Poor operating procedures are one hazard that can be avoided.

10.3 Handling and Sampling of Core Recovery

The ultimate purpose of coring is usually the quantitative analysis of the physical properties and fluid content of the recovered core. (Some samples, however, may be evaluated by visual inspection — such as solid shale — while some are cut for lithological information only.) Consequently, it is extremely important that proper care be exercised to insure that the core reaches the laboratory in the best possible condition. The following are recommended handling and sampling procedures which have been furnished by Core Laboratories, Inc.[3]

Field Sampling

1. Field check list

   A log sheet is provided for the field engineer on which a record is kept of the depths of the interval cored, coring time for each foot, lithologic description of the core, sample number and depth, fractures and any other notable features of the core, and type and properties of drilling fluid.

2. Removal of core from core barrel and handling prior to preservation

a. Whole core analysis

   The core should be removed from the barrel in segments as long as possible and care should be taken to prevent excessive breaking up of the core. Jarring and hammering on the core barrel is often necessary, but this should be done as carefully as possible to avoid crushing the core or opening fractures. Each piece should be wiped clean with dry rags (not washed) as soon as it is removed from the barrel, and laid out on the pipe rack and marked as to top and bottom. After all the core is removed from the barrel, the core is measured with a tape and marked off into feet. Any lost core is logged at the bottom of the cored interval. If more core is measured than was supposedly cut, the discrepancy is resolved by the operator.

b. For conventional *plug type* core analysis

   The procedure for removal of core from the core barrel for conventional analysis is generally the same as for whole core analysis. In this case short pieces of core are used for analysis, and the extra precautions to recover long pieces are not necessary.

c. For sidewall core analysis

   Due to the normally fragile condition of sidewall cores, care should be exercised in removing them from the coring instrument. It is recommended that they be securely sealed in small containers immediately upon removal from the coring instrument.

3. Frequency of sampling

a. For whole core analysis

   Frequency of sampling is no problem in whole core analysis; all the core recovered from any section to be studied should be analyzed.

b. For conventional core analysis

   In plug type analysis, one sample per foot is ordinarily taken. Formation which is obviously nonproductive, such as solid shale, is not sampled. If the section to be analyzed is heterogeneous, samples may be taken closer than one per foot. Taking fewer samples than one per foot is not recommended. Sufficient samples should always be taken to define net productive thickness, transitional zones, and contacts.

c. For sidewall core analysis

   Sample frequency in sidewall cores is normally beyond the control of the core analyst.

4. Preservation of core

If the core samples selected for analysis are to be analyzed for fluid content, it is necessary that the samples be preserved for transportation to the laboratory to prevent the evaporation of the liquids. This is

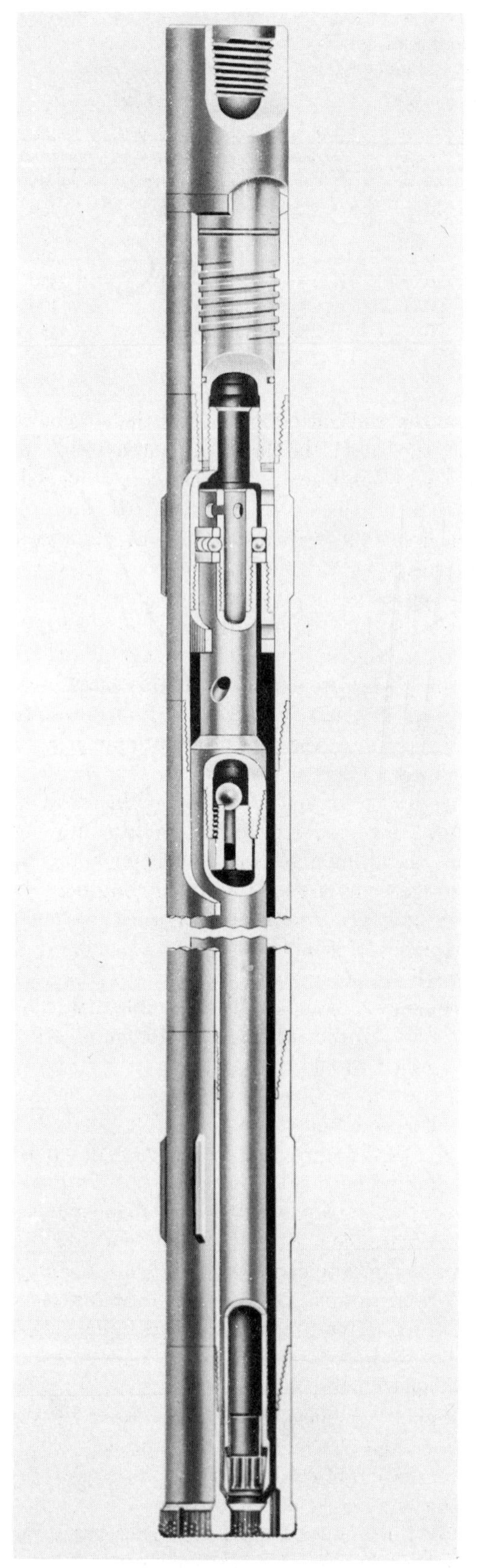

Fig. 10.3. Conventional diamond core barrel. Courtesy Christensen Diamond Products.

normally done by freezing with dry ice. It has been shown that samples frozen with dry ice can be stored for long periods of time without their fluid content or other properties being affected. Alternatively, one can wrap the samples tightly in plastic bags so as to exclude air. Care must be taken to prevent puncturing the bag or

Fig. 10.4. Sidewell coring device. Courtesy Schlumberger Well Surveying Corporation.

exposing it to extremes of temperature. Samples of both whole core and conventional analysis are preserved by the above procedures. A third method that is sometimes used for conventional samples is to wrap them tightly with foil and coat them with paraffin. Sidewall samples are usually kept in the bottles supplied by the coring service.

5. Core handling in laboratory

After the samples arrive in the laboratory, they are placed in order of depth and sample number. If frozen, they are allowed to thaw until they can be handled. They are wiped clean again and an ultraviolet examination and a visual (microscopic) description are made and recorded. A detailed notation of fractures and vugs is also made at this time. Whole diamond core sections are frequently photographed to permit later detailed study of fractures and vugs.

10.4 Routine Core Analysis

The core which reaches the laboratory has undergone an extreme environmental change from its original undisturbed state in the reservoir. First it has been subjected to flushing and contamination by the drilling fluid; then the confining pressure and temperature have

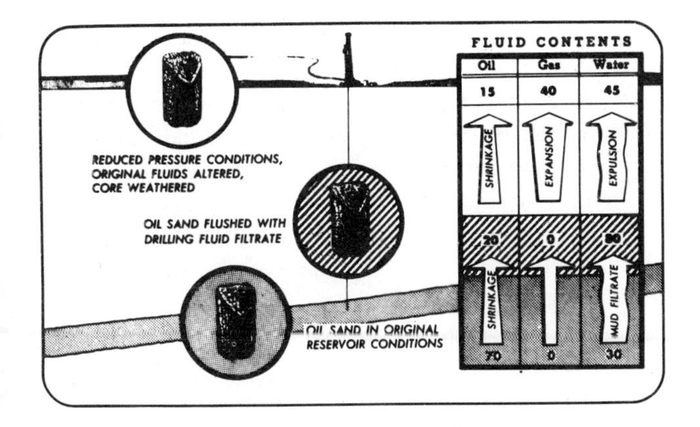

Fig. 10.5. Environmental changes undergone by core between the reservoir and laboratory. After Clark and Shearin,[4] courtesy AIME.

been reduced when it was brought to the surface. Thus gas has evolved from solution in the oil. The expansion of this gas expels some part of the liquid content. Evaporative or weathering losses occur during surface handling before the core is finally frozen or sealed. Therefore it is obvious that the fluid content as determined in the laboratory cannot be the original content.

Fortunately, the porosity and absolute permeability are not appreciably altered by these factors in most cases, provided the sample is properly cleaned. It is possible, however, that invasion of the core by drilling fluid solids may affect these values to a measurable extent. Also, the mud filtrate may, in some cases, permanently alter or cause movement of interstitial clay particles within the rock. Normally, no specific attention is given to these possibilities in routine analyses. These factors will be discussed further in the next chapter when we consider the general topic of formation damage. Figure 10.5 illustrates the conditions and factors which affect the reservoir to laboratory alteration of core samples.[4]

Routine core analysis, as discussed here, will include the measurement of porosity, absolute permeability, and fluid saturations. These quantities were defined in Chapter 2; however, a brief summary is in order.

1. Porosity: the fractional void space within a rock.

$$\phi = \frac{V_b - V_s}{V_b} = \frac{V_p}{V_b} \tag{2.1}$$

where ϕ = porosity
V_b = bulk volume
V_s = solid or grain volume
V_p = pore volume

2. Permeability: the ability of a rock to transmit fluids. The discussion to this time has considered only the absolute permeability which is purely a rock property and independent of fluid content. This was defined analytically by the Darcy equations.

3. Fluid saturation: the fraction of the pore volume occupied by a particular fluid:

$$S_w = \frac{V_w}{V_p}$$

$$S_o = \frac{V_o}{V_p}$$

$$S_g = \frac{V_g}{V_p}$$

where S_w, S_o, S_g = water, oil, and gas saturations, respectively
V_w, V_o, V_g = water, oil, and gas volumes in the rock

Since many measurements of rock properties require a clean dry sample, let us first consider the cleaning, drying, and other preparational procedures.

10.41 Sample Preparation

It is generally desirable to perform tests on samples as large as possible. This is particularly important when the formation is quite heterogeneous, as is the case with many limes and dolomites, or any fractured rocks. These are best analyzed by using the entire core, hence the common name, *whole core* analysis. Reasonably homogeneous rocks having intergranular porosity are normally analyzed by selection of small, representative samples at frequent intervals through the section of interest. This is commonly referred to as conventional or *plug* analysis. Porosity and permeability values are measured from small cylindrical or cubical plugs in conventional work, or from a length of the full diameter core in whole core analysis.

In plug analysis, it is customary to cut samples parallel to the bedding planes, so that measurements of *horizontal permeability* may be made. It is often desirable to measure the permeability perpendicular to bedding planes (vertical permeability) as well. Cubical samples are used when directional measurements are desired. This also requires that the individual cubes be carefully marked when cut so that no orientation mixup results.

After the method of sampling has been decided upon, the samples to be used for permeability (and, normally, for porosity measurements also) are thoroughly cleaned of all interstitial fluids and dried. The cleaning process is commonly performed in extraction apparatus such as that indicated in Figure 10.6. The samples are placed in the extractor with a solvent (pentane, naphtha, toluene, carbon tetrachloride, etc.) and boiled for several hours. This method of cleaning is not entirely satisfactory, and other methods are often employed. Centrifugal extractors which cycle clean solvent through the sample have several obvious advantages in that they produce a cleaner sample in much less time. After extraction, the samples are oven-dried at 250°F, or less if alteration of

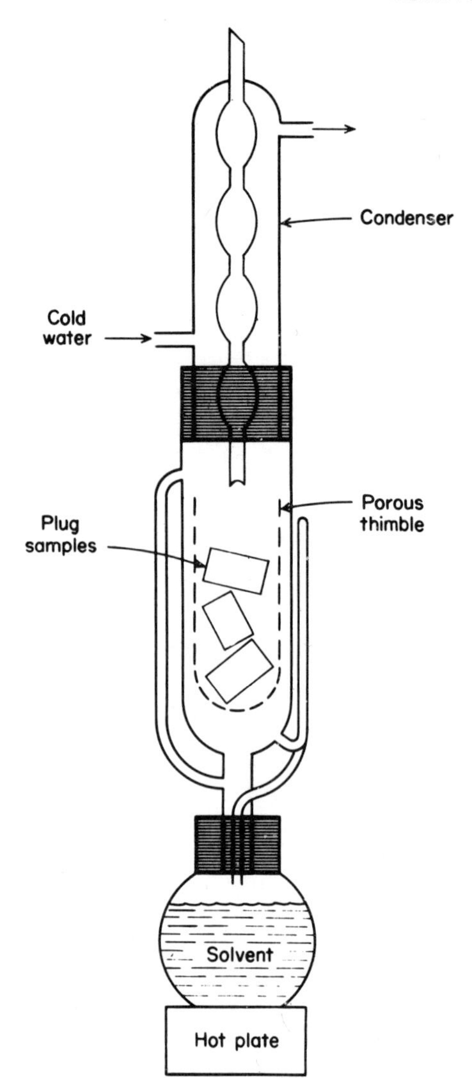

Fig. 10.6. Soxhlet extractor for plug cleaning.

interstitial clays is anticipated. After the sample has cooled, it is ready for the necessary testing.

A special method of cleaning whole core samples is shown in Figure 10.7.[5] The cores are placed in the core chamber through the open end. An O-ring cap is screwed in place and the chamber is then filled with gas to a pressure equal to that of the gas dissolved in the solvent. Next, this gas is displaced at constant pressure by the solvent-gas mixture. Then, the chamber is pressured up by means of the hydraulic pump to approximately four or five times the solvent-gas pressure. When liquid flow into the cores ceases, the core chamber is depressured rapidly to atmospheric pressure; the cores are left submerged in solvent until most of the gas has flowed from the cores. The solvent is then drained off and the cycle repeated. Data showing the number of cycles necessary to clean four different types of formations are shown here. The core samples reported were cleaned and dried and their porosity determined. They were then subjected to additional cleaning cycles, and dried each time until the porosity showed no increase.

10.42 Porosity Measurements

It may be deduced from Eq. (2.1) that the porosity of a sample can be determined from a knowledge of any two of the factors V_b, V_s, or V_p. There are several ways by which these may be measured with satisfactory accuracy.[6] The choice of method depends largely on the type of sample, subsequent tests to be made, and the preference of the individual laboratory. Only a few techniques will be discussed here.

Grain and/or Pore Volume Measurements

1. Boyle's law porosimeters: the operation of these devices is based on the gas law. The two-cell type

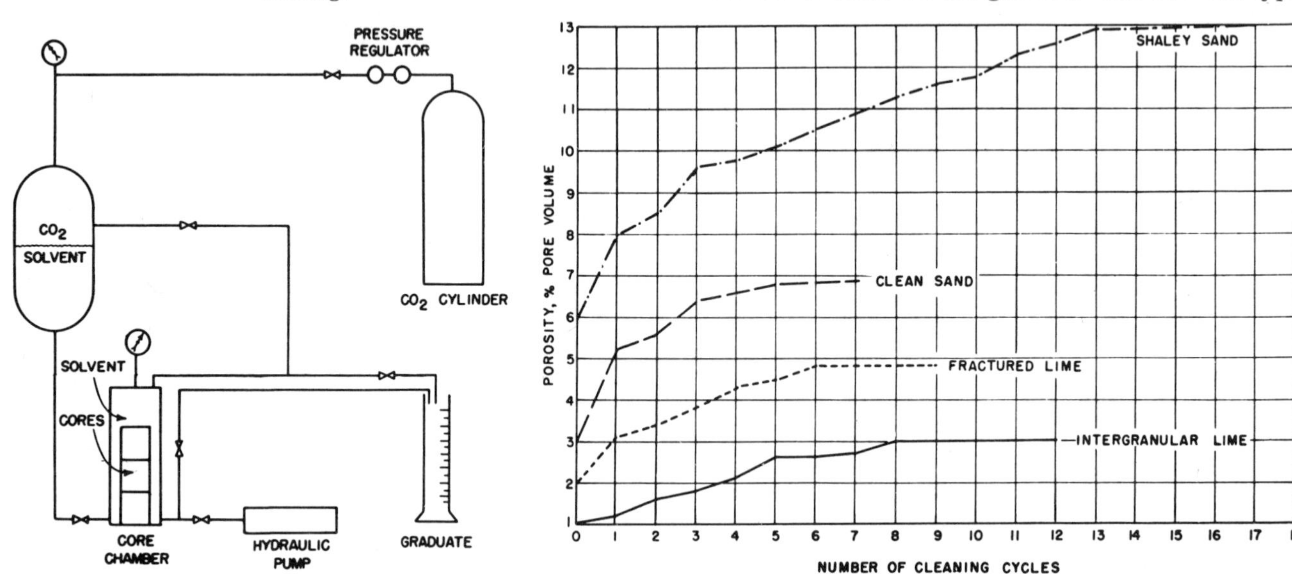

Fig. 10.7. Special apparatus for cleaning whole core samples. After Stewart and Spurlock,[5] courtesy *Oil and Gas Journal.*

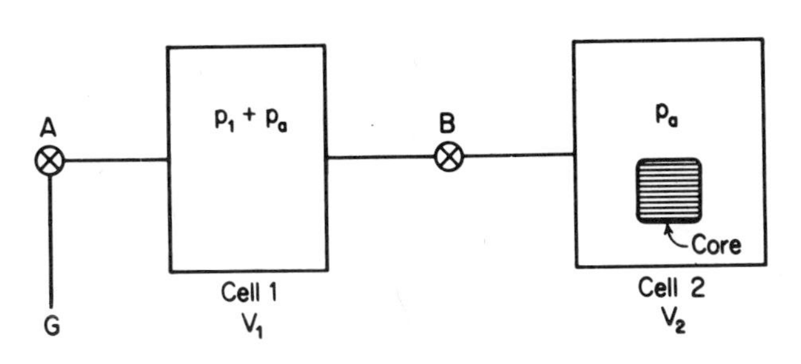

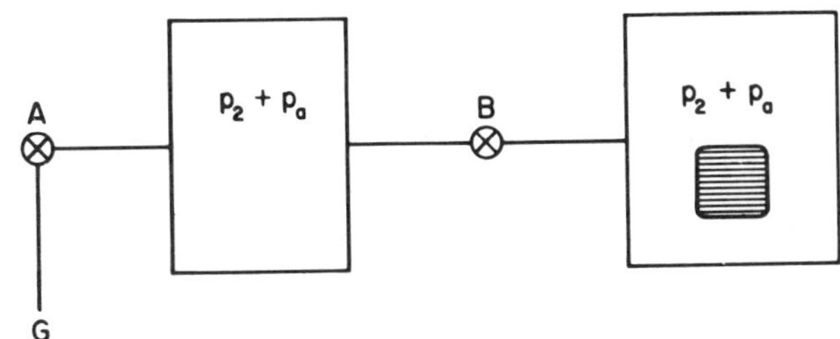

Fig. 10.8. Operation of two cell Boyle's law porosimeter.

shown in Figure 10.8 is well known and illustrates the principle involved.[7,8]

Two cells of known volumes, V_1 and V_2, are connected to a source of gas pressure G as shown in the diagram. The connecting valve B between the cells is closed and cell 1 is filled to an initial absolute pressure $p_1 + p_a$. Valve A is then closed. The clean, dry core sample is placed in cell 2 at atmospheric pressure p_a. Valve B is then opened, allowing the pressure in both cells to equalize at $p_2 + p_a$. Assuming an isothermal expansion of a perfect gas, the following equation for the grain volume of the sample is derived. First, by a material balance between conditions I and II:

$$n_1 + n_2 = n_3 + n_4 \tag{1}$$

where n_1, n_2 = mols of gas in cells 1 and 2 at condition I

n_3, n_4 = mols of gas in cells 1 and 2 at condition II

From the ideal gas law, $n = pV/RT$; therefore,

$$\frac{(p_1 + p_a)V_1}{RT} + \frac{p_a(V_2 - V_s)}{RT} = \frac{(p_2 + p_a)V_1}{RT} + \frac{(p_2 + p_a)(V_2 - V_s)}{RT} \tag{2}$$

from which:

$$V_s = V_1 + V_2 - \frac{p_1}{p_2} V_1 \tag{10.1}$$

where p_1, p_2 = *gage* pressures at conditions I and II, respectively

V_1, V_2 = volumes of cells (1) and (2)

V_s = grain volume of core sample

It is not desirable to calculate every measurement from Eq. (10.1), due to the excessive time and possible errors involved. Further, the cell volumes V_1 and V_2 are difficult to measure with the desired accuracy. This instrument is, however, easily calibrated with precisely known solid volumes such as steel balls. If all measurements are then started at the same p_1, it is a simple matter to obtain V_s from a previously determined calibration plot of V_s vs p_2. Considerable attention must also be given to the gas used. Air is generally unsatisfactory for precise work due to its adsorption on the sand grains at high pressures. Helium does not exhibit this behavior and is therefore an excellent choice for Boyle's law instruments;[7] also, its deviation from the perfect gas law is negligible at the pressures involved (generally from 50 to 100 psig). Numerous other Boyle's law porosimeters with different design features are used for grain volume measurements.

2. Saturation methods: The pore volume of a sample may be measured gravimetrically by completely saturating a sample with a liquid of known density, and noting the weight increase. A simple apparatus for this procedure is shown in Figure 10.9. A high vacuum is created inside the flask which contains the clean, dry, previously weighed sample. A liquid (commonly kerosene or a similar hydrocarbon) is allowed to enter the vessel (and the pores of the sample) until it covers the sample. The saturated sample is then removed from the flask with forceps, dried of excess liquid, and quickly weighed. The pore volume is obtained as:

$$V_p = \frac{W_s - W_d}{\rho_l} \tag{10.2}$$

where W_s = saturated sample weight

W_d = dry sample weight

ρ_l = density of saturating liquid.

Grain volume may also be calculated from:

$$V_s = \frac{W_d}{\rho_s} \tag{10.3}$$

where ρ_s = sand grain density.

Equation (10.3) is often used with the typical value for ρ_s of 2.65 gm/cc. This is generally quite accurate for sands; however, small errors in ρ_s become appreciable in low porosity samples. Grain density may be measured by pulverizing a sample to grain size, weighing the prescribed amount, and measuring the volume of liquid it displaces in a volumetric flask. It should be pointed out that measurements using grain density measure total, not just effective, porosity.

Another widely used porosity determination method utilizes the fluid contents of the core as it comes from the field. A sample of approximately 20 to 30 cc bulk volume is used. Mercury is injected into the sample at relatively high pressure, the injected volume being taken as the volume of gas space V_g existing in the core. Water and oil contents are then determined by distillation. The summation of these three fluid volumes, $V_g + V_o + V_w$, is the pore volume of the sample. Bulk volume is obtained by separate measurement, and is required for computation of porosity.

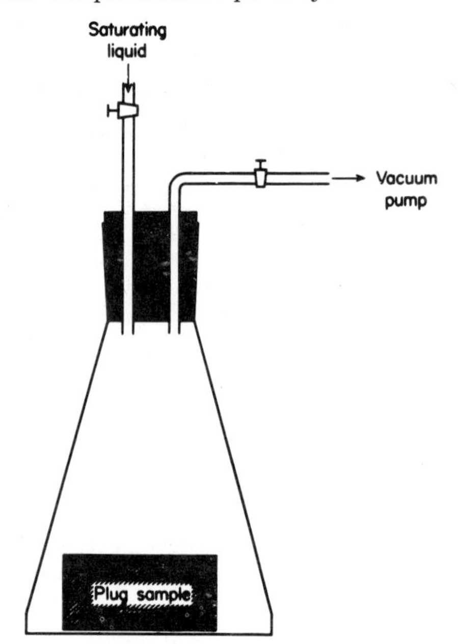

Fig. 10.9. Saturation apparatus for pore volume determination.

Bulk Volume Determinations

The bulk volume of a sample is normally found by measurement of the volume of liquid it displaces. If the liquid is one which readily penetrates the pores, the sample must first be saturated so that the true bulk volume is displaced. To avoid this problem, mercury is commonly used as the displaced liquid since it will not appreciably penetrate normal pore sizes at low pressures. To keep the pressure low requires, of course, that submergence of the sample be restricted to small depths (2–5 cm). If the sample is a uniform cube or cylinder, its volume may be calculated from dimensional measurements.

Example 10.1

Given the following data, compute the porosity of a cylindrical sample. The grain volume was measured in a two-cell Boyle's law porosimeter.

Sample Dimensions	*Porosimeter Data*
Length = 4.00 cm	V_1 = 25.0 cc
Diameter = 2.50 cm	V_2 = 50.0 cc
	p_1 = 100.0 psig
	p_2 = 50.0 psig

Solution:

$$V_B = \frac{\pi(2.50)^2}{4} \times 4.00 = 19.6\text{cc}$$

$$V_s = 25 + 50 - \left(\frac{100}{50}\right)25 = 25.0 \text{ cc [Eq. (10.1)]}$$

$$\phi = \frac{25.0 - 19.6}{25.0} = 0.216 \text{ or } 21.6\%$$

Example 10.2

The bulk volume of a core sample was measured by mercury displacement as 25.0 cc. Pore volume was obtained by saturating the sample with a hydrocarbon solvent as shown by the following data:

$$W_d = 50.25 \text{ gm}$$
$$W_s = 54.50 \text{ gm}$$
$$\rho_l = 0.701 \text{ gm/cc}$$

(a) What is the sample's porosity?

$$V_p = \frac{54.50 - 50.25}{0.701} = 6.06 \text{ cc[Eq. (10.2)]}$$

$$\phi = \frac{6.06}{25.0} = 0.242 \text{ or } 24.2\%$$

(b) What is the grain density of this rock?

$$\rho_s = \frac{W_d}{V_g} = \frac{W_d}{V_b(1-\phi)} = \frac{50.25}{25.0(1-0.242)} = 2.65 \text{ gm/cc}$$

10.42 Permeability Measurements

The absolute permeability of a core sample is determined from flow test data. A gas (usually air) is used as the flowing fluid for several reasons: (1) steady state flow is quickly obtained, which allows rapid determinations; (2) dry air does not alter the mineral constituents of the rock; (3) 100% saturation to the flowing fluid is easily obtained. Specific instructions for permeability measurements may be found in the *API Code No. 27;* however, the apparatus shown schematically in Figure 10.10 is typical for small sample or plug measurements. The core is placed in a suitable holder such as either the Fancher or Hassler type which seals the sides, allowing linear flow only. The pressure differential is measured by suitable manometers. The flow volume may be obtained with a calibrated capillary or orifice, a gas (wet test) meter, or by liquid displacement, depending on individual preference.

Air or gas permeability is calculated from the suitable form of Darcy's equation as developed in Chapter 2. For linear flow tests:

$$(2.14) \qquad k = \frac{2q_2\mu L p_2}{A(p_1^2 - p_2^2)}$$

or

$$(2.16) \qquad k = \frac{q_m \mu L}{A\Delta p}$$

where k = permeability, darcys

q_2 = flow rate at exit conditions, cc/sec

q_m = flow rate at mean conditions, $\frac{p_1 + p_2}{2}$, cc/sec

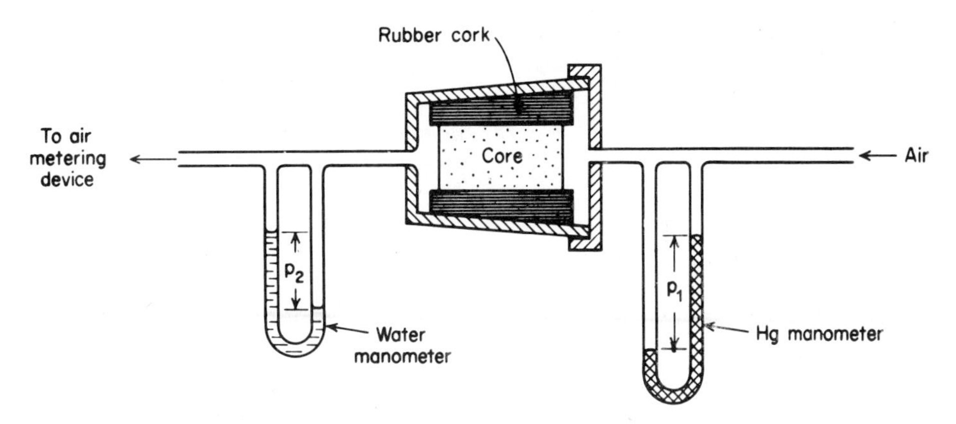

Fig. 10.10. Apparatus for routine air permeability determinations (Fancher core holder).

μ = gas viscosity at test temperature, cp
L = sample length, cm
A = sample area, cm^2
Δp = pressure differential across sample, atm
p_1 = inlet pressure, atm (absolute)
p_2 = exit pressure, atm (absolute)

In whole core analysis, the procedure for permeability measurement must be altered. Normally, the horizontal permeability is of principal interest; however, vertical permeability may also be desired. Radial flow tests are sometimes performed by boring a vertical hole down the center of the core through which gas is then allowed to flow. Equation (2.23) is then applied (if a gas is the flowing fluid):

$$k = \frac{q_w \mu p_w \ln(r_e/r_w)}{\pi h(p_e^2 - p_w^2)} \tag{2.23}$$

where q_w = rate of flow at the sample's well bore conditions (commonly atmospheric), cc/sec
p_w = pressure at well bore, atm (absolute)
r_e, r_w = outer and *well bore* radii of sample, cm
h = sample thickness, cm

Radial flow tests require considerable sample preparation and are not well adapted to routine, large volume work. Consequently other methods have been devised.[5,9] Since horizontal permeabilities are usually desired, the measuring fluid must flow from one area along the side of a cylindrical specimen to a similar area directly opposite. This procedure introduces the problem of determining the true flow path length L for use in Darcy's equation. A simple solution to this problem is the method presented by Stewart and Spurlock.[5] Their multiple core apparatus is shown in Figure 10.11. This particular permeameter can accommodate 24 in. of total core which may consist of 4-, 6-, 8-, 12- or 24-in. samples. Flow in any horizontal direction is obtained by orienting each sample. Vertical permeability may also be measured by using separate, easily inserted ram heads and spacers. Permeameter operation is as follows:

1. Vacuum is applied by means of the aspirator, and the ram is moved to the top of the instrument.
2. Preselected cores with screens attached are placed on top of the ram, and the ram and cores lowered into position.
3. Ram pressure and air pressure to the diaphragm are then applied.
4. Air under pressure is then admitted to the upstream manifold and flows through each core individually. After stabilized flow is reached, the pressure drop and air flow rates for each core are measured and recorded.

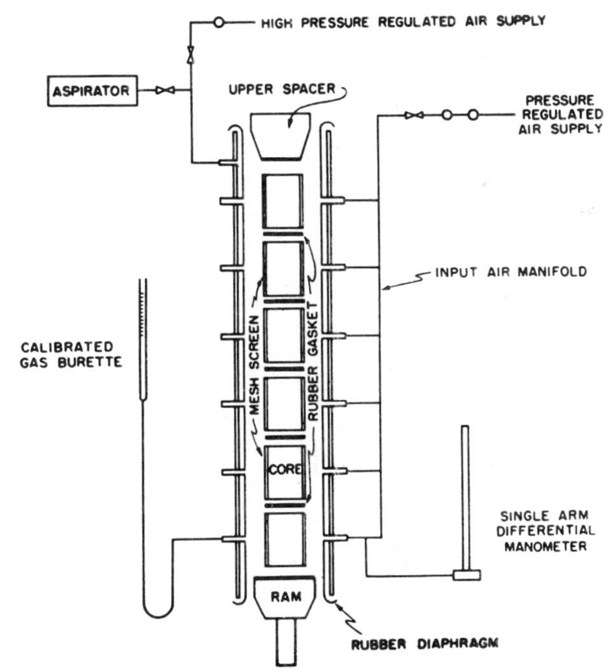

Fig. 10.11. Whole core permeameter. After Stewart and Spurlock,[5] courtesy *Oil and Gas Journal.*

This apparatus divides the circumference of the core into quarters, with flow taking place between two diametrically opposed, full length sections. Under these conditions the use of a shape (or flow path distortion) factor is unnecessary. This may be shown as follows:

$$k = \frac{q}{A} \times \frac{L_m}{\Delta p} \mu \qquad (1)$$

where L_m is the mean flow path length, and other notations are standard Darcy equation variables. However, the area A through which flow occurs may be taken as the projected area covered by the 90° screen:

$$A \cong \sqrt{2}rh \qquad (2)$$

where r = core radius

h = vertical length of the core

Also

$$L_m \cong \sqrt{2}r \qquad (3)$$

Substituting Eqs. (2) and (3) in Eq. (1),

$$k = \frac{q}{\sqrt{2}rh} \times \frac{\sqrt{2}r}{\Delta p} \times \mu = \frac{q\mu}{h\Delta p} \qquad (10.4)$$

Equation (10.4) is then a simplified permeability formula for the operation of a whole core permeameter of this type for these conditions:

(1) The screens (or flow openings) are diametrically opposed and each covers $\frac{1}{4}$ of the core perimeter.

(2) All other surfaces of the core, including the ends, are sealed.

Normally, two horizontal permeabilities are measured. One is parallel to any fractures or other highly permeable pores, k_{max}; and another is perpendicular to this direction, $k_{90°}$. Vertical permeability is also frequently measured.

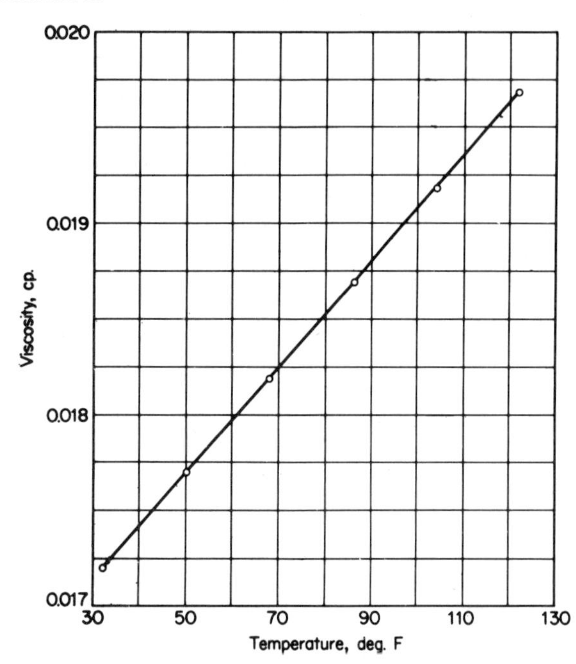

Fig. 10.12. Viscosity of air at one atmosphere pressure. Data of Bearden, from API Code 27.

Example 10.3

A whole core specimen $3\frac{1}{2}$ in. in diameter × 12 in. in length is subjected to a flow test in the whole core permeameter just described. The following data are given:

Flowing temperature = 70°F

Δp = 0.05 atm

Flow rate at mean flowing pressure $(p_1 + p_2)/2$ = 1.50 cc/sec

Compute the permeability of the sample at the test conditions.

Solution:

From Figure 10.12, μ = 0.0182 cp

Using Eq. (10.4)

$$k = \frac{(1.50)(0.0182)}{(12 \times 2.54)(0.05)} = 0.018 \text{ darcys or } 18 \text{ md}$$

In order to obtain the absolute permeability of a rock from gas flow tests, it is necessary that an anomaly caused by the nature of a gas be accounted for. This was first recognized by Klinkenberg[10] and is known as the Klinkenberg effect or correction. This principle states that permeability to gas is a function of the mean free path of the molecules, and is therefore dependent on the mean pressure at which the test is performed. This is expressed by Eq. (10.5):

$$k_a = k\left(1 + \frac{b}{\bar{p}}\right) \qquad (10.5)$$

where k_a = the measured air permeability at the pressure $\bar{p}$

k = the true or absolute permeability of the rock (also called equivalent liquid permeability)

b = a constant primarily dependent on pore size (hence on permeability) which increases in value as pore size decreases

$\bar{p}$ = mean pressure at which test is conducted

It is apparent that as $\bar{p}$ becomes large, $k_a \rightarrow k$. This is the basis for making the Klinkenberg correction. Air permeabilities are measured at several values of $\bar{p}$, and a plot of k_a vs $1/\bar{p}$ is made. Extrapolation to $1/\bar{p} = 0$ (or $\bar{p} = \infty$) yields the true or equivalent liquid value k.

Example 10.4

The air permeability of a conventional plug sample was measured at several mean pressures with the following results. Determine the absolute or equivalent liquid permeability of the sample.

Sample area = 3.00 cm²

Sample length = 3.00 cm

Flowing temp. = 70°F

Atmos. Pressure = 29.30 in. Hg = 29.30/29.92 = 0.979 atmos.

μ = 0.0182 cp from Figure 10.12. No correction for pressure is necessary over the normal ranges involved.

Calculation Steps

1. Δp = inlet pressure, since discharge is to atmosphere. To convert to atmospheres: column (3) = column (1)/29.92 where 29.92 inches Hg = 1 atm.

Run	(1) Inlet pressure, in. Hg	(2) Flow rate at discharge q_2, cc/sec	(3) Δp, atm	(4) Mean core pressure $\bar{p}$, atm	(5) Flow rate at mean pressure q_m, cc/sec	(6) Calculated air perm. k_a, md	(7) $1/\bar{p}$
1	5.0	0.348	0.167	1.062	0.321	35	0.941
2	15.0	1.17	0.501	1.229	0.930	33.7	0.814
3	25.0	2.17	0.836	1.397	1.52	33.1	0.716
4	75.0	9.73	2.50	2.229	4.27	31.1	0.448

2. $\bar{p} = p_{atm} + \Delta p/2$, or (4) = 0.979 + (3)/2.
3. $q_m = q$ at discharge condition (q_2) converted to mean core pressure.

$$\therefore q_m = q_2 \times p_2/\bar{p}, \text{ or } (5) = (2) \times 0.979/(4)$$

4. Uncorrected air permeability is then calculated from Eq. (2.14) or Eq. (2.16), as desired. For run no. 1: Note that if Eq. (2.14) is used, q_m and $\bar{p}$ need not be calculated.

$$k_{a1} = \frac{(0.321)(0.0182)(3.00)}{(3.00)(0.167)} = 0.035\ d \text{ or } 35 \text{ md [Eq. (2.16)]}$$

or

$$k_{a1} = \frac{(2)(0.348)(0.0182)(3)(0.979)}{(3.00)[(1.146)^2 - (0.979)^2]} = 0.035\ d \text{ or } 35 \text{ md [Eq. (2.14)]}$$

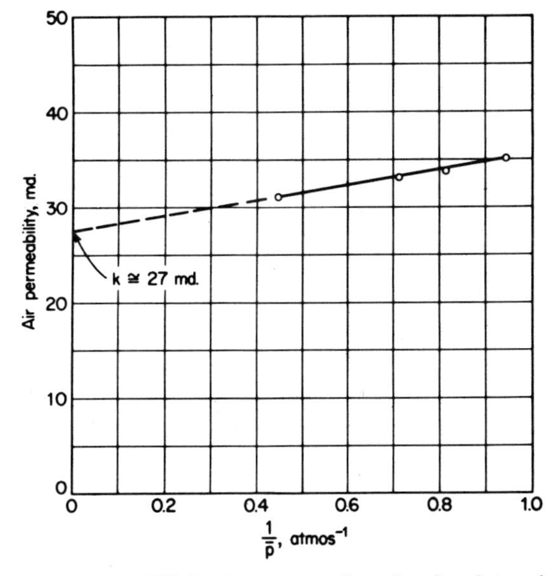

Fig. 10.13. Klinkenberg correction plot for data of Example 10.4.

5. Similar calculations for k_a are made for runs 2, 3, and 4. These values are then plotted vs $1/\bar{p}$ to obtain the equivalent liquid or absolute k. This is shown in Figure 10.13. *Ans.* $k = 27$ md.

The Klinkenberg correction is not particularly critical in highly permeable samples or if high mean pressures are used. For low permeability rocks, the error may be quite large (20 to 100% high) if the correction is not made. Routine measurements often ignore this correction.

Another source of error in gas permeability measurements is that the flow may not be laminar as demanded by Darcy's law. This may be checked by plotting q_m/A vs $\Delta p/L$, as suggested by Eq. (2.16). Conformance with Darcy's law requires that a straight line through the origin of slope k/μ be obtained. The critical value of pressure gradient $\Delta p/\Delta L$ may therefore be established. Routine laboratory procedures are normally set up to meet this requirement.

Liquid permeability measurements are infrequently made, due to the time required to fully saturate the specimen and establish steady state flow, and the difficulty of finding a liquid which is completely inert with respect to the rock. Occasionally, such data are desired for special purposes, in which case the tests are performed. Calculations are made using the proper form of Darcy's equation for incompressible fluid flow.

10.43 Liquid Saturations

As was mentioned at the beginning of this section, the fluid saturations determined in the laboratory are greatly altered from their original, underground values. However, much valuable information may be derived from a knowledge of the fluid contents of a core. Saturation determinations require a knowledge of the individual fluid volumes contained in a known sample pore volume. Two methods are in common use: the retort, and the distillation methods.

1. Retort method: a fresh, relatively large sample (100 to 200 gm) is placed in the retort and heated at approximately 400°F for 20 minutes to an hour. During this period the water and light oil fractions are recovered and their volumes noted. The mixture may be centrifuged to obtain total separation of the oil and water. The temperature is then raised to approximately 1200°F to distill any heavy residual hydrocarbons not previously recovered (many labs employ high vacuums and operate at lower temperature). Any water recovered during this latter period is considered to be of intercrystalline origin and is not included in the computed water saturation. The oil volume collected during the 1200°F period is corrected for volume changes due to cracking* by an experimentally derived retort calibration curve. The bulk volume of the sample may be determined by the standard displacement method or it may be

*The breakdown of a long hydrocarbon chain to several smaller molecules when the material is heated above some critical value. This is a widely applied principle in petroleum refining.

computed from grain density. Porosity is normally obtained from a small adjacent, or companion sample. A typical retort apparatus is shown in Figure 10.14.

The retort method has several advantages. It is rapid and well adapted to routine work. The relatively large sample volume involved improves accuracy. Both water and oil volumes are measured directly. The principal disadvantages are the possible errors involved due to hydrocarbon cracking and the driving off of waters of crystallization, as mentioned previously. These errors are not serious in routine work, and the retort method is widely used. Properly corrected oil saturation values are reportedly accurate to ± 5%, with reproducibility within 2%.[3]

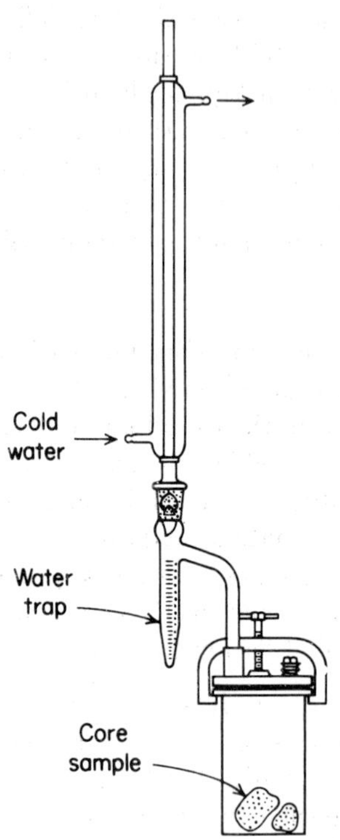

Fig. 10.14. Retort apparatus for fluid saturation determination.

2. Distillation methods: the Stark-Dean apparatus of Figure 10.15 is widely used and serves to illustrate the procedure. About 50 to 75 gm of sample (this may be the cylindrical or cubical plugs for ϕ and k determinations) are placed in a weighed, porous thimble. The thimble is then suspended in the flask which contains a suitable solvent having a boiling point near that of water. Toluene, $C_6H_5CH_3$, with a boiling point of 112°C is commonly used. The solvent is then boiled; this vaporizes both water and solvent. The vapors are condensed and fall into the calibrated trap where the water collects at the bottom. The distillation is continued until the collected water volume remains constant. The thimble and sample are then removed from the flask, dried, and reweighed. The liquid saturations may be calculated from data on the total weight loss, collected water volume, and oil density. Gas saturation is taken as the fraction of pore volume not occupied by liquids.

The primary advantages of the distillation procedure are that the disadvantages of the retort are eliminated; however, this method also has certain disadvantages. The time required for complete water distillation and oil leaching is usually several hours (2 to 24+). Also, the samples used are generally smaller than retort samples. Since the oil volume must be calculated from the weight loss not attributed to water, it is necessary to know or accurately measure the oil density. Salt deposition from the core water during distillation may also introduce appreciable errors when small, low oil content cores are

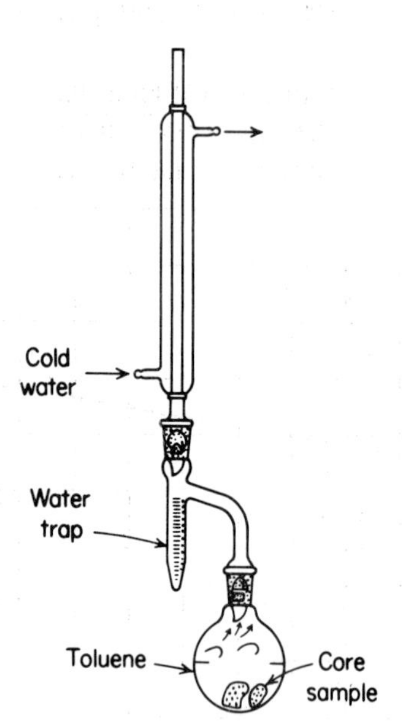

Fig. 10.15. Stark-Dean distillation apparatus for water saturation determination.

analyzed. Consequently, the choice of retort vs. distillation methods is largely a matter of personal preference. Comparable accuracy is generally obtained with either method, provided proper care and the necessary corrections are applied. Other methods are also available for special purposes; however, the adaptability of the two discussed above has caused them to be widely used by commercial and company laboratories engaged in large volume, routine work.

Fluid saturations of whole core samples may be determined by similar methods, except that the apparatus has to be much larger. Vacuum distillation or retorting is often used to expedite the measurements and improve accuracy, both in whole core and small sample analysis.

A typical saturation determination is illustrated by the following example.

Example 10.5

Given the following data on a core sample, compute the porosity, and the oil, water, and gas saturations.

1. Sample weight as received from field = 53.50 gm
2. Water volume recovered during extraction = 1.50 cc
3. Sample weight after extracting and drying = 51.05 gm
4. Density of core oil = 0.850 gm/cc
5. Bulk volume of sample = 23.60 cc
6. Grain density of sample = 2.63 gm/cc

Solution:

$$(1)\ \phi = \frac{V_b - V_s}{V_b}$$

$$V_s = \frac{W_s}{\rho_s} = \frac{51.05}{2.63} = 19.4 \text{ cc}$$

$$\therefore \phi = \frac{23.6 - 19.4}{23.6} = 0.178 \text{ or } 17.8\%$$

$$(2)\ S_w = \frac{V_w}{V_p} = \frac{1.50}{4.2} = 0.357 \text{ or } 36\%$$

$$S_o = \frac{V_o}{V_p}$$

$$V_o = \frac{W_o}{\rho_o} = \frac{53.50 - (51.05 + 1.50)}{0.850} = 1.12 \text{ cc}$$

$$\therefore S_o = \frac{1.12}{4.2} = 0.267 \text{ or } 27\%$$

$$S_g = 1 - (S_o + S_w) = 1 - (0.27 + 0.36) = 0.37 \text{ or } 37\%$$

Numerous other miscellaneous core tests are performed by commercial laboratories. Some of these will be discussed in a later section on special core testing; some, however, such as chloride content, core oil density, and grain size distribution tests will not be considered. Other texts which devote more space to these topics are listed as recommended outside readings and should be consulted. Also, the references at the end of this chapter contain specific information on more detailed analysis.

Before proceeding into more involved core testing, it is necessary to introduce certain fundamental concepts regarding the spatial distribution of fluids within a rock and the resulting effects of such distribution on flow behavior.

10.5 Fundamental Fluid Distribution Concepts — Multiphase Systems

In Chapter 2, mention was made of the fact that petroleum reservoir rocks always contain at least two, and sometimes three, separate or immiscible fluids. Water and oil; water and gas; or water, oil, and gas are the combinations of interest. The portion of the pore space which each occupies depends on the amount of each present and the wettability of the system. Wettability refers to the relative affinity between the rock and each fluid present, i.e., which fluid is preferentially adsorbed on the rock surface and is held in the most minute portions of the interstices. The determination of a particular reservoir's wettability is a difficult problem and lies beyond the scope of present treatment. In general, most rocks are wetted by water, a few are wetted by oil, and some deep high pressure reservoirs may be wetted by gas, although the latter occurrence is not definitely established. An oil-wet condition is believed to be due to the presence of polar impurities or surface active materials in the oil which, over geologic time, have been adsorbed by the rock thereby increasing its surface affinity for oil and rendering it oil-wet. The significance of fluid distribution in a porous network will become apparent as we discuss multiphase flow concepts.

Up to this time, our discussion of permeability has been restricted to the *absolute* permeabilities, which are relevant to rock completely saturated with flowing fluid. It is now necessary to expand the permeability concept to include cases where fractional saturations to two or three fluids exist. Two basic definitions must be introduced:

1. *Effective permeability:* the permeability of a rock to a particular fluid at saturations less than 100%, i.e., when other fluid(s) is present. This is measured in darcys or millidarcys and is therefore the dimensional equivalent of absolute permeability, hence:

   k_o = effective permeability to oil, darcys or md
   k_w = effective permeability to water, darcys or md
   k_g = effective permeability to gas, darcys or md
   Individual values of k_o, k_g, k_w may vary from zero up to the absolute value, k:

   $$0 \leqq k_w, k_o, k_g \leqq k$$

2. *Relative permeability:* this is merely a convenient, dimensionless quantity defined by:

   $$k_{rw} = \frac{k_w}{k} \qquad k_{ro} = \frac{k_o}{k} \qquad k_{rg} = \frac{k_g}{k}$$

   where k_{rw}, k_{ro}, k_{rg} = relative permeability to water, oil, and gas, respectively.

   Since the effective permeabilities may range from zero to k, the relative permeabilities may have any value between zero and one:

   $$0 \leqq k_{rw}, k_{ro}, k_{rg} \leqq 1$$

   Another widely used parameter is the ratio of the effective (or relative) permeabilities of water and oil, and gas and oil:

   $$\frac{k_w}{k_o} \text{ or } \frac{k_{rw}}{k_{ro}} \quad \text{and} \quad \frac{k_g}{k_o} \text{ or } \frac{k_{rg}}{k_{ro}}$$

These ratios are dimensionless and may vary from zero to infinity.

Consider the two-phase flow behavior depicted in Figure 10.16. The entire pore space is filled with water and oil so that $S_w + S_o = 100\%$ at all times. To visualize what is happening, assume that the rock is originally 100% saturated with oil. Further, assume that we introduce water into every pore simultaneously and that a *water-wet* equilibrium is instantaneously established. This, of course, we cannot do, except mentally to visualize the mechanism involved. When water is first introduced, it is adsorbed by the rock and held immobile both on the rock surfaces and in the small corners around the junctions of the individual grains. This immobility is indicated by $k_{rw} = 0$ in region A. Note, however, that k_{ro} is essentially constant at 1.0 over the same saturation range. As this process continues, the water saturation reaches some critical value S_{wc} at which water becomes mobile, ($k_{rw} > 0$). At this time, both oil and water flow; as water saturation is increased (and oil saturation is decreased), however, k_{ro} decreases and k_{rw} increases, as shown in region B. Continued increase of S_w causes the oil saturation to reach a residual value S_{or} at which oil becomes immobile ($k_{ro} = 0$) and only water flows. This is the minimum saturation to which oil may be reduced by injecting water. If it were possible to remove the oil by some other means, k_{rw} would continue to increase and finally reach the value of one as shown. This process could have been visualized in reverse just as well. It should be noted that this example portrays oil as non-wetting and water as wetting. The curve shapes shown are typical for wetting and non-wetting phases and may be mentally reversed to visualize the behavior of an oil-wet system. Note also that the total permeability to both phases, $k_{rw} + k_{ro}$, is less than 1, in regions B and C.

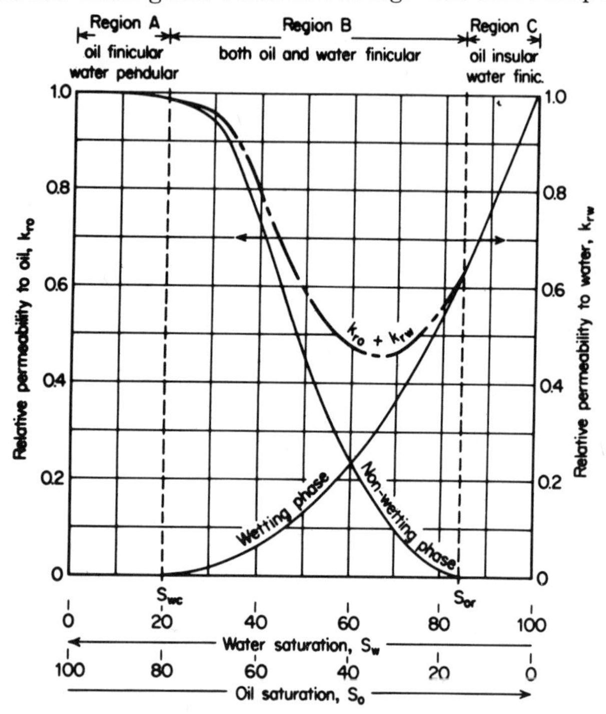

Fig. 10.16. Typical two-phase flow behavior.

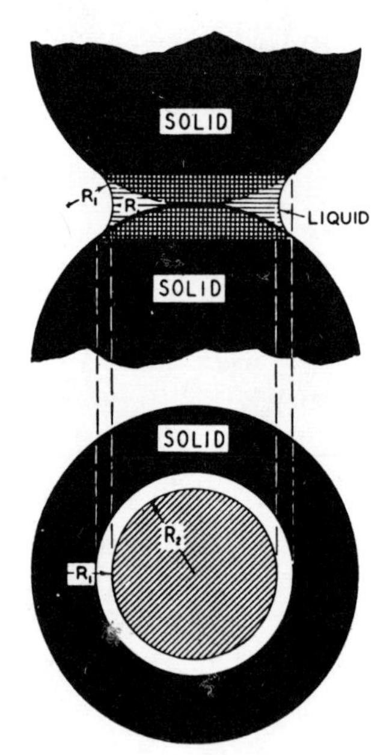

Fig. 10.17. Idealized conception of pendular rings around sand grain junctions. After Leverett,[11] courtesy AIME.

The distribution of the wetting and non-wetting phases is commonly classified as pendular, finicular, or insular, depending on their saturations. In region A, the aqueous phase exists mainly as pendular rings around the grain junctions which may only contact each other via an extremely thin adsorbed layer on the rock surface (Figure 10.17). In region B, both phases exist in continuous flow paths through their own pore networks, and both are said to be in finicular saturation. As water saturation continues to increase, the oil saturation is finally reduced to the point where the connecting threads break and oil becomes discontinuous at the value S_{or}. Thus in region C the oil exists in small, isolated groups of pores (islands) or in a state of insular saturation. Again, this discussion has considered water as the wetting phase and oil as non-wetting; however, the general concepts apply to any system of wetting and non-wetting fluids.

Therefore, in summary:

Region	*Wetting phase saturation*	*Non-wetting phase saturation*
A	pendular	finicular
B	finicular	finicular
C	finicular	insular

The practical significance of this behavior is extremely important. First, it becomes obvious that the mere presence of oil in a rock is not proof that oil will be produced. Actually, many rocks which show traces of oil in cores and cuttings will produce 100% water. This means that $S_o \leqq S_{or}$. Likewise, water will not be produced if $S_w \leqq S_{wc}$. Predictions of the future producing behavior of entire fields are based on the calculation of future production rates at steadily decreasing oil saturations; these predictions require that relative (or effective) permeabilities to oil, gas, and/or water at the proper saturations be known.

The Darcy equations derived in Chapter 2 may now be altered such that the effective permeability to the phase of interest replaces the absolute permeability. For example, the linear incompressible fluid equation

$$q = \frac{kA\Delta p}{\mu L} \tag{2.10}$$

becomes for multiphase flow systems:

$$q_o = \frac{k_o A\Delta p}{\mu_o L}$$

or

$$q_w = \frac{k_w A\Delta p}{\mu_w L}$$

where q_o, q_w = oil and water flow rates, respectively

μ_o, μ_w = oil and water viscosities, respectively

Example 10.6

A well is producing from a reservoir having the relative permeability characteristics of Figure 10.16. The following data are available:

$p_e = 2500$ psi $\quad \mu_o = 5.0$ cp

$p_w = 1000$ psi $\quad \mu_w = 0.6$ cp

$r_e = 700$ ft $\quad h = 25$ ft

$r_w = 0.33$ ft $\quad k = 50$ md (absolute)

$B_o = 1.30$ (formation volume factor)

(a) What will the steady state tank oil production rate be if the water saturation is at the critical value?

Solution:

Recall equation (2.27)

$$q = \frac{7.07hk(p_e - p_w)}{\mu \ln (r_e/r_w)}$$

where q = bbl/day

h = ft

k = darcys

p_e, p_w = psi

μ = cp

Altering the above to the problem conditions:

$$q_o = \frac{(7.07)(25)(0.050)(2500 - 1000)}{(5.0) \times \ln (2100)}$$

$= 350$ bbl/day of reservoir oil

However,

$$q_{ot} = \frac{q_o}{B_o} = \frac{350}{1.30} = 270$$ bbl/day of tank oil since the flowing volume q_o will shrink prior to reaching the tanks

(b) What will the tank oil flow rate be when $S_o = 0.50$, $p_e - p_w = 1000$ psi, $\mu_o = 7.0$ cp, and $B_o = 1.20$?

Solution:

From Figure 10.16, $k_{ro} = 0.45$ at $S_o = 50\%$

$$k_o = (k_{ro})(k) = (0.45)(50) = 22.5 \text{ md}$$

Combining B_o in the flow equation,

$$q_{ot} = \frac{(7.07)(0.0225)(25)(1000)}{(7.0)(7.64)(1.20)} = 62 \text{ bbl/day}$$

or:

$$q_{ot} = 270 \times \frac{1.30}{1.20} \times \frac{1000}{1500} \times \frac{.0225}{.050} \times \frac{5.0}{7.0} = 62 \text{ bbl/day}$$

(c) What will the producing water-oil ratio be when $S_o = 0.40$ and $\mu_o = 7.5$ cp?

Assume same Δp and B_o as part (b).

Solution:

Since:

$$q_{ot} = \frac{q_o}{B_o} = \frac{7.07k_o h(p_e - p_w)}{\mu_o B_o \ln (r_e/r_w)}$$

and

$$q_w = \frac{7.07k_w h(p_e - p_w)}{\mu_w \ln (r_e/r_w)}$$

the water-oil ratio is then:

$$\frac{q_w}{q_{ot}} = \frac{k_w \mu_o}{k_o \mu_w} \times B_o = \frac{k_{rw}\mu_o}{k_{ro}\mu_w} \times B_o \tag{10.6}$$

Note: The term $k_w\mu_o/k_o\mu_w$ is called the mobility ratio of water to oil, λ_w/λ_o.

From Figure 10.16, $k_{rw} = 0.23$ and $k_{ro} = 0.23$ at $S_o = 0.40$.

$$\therefore \frac{q_w}{q_{ot}} = \frac{(0.23)(7.5)}{(0.23)(0.6)} \times 1.20 = 15$$

Similar problems involving gas vs. oil, or gas vs. water behavior may be solved in the same manner.

This has been a brief but necessary discussion of a basic concept. Later sections and chapters require some fundamental knowledge of multiphase flow. It might be mentioned that three-phase flow sometimes occurs, further complicating the picture. Fortunately, most practical calculations may be resolved as two-phase problems with the third phase (if present) being considered immobile or constant. Oil and gas flow commonly occurs with water present in pendular saturation. Relative permeability curves for oil and gas may then be used, assuming water saturation constant. Similarly, occasions arise where simultaneous water and oil flow occur at some constant, insular gas saturation. Three-phase relative permeability data are scarce; however, the classic work of Leverett and Lewis[12] is suggested as a basic reference.

10.6 Special Core Analysis Procedures

If the literature on special core tests were bound in a single volume, its size would probably approach that of an unabridged dictionary. Consequently, this section will merely introduce the subject and cite a few basic references which may be used as a starting point. Since relative permeability has just been discussed let us start with its measurement.

10.61 Relative Permeability Measurements

A common method for measuring two phase relative permeability utilizes the apparatus shown in Figure 10.18. This is a slight modification of the Penn State method developed by Morse et al.[13] The test sample is confined at the ends between samples having similar properties. Intimate contact is maintained between the three cores to eliminate any capillary effects at the ends (particularly the downstream end) of the test sample. This insures that the saturation distribution of each fluid will be uniform during a steady state flow test. The upstream plug also serves as a mixing head for the injected fluids. The cores are first saturated with the fluid to be displaced, (which is commonly oil), and the weight of the test section is recorded. A constant oil flow rate is then established such that the desired pressure drop occurs. The oil flow rate is then reduced slightly and the displacing fluid (gas or water) is simultaneously injected at a rate sufficient to maintain the originally established pressure drop. Equilibrium is established when the respective input and outflow volumes are equal. Saturations are determined either gravimetrically by removing and weighing the test section, or electrically by measuring resistivity. The oil rate is then decreased further and the gas or water flow rate increased proportionally. Repetition of this procedure in sufficiently small steps allows calculation of the permeability to each phase at various saturations. Saturations are, of course, measured at each step. The porosity and absolute permeability of the test core are measured prior to the test.

There are numerous other methods for measuring relative permeability which are also capable of defining the flow behavior of the rock-fluid system used. Papers by Osoba et al.,[14] and Richardson et al.,[15] have summarized existing techniques and compared the results obtained from each; a discussion of the basic factors affecting such measurements has been presented by Geffen et al.[16] While space does not permit anything like a complete discussion of the subject, it seems pertinent to mention some precautions regarding the validity of basing field or well behavior predictions on core test data.

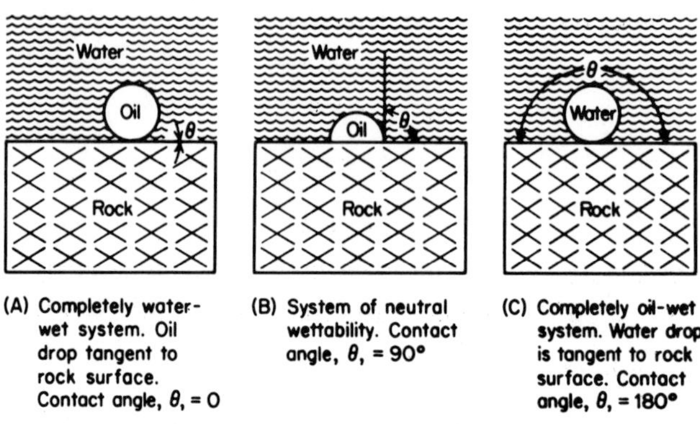

Fig. 10.19. Idealized oil-water-rock systems showing three stages of wettability as defined by contact angle.

The first rather obvious question to arise is whether or not a small core sample can represent the average behavior of a reservoir. This is a problem in all core analysis work. The apparent answer is that sufficient, properly selected cores must be analyzed to obtain a reasonable

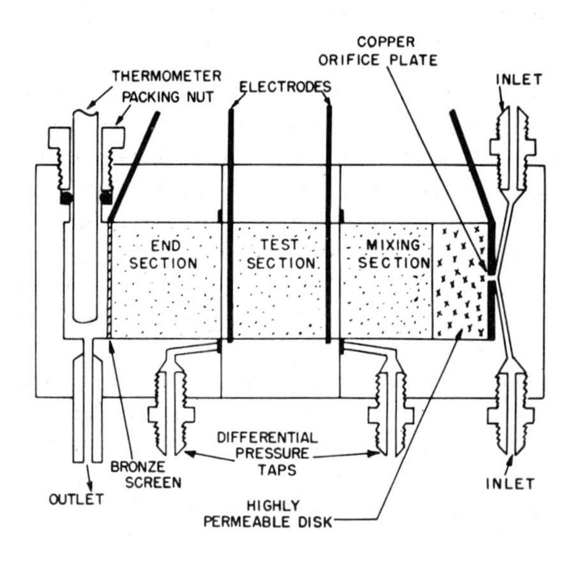

Fig. 10.18. Modified Penn State permeability apparatus. After Geffen, *et al.*,[16] courtesy AIME.

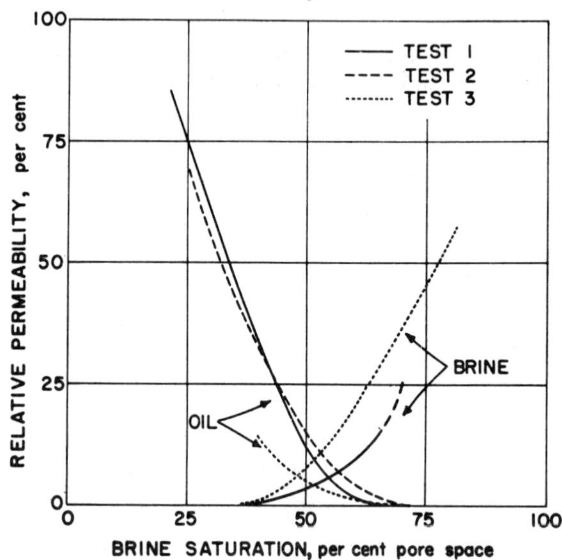

Fig. 10.20. Effect of wettability on flow behavior of a sandstone core. After Geffen, *et al.*,[16] courtesy AIME.

statistical sampling. The reservoir behavior will then be the properly weighted average of the individual observations. Aside from the problems of sampling or laboratory technique, there are two other factors of great fundamental importance.

1. Wettability alterations: Laboratory flow tests are normally conducted on core samples which have been thoroughly cleaned and dried. The test fluids used are usually synthetic brines, close hydrocarbon cuts (C_{10} to C_{12} for example), and air or nitrogen. Use of the actual reservoir fluids introduces severe problems in technique and handling and is not generally practiced. Also the reservoir temperature and pressure are usually not simulated. Therefore the wettability of the normal laboratory system is the same as that of the reservoir only by accident. In our earlier discussion of interstitial fluid distribution, it was shown that the shapes of the individual relative permeability curves were functions of the fluid which wetted the rock surface. Consequently, it may be expected that alterations in wettability will change the relative permeability behavior.

Wettability may be visualized in terms of contact angle, as shown by Figure 10.19. A zero contact angle implies complete wettability by the water as shown in *A*. A contact angle of 180° denotes complete wetting by the oil (*C*). Intermediate wettabilities are indicated by angles between these extremes. In oil-water-rock systems it is the convention to measure contact angle through the aqueous phase as shown.[17]

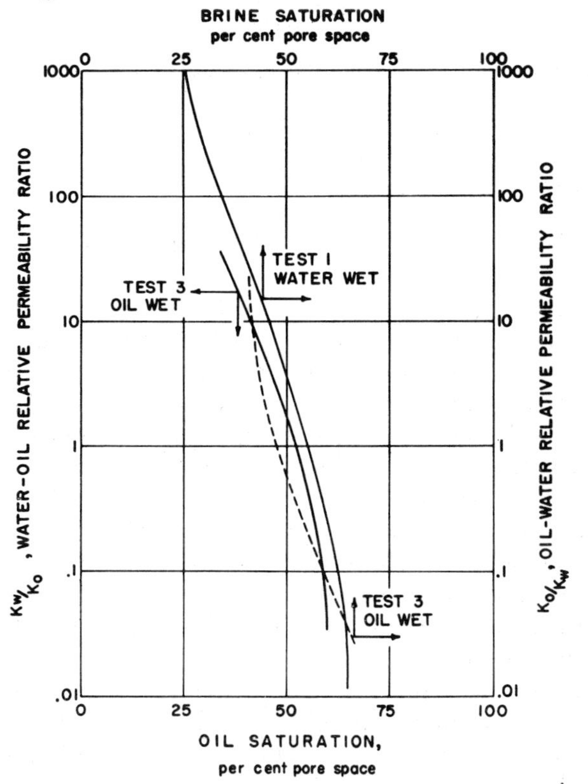

Fig. 10.21. Comparison of data from Fig. 10.20 on effective permeability ratio bases. After Geffen, *et al.*,[16] courtesy AIME.

Experiments have been conducted to show the effect of wettability on flow behavior, the sandstone sample's wettability being altered by a surface active material (Dri-film). The relative permeability behavior is shown in Figure 10.20. Figure 10.21 shows the data of tests 1 and 2 replotted in terms of relative permeability ratio. Note that the curves of k_w/k_o for test 3 (oil-wet) and k_o/k_w for test 1 (water-wet) are almost identical, which implies that the fluids merely changed positions in the core.

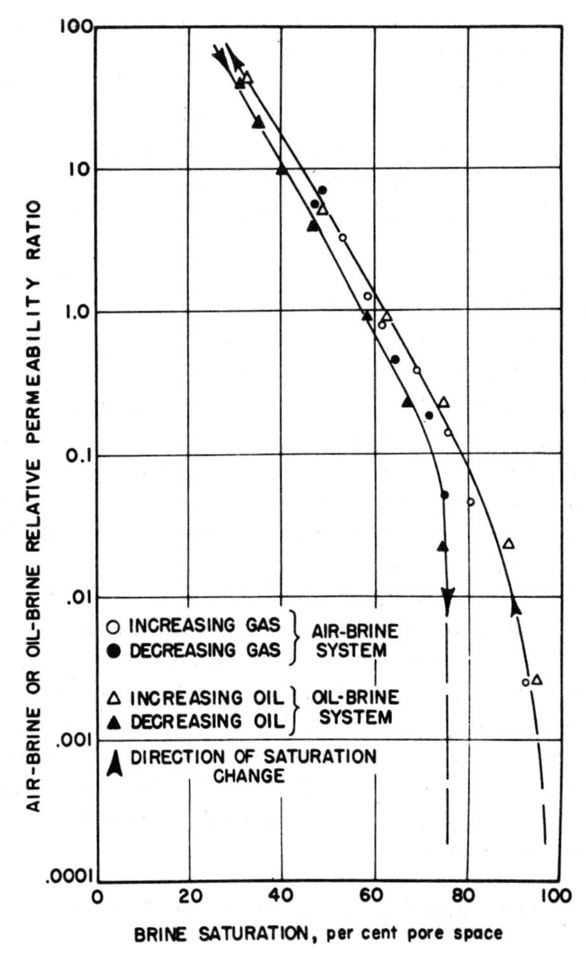

Fig. 10.22. Typical effect of saturation history on relative permeability behavior. After Geffen, *et al.*,[16] courtesy AIME.

Thus it is apparent that wettability alterations may cast considerable doubt on the validity of laboratory flow test data. Apparently, most laboratory tests are made on the assumption of water-wet conditions which, fortunately, holds true in most cases. However, it should be realized that wettability is a gradational phenomenon, and a particular system may be completely water-wet, oil-wet, or any condition between. A further complication of this problem is the lack of any completely satisfactory means of measuring wettability. Temperature and pressure also affect the magnitude of surface forces, thereby altering wettability.[18] The solution to the wettability problem has not been attained; however, its possible effects on flow behavior must be recognized.

2. Saturation history effects: It has also been demonstrated that relative permeability is not a unique function of saturation but depends on the direction from which the saturation is approached. This means that the curves obtained by displacing oil with water will not be the same as those from the reverse process. Practically speaking, there are two saturation histories or directional changes which are of interest. These are:

(a) Gas drive process: displacement of oil by gas. Oil is assumed to be the wetting phase with respect to gas. This is also called the drainage process.

(b) Water drive process: displacement of oil by water, where water is the wetting phase. This is an imbibition process.

The typical behavior of these processes is shown in Figure 10.22, in which the arrows denote the direction of saturation change. Note that in the gas drive, (in which water, rather than oil, was used as wetting phase) permeability to gas exists at a very high water saturation. In the reverse procedure, however, gas permeability approaches zero at a much lower water saturation. Therefore, laboratory measurements must have the proper saturation history which applies to the field problem at hand.

The relative permeability concept is one which is sometimes difficult to grasp. It is often helpful to consider it as a correction factor which must be applied to the absolute permeability to account for the presence of other immiscible fluids. This is precisely what it amounts to, although the determination of its correct magnitude may be extremely difficult. For a particular rock, it is a function of saturation, wettability, and saturation history.

10.62 Determination of Connate Water Saturation

The determination of the actual (or connate) water saturation in the reservoir rock is not obtainable from routine analysis, due to the environmental factors mentioned earlier, except when the rock exists at its critical (minimum or irreducible) water saturation and is cored with oil or an oil base mud.* In this case the water, being immobile, is not disturbed by the mud filtrate. If the cores are expediently handled and properly preserved, the routine water saturation determinations will be indicative of the original reservoir. In most cases, however, this procedure is not followed, and many special laboratory methods have been used to procure an irreducible water saturation value. In order to discuss this topic it is first necessary to review the elementary concept of capillary behavior.

The familiar rise of a liquid in a capillary tube is shown in Figure 10.23. Equating of the upward and downward forces on the elevated column results in the well known expression for the surface tension of the liquid:

$$F_u = 2\pi r\sigma \cos\theta \tag{1}$$

and

$$F_d = \pi r^2 l \rho g \tag{2}$$

where F_u = upward force on the elevated liquid column
F_d = downward force
r = capillary radius
σ = surface tension of the liquid or interfacial tension if two liquids are involved.
θ = contact angle of the system
l = height of the column
ρ = liquid density
g = gravitational constant

At equilibrium, $F_u = F_d$, hence:

$$\sigma = \frac{rl\rho g}{2\cos\theta} \tag{10.7}$$

If the downward force F_d is expressed as a pressure, then

$$P_c = \frac{F_d}{\pi r^2} = l\rho g$$

or from Eq. (10.7)

$$P_c = \frac{2\sigma\cos\theta}{r} = \text{capillary pressure} \tag{10.8}$$

Capillary pressure may also be defined as the pressure necessary to displace a wetting fluid from a capillary opening. Obviously, this definition will result in a negative value if the fluid does not wet the tube, i.e., if $\theta > 90°$. The factors which govern this capillary or displacement pressure are the surface or interfacial tension of the fluids involved, the system wettability θ, and the size (radius) of the capillary. The significance of capillary pressure becomes apparent when one considers that most oil-containing porous media may be visualized as a heterogeneous and tortuous bundle of capillary tubes.

Nearly all petroleum reservoirs occur in marine sediments which were originally saturated with water. As oil was formed and it migrated into the rock, it displaced water to an extent dependent on its driving pressure opposed to the rock's capillary pressure. Over geologic time, the oil accumulated in traps where

*The term connate water is widely used as a synonym for critical water saturation. Here we will use it as the actual reservoir water saturation, which may or may not be the irreducible value.

differential pressure became sufficient to reduce the water saturation to its minimum value. This minimum or critical value will normally exist in formations which produce clean (water-free) oil. Oil reservoirs underlaid by, and in intimate contact with, water have a transition

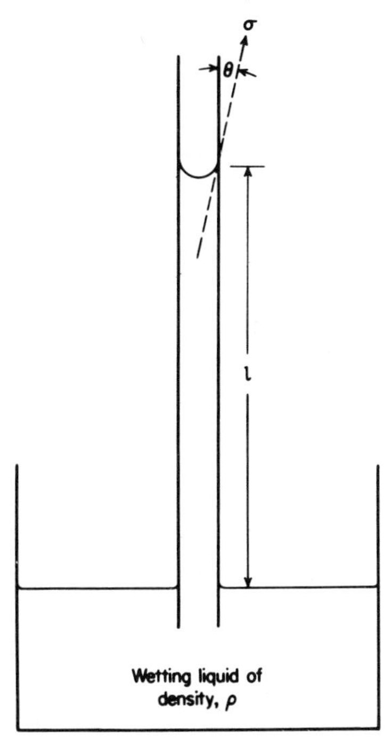

Fig. 10.23. Capillary rise of a liquid which wets the walls of the capillary tube.

zone through which water saturation decreases from 100% in the water zone to some irreducible value in the oil zone. Wells completed in the transition interval produce both oil and water. Similarly an oil-gas transition zone will exist in fields having a gas cap. The thickness of these zones depends on the same factors appearing in the capillary pressure equation: interfacial tension, wettability, and average pore radius. Equation (10.8) is normally altered to Eq. (10.9) when applied to porous media, as indicated by Figure 10.24(A), (B).

$$(10.9)\quad P_c = l(\rho_w - \rho_o)g = \sigma_{ow}\left(\frac{1}{r_1} + \frac{1}{r_2}\right)\cos\theta$$

where r_1, r_2 are the radii of curvature of the oil water interface measured as was shown in Figure 10.17. The term $1/r_1 + 1/r_2$ may be considered as the mean curvature of the surface.[19]

With these concepts in mind, let us consider some methods of measuring critical water saturation.

1. Restored-state method: This is a well known, and widely used method which was proposed in 1947 by Bruce and Welge.[20] Its name derives from its similarity to the original reservoir process. A typical apparatus is shown in Figure 10.25. A reservoir core of known pore volume is saturated 100% with water

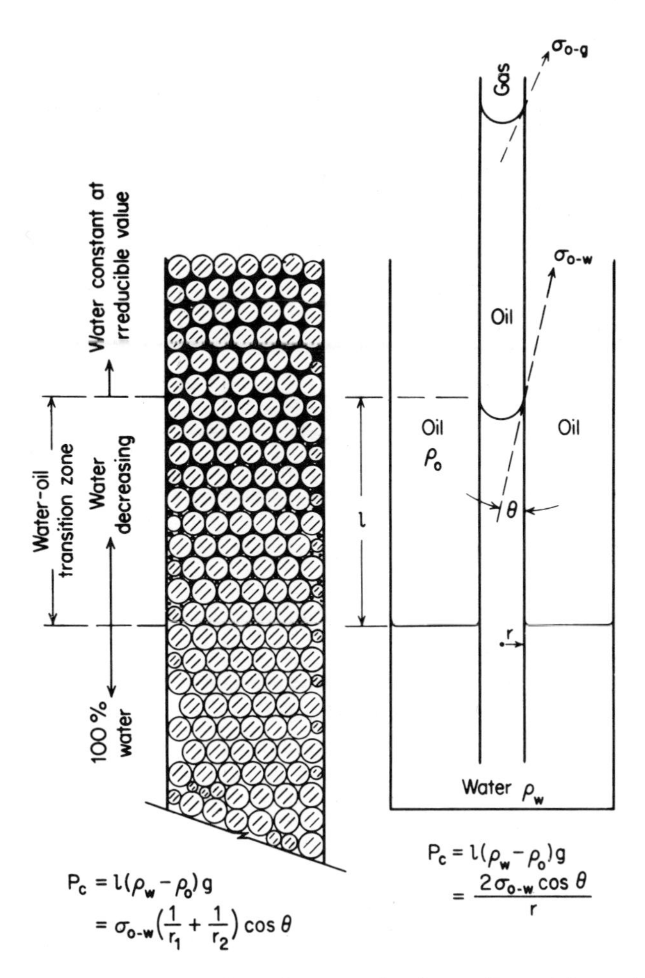

Fig. 10.24. Analogy between simple capillary tube and porous media.

and placed in contact with the water-wet membrane, as shown. The membrane has extremely small pores and will not allow a non-wetting fluid to enter it at the pressures to be used in the test. A non-wetting fluid (oil, air, nitrogen, etc.) is then introduced into the cell at slightly elevated pressure. The air (or non-wetting fluid) will enter all the pores in the core sample having a capillary pressure less than that applied. The water displaced from the core is forced through the membrane and collected in a suitable graduate. When the displaced water volume becomes constant at a given pressure, it is recorded. The saturation may then be computed at that particular pressure. This procedure is repeated in subsequent, higher pressure, steps until an increase in pressure forces no more water from the core. This is then the irreducible value. The resultant plot of capillary pressure vs. saturation is shown in Figure 10.26. The minimum pressure which will displace water from the largest pore is called the displacement or entrance pressure. Note also the analogy between the curve obtained and the reservoir condition of Figure 10.24.

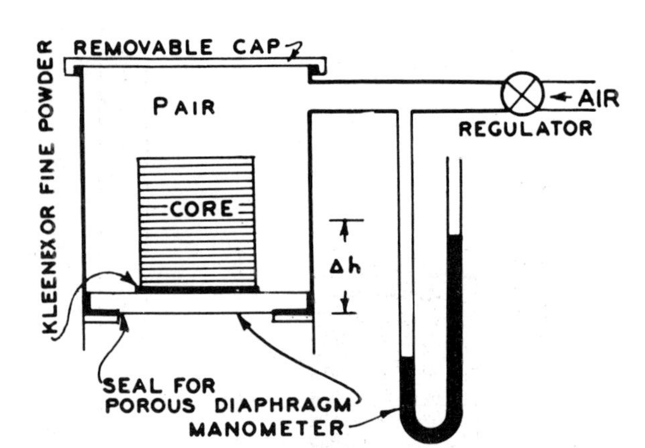

Fig. 10.25. Restored state apparatus for determination of capillary pressure curve.

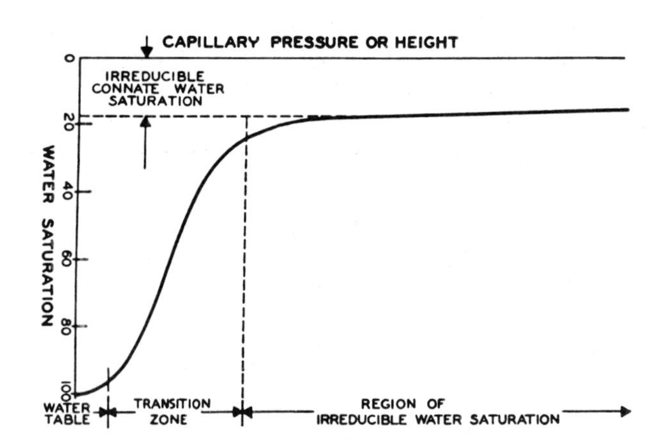

Fig. 10.26. Typical capillary pressure curve showing significant features.

The principal disadvantage of this technique is the time required to carry it out. Several days may be required to reach a satisfactory equilibrium at each step. However, by installing batteries of cells, one may conduct numerous tests simultaneously.

2. Mercury injection method: This is similar in principle to the restored state method. The dry sample is placed in a mercury cell. Pressure is then applied incrementally as before, with the volume injected into the rock pores being noted at each pressure. A curve of capillary pressure vs mercury saturation is thus obtained. This must be corrected as shown by Purcell:[21]

$$\frac{P_{c\mathrm{Hg}}}{P_{cw}} = \frac{\sigma_{\mathrm{Hg}} \cos \theta_{\mathrm{Hg}}}{\sigma_w \cos \theta_w} \cong 5$$

where $\sigma_{\mathrm{Hg}} = 480$ dynes/cm

$\sigma_w = 70$ dynes/cm

$\theta_{\mathrm{Hg}} = 140°$

$\theta_w = 0°$

Reasonable checks are obtained between this and the previous method. The primary advantage of the method is speed, since only a few hours are required to obtain a complete curve. A disadvantage is that the sample is ruined for subsequent testing.

Much may be determined from the capillary pressure curve, besides the irreducible water saturation. Certainly, the curve affords some measure of pore size distribution. If all the pores were essentially one size, a very flat curve would result. A steep slope implies that many pore sizes exist. Stepwise calculations of pore radii may be made with Eq. (10.8) at various pressures along the curve, providing σ and θ are known. Methods for computing both absolute and relative permeability from such data are in use.[21–26] Such calculations are based on the fundamental relationship between permeability and pore size.

In this regard, it is also logical to expect that the critical water saturation should correlate with absolute permeability, other factors being reasonably constant. A correlation of this type is given as Figure 10.27. Note that height above the water table is also a parameter. Such curves may be developed for a particular formation, providing data over a sufficient permeability range are available. This is a further illustration of the need for proper sampling.

3. Evaporation method: This is a simple and unique method which was proposed by Messer[27] in 1951. A sample of known porosity is completely saturated with water. It is then placed in a suitable oven and dried under constant conditions. The weight loss is recorded, either continuously or in increments, and plotted against time. Adsorbed water evaporates at a slower rate than the free or mobile water, due to the capillary and surface forces which oppose its escape. This difference in drying rate exhibits a break in the drying curve which may be taken as the critical value. Other liquids such as toluene, benzene, or tetrachloroethane may be used instead of water, provided that the proper volume correction is made. Satisfactory correlation between this method and the restored state or capillary pressure technique was obtained by Messer.

The advantage of the evaporation method is the speed of measurement; usually, a test can be completed in twenty minutes to an hour. Some theoretical objections have been raised regarding the validity of the method; however, it is a fast, cheap, and reasonably accurate means of obtaining an irreducible water saturation value.

All methods used for connate water determination may be criticized on one basis or another. A serious drawback to capillary pressure measurements lies in the general use of synthetic fluids which may not reproduce the proper values of σ and/or θ. The seriousness of these procedural shortcuts may be considerable, in some instances. It has also been shown that core weathering and aging can affect the obtained values.

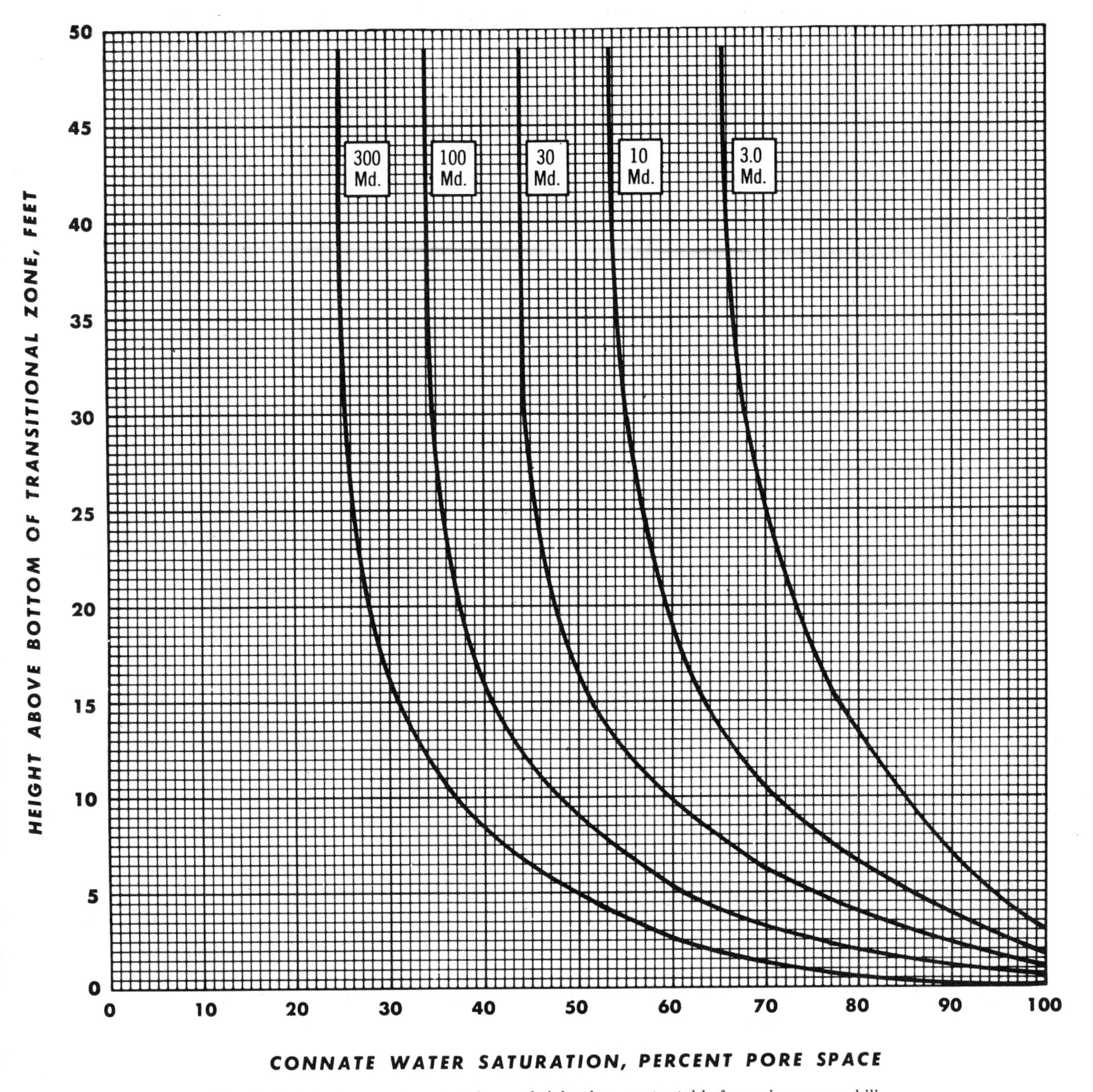

Fig. 10.27. Connate water saturation vs. height above water table for various permeabilities. Courtesy Core Laboratories, Inc.

10.63 Water Flood Tests

Freshly cut cores may be flooded with water (usually a synthetic brine) to determine the residual oil content S_{or} after flooding. Such data are often useful in estimating the quantity of oil which may be recovered from an actual field water flood. These tests are often referred to as flood pot tests and are performed under radial flow conditions with a full diameter section. In some cases the cores are cleaned, resaturated to simulate initial reservoir conditions, and flooded in a manner similar to that mentioned with regard to relative permeability measurements.

10.64 Miscellaneous Special Tests

These are, of course, numerous and varied. Water permeabilities are often measured to determine the compatibility of a particular water-sand system. The principal factor involved in such tests is the hydration

DEPTH	LITHO.	K	ø	% PORE SPACE OIL	% PORE SPACE WATER	PPM CHLORIDE	INTERP.
7201		Shale, No Analysis		—	—	—	—
7202		Shale, No Analysis		—	—	—	—
7203		3000	31	0.0	45.0	4,000	COND
7204		2800	31	2.0	47.0	8,000	COND
7205		3500	33	1.0	42.0	6,000	COND
7206		2500	30	3.0	45.0	10,000	COND
7207		3800	32	10.0	47.0	12,000	OIL
7208		3200	31	12.0	48.0	15,000	OIL
7209		155	27	3.0	70.0	18,000	OIL
7210		4000	33	15.0	40.0	10,000	OIL
7211		5500	32	18.0	41.0	3,000	OIL
7212		Shale, No Analysis		—	—	—	—
7213		3000	31	40.0	20.0	60,000	OIL
7214		4500	34	20.0	48.0	8,000	OIL
7215		2500	29	15.0	52.0	5,000	OIL
7216		5000	33	7.0	50.0	8,000	WATER
7217		3000	31	3.0	65.0	10,000	WATER
7218		100	26	3.0	79.0	8,000	WATER
7219		4500	30	0.0	80.0	60,000	WATER
7220		3000	29	0.0	78.0	14,000	WATER
7221		Shale, No Analysis		—	—	—	—

Fig. 10.28. Core analysis and productivity interpretation for a predominantly clean, Frio sand section. After Elmdahl,[28] courtesy AIME.

behavior of interstitial clays when brought into contact with various waters. Electrical measurements are also made to facilitate electric log interpretation. Mineralogical composition may be studied for geological purposes. Wettability studies are also conducted in specific instances, particularly when an oil-wet condition is indicated.

10.7 Practical Uses of Core Analysis Data

In exploratory drilling, the first question which core analysis is expected to answer is (1) what fluid(s) will be produced. If the answer to the above is oil and/or gas, the next questions are (2) what is the possible rate of production, and (3) what total amount will ultimately be produced. It would indeed be satisfying if core analysis, with no other substantiating data, could answer these queries. Unfortunately, such an abundance of information is not so easily obtained; however, the experienced engineer or core analyst is able to make reasonable estimates of productivity from core analysis data. Let us, then, consider the extent to which core analysis provides answers to these basic questions.

10.71 Probable Produced Fluid

It was previously stated that the fluid content of a laboratory core is vastly different from its original reservoir condition, due to the flushing pressure reduction and weathering undergone in transit. The laboratory saturations are therefore dependent on the following factors:

1. Original fluid content: if oil and/or gas were originally present, some should still exist in the captured core.
2. Properties of the reservoir fluids:
   a. oil viscosity: Low viscosity oils are more easily flushed by the drilling fluid; therefore, if all other factors are constant, the higher the viscosity, the higher the observed oil saturation.
   b. formation volume factor (liquid shrinkage): An oil which has a high volume factor B_o will naturally show a lower laboratory oil saturation than one which undergoes little shrinkage.
   c. oil volatility: Highly volatile oils (oils having high vapor pressures) evaporate easily; this may greatly reduce their measured saturations.
   d. gas sands: In the case of a dry gas sand, only gas and water saturations will be noted. Gas condensate systems will normally have low oil saturations (1 to 5%). The low viscosity and high compressibility of gas are conducive to severe flushing.
3. Rock permeability: Low permeability and/or shaly rocks have high irreducible water saturations to start with, and will therefore indicate high laboratory water saturations. However, low permeability rocks are not as susceptible to flushing. Many tight cores

DEPTH	LITHO.	K	Ø	% PORE SPACE		PPM	INTERP.
				OIL	WATER	CHLORIDE	
6401		Shale, No Analysis		—	—	—	—
6402		Shale, No Analysis		—	—	—	—
6403		80	27	0.0	68.0	8,000	COND
6404		200	32	0.0	64.0	10,000	COND
6405		1500	35	2.2	50.0	6,000	COND
6406		500	33	3.4	52.0	12,000	COND
6407		50	28	4.0	70.0	16,000	OIL
6408		300	31	10.0	65.0	14,000	OIL
6409		20	25	3.4	75.0	16,000	OIL
6410		30	27	6.0	70.0	20,000	OIL
6411		100	32	4.0	71.0	18,000	TRANS
6412		200	34	2.0	78.0	5,000	WATER
6413		1500	36	3.0	75.0	6,000	WATER
6414		100	26	0.0	78.0	7,500	WATER
6415		Shale, No Analysis		—	—	—	—

Fig. 10.29. Core analysis and productivity interpretation for a predominately silty Cockfield sand. After Elmdahl,[28] courtesy AIME.

will bleed oil and gas for several hours after being recovered. The saturations obtained from samples taken near the center of such cores may have undergone relatively little flushing. Highly permeable cores are readily flushed and may also exhibit high water saturations. The salinity of the core water is a good qualitative indicator of the degree of flushing, provided there is a sharp salinity contrast between the mud filtrate and the connate water.

4. Drilling fluid properties:
   a. water or oil base: This effect is rather obvious. The water saturation of oil-cut cores is often taken as the critical or irréducible saturation.
   b. filtration properties: The degree of flushing varies directly with the fluid loss of the drilling fluid used. The API filter press test is not necessarily indicative of this behavior, but is commonly used in lieu of something better.
   c. density: Since the differential pressure between the formation and bore hole depends on mud density, it may be expected that flushing increases with mud density.
5. Coring rate: The major part of flushing occurs when the core is cut, before it enters the barrel. Thus exposure time varies with coring rate.
6. Care of handling: This effect is quite obvious. Evaporative losses depend on the prevailing weather conditions, as well as on time. A permeable core which is left lying on the walk for an hour or two during a West Texas afternoon may be completely stripped of any light hydrocarbon content. Heavy oils and water are not as easily evaporated; however, their saturations will also be reduced if the core is not quickly handled and properly preserved.

One cannot make a quantitative appraisal of the effects of all these factors. When these uncertainties are coupled with the lack of relative permeability data, it becomes apparent that precise definition of the produced fluid types and ratios from routine core analysis is not possible. Any predictions based solely on such data with no other substantiating information must be treated as very qualitative estimates. Fortunately, other data are generally available which can supplement the over-all picture. Experience in the same or similar areas is invaluable in this regard. Generally, the producible fluid type may be predicted, but there are always borderline cases where data are misinterpreted.

The observed saturations in cases of interest will generally lie somewhere between a completely flushed and a slightly flushed stage. This allows the application of certain saturation limits between which the observed saturations must lie to prognosticate commercial production.[28,29] Correlations of this type are reasonably valid for the specific geologic areas and formations on which they are based, provided that similar coring and handling procedures have been followed.

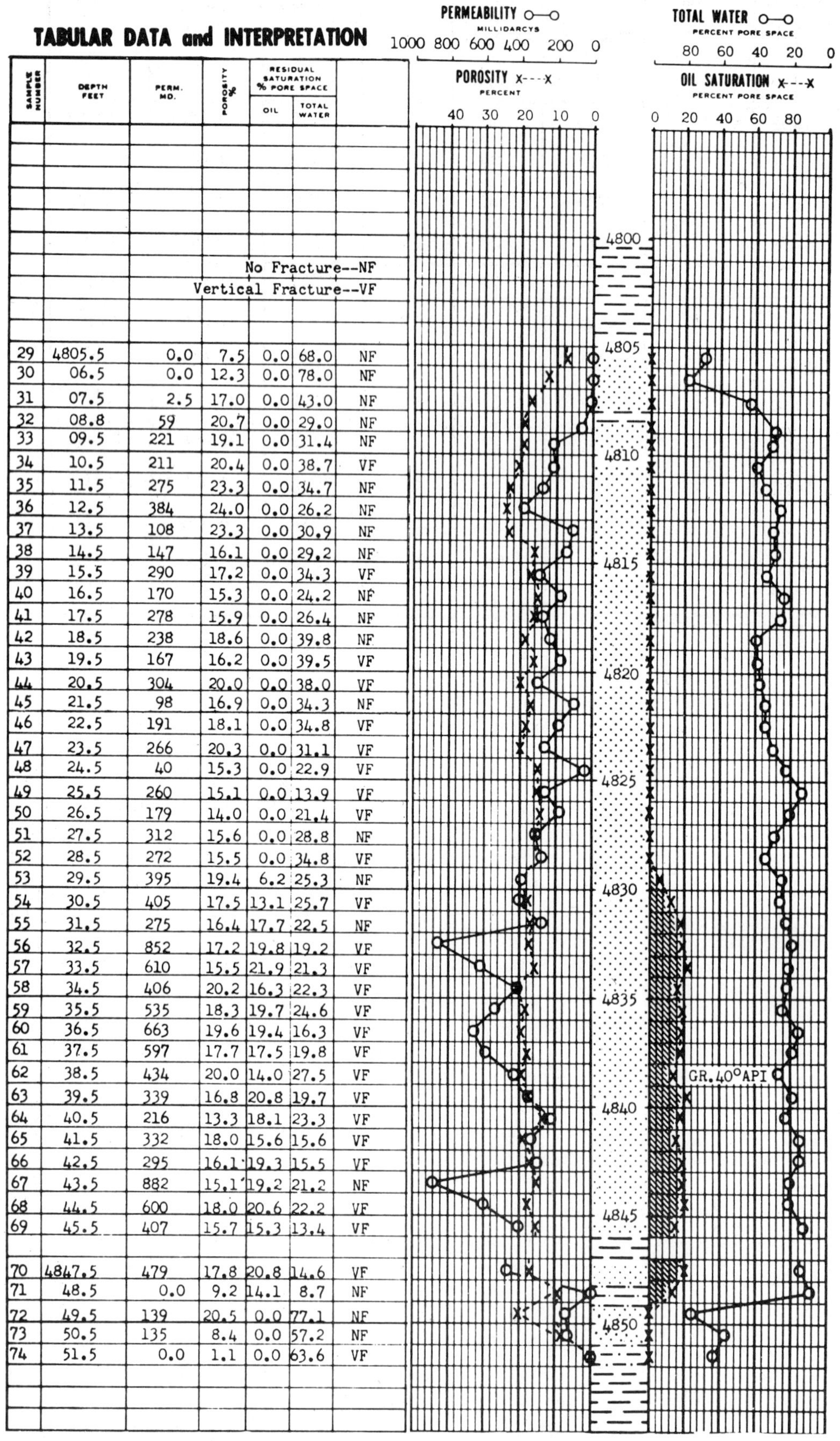

TABULAR DATA and INTERPRETATION

No Fracture--NF
Vertical Fracture--VF

SAMPLE NUMBER	DEPTH FEET	PERM. MD.	POROSITY %	RESIDUAL SATURATION % PORE SPACE OIL	RESIDUAL SATURATION % PORE SPACE TOTAL WATER	
29	4805.5	0.0	7.5	0.0	68.0	NF
30	06.5	0.0	12.3	0.0	78.0	NF
31	07.5	2.5	17.0	0.0	43.0	NF
32	08.8	59	20.7	0.0	29.0	NF
33	09.5	221	19.1	0.0	31.4	NF
34	10.5	211	20.4	0.0	38.7	VF
35	11.5	275	23.3	0.0	34.7	NF
36	12.5	384	24.0	0.0	26.2	NF
37	13.5	108	23.3	0.0	30.9	NF
38	14.5	147	16.1	0.0	29.2	NF
39	15.5	290	17.2	0.0	34.3	VF
40	16.5	170	15.3	0.0	24.2	NF
41	17.5	278	15.9	0.0	26.4	NF
42	18.5	238	18.6	0.0	39.8	NF
43	19.5	167	16.2	0.0	39.5	VF
44	20.5	304	20.0	0.0	38.0	VF
45	21.5	98	16.9	0.0	34.3	NF
46	22.5	191	18.1	0.0	34.8	VF
47	23.5	266	20.3	0.0	31.1	VF
48	24.5	40	15.3	0.0	22.9	VF
49	25.5	260	15.1	0.0	13.9	VF
50	26.5	179	14.0	0.0	21.4	VF
51	27.5	312	15.6	0.0	28.8	NF
52	28.5	272	15.5	0.0	34.8	VF
53	29.5	395	19.4	6.2	25.3	NF
54	30.5	405	17.5	13.1	25.7	VF
55	31.5	275	16.4	17.7	22.5	NF
56	32.5	852	17.2	19.8	19.2	VF
57	33.5	610	15.5	21.9	21.3	VF
58	34.5	406	20.2	16.3	22.3	VF
59	35.5	535	18.3	19.7	24.6	VF
60	36.5	663	19.6	19.4	16.3	VF
61	37.5	597	17.7	17.5	19.8	VF
62	38.5	434	20.0	14.0	27.5	VF
63	39.5	339	16.8	20.8	19.7	VF
64	40.5	216	13.3	18.1	23.3	VF
65	41.5	332	18.0	15.6	15.6	VF
66	42.5	295	16.1	19.3	15.5	VF
67	43.5	882	15.1	19.2	21.2	NF
68	44.5	600	18.0	20.6	22.2	VF
69	45.5	407	15.7	15.3	13.4	VF
70	4847.5	479	17.8	20.8	14.6	VF
71	48.5	0.0	9.2	14.1	8.7	NF
72	49.5	139	20.5	0.0	77.1	NF
73	50.5	135	8.4	0.0	57.2	NF
74	51.5	0.0	1.1	0.0	63.6	VF

Fig. 10.30. Sample core analysis report. Courtesy Core Laboratories, Inc.

In any given reservoir, the sequence of fluids from the top down must be free gas (if present), oil, and water. It must also be remembered that the contacts between these fluids are zones and not sharp interfaces, unless permeability is very high. It is therefore possible to predict such contacts from saturation changes, providing other factors which obscure the picture are not present. This is illustrated by the Gulf Coast sample analyses of Figures 10.28 and 10.29.

PROBLEMS

1. A two-cell Boyle's law porosimeter was calibrated using steel spheres, and the following data were obtained:

Sphere diameter, cm	p_1, *psig*	p_2, *psig*
1.0	100	33.6
2.0	100	35.7
4.0	100	60.2

(a) Plot a calibration curve for the instrument as V_s vs. p_1/p_2.
(b) Plot a calibration curve as V_s vs. p_2, with $p_1 = 100$ psi.
(c) Determine the cell volumes of the instrument from the plot in (a). *Ans.* (c) $V_1 + V_2 = 75.0$ cc.

2. The following data were obtained during linear flow tests using air. Determine the absolute permeability of the sample by making the Klinkenberg correction.

p_1, *atm*	p_2, *atm*	k_a, *md*
5	1	135
3	1	140
2	1	142
1.5	1	150

Ans. $k \cong 125$ md

3. Using Eq. (10.5) and the plot from Problem 2, determine the value of kb for the above sample. Would you expect k and b to be interrelated in a predictable manner? For a discussion of this see reference 30.

4. A core sample has an absolute permeability of 100 md. It was subjected to a relative permeability test using water and oil, and the following data were obtained.

	S_w	q_o, *cc/sec*	q_w, *cc/sec*	k_o, *md*
	.30	1.0	0.03	62
	.40	1.0	0.20	37
$\mu_o = 2$ cp	.50	1.0	0.80	25
$\mu_w = 1$ cp	.60	0.5	1.0	18
	.80	.03	1.0	4

(a) Calculate and plot relative permeability curves for water and oil vs. water saturation.
(b) Plot k_w/k_o vs S_w. (Use semi-log paper.)
(c) From inspection of the curves in part (a), which is the wetting phase?
(d) Estimate S_{wc} and S_{or} from the same plot.

5. Figure 10.30 is a core analysis report for a productive Dakota "J" Sand Zone as obtained from a wildcat well in Nebraska.
(a) Estimate the probable produced fluid throughout the interval.
(b) Determine the net feet of gas and/or oil zone present.
(c) Determine the average ϕ and k for the zones in (b).

REFERENCES

1. Hughes Tool Company, *Rotary Core Drilling,* Houston, Texas.
2. Christensen Diamond Products, *Diamond Coring Manual.* Salt Lake City, Utah.
3. Core Laboratories, Inc., *Summary of Core Analysis Procedures.* Dallas, Texas.
4. Clark, N. J., and H. M. Shearin, "Formation Evaluation of Oil and Gas Reservoirs," *The Petroleum Engineer,* Apr. 1956, p. B-21.
5. Stewart, C. R., and J. W. Spurlock, "How to Analyze Large Core Samples," *Oil and Gas Journal,* Sept. 15, 1952, p. 89.
6. Dotson, B. J., Slobod, R. L., McCreery, P. N., and J. W. Spurlock, "Porosity-Measurement Comparisons by Five Laboratories," *Trans. AIME,* Vol. 192, (1951), p. 341.
7. Coberly, C. J., and A. B. Stevens, "Development of Hydrogen Porosimeter," *Trans. AIME,* Vol. 103, (1933), p. 261.
8. Beeson, C. M., "The Kobe Porosimeter and The Oilwell Research Porosimeter," *Trans. AIME,* Vol. 189, (1950), p. 313.
9. Kelton, F. C., "Analysis of Fractured Limestone Cores," *Trans. AIME,* Vol. 189, (1950), p. 225.
10. Klinkenberg, L. J., "The Permeability of Porous Media to Liquids and Gases," *API Drilling and Production Practices,* 1941, p. 200.
11. Leverett, M. C., "Capillary Behavior in Porous Solids," *Trans. AIME,* Vol. 142, (1941), p. 152.
12. Leverett, M. C., and W. B. Lewis, "Steady Flow of Gas-Oil-Water Mixtures through Unconsolidated Sands," *Trans. AIME,* Vol. 142, (1941), p. 107.
13. Morse, R. A., Terwilliger, P. L., and S. T. Yuster, "Relative Permeability Measurements on Small Core Samples," *Oil and Gas Journal,* Aug. 23, 1947, p. 109.
14. Osoba, J. S., Richardson, J. G., Kerver, J. K., Hafford, J. A., and P. M. Blair, "Laboratory Measurements of Relative Permeability," *Trans. AIME,* Vol. 192, (1951), p. 47.
15. Richardson, J. G., Kerver, J. K., Hafford, J. A., and J. S. Osoba, "Laboratory Determination of Relative Permeability," *Trans. AIME,* Vol. 195, (1952), p. 187.
16. Geffen, T. M., Owens, W. W., Parrish, D. R., and R. A. Morse, "Experimental Investigation of Factors Affecting Relative Permeability Measurements," *Trans. AIME,* Vol. 192, (1951), p. 99.
17. Benner, F. C., and F. E. Bartell, "The Effect of Polar Impurities Upon Capillary and Surface Phenomena in Petroleum Production," *API Drilling and Production Practices,* 1941, p. 341.

18. Hough, E. W., Rzasa, M. J., and B. B. Wood, "Interfacial Tensions at Reservoir Pressures and Temperatures; Apparatus and the Water-Methane System," *Trans. AIME*, Vol. 192, (1951), p. 57.

19. Leverett, M. C., "Capillary Behavior in Porous Solids," *Trans. AIME*, Vol. 142, (1941), p. 152.

20. Bruce, W. A., and H. J. Welge, "The Restored-State Method for Determination of Oil in Place and Connate Water," *API Drilling and Production Practices*, 1947, p. 166.

21. Purcell, W. R., "Capillary Pressures — Their Measurement Using Mercury and the Calculation of Permeability Therefrom," *Trans. AIME*, Vol. 186, (1949), p. 39.

22. Rose, W., "Theoretical Generalizations Leading to the Evaluation of Relative Permeability," *Trans. AIME*, Vol. 186, (1949), p. 111.

23. Rose, W., and W. A. Bruce, "Evaluation of Capillary Character in Petroleum Reservoir Rock," *Trans. AIME*, Vol. 186, (1949), p. 127.

24. Purcell, W. R., "Interpretation of Capillary Pressure Data," *Trans. AIME*, Vol. 189, (1950), p. 369.

25. Gates, J. I., and W. Templaar Lietz, "Relative Permeabilities of California Cores," *API Drilling and Production Practices*, 1950, p. 285.

26. Burdine, N. R., "Relative Permeability Calculations from Pore Size Distribution Data," *Trans. AIME*, Vol. 198, (1953), p. 71.

27. Messer, E. S., "Interstitial Water Determination by an Evaporation Method," *Trans. AIME*, Vol. 192, (1951), p. 269.

28. Elmdahl, B. A., "The Fundamental Principles of Core Analysis and Their Application to Gulf Coast Formations," AIME Tech. Paper 588-G, presented at Formation Evaluation Symposium, Univ. of Houston, Oct. 27-28, 1955.

29. Pyle, H. G., and John E. Sherborne, "Core Analysis," *Trans. AIME*, Vol. 132, (1939), p. 33.

30. Heid, J. G., McMahon, J. J., Nielsen, R. F., and S. T. Yuster, "Study of the Permeability of Rocks to Homogeneous Fluids," *API Drilling and Production Practices*, 1950, p. 230.

SUPPLEMENTARY READINGS

Calhoun, J. C., Jr., *Fundamentals of Reservoir Engineering*. Norman, Oklahoma: University of Oklahoma Press, 1953.

Muskat, M., *Physical Principles of Oil Production*, 1st ed. New York: McGraw-Hill Book Co., Inc., 1949, pp. 114–169.

Pirson, S. J., *Oil Reservoir Engineering*, 2nd ed. New York: McGraw-Hill Book Co., Inc., 1958, pp. 30–97.

Chapter 11 . . .

Well Logging

A well log may be defined as a tabular or graphical portrayal of any drilling condition(s) or subsurface feature(s) encountered which relate to either the progress or evaluation of an individual well. The records of core analysis data vs. depth, shown in the last chapter, are often called core logs.

This chapter will discuss the following common logs and introduce the general subject of electric and radioactivity log interpretation.

1. Driller's logs (including drilling time)
2. Sample logs
3. Mud logs
4. Electric logs
5. Radioactivity logs
6. Miscellaneous logs

11.1 Driller's Logs

In the early days of drilling, particularly in the cable tool era, the driller's log was the principal well record kept. It recorded the types of formations encountered, any pertinent fluid flows or oil and gas shows observed, and other related operational remarks. While these records appear crude by present standards, they were considered to be very informative at the time. Such logs are still frequently encountered as the only available source of data in old areas. The geologic descriptions of various formations may be quite colorful and full of expressions unique to either or both the particular area and driller involved.

The current rotary driller's log is contained under the more common heading of the daily tour report. This is filled out daily by each driller (or the toolpusher) as a record of the operations and progress which occurred during his working hours (tour). It is largely used to inform office personnel of daily occurrences, to provide operational data, and to serve as a legal record of the contractor's compliance with the operator's instructions as set forth in their agreement or contract. The hourly breakdown of time spent on various operations is also used to compute the amount of the contractor's invoice. Ordinarily, the formation type (such as sand, shale, lime, etc.) is the only geological information recorded.

A drilling time log is often kept by the driller when hole depth approaches a zone of particular interest. This is done manually by marking the kelly joint at the prescribed intervals (1 ft, 5 ft, etc.) and recording the drilling time for the increment. Such a record is quite useful for locating precisely formations or porous zones which are anticipated as productive possibilities. Abrupt changes in drilling rate will immediately indicate a change in lithology although the cuttings may not reach the surface for some time. Automatic devices which furnish a continuous record of drilling progress are also in common use. These instruments consist basically of a spring-actuated drum containing a flexible steel cable whose other end is fastened to the gooseneck of the swivel above the kelly joint, via a pulley at the crown of the derrick. As drilling progresses, the downward movement of the kelly rotates the recording chart so that a continuous record of its position is made. The record obtained is quite accurate and includes an accounting of all non-drilling time such as passes in making trips, connections, and repairs to equipment. Two lines are obtained, as shown in the sample chart of Figure 11.1. The left hand tract furnishes the foot-by-foot drilling rate by recording a diagonal line (to the left and upward) as each foot is drilled. The offsets to the right occur at intervals of five feet. Non-drilling time is shown by the deflections to the right on the right hand track. The net drilling time is obtained by subtracting any non-drilling time from the total interval.

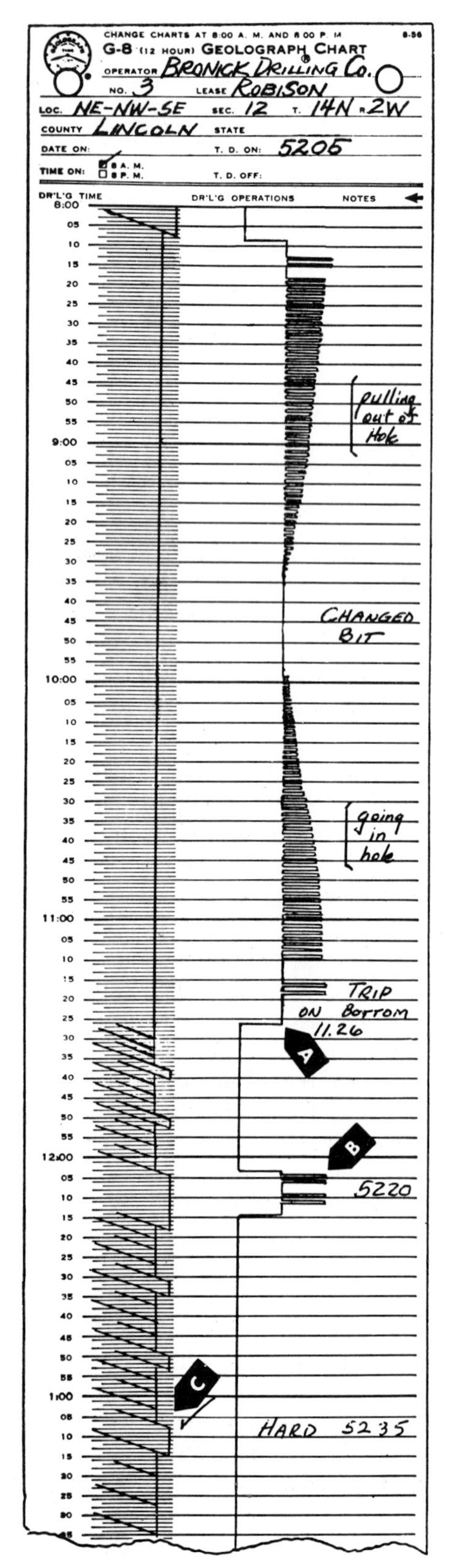

A Line in drilling operations column moves to the left indicating that driller got on bottom with new bit and started drilling at 11:26. Total trip time, as indicated by "Trip Action", 3 hours and 17 minutes.

B This is the way a connection looks on the Geolograph chart. The driller raised the drill pipe from bottom at 12:03, broke out the kelly, picked up a single pipe (adding it to the drilling string), picked up the kelly and resumed drilling. This operation required 11 minutes, and the driller has written the depth of the hole, at that time, on the chart. Thus, every connection is a convenient datum for determining the depth of any drilling or down-time break, either immediately above or below.

C A 4-foot hard streak was encountered at 5,235 feet, as indicated by the increased spacing of the foot marks on this time chart.

D A connection was made at 5,259 feet and a vertical test was run at this point to determine the vertical deviation of the hole. The driller has noted on the chart that the test was actually taken at 5,250 feet and the deviation was ½ degree. The vertical test and connection required 34 minutes.

E Soft bed was drilled from 5,266 to 5,269 feet. Because of the thinness of this bed, no core or drill stem test was attempted.

F This section represents 5 feet of drilling. Note that every 5 feet the base line is offset for 1 foot, making a convenient marker for determining the depth of significant drilling changes.

G Connection was made at 5,287 feet. Note similarity to the record at "B".

H A hard streak was encountered from 5,288 to 5,290 feet.

I At 5,290 feet, the formation softened, drilling continued to 5,300 feet where the driller was given orders to cease drilling and circulate for samples.

J Circulating for samples started at 6:39 as indicated by movement of the line to the right. After circulating for 35 minutes, samples showed stain and odor, and a drill stem test was ordered.

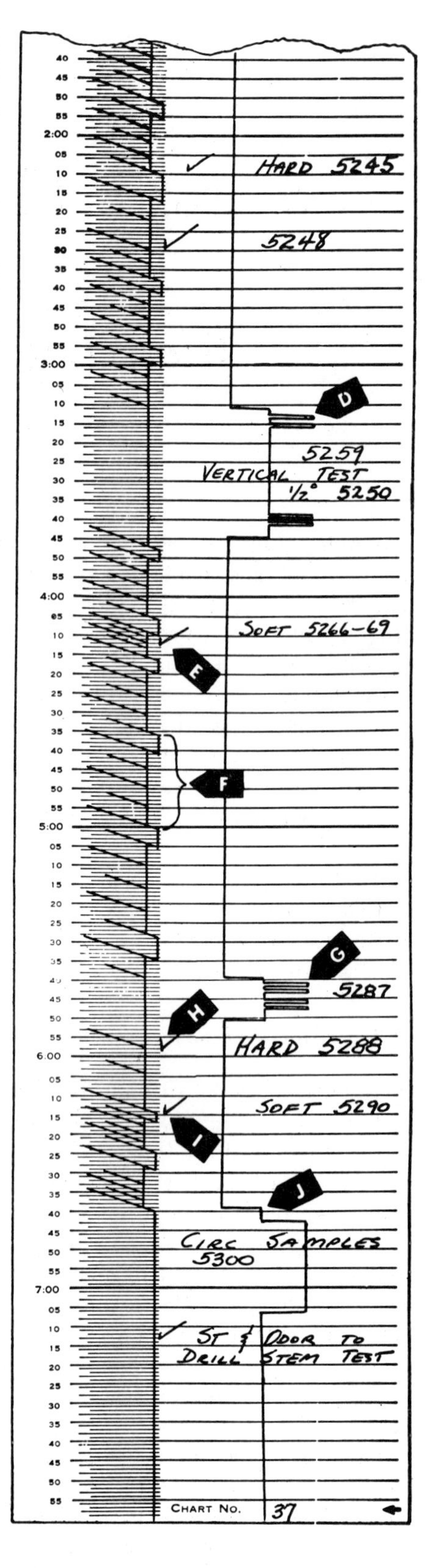

Fig. 11.1. Typical mechanical drilling log record. Courtesy Geolograph Mechanical Well Logging Service.

Accurate drilling time data are of considerable use in geologic interpretations and are widely used by geologists in exploratory or *well-sitting* work. The unique advantage of such data is that the information is immediately available. Engineering studies of penetration rate, bit performance, and operating procedures are also improved by the accuracy and completeness of automatically recorded data.

11.2 Sample Logs

The well cuttings are a source of considerable information, provided that they are properly procured. In an exploratory well, the driller's instructions may call for samples to be caught at a definite footage interval over the entire hole. In field development wells, such complete coverage is often unnecessary and only the interval(s) of interest may be sampled. A typical procedure for catching rotary samples runs as follows. A portion of the returning mud stream is diverted into a sample box where the reduced velocity allows the cuttings to be deposited. At regular intervals (10 ft, for example), a crew member removes a representative portion or sample from the box and then cleans the trough. The end gate or baffle is then replaced in order to catch fresh cuttings from the next interval. The sample is washed with water (or a solvent in the case of oil base mud) and decanted several times, after which it is dried and placed in a cloth bag (sample sack). Each sack is tagged according to the interval which was drilled while it was accumulating. The bags are then stored until called for by the company representative (commonly the geologist). This is, of course, a general procedure and is not always followed. As the well depth approaches promising formations, the geologist may be on continuous duty at the well, in which case the procedure is varied to suit the circumstances.

From examination of the cuttings, the skilled observer is able to determine the rock type, the specific formation being drilled, the depth at which a certain formation was encountered (called the formation's top), and qualitative indications of porosity and oil content. Other features such as the texture, fossil content, and mineral composition may be noted and recorded in some cases. Normally, such fine points are merely used as mental observations which aid in identifying the formation.

In sample analysis, there are many complicating factors which can obscure the true subsurface relationships. There is a discrepancy between the time the rock was drilled and the time it reached the sample box. This is particularly troublesome in deep wells, where a cutting may take two or more hours to reach the surface. Consequently, the label on the sack, which gives the depth interval at the time the sample was accumulating, may differ greatly from the true depth at which it was drilled. The geologist must adjust this to approximate more nearly the true condition. Our previous calculations of cutting slip velocity may be used as a guide to such corrections; however, a common rule of thumb is to assume a cutting lag of 10 min/1000 ft, which amounts to a net rising velocity of 100 ft/min. This may be considerably in error in some cases.

Another problem in sample inspection is that the cuttings obtained over a given interval usually contain fragments from upper beds which have sloughed into the annulus. Shales are particularly prone to do this and may constitute half the sample, even though solid limestone is being drilled. The practiced observer is usually able to eliminate such cuttings from the picture; however, it is not always easy. Data on the drilling time are of great help in overcoming both the lag and sloughing problems so that the true picture is obtained.

Despite the smallness of cutting samples, a good qualitative picture of formation porosity may be obtained from a microscopic examination of them. The microscope commonly used is a low-power (generally 12 to 24) binocular type which is adequate, and in fact, desirable, for routine cutting examination. Oil stains may be detected, despite the severe flushing of the samples. An odor of oil may be noted in many cases and is, of course, an excellent qualitative indicator. The fluorescent lamp is also a great aid in detecting oil shows, since crude oils exhibit some, although highly variable, fluorescence. Pieces of the sample which show apparent oil staining may be placed in carbon tetrachloride, which will leach out minute traces of oil, yielding a fluorescent solution. The use of fluoroscopy is hampered by oil-base or oil-emulsion muds; however, crude oil fluorescence may usually be distinguished from the known, and previously observed, fluorescence of the oil in the mud system. Certain minerals also exhibit fluorescence which is often difficult to distinguish from that of crude oil. The carbon tetrachloride test mentioned above will indicate whether the fluorescence is of oil or mineral origin.

These are but a few of the more common observations obtained from cutting analysis. Such data are normally presented in a sample or strip log. A percentage log, which represents all formations noted in the cuttings according to their proportion in the total sample, is often used in areas where the section is not well known and the geologist is hesitant to make precise distinctions. Specific formations are normally shown on the logs of well-known areas. The drilling time may be plotted beside the sample log as an aid to interpretation. As mentioned previously, a depth correction is often required to make the samples jibe with later data.

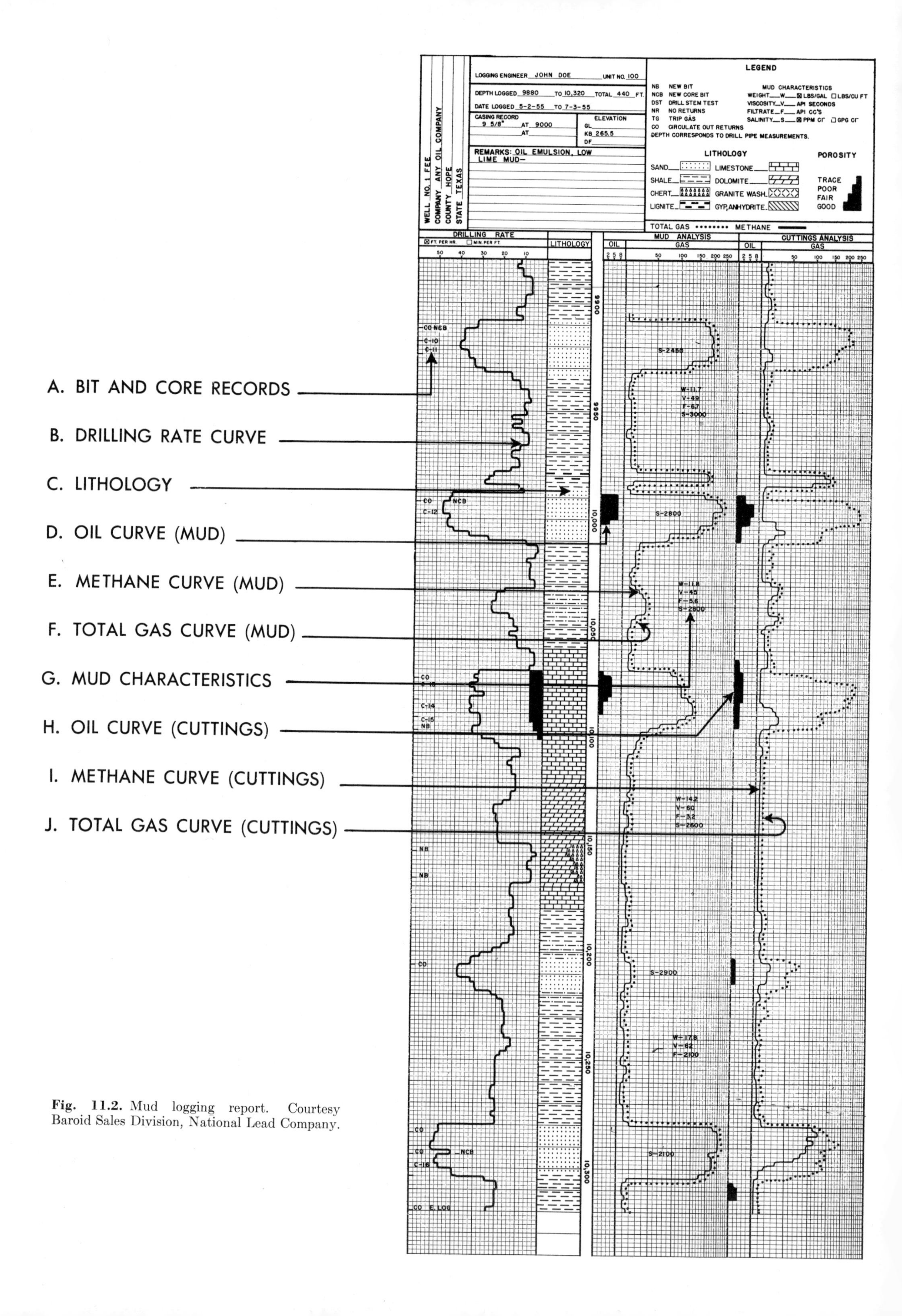

Fig. 11.2. Mud logging report. Courtesy Baroid Sales Division, National Lead Company.

11.3 Mud Logging

Mud logging, as the term is used here, refers to the continuous analysis of the drilling mud for oil and gas content. Cutting inspection and analysis are also included in the complete service. This procedure is widely used in exploratory drilling, and affords an extra tool for detecting the presence of oil and gas.[1] The basic oil or gas detection equipment consists of a mud-gas separator and an electric filament or hot wire. This wire is one component of a bridge circuit similar to the well known Wheatstone bridge.

A portion of the mud is diverted from the return flow line into the gas-mud separator. Air is injected into the mud to agitate it thoroughly and liberate a portion of any entrained gas. The air-gas mixture (if volatile hydrocarbons are present) is ignited by the hot wire. This raises the wire's temperature, thereby increasing its electrical resistance and unbalancing the bridge circuit. The resistance increase depends on the amount of gas present to combust. Hot wire cutting analyzers are essentially the same device. These use a grinder, such as a Waring Blendor, which is furnished with a vacuum cap. The cuttings are placed in the grinder and a vacuum applied. As the particles are pulverized, minute traces of gas are liberated and sucked into contact with the hot wire, with the same effect as that just described. This equipment is not unique to mud logging services, and small, battery powered units are available as an aid to routine sample logging.

Quantitative interpretation of formation content is not obtainable by mud or sample logging for the same reasons that flushed cores do not yield original saturations. Continuous mud logging is, however, an excellent exploratory tool. A typical mud log is shown as Figure 11.2. Note that other data, such as drilling rate and lithology, are included in the presentation.

11.4 Electric Logging

This method of formation evaluation was developed by Conrad and Marcel Schlumberger and was introduced to the United States in 1929. By 1935, its use was widespread and electric logging has since become a standard practice. The literature on this subject is voluminous and the entire topic is one of considerable complexity. Indeed, there are many technical experts who devote essentially all of their time to this subject. Consequently, this section will present only a brief introduction to the basic concepts and ideas on which these electrical measurements and their interpretations are based. No attempt will be made to cite all the pertinent references; however, the books, manuals, and article series listed at the end of this chapter will serve as adequate material for those wishing to delve deeper at this time.

For our purposes, an electric log will be considered as a plot of certain electrical properties of the strata in contact with the well bore. These properties are measured by various electrode configurations which are lowered into the borehole on electric cables. The standard electric log normally presents two different sets of graphs. The left-hand side shows the spontaneous potential (called the SP), while the *resistivity* measurements are recorded on the right. Several of these measurements may be recorded simultaneously with one run of the instrument. Before discussing the individual curves, let us first consider a number of basic concepts and definitions.

11.41 Basic Concepts

The resistivity of a material is the specific resistance which it offers to the flow of electrical current. In this regard it is much like specific gravity or density as opposed to weight.

$$R = \frac{rA}{L} \tag{11.1}$$

where R = resistivity of media or conductor through which current is flowing

r = resistance of conductor

A = cross-sectional area of conductor

L = length of conductor

The practical units of R as used in electric logging are ohms $\times \frac{\text{meters}^2}{\text{meters}}$, or simply ohm-meters. A conductor having an area of one square meter and a length of one meter and offering one ohm resistance to current flow is said to have a resistivity of one ohm-meter.

Fluid Resistivities

With but a very few and exceedingly minor exceptions, dry sedimentary rocks are non-conductive, which is to say that their resistivity is extremely high. Shales are often considered as conductors; however, their low resistivities are due to a high interstitial water content rather than to any conduction by the dry clay minerals. Oil and gas are also insulators and will not conduct an electric current. The conductance of subsurface strata is then normally due to contained water. The resistivity of water depends on its salinity and temperature as shown in Figure 11.3. Note that salinity is expressed as parts per million or grains per gallon of sodium chloride. Formation water resistivity may be either measured from a sample or calculated from a chemical analysis used in conjunction with Figure 11.3. Concentrations of other ions may be converted to electrically equivalent sodium chloride concentrations by the following factors:[2]

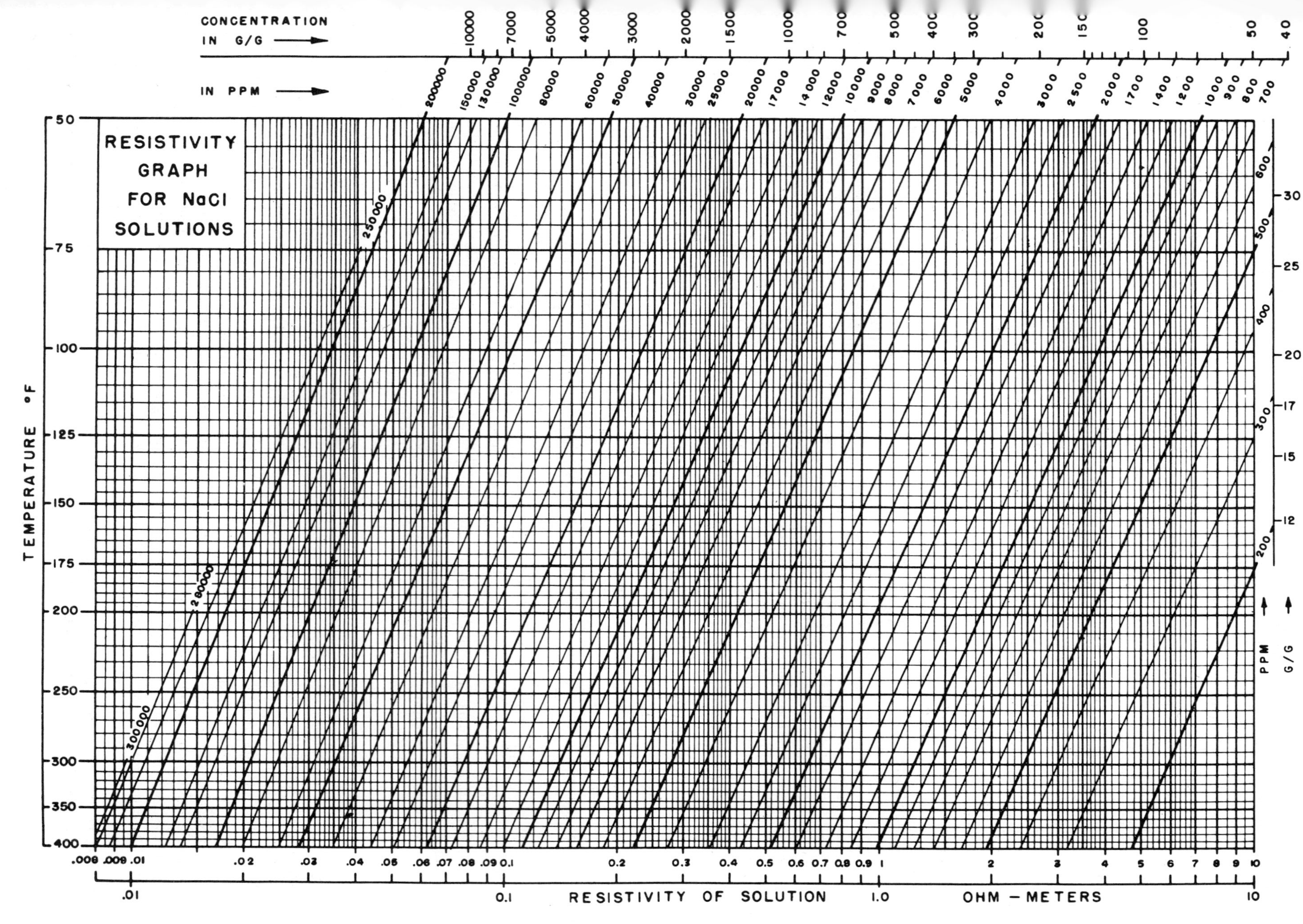

Fig. 11.3. Resistivity of water as a function of salinity and temperature. Salinities are in terms of NaCl concentration. Courtesy Schlumberger Well Surveying Corporation.

TABLE 11.1

FACTORS FOR CONVERTING VARIOUS IONS TO ELECTRICALLY EQUIVALENT SODIUM CHLORIDE CONCENTRATIONS

Ion	*Multiplier*
Na^+	1.0
Ca^{++}	0.95
Mg^{++}	2.0
K^+	1.0
SO_4^{--}	0.5
Cl^-	1.0
HCO_3^-	0.27
CO_3^{--}	1.26

For most practical purposes the total solids may be summed directly from a chemical analysis without making the above corrections.

Example 11.1

Given the following chemical analysis of an oil field brine, compute its equivalent salinity in ppm sodium chloride and its resistivity at 100, 150, and 200°F.

(1) *Ion*	(2) *Concentration, ppm*	(3) *Factor*	(4) (2) × (3)
Na	50,000	1.0	50,000
Mg	10,000	2.0	20,000
Cl	52,000	1.0	52,000
SO_4	6,000	0.5	3,000
HCO_3	4,000	0.27	1,080
Total	122,000		126,080

Solution:

The equivalent NaCl salinity is the total of column (4), or, 126,000 ppm which, practically speaking, is the same as the sum of column (2). From Figure 11.3 the formation water resistivities are obtained as:

Temperature, °F	*R_w, ohm-meters*
100	0.05
150	0.034
200	0.025

A rule of thumb for estimating the effect of temperature on water resistivity which is useful and quite accurate for all except very high salinities, is the following:

$$R_{w2} = R_{w1} \times \frac{T_1}{T_2} \tag{11.2}$$

where R_{w1}, R_{w2} = water resistivities at temperatures T_1 and T_2, respectively

Applying this rule of thumb to this example and assuming that R_{w1} is 0.05 ohm-meters at 100°F,

$$R_{w150} = 0.05 \times \frac{100}{150} = 0.033$$

$$R_{w200} = 0.05 \times \frac{100}{200} = 0.025$$

Note the close check with the previous values.

A knowledge of mud resistivity is essential for electric log interpretation. This property is always measured by the logging crew, either on a surface sample or in the borehole, and appears on the log heading. Temperature corrections may be made using Figure 11.3. The resistivity of a mud is greater than that of its filtrate, due to the presence of non-conductive solids. Charts are available which make possible the estimation of the values of any two of the factors, mud resistivity, mud filtrate resistivity, or the resistivity of the mud cake, from a knowledge of the third. Useful and quite accurate approximations are afforded by Eqs. (11.3) and (11.4):

$$R_{mf} = 0.75R_m \tag{11.3}$$

where R_{mf} = mud filtrate resistivity at a particular temperature, ohm-meters

R_m = mud resistivity at the same temperature, ohm-meters

$$R_{mc} = 1.5R_m \tag{11.4}$$

where R_{mc} = resistivity of the mud cake

Formation Resistivities

Any conductivity exhibited by sedimentary rock strata is attributed to interstitial fluid content. The only exceptions to this rule are a few sands which contain appreciable quantities of glauconite and pyrite, both of which are conductors. It would then appear logical that for a particular porous medium:

(1) the greater the water content, the lower will be the formation resistivity.

(2) a rock which contains an oil and/or gas saturation will have a higher resistivity than the same rock completely saturated with formation water.

Conclusions (1) and (2) form the principal basis for electric log interpretation and are completely valid except for those cases where the formation water is relatively fresh (only slightly saline). The complicating factor in such cases is the small contrast in resistivity between the water and oil or gas.

A fundamental definition is the following:

R_0 = the resistivity of a rock which is 100% saturated with formation water

The value of R_0 is, of course, greater than the resistivity of the formation water with which the rock is saturated. However, these two quantities are intimately related by:

$$R_0 = FR_w \tag{11.5}$$

where F = formation factor.

Consider a clean, porous sandstone sample which is 100% saturated with a brine of resistivity R_w. Let a voltage E be impressed across the sample. Since current can flow only through the water, it would appear that the value of F should depend on:

(1) The amount of water present, hence the porosity.

(2) The pore geometry of the particular rock, the main factor being, probably, the tortuosity τ which is the square of the ratio of the actual path length of the current to the length of the sample.

(3) The degree of the rock's consolidation: the extent to which the individual grains are cemented together.

Experimental evidence has shown that porosity, ϕ, and F are related by Eq. (11.6):

$$F = \phi^{-m} \tag{11.6}$$

where m = the cementation factor.

Equations (11.5) and (11.6) are from the work of G. E. Archie and are among the expressions commonly referred to as Archie's equations.[3] The value of m varies with the rock and may be determined from laboratory tests; however, Table 11.2 shows its typical range of magnitude.

TABLE 11.2

VALUES OF CEMENTATION FACTOR m (AFTER GUYOD[4])

Rock type	m
Highly cemented: limestone, dolomite, quartzite	2.0–2.2
Moderately cemented: consolidated sands	1.8–2.0
Slightly cemented: friable, crumbly sands	1.4–1.7
Unconsolidated sands	1.3

A later empirical expression which relates F and ϕ is the Humble equation:[5]

$$F = 0.62\phi^{-2.15} \tag{11.7}$$

Example 11.2 serves to illustrate the use of these equations.

Example 11.2

A cylindrical core sample of a well consolidated sand is completely saturated with a synthetic brine of 50,000 ppm salinity. At 104°F the resistance of the core (from end to end) is 980 ohms. The core is 3½ in. in diameter and 12 in. long. Compute its porosity by both the Archie and Humble equations.

Solution:

From Eq. (11.1):

$$R_0 = \frac{(980)(3.5^2\pi/4)}{12} \times \frac{1}{39.4} = 2.0 \text{ ohm-meters}$$

From Figure 11.3, R_w = 0.10 ohm-meter at 104° F.

From Eq. (11.5),

$$F = \frac{2.0}{0.10} = 20$$

Using Eq. (11.6) for m = 2,

$$\phi = \frac{1}{F^{1/2}} = \frac{1}{4.47} = 0.22 \text{ or } 22\%$$

Using Eq. (11.7):

$$\phi = \frac{1}{(F/.62)^{0.465}} = \frac{1}{(32.3)^{0.465}} = \frac{1}{5.0} = 0.20 \text{ or } 20\%$$

A rock which has an oil or gas saturation will naturally exhibit a higher resistivity than the same rock with 100% water saturation. Further, the greater the hydrocarbon saturation, the greater will be the resistivity. This behavior is expressed by the empirical relationship of Eq. (11.8) which is another of the Archie equations:

$$S_w = \left(\frac{R_0}{R_t}\right)^{1/n} = \left(\frac{FR_w}{R_t}\right)^{1/n} = \left(\frac{\phi^{-m}R_w}{R_t}\right)^{1/n} \tag{11.8}$$

where S_w = water saturation of the rock in question.

R_t = true resistivity of the formation, ohm-meters. (The designation "true" is used to distinguish between this and the apparent value read from a log. Apparent values may or may not require corrections to convert them to R_t.)

n = saturation exponent. For clean, water-wet rocks n = 2 is commonly used. The precise value of n for shaley or oil-wet rocks is difficult to obtain and discussion of it is beyond our treatment. Its range *for shaley sands* is between 1.0 and 1.7. For oil wet rocks, n = 2 to 10.

Example 11.3

Suppose that the core sample of Example 11.2 is now partially saturated with oil so that its resistivity is 20 ohm-meters. What is its oil saturation?

Solution:

$$S_w = \left(\frac{2.0}{20}\right)^{1/2} = 0.316 \text{ or } 32\%$$

and

$$S_o = 1 - S_w = 68\%$$

Question: If these values represented the original reservoir condition, would such a sand produce water, oil, or both?

Answer: Since the precise value of the critical water saturation of this sand is not known, the question cannot be definitely answered. However, 32% is a reasonable value for the irreducible water saturation of a water-wet sand, hence clean oil production is probable.

Equations (11.5), (11.6) or (11.7) and (11.8) form the backbone of quantitative electric log interpretation. The main problem involved is to obtain reasonably accurate values of the necessary parameters. The many logging devices in use have been designed primarily to overcome the problems of obtaining such data under various operating and lithological conditions.

Before discussing resistivity measurements as they are obtained in boreholes, let us consider another type of measurement.

11.42 Spontaneous or Self Potential Measurements[6]

The circuit for SP measurements is shown schematically in Figure 11.4. An electrode M is lowered into the hole on an insulated cable. Another electrode N is grounded in a shallow hole dug at the surface which has been filled with the drilling mud. All potential (or voltage) changes which occur between M and N as the electrode is raised from the hole are recorded as a plot of voltage difference vs. depth.

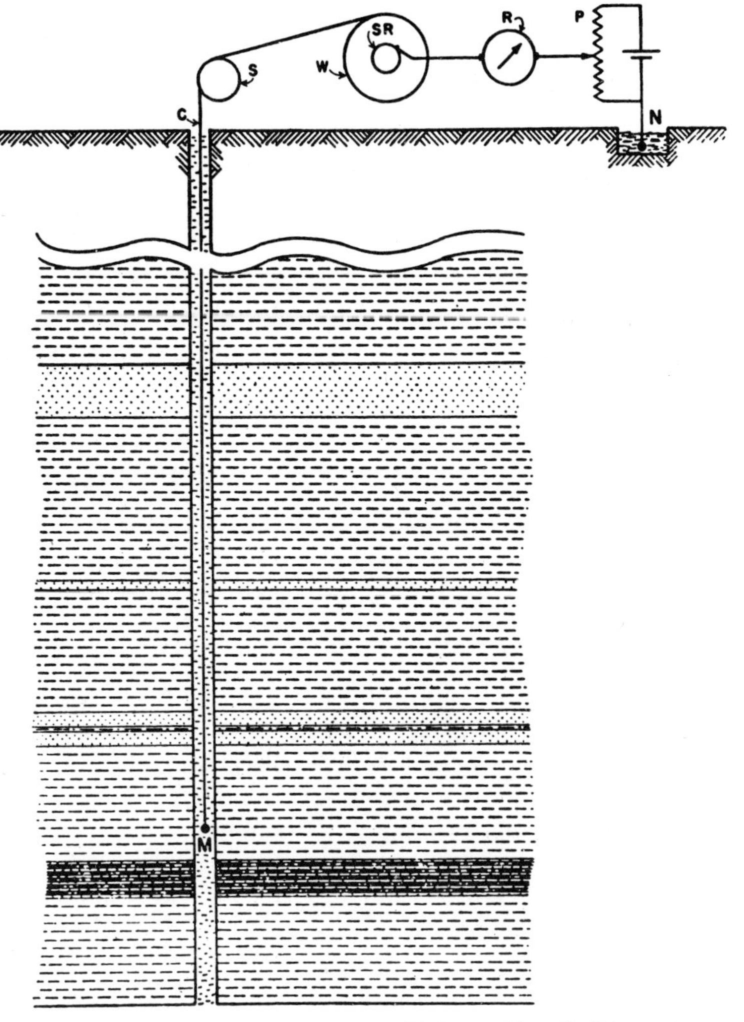

Fig. 11.4. Schematic circuit for recording SP logs. After Doll,[6] courtesy AIME.

The various factors which give rise to the SP behavior are quite complex. A qualitative picture of the main contributory subsurface occurrences is shown in Figure 11.5. The presence of the mud filled borehole causes currents to flow from zones of high electrical potential (shales) into permeable zones of lower potential (such as permeable sands). Further, ions will move from the formation water into the more dilute mud, or in the opposite direction, if the mud is the more saline fluid. Thus there are two principal sources of SP deflection:

(1) the shale potential (also called the Mounce potential).
(2) the diffusion cell effect caused by the salinity contrast between the drilling mud and formation waters.

The magnitude of the potential difference between M and N is then dependent on the lithological character (shaliness) of the permeable zone and the relative resistivities of the mud filtrate and the formation waters. If the zone contained between the shale bodies in Figure 11.5 were impermeable, no change in potential would occur at the bed boundaries. Thus it would appear that the SP curve is indicative of permeability.

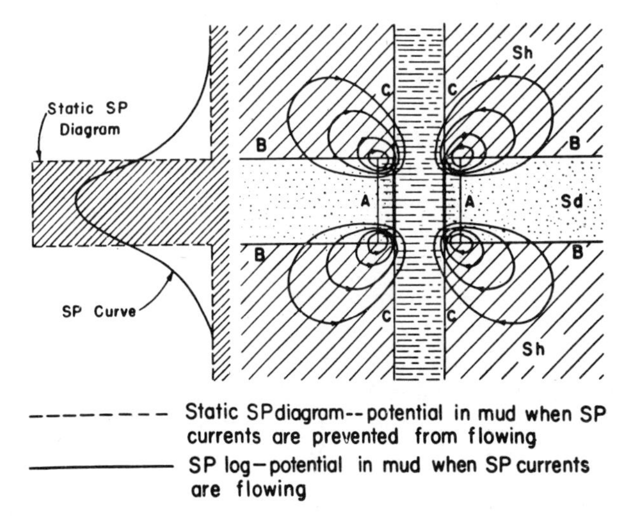

Fig. 11.5. Schematic diagram of borehole currents which cause SP behavior. Courtesy Schlumberger Well Surveying Corporation.

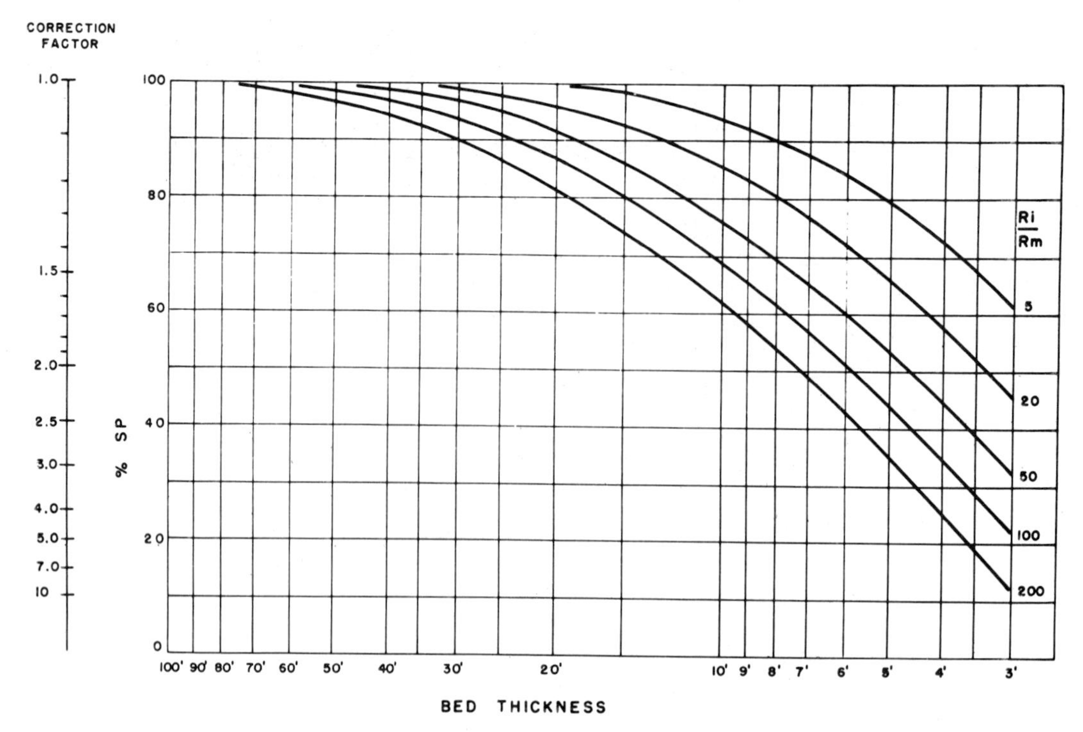

Fig. 11.6. Empirical correction chart for the SP curve. Courtesy Schlumberger Well Surveying Corporation.

This is qualitatively true; however, there is absolutely no known quantitative relationship involved. A bed with any permeability may exhibit an SP even though the magnitude of its permeability is too small to be of commercial interest.

The equation usually used for quantitative interpretation of self-potential behavior is

$$SSP = -K \log \frac{R_{mf}}{R_w} \tag{11.9}$$

where SSP = static self-potential value, mv.

K = lithological factor for the particular formation which is also dependent on temperature, mv.

The apparent value of the SP (the ASP) as read from the log is greatly influenced by bed thickness and will not reach its maximum (or static) value except in thick beds. One method of correcting the ASP to the SSP is the empirical chart of Figure 11.6.

The principal uses of the SP curve are:

(1) Delineation of clean permeable beds from shaly and/or impermeable beds
(2) Determination of bed thickness
(3) Calculation of formation water resistivity R_w—this calculation from SP data is, of course, not as desirable as having either a physical measurement or a chemical analysis
(4) Determination of net pay
(5) Correlation, for subsurface mapping

Example 11.4

Figure 11.8 shows a short section of an SP log from a Texas well. The following steps illustrate a typical SP analysis:

1. The shale base line is drawn through the extreme right hand excursions of the curve. This represents the maximum potential observed (note that positive is to the right) and is recorded opposite impermeable beds. An exception to this occurs when the formation water is more resistive than the mud filtrate, i.e., opposite fresh water zones or when very saline mud is used. This effect is evident from Eq. (11.9) since when $R_{mf} < R_w$, $\log(R_{mf}/R_w)$ is negative and the deflection SSP will be positive (to the right).
2. The only permeable (and porous) zone of any consequence is from 5824 to 5856 ft, as shown. The top and bottom of the zones are taken at the inflection points (not the midpoint) of the curve. This is difficult to determine at the base of the sand, due to the slight irregularity probably caused by a thin impermeable streak at 5858. Therefore the gross thickness of this zone is 32 ft. Note also the presence of a shaly streak at 5835 to 5837 ft. Hence the net section (permeable strata only) is 30 ft.
3. The maximum deflection observed opposite the bed is approximately ASP = -100 mv.
4. From the log heading (not shown): $R_m = 0.80$ ohm-meters at 192°F, the temperature at 7000 ft. Assuming a mean surface temperature of 70°F, the temperature at 5850 ft is:

$$T_{5850} = (192 - 70)\frac{5850}{7000} + 70 = 172°\text{F}$$

Therefore, R_m at 172°F $= \frac{192}{172} \times 0.80 = 0.89$, which checks a value of 0.90 from Figure 11.3.

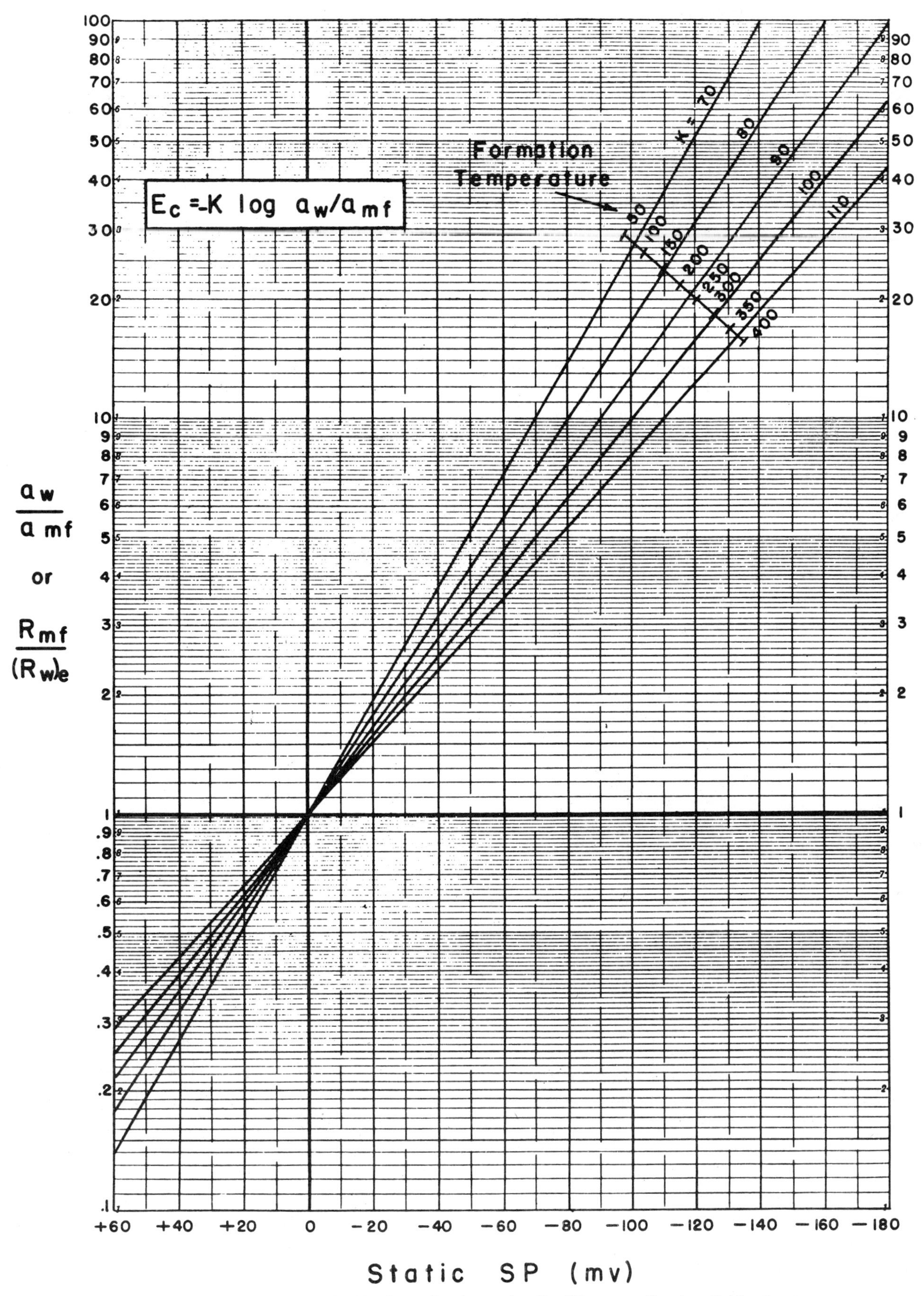

Fig. 11.7(A). Charts for R_w determinations using the SP curve. Courtesy Schlumberger Well Surveying Corporation.

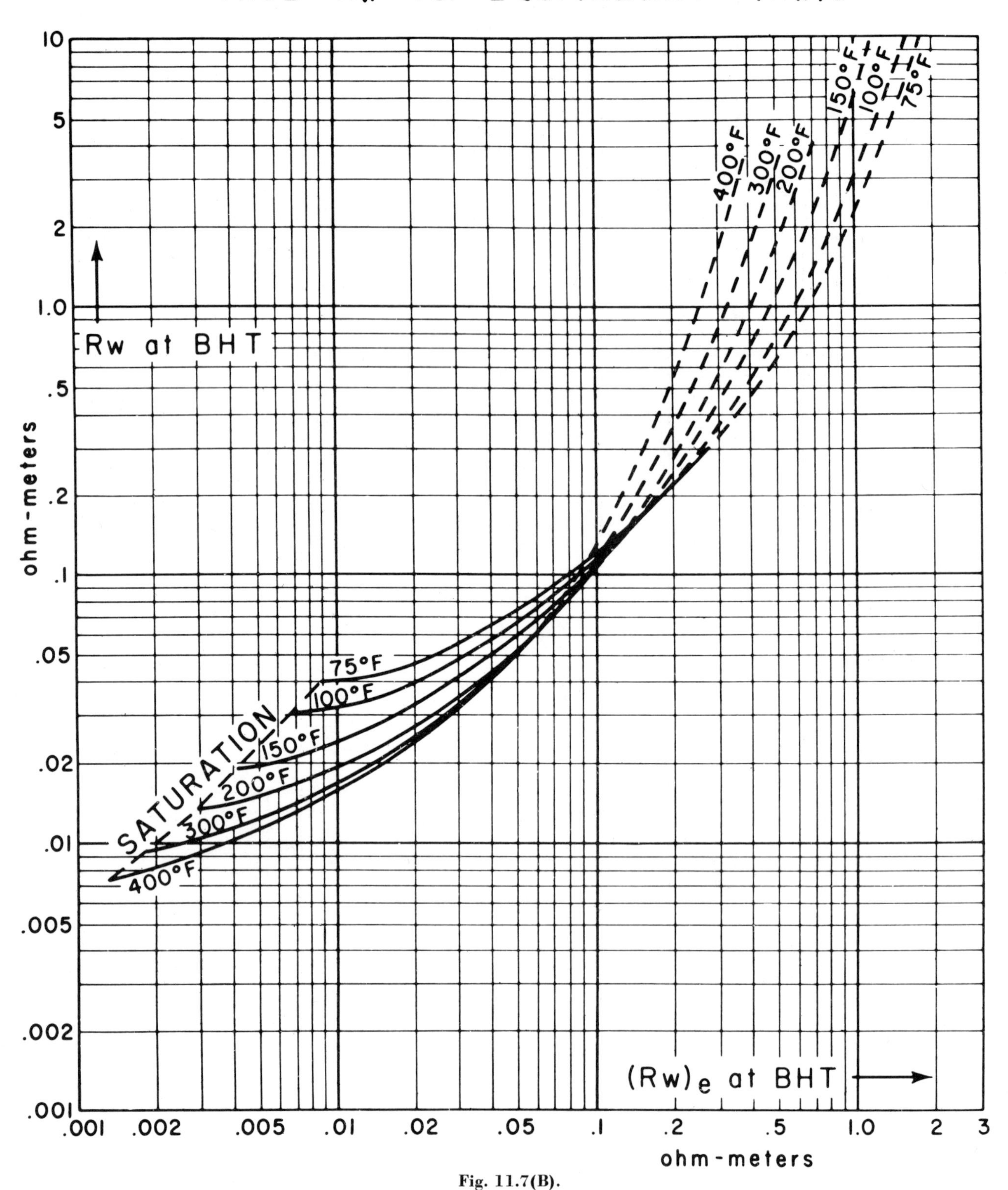

Fig. 11.7(B).

5. In order that Figure 11.6 may be used to evaluate the SSP, a resistivity R_i must be known. This determination will be explained in the next section. For now let us assume it is given as 30 ohm-meters. Therefore,

$$R_i/R_m = 30/0.90 = 33$$

Entering Figure 11.6 at $h = 32$, proceeding upward to $R_i/R_m = 33$, and extrapolating horizontally to the left ordinate, we obtain a value of %SP $= 99$ and a correction factor of 1.01. Practically speaking, these corrections are insignificant and the SSP = ASP = -100 mv.

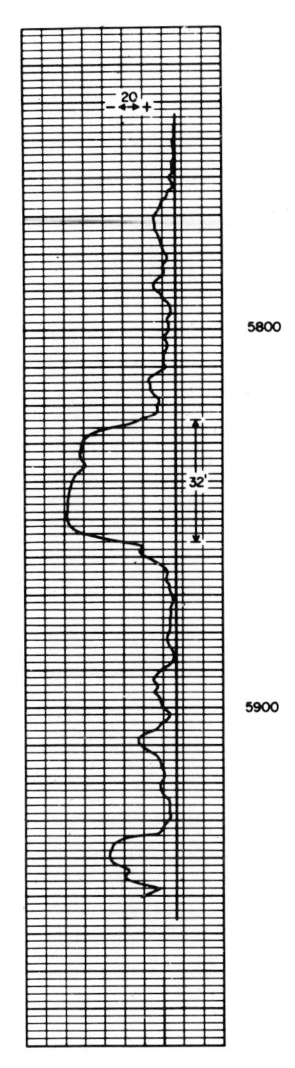

Fig. 11.8. SP curve for Example 11.4.

6. It is now possible to estimate the formation water resistivity R_w which we will assume is the unknown in this case. Figure 11.7(A) is entered at SSP $= -100$ mv; a vertically upward projection interpolated to the temperature line $T = 172°F$ is drawn. It may be noted that $K \cong 83$ is also obtained as either K or T may be used on this chart. Horizontal extrapolation to the left gives $R_{mf}/R_{we} = 16$. Since R_{mf} was not measured, it may be approximated by

$$R_{mf} = 0.75R_m = (0.75)(0.90) = 0.67 \text{ ohm-meters}$$

$$R_{we} = \frac{R_{mf}}{R_{mf}/R_w} = \frac{0.67}{16} = 0.042 \text{ ohm-meters}$$

Finally, from Figure 11.7(B), $R_w = 0.047$ ohm-meters. Checking this on Figure 11.3 at 172°F shows that the salinity of such a water is approximately 70,000 ppm. Note that all calculations are made at the temperature opposite the zone under analysis.

All SP curves are not so simple as Figure 11.8 and the interpretative measures may be much more complex. The interpretation of Example 11.4 would have been enhanced by the resistivity curves for the same section, since these yield complementary data. A few more general points should be mentioned:

(1) The SP curve is very reproducible and accurate, providing that instrumentation errors are not present.
(2) From Eq. (11.9) it is apparent that if $R_{mf} \cong R_w$, no SP definition is obtained, hence the SP curve is normally of little value in salty muds.
(3) Much interpretive information may be obtained from a detailed study of the small irregularities and features of the SP curve. The reader is referred to Doll's paper already cited,[6] and to other references at the end of this chapter, for a complete treatment of this subject.

11.43 Resistivity Measurements

Resistivity measuring devices are more complicated than the SP circuit, since multiple electrode arrangements and externally applied currents are used. Such instruments are numerous in type and only two of the simpler configurations will be discussed in any detail. As was shown in Section 11.41, the Archie equations require a knowledge of the true formation resistivity R_t. Figure 11.9 shows the typical condition around the borehole in an oil zone and the nomenclature to be used in this discussion. The term *annulus* pertains to the zone of relatively high formation water saturation, banked up by the displacing mud filtrate.

The extent of contamination around the borehole depends on the filtration characteristics of the mud, the pressure differential between the well bore and the formation, and the time of exposure, as well as the nature of the rocks. The resistivities which may influence measurements are those shown as R_m, R_{mc}, R_{xo}, R_i, R_{an}, and, finally, R_t. In general, it is the latter which is desired. A standard set of resistivity curves, designed to obtain R_t under certain conditions, consists of two normal curves (2 electrode) and one lateral curve (3 electrode). A further description of the circuits for these devices will serve to illustrate the basic principles of well bore resistivity measurements.

Normal Curves (2 electrode)

The simplified equivalent circuit for this device is shown in Figure 11.10. Assume that the medium surrounding the electrodes is completely homogeneous and of resistivity R. As current is transmitted from the current electrode A, it will flow spherically outward as shown by Figure 11.11.

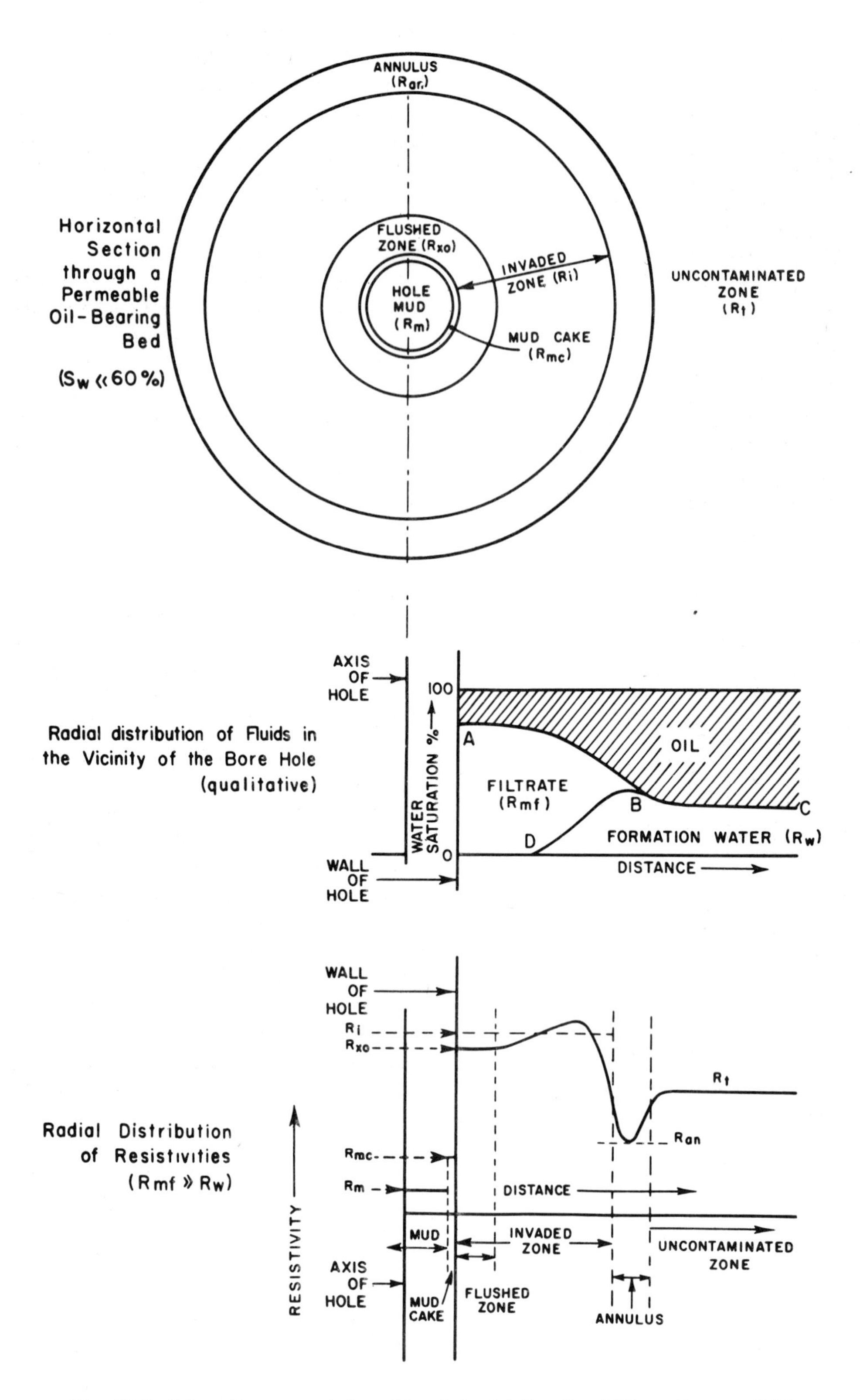

Fig. 11.9. Schematic representation of borehole vicinity after drilling. Courtesy Schlumberger Well Surveying Corporation.

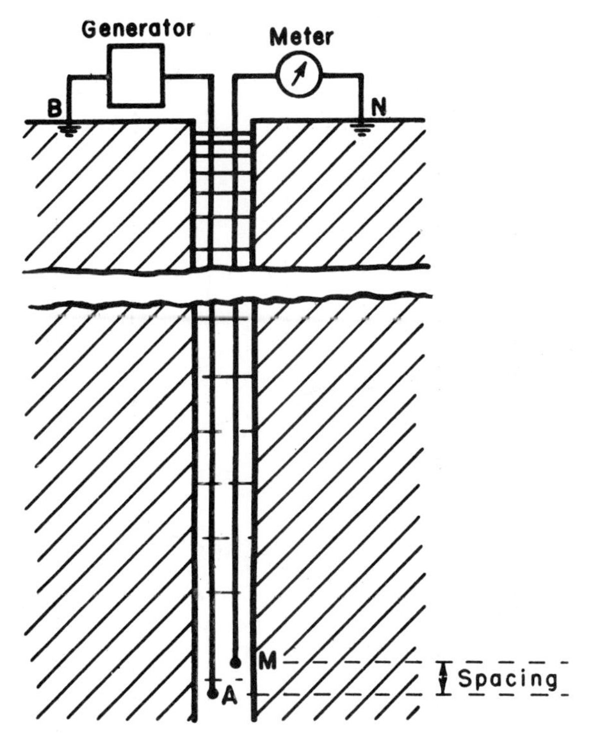

Fig. 11.10. Simplified two-electrode circuit for measuring resistivity. Courtesy Schlumberger Well Surveying Corporation.

The potential difference between any two radii from A is defined by Ohm's law:

$$E = ir = iR\frac{L}{A} \qquad (1)$$

where E = electromotive force or voltage
i = current
r = resistance
R = resistivity
L = length of conductor over which E is measured
A = cross-sectional area of conductor

However, for the spherical system of Figure 11.11,

$L = dx$

$A = 4\pi x^2$ (the surface area of a sphere of radius x)

Hence:

$$dE = \frac{iR}{4\pi x^2}\,dx \qquad (2)$$

Integration of (2) from $x = M$ to $x = \infty$ (the potential is zero at $x = \infty$) yields:

$$E = \frac{Ri}{4\pi x} \qquad (11.10)$$

where E = potential at any radii x due to the current i transmitted at A.

Considering then the two electrode system of Figure 11.10, it is apparent that the potential measured at M due to A is:

$$E_{MA} = \frac{Ri}{4\pi(AM)}$$

where AM is the distance between A and M, which is the spacing of the device. Since it is the resistivity that is desired,

$$R = 4\pi(AM)\frac{E_{MA}}{i} \qquad (11.11)$$

The quantity $4\,\pi\,(AM)$ is then the constant for the particular sonde (logging tool assembly) used. In practice, the medium surrounding the sonde is not homogeneous, hence the current flow will be distorted, and will not be spherical.

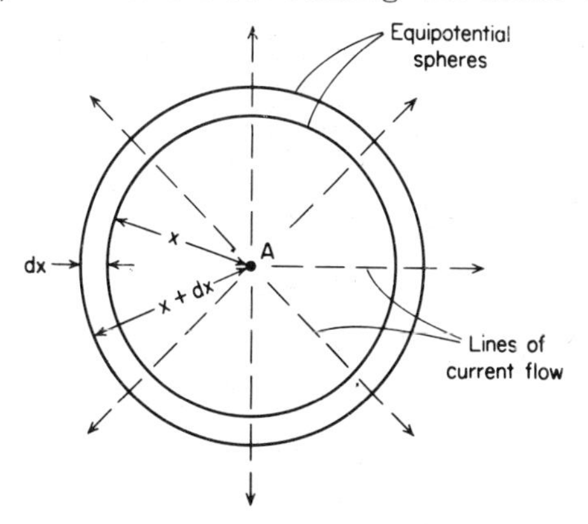

Fig. 11.11. Current flow pattern in isotropic and homogeneous medium. Courtesy Schlumberger Well Surveying Corporation.

Two normal curves of different spacing, called the short normal and long normal, are usually obtained. Logging companies use different spacings; however, 16 and 64 in. are common combinations. The spacing governs the distance of investigation around the borehole — the greater the spacing, the greater the radius of investigation. For example, a very short normal might investigate only the mud in the borehole.

The typical curve shapes recorded by two electrode devices are shown in Figures 11.12 and 11.13. These can be interpreted as follows:

(1) The curves are symmetrical with respect to the bed centers.
(2) Bed thickness is not sharply defined, due to the irregularity of current flow when A and M are on opposite sides of the boundaries. A short normal records boundaries more accurately than one of longer spacing; however, the shorter the spacing, the greater the effect of the mud filled borehole; this will flatten the curve features. In practice, the 16-in. curve normally provides a satisfactory balance between these factors.
(3) Beds having a thickness less than the spacing will indicate a low resistivity or crater, as shown. The width of the crater is the sum of the spacing and bed thickness; therefore, no true measure of the resistivity of such beds is possible from these devices.

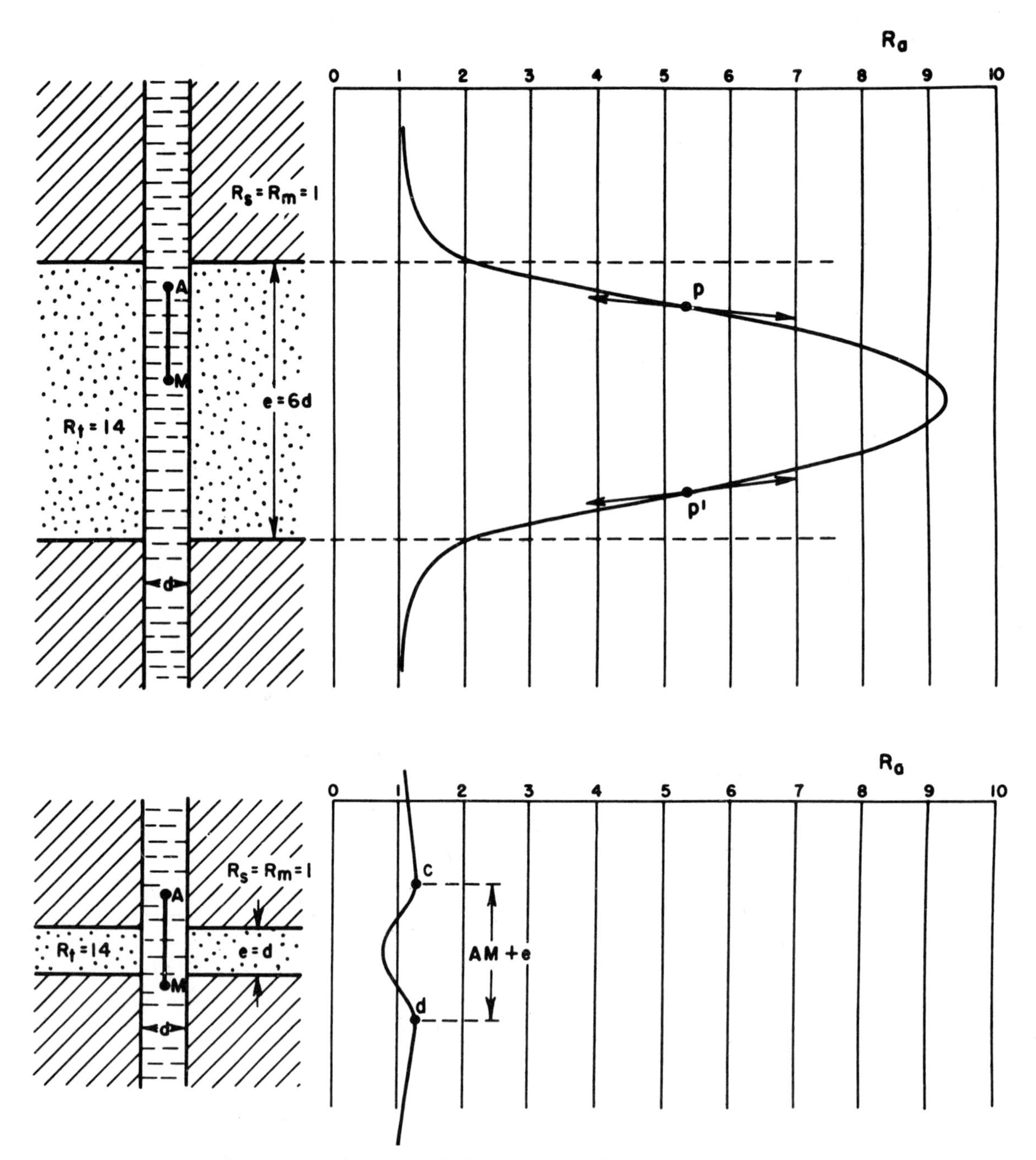

Fig. 11.12. Curve shapes for normal device opposite bed more resistive than adjacent beds. (R_s = resistivity of shale, e = bed thickness, d = hole diameter). Courtesy Schlumberger Well Surveying Corporation.

Also, the resistivity curve opposite a bed of thickness equal to the spacing will be almost flat and indistinguishable.

(4) Due to their relatively short spacing and consequent susceptibility to borehole and invaded zone effects, the normal curves are not well adapted for direct measurement of R_t. Under certain conditions, however, they may be used for obtaining R_t, provided that necessary corrections are made. The resistivity reading from the short normal is often taken as R_i (see Figure 11.6).

The Lateral Curve (3 electrode)

This device is designed to overshoot the borehole and invaded zone, thereby measuring R_t. The equivalent circuit is shown in Figure 11.14. Two pickup electrodes M and N are relatively close together with A being some distance removed. The distance from A to the midpoint of M and N, point O, is considered the spacing, and will

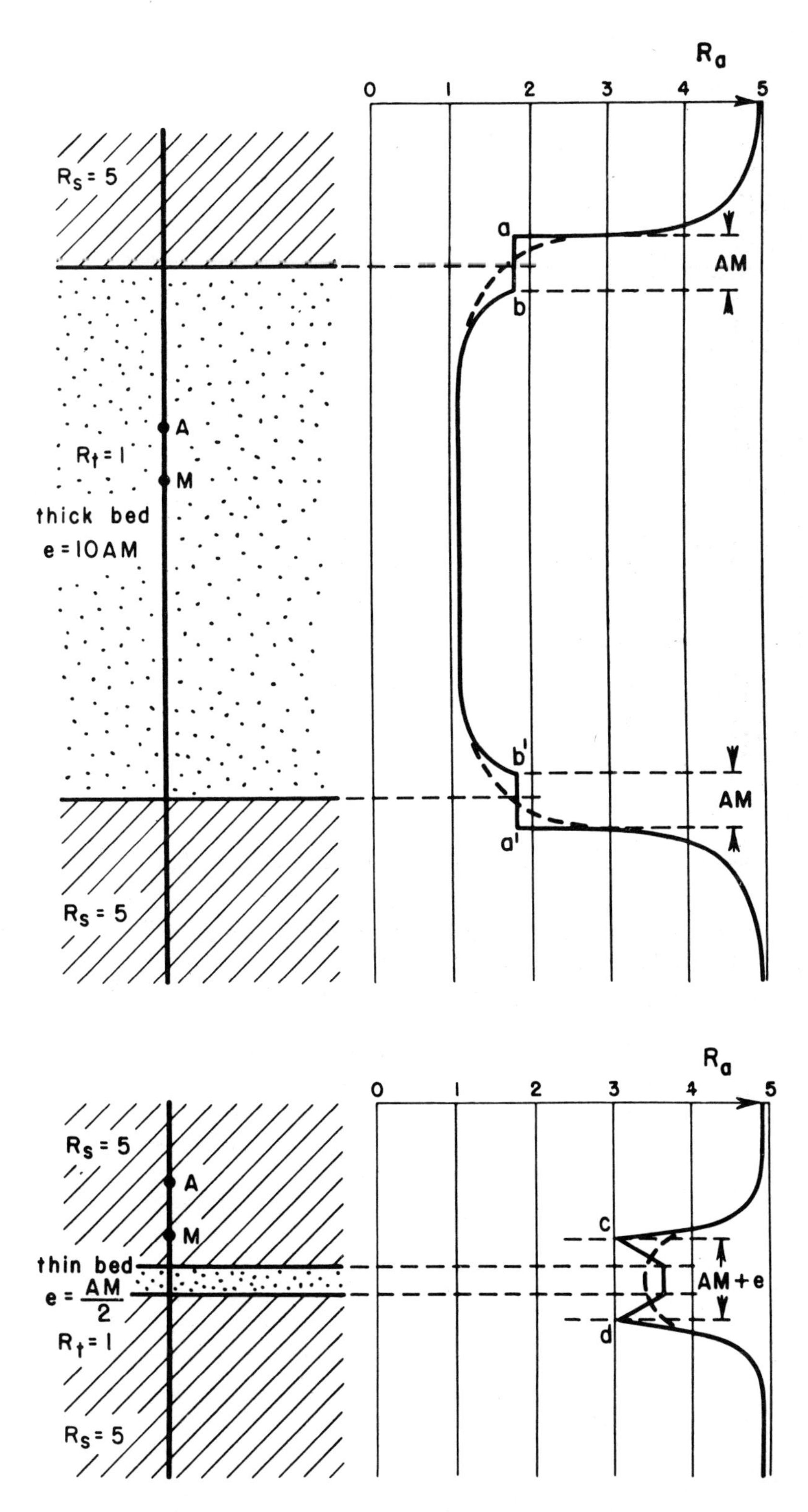

Fig. 11.13. Normal device curve shapes opposite bed less resistive than adjacent formations. Solid lines — theoretical, borehole effect neglected. Dashed lines — shows qualitatively the rounding effect of mud filled borehole. Courtesy Schlumberger Well Surveying Corporation.

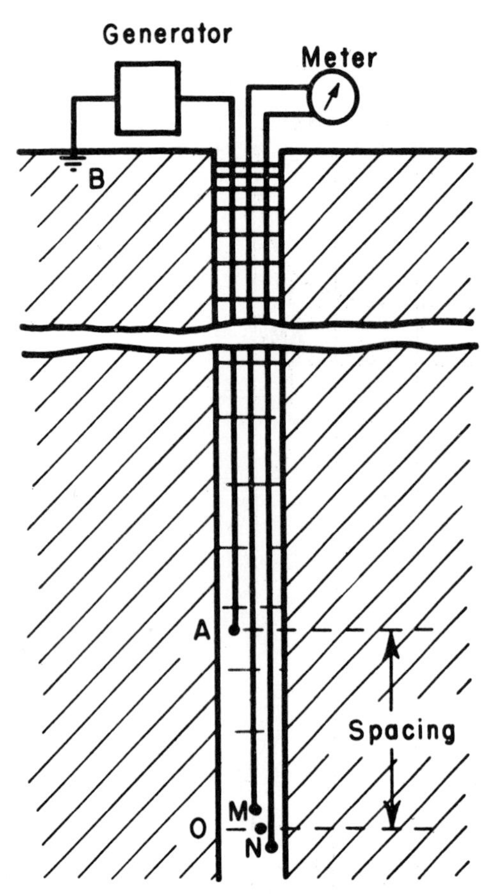

Fig. 11.14. Equivalent circuit for the lateral (three-electrode) device. Courtesy Schlumberger Well Surveying Corporation.

be referred to as AO. With this instrumentation, the potential between M and N due to the current from A is measured. This may be developed in the same manner as before:

$$E_{MA} = \frac{R_i}{4\pi(AM)} = \frac{R_i}{4\pi(x_1)} \tag{1}$$

and

$$E_{NA} = \frac{R_i}{4\pi(AN)} = \frac{R_i}{4\pi(x_1 + x_2)} \tag{2}$$

hence

$$\Delta E_{MN} = E_{MA} - E_{NA} = \frac{R_i}{4\pi}\left(\frac{1}{x_1} - \frac{1}{x_1 + x_2}\right) = \frac{R_i}{4\pi}\left[\frac{x_2}{(x_1)(x_1 + x_2)}\right] \tag{3}$$

or in terms of the desired quantity, resistivity:

$$R = \frac{4\pi(x_1)(x_1 + x_2)}{x_2} \times \frac{\Delta E_{MN}}{i} \tag{11.12}$$

The term $4\pi(x_1)(x_1+x_2)/x_2$ is the instrument constant for the particular sonde. The resistivity measured is that of a thin shell of material surrounding A at a distance AO. This may be visualized by considering the system as a thin-skinned orange with A at its center. The radius of the orange is approximately AO, the thickness of the skin being MN. The resistivity measured is that of the orange's skin. If AO is sufficiently large, the borehole and invaded zone effects will become negligible so that the resistivity measured is that of the undisturbed formation, R_t. The reading taken from the log is, however, considered as an apparent value R_a, until any necessary corrections are applied to convert it to R_t.

In practice, the spacing of the lateral device is usually 19 ft (18 ft, 8 in., to be exact). While this gives a considerable radius of investigation, it also causes a loss of detail in thin beds, as well as certain curve distortions which occur when the electrodes are on opposite sides of bed boundaries.

A few of these are shown in Figures 11.15 and 11.16. A study of these figures reveals that considerable care must be exercised in determining where to read the proper resistivity value. A detailed study of curve shapes is beyond our scope; however, the rules of thumb in Table 11.3 are applicable to lateral and normal curve readings. The use of these rules will be demonstrated in later examples.

Departure Curves

It is apparent that the resistivity curves just described are, of necessity, influenced by the effects of the borehole and the invaded zone, namely, R_m, R_{mc}, R_{xo}, and R_i. The thickness of the bed with relation to the electrode spacing is a further complicating factor. Sets of curves have been prepared which allow log readings (apparent resistivities) to be corrected for these effects. These are called departure curves. A simplified chart of this type for the short normal (16-in.) and lateral (18-ft 8-in.) curves is shown in Figure 11.17. Complete and detailed sets of curves for these devices are available for more rigorous analyses;[7] however, the single chart based on average conditions will be adequate for our purposes.

Use of Figure 11.17 is described as follows:

(1) Required information
  (a) R_m at formation temperature.
  (b) Hole diameter. (This is normally taken as bit size, unless a caliper log is available.)
  (c) Apparent resistivity readings from short normal ($R_{16''}$) and lateral ($R_{18'8''}$) curves.

(2) For the 16 in. normal: Calculate $R_{16''}/R_m$. Enter at left margin and proceed to appropriate hole size (solid lines) and read R_i/R_m at the bottom.

(3) For the 18 ft 8 in. lateral: Calculate $R_{18'8''}/R_m$ and proceed as before, using dashed hole size lines.

TABLE 11.3

RULES FOR ESTIMATING R_t FROM NORMAL AND LATERAL CURVES UNDER VARIOUS CONDITIONS (COURTESY SCHLUMBERGER WELL SURVEYING CORP.)

Bed thickness e, ft	*Qualifications*	*Device*	*Response*
A. *In low resistivity when $R_{16''}/R_m < 10$ (invasion up to 2d)*			
$e > 20$ $(> 4\,AM)$		Long normal	$R_{64''} = R_t$
$e \simeq 15$ $(> 3\,AM)$	$R_m \simeq R_s$ $R_{64''}/R_s \geq 2.5$	Long normal	$R_{64''} = 2/3\, R_t$
$e \simeq 15$ $(> 3\,AM)$	$R_m \simeq R_s$ $R_{64''}/R_s \leq 1.5$	Long normal	$R_{64''} = R_t$
$e \simeq 10$ $(> 2\,AM)$	$R_m \simeq R_s$ $R_{64''}/R_s \geq 2.5$	Long normal	$R_{64''} = 1/2\, R_t$
$e \simeq 10$ $(> 2\,AM)$	$R_m \simeq R_s$ $R_{64''}/R_s = 1.5$	Long normal	$R_{64''} = 2/3\, R_t$
$5 < e < 10$	When oil bearing and SP is $-50 - 80$ MV	Short normal	$R_{16''} \simeq R_t$
$5 < e < 10$	Surrounding beds homogeneous	Lateral in resistive bed	$R_t \overline{>} R_{max} \times R_s/R_{min}$
Thin beds (in general)	Surrounding beds homogeneous	Lateral in conductive bed	$R_{19''} \simeq R_t$
$e > 3$		Induction 27″	$R_{ind} \simeq R_t$
$e > 5$		Induction 40″	

B. *Rules for using lateral curves* (AO = 18′ 8″)

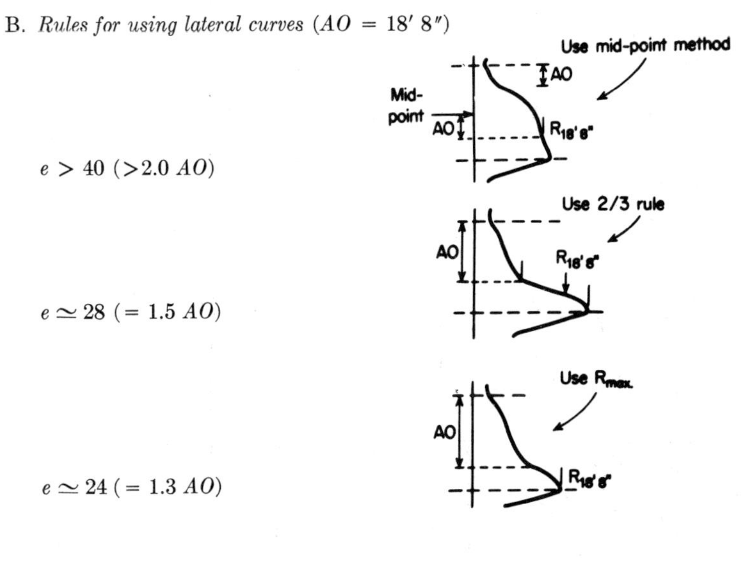

$e > 40$ $(> 2.0\,AO)$

$e \simeq 28$ $(= 1.5\,AO)$

$e \simeq 24$ $(= 1.3\,AO)$

$5 < e < 10$ (Resistive bed and surrounding beds homogeneous)

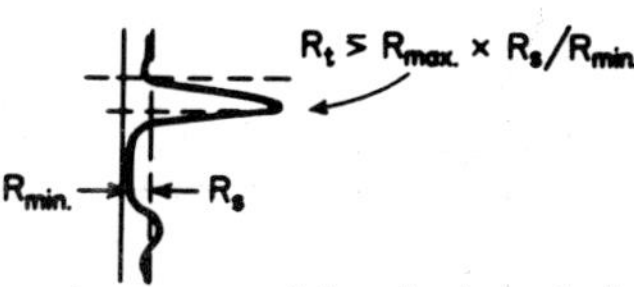

(When $R_{16''}/R_m > 50$, these values must then be corrected for the bore hole.)

Example 11.5

Given the following log readings and other data, compute R_i and R_t.

$R_{16''}$ = 50 ohm-meters

$R_{18'8''}$ = 250 ohm-meters

R_m = 0.50 ohm-meters at formation temperature

Hole size = 9 in.

Solution:

$$R_{16''}/R_m = \frac{50}{0.5} = 100$$

$$R_{18'8''}/R_m = \frac{250}{0.5} = 500$$

From Figure 11.17,

$$R_i/R_m = 150$$

$$R_t/R_m = 300$$

Therefore, the true or corrected values are:

$$R_i = 150 \times 0.5 = 75 \text{ ohm-meters}$$

$$R_t = 300 \times 0.5 = 150 \text{ ohm-meters}$$

With these factors in mind let us consider a simple example of an electric log interpretation.

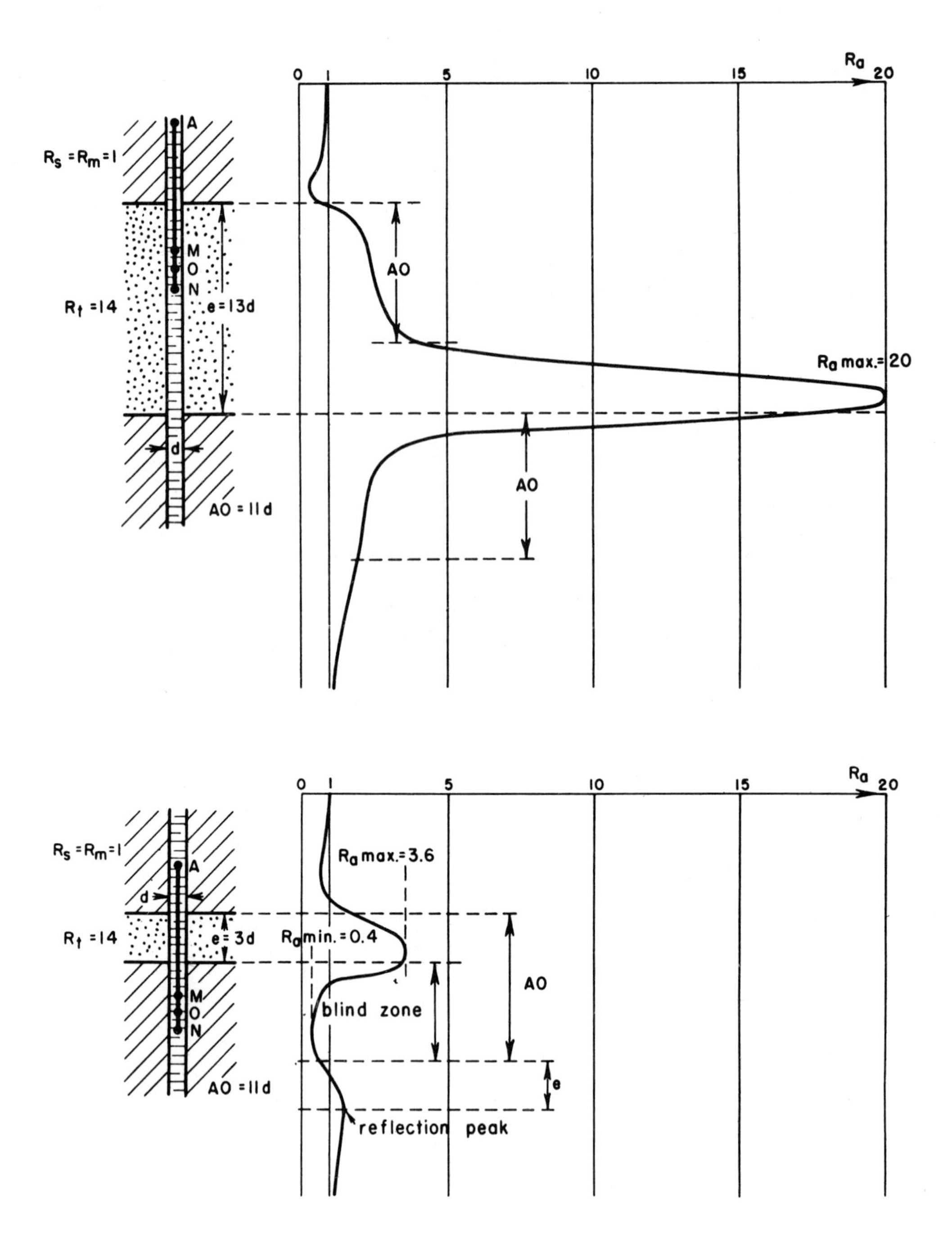

Fig. 11.15. Lateral device curve shapes for beds more resistive than adjacent beds. Note effect of bed thickness. Courtesy Schlumberger Well Surveying Corporation.

11.44 Interpretation Example

Before illustrating a rather simplified interpretation procedure, it will be proper to insert a few general words of caution regarding electric log interpretation. The electric log is an excellent and valuable tool in the hands of the skilled interpreter. For the amateur, however, such may not be the case. The purpose of this treatment is merely to introduce the subject with enough application to stimulate the reader's interest. The following example has been chosen to illustrate the principles involved, and the procedure illustrated must not be construed as a general method applicable under all conditions.

Example 11.6

The electric log shown in Figure 11.18 is from a North Louisiana well. The section from 7014 ft to 7064 ft is the Pettit Limestone. The curves shown are the SP, short normal,

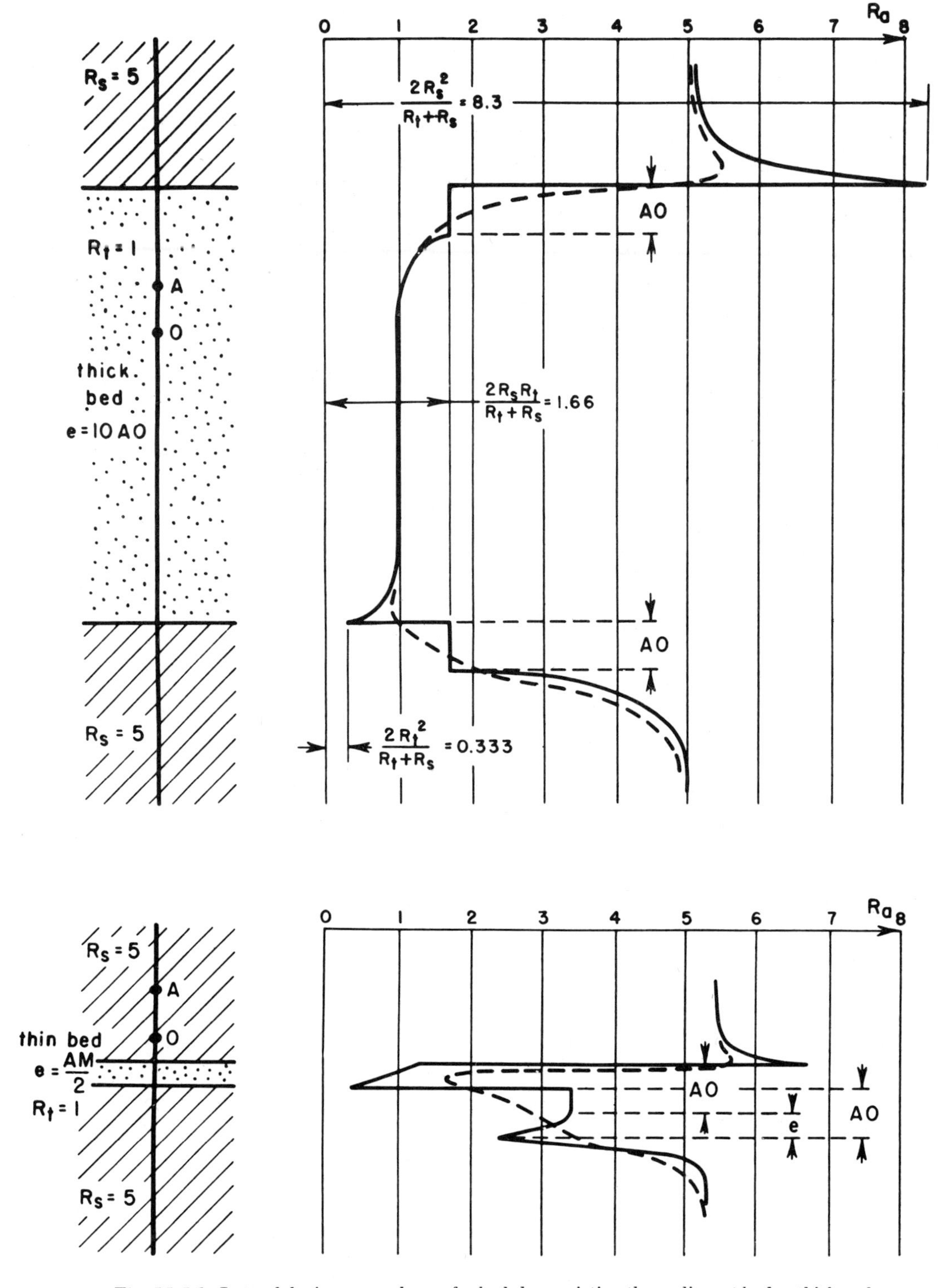

Fig. 11.16. Lateral device curve shapes for beds less resistive than adjacent beds: thick and thin beds. Courtesy Schlumberger Well Surveying Corporation.

long normal, and the lateral. This combination constitutes what is commonly referred to as the standard ES (electrical survey). Note that the long normal is the dotted line on the same track as the short normal, although usually the lateral curve is presented in this position. The log heading must always be checked to determine (1) the curve types shown, (2) the resistivity scales, (3) other pertinent data such as the hole size, mud type, mud resistivity, temperature, etc.

Qualitative inspection of Figure 11.18 would immediately suggest this section as a productive zone due to the relatively high resistivities and prominent SP curve. A quantitative appraisal might proceed as follows:

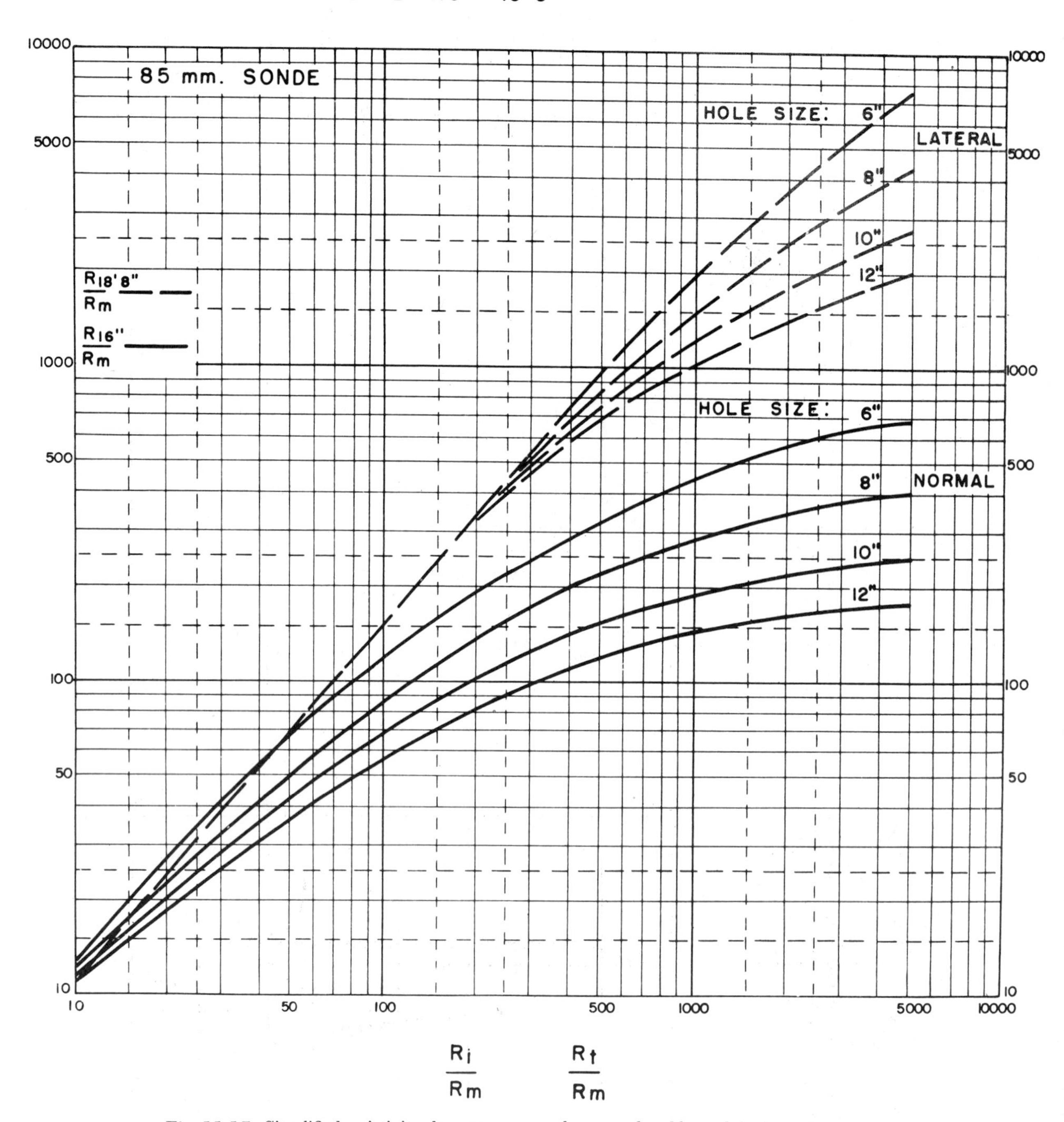

Fig. 11.17. Simplified resistivity departure curves for normal and lateral devices. Courtesy Schlumberger Well Surveying Corporation.

1. Water Saturation Estimate: Since the bed thickness is about 50 ft, or > 2 AO, we will use the lateral curve to determine R_t. The apparent value is taken as shown; R_a = 55 ohm-meters. This, however, must be corrected for borehole effects.

$$R_{18'8''} = 55$$

$$R_m = 0.40$$

$$R_{18'8''}/R_m = 137$$

From Figure 11.17, $R_t/R_m = 90$

$$R_t = (90)(0.40) = 36 \text{ ohm-meters}$$

Application of Eq. (11.8) requires a knowledge of R_0. Since no water zone exists on our log we will assume that the log of a nearby dry hole in the same formation is available. From this we obtain $R_0 = 0.9$ ohm-meters, hence:

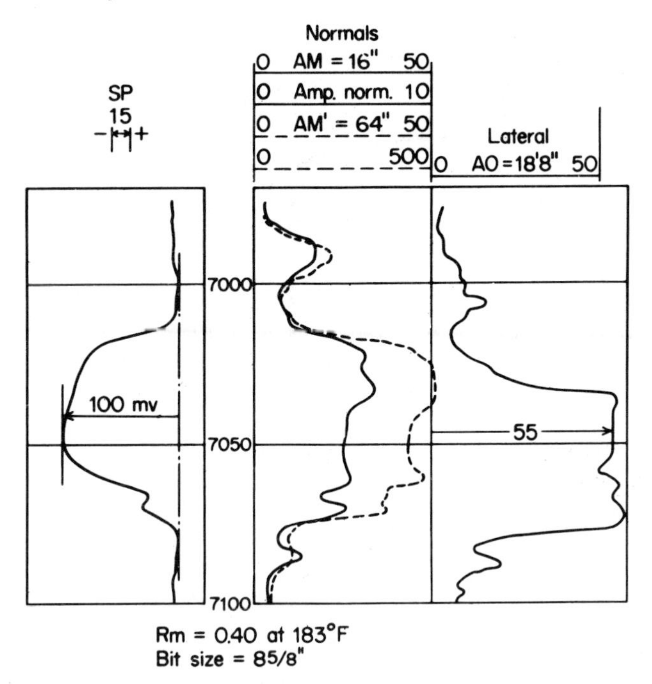

Fig. 11.18. Portion of electric log showing possible productive interval in the Pettit Limestone. Courtesy Schlumberger Well Surveying Corporation.

$$S_w = \left(\frac{R_0}{R_t}\right)^{1/2} = \left(\frac{0.9}{36}\right)^{1/2} = 0.16 \text{ or } 16\%$$

2. Porosity Estimation: Here we will apply either Eq. (11.6) or (11.7), both of which require values of formation factor F. Since $F = R_0/R_w$, we require R_w. In lieu of a better method, we may calculate this from the SP curve. From the log:

$$\text{ASP} \cong -100 \text{ mv}$$

(a) Since this bed is fairly thick ($h \cong 50$ ft) no SSP correction is necessary (Figure 11.6) and

$$\text{SSP} = \text{ASP} \cong -100 \text{ mv}$$

(b) From Figure 11.7(A) at $T = 183°$F,

$$\frac{R_{mf}}{R_{we}} = 16$$

Since $R_{mf} = 0.75\ R_m = (0.75)(0.40) = 0.30$ ohm-meters,

$$R_{we} = \frac{0.30}{16} = 0.019 \text{ ohm-meters}$$

And finally, from Figure 11.7(B),

$$R_w = 0.03 \text{ ohm-meters}$$

(c) $$F = R_0/R_w = \frac{0.9}{0.03} = 30$$

(d) $$\phi = \frac{1}{30^{1/2}} = 0.18 \text{ or } 18\%, \text{ by Archie's formula}$$

Thus our qualitative observation is checked by calculation, and we would definitely predict clean oil production for this zone. Naturally, core analysis, sample inspection, and/or drill stem test data may also be available to substantiate this opinion.

11.45 Other Resistivity Logging Devices

While the two and three electrode curves just described are quite basic and furnish reliable information in many instances, they also have certain drawbacks, under various conditions. In particular, the definition of thin, interbedded zones is not always clear. In limestone areas where the dense sections are essentially non-conductive, it is difficult to locate the conductive, and hence porous and permeable zones. The measurement of R_t is also difficult to obtain in thin beds due to the long spacing, the geometric distortions, and the averaging of resistivities which are characteristic of the lateral device. Estimations of R_t using the long normal are often in error, due to the uncertainty of the invasion depth of the mud filtrate. Further, the standard resistivity curves will not function in boreholes containing non-conductive fluids such as air, gas, and oil. To solve these and other problems, a number of newer resistivity logs have been developed. Some of these have practically replaced the standard survey in many areas, due to their superiority under a particular set of operating conditions. The following discussions will present only the main features of each device.

1. The limestone sonde: This device was designed to give a clear and easily interpreted log in highly resistive formations such as limestones and dolomites. It consists of a five-electrode system arranged as in Figure 11.19, and may be considered as a combination of two lateral devices. It has the advantage that a symmetrical curve is recorded opposite thin, conductive streaks in an otherwise dense and highly resistive section. It is these relatively conductive sections which may be oil productive. The spacing of the limestone sonde is normally 32 in. and it is generally run in combination with a special 10-in. normal curve. Excellent correlations between the response of this device and porosity are obtained in many cases; this further enhances its use. A sample log is shown as Figure 11.20. Note that many thin conductive streaks are shown on the limestone curve (solid line) which are not so clearly defined on the normal curve (dashed). Also note the very high resistivities recorded by the normal, which are much greater than those shown previously.
2. The Microlog:[8] This device was designed to furnish a means of accurately locating thin, permeable zones. Such determinations allow precise measurement of the net pay (permeable section) in a total interval. The SP curve, while useful in this respect as mentioned before, does not furnish adequate definition in many areas. The Microlog sonde is shown in Figure 11.21. Three electrodes spaced 1 in. apart, (A, the current electrode, and M_1 and M_2, the pickup electrodes), are placed in a vertical line in an

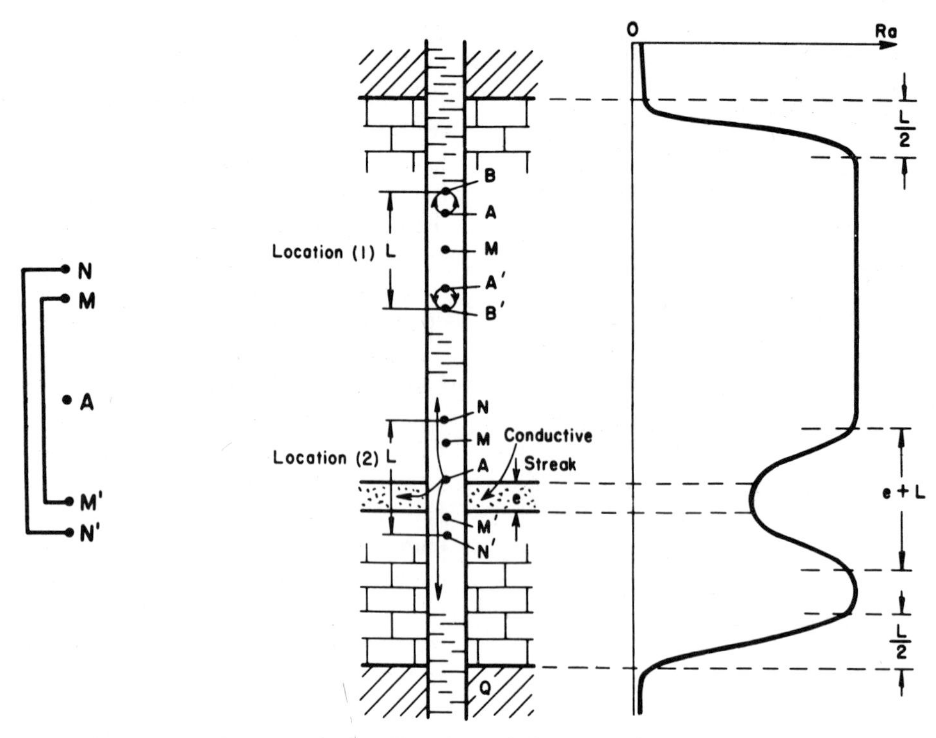

Fig. 11.19. Limestone device (five electrodes) and qualitative diagram opposite thin, relatively conductive streaks. Courtesy Schlumberger Well Surveying Corporation.

insulated pad as shown. The potential differences between M_1 and M_2, and between M_2 and a reference electrode at the surface, due to a constant current transmitted at A, is measured. Hence the device may be thought of as a $1\frac{1}{2}$-in. lateral and a 2-in. normal arrangement. Since the pad is pressed tightly against the hole wall, these electrode systems measure the resistivity of a very small volume of material adjacent to the borehole. The system AM_1M_2 is generally called the 1 × 1-in. microinverse, and AM_2 the micronormal; their measured resistivities are denoted by $R_{1\times1''}$ and $R_{2''}$, respectively. A second pad is often run opposite the Microlog sonde for recording hole diameter; this is a caliper log known as the microcaliper.

Opposite permeable beds, the microinverse system essentially measures the mud cake resistivity R_{mc}, while the micronormal is more affected by the resistivity of the flushed zone, R_{xo}.

The mud cake is generally about twice as resistive as the mud for water base muds, but may be more than this for oil emulsions. The apparent resistivity of the flushed zone R_{xo} which is estimated as extending about three inches around the hole, is normally a few (3 to 10) times R_m depending on the thickness of the mud cake and the formation factor of the porous bed. Generally then, $R_{2''} > R_{1\times1''}$. Consequently, a separation between the curves will nearly always occur opposite a permeable zone. Opposite impermeable beds, $R_{1\times1''} = R_{2''}$, since no mud cake or filtrate invasion occurs. Exceptions do occur when $R_{1\times1''} > R_{2''}$, e.g., in the measurement obtained opposite a salt water sand when little filtrate invasion has occurred. Here the formation resistivity may be less than R_{mc} or R_{xo}; this causes the reversed behavior.

Figure 11.22 shows a Microlog and microcaliper survey in combination with the standard SP, normal, and lateral curves. Note the separation of the Microlog curves and the obvious shale breaks between 7028 and 7088 ft. Note also that the microcaliper shows a reduced hole size (due to mud cake) opposite the permeable zones and a gauge hole opposite the shale breaks.

The Microlog is also useful in estimating porosity, calculating the formation factor F, locating permeable beds, and defining water-oil contacts under certain conditions. Accurate definition of bed boundaries and delineation of pay from non-pay zones are available from Microlog readings, as was mentioned previously.

Focused Current Devices

3. The laterolog (guard logging):[9,10] This is the first of the focused current devices to be discussed. The circuit of a seven-electrode system of this type (called laterolog 7) is shown in Figure 11.23. Three pairs of electrodes, M_1M_2, $M_1'M_2'$, and A_1A_2, are positioned symmetrically with respect to a seventh electrode, A_0. Each of the pairs is short circuited as shown. A constant current is fed through A_0. At the same time, a current of the same polarity is transmitted through the auxiliary electrodes A_1 and A_2

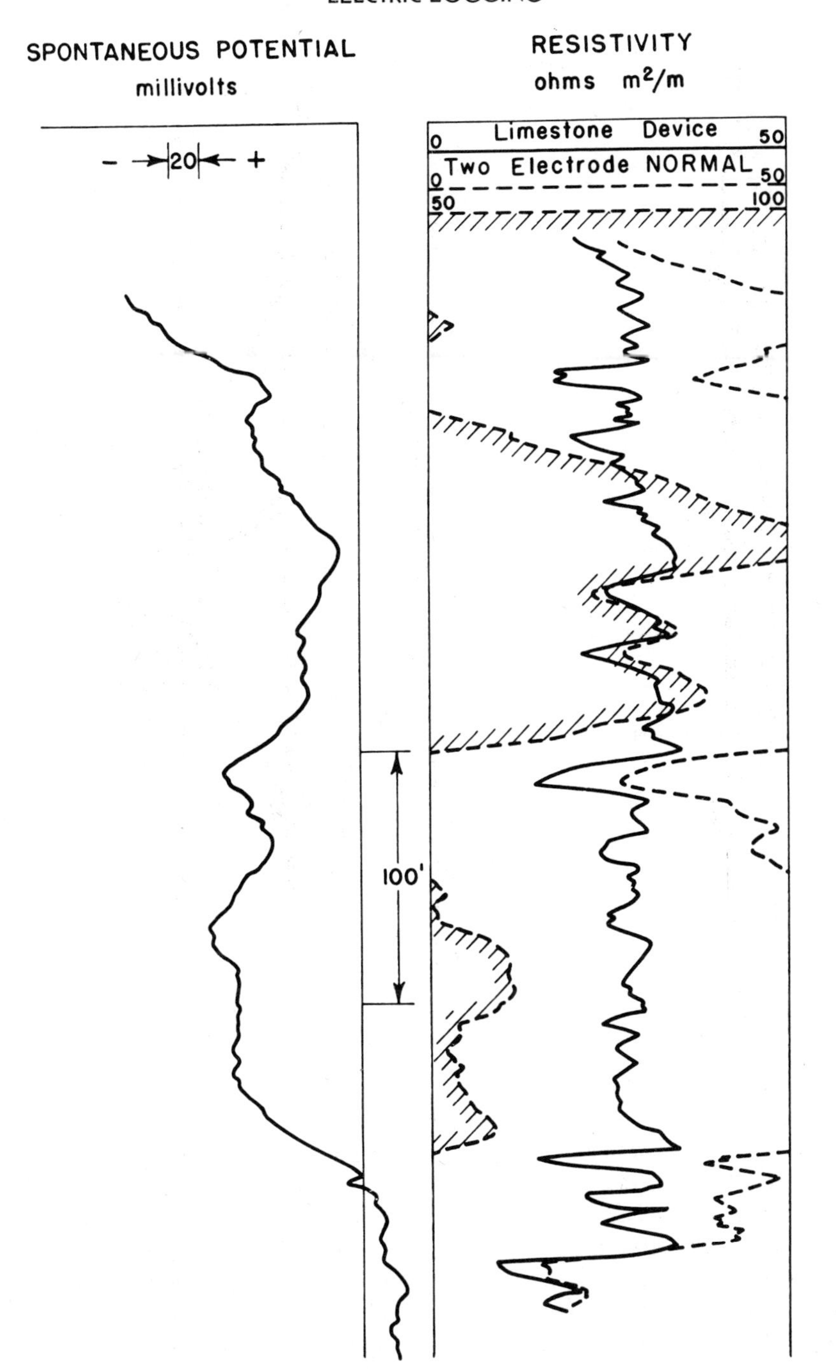

Fig. 11.20. Example of limestone curve application. Note numerous conductive streaks not clearly defined by normal curve. Courtesy Schlumberger Well Surveying Corporation.

which is automatically adjusted by a control apparatus so that the potential between M_1M_2 and $M_1'M_2'$ is maintained at zero. As a result of this arrangement, the current leaving A_0 is prevented from flowing upward or downward just as though insulating plugs were placed at the pickup electrodes M_1M_2, $M_1'M_2'$. Hence, the current flows horizontally outward from A_0; this, in effect, places R_m, R_{mc}, R_{xo}, R_{an}, R_i, and R_t in series. If R_t is very large compared to the other resistivities, as is often the case when salty muds are used, the apparent resistivity value is essentially R_t. Departure curves are available for checking this and making any necessary corrections for borehole effects.[11]

Because measurements with this type of device are little affected by adjacent beds, good thin-bed definition and quantitative resistivity information are provided under conditions where other logs are often unsatisfactory. This is illustrated by Figure 11.24 which shows a comparison between this and the

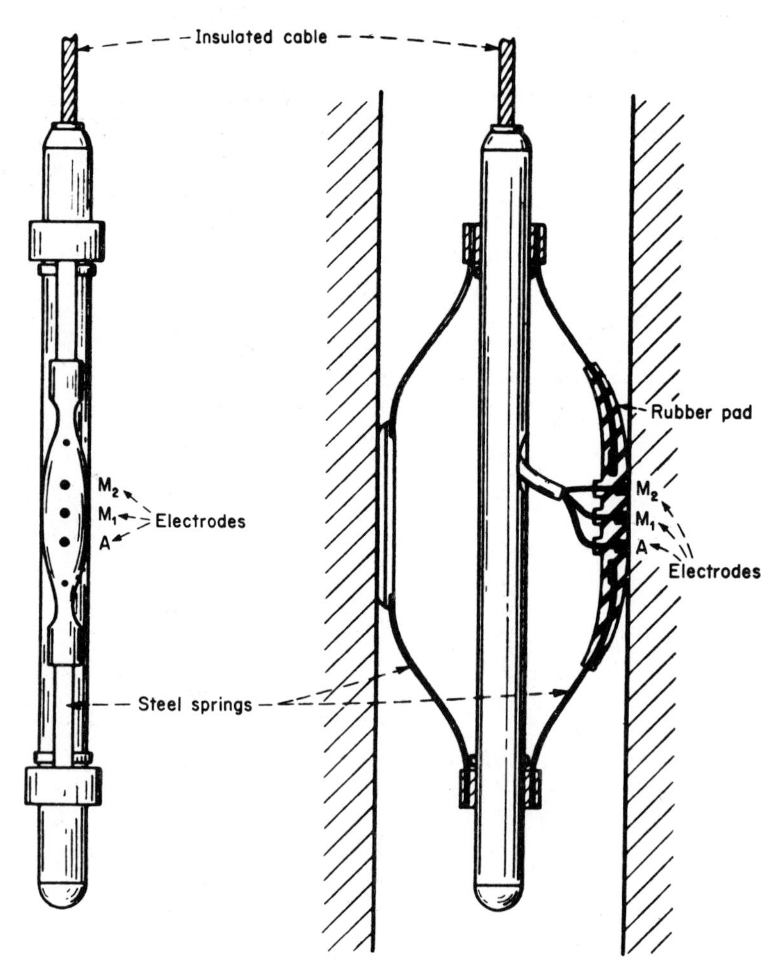

Fig. 11.21. Common type of Microlog sonde. To obtain the microcaliper log, a second pad is added. Courtesy Schlumberger Well Surveying Corporation.

standard survey in a hard rock area. Note the featureless appearance of the SP which is caused by the lack of contrast between R_{mf} and R_w.

4. Microlaterolog:[12] This is also a focused current device and has the same circuit as the laterolog 7 just described. The electrodes are placed in a micro-pad, as shown in Figure 11.25, and cause the current to flow outward in a trumpet-like ray. The resistivity measured is approximately that of a cylindrical plug (much like a sidewall core) of material some 3 or 4 in. long. Hence, under proper conditions, it may be used to evaluate R_{xo}, a parameter useful in more involved quantitative interpretation. A log showing the curve obtained, as compared with other types, is also shown in Figure 11.25
5. Induction Logging:[13] All of the devices described so far require the presence of a reasonably conductive fluid in the borehole, so that current may flow from the power electrode(s). Consequently, none of them can be used in oil, oil base mud, gas, or air. This shortcoming of ordinary resistivity devices led to the development of another focused current device which could function under such conditions. This is the induction log. The circuit for the induction log is shown in Figure 11.26. A constant alternating current is made to flow through a transmitter coil supported by an insulating mandrel. The magnetic field caused by this current generates eddy currents which induce a voltage in a second coil (receiver) mounted on the same mandrel at a certain distance (the instrument spacing) from the transmitter. The signal generated by the receiver, which is measured and recorded at the surface, is dependent on the conductivity of the formation between the coils. The quantity measured is conductivity, which is the reciprocal of resistivity.

Although designed for the purpose mentioned, the induction log has proved to offer considerable advantage over conventional devices in other aspects as well. Bed boundaries are well delineated and measured resistivities are not appreciably affected by nearby strata. In fresh muds, measurements of R_t may be obtained which require simple corrections or none at all. In salty muds, the response is

The example illustrates the ability of the MicroLog to locate porous permeable zones. The movement of the pads against the borehole face also gives a MicroCaliper that confirms mud cake deposition in permeable zones and indicates caving sections such as may be present in shales and other hole conditions important to the completion of the well.

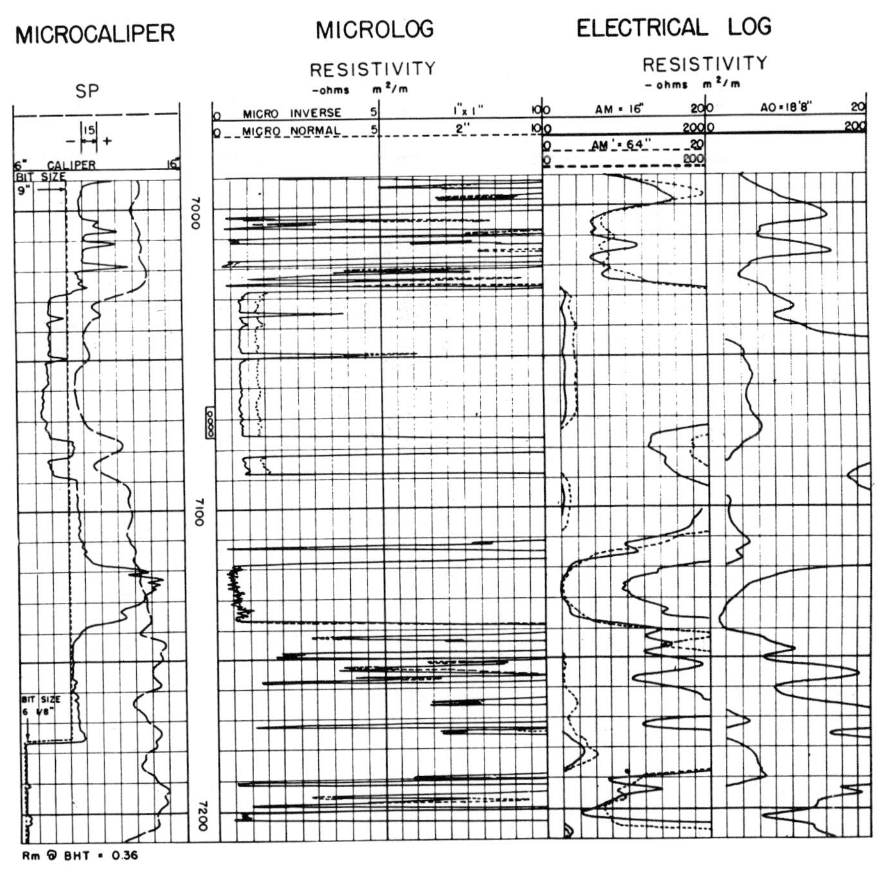

Fig. 11.22. Comparison of Microlog and microcaliper with the standard electrical survey. Note the separation of the microinverse and micronormal curves. Courtesy Schlumberger Well Surveying Corporation.

greatly influenced by the mud and invaded zone resistivity; other devices must be used in combination with it. As a consequence of these features, the induction log has partially replaced other resistivity logs in many *fresh mud* areas. When used in conventional muds it is usually combined with the SP and short normal curves which supplement the information obtained and help to furnish a complete survey. Such a log is shown in Figure 11.27. Note that the induction log curve is presented as both a conductivity (with scale from right to left) curve and its reciprocal, a resistivity curve. The standard resistivity curves and the Microlog are also shown.

11.5 Radioactivity Logging

All electric logs must be run in an open hole, to avoid short circuits through the steel casing. This restriction does not apply to radioactivity logs, which may be run in either open or cased holes. This advantage accounts for many applications of radioactivity logging. Two curves are included in a complete log of this type: the gamma ray curve, and the neutron curve. The gamma ray curve is presented on the left hand side and is similar to the SP curve. The neutron curve appears on the right hand track and is somewhat analogous to the resistivity curves.

11.51 The Gamma Ray Curve

Certain elements exhibit nuclear disintegration by emitting energy in the form of alpha, beta, and gamma particles. Alpha particles are helium nucleii and beta particles are electrons. Both of these have relatively low penetrating power and may be effectively stopped by small thicknesses of solid material. Gamma rays are similar to X-rays (electromagnetic waves) and are able to penetrate several inches of rock or steel. The relative penetrating power of the three particles is about 1, 100, 10,000, respectively. Consequently, a properly shielded device can be made to respond only to gamma radiation.

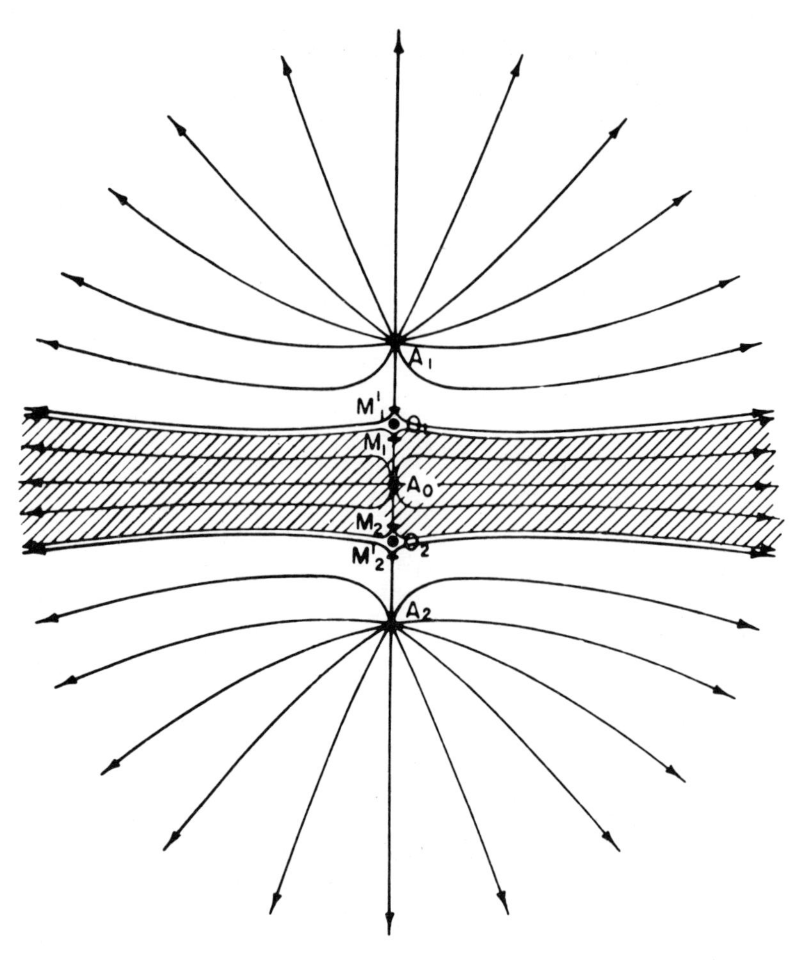

Fig. 11.23. Schematic electrode arrangement for seven-electrode laterolog device. Current flow is as shown. Courtesy Schlumberger Well Surveying Corporation.

The principal radioactive families of elements are the uranium-radium series, the thorium series, the actinium series, and the potassium isotope K^{40}. Nearly all sedimentary rocks contain traces of radioactive salts and, as a consequence, emit measurable radiation. The magnitude of these emissions is highly variable in different geologic provinces but is quite uniform over wide areas of a single province. Figure 11.28 shows the relative intensity of gamma radiation in typical sedimentary rocks.

A gamma ray logging device of the ionization chamber (Geiger-Muller Counter) type is shown in Figure 11.29. Scintillation counter equipment is also in common use; the ionization chamber, however, provides the simplest example of the principles involved. An ionization chamber containing an inert gas at high pressure is penetrated by gamma rays (α and β particles do not penetrate the case). Some of these rays collide with gas atoms, liberate electrons from the gas, and thereby ionize the gas. The current resulting from this liberation of electrons is automatically amplified at the surface and recorded as a function of depth. Its magnitude is, of course, directly related to the intensity of the gamma radiation at any level. Gamma radiation is not emitted at a constant rate. Consequently, for a given length of ionization chamber (commonly 3 to 4 ft) one has to adjust the logging speed to obtain a true statistical picture of radiation in different localities. However, a few logs will normally establish the proper logging speed range as a standard practice for each particular area.

Since the gamma ray curve measures the natural radioactivity of sediments, and since the latter varies with the rock type, it is apparent that gamma ray logging may be used to define the lithology of a section. This is illustrated by the idealized log shown in Figure 11.30. Note that the left-hand excursions of the gamma ray curve denote decreases in radioactivity. Shaly rocks normally have the greatest radioactivity and are indicated by right-hand deflections. Igneous rocks are more radioactive than sediments, as is indicated by the response to granite at the bottom of the section. Hence the gamma ray curve is quite similar in appearance to

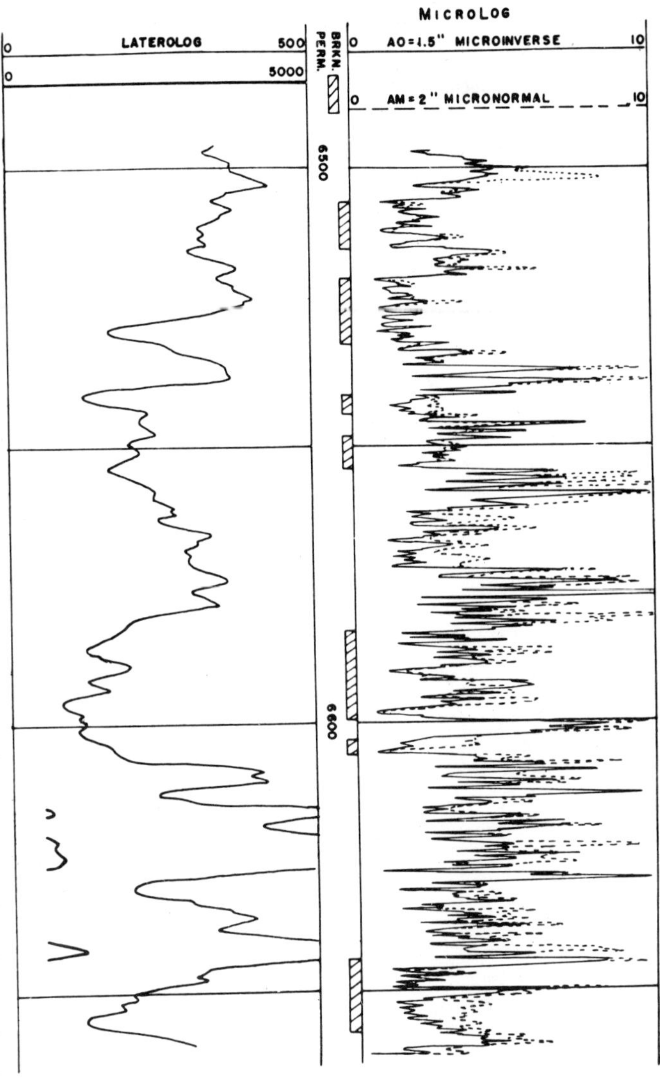

Fig. 11.24. Comparison of laterolog 7 with other devices in salty mud. Log taken in West Texas. Courtesy Schlumberger Well Surveying Corporation.

the SP curve; however, it is insensitive to permeable or porous conditions, distinguishing only between clean and shaly rocks.

11.52 The Neutron Log

Neutrons exist in the nucleii of all elements except hydrogen. They are of approximately the same mass as a hydrogen atom but have no charge. When emitted from fissionable material, they possess very high velocities but are rapidly slowed down by collisions with other atoms. Atoms of nearly the same mass as the neutron are most effective in reducing neutron velocity. To visualize this, consider an ordinary golf ball hurled against another ball of the same size but made of lead. The golf ball will rebound with a velocity almost as great as its original value, while the lead ball will move very little. If, however, two golf balls are used in the same manner, each may leave the collision with about half the speed of the first ball. Similarly, the neutron is greatly slowed by collisions with hydrogen atoms. Since fluids (water, oil, and gas) contain a much higher hydrogen content than rocks, it is apparent that the behavior of emitted neutrons affords a means of evaluating the hydrogen (and hence fluid) content of a formation.

A typical neutron logging arrangement utilizes a modification of the apparatus shown in Figure 11.29. Other instrumentations are also used which yield equivalent results. Neutrons are emitted from a highly radioactive source which is shielded from the ionization chamber. These particles enter adjacent formations and collide with various atomic nucleii. If the rocks entered are dry, the neutrons will not immediately be slowed and will continue to penetrate the rock until they are captured by some element. When this capture takes place, secondary gamma rays are emitted from the capturing element. These are of greater intensity than the natural radiation recorded by the gamma log instrument; thus, they can be measured separately. The ionization chamber is also shortened (to about one foot) to diminish further the effects of natural gamma rays. When the source is opposite fluid-bearing rocks, the emitted neutrons are quickly slowed by collisions with hydrogen nucleii, their penetration is reduced, and consequently, the gamma rays of capture occur relatively close to the neutron source. Hence the magnitude of the ionization current is lower opposite porous zones.

The typical response of a neutron device is shown in the log of Figure 11.30. Note that radioactivity increases to the right, and that left-hand deflections indicate high fluid content. Again, the shales furnish the greatest left-hand excursion, because of their high water content. Hence the neutron response curve has an appearance similar to that of a normal resistivity curve. Let us consider, now, the applications of radioactivity logging and some of the interpretation techniques.

11.53 Applications of Radioactivity Logs

Radioactive techniques have been extremely useful in logging old wells which were cased and completed in the era when the driller's log was the only recorded source of information. Reappraisal and recompletion of such wells based on gamma ray-neutron data has often resulted in substantial production increases. In this regard, the collar log should be mentioned as an aid to accurate downhole depth measurements. When the radioactivity survey is run inside the casing, it is customary to record simultaneously the position of casing collars with respect to the regular curves. These marks are recorded as small horizontal lines at the side of the log, and furnish a useful and accurate bench mark

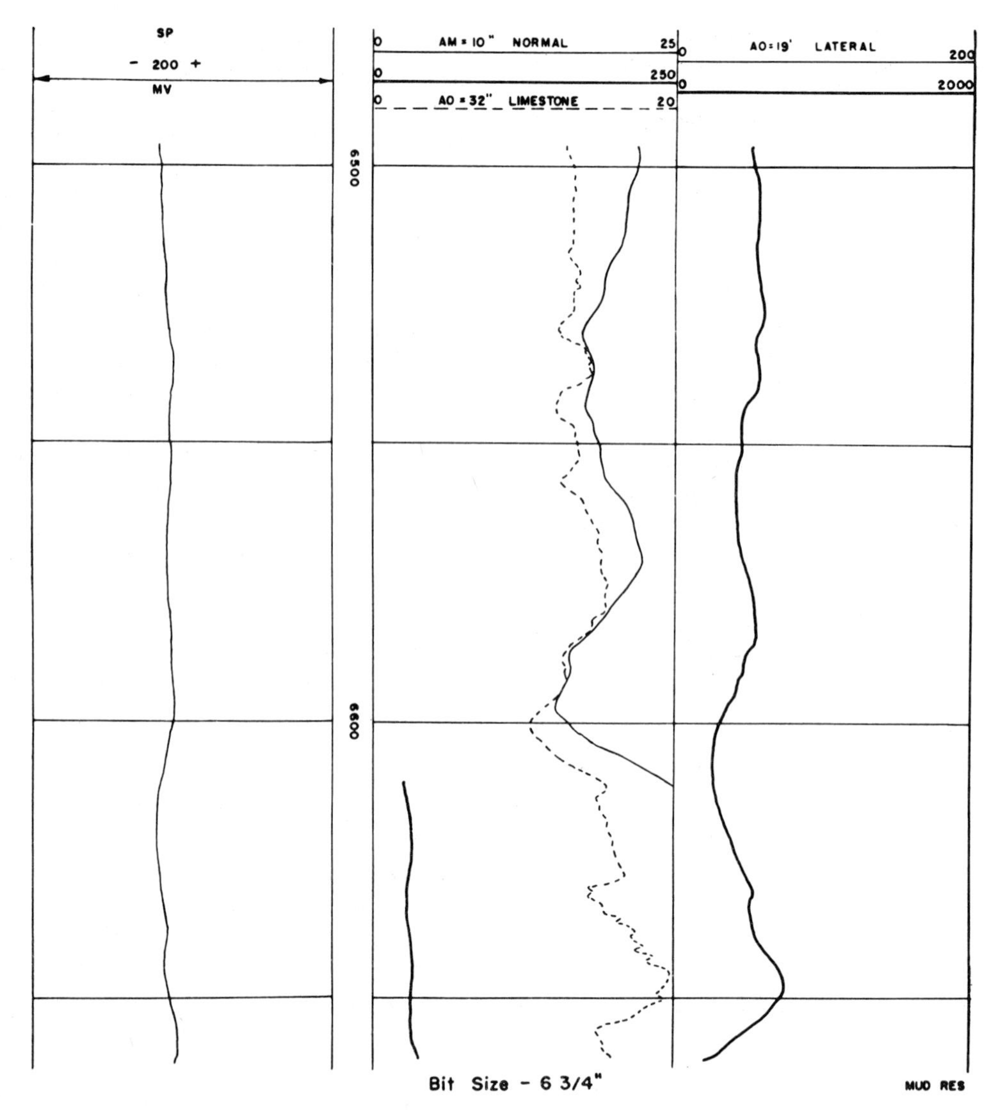

Fig. 11.24. (cont.).

which may be used in later operations. This eliminates inaccuracies in perforating operations, as will be mentioned in a later section.

By virtue of the small length of the ionization chambers (which results in small spacing between measurements), both the gamma ray and neutron curves give good definition of formation and porous zone boundaries. The neutron curve gives the sharper record, due to the shorter counter length. The tops or bottoms of the curve may be taken as the midpoint of the deflections. Tight or shaly zones within a pay section are shown by the neutron log; this definition allows accurate calculation of net pay, in most instances.

The gamma ray log is often substituted for the SP curve where operating conditions make the latter unsatisfactory. This usage is particularly widespread in holes drilled with salt water muds which exhibit a featureless SP (see Figure 11.24). The gamma ray-laterolog-microlaterolog-neutron combination of measurements is often called a salt mud survey and is widely employed in many areas.

Although radioactivity logs may be run inside casing and/or tubing, in the open hole, and in mud, oil, or gas, these borehole conditions do cause uniform shifts in the curves, which must be recognized. Figure 11.31 shows the typical responses to these conditions.

Porosity Determinations

Since the neutron curve responds to changes in hydrogen content, it may also be used for both qualitative and quantitative porosity determinations. This follows from the fact that all formation voids are filled with a fluid which is always oil, water, or gas. It might be noted that the gas is sometimes air; such occurrences are, however, relatively rare.

The qualitative detection of porous zones is illustrated by the following stepwise example. This procedure is a reasonably accurate and useful technique

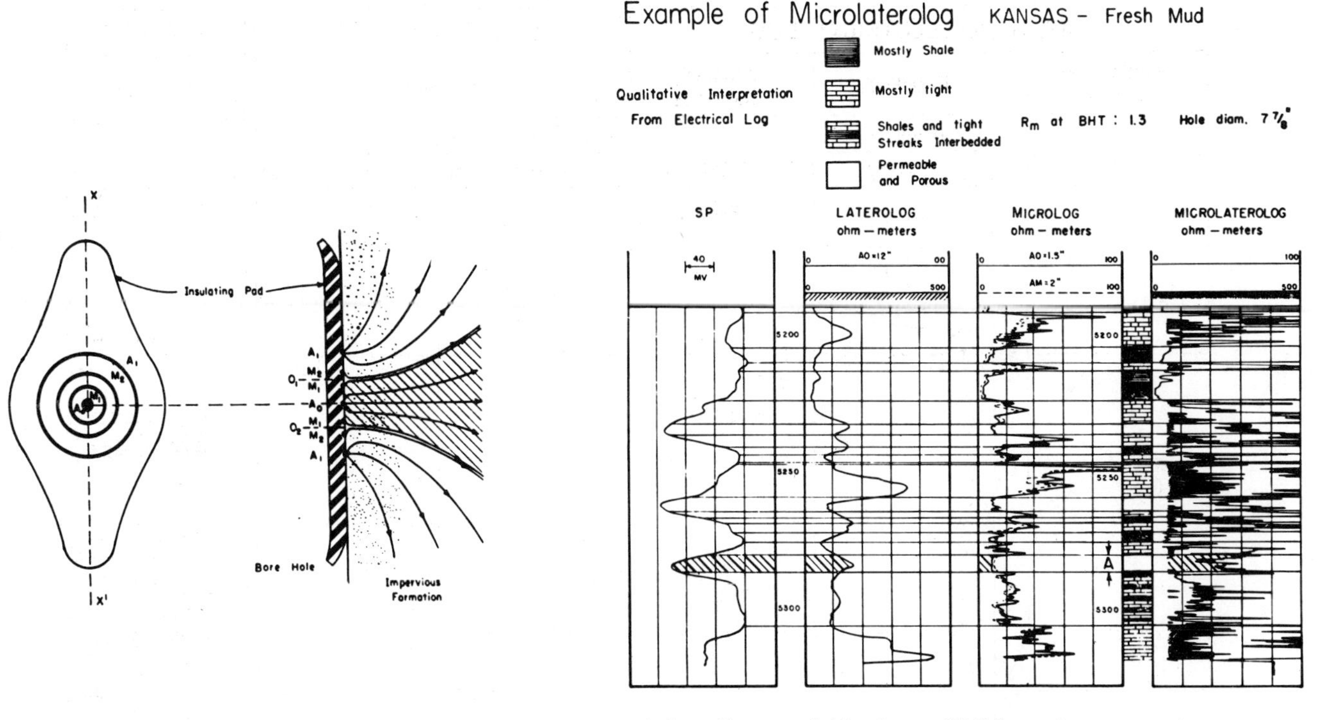

Fig. 11.25. Microlaterolog circuit and example log. Courtesy Schlumberger Well Surveying Corporation.

for determining net pay, where to perforate, and other similar factors.

Example 11.7

The following steps may be considered as a set of general rules for marking porous zones. Each step is shown on the log of Figure 11.32.

1. Establish a shale reference on the neutron curve by using the average minimum shale value. This will be called the minimum neutron shale line.
2. Establish a maximum reference by drawing a line through the average of the maximum curve values, as shown on the neutron curve. This line will be used as the 100% neutron line. Care should be exercised in determining this maximum neutron line, and thorough knowledge of the territory will help in determining its position.

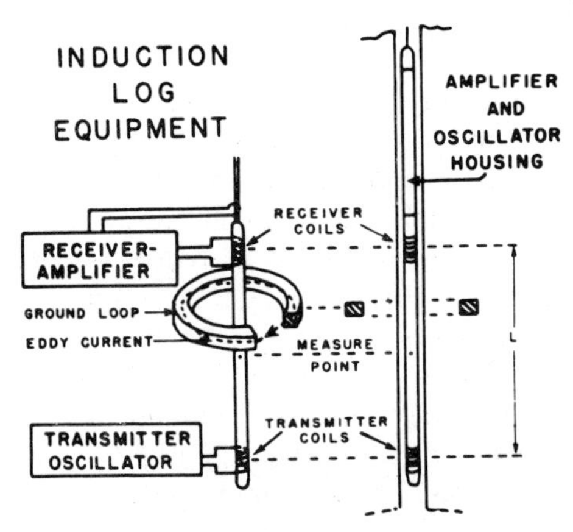

Fig. 11.26. Schematic drawing of induction logging apparatus. After Doll, courtesy AIME.

3. On the neutron curve, draw a line which is $\frac{3}{5}$ of the distance from the minimum neutron shale line toward the 100% neutron line. This will be known as the 60% neutron line. Draw another line midway between the minimum neutron shale line and the 100% neutron line. This will be known as the 50% neutron line.
4. Establish a shale reference on the gamma ray curve by drawing a line through the average shale value. This will be known as the average or 100% gamma ray shale reference line.
5. Draw a line through the average minimum gamma ray curve value in a clean limestone or sandstone. This will be known as the average minimum gamma ray line.
6. Draw a line on the gamma ray curve $\frac{1}{5}$ of the distance between the minimum line and the 100% shale line; this will be known as the 20% gamma ray line. Do the same with another line $\frac{2}{5}$ of the distance, calling this a 40% gamma ray line.
7. For all values on the gamma ray curve between the zero (or minimum line) and the 20% line, pick all porous zones on the neutron curve that extend to the left of the 60% line.
8. For all values on the gamma ray curve between the 20% line and 40% line, pick all porous zones on the neutron curve that extend to the left of the 50% neutron value.
9. Any zone that lies on the gamma ray curve beyond or to the right of the 40% value should not be picked as a porosity zone, even though the neutron curve indicates a very low neutron value.

Note: All porous zones are shown as the hachured intervals of Figure 11.32.

These values have been determined by experience from the quantitative calculation of numerous radioactivity logs, and will in general lead to designation of all zones of porosity

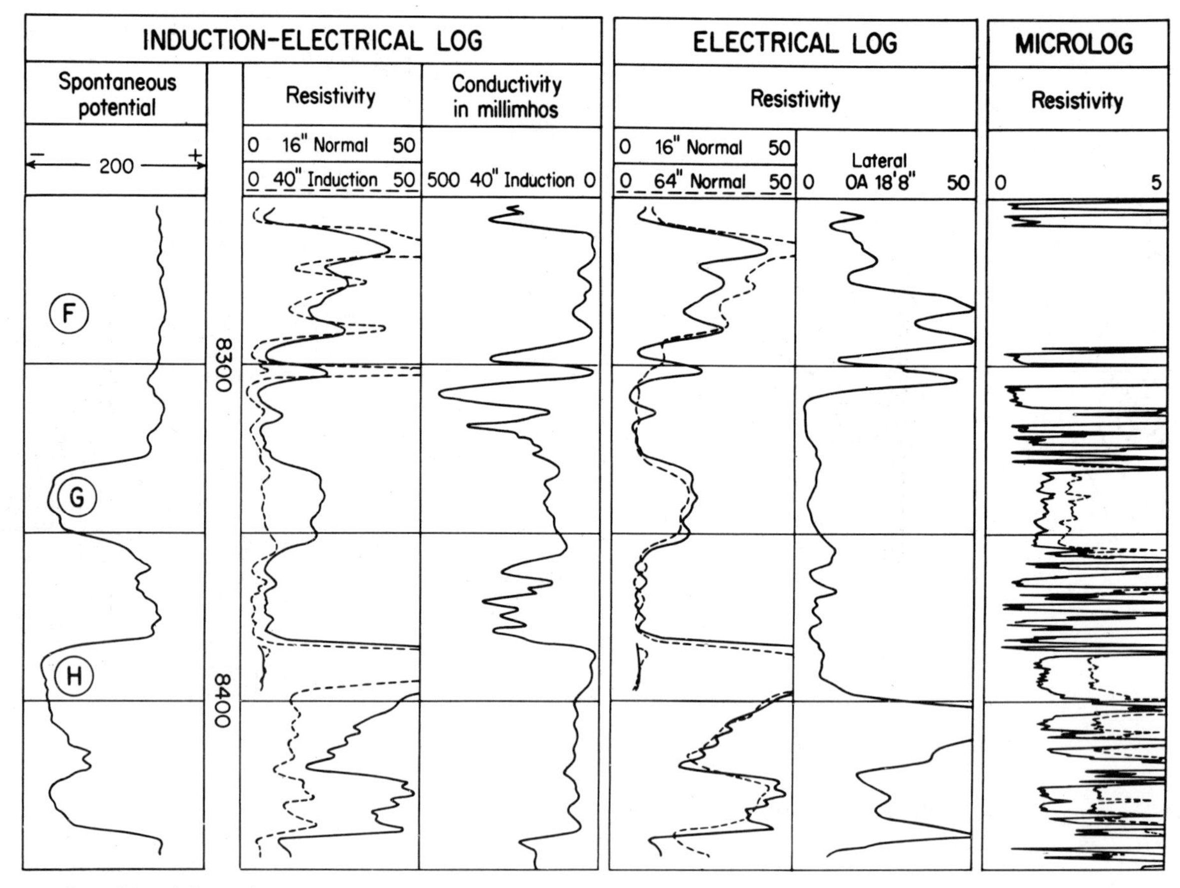

(F) The superior ability of the Induction Resistivity Curve to define thin resistive formations is illustrated here. Often the detail is greater than that of all other resistivity curves, including the Short Normal.

(G) This section is deeply invaded water sand. Only two curves, the Induction and the Lateral, have sufficient depth of investigation to reach values near true resistivity.

(H) This is an outstanding example of the superiority of the Induction-Electrical Log in consolidated formations. The top part of the section is marked by high resistivity readings on the Short Normal, Long Normal and Induction Resistivity Curves. The Lateral Curve cannot be properly interpreted in this zone due to its characteristic lag. In the middle and lower sections relatively high resistivity values are given by the Short Normal, Long Normal, and Lateral Curves. These values are not due to presence of hydrocarbons. The high reading of Short Normal and Long Normal Curves is the result of invasion. The Lateral Curve sustains a high reading due to the numerous hard streaks as shown by the MicroLog. Only the Induction Resistivity Curve shows clearly that the middle and lower sections of the sand are water bearing. The water level is dramatically marked by the abrupt drop on the Induction Resistivity Curve. The upper 10 feet of the sand produces 2.3 million cubic feet of gas per day with 20 barrels of distillate per million cubic feet of gas.

Fig. 11.27. Comparison of induction log with other devices. Courtesy Schlumberger Well Surveying Corporation.

above 5% core analysis as valuable and, conversely, will lead to rejection of zones of lower than 5% porosity. Since 5% seems to be the value where many wells are deemed commercial or non-commercial, this seems to be a reasonable dividing point for interpretation of the neutron log.

Quantitative Porosity Determination

Various techniques are employed for the determination of porosity from the neutron response. Only one general method and the qualitative basis for its application will be presented. It is assumed that:

(1) The lowest neutron deflection will be observed opposite shales. In general, shales contain from 35 to 40% water; therefore, their porosity may be taken as 35 to 40%. In lieu of more precise data, a value of 40% is usually satisfactory.

(2) The highest neutron deflection will occur opposite nonporous (low fluid content) rocks such as dense limestone, dolomite, or anhydrite sections. The porosities of these rocks normally range from 1 to 4%. In lieu of actual data, values of 3% for lime and dolomite, or 1% for anhydrite may be chosen. Specific values for the maximum and minimum values in various areas are given in Table 11.4.

After minimum and maximum porosity values have been assigned, a logarithmic scale is constructed between the limits as shown in Figure 11.33. The porosities of individual sections may then be taken as the

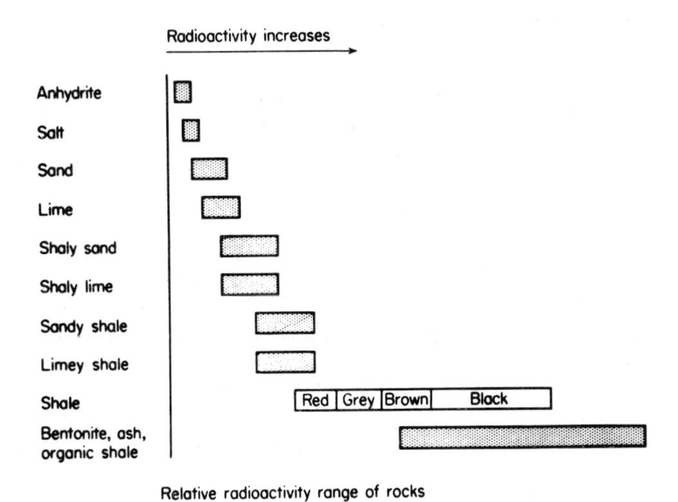

Fig. 11.28. Relative intensity of gamma radiation of sedimentary rocks. Courtesy Lane-Wells Company.

TABLE 11.4

NEUTRON CALIBRATION VALUES IN SOME AREAS (COURTESY LANE-WELLS CO.)

Locality	*Shale value, %*	*Dense value, %*
Oklahoma		
southern	35	7 limestone
northern	40	3 Oswego lime
Williston Basin	40 (Pierre shale)	1 anhydrite
Montana	40	1 dense Madison
Big Horn Basin	35 (Chugwater shale)	1 Gypsum Springs or dense zone in Madison
Illinois Basin	40	3 Glen Dean lime
Panhandle area, Texas	40	1 anhydrite
Panhandle area, Kansas and Oklahoma	40	3 limestone
Kansas	40	3 limestone
Scurry Reef (Texas)	40	1 dense limestone
Smackover Lime (Claiborne)	40	4 dense limestone
Cambrian Sand	40	2 dense limestone
Caddo Limestone (Nolan)	40	2 dense limestone

values corresponding to the observed deflections. The gamma ray curve is useful in defining lithological changes as mentioned previously. The accuracy of porosities obtained by this general method is improved by a previously derived calibration curve for the specific formation in question, which ties the neutron deflection to porosities obtained from core analysis. The latter method should be applied where data are available. It should also be realized that these procedures must be applied over intervals which include no curve shifts due to casing seats, changes in hole diameter, fluid levels, etc.

Fluid Contacts

Gas-oil and/or water-oil contacts in a pay section are always of extreme interest. Under proper conditions,

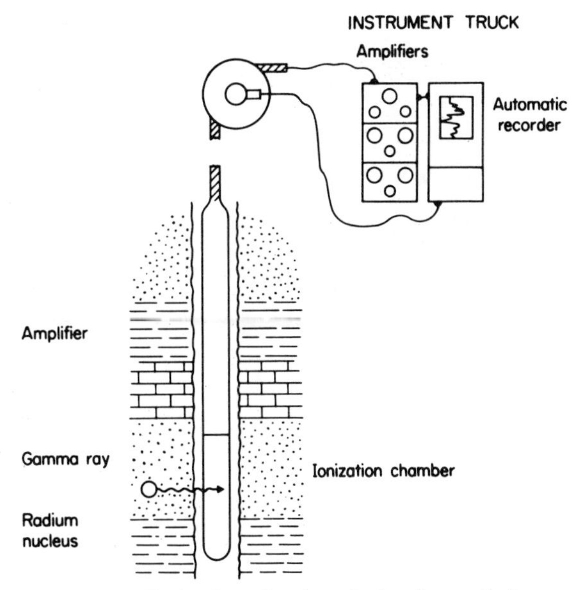

Fig. 11.29. Ionization chamber device for radiation logging. Courtesy Lane-Wells Company.

the neutron curve is able to distinguish gas from oil or water. This is due to the lower hydrogen density of gas at low pressures. Water-oil contacts are generally not apparent, since their hydrogen contents are so nearly the same magnitude. An excellent example of a gas-oil contact is shown in Figure 11.34 (page 231). Not all cases are so apparent; however, the neutron curve is useful for this purpose in some instances, though it should not be applied indiscriminately.

11.6 Miscellaneous Logging Devices

To be included in this category are a number of special well logs, namely:

(1) Acoustic logs
(2) Caliper logs
(3) Temperate logs
(4) Dipmeter surveys

11.61 Continuous Velocity or Acoustic Logging

This method is in its infancy insofar as formation evaluation is concerned; however, its future application seems quite promising. Although originally designed to furnish geophysicists with velocity information for use in seismic interpretation, the acoustic log has been shown to have considerable value for other purposes. A simplified sketch of one device is shown in Figure 11.35. Ultrasonic signals are generated which pass through the mud to the adjacent formations, are refracted parallel to the borehole, and finally arrive at the receivers. The travel time depends on the character of the rocks through which the signal travels and the spacing between the receivers (spacings of 1 ft are common).

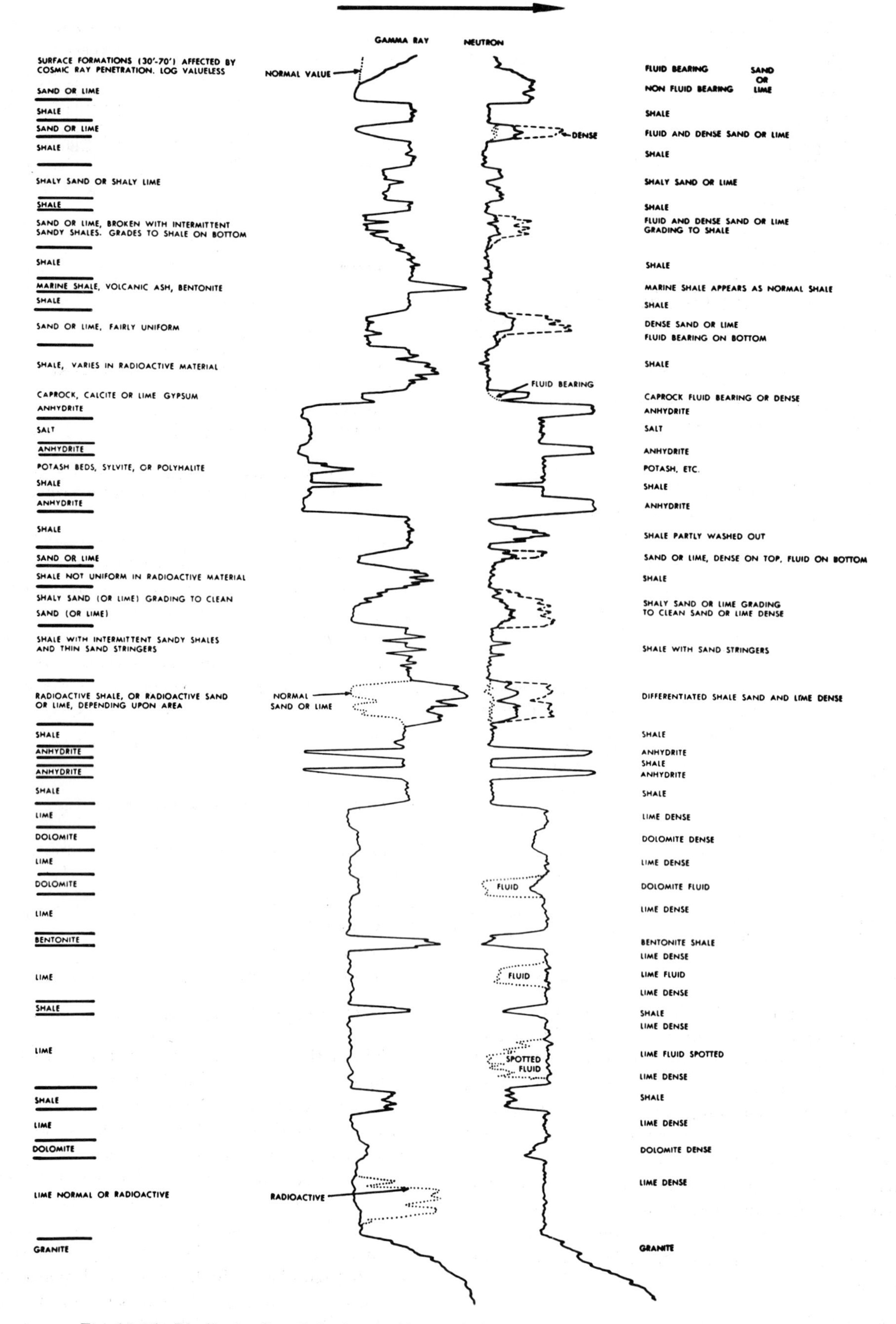

Fig. 11.30. Idealized radioactivity log showing typical responses of various rocks. Courtesy Lane-Wells Company.

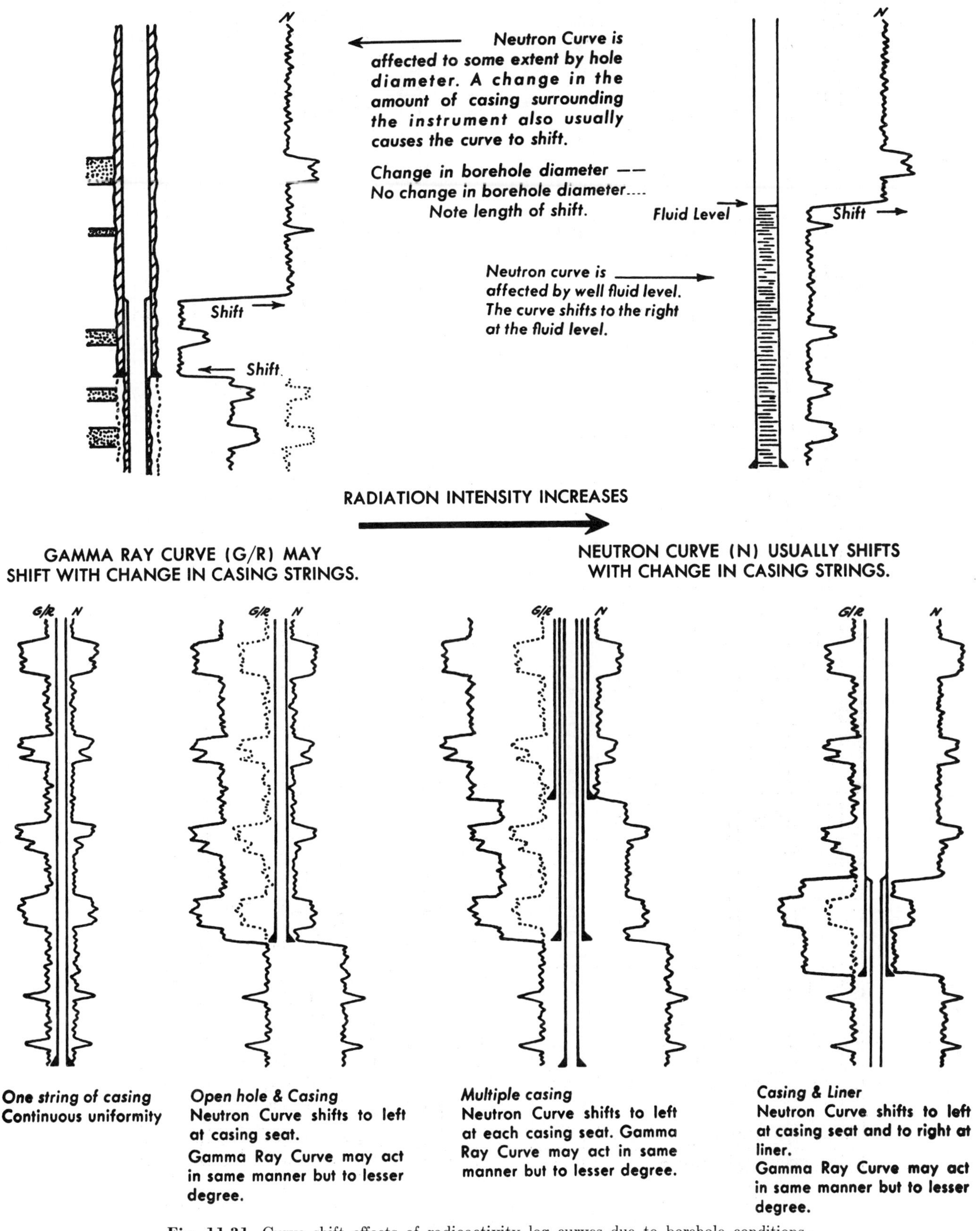

Fig. 11.31. Curve shift effects of radioactivity log curves due to borehole conditions. Courtesy Lane-Wells Company.

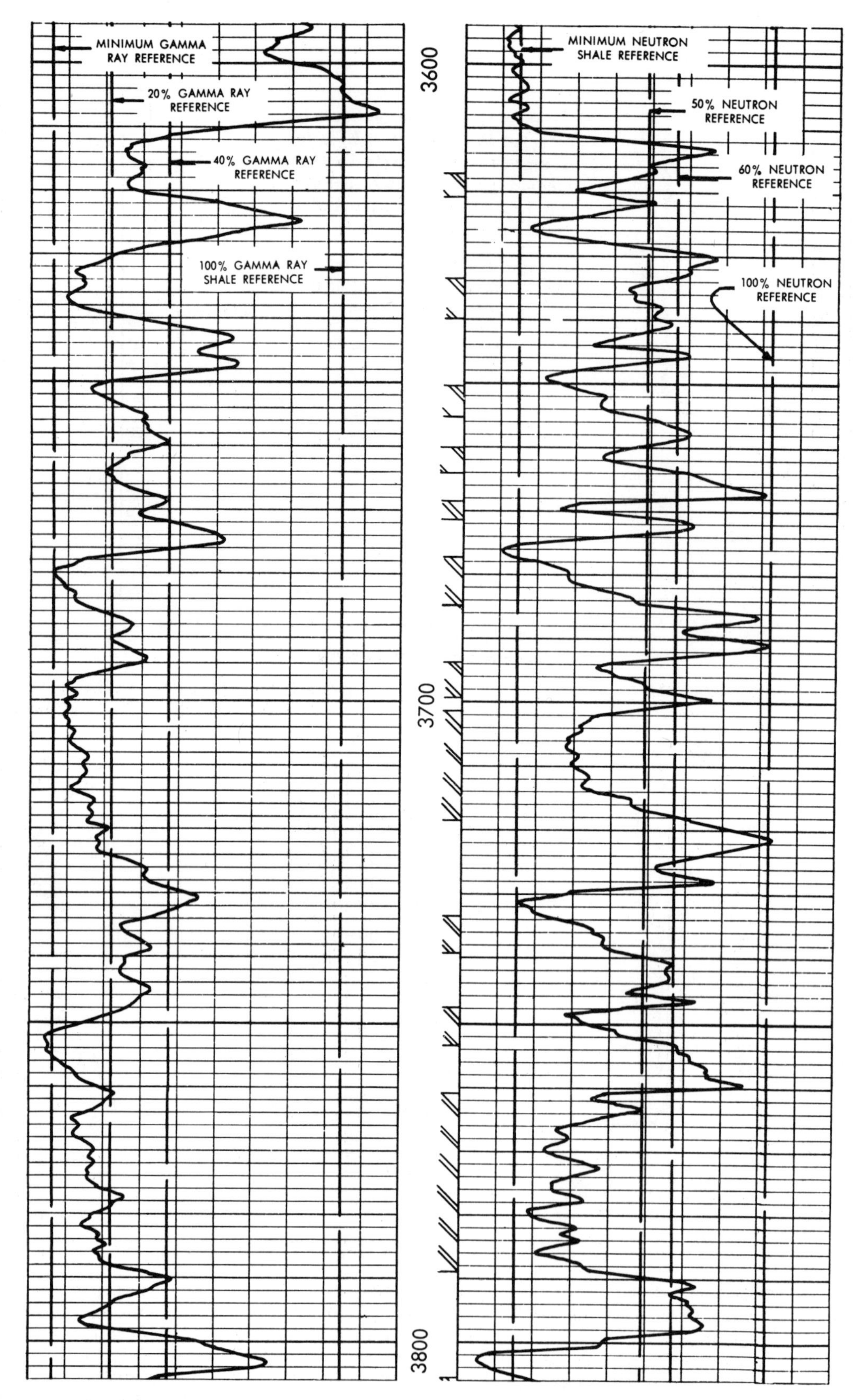

Fig. 11.32. Sample log illustrating qualitative interpretation procedure. Courtesy Lane-Wells Company.

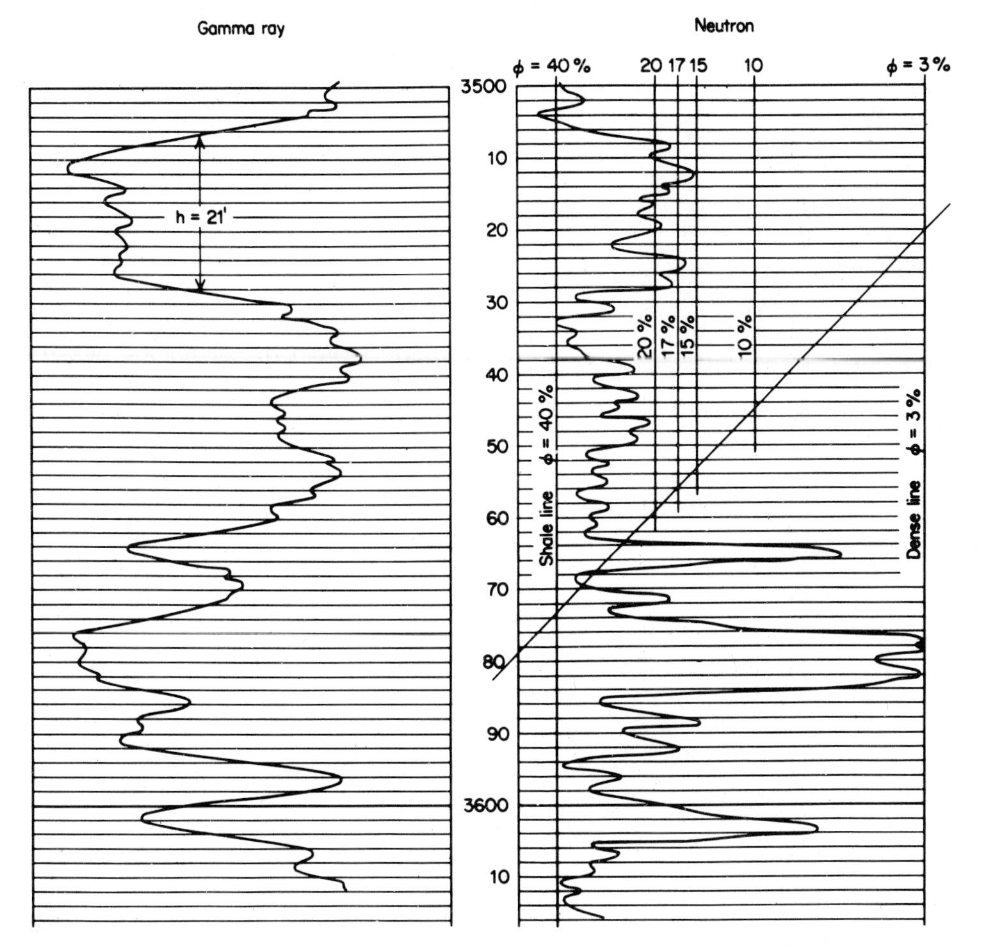

Fig. 11.33. Radioactivity log illustrating porosity determination from neutron curve. Courtesy Lane-Wells Company.

A rather complicated electrical circuit furnishes the log as a plot of velocity vs. depth. The recorded velocity is then the average for the material between the receivers, and is not affected by adjacent beds.

Qualitatively, the interpretation of acoustic logs furnishes useful lithological information, since various rock types exhibit different sound velocities. Future work may achieve quantitative evaluation of fluid content, since sound travels more slowly in both oil and gas than in water; at the present time, however, the quantitative use of acoustic logs is restricted to porosity estimation. Those readers desiring a more detailed discussion of this logging technique are referred to Wiley's text.[14]

11.62 Caliper Logging

The caliper log is a presentation of hole size, either diameter or area, vs. depth. This log is recorded by a device such as that shown in Figure 11.36. The flexible springs expand to conform with the hole as its diameter varies. The lower ends of the springs are connected to a rod which telescopes into a cavity as shown. The position of this rod depends on the compression of the springs and hence on the hole size. A currrent and a pickup coil form an inductive coupling such that the voltage induced in the pickup coil depends on the position of the rod. This allows the recorded voltage to

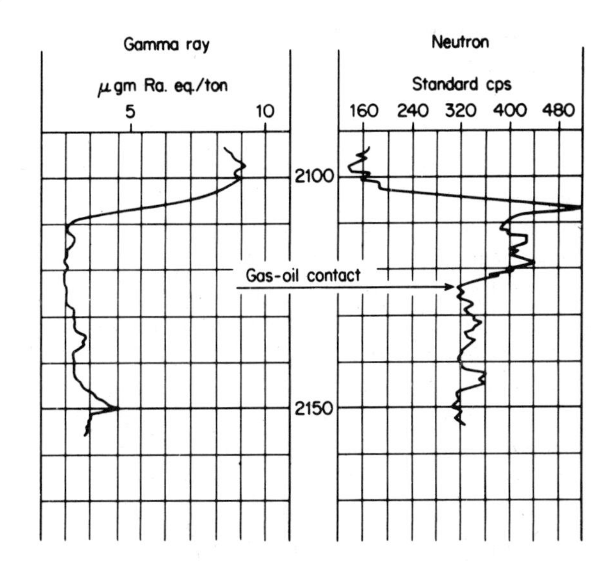

Fig. 11.34. Example of gas-oil contact determination from neutron curve. Courtesy Schlumberger Well Surveying Corporation.

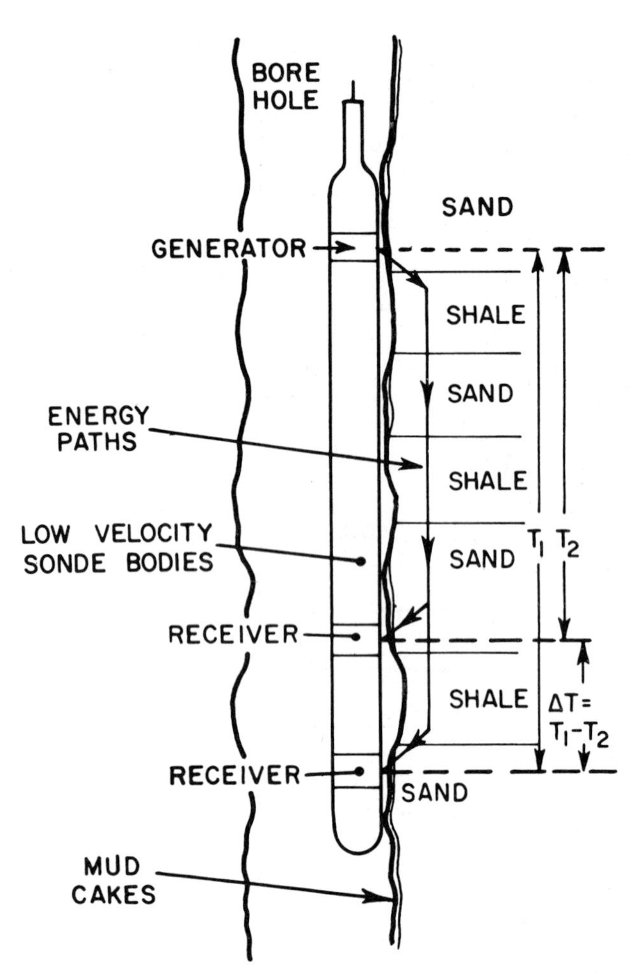

Fig. 11.35. Two-receiver type acoustic logging sonde. Courtesy Schlumberger Well Surveying Corporation.

vary as hole size, the latter being recorded by proper instrumentation.

The uses of a caliper log are many. The most obvious is computation of hole volume for cementing purposes. Other uses are:

(1) Selection of true gauge sections for packer settings (as in drill stem testing).
(2) Interpretation of electric logs: The assumption that hole size equals bit size may introduce considerable error in some cases. Also, the estimation of mud cake thickness opposite permeable zones is of interest in more advanced quantitative log analysis.
(3) Calculation of annular mud velocity with regard to the lifting of cuttings. While this seems rather academic, it could conceivably be of value in establishing a safe, minimum circulation volume which would provide adequate velocity in the largest section of the hole. One or two such logs could be taken as being indicative of the area, allowing hydraulic practices in subsequent wells to be standardized. This presupposes that similar mud types are used.
(4) Lithologic correlations based on caliper logs are sometimes possible. This usage may be visualized from the log shown in Figure 11.37. Note that many formation changes indicated by the electric log are also apparent on the caliper log.

11.63 Temperature Logging

As might be anticipated, this category refers to plots of borehole temperature vs. depth. Such measurements may be obtained by either electrical devices or self-contained temperature bombs. Electrical instruments utilize the resistivity variation of a conductor with temperature. The voltage change due to this variance is recorded as a temperature change, by proper instrumentation. Self-contained instruments actually record temperature vs. time. Time is correlated with depth by making timed stops at desirable depth intervals. These stops appear on the chart as constant temperature-time intervals or steps. Since the depths of the stops are known, a plot of temperature vs. depth is obtained. Electrical measurements are more detailed and accurate, since a continuous record is obtained. However, the self-contained instrument may be run on a steel measuring line; it requires a minimum of equipment, and is therefore cheaper. The choice of instrument depends on the application and the accuracy desired.

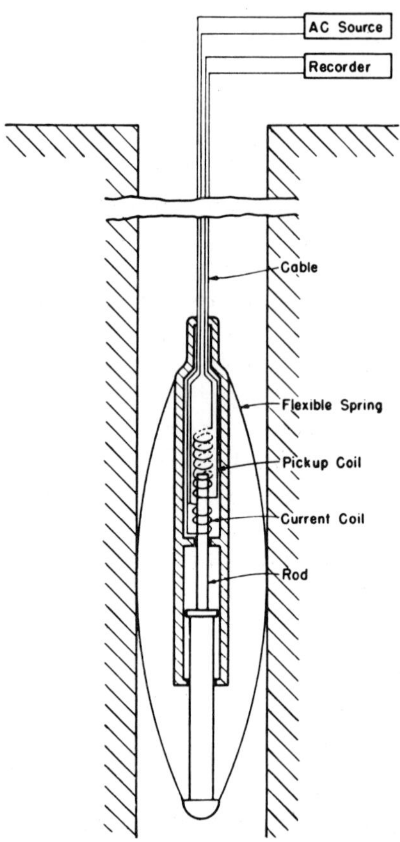

Fig. 11.36. Schematic drawing of the section-gauge or hole caliper. Courtesy Schlumberger Well Surveying Corporation.

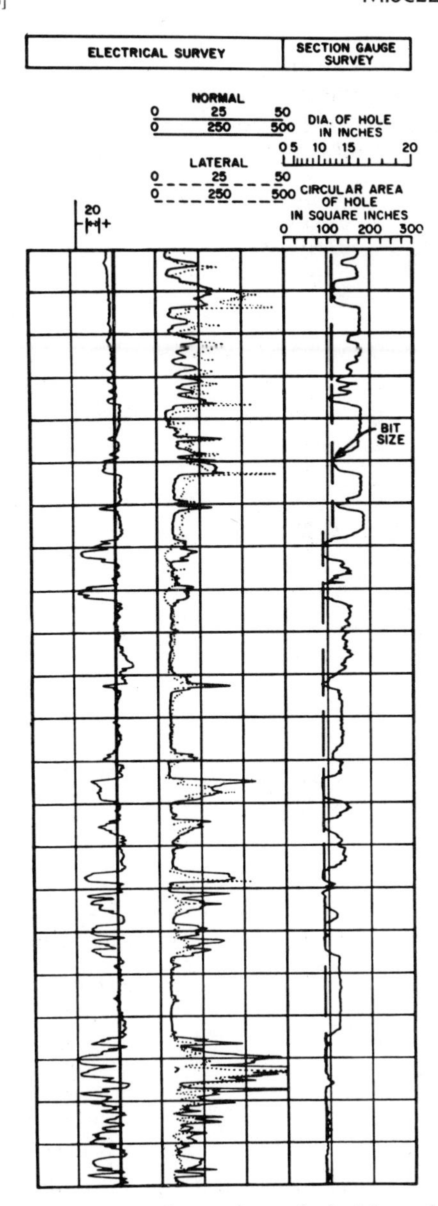

Fig. 11.37. Comparison of electric and caliper logs. Courtesy Schlumberger Well Surveying Corporation.

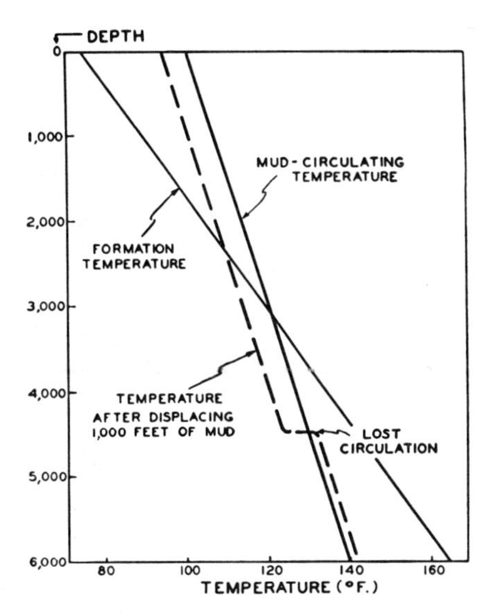

Fig. 11.38. Use of temperature survey to locate zone of lost circulation. After Goins and Dawson,[15] courtesy *Oil and Gas Journal.*

The usefulness of temperature logs or surveys is due to the normally uniform temperature vs. depth behavior observed in sedimentary basins. Although the gradients vary in different areas, any given area typically exhibits linear behavior. Indications of abrupt deviations from linearity, caused by gas expansion or other fluid movement, may be used for many purposes. Among these are:

(1) Determination of cement fillup (this is a principal use of temperature surveys and will be discussed later.)

(2) Location of thief zone(s) causing lost circulation. This procedure is illustrated by Figure 11.38. Under normal circulating conditions, the drilling mud is not in equilibrium with the normal gradient for the area and will exhibit a gradient as shown. Suppose that lost circulation occurs in a way such that a considerable volume of mud from above the loss zone is suddenly displaced downward. The mud below the loss zone, however, is not displaced and remains in its normal position. A temperature survey run shortly after such an occurrence would show the anomaly indicated by the dashed line. The position of the break locates the thief zone. Specific field methods utilizing this principle have been presented by Goins and Dawson.[15]

(3) Location of gas bearing zones: Since gases cool on expansion (Joule-Thompson effect), the temperature opposite a gas producing zone will be lower than that predicted by the normal gradient. Thus, points of gas entry can be determined; such information is extremely useful in various instances. Quantitative determinations of produced gas volumes from various zones may be made by utilizing temperature survey data.[16]

(4) Location of casing leaks: Again, a temperature anomaly may provide the data for diagnosis of the problem. A hole in the casing may allow external fluids (often water) to enter the well bore where it mingles with the produced fluids. If the internal pressure is higher than that outside the casing, the reverse occurs.

Detailed discussion of these topics is beyond the present scope; however, the reader is referred to a paper by C. V. Millikan for an interesting discussion of these and other applications.[17]

11.64 Dip Logging

This refers to a record of formation dip, both angle and direction, vs. depth. The devices used to obtain

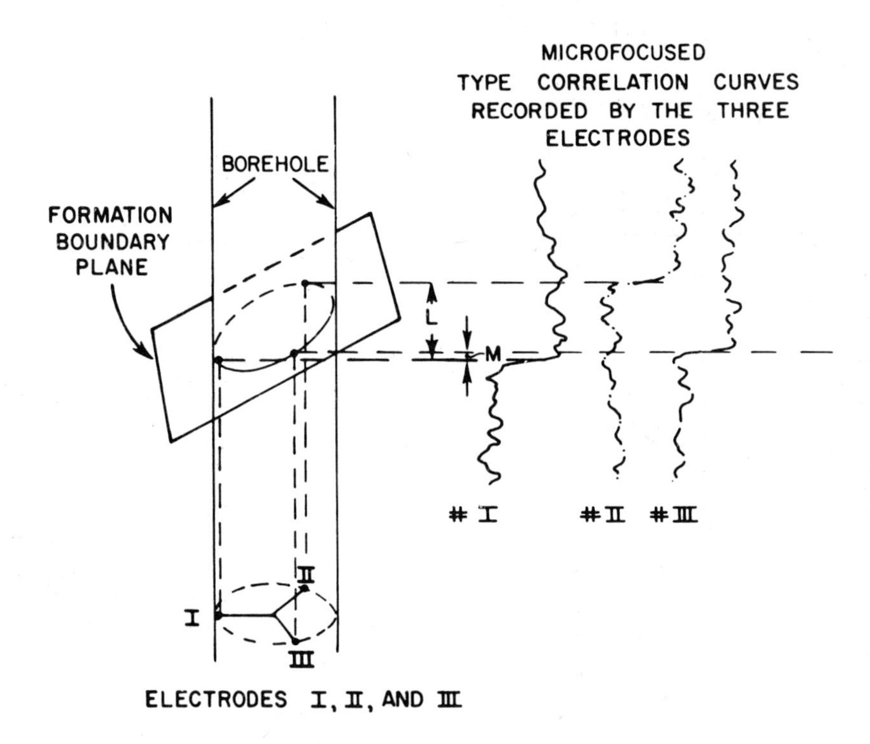

Fig. 11.39. Schematic drawing of Microlog dipmeter curves as instrument passes bed boundary, and consequent determination of formation dip. Courtesy Schlumberger Well Surveying Corporation.

these measurements are called the SP, resistivity, or Microlog continuous dipmeter, depending on which electrode system is used. The latter is a considerable improvement over the earlier SP and resistivity instruments, in that a great number of intervals are available for dip measurements, and therefore the precision of the data is increased. By using three Microlog electrode systems and simultaneously recording three curves, one can determine the variation in depth at which each curve records a bed or zone boundary. The directional orientation of the electrodes as well as the hole deviation and direction are also recorded simultaneously. These data make it possible to establish the dip at any level. (See Figure 11.39.) Some uses of dip information are:

(1) Hole deviation problems, as shown in Chapter 9.
(2) Geological purposes
  (a) Subsurface mapping of nearly any type
  (b) Determination of proper direction in which to offset discovery wells, dry holes, etc.

In later chapters we will have occasion to apply certain well log information to a few specific problems; for the most part, however, this concludes our discussion of the general subject. Those desiring to pursue the topic are referred to the references and recommended outside readings of this chapter.

PROBLEMS

1. The analysis of a formation water is as follows:

Ion	*Concentration, ppm*
Na	40,000
Mg	15,000
K	5,000
Cl	45,000
SO_4	10,000
HCO_3	8,000

(a) Determine the equivalent NaCl salinity of this water and its resistivity at 75, 100, 150, and 200°F.
(b) Check the resistivities of (a) by Eq. (11.2) using the 75°F value as R_{w1}. Is the agreement satisfactory?
(c) Determine the resistivities at the same temperatures as in (a) by ignoring the ion multipliers (Table 11.1) and directly summing the concentrations. Do these check reasonably well?

2. A drilling mud has a resistivity of 1 ohm-meter at 80°F. Estimate R_{mf}, R_{mc}, and R_m at 150°F.

3. A drilling mud has a resistivity of 0.30 ohm-meters at 200°F which is the bottom hole temperature at 10,000 ft. In interpreting the log at 7000 ft, what R_m should be used? The mean surface temperature of the area is 80°F. ($T_{7000} = 80° + ad$; a = geothermal gradient, d = depth.)

4. Given: R_o = 1.5 ohm-meters at 150°F
R_w = 0.05 ohm-meters at 80°F
R_t = 30 ohm-meters at 150°F

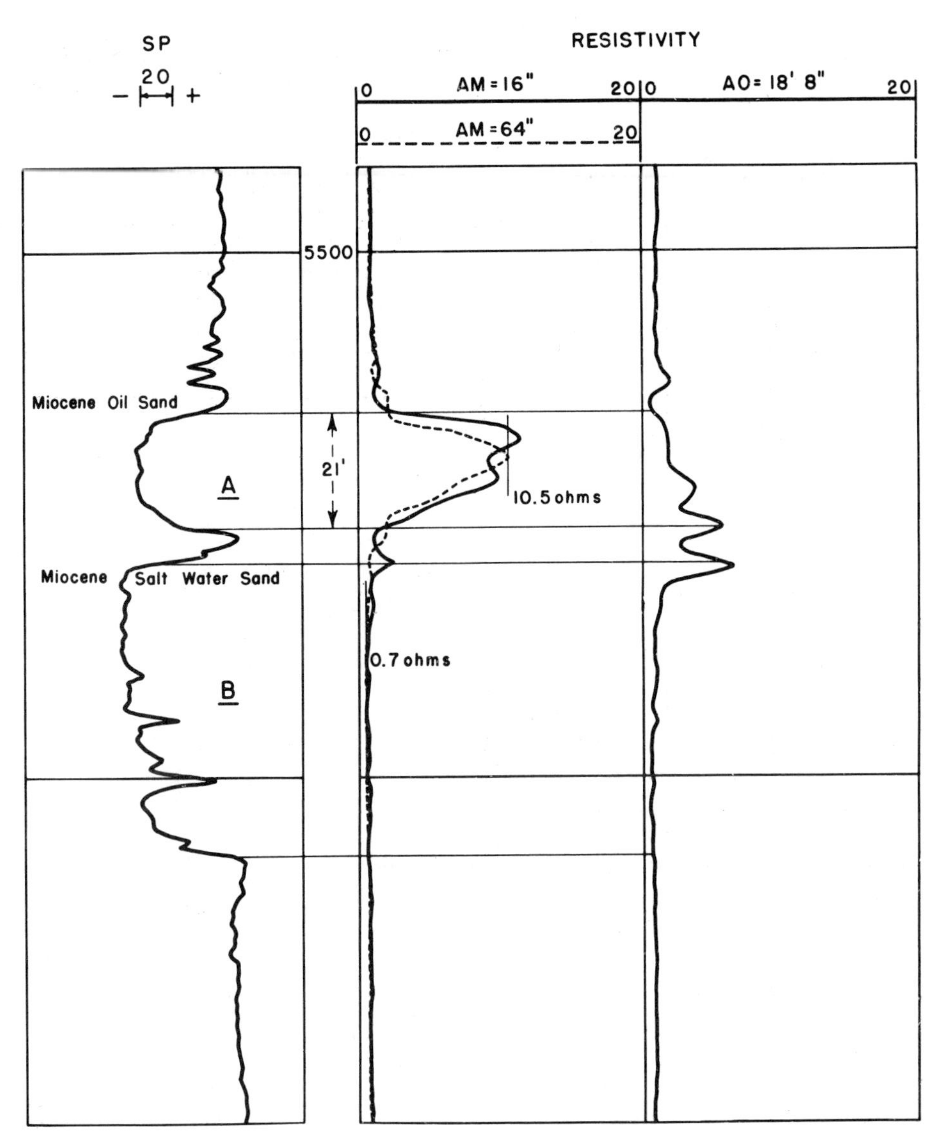

Fig. 11.40. Electric log for Problem 9. Courtesy Schlumberger Well Surveying Corporation.

(a) Compute the formation factor for this zone, and its porosity by both the Archie and Humble equations.
(b) Compute S_w by Archie's equation for $n = 2$. *Ans.* (b) $S_w = 22\%$.

5. From core analysis, the porosity of a well consolidated sand is known to be 20%. Formation water resistivity has been measured from an actual sample as 0.04 ohm-meters at formation temperature. From electric logs, $R_t = 80$ ohm-meters. Estimate S_w for this zone. *Ans.* $S_w = 35\%$ for $m = 2$ (Archie's equation).

6. Given the following electric log data, determine R_i and R_t by making departure curve corrections.

$R_{16''} = 40$ ohm-meters

$R_{18'8''} = 150$ ohm-meters

$R_m = 0.5$ ohm-meters at formation temperature

Hole size = 9 in.

Ans. $R_i = 55$ ohm-meters.

7. Given the following data, determine the formation water resistivity.

SP = -70 mv

$R_m = 0.4$ ohm-meters (at formation temperature)

$h = 15$ ft (bed thickness)

$R_i = 20$ ohm-meters

$T = 150°$F *Ans.* $R_w \cong 0.040$ ohm-meters.

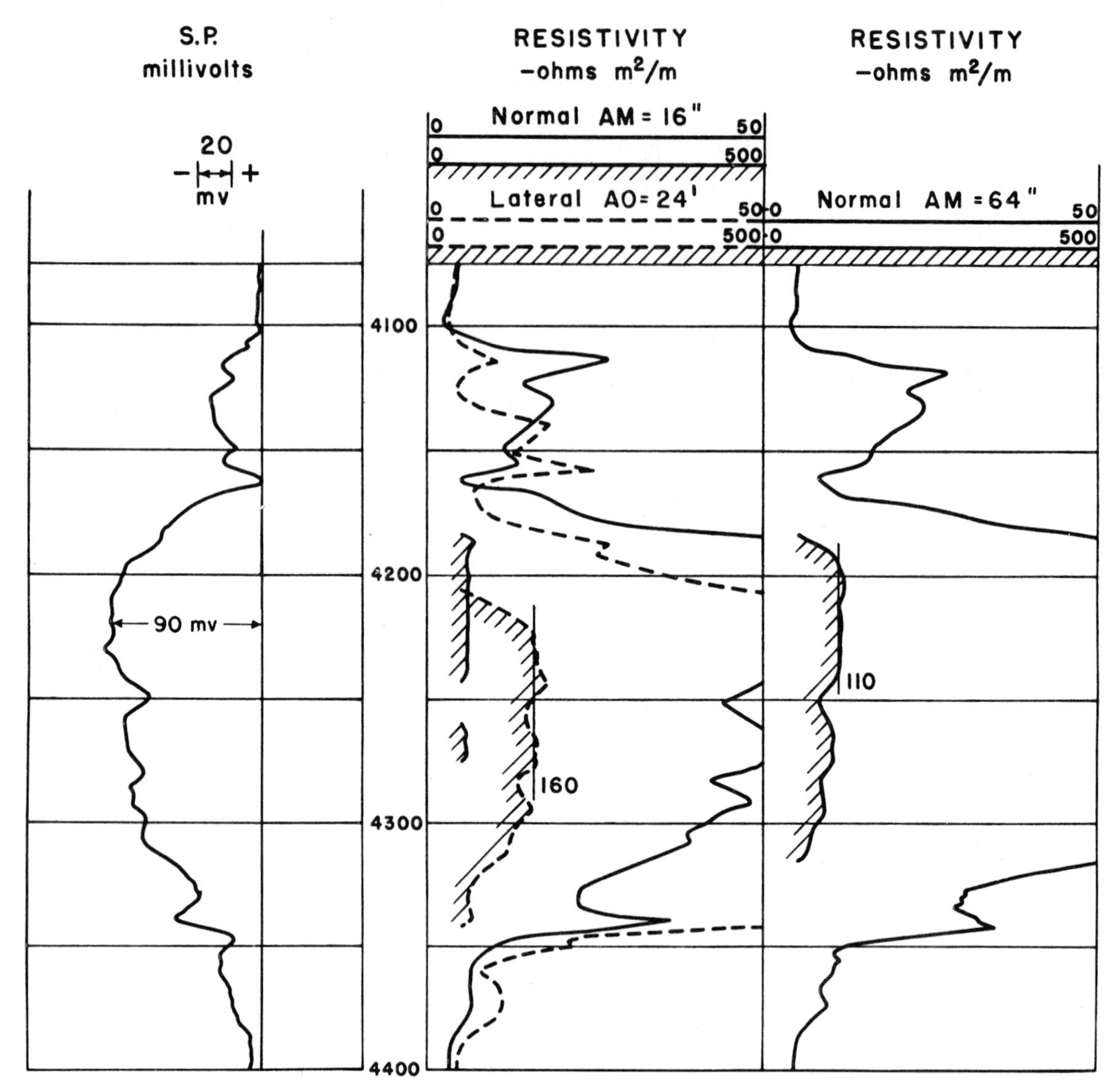

Fig. 11.41. Electric log for Problem 10. Courtesy Schlumberger Well Surveying Corporation.

8. Make a foot-by-foot porosity determination for the 3507–3528 ft zone of Figure 11.33. (Delete any shale sections.) Compute the initial reservoir oil in place per acre-foot, assuming an oil zone and $S_w = 30\%$.

9. Given the log shown as Figure 11.40. Estimate the water saturation of zone A.

(a) Assume sands A & B are similar, contain the same water, and that R_o for zone A is the same as that of zone B.

(b) Take R_t for zone A as the long normal reading, $R_{64''} = 10.5$ ohm-meters, due to the thinness of the bed. *Ans.* $S_w \cong 26\%$.

10. Interpret the log of Figure 11.41. The following data are available:

$T = 130°F$

$R_o = 2.4$ ohm-meters from neighboring well

(Note that lateral curve used had 24 ft spacing; in lieu of other charts, use the departure curve for the 18 ft 8 in. device to correct $R_a = 160$ ohm-meters to R_t. Compare answer to $R_{64''}$ on log.) *Ans.* $\phi \cong 13\%$ (Archie); $S_w \cong 15\%$.

REFERENCES

1. Wilson, R. W., "Mud Analysis Logging and its Use in Formation Evaluation," AIME Tech. Paper 587-G, Presented University of Houston, Oct. 1955.

2. Dunlap, H. F., and R. R. Hawthorne, "The Calculation of Water Resistivities From Chemical Analyses," *Trans. AIME*, Vol. 192, (1951), p. 373.

3. Archie, G. E., "The Electrical Resistivity Log as an Aid in Determining Some Reservoir Characteristics," *Trans. AIME*, Vol. 146, (1942), p. 54.

4. Guyod, H., "Fundamental Data for the Interpretation of Electric Logs," *Oil Weekly*, Oct. 30, 1944.

5. Winsauer, W. O., H. M. Shearin, P. H. Masson, and M. Williams, "Resistivity of Brine Saturated Sands in Relation to Pore Geometry," *AAPG Bulletin*, Vol. 36. No. 2, Feb. 1952.

6. Doll, H. G., "The S.P. Log: Theoretical Analysis and Principles of Interpretation," *Trans. AIME*, Vol. 179, (1949), p. 146.

7. *Resistivity Departure Curves*, Schlumberger Document 3, (1949), Schlumberger Well Surveying Corporation, Houston, Texas.

8. Doll, H. G., "The Microlog — A New Electrical Logging Method for Detailed Determination of Permeable Beds," *Trans. AIME*, Vol. 189, (1950), p. 155.

9. Doll, H. G., "The Laterolog: A New Resistivity Logging Method with Electrodes Using an Automatic Focusing System," *Trans. AIME*, Vol. 192, (1951), p. 305.

10. Owen, J. E., and W. J. Greer, "The Guard Electrode Logging System," *Trans. AIME*, Vol. 192, (1951), p. 347.

11. *Departure Curves for Laterolog*, Schlumberger Document 6, Schlumberger Well Surveying Corporation, Houston, Texas.

12. Doll, H. G., "The Micro-Laterolog," *Trans. AIME*, Vol. 198, (1953), p. 17.

13. Doll, H. G., "Introduction to Induction Logging and Application to Logging of Wells Drilled with Oil Base Mud," *Trans. AIME*, Vol. 186, (1949), p. 148.

14. Wyllie, M. R. J., *The Fundamentals of Electric Log Interpretation*, 2nd ed. New York: Academic Press, Inc., 1957, pp. 112–126.

15. Goins, W. C., Jr., and D. D. Dawson, Jr., "Temperature Surveys to Locate Zone of Lost Circulation," *Oil and Gas Journal*, June 22, 1953, p. 169.

16. Kunz, K. S., and M. P. Tixier, "Temperature Surveys in Gas Producing Wells," *Trans. AIME*, Vol. 204, (1955), p. 111.

17. C. V. Millikan, "Temperature Surveys in Oil Wells," *Trans. AIME*, Vol. 142, (1941), p. 15.

SUPPLEMENTARY READINGS

Wyllie, M. R. J., *The Fundamentals of Electric Log Interpretation*, 2nd ed. New York: Academic Press, Inc., 1957.

Pirson, S. J., *Oil Reservoir Engineering*, 2nd ed. New York: McGraw-Hill Book Co., Inc., 1958, pp. 137–302.

LeRoy, L. W., *Subsurface Geologic Methods*, 2nd ed. Golden, Colorado: Colorado School of Mines, 1951. This is a symposium and contains articles on all phases of logging.

Guyod, H., *Electrical Well Logging Fundamentals*, (Houston, Texas, 1952). A series of 27 papers, plus charts.

Fundamentals of Quantitative Analysis of Electric Logs. (Fort Worth, Texas: Welex, Inc.), Bull. A120.

Interpretation Charts for Electric Logs and Contact Logs. Fort Worth, Texas: Welex, Inc., Bull. A101.

Radioactivity Well Logging. Lane-Wells Company, 1952, Bull. RA-47-B-4.

Nuclear Well Logging. Houston, Texas: Perforating Guns Atlas Corporation. A manual including interpretation charts.

Introduction to Well Logging. Schlumberger Document 8, Schlumberger Well Surveying Corporation, Houston, Texas, 1958.

Log Interpretation Charts. Schlumberger Well Surveying Corporation, Houston, Texas, 1958.

Chapter 12 . . .

Formation Damage

Before proceeding into the topics of drill stem testing and well completions, it is necessary to discuss the possibilities and important consequences of permeability alteration which may occur around the well bore as a result of drilling and completion operations. In general these alterations are unfavorable, i.e., the permeability of the altered zone is less than that of the virgin reservoir rock; hence the term formation damage. As we shall see, this relatively thin, altered veneer (or skin) has a considerable effect on the productivity of a well. However, before attempting a quantitative appraisal, we will first consider the basic causes of such damage.

12.1 Causes

Formation damage is caused by the invasion of foreign fluids and/or solids into the exposed section adjacent to the well bore. Generally, the drilling mud is the main source of such contaminants. Fluids used in stimulation treatments (acidizing, hydraulic fracturing, etc.) may also have some undesirable effects which partially nullify their beneficial actions. Our present discussion will, however, be primarily concerned with drilling mud damage, although the principles involved may be applied to any fluid.

In our previous discussion of drilling fluids, it was stated that in many areas a necessary function of the drilling mud is the control of encountered subsurface pressures. To carry out this function, the mud column pressure must exceed that of the formation. Hence the mud filtrate flows radially outward in accordance with the filtration characteristics of the particular mud in use. This filtration effect was also recognized in our discussion of electric logging in terms of the flushed and invaded zones. Since all emphasis up to this point has been on filtrate invasion, let us consider the possible injurious effects that various liquids may have on invaded zone permeability.

12.11 Liquid Invasion

The susceptibility of a particular formation to damage by foreign fluids is largely dependent on its clay content. Dirty sands (those with high clay content) are generally quite sensitive to the filtrate from fresh water base muds which brings about the hydration and swelling of interstitial clay particles. Saline filtrates cause less of this kind of trouble, and may in fact reduce particle size and increase oil permeability in some cases. This latter effect was noted by Nowak and Krueger[1] in their studies of the effects of various mud filtrates on the permeability of sandstone cores. It was shown in their tests that polyvalent salts, such as calcium chloride and aluminum chloride, did not permanently damage permeability as much as did sodium chloride solutions. These results are shown in Figure 12.1. Note that in the cases of the 3% calcium chloride solution and the calcium chloride-starch mud (tests 4 and 9), the final permeability after oil backflow was equal to or greater than the original. Both of these effects were explained in terms of ion exchange as follows: The clay in the test cores was, presumably, a sodium clay; sodium ions were initially present to neutralize the surface charge of the individual clay particles. When brought into contact with the calcium chloride filtrate, some of the sodium ions were replaced by calcium in the ratio of two sodium for one calcium; this exchange resulted in clay particle shrinkage and hence less, if any, reduction in permeability. Therefore it may be surmised that clay swelling problems depend on the interaction between the particular clay and filtrate involved.

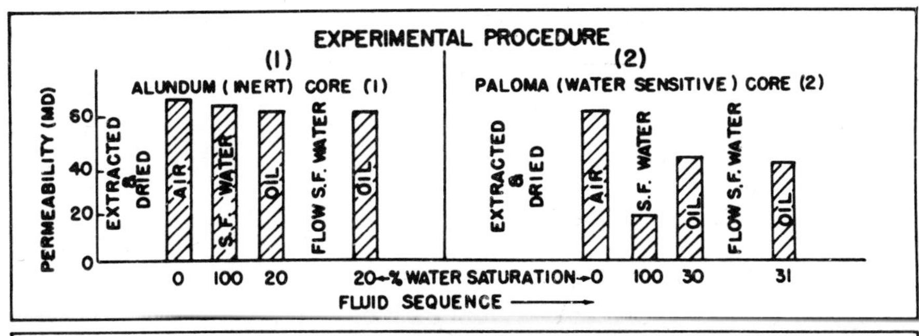

EXPERIMENTAL DATA (PALOMA CORES)
(3)

TEST NO.	TEST SOLUTION % BY WT.	% INTERSTITIAL WATER		% OF ORIGINAL OIL PERMEABILITY RECOVERED (0 20 40 60 80 100 120)
		BEFORE	AFTER	
1	DISTILLED WATER	32.5	45.8	
2	1 % SODIUM CHLORIDE	33.2	41.4	
3	1 % CALCIUM CHLORIDE	38.4	38.7	
4	3 % CALCIUM CHLORIDE	29.9	29.1	
5	1 % ALUMINUM CHLORIDE	33.4	34.3	

Fig. 12.1(A). Effect of aqueous solutions on oil permeability of Paloma field (Stevens sand) cores.

Secondly, as the aqueous filtrate invades dirty sands, the interstitial clay particles may shift position as the increased water saturation gives them freedom of movement. This hypothesis, based on the assumption that clay fines are water-wet in the reservoir, was proposed by Bertness.[2] The movement of such particles enables them to relocate within the pore network, thereby obstructing flow and causing a reduction in permeability.

Other possible effects of foreign fluid invasion are:

(1) emulsification with formation fluids, resulting in highly viscous mixtures, and capillary blocking by insular bubbles (the latter is commonly called the Jamin effect).

(2) precipitation of solids: insoluble salts and asphaltic or wax particles.

(3) reduction of relative permeability to gas by the presence of a third immiscible fluid. For example, consider a gas sand which originally contains only gas and water. If this section is drilled or stimulated with an oil base fluid, some filtrate invasion will occur. When gas production begins, some of this oil will backflow and clean up; however, some will remain as an irreducible or immobile saturation. This reduces the permeability to gas in the affected zone and hence lowers the productivity of the well.

(4) reduction of relative permeability to oil due to an increased irreducible water saturation. This is a common, auxiliary consequence of clay swelling and results from pore size reduction. Note, for example, in Figure 12.1, the interstitial water values before and after filtrate contamination.

TEST NO.	TYPE OF MUD FILTRATE	% INTERSTITIAL WATER		% OF ORIGINAL OIL PERMEABILITY RECOVERED (0 20 40 60 80 100 120)
		BEFORE	AFTER	
6	CLAY-WATER BASE	34.3	37.6	
7	FRESH WATER-STARCH	34.6	45.3	
8	FRESH WATER-DRISCOSE	32.2	36.2	
9	CALCIUM CHLORIDE-STARCH	32.3	25.7	
10	LIME-STARCH	28.5	27.4	
11	LIME-TANNATE	36.2	43.3	
12	FRESH WATER EMULSION	32.0	37.7	
13	SALT WATER EMULSION	28.8	26.6	
14	OIL BASE	25.2	24.9	

Fig. 12.1(B). Effect of field mud filtrates on oil permeability of Paloma field (Stevens sand) cores.

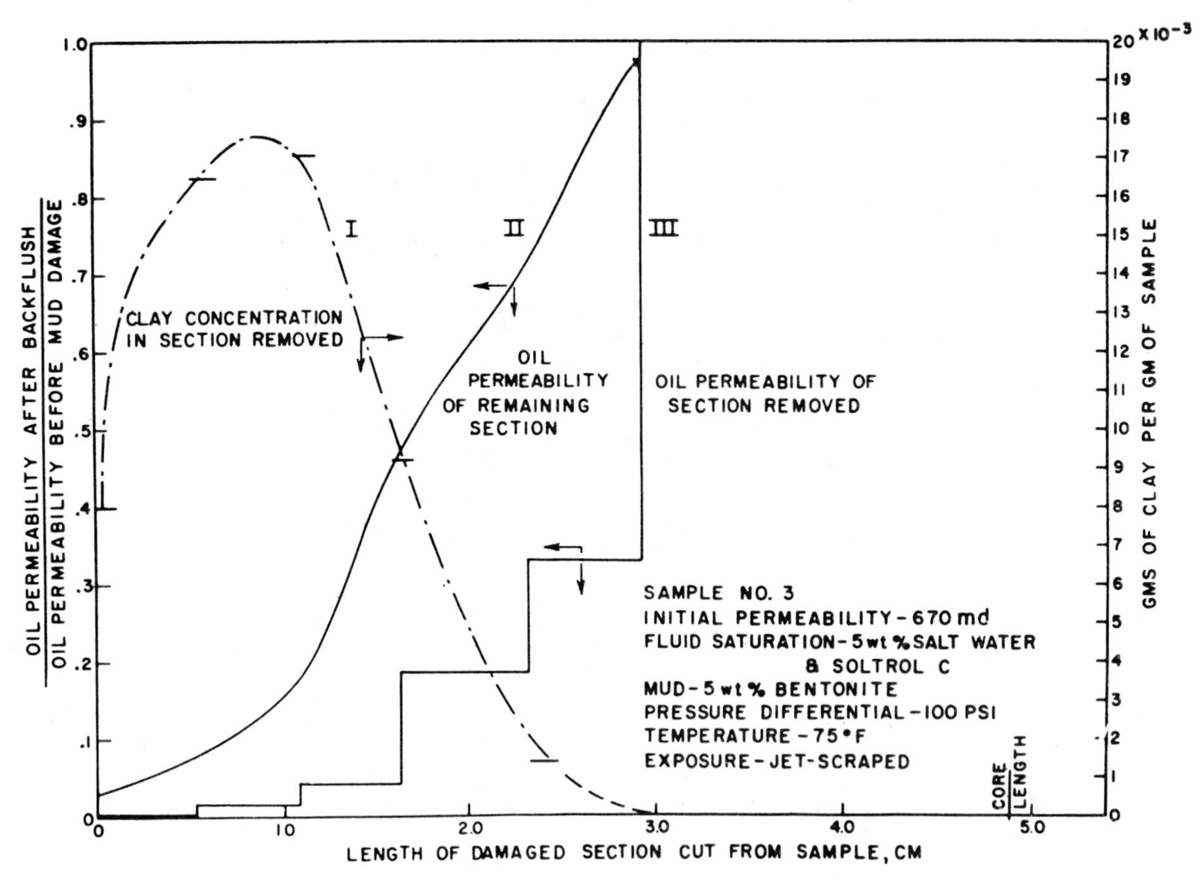

Fig. 12.2. Ratio of damaged permeability to original permeability of an alundum sample exposed to a 5 per cent Bentonite mud. After Glenn and Slusser,[5] courtesy AIME.

The effect of a particular fluid on a specific formation is difficult to evaluate, since each case is somewhat unique. From a formation damage standpoint, the best damage prevention rule is to use drilling and completion fluids as much like the formation fluids as possible. In general, it is best to use lease crude oil, other oils, salt water, and finally fresh water, in that order. The use of gas or aerated liquid muds often alleviates the problem in those areas where their usage is feasible.

12.12 Solid Invasion

It has been recognized that the invasion of solid particles may also be a considerable source of formation damage. For solids to enter into a rock the solid particles must, naturally, be smaller than the pore openings. An extreme case of whole mud invasion is the loss of circulation.

Various publications have analyzed the extent of mud solid invasion and its effect on formation permeability.[1,3-5] The observed permeability decreases are believed due to two basic causes:

(1) plugging of internal pores by solid particles.
(2) reduction of effective pore radius with consequent increase in interstitial water content and reduction in effective permeability to oil or gas. (This same effect was also attributed to clay swelling.)

The depth of solid particle invasion is of course less than that of the filtrate. Krueger and Vogel[4] report invasion depths up to 12 in. or more in 350 to 550 md cores exposed to mud for five days. Also, the severity of damage decreases with distance from the well bore, as was shown by the Glenn and Slusser[5] experiment depicted in Figure 12.2.

Up to this point, we have considered only what may happen within the damaged zone after its invasion by the contaminating materials. Further appraisal of the cause of such invasion requires a basic look at the general filtration behavior of fluid-rock systems.

The most widely used indicator of a mud's filtration properties is its water loss as determined by the standard API test described in Chapter 6. This is a static filter press test using a standard filter paper of specified mesh. The initial surge or spurt, which occurs before the mud solids bridge and begin building a filter cake, is generally not considered, and the corrected 30 minute water loss is taken as the basic measurement. This is shown as V_c in Figure 12.3. While this procedure may be reasonably satisfactory for mud control purposes, it does not tell the complete story insofar as formation damage is concerned.

Beeson and Wright[3] studied the filtration behavior of numerous muds against sands and filter papers of

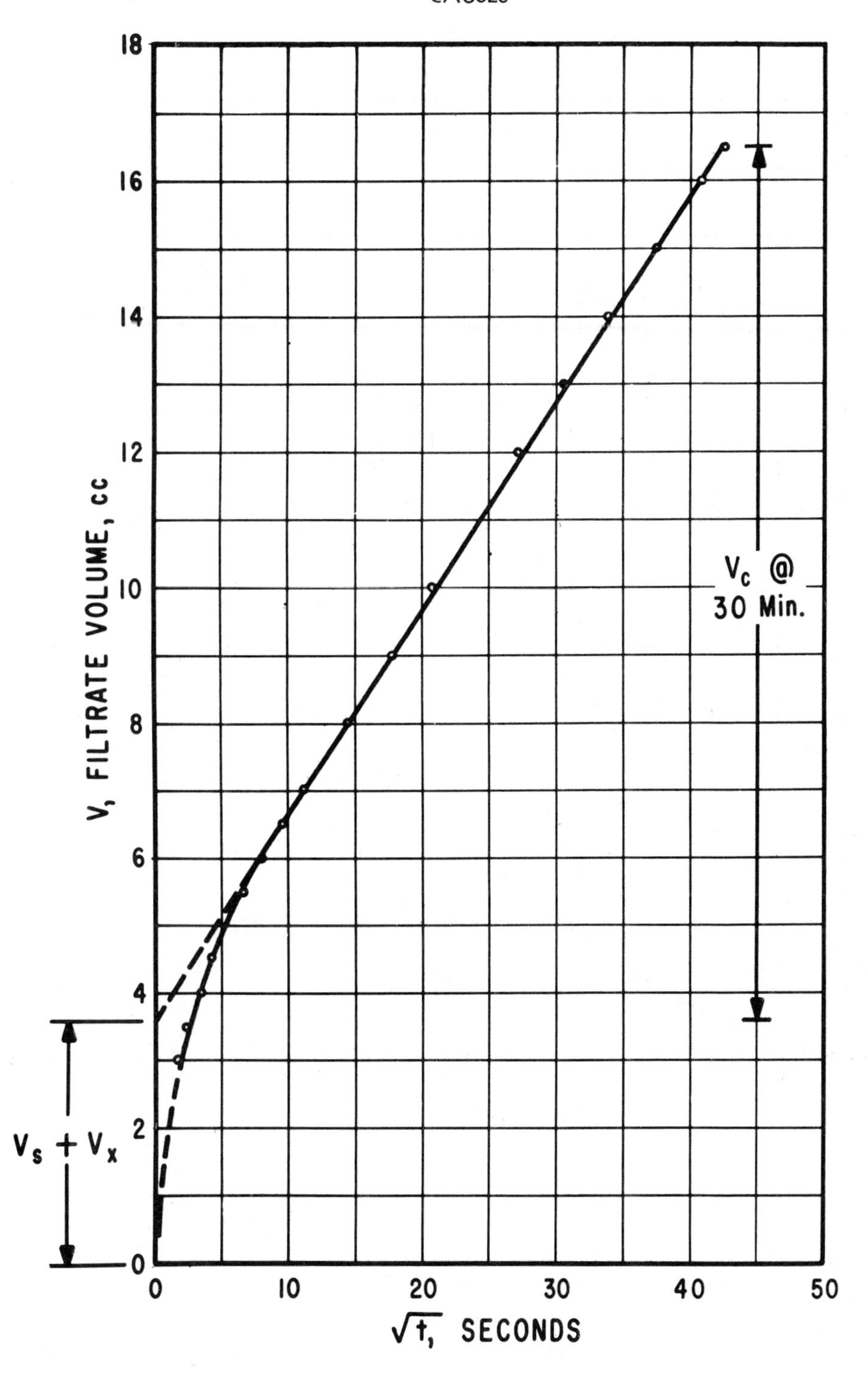

Fig. 12.3. Standard method of obtaining API water loss. Corrected volume, V_c, is commonly reported. $V_s + V_x$ = initial spurt or surge correction.

varying permeability. Their results indicated that considerable solid invasion occurred and that an internal filter cake was formed within the sand. The factors governing the entry of mud filtrate and solids into a particular sand were the pore size distribution of the sand, the particle size distribution of the mud solids, and the *plastering* ability of the mud. The standard API test provides data on the latter factor only. Beeson and Wright also suggested that the generally disregarded initial surge was indicative of the invasion which occurred prior to bridging. The addition of sized particles was effective in reducing the surge loss.

Ferguson and Klotz[6] have shown that downhole filtration should be considered as three separate phenomena, namely, beneath bit, dynamic, and static filtration. No appreciable filter cake can form on the

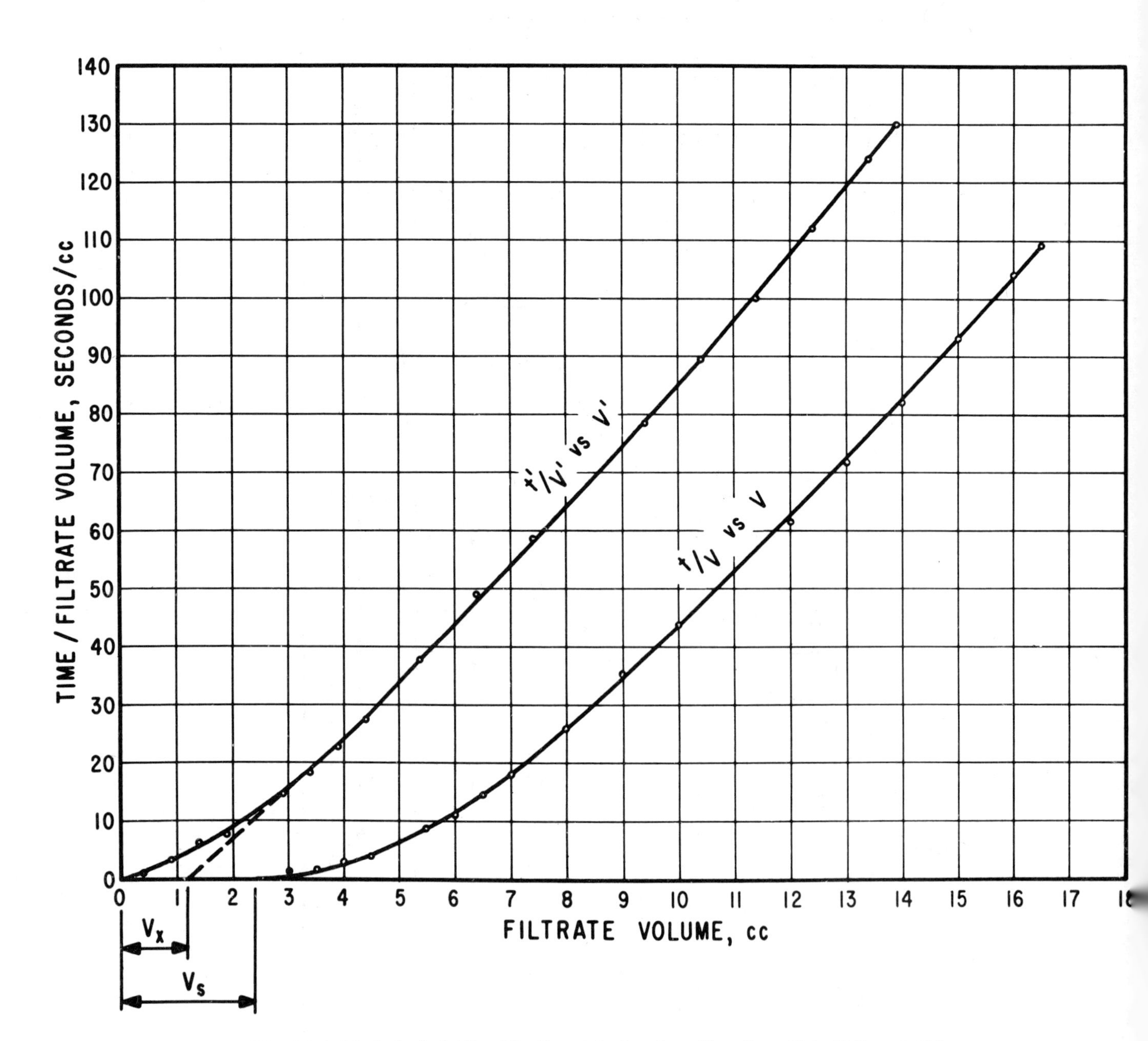

Fig. 12.4. Method of plotting filtration data based on Equations 12.1, 12.2, and 12.3. V_s, V_x, and V_c are obtained as shown. After Slusser, Glenn, and Huitt,[7] courtesy AIME.

surface beneath the bit; however, experiments have shown that pore plugging can occur some distance ahead of the bit. This finding is in agreement with the internal bridging proposed earlier. Dynamic filtration occurs above the bit while mud is circulating. The rate of fluid loss during this period is higher than the static rate and eventually falls to a constant value when the filter cake reaches its maximum thickness. Further cake growth is prevented by erosion of the gelatinous, low strength, outer layer by the circulating mud and drill pipe action. Static filtration occurs during periods of non-circulation.

A more fundamental method of analyzing water loss data has been presented by Slusser, Glenn, and Huitt.[7] This procedure utilizes standard filter press data in such a manner that three separate stages of filtration may be evaluated:

1. The initial surge period: before any appreciable cake is formed.
2. The transition period: after a filter cake is initiated

but before it becomes uniform, i.e., the period during which the cake surface is irregular and is under unequal pressure gradients at different points.

3. The constant pressure gradient period: filtrate volume varies linearly with the square root of time.

Quantitative analysis of the water loss in each stage is based on the Carman equation:

$$t = mV^2 + nV \tag{12.1}$$

where t = time

V = cumulative filtrate volume

m = a number which defines the filtration characteristics of the mud cake. It increases from zero during stage 1 to some constant value in stage 3.

n = constant depending essentially on the permeability of the septum (filter paper or porous formation)

Rearrangement of Eq. (12.1) is useful for plotting purposes:

$$\frac{t}{V} = mV + n \tag{12.2}$$

A plot of t/V vs. V becomes linear when m becomes constant in stage 3 of the filtration process. The surge loss V_s is obtained as the magnitude of V when $t/V = 0$. Careful analysis of the original data will also allow determination of the corresponding surge time t_s. Since stage 3 is completely governed by the permeability of the filter cake, Eq. (12.2) may be rewritten in terms of the surge corrected volume V' and time t':

$$\frac{t'}{V'} = mV' \tag{12.3}$$

where $t' = t - t_s$

$V' = V - V_s$

A plot of t'/V' vs. V' becomes linear when m = constant. Hence the filtration loss during stage 2, V_x, may be obtained as the linearly extrapolated intercept at $t'/V' = 0$.

This method is illustrated by Figure 12.4 which is plotted from the data in Table 12.1. The standard API procedure for treating the same data was shown in Figure 12.3. Note that the intercept in Figure 12.3 is the sum $V_s + V_x$.

This analysis requires that mud filtration loss be reported as three separate values, V_s, V_x, and V_c, which are indicative of behavior during three separate periods. The importance of this procedure lies in the fact that additives which improve stage 3 behavior do not necessarily improve losses during other stages. The surge loss of stage 1 is dependent on the rapidity with which the mud solids bridge surface and internal pore openings. A mud which contains properly sized solids for a particular sand may be expected to bridge more quickly and efficiently than one which is deficient in some critical sizes. This expectation is substantiated by Figure 12.5, which shows the difference in surge loss when various sized particles were added. Also note the per cent permeability recovery.

TABLE 12.1

TYPICAL FILTRATION LOSS DATA FROM STANDARD FILTER PRESS TEST (AFTER SLUSSER, GLENN, AND HUITT[7])

(1) V, cc	(2) t, sec	(3) $\sqrt{t}$	(4) (2) ÷ (1): t/v, sec/cc	(5) (1) − 2.6 cc: v', cc	(6) (2) − 3.0 sec: t', sec	(7) (6) ÷ (5): t'/v', sec/cc
3.0	3.5	1.87	1.17	0.4	0.5	1.25
3.5	6.0	2.45	1.72	0.9	3.0	3.34
4.0	12	3.46	3.00	1.4	9.0	6.41
4.5	18	4.24	4.00	1.9	15.0	7.89
5.5	46	6.78	8.37	2.9	43	14.8
6.0	65	8.06	10.8	3.4	62	18.2
6.5	92	9.59	14.2	3.9	89	22.8
7.0	125	11.2	17.9	4.4	122	27.7
8.0	206	14.3	25.8	5.4	203	37.6
9.0	317	17.8	35.2	6.4	314	49.0
10.0	436	20.9	43.6	7.4	433	58.5
12.0	737	27.1	61.4	9.4	734	78.1
13.0	932	30.5	71.6	10.4	929	89.3
14.0	1150	33.9	82.0	11.4	1147	100
15.0	1392	37.3	93.0	12.4	1389	112
16.0	1668	40.8	104	13.4	1665	124
16.5	1800 (30 min.)	42.4	109	13.9	1797	130

Note: $V_c = 16.5 - (V_x + V_s) = 12.9$ cc (See Figure 12.3.)

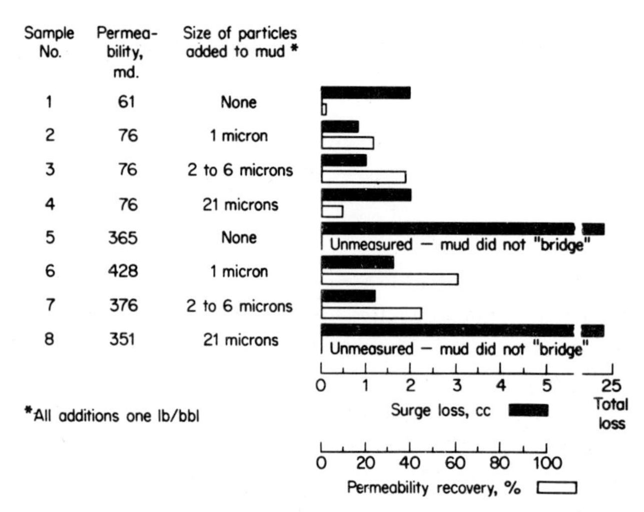

Fig. 12.5. Dependance of surge loss on mud particle sizes. After Slusser, Glenn, and Huitt,[7] courtesy AIME.

The question immediately arises, how does one predict the necessary mud particle sizes for each sand? While pore size distribution can be obtained from capillary pressure data for a particular sample, it must also vary with each permeability in the zone, as well as in all other sections up and down the hole. Consequently, the logical solution is to use colloidal mud materials in proper size gradations which will quickly bridge any encountered openings. Just what this optimum size distribution should be is open to conjecture at this time. Some general relationships are, however, available in the work of Gates and Bowie.[8]

A bridge may be initiated when two large particles start into an opening at the same time and lodge against each other. Other smaller particles may then bridge the openings between the larger, previously bridged particles. If the proper particle sizes are present, this process may continue until the openings become too small for any contained solid to penetrate. It is at this time that only the filtrate flows through the filter cake. Theoretically, if enough successively smaller particles were present, the voids in the filter cake could become so small that even water molecules could not pass.

Lost circulation problems are analogous to those of filtration except that in the former case the scale is larger, i.e., the initial surge may often be measured in thousands of barrels. A common solution to lost circulation is the addition of properly sized granular materials which can initiate bridging and allow the normal mud solids to wall the hole. Considerable reference to proper or optimum particle size distribution for efficient bridging is also found in lost circulation literature. However, as is the case in filtration studies, no precise definition is available. Perhaps the work of Furnas cited in the parallel discussion of Chapter 6 may be a solution to both problems.

It should also be mentioned that results obtained at laboratory conditions are not necessarliy indicative of a mud's bottom hole filtration behavior. Schremp and Johnson[9] have observed that some premium muds may have poorer filtration characteristics at high temperature

TABLE 12.2

SUMMARY OF METHODS FOR APPRAISING AND PREVENTING FORMATION DAMAGE

Damage Effects	*Appraisal Methods*	*Practical Preventive Measures*
1. *Foreign Fluid Invasion*		
(a) clay swelling (b) emulsification (c) precipitation of solids (d) rock alteration, movement of interstitial fines (e) reduction in relative permeability due to introduction of third phase	(1) Special contamination and flow tests on core samples to determine compatibility of fluids involved (2) 3 stage filtration test based on Carman equation (3) API test which includes initial correction $V_s + V_x$ (4) Standard API test (5) Conduct the above at the temperature and pressures involved	(1) Use of additives which will reduce filtration losses (2) Reduce pressure differential against formation to lowest safe value (3) Air and/or gas drilling may eliminate the problem in areas where applicable (4) Use fluids which are compatible with formation and its contents if possible, i.e., lease crude, formation water, saline or oil base muds (5) Minimize exposure time as much as possible
2. *Foreign Solid Invasion*		
(a) size reduction or plugging of internal pores by intruding solids (b) increased interstitial water content and consequent reduction in oil or gas permeability	(1) Analysis of initial surge filtration data from filter press tests (2) Actual surge data against core sample of rock in question	(1) Addition of properly sized colloidal solids which rapidly form an efficient bridge Also items (2), (3), (5) above

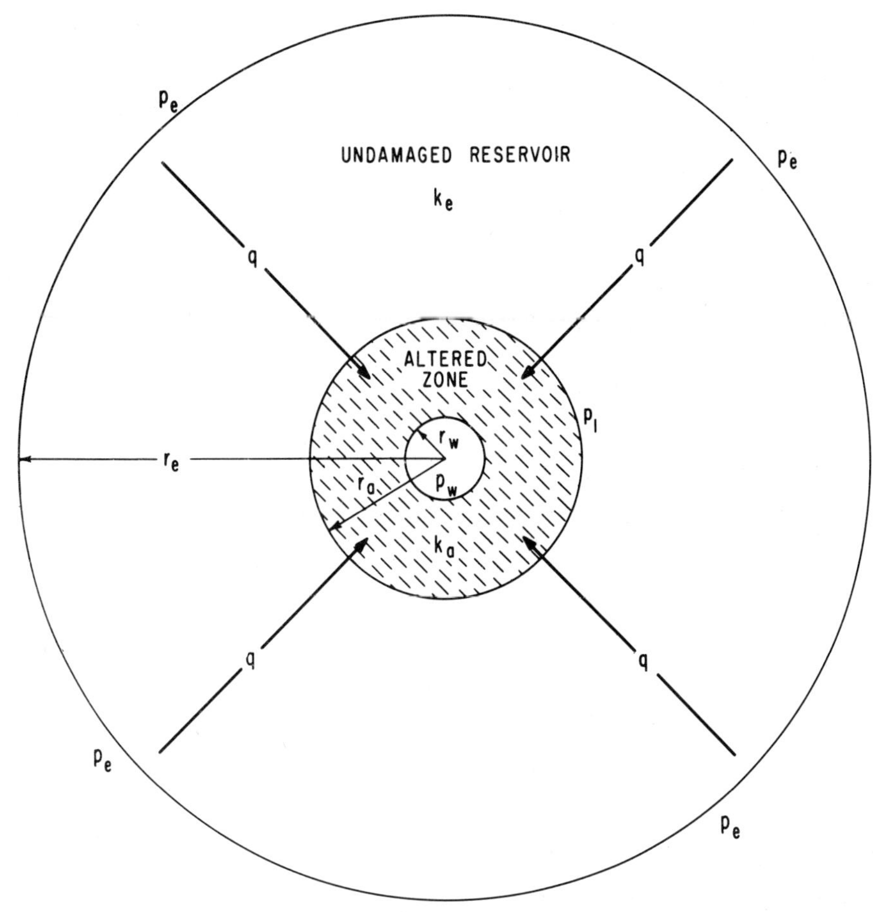

Fig. 12.6. Idealized sketch of drainage area around a well. Reservoir permeability $= k_e$, altered zone $= k_a$. Flow rate, q, at a given drawdown, $p_e = p_w$, depends on k_a and r_a/r_w as well as k_e and r_e/r_w.

and pressure than some cheaper, supposedly lower quality types of muds. They very logically conclude that muds should be tested under the conditions of use. At present, however, this is not the normal procedure.

While this section has dealt primarily with drilling fluid contamination, the principles involved are applicable to any fluid. Cement slurries and stimulation fluids may be subjected to the same appraisal where applicable; however, these subjects will be dealt with in later chapters.

12.2 Prevention of Formation Damage

The discussion of causes might also serve as a coverage of preventive measures, since the obvious prevention is removal of the cause. It might be well, however, to re-emphasize certain practical measures which may be taken. Table 12.2 attempts to list the main points previously discussed along with practical preventive measures. Considerable overlapping occurs both in appraisal and prevention methods.

Practically speaking, the cost of the preventive measures should not exceed the benefits obtained. As will be discussed later, stimulation techniques are successful in overcoming formation damage effects in most cases. It is worthwhile to avoid stimulation if productivity is satisfactory without it; however, the cost of the preventive measures must be balanced against the cost and possible results of stimulation. This is not to say that both preventive measures and stimulation may not be desirable in many instances, but rather that each situation requires a separate economic analysis. Formation damage is of extreme importance in some cases and of little consequence in others. Some formation damage will almost always occur during well completion, regardless of the preventive measures applied.

12.3 Quantitative Analysis of Formation Damage

12.31 Introductory Concepts

Figure 12.6 shows schematically the typical permeability distribution around a borehole with the altered zone of permeability k_a extending a distance r_a from the well bore. This picture is, of course, idealized, since the alteration is greatest near the well and diminishes rather rapidly as the radius increases. It is difficult to estimate the radial extent of normal damage; however, in most cases the invasion depth is probably on the order of a few feet, with most of the damage being restricted to the first few inches (for an example, see Figure 12.2).

The effect of a damaged zone on well productivity may be illustrated by considering the system of Figure 12.6 as a case of two permeabilities in radial series. The productivity of such a well will be governed by a properly computed average permeability, $\bar{k}$. Applying Darcy's radial flow equation (steady state):

$$(12.4)\quad q = \frac{2\pi h k_a (p_1 - p_w)}{\mu \ln (r_a/r_w)} = \frac{2\pi h k_e (p_e - p_1)}{\mu \ln (r_e/r_a)} = \frac{2\pi h \bar{k} (p_e - p_w)}{\mu \ln (r_e/r_w)}$$

Note that Eq. (12.4) merely divides the system into concentric hollow cylinders of radii r_a and r_e and states that, for steady state conditions, the flow rate q is the same at any radius. Also:

$$(12.5)\quad p_e - p_w = (p_1 - p_w) + (p_e - p_1)$$

Substitution of Eq. (12.4) in Eq. (12.5) and simplification yields:

$$(12.6)\quad \bar{k} = \frac{\ln(r_e/r_w)}{\frac{1}{k_e}\ln(r_e/r_a) + \frac{1}{k_a}\ln(r_a/r_w)}$$

where $\bar{k}$ = average or equivalent permeability of the altered system.

Examination of Eq. (12.6) shows the logical result that as $k_a \to 0$, $\bar{k} \to 0$. Further, as r_a increases, $\bar{k} \to k_a$. Example 12.1 illustrates the strong influence of k_a on $\bar{k}$.

Example 12.1

What is the average or equivalent permeability which will govern the productivity of the following well?

r_e = 700 ft k_e = 500 md

r_w = 4 in. k_a = 5 md

r_a = 2 ft

Applying Eq. (12.6):

$$\bar{k} = \frac{\ln \frac{700}{0.33}}{\frac{1}{0.5}\ln\frac{700}{2} + \frac{1}{.005}\ln\frac{2}{0.33}} \cong 0.021 \text{ d} = 21 \text{ md}$$

Hence the steady state productivity of such a well is only 21/500, or approximately 4% of what it could be if $k_a = k_e$.

It must be realized that Eq. (12.6) is based on Darcy's equation and is therefore subject to the principal restriction that steady state conditions prevail. The average permeability $\bar{k}$ may best be evaluated by flow testing of the specific well in question; however, caution must be exercised to insure that essentially steady state conditions exist. Equation (12.7) applies for oil wells:

$$(12.7)\quad \bar{k}_o = \frac{B_o q_o \mu_o \ln (r_e/r_w)}{7.07h(p_e - p_w)}$$

For gas wells, Eq. (12.8) may be used:

$$(12.8)\quad \bar{k}_g = \frac{q_g \mu_g T \ln(r_e/r_w)}{0.704h(p_e^2 - p_w^2)} \quad \text{(for an ideal gas)}$$

where $\bar{k}_o$, $\bar{k}_g$ = effective permeabilities to oil and gas, respectively, darcys

q_o = tank oil flow rate, bbl/day

q_g = gas flow rate, MCF/day at 14.7 psia and 60°F.

B_o = oil formation volume factor, reservoir oil volume/tank oil volume

μ_o, μ_g = oil and gas viscosities at reservoir conditions, cp

r_e, r_w = drainage and well bore radii, any consistent units

ln = base e

h = net producing pay thickness, ft

T = reservoir temperature, °R

The productivity of oil wells is often expressed in terms of productivity index (or PI). This is a useful parameter defined by:

$$(12.9)\quad J = \frac{B_o q_o}{p_e - p_w} = \text{productivity index, bbl reservoir oil/day/psi}$$

The PI is sometimes stated in terms of tank oil flow rate, in which case the factor B_o must be applied before any reservoir flow calculation is made. Throughout our discussion the PI will be defined in terms of reservoir flow rate. Note that the direct comparison of the productivity indices of two wells does not account for pay thickness differences; however, a similar quantity which does correct for this is the specific productivity index J_s where:

$$(12.10)\quad J_s = \frac{J}{h} = \frac{B_o q_o}{h(p_e - p_w)} = \text{bbl/day/psi/ft of pay}$$

Therefore, in terms of the two productivity indices just defined,

$$(12.11)\quad \bar{k}_o = \frac{J \mu_o \ln(r_e/r_w)}{7.07h} = \frac{J_s \mu_o \ln(r_e/r_w)}{7.07}$$

The value of $\bar{k}_{o,g}$ does not in itself furnish sufficient data to indicate whether a well's low productivity is due to low natural permeability $(k_e)_{o,g}$, or to severely damaged, altered zone permeability k_a. Some independent means of determining $(k_e)_{o,g}$ is required in order to compare the two values. One of the first approaches of this type was proposed by Wade[10] and consisted of comparing the computed $\bar{k}_o$ with absolute permeabilities from core analysis, the latter being corrected to effective permeabilities by relative permeability data where applicable. The result of this method is an efficiency factor E defined as:

$$E = \frac{\bar{k}_o}{k_{ro}\bar{k}_c} = \frac{q_o}{q_{oI}} \tag{12.12}$$

where k_{ro} = relative permeability to oil

$\bar{k}_c$ = average absolute (or air) permeability of the section as computed from core analysis

q_{oI} = ideal or theoretical steady state production rate for an undamaged system

From our previous discussion of relative permeability curve shapes, the assumption that $k_{ro} = 1$ for oil reservoirs above the bubble point pressure (no gas saturation) is a reasonable approximation, provided that oil is the non-wetting phase and water saturation is at the irreducible value. Hence this approach affords a practical means of evaluating formation damage in many cases where only routine core analysis data are available.

Example 12.2

Given the following well data, compute the productivity indices and the completion efficiency of the well.

Core Analysis

Depth	*Air Permeability*
6105–10	50 md
6110–15	0 (shale)
6115–20	150
6120–30	100

Fluid properties from PVT analysis:

$B_o = 1.20$

$\mu_o = 3.0$ cp

Bubble point pressure = 2000 psia

Steady state flow test data:

Static (shut-in) reservoir pressure = 3000 psia

Bottom hole producing pressure = 2500 psia

Tank oil producing rate = 100 bbl/day

$\ln(r_e/r_w) = 7.0$

Solution:

The productivity indices are readily calculated from Eqs. (12.9) and (12.10):

(a) $J = \dfrac{(100)(1.20)}{3000 - 2500} = 0.24$ bbl/day/psi

and

(b) $J_s = \dfrac{0.24}{20} = 0.012$ bbl/day/psi/ft

The average effective permeability to oil is obtained from Eq. (12.11):

(c) $\bar{k}_o = \dfrac{(0.012)(3.0)(7.0)}{7.07} = 0.036$ d $= 36$ md

The average air permeability from core analysis is:

(d) $\bar{k}_c = \dfrac{(5)(50) + (10)(100) + (5)(150)}{20} = 100$ md

Since the reservoir pressure is well above the bubble point, we may assume in lieu of other data, $k_{ro} = 1$, hence:

(e) $E = \dfrac{36}{100} = 36\%$, i.e., the well is producing only 36% of its theoretical, undamaged rate.

While this method of evaluating formation damage has largely been replaced by the pressure build-up method, it is still a valuable tool in many instances.

12.32 Pressure Build-up Method

Thus far, we have limited our discussion of flow problems to steady state conditions. Certainly, this is an approximation in almost every actual case, since steady state requires that pressure at all points in the system remain constant with time; for the radial case, flow into the system across the external boundary r_e must equal the amount removed at the well boundary r_w. Also, we assumed that the liquids were completely incompressible, which is again an approximation. In our forthcoming, rather limited, treatment of pressure build-up it will be necessary to note that liquid compressibilities, while small, must be reckoned with and that unsteady state flow is actually the normal reservoir situation. This does not, however, invalidate our previous discussions, since the errors involved in the steady state assumption are quite negligible in many applications.

Unsteady state flow is mathematically quite complex; it is a principal topic of various reservoir engineering texts as well as numerous other technical publications. Consequently, we will only introduce the general subject as it specifically deals with the pressure build-up method of evaluating formation damage. Since all unsteady state treatments begin with some form of the so-called diffusivity equation, we will begin with its development. We will restrict our discussion to homogeneous liquid flow and assume that the liquid density may be expressed as:

$$\rho = \rho_i e^{-c(p_i - p)} \tag{12.13}$$

where ρ = density of reservoir fluid at pressure p

ρ_i = density of reservoir fluid at initial reservoir pressure p_i

e = natural logarithm base

c = compressibility of reservoir liquid, volume per volume per psi, or psi^{-1}. (In some instances of heterogeneous flow, c is given a gross value combining gas, oil, and water compressibilities and solubilities.)

Considering next Figure 12.7, which depicts a plane radial system, we develop the basic continuity equation which expresses the law of conservation of matter.[11] Since radial symmetry is assumed, the use of plane polar coordinates is expedient. Visualize Figure 12.7 as a homogeneous cylinder of a porous medium h cm thick,

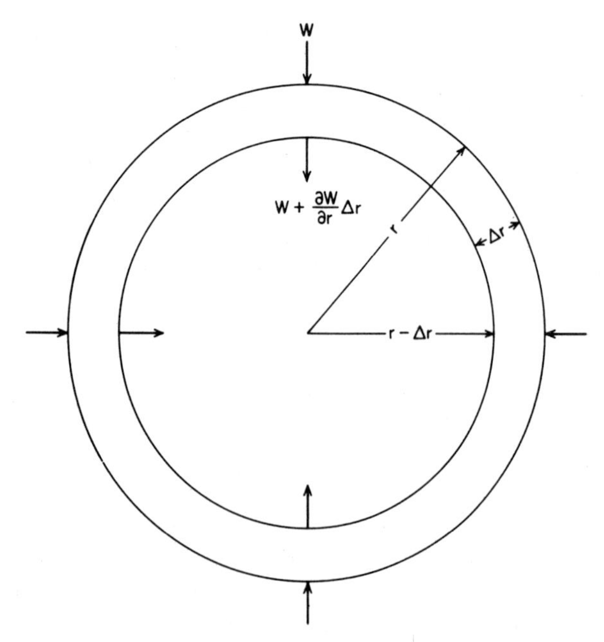

Fig. 12.7. Conditions of continuity equation for radial flow.

having a porosity ϕ. At some distance r from the center, the inflow rate is W gm/sec, while at distance $(r - \Delta r)$ the mass inflow rate is $W + (\partial W/\partial r)\Delta r$. The difference between these, $-(\partial W/\partial r)\Delta r$, is then the rate of accumulation of matter within the ring of width Δr. The rate of density change within this ring is the change in mass ÷ the total pore volume of the element, or:

$$\frac{\partial \rho}{\partial t} = -\frac{\partial W}{\partial r}\Delta r \div (2\pi r h \phi \, \Delta r) = -\frac{\partial W}{\partial r}\left(\frac{1}{2\pi r h \phi}\right) \tag{12.14}$$

Equation (12.14) is a form of the continuity equation for plane radial flow in porous media. Expressing the flow rate W in terms of Darcy's law:

$$W = q\rho = -\frac{2\pi r h k \rho}{\mu}\frac{\partial p}{\partial r} \tag{12.15}$$

or

$$\frac{\partial W}{\partial r} = -\frac{2\pi h k \rho}{\mu}\left(\frac{\partial p}{\partial r} + r\frac{\partial^2 p}{\partial r^2}\right) \tag{12.16}$$

Substitution of Eq. (12.16) in Eq. (12.14) results in:

$$\frac{\partial^2 p}{\partial r^2} + \frac{1}{r}\frac{\partial p}{\partial r} = \frac{\phi\mu}{k\rho}\frac{\partial \rho}{\partial t} \tag{12.17}$$

From Eq. (12.13),

$$\frac{\partial \rho}{\partial t} = c\rho_i e^{-c(p_i - p)}\frac{\partial p}{\partial t}$$

or

$$\frac{\partial \rho}{\partial t} = c\rho\frac{\partial p}{\partial t} \tag{12.18}$$

Substitution for $\partial\rho/\partial t$ in Eq. (12.17) gives

$$\frac{\partial^2 p}{\partial r^2} + \frac{1}{r}\frac{\partial p}{\partial r} = \frac{\phi c \mu}{k}\frac{\partial p}{\partial t} \tag{12.19}$$

Equation (12.19) is generally known as the diffusivity equation for radial flow. Its solution may be either simple or complex, depending on the boundary conditions assumed. For example, let us assume steady state conditions, or $\partial p/\partial t = 0$. Note that this automatically eliminates the right-hand side of Eq. (12.19). We could just as well have assumed an incompressible liquid, $c = 0$, in which case the same result would have been obtained. Hence, these statements amount to the same thing. Integration of Eq. (12.19) for $c = 0$ or $\partial p/\partial t = 0$, gives the standard Darcy law pressure distribution.

Pressure build-up curves are representations of the bottom hole pressure vs. time behavior of a well immediately, and for some time after shut-in. Analytical description of such curves has been developed from solutions of the diffusivity equation for appropriate boundary conditions. Among those concerned with this general topic have been Muskat,[12] Miller, Dyes, and Hutchinson,[13] Horner,[14] Van Everdingen,[15] and Hurst,[16] Other authors, including Thomas,[17] Arps,[18] and Gladfelter, Wilsey, and Tracy[19] have been instrumental in extending the theory and providing practical means of applying the method to evaluation of formation damage. For a summary of this general topic the reader is further referred to Perrine's comprehensive treatment.[20]

Horner derived a simplified expression for the bottom hole pressure of a newly completed oil well after shut-in:

$$p_{ws} = p_e - \frac{0.163\, q_o \mu_o B_o}{k_o h}\log\frac{t_o + \Delta t}{\Delta t} = p_e - \frac{0.0707\, q_o \mu_o B_o}{k_o h}\ln\frac{t_o + \Delta t}{\Delta t} \tag{12.20}$$

where p_{ws} = bottom hole pressure at any shut-in time, $t_o + \Delta t$, psi
p_e = static bottom hole pressure
t_o = total producing life of well
Δt = shut-in time, same units as t_o (usually hours)
log = base 10
ln = base e
q_o = stabilized tank oil production rate prior to shut-in, bbl/day

For older wells which have produced at various rates during their life, a corrected time t_c is often substituted for t_o.

$$t_c = \frac{N_p}{q_o} \tag{12.21}$$

where N_p = cumulative oil production from the well
q_o = last stabilized oil flow rate.

Equation (12.20) is based on a number of assumptions, some of which are:
(1) The well is considered as a point sink producing

from the center of an infinite reservoir having a constant p_e.

(2) The flowing fluid is homogeneous, of essentially constant viscosity and compressibility over the pressure and temperature range of interest.

(3) The well is shut-in at the sand face and no after production enters the borehole after shut-in.

(4) Formation permeability is homogeneous in the direction of flow.

Despite these and other simplifications, the method has been shown to work amazingly well, even for multiphase and gas flow. The most serious restriction seems to be assumption (3), since wells are normally shut-in at the surface and some *after production* (fluid entry into the well after shut-in) always occurs. This effect greatly distorts the early portion of many buildup curves requiring either long shut-in times or further manipulation of early data. We will not be concerned with these problems; however, reference 19 presents a useful means of eliminating the after production effect. Assuming, then, a sufficiently long shut-in period during which the linear behavior has developed, we readily see that the slope of the straight line section m is defined as

$$m = -\frac{0.0707 q_o \mu_o B_o}{k_o h} \tag{12.22}$$

Rearrangement allows calculation of k_o:

$$k_o = -\frac{0.0707 q_o \mu_o B_o}{mh} \tag{12.23}$$

This value of k_o represents the average oil permeability of the reservoir beyond the altered zone. The expression is valid, since the damaged zone radius r_a is quite small compared to r_e. Therefore, any transient effects caused by the damaged zone are of short duration and do not affect the shape of the build-up curve after an appreciable Δt.[15,16]

Application of Eq. (12.23) requires a knowledge of fluid properties μ_o and B_o at reservoir conditions. If these are not known from PVT analysis, they may often be satisfactorily estimated from correlations; however, direct computation of k_o is not necessarily required for determining formation damage. Instead, the ratio of the actual average permeability $\bar{k}_o$ to the unaltered value k_o may be computed from Eqs. (12.7) and (12.23) as:

$$PR = \frac{\bar{k}_o}{k_o} = -\frac{2m\ln(r_e/r_w)}{p_e - p_w} \tag{12.24}$$

where PR = productivity ratio

The PR is then a measure of permeability alteration around the well bore; in this respect, it is similar to the efficiency factor E presented earlier. If $\bar{k}_o = k_o$, no alteration exists and the formation is in the virgin state; $PR = 1$. If $\bar{k}_o < k_o$, the zone has been damaged and is not producing up to its natural capacity. Similarly, if $\bar{k}_o > k_o$, or $PR > 1$, the permeability around the well has been improved. Normally, the latter situation arises only after stimulation such as acidizing or hydraulic fracturing.

The following example will serve to illustrate these calculations.[21]

Example 12.3

Given the following data on a newly completed well, compute its productivity ratio.

Figure 12.8 is the prescribed plot of p_{ws} vs. log $(t_o + \Delta t)/\Delta t$. The last five points exhibit linearity and allow extrapolation to $(t_o + \Delta t)/\Delta t = 1$, or log $(t_o + \Delta t)/\Delta t = 0$, which is the value when Δt becomes infinite. The pressure intercept may be taken as the final shut-in or boundary pressure p_e. This procedure of obtaining p_e may not always be satisfactory, and Horner's paper should be consulted for corrections when the infinite reservoir assumption is invalid. The latter is a valid premise only if the fluid withdrawals from the well have been quite small with respect to the total content of its drainage

Build-up Data

Shut-in time, Δt hrs	Shut-in pressure, p_{ws}	$(t_o + \Delta t)/\Delta t$
0.5	1480	1441
1	1645	721
2	1750	361
4	1970	181
6	2020	121
8	2050	91
10	2070	73
12	2090	61
14	2105	52.3
16	2115	46
20	2120	37
24	2125	31
30	2129	25
36	2134	21

Flow Test Data

q_o = 120 bbl/day
h = 20 ft
p_{wf} = 1250 psi (flowing bottom hole pressure)
r_e = 700 ft (taken as half the distance to adjacent wells)
r_w = 0.33 ft

Fluid Properties

μ_o = 5.0 cp
B_o = 1.15
N_p = 3600 bbl, cumulative production
t_o = 30 days (720 hr) producing life

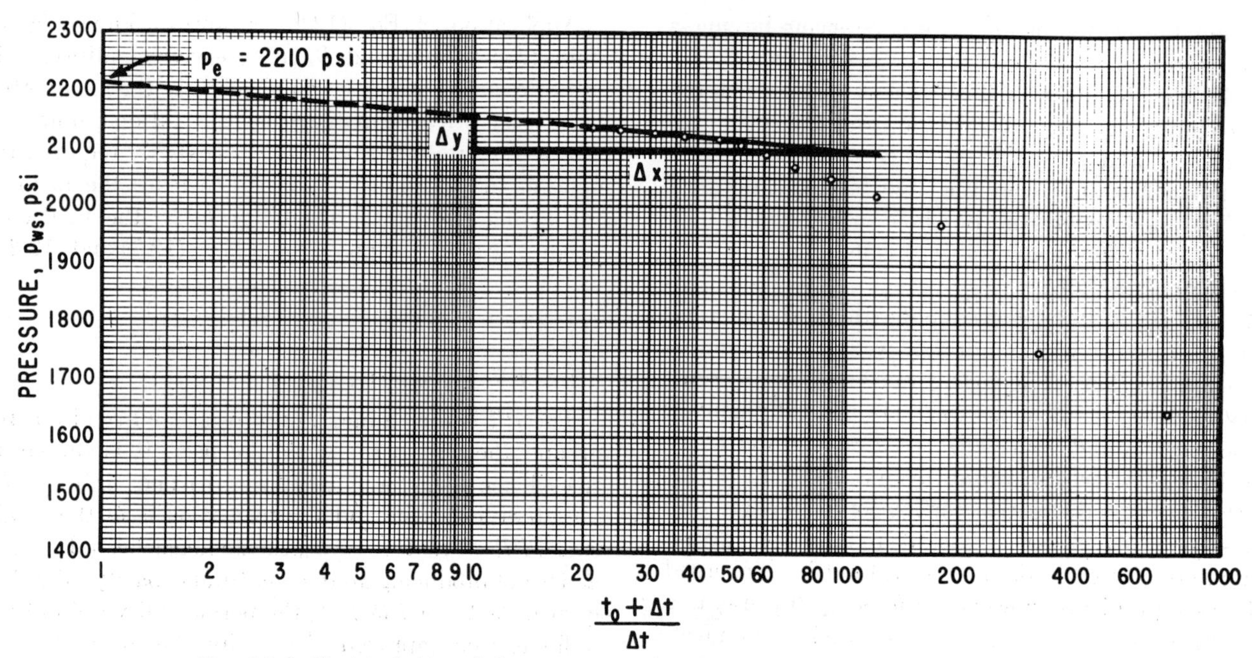

Fig. 12.8. Shut-in bottom hole pressure vs. log $(t_o + \Delta t)/\Delta t$. Data of Example 12.3.

volume. Since we will be considering only completion or precompletion problems, this technique will be used, the extrapolated value of p_e being taken as the static reservoir pressure at the time of the well's completion. This is, of course, the initial pressure as far as the well under study is concerned.

The slope m is conveniently calculated over an abscissa increment of one cycle, as indicated. (For a discussion on taking slopes from various types of plots, see Appendix B.)

$$m = \frac{\Delta y}{\Delta x} = \frac{2153 - 2092}{2.3(\log 10 - \log 100)} = \frac{61}{2.3(1 - 2)} = -26.5$$

Applying Eq. (12.24) next:

$$PR = \frac{(-2)(-26.5) \ln (700/0.33)}{2210 - 1250} = 0.42$$

Since 0.42 < 1, a considerable damaged zone exists. For further illustration, $\bar{k}_o$ and k_o could have been computed from Eqs. (12.7) and (12.23):

$$\bar{k}_o = \frac{(1.15)(120)(5)(7.6)}{(7.07)(20)(960)} = 0.039\ d \text{ or } 39 \text{ md}$$

$$k_o = -\frac{(0.0707)(120)(5)(1.15)}{(-26.5)(20)} = 0.092\ d \text{ or } 92 \text{ md}$$

and

$$PR = \frac{39}{92} = 0.42, \text{ as before}$$

To avoid confusion, natural logarithms have been used throughout; however, if use of the factor 2.3 in computing the slope m is confusing, it may be omitted, provided that the 7.07 in Eq. (12.7) is changed to 3.07 and $\ln(r_e/r_w)$ is changed to $\log (r_e/r_w)$.

Although the pressure build-up theory was advanced for liquid saturated systems (low compressibility) it has been successfully applied to gas wells through use of Eq. (12.25):[22]

$$(12.25) \quad PR = \frac{4mp\ln(r_e/r_w)}{p_e^2 - p_w^2} \quad \text{(for gas wells)}$$

where p = average reservoir pressure at the time of the test.

In the Van Everdingen[15] and Hurst[16] studies, the altered zone was treated as a very thin veneer or skin which was considered to exert a *skin effect* on well productivity. Thus the skin effect is defined as the additional pressure drop between r_e and r_w caused by the presence of the skin, and is normally expressed in dimensionless units. A positive skin effect denotes formation damage, a negative effect denotes improvement, and zero effect indicates no alteration. The skin effect depends (as does the PR) on both the permeability k_a and extent r_a of altered zone. As shown by Hawkins:[23]

$$(12.26) \quad S = \left(\frac{k_e}{k_a} - 1\right) \ln \frac{r_a}{r_w}$$

where the subscripts refer to the general condition of Figure 12.6, and S = skin effect. Figure 12.9 shows the general relationship between these factors. Consequently, it is apparent that some relationship between S and PR exists. This was also given by Hawkins as:

$$(12.27) \quad S = \frac{\ln(r_e/r_w)}{PR} - \ln(r_e/r_w)$$

Hence, for the well of Example 12.3:

$$S = \frac{7.6}{0.42} - 7.6 = 10.5$$

This could then be due to any combination of r_a and k_e/k_a values giving S = 10.5 on Figure 12.9. While

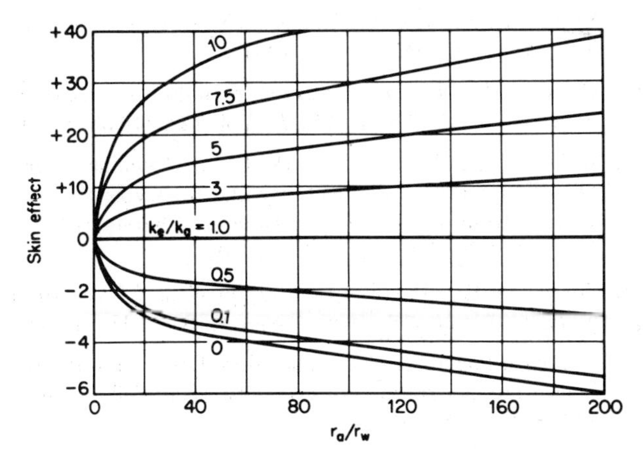

Fig. 12.9. Magnitude of the skin effect as a function of altered zone permeability, k_a, and radius, r_a. After Hawkins,[23] courtesy AIME.

we will not place particular emphasis on the skin effect as such, it is another widely used parameter in the analysis of formation damage. In our later work the *PR* approach will be utilized.

The main purpose of this chapter has been to point out the factors attributing to formation damage, and to cite current methods for diagnosing the extent of such damage. Bulletin D-6 of the API[22] contains a large amount of statistical data concerning typical *PR* values for newly completed wells. We will refer to this later in our discussion of well completion practices.

PROBLEMS

1. Given the following water loss data, analyze the filtration behavior of the mud by both the standard API and the Slusser, Glenn, and Huitt methods. Make the necessary plots and determine V_c, V_s, V_x. Compare this mud with that of Figures 12.3 and 12.4, discussing any differences.

V, cc	*t, sec*
1.3	1
2.4	4
4.2	16
4.9	25
6.3	64
7.1	100
9.1	225
11.1	400
13.1	625

2. Suppose that two productive zones having porosities of 15 and 20%, respectively, were drilled with the same water base mud. Assume equal exposure time. Considering only the stage 3 filtration behavior, in which zone would you expect the deeper filtrate invasion (greater r_a)? Why? Resolve this observation into a general statement, clearly indicating any limitations.

3. A well is completed on 40 acre spacing. Given that $k_e = 200$ md, $r_w = 4$ in., $r_a = 1$ ft, $k_a = 10$ md, compute the well's *PR*.

4. Show that, under certain conditions, Wade's efficiency factor E may be expressed as:

$$E = \frac{J\mu_o}{k_c h}$$

List the assumptions involved. Do they check with those in Example 12.2?

5. Given the following data on a newly completed well, compute the completion efficiency E.

Producing rate = 200 bbl/day (tank oil)
p_e = 3000 psig
p_w = 2800 psig
$\bar{k}_c$ = 1 darcy
h = 100 ft
k_{ro} = 0.25
B_o = 1.25
μ_o = 4 cp
Spacing = 40 acres

6. Given the following pressure build-up data,[14] estimate the formation permeability to oil, k_o, and the static pressure p_e.

Cumulative tank oil production before shut-in = 5847 bbl

Last stabilized producing rate = 641 bbl/day

Pay section thickness h = 349 ft

PVT data: μ_o = 40 cp, B_o = 1.0

Build-up Data

Δt, *hr*	p_{ws}, *psig*
19	1192
25	1200
31	1206
37	1212
43	1216
49	1220
55	1223
61	1227

Ans. p_e = 1280 psig; k_o = 156 md.

7. Given the following data on a newly completed well, determine:

(a) Undamaged oil permeability k_o
(b) Average oil permeability $\bar{k}_o$
(c) *PR* of the well
(d) Static formation pressure, p_e. *Ans.* (d) $p_e \cong 2445$ psig
(e) What q_o could be obtained if the *PR* of this well were increased to 2?

Build-up Data

Δt, *hrs*	p_{ws}, *psig*
1.0	1830
2.0	2005
3.0	2130
4.2	2200
8.5	2265
32.0	2315
73.2	2345

Other Data

Spacing = 40 acres
B_o = 1.10
μ_o = 5.0 cp
h = 10 ft
N_p = 2600 bbl
r_w = 4 in.

Flow data Prior to Shut-in:

q_o = 65 bbl/d (stabilized rate over life of well)
p_w = 400 psig (producing bottom hole pressure)

REFERENCES

1. Nowak, T. J., R. F. Krueger, "The Effect of Mud Filtrate and Mud Particles upon the Permeabilities of Cores," *API Drilling and Production Practices*, 1951, p. 164.

2. Bertness, T. A., "Observations of Water Damage to Oil Productivity," *API Drilling and Production Practices*, 1953, p. 287.

3. Beeson, C. M., and Charles C. Wright, "Mud Loss to Formation Pores," AIME Tech. Paper No. 79-G, presented Los Angeles, Oct. 1950.

4. Krueger, R. F., and L. C. Vogel, "Damage to Sandstone Cores by Particles from Drilling Muds," *API Drilling and Production Practices*, 1954, p. 158.

5. Glenn, E. E., and M. L. Slusser, "Factors Affecting Well Productivity: II. Drilling Fluid Particle Invasion into Porous Media," *Journal of Petroleum Technology*, May, 1957, p. 132.

6. Ferguson, C. K., and J. A. Klotz, "Filtration from Mud During Drilling," *Trans. AIME*, Vol. 201, (1954), p. 29.

7. Slusser, M. L., E. E. Glenn, and J. L. Huitt, "Factors Affecting Well Productivity: I. Drilling Fluid Filtration," *Journal of Petroleum Technology*, May, 1957, p. 126.

8. Gates, G. L., and C. P. Bowie, "Correlation of Certain Properties of Oil-Well Drilling-Mud Fluids with Particle-Size Distribution," U. S. Bureau of Mines R. I. 3645, May 1942.

9. Schremp, F. W., and V. L. Johnson, "Drilling Fluid Filter Loss at High Temperatures and Pressures," *Trans. AIME*, Vol. 195, (1952), p. 157.

10. Wade, F. R., "The Evaluation of Completion Practice from Productivity Index and Permeability Data," *API Drilling and Production Practices*, 1947, p. 186.

11. Nielsen, R. F., "How to Calculate Unsteady State Flow," *Oil and Gas Journal*, July 26, 1954.

12. Muskat, M., "Use of Data on the Build-up of Bottom Hole Pressures," *Trans. AIME*, Vol. 123, (1937), p. 44.

13. Miller, C. C., A. B. Dyes, and C. A. Hutchinson, Jr., "The Estimation of Permeability and Reservoir Pressure from Bottom Hole Pressure Buildup Characteristics," *Trans. AIME*, Vol. 189, (1950), p. 91.

14. Horner, D. R., "Pressure Build-up in Wells," *Proceedings of Third World Petroleum Congress*, Sect. II, p. 503.

15. Van Everdingen, A. F., "The Skin Effect and Its Influence on the Productive Capacity of a Well," *Trans. AIME*, Vol. 198, (1953), p. 171.

16. Hurst, W., "Establishment of the Skin Effect and Its Impediment to Fluid Flow into a Well Bore," *The Petroleum Engineer*, Oct. 1953, p. B–6.

17. Thomas, G. B., "Analysis of Pressure Build-up Data," *Trans. AIME*, Vol. 198, (1953), p. 125.

18. Arps, J. J., "How Well Completion Damage Can be Determined Graphically," *World Oil*, Apr. 1955, p. 225.

19. Gladfelter, R. E., G. W. Tracy, and L. E. Wilsey, "Selection of Wells Which Will Respond to Production-stimulation Treatment," *API Drilling and Production Practices*, 1955, p. 117.

20. Perrine, R. L., "Analysis of Pressure Build-up Curves," *API Drilling and Production Practices*, 1956, p. 482.

21. Gatlin, C., "Formation Damage," *The Petroleum Engineer*, Nov. 1957, p. B–102.

22. *Selection and Evaluation of Well-Completion Methods*, API Bulletin D-6, July 1955. Also published in *API Drilling and Production Practices*, 1955, p. 421.

23. Hawkins, M. F., Jr., "A Note on the Skin Effect," *Journal of Petroleum Technology*, Dec. 1956, p. 65.

Chapter 13 . . .

Drill Stem Testing

Up to this point, we have discussed two methods of evaluating formation productivity, core analysis and well logging. In this chapter a third technique, namely, formation or drill stem testing will be presented. Despite the tremendous value of core analysis and logging, some shadow of doubt always remains concerning the potential productivity of an exploratory well, and this doubt is not dispelled until a sizeable sample of oil has been delivered to the surface. This drawback is not inherent in drill stem testing.

The decision to run a drill stem test on a zone is often based on shows of oil in the cuttings which, in the opinion of the geologist or engineer in charge, deserve detailed investigation. This may happen many times in the course of drilling a wildcat, with as many as 20 or 30 tests being conducted on a single well. Although the cost of such detailed testing is quite high, it is much better to test and be sure, rather than miss a productive zone.

13.1 General Procedure

A drill stem test is a temporary completion whereby the desired section of the open hole is isolated, relieved of the mud column pressure, and allowed to produce through the drill pipe (drill stem).

The basic test tool assembly consists of:

(1) a rubber packing element or packer which can be expanded against the hole to segregate the annular sections above and below the element

(2) a tester valve to (a) control flow into the drill pipe, that is, to exclude mud during entry into the hole and to allow formation fluids to enter during the test, and an equalizing or by-pass valve to (b) allow pressure equalization across the packer(s) after completion of the flow test.

Figure 13.1 illustrates the procedure for testing the bottom section of a hole. While going in the hole, the packer is collapsed, allowing the displaced mud to rise as shown by the arrows. After the pipe reaches bottom and the necessary surface preparations have been made, the packer is set (compressed and expanded); this isolates the lower zone from the rest of the open hole. The compressive load is furnished by a slacking off of the desired amount of drill string weight, which is transferred to the anchor pipe below the packer.

The tester valve is then opened and thus the isolated section is exposed to the low pressure inside the empty, or nearly empty, drill pipe. Formation fluids can then enter the pipe, as shown in the second picture. At the end of the test, the tester valve is closed, trapping any fluid above it, and the by-pass valve is opened to equalize the pressure across the packer. Finally, the setting weight is taken off and the packer is pulled free. The pipe is then pulled from the hole until the fluid-containing section reaches the surface. As each successive stand is then broken (unscrewed), its fluid content may be examined. Frequently such stand-by-stand sampling is neither necessary nor desirable and the test is reversed (circulated opposite to the normal direction) as shown. This reversal is performed by closing the blowout-preventers and pumping mud down the annulus; the mud then enters the drill pipe through the reversing ports, thereby displacing any formation fluids in the pipe. The recovered fluids may be sampled as they are discharged at the surface.

Although the above is a very common type of test, there are many other variations of procedure, as indicated in Figure 13.2. The straddle packer test is necessary when isolation from formations both above and below the test zone is necessary. Such a situation commonly arises when evidence (electric logs, radioactiv-

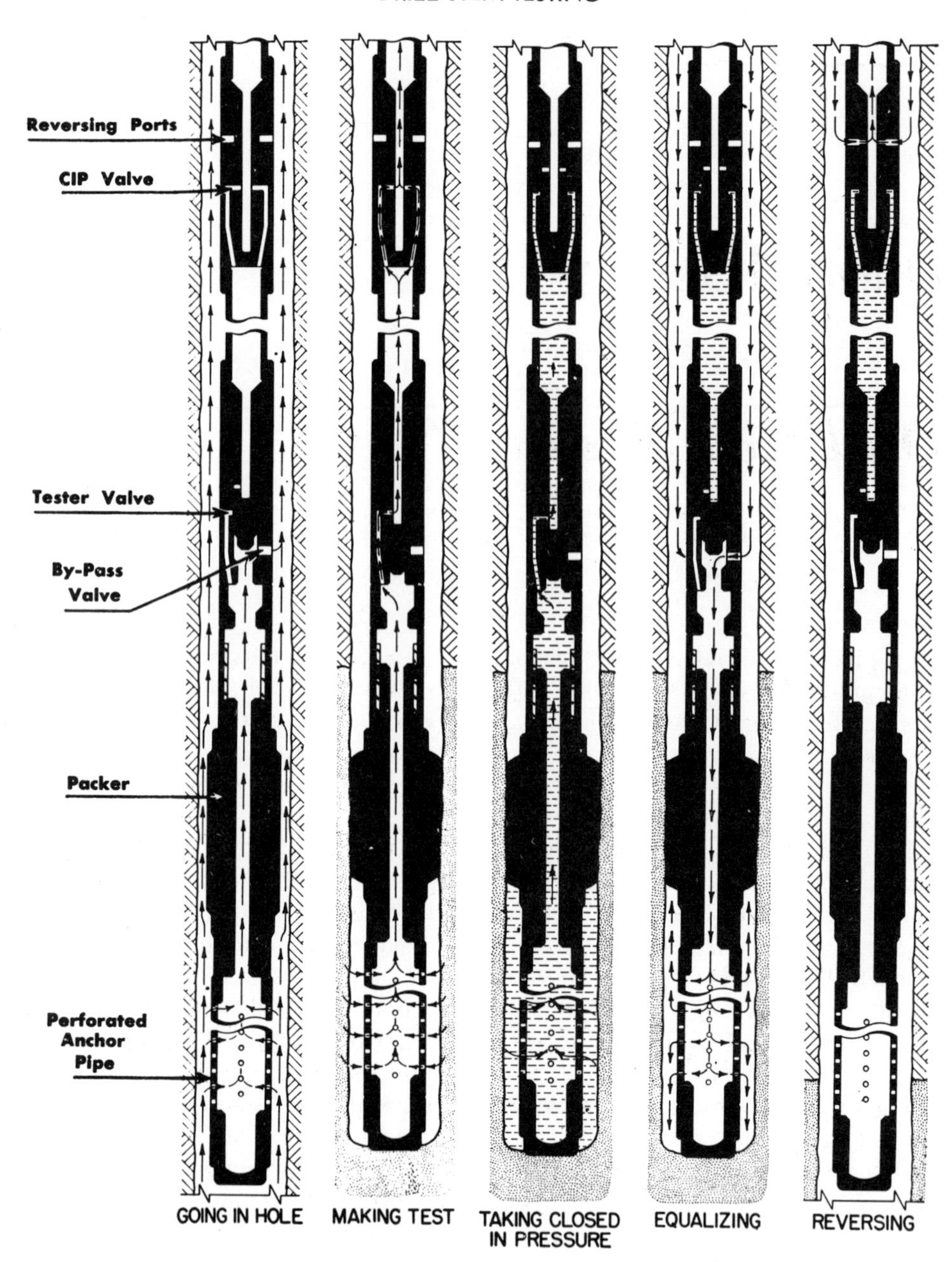

Fig. 13.1. Fluid passage diagram for a conventional bottom section, drill stem test. Courtesy Halliburton Oil Well Cementing Company.

ity logs, detailed sample analysis, etc.) indicates that a zone previously passed by has productive possibilities. Straddle testing is less desirable than conventional testing, from both a cost and an operational hazard standpoint. Two packers are more apt to become stuck than one, since any material which sloughs or caves from the test zone may accumulate between the packers. Also, two positive, pressure-tight packer–formation seals are required for a successful test. Consequently this procedure is not preferred, and is applied only when necessary. This should not be construed to mean that these disadvantages prevent one from making such tests but rather that the additional problems the tests entail should be recognized.

The cone packer or *rat-hole* method is used when the test section is smaller in diameter than the hole above. This situation commonly occurs as a consequence of coring operations in which the corehead used was smaller than the regular bits. The cone-shaped packer is compressed against the shoulder, forming the necessary seal. Note that the anchor pipe does not touch bottom. If the formation opposite the shoulder is soft, an additional conventional wall packer placed above the cone packer may be necessary to provide the desired seal.

Zones behind casing may be tested through perforations by the same basic procedures except that the packer used has slips which engage or grab the casing

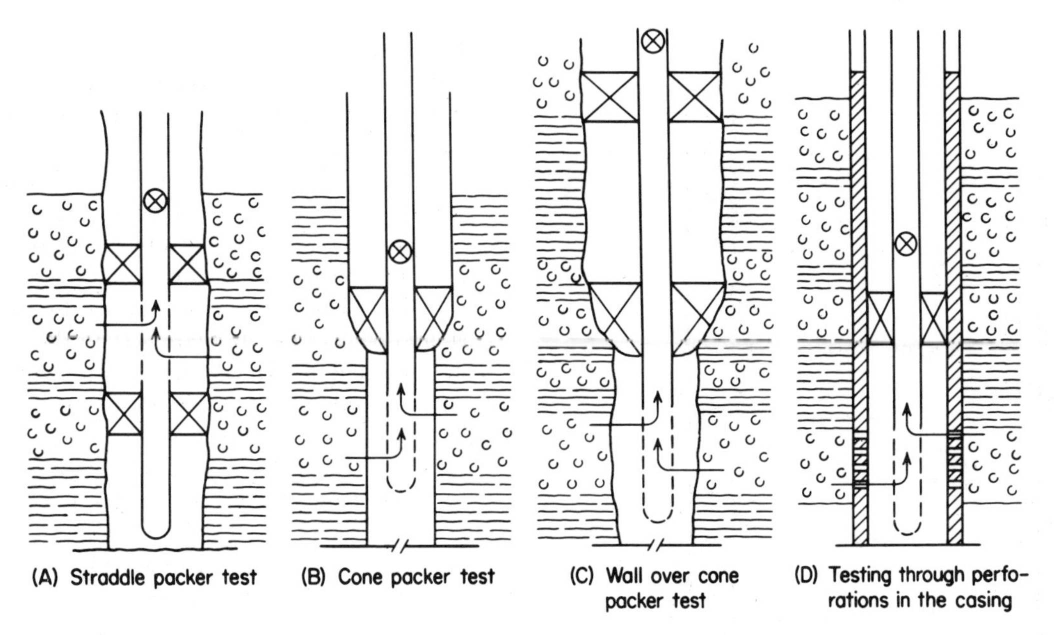

Fig. 13.2. Schematic illustration of various drill stem test conditions. After Kirkpatrick,[1] courtesy *Petroleum Engineer*.

wall. These are commonly called hook-wall packers. Since the slips support the compressive load used to expand the packer element, no anchor pipe is required, and the packer may be reset several times at different depths if necessary. Testing inside casing is widely used in soft rock areas where open hole testing is particularly hazardous. Such testing departs somewhat from our original definition but may be included in this general category. Drill stem testing usually refers to open hole operations.

13.2 General Considerations

Proper planning and consideration of the factors involved are essential to successful testing. The following points are of particular importance.

1. Condition of the hole: The close tolerance between the hole and the tool assembly requires a full gage, clean, well bore if the tool is to reach bottom in an undamaged, unplugged condition. Wall cake and cavings shoved ahead of the packer may plug the perforations and/or choke when the valve is opened. It is common practice to circulate for some time prior to testing, so that all cuttings are removed from the hole. The drilling mud should be conditioned to the desired density and viscosity before the test is started.
2. Pressure surges: the pressure effects of pipe movement were discussed previously in terms of running pressure and pulling suction. The drill stem test conditions represent a severe case of pressure surge, because the lower end of the pipe is closed, necessitating displacement of the total drill pipe volume. Special consideration should be given to pipe running and pulling speeds to avoid undue bottom hole pressure variations.
3. Operating conditions: (a) The length and location of test section govern the amount of tail pipe required and the choice of a conventional or straddle test. The testing of short sections is more conclusive, and is generally preferable. Also, the volume of drilling fluid below the packer should not fill the pipe to such an extent that its back pressure interferes with the test. If a long anchor is necessary, it should be made up of drill collars to avoid excessive bending and permanent kinking, which may occur if drill pipe is used.

   (b) The packer seat location, while of no particular importance for tests run inside casing, is critical for a successful open hole test. The seat should be placed in a true gage section of the hole opposite as dense and consolidated a formation as possible. Limestone or dolomite, anhydrite, or hard dense shales are all satisfactory. Fractured zones of any type are undesirable and may cause either or both packer failure and fluid by-passing. Packer location is also governed by the precision with which the test zone must be isolated. If no other permeable sections are close by, the packer may be set some distance above the section of interest. As a general rule, electric or caliper logs are not taken prior to open hole testing, and the main guides to packer seat location are the drilling time and sample logs.

   (c) Size and number of packers: The pressure differential which a packer can stand depends on the amount it must expand to furnish the desired seal. The service companies which supply these tools have a standard range of sizes for various hole diameters.

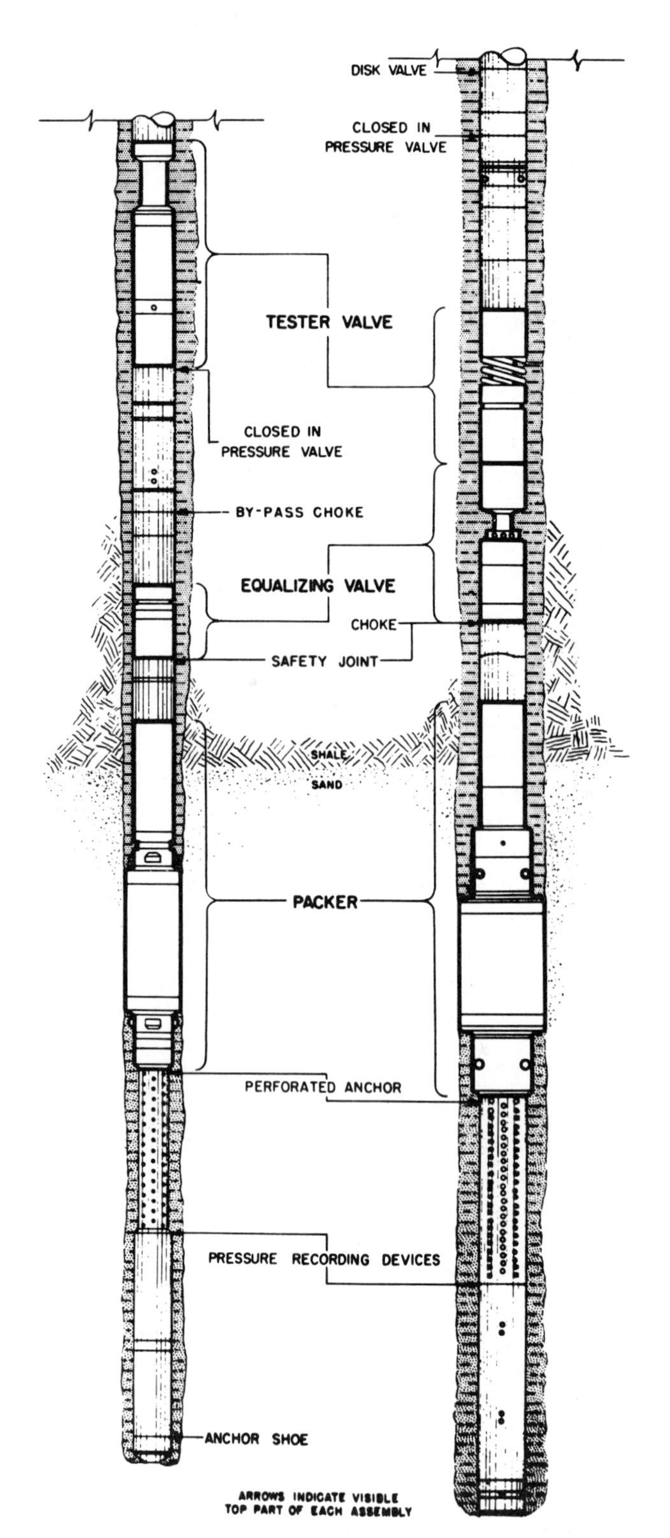

Fig. 13.3. Typical conventional drill-stem test tools (schematic drawing not to scale). After Black,[2] courtesy AIME.

The ratio of hole diameter to unexpanded packer diameter is kept as low as possible, and commonly ranges from 1.1 to 1.2. In deep, open hole tests, two closely spaced packers are often run as a precautionary measure. This often eliminates test failures due to by-passing in fractured zones; also, if one packer fails, the other may carry the load. The cost of the second packing element is quite nominal compared to the cost of a misrun.

(d) Choke sizes: The size of the bottom hole and surface orifices selected depends on the anticipated test conditions. The bottom choke is of prime importance and is used to govern the flow rate. The top choke is used primarily as a safety measure and should be considerably larger than the bottom choke in order to minimize surface pressure in case a flowing test is obtained. It is often omitted in many areas. The producing pressure drop around the borehole depends on the flow rate, which is in turn governed by the bottom choke size. Large chokes allow build-up of a large pressure differential between formation and well bore; this may be undesirable in tests of thin zones underlain by water and/or overlain by gas. Also, wall caving and/or packer failures may result from high pressure differentials during the test period. Bottom chokes of $\frac{3}{16}$ to $\frac{3}{8}$ in. diameter in combination with $\frac{5}{8}$ to 1 in. diameter surface chokes are commonly used in the Gulf Coast area where high pressure, flowing tests are often obtained. In the mid-continent area, larger sizes are employed, the $\frac{3}{4}$ in. choke being the most common. Chokes less than $\frac{3}{16}$ in. in diameter are too easily plugged and are not in general use.

(e) Use of cushions: This refers to the practice of placing a certain length or head of liquid inside the drill pipe, rather than running it dry. This is commonly done for two reasons: (1) To reduce the collapse (external) pressure on the drill pipe in deep holes; and/or (2) to reduce the pressure drop on the formation and across the packer(s) when the tool is first opened. While this is a safe and necessary practice in many cases, it should not be carried to the extreme that back pressure becomes large enough to prevent formation fluid flow. Some types of formation test tools offer a considerable advantage in this respect. These have a slow, staged tool opening which is analogous to slowly cracking or opening a valve in a high pressure pipe line. This greatly reduces the initial shock across the packers and on the formation when the tool is first opened. This type of tester may eliminate the necessity of a water cushion in some instances.

(f) Length of test: This is difficult if not impossible to predict until after the test has commenced and some observations are available. In hard rock areas where the pipe is not in too great danger of sticking, the flow test is often several hours long. If the fluid has not reached the surface within this period of time, it is usually desirable to leave the tool open, as long as appreciable entry into the pipe is taking place. The shut-in period after the flow test should

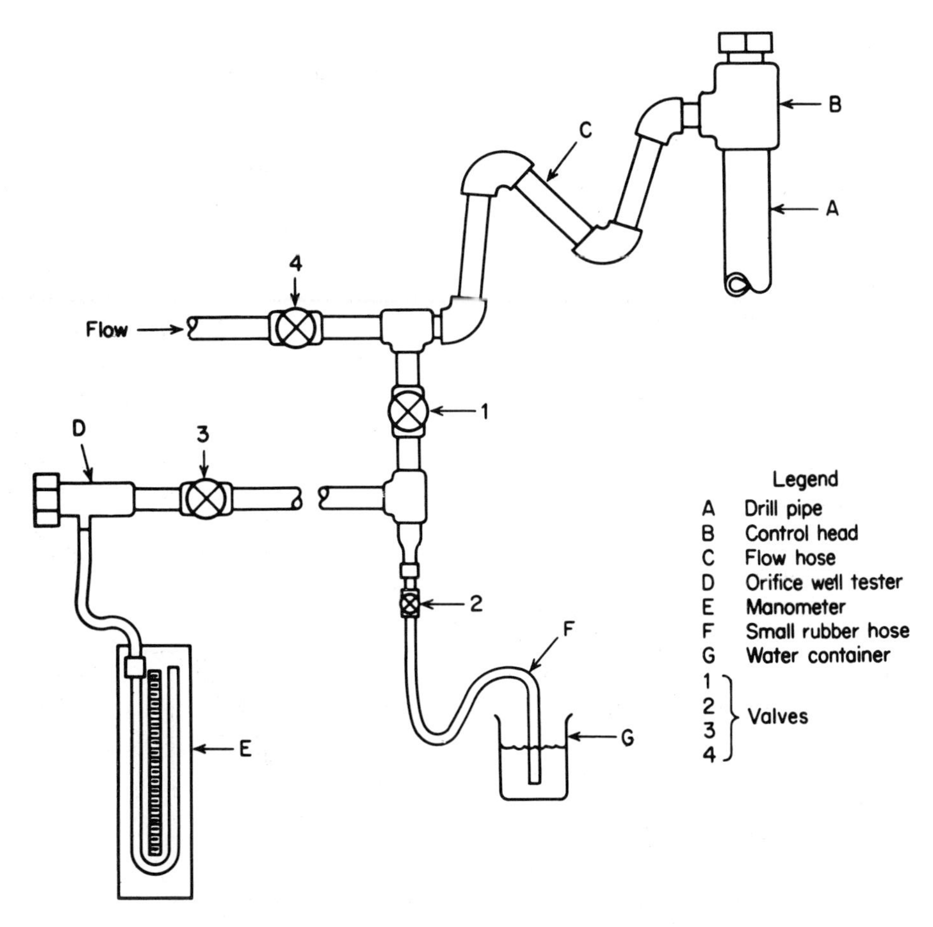

Fig. 13.4. Schematic arrangement of surface equipment for a drill stem test. When tool is opened, the entry of fluid into the pipe displaces air so that a blow is obtained through the hose, F (valves 1, 2 open; valves 3, 4 closed). If the blow is excessive, valves 3 or 4 may be opened. If the produced fluid surfaces, it is diverted through valve 4 if liquid, or metered through D if gas.

be long enough to permit establishment of a stabilized static pressure, if possible. The time required for this depends primarily on formation and damaged zone permeability, fluid properties, the length of the flow period, and the rate of flow. In many cases, it is not practical to obtain the static formation pressure in this manner; however, use of a shut-in pressure valve arrangement permits measurement of the static pressure prior to the flow test. This technique will be discussed later.

(g) Handling of test production at the surface: Safety considerations dictate that some arrangement should always be made for disposing of any produced fluids at some safe distance from the rig. Commonly, a flow line from the rig floor to the reserve mud pit is adequate. If accurate gauging of oil and gas is desired, a test tank and/or separator may be set. The advance knowledge and anticipated results will usually govern the surface preparations made in a specific instance.

13.3 Test Tool Components and Arrangement

The specific selection and arrangement of downhole equipment depends on the conditions and purpose of the particular test. Different mechanical features are available in the tools of different service companies; however, the basic assemblies shown in Figure 13.3 are typical and in common use.[2] Further consideration of the individual components and their functions is in order.

The functions of the packer(s), tester valve, and equalizing valve were previously stated. The functions of other items shown in Figure 13.3 are as follows:

(1) Anchor: This is merely the extension below the tool which supports the weight applied to set the packer. On normal open hole tests, it rests either on the bottom of the hole or on cement plugs which have been spotted at the desired location. However, straddle tests conducted a considerable distance off bottom may employ a specially designed anchoring

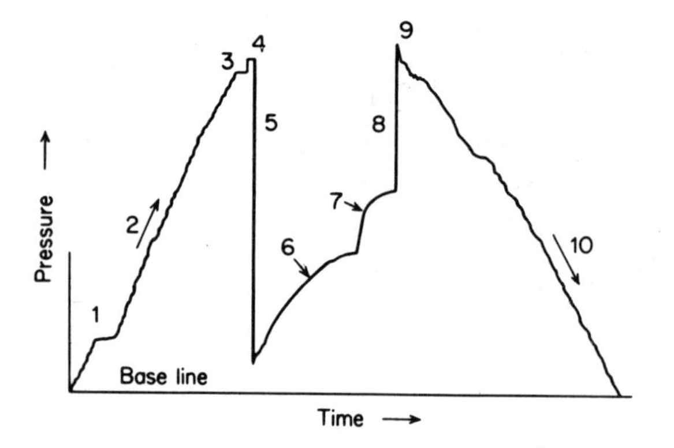

1. Putting water cushion in drill pipe
2. Running in hole
3. Hydrostatic pressure (weight of mud column)
4. Squeeze created by setting packer
5. Opened tester, releasing pressure below packer
6. Flow period, test zone producing into drill pipe
7. Shut in pressure, tester closed immediately above packer
8. Equalizing hydrostatic pressure below packer
9. Released packer
10. Pulling out of hole

Fig. 13.5. Normal sequence of events in successful drill stem test. After Kirkpatrick,[1] courtesy *Petroleum Engineer*.

device employing steel prongs or *dogs* which may be expanded to dig into the formation at the desired depth.

(2) Pressure recorders: These furnish a complete record of all events which may occur during a particular test. This record is in the form of a graph of pressure vs time. These charts are absolutely essential to the accurate interpretation of test results, and every effort should be made to insure that all available data are obtained. Two pressure recorders are usually desirable and should be located so that one will measure the pressure inside, and the other the pressure outside the anchor. These two measurements allow accurate determination of whether or not the perforations have become plugged during the test. This will be discussed further in conjunction with pressure chart analysis.

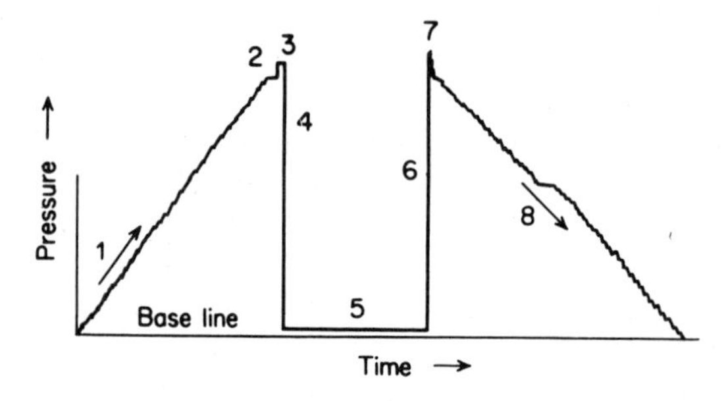

1. Running in hole
2. Hydrostatic pressure (weight of mud column)
3. Squeeze created by setting packer
4. Opened tester, releasing pressure below packer
5. Flow period, test zone open to atmosphere
6. Closed tester and equalizing hyd. pressure below packer
7. Pulled packer loose
8. Pulling out of hole

Fig. 13.6. *Dry* test. After Kirkpatrick,[1] courtesy *Petroleum Engineer*.

(3) Safety joints: These merely afford a means of unscrewing the drill string at a point convenient for fishing operations, should the packers become stuck.

(4) Closed-in pressure valve: This makes possible measurement of the static formation pressure prior to the test. An air volume is trapped between this valve and the disk valve above it. When the closed-in valve is opened, the pressure below it is relieved and the drilling fluid expands, allowing a small volume of formation fluid to enter the tool below the disk valve. Since this volume is quite small, the pressure will quickly rebuild to what is essentially the undisturbed or static formation pressure. The disk valve is opened later to start the flow test.

The surface control head should also be mentioned. This is placed on the top of the drill pipe before the test is started. There is no particularly standard hook-up

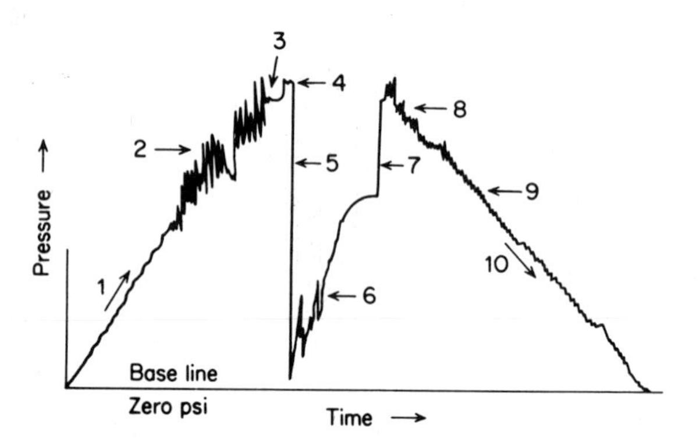

1. Running in hole
2. Indicates bad hole condition, scraping wall cake
3. Hydrostatic pressure (weight of mud column)
4. Squeeze created by setting packer
5. Opened tester, releasing pressure below packer
6. Indicates plugging of perf. anchor or tester
7. Closed tester and equalizing hyd. pressure below packer
8. Indicates swabbing due to bad hole condition
9. Indicates less swabbing, better hole condition
10. Pulling out of hole

Fig. 13.7. Effect of poor hole condition. After Kirkpatrick,[1] courtesy *Petroleum Engineer*.

which is widely used for it; however, all fittings and valves should have sufficient pressure rating to handle any condition which might arise. Figure 13.4 is a schematic sketch of a typical surface control arrangement.

13.4 Qualitative Pressure Chart Analysis

While considerable information is obtained from surface observations during the test, it is not until the tool is retrieved and the pressure charts are examined that a complete analysis is possible. The appearance of these charts is quite variable; nevertheless, a number of typical cases will be illustrated. Figure 13.5 shows the normal sequence of events in a successful flowing test. Actual pressure measurements require a calibration curve (pressure vs stylus deflection) for the particular pressure element used. The base line denotes atmospheric pressure and is drawn on the chart at the surface when it is first inserted into the recorder. Figures 13.6 to 13.9 show other common types of charts.[1]

Figure 13.10 illustrates the method and results of recording the static or shut-in formation pressure before flow testing. After the packer is set, the main tester valve is opened, allowing fluid to enter the air chamber between it and the upper (disk) valve. Since the volume of the air chamber is small, the pressure builds up rapidly, so that little disturbance is imposed on the formation. This allows an accurate measurement of static formation pressure p_e to be quickly obtained, in many cases.

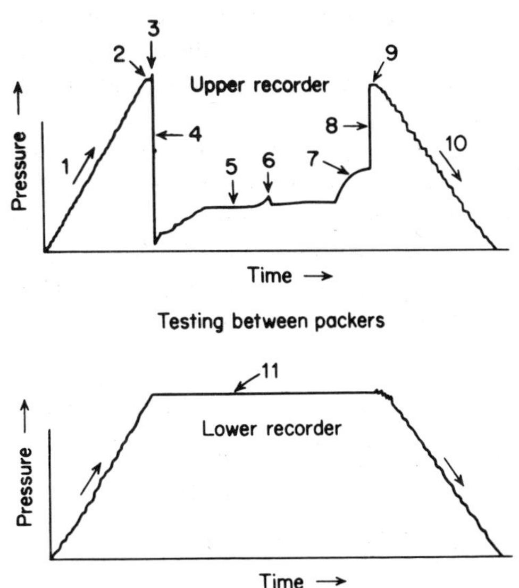

1. Running in hole
2. Hydrostatic pressure (weight of mud column)
3. Set packer
4. Opened tester, releasing pressure between packers
5. Test zone flowing
6. Rise in pressure due to closing in to change chokes
7. Shut in bottom hole pressure
8. Equalizing hydrostatic pressure between packers
9. Hyd. pressure at conclusion of test
10. Pulling out of hole
11. Lower recorder, below bottom packer shows no drop in pressure, proves bottom packer is holding

Fig. 13.8. Straddle packer test. After Kirkpatrick,[1] courtesy *Petroleum Engineer*.

Figure 13.11 illustrates the advantage of using two pressure recorders in a conventional test. The inside recorder measures the pressure inside the anchor—downstream from the perforated anchor but upstream of the bottom hole choke. The outside recorder measures the pressure upstream, or outside the perforated anchor. The time scale shown progresses from right to left. Part A shows that the anchor plugged, since no pressure increase was recorded by the inside gauge and a buildup did occur outside. If only the top chart were available, a *dry test* would be indicated; hence the zone could possibly be overlooked. Part B

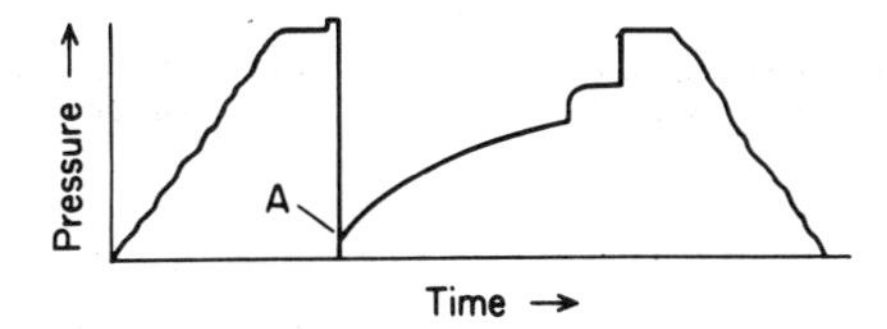

It will be noted from point A on the chart above that the pressure line indicating the opening of the tool, coincides with the start of the build-up in flow pressure. The exact initial flow pressure point cannot be determined. This condition is caused by the successive pressure events happening too rapidly to allow the clock to turn the chart and separate the lines.

Fig. 13.9. Effect of pressure recorder inertia. After Kirkpatrick,[1] courtesy *Petroleum Engineer*.

shows the identical behavior of both charts, which indicates that the anchor did not plug. Note, however, that pressure change due to the recovery of 15 ft of drilling mud is far too little to account for the observed pressure build-up on the charts. Therefore, the choke in the tool itself plugged, and the charts merely recorded the shut-in pressure build-up of the formation. Hence it is apparent that the use of two pressure recorders may be well justified.

Three charts are often used in straddle testing, two to check possible tool or anchor plugging and the third to check the pressure below the lower packer. Sample charts for this type of test are shown in Figure 13.8.

13.5 Analysis of Test Data

The sample charts shown above are typical of various test conditions which may be recognized by qualitative inspection. Certain quantitative estimates are also obtainable from formation tests.

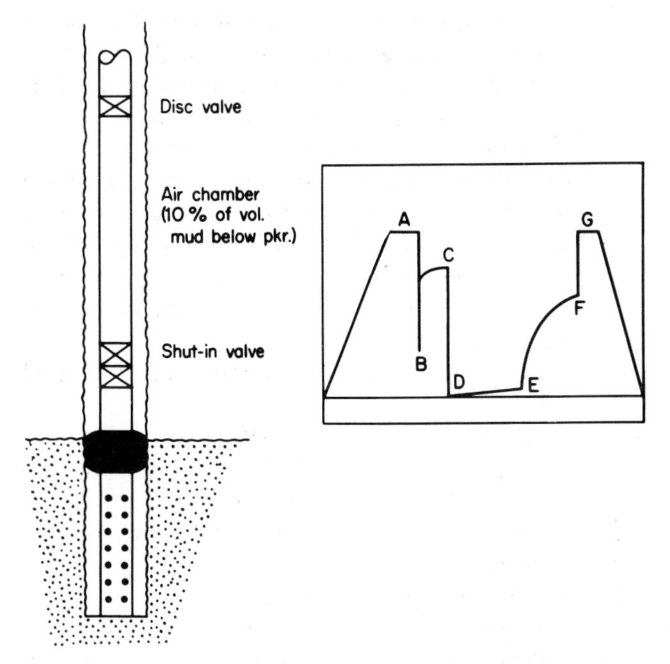

Fig. 13.10. Schematic tool arrangement for procuring initial closed-in pressure. Point C on chart is taken as reservoir pressure. Courtesy Johnson Testers.

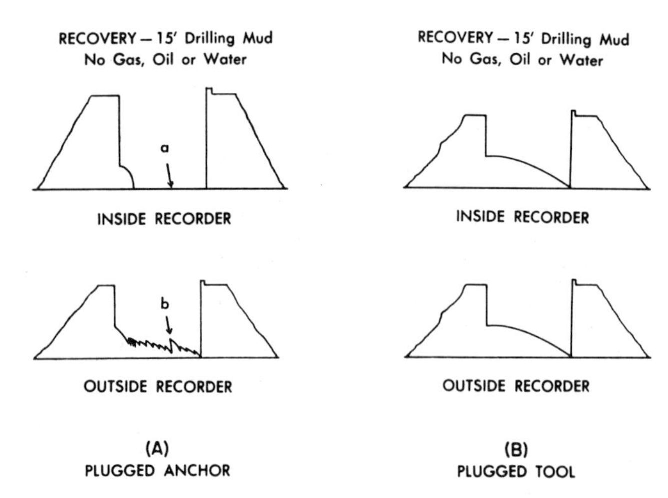

Fig. 13.11. Illustration of double pressure chart analysis. Courtesy Johnson Testers.

13.51 Estimation of Formation Productivity

Quantitative analysis of drill stem test data with regard to formation productivity is not highly accurate. This is a natural consequence of the relatively short flow test period, which does not permit attainment of steady state conditions. In many cases, however, the reservoir pressure, formation permeability, productivity index, and extent of formation damage may be calculated with reasonable accuracy from drill stem test data. The following sequence illustrates a general analysis procedure.

A. A qualitative inspection of the pressure chart should be made to insure that the sequence of events is understood and that no irregularities or tool failures occurred. The pressures listed below should be taken from the chart and recorded. They are computed by measuring the extension (deflection) above the base line and obtaining the corresponding pressure from a calibration curve for the recorder used. Drill stem test pressures were considered rather inaccurate at one time; however, current instruments are reportedly accurate to 1% and can detect changes of $\frac{1}{2}$ psi. Various chart magnification techniques are used to obtain more accurate chart deflections.

(1) Initial hydrostatic pressure (IHP), exerted by mud column
(2) Initial closed-in pressure (ICIP)
(3) Initial flowing pressure (IFP), the lowest pressure recorded just after the tool is opened
(4) Final flowing pressure (FFP), the pressure just before the tool is closed
(5) Final closed-in pressure (FCIP)
(6) Final hydrostatic pressure (FHP)

These points are shown in Figure 13.12. The mud column presures IHP and FHP should be compared with those calculated from the mud density. This serves as a rough check on the pressure recorder.

B. The quantity of fluid recovered during the test should be computed. In cases where flowing production is obtained, the producing rate may be measured at the surface by metering through a test separator and/or tank. If only gas is recovered, its volume may be measured with an orifice well tester or pitot tube. In most cases, however, the amount of recovered liquid is measured in terms of feet of fillup in the drill pipe. This procedure often leads to rather ambiguous descriptions when the recovered liquids are mixtures, such as slightly oil-cut mud, oil-cut mud, heavily oil-cut mud, gas-cut mud, oily gas-cut mud, etc. Unless such descriptions are accompanied by a percentage estimate or unless the company involved has some standard code defining these terms, it is virtually impossible to know just how much of each fluid is involved. A portable hand crank centrifuge may be used to check the liquid samples as the drill pipe is withdrawn, allowing percentage compositions to be reported. In highly successful tests the quantity of these mixtures may be negligible with respect to clean oil recovery. The volume of liquid recovered in the pipe is

$$V = bL \tag{13.1}$$

where V = volume recovery of a particular liquid, bbl
b = capacity of pipe, bbl/ft
L = length of pipe filled, ft

A common rule of thumb for b is:

$$b = \frac{d^2}{1000} \tag{13.2}$$

where d = inside diameter of pipe, in.

This is often a satisfactory (3% error) approximation of

$$b = \frac{(\pi d^2/4)12 \text{ in.}^3/\text{ft}}{9702 \text{ in.}^3/\text{bbl}} = 0.00097\, d^2$$

The average oil flow rate over the test time interval is then:

$$q_o = \frac{1440 V_o}{t}$$

where q_o = bbl/day
t = test time, min
V_o = oil volume recovered, bbl

C. Estimates of formation permeability, reservoir pressure, and extent of formation damage may now be made. The effective permeability to the flowing fluid is obtained from the pressure build-up curve. Recall equation

$$p_e - p_{ws} = \frac{0.163 q_o \mu_o B_o}{k_o h} \log \frac{t_o + \Delta t}{\Delta t} \tag{12.20}$$

where $0.163q_o\mu_o B_o/k_o h = m$, the slope of the linear portion of the buildup curve

Equation (12.20) may be expressed in fundamental units as

$$p_e' - p_{ws}' = \frac{q_o' B_o \mu_o}{4\pi k_o h'} \ln \frac{t + \Delta t}{\Delta t} \tag{12.20a}$$

where p_e', p_{ws}' = atm
q_o' = tank oil flow rate, cc/sec
h' = cm

If the shut-in period following the flow period is sufficiently long that after-production effects are eliminated, the pressure build-up curve may be plotted. Since the volume below the packer and tester valve is quite small, the time required to eliminate the after-production effect is, in general, not excessive. It is commonly recommended that the shut-in period should equal the flow period. The initial closed-in pressure may normally be taken as an accurate value of p_e; this value may serve to guide the extrapolation beyond the final shut-in pressure, allowing a more accurate slope to be obtained. In order to estimate formation damage, it is again necessary to compare the unaltered flow capacity from pressure build-up analysis with some measure of the average capacity including that of the altered zone. Again, the productivity index based on the average flow rate during the test may be used. This, however, requires that certain assumptions be made concerning the effective drainage radius during the flow test. Recall that the oil *PI* was defined in Chapter 12 as

$$J_o = \frac{q_o B_o}{p_e - p_w} = \frac{7.07\bar{k}_o h}{\mu_o \ln(r_e/r_w)} \tag{12.9}$$

or in fundamental units as

$$J_o = \frac{q_o' B_o}{p_e' - p_w'} = \frac{2\pi \bar{k}_o h'}{\mu_o \ln(r_e/r_w)} \tag{12.9a}$$

The term kh/μ is commonly called the transmissibility factor. Solving Eq. (12.9a) for this,

$$\frac{k_o h'}{\mu_o} = \left(\frac{q_o' B_o}{p_e' - p_w'}\right)\frac{\ln(r_e/r_w)}{2\pi} \tag{13.3}$$

For drill stem test purposes, it is often assumed that $\ln(r_e/r_w) \cong 2\pi$; hence Eq. (13.3) becomes:

$$\left(\frac{\bar{k}_o h'}{\mu_o}\right)_J = \frac{q_o' B_o}{p_e' - p_w'} \tag{13.4}$$

This amounts to assuming that $r_e/r_w \cong 500$. As we have noted before, radial flow computations are not particularly sensitive to r_e/r_w, since the term appears as a logarithm. From Eq. (12.20a) it is seen that

$$\left(\frac{k_o h'}{\mu_o}\right)_{BU} = \frac{q_o' B_o}{4\pi(p_e' - p_w')} \ln \frac{t + \Delta t}{\Delta t} \tag{13.5}$$

or

$$\left(\frac{k_o h'}{\mu_o}\right)_{BU} = \frac{q_o' B_o}{4\pi}\left(\frac{1}{\mathrm{m}}\right) \tag{13.5a}$$

Note: Subscripts *J* and *BU* designate values obtained from productivity index and build-up tests, respectively.

However, m may be expressed as

$$m = \frac{\Delta p_c}{\ln 10} = \frac{\Delta p_c}{2.3}$$

where Δp_c = pressure drop per common (base 10) logarithm cycle from the buildup curve

If we now equate q_o' in Eqs. (13.4)and (13.5), i.e., the average producing rate for the flow test interval, then our standard definition of productivity ratio becomes, conveniently,

$$PR = \frac{\bar{k}_o}{k_o} = \frac{(\bar{k}_o h'/\mu_o)_J}{(k_o h'/\mu_o)_{BU}} \tag{12.24}$$

or

$$PR \cong 5.5 \frac{\Delta p_c}{p_e - p_w} \tag{13.6}$$

Equation (13.6) is the same as that presented by Dolan et al.,[3] except that the inverse of the *PR* was used and denoted as a damage factor:

$$DF = \frac{1}{PR} = 0.183 \frac{p_e - p_w}{\Delta p_c} \tag{13.7}$$

We will use Eq. (13.6), as it is consistent with Chapter 12. The validity of this rather empirical approach has been verified by electric analyzer studies and field experience. It is not as precise as data based on postcompletional flow tests which are carried out under more stabilized conditions. Furthermore, the *PR* value obtained from drill stem test data cannot be expected to correspond with later results, owing to possible permeability alterations during the completion process. However, quantitative utilization of pressure chart data has proved to be a useful tool. Example 13.1 illustrates the calculations involved.

Example 13.1

The pressure chart for a DST is given in Figure 13.12. Other available data are listed below.

Interval tested = 9990–10,010 ft

Mud density = 9.8 lb/gal

Recovery = 200 ft mud, 7500 ft clean 40°API oil, no water

No water blanket run

Drill pipe = $4\frac{1}{2}$-in., 16.6 lb/ft (i.d. = 3.826 in.)

1. The sequence of events is clear from the chart. Note that the pressure apparently stabilized during the initial closed-in period, hence the static reservoir pressure is probably very closely approximated by 4500 psig, the ICIP. This

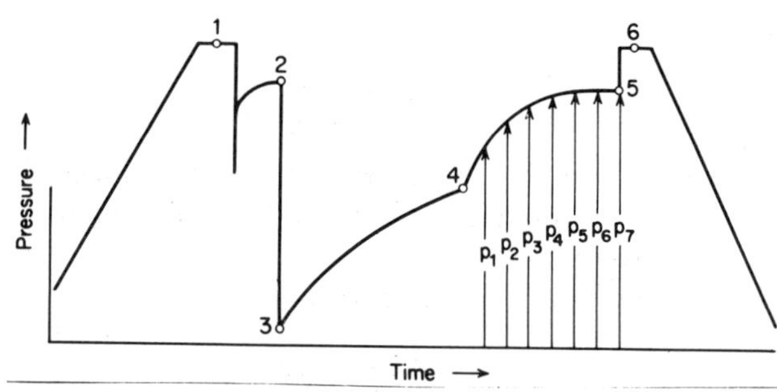

Fig. 13.12. Sketch of DST chart for Example 13.1.

(1) IHP = 5120 psig	(4) FFP = 2700 psig
(2) ICIP = 4500 psig	(5) FCIP = 4325 psig
(3) IFP = 250 psig	(6) FHP = 5140 psig

will be checked by the build-up curve extrapolation.

2. The initial and final hydrostatic pressures, 5120 and 5140 psig, may be compared with the value calculated from the mud density:

$$p_m = \frac{9.8}{8.33} \times .433 \times 10{,}000 = 5100 \text{ psig}$$

This is excellent agreement.

3. The oil recovery during the 80-min flow test was:

$$V_o \cong \frac{(3.83)^2}{1000} \times 7500 = 110 \text{ bbl}$$

The corresponding average daily oil producing rate is:

$$q_o = \frac{1440 \times 110}{80} = 1980 \text{ bbl/day}$$

4. Next, the pressure build-up section is divided into convenient Δt increments and the pressures at successive intervals are tabulated.[4] These are shown as p_1, p_2, etc., on Figure 13.12. Note that the producing time t was 80 min.

Pressure, atm	Δt, *min*	$(t + \Delta t/\Delta t)$
$p_1 = 3460$	10	9
$p_2 = 3900$	20	5
$p_3 = 4190$	30	3.67
$p_4 = 4300$	40	3
$p_5 = 4375$	50	2.60
$p_6 = 4400$	60	2.33
$p_7 = 4410$	70	2.14

The build-up curve is plotted as Figure 13.13. Note that only the last two points fall on the extrapolation to p_e = ICIP = 4500 psig. This illustrates the advantage of obtaining an ICIP for extrapolation purposes.

(a) The productivity ratio may be calculated from Eq. (13.6).

$$\Delta p_c = 4500 - 4200 = 300 \text{ psi per cycle} = m$$

$$p_e = 4500$$

$$p_w = \text{FFP} = 2700 \text{ psig}$$

and

$$PR = 5.5\left(\frac{300}{1800}\right) = 0.92$$

(b) The estimated productivity index is:

$$J_o = \frac{1980}{4500 - 2700}$$

$$= 1.1 \text{ bbl tank oil/day/psi at a } PR = 0.92$$

or,

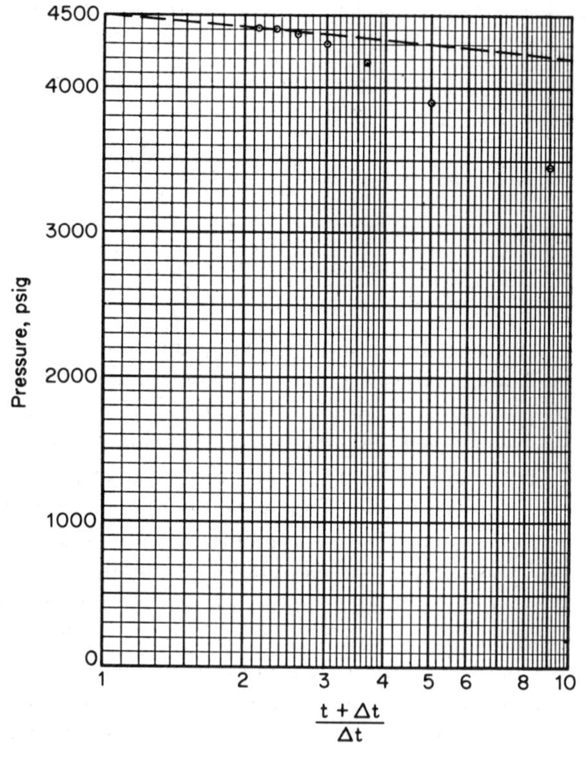

Fig. 13.13. Pressure buildup curve for Example 13.1.

$$J_o = \frac{1.1}{.92} = 1.2 \text{ at a } PR = 1$$

Note that the formation volume factor B_o is omitted. Actually, the oil recovered in the drill pipe is neither tank oil nor reservoir oil, but something in between. If *PVT* or other data are available so that B_o may be estimated, the calculations may be refined. The seriousness of this omission is, however, minor as far as DST estimates are concerned.

(c) Calculation of unaltered reservoir permeability requires a knowledge of fluid viscosity. Since it is unusual to have such data at the time of a DST, Black has presented the correlation of Figure 13.14, which may be used for oil viscosity estimates.

Using $\mu_o = 0.4$ cp for 40°API oil

$$k_o = \frac{.163 q \mu_o}{mh} = \frac{(.163)(1980)(0.4)}{(300)(20)} = 0.022 \text{ d or } 22 \text{ md}$$

(d) An estimate of the producing rate and possibility of flowing production completion may be made. For oil to flow naturally, the bottom hole producing pressure must be high enough to support the flowing column to the surface, while compensating for the frictional losses enroute. The estimation of the gradient in a vertical flow string is complicated by the following principal factors:

(1) As the fluid moves upward, pressure is steadily decreased, allowing gas to evolve continuously from solution.

(2) Any gas in the flow string expands as the pressure is reduced, hence the average density of the oil-gas (and possibly water) mixture decreases as depth decreases.

(3) The effective viscosity of this mixture also changes with gas evolution and expansion, making friction loss calculations difficult.

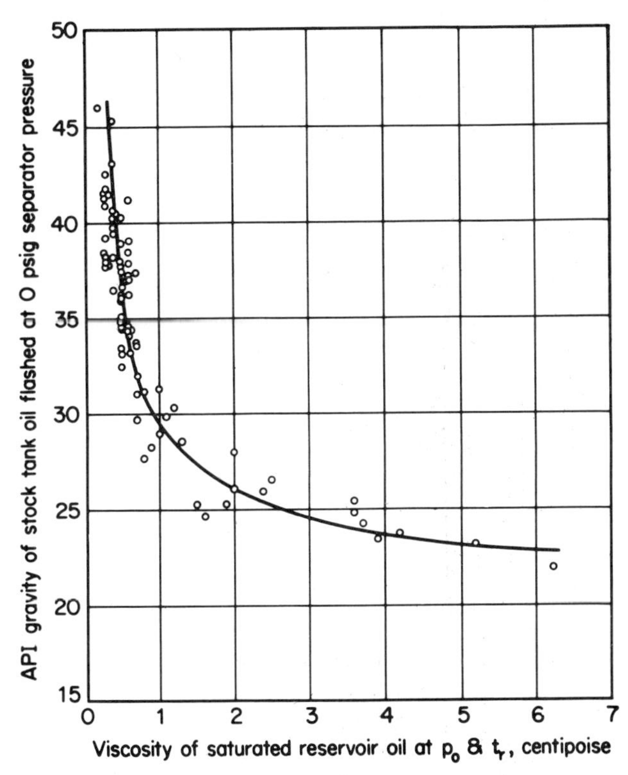

Fig. 13.14. Correlation of API gravity with reservoir oil viscosity. After Black,[2] courtesy AIME.

A reasonably accurate method of solving such problems is available;[5] however, some of the data required are not generally available at the time of testing. A usual, although generally pessimistic approach, is to assume a flow string gradient equal to the static stock tank oil gradient. This is equivalent to assuming that the lightening effect of free and/or dissolved gas compensates for frictional losses in the tubing and surface piping. Therefore, for 40°API oil, the necessary bottom hole producing pressure for flowing production is:

$$p_w = 0.358 \times 10{,}000 = 3580 \text{ psig}$$

The corresponding flow rate based on the DST productivity index is:

$$q_o = J_o(p_e - p_w) = (1.1)(4500 - 3580) = 1000 \text{ bbl/day}$$

Certainly this well could be expected to flow naturally.

13.52 Formation Water Analysis

Chemical analysis of formation water samples are valuable for numerous purposes. Among these are electric log interpretation and subsequent determinations of the source of produced water. However, there is always considerable question as to whether drill stem test samples are representative of actual formation water. Consider a test which recovers only salt water. When the test tool is opened, mud, mud filtrate, diluted formation water, and perhaps finally formation water will enter the pipe, if the tool is left open long enough. A single sample could not be considered representative of the unaltered formation water. However, if continuous sampling is performed and the results are plotted as shown in Figure 13.15(A), the constant salinity section may be taken as reasonably representative of undiluted formation water. If, however, the results are as shown in 13.15(B), it is probable that a representative sample was not obtained.

13.6 Wire Line Formation Testing

A wire line formation tester is available which has found considerable application in areas where conventional open hole testing is undesirable.[6] Proper application of this device has, in many cases, simplified precompletion testing and made it feasible to evaluate possibly pro

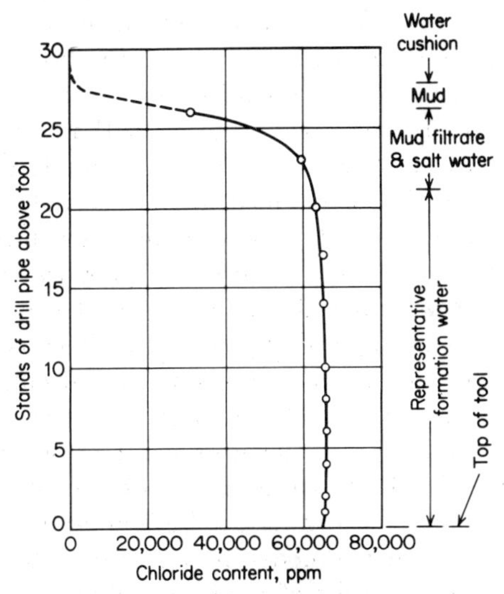

(A) Example of usual variation of salinity with depth in recovered water column

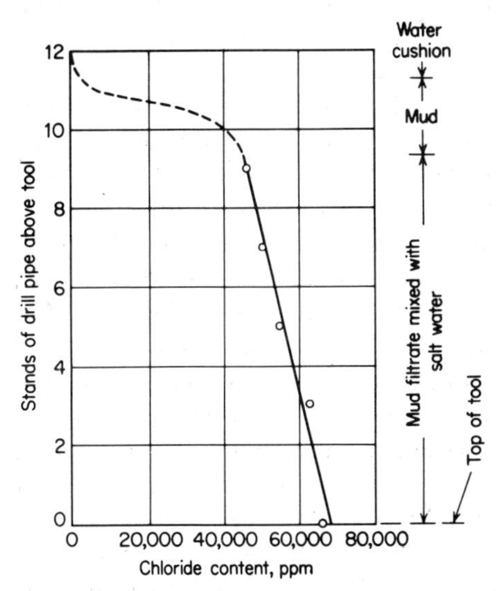

(B) Example of unsufficient yield to recover representative water

Fig. 13.15. Method of plotting water analysis data to determine if sample was representative. After Black,[2] courtesy AIME.

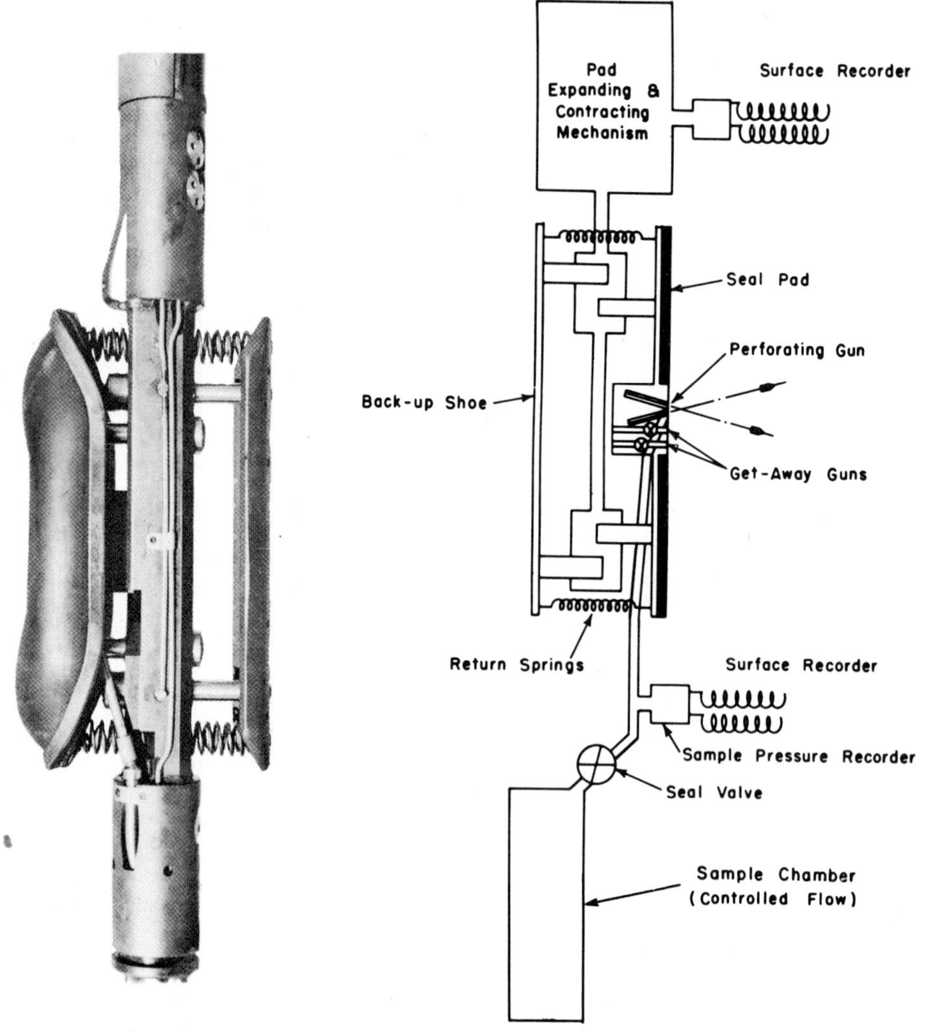

Fig. 13.16. Photograph and schematic diagram of wire line formation tester. After Lebourg, Fields, and Doh,[6] courtesy AIME.

ductive formations with a considerable saving of time and money. This tool, called the formation tester, is shown in Figure 13.16.

The tester is run into the hole on electric logging cable and may be positioned at any desired level. An *SP* curve is recorded simultaneously and is used to position the tester at the precise depth desired (± 6 in.). Positioning is accomplished by comparison of the *SP* from the tester run with the original electric log from which the desirable test zones are often selected. After the tool is positioned, the sealing pad is expanded against the hole wall to form a seal between the formation and the mud column. Two perforating bullets are then fired from within the pad (Figure 13.16) which establish flow channels between the formation and sample chamber. When the recording chart at the surface indicates that the sample container is filled, the seal valve is closed and the tool is automatically collapsed. Generally, the seal pad remains stuck to the hole wall owing to the pressure differential between the mud column and the formation. To alleviate this problem, two additional perforating bullets are included in the tool. When these are fired through the pad, their recoil plus the two new holes in the seal pad is generally sufficient to bring about pressure equalization that frees the tool and allows it to be withdrawn. A complete record of these events is made simultaneously at the surface. Also obtained are the sampling and shut-in formation pressure, as shown in Figure 13.17. Note that time is plotted vertically and pressure is the abscissa.

The principal advantages of this device, other than its particular applicability to soft rock areas, are its speed and economy. It requires no special hole conditioning and may be run day or night. Successive tests over thick intervals define gas-oil and/or oil water contacts. Undoubtedly it will find use in hard rock areas as well.

Its disadvantage is the limited size of the sample it draws. Current models procure from 1 to $5\frac{1}{2}$ gal of fluid. This may consist of mud, mud filtrate, salt water, oil, and gas, or any mixture of these. Consequently, a test recovering oil, gas, and salt water will neither prove nor completely disprove the formation potential, although

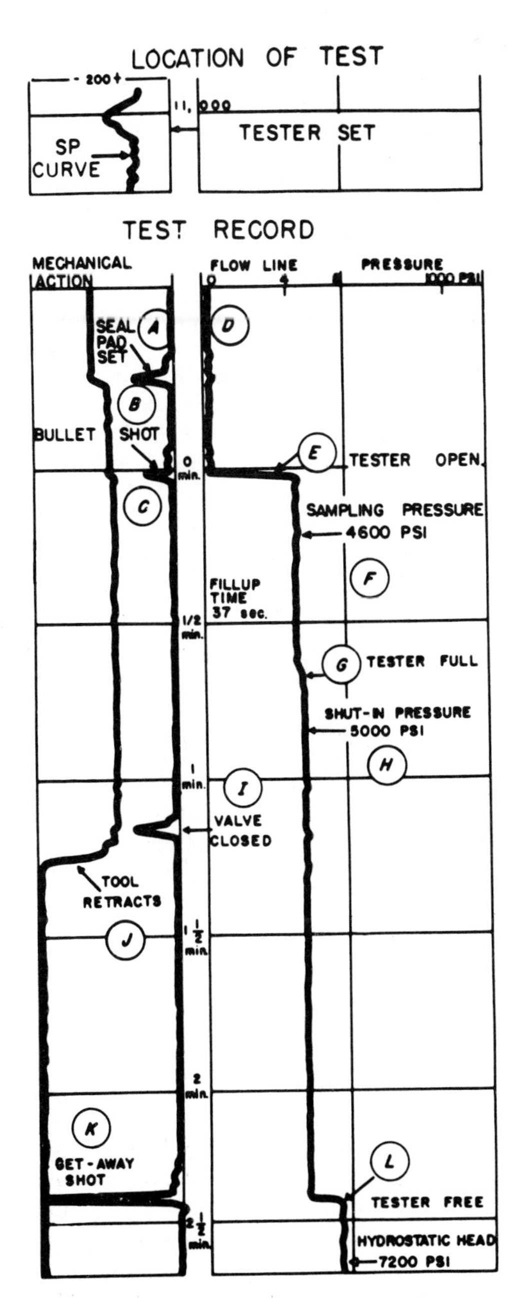

Fig. 13.17. Surface record (or log) of wire line formation test at 11,000 ft.[6] Courtesy AIME.

Sampling pressure: 4600 psi
Shut in pressure: 5000 psi
Hydrostatic (mud) pressure: 7200 psi
Recovery: 3.9 SCF gas, 1500 cc oil, 50 cc sand, 2000 cc mud filtrate
Gas/oil ratio of sample: 414 SCF/bbl
Production test: perforated 6 ft, produced 244 bbl/day, 34.6° API oil at gas/oil ratio of 848 SCF/bbl.

knowledge of the relative amounts of each will be of some value. If only gas, oil, and mud filtrate are recovered, clean oil recovery is indicated. Analysis of the sample will yield an estimate of producing gas/oil ratio, oil gravity, and gas richness. In all cases, the formation pressure is of interest and may generally be taken as a reasonable estimate of static pressure p_e.

The pressure charts of Figure 13.18 are presented as being typical of eight general categories into which all tests fall. The following examples illustrate the application of the wire line formation tester and its use in conjunction with electric logging. These have been taken from the paper of Lebourg, Fields, and Doh.[6]

Example 13.2

One of the most difficult problems on the Gulf Coast, both in logging and, to a lesser extent, in core analysis, is the differentiation between oil- and gas-bearing formations. The difficulty is increased when the sands are unconsolidated and the gravity of the oil is greater than 40°API.

The electrical log and microlog covering a sand in the second well of a new field are shown in Figure 13.19. Analysis of side-wall cores in this same sand in the discovery well indicated zero oil saturation at the equivalent depth of 9300 ft. Three side-wall cores below 9300 ft gave an oil saturation varying from 2% to 6.3%. It was concluded on the basis of these cores that the sand was gas-bearing; for this reason, the well was completed in another horizon.

Core analysis of the section in the second well, shown here, gave an average oil saturation of only 6%; however, at 9294 ft the formation tester gave 1.75 pints of oil, 7.5 CF of gas, and 3 pints of filtrate. The calculated GOR from this test was 1440, with a flowing pressure of 4300 PSI and a shut-in pressure of 4400 PSI. A second test at 9290 ft (4 ft above the first test) gave a GOR of 3494.

It was then concluded that an oil completion should be made. After perforating from 9296 ft to 9300 ft, the well produced 211 barrels of oil with a normal gas-oil ratio of 1284, an FFBHP of 4275, and an SIP of 4375. The excellent correspondence between the pressures given by the formation tester and the pressures obtained on the final production test was very encouraging. It should be noted that successive measurements of GOR on the same test, when made by methods now in common practice, will vary by 200 ft^3/bbl. It was later established that the upper formation test which yielded a GOR of 3494 had been taken at the exact gas-oil contact of the field.

Example 13.3

An electrical log and microlog (Figure 13.20) from the *Hard Rock* country of Central Mississippi indicated a potentially productive sand in the Glen Rose. Side-wall core analysis showed porosities of 19% to 29% with oil saturations of 5% to 20%. Two formation tests were made at 8421 ft, recovering traces of oil and 3.8 quarts of filtrate, each. Flowing pressures in the two tests were 600 and 500 psi; shut-in pressures were identical at 3800 lb.

A third formation test at 8405 ft recovered one quart of filtrate, with 500 psi flowing pressure, and 3700 psi shut-in pressure.

The recoveries of filtrate indicated extreme invasion, which explained, partially, the high resistivities measured on the electrical log. The very low flowing pressures were also indicative of extremely low permeabilities. The traces of black oil

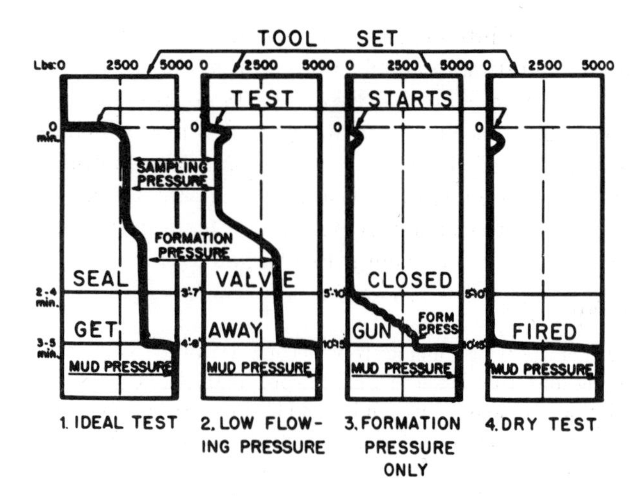

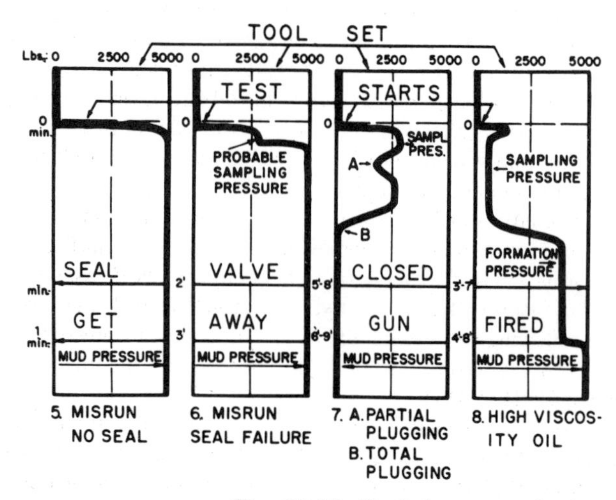

Chart No. 1: This type of pressure curve is obtained from a highly permeable formation.

Chart No. 2: This type of pressure curve is obtained from a formation with a somewhat lower permeability than that of Chart No. 1. This is indicated by the lower sampling pressure and a slightly longer testing time.

Chart No. 3: This pressure recording is obtained from a formation with very low permeability. The flow was insufficient to indicate a pressure in the sample container; however, it was possible to obtain the formation pressure in the flow line.

Chart No. 4: This illustrates the type of chart obtained from formations with extremely low permeability. The flow was too small to indicate a pressure in either the sample container or the flow line.

Chart No. 5: In this test, the pressure curve shows clearly that a seal was not affected; therefore, this is a misrun.

Chart No. 6: The pressure curve here indicates that the seal was obtained but failed before completion of the test.

Chart No. 7: This is an example of plugging of the flow line before completion of the sampling. Point A indicates partial plugging. Point B indicates total plugging, making it impossible to obtain the formation pressure. If only partial plugging occurs, formation pressure can be obtained.

Chart No. 8: This chart is similar to No. 2. In this case, the high viscosity of the oil contributes to a low sampling pressure and a slightly longer testing time. This is caused by the low relative permeability of this sand to the viscous fluid.

Fig. 13.18. Typical pressure charts and their analysis. After Lebourg, *et al.*,[6] courtesy AIME.

recovered further contributed to a noncommercial evaluation of the reservoir. Nevertheless, an open hole DST of the entire sand body was made; it recovered 9.4 barrels of salty gas-cut mud and 9.7 barrels of slightly oily salt water, with a flowing pressure of 825 PSI. The well was abandoned.

Example 13.4

High invasion is indicated by the large separation of the resistivity curves in Figure 13.21. The very low reading on the lateral curve is due to the anisotropy of the shale above 7350 ft. Standard quantitative evaluation of the electric log would have condemned the sand. Due to the high invasion, the 1 gallon tester would probably have recovered only filtrate.

This concludes our discussion of not only drill stem and formation testing, but of formation evaluation methods as a whole. It has been shown that each technique yields valuable data which guides the interpreter's decision as to whether or not a given formation is capable of yielding commercial oil and gas production. No one technique furnishes all the answers, and each has its own advantages under certain circumstances. The usefulness of these data, i.e., core analysis, well logging, formation testing, etc., does not end with the completion of the well but may be used years later for other production and reservoir engineering purposes. For example, the estimation of reserves, predictions of future producing behavior, planning of remedial and recompletion operations, and many other postcompletional problems rely heavily on these data obtained at the time of drilling. Consequently, it is desirable that formation evaluation programs be planned in such a way that the desired information is available when needed.[7]

PROBLEMS

1. Several simplifications of the pressure build-up equations have been developed for well site analysis of DST charts. H. K. van Poolen has presented the following in conjunction with a convenient nomograph:[8]

$$DR = 0.183 \frac{\text{ICIP} - \text{FFP}}{\text{ICIP} - \text{FCIP}} \log \frac{t + \Delta t}{\Delta t}$$

where DR = Damage Ratio $= \dfrac{1}{PR}$

ELECTRICAL LOG MICROLOG

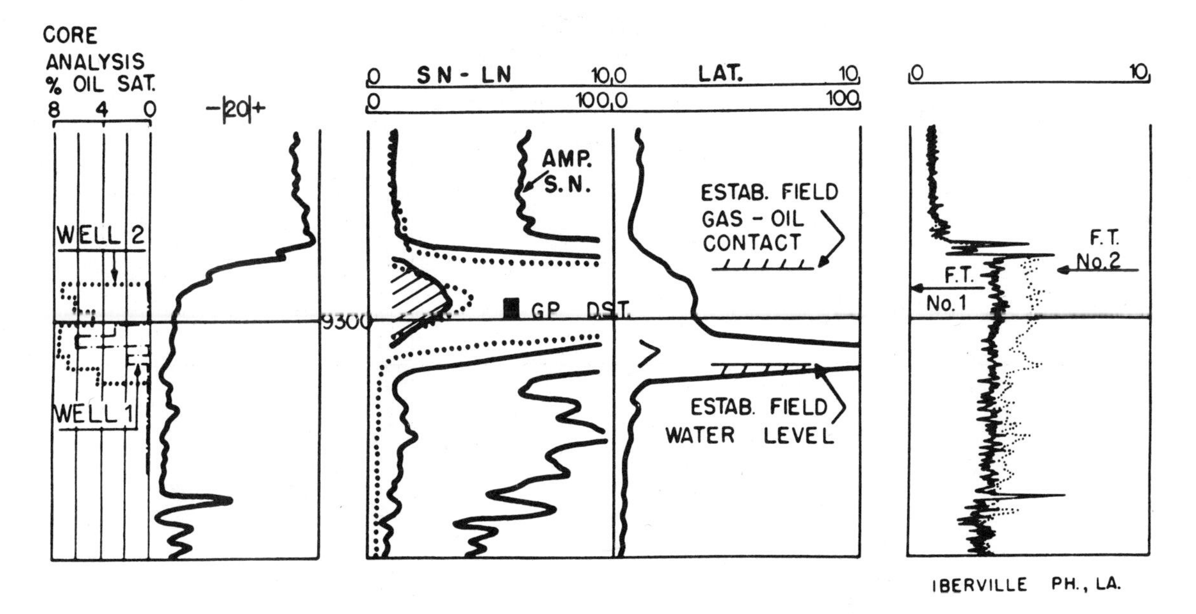

FORMATION TEST No. 1 at 9294'
REC. 1¾ PTS. OIL; 7.5 CU. FT. GAS; 3 PTS. FILTRATE & SAND; G.O.R. 1440; F.P. 4300#; S.I.P. 4400#

FORMATION TEST No. 2 at 9290'
REC. 1 PT. OIL; 10.4 CU. FT. GAS; 1 PT. SAND; 4 PTS. FILTRATE; G.O.R. 3494; F.P. 4300#; S.I.P. 4400#

DRILL STEM TEST — G.P. 9296' - 9300'
OPN. 5¾ HRS. T.P. 1980#; FLWD. 211 OIL; 1/8 CK.; G.O.R. 1284; F.F.B.H.P. 4375#

Fig. 13.19. Data for Example 13.2.

ELECTRICAL LOG MICROLOG

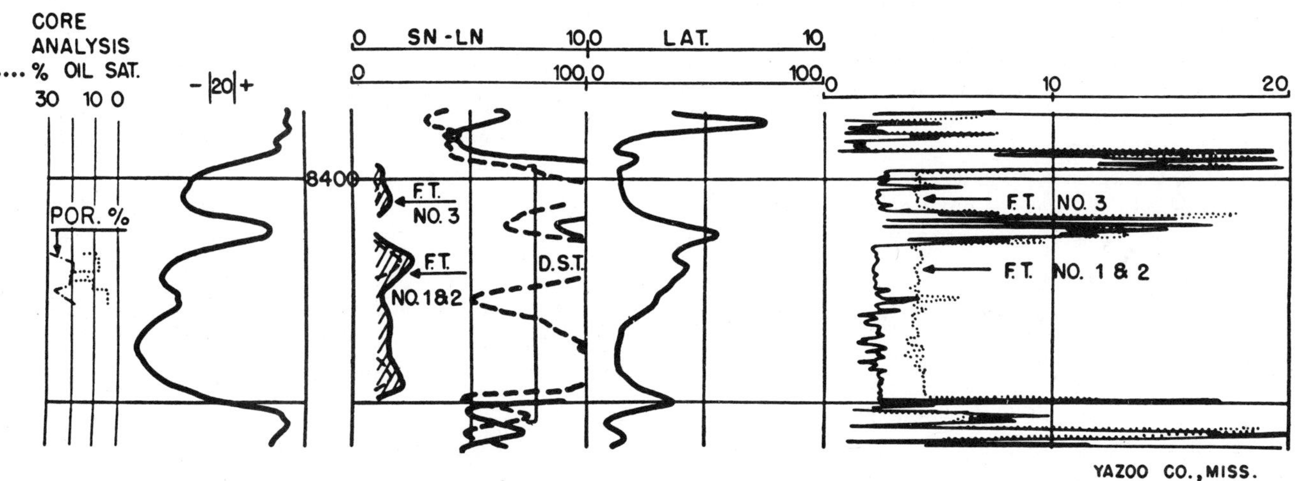

FORMATION TEST No. 1 at 8421'
REC. 3.8 QTS. FILTRATE, TRACE OIL, F.P. 600#; S.I.P. 3800#

FORMATION TEST No. 2 at 8421'
REC. 3.8 QTS. FILTRATE, TRACE OIL, F.P. 500#; S.I.P. 3800#

FORMATION TEST No. 3 at 8405'
REC. 1 QT. FILTRATE, F.P. 500#; S.I.P. 3700#

D.S.T. 8397 - 8454'
O.P. 25 MIN., PRESS. ¾#, REC. 9.4 BBLS. SALTY GAS CUT MUD & 9.7 BBLS. GASSY, SLIGHTLY OILY SALT WTR. (16) F.F.P. 825

Fig. 13.20. Data for Example 13.3.

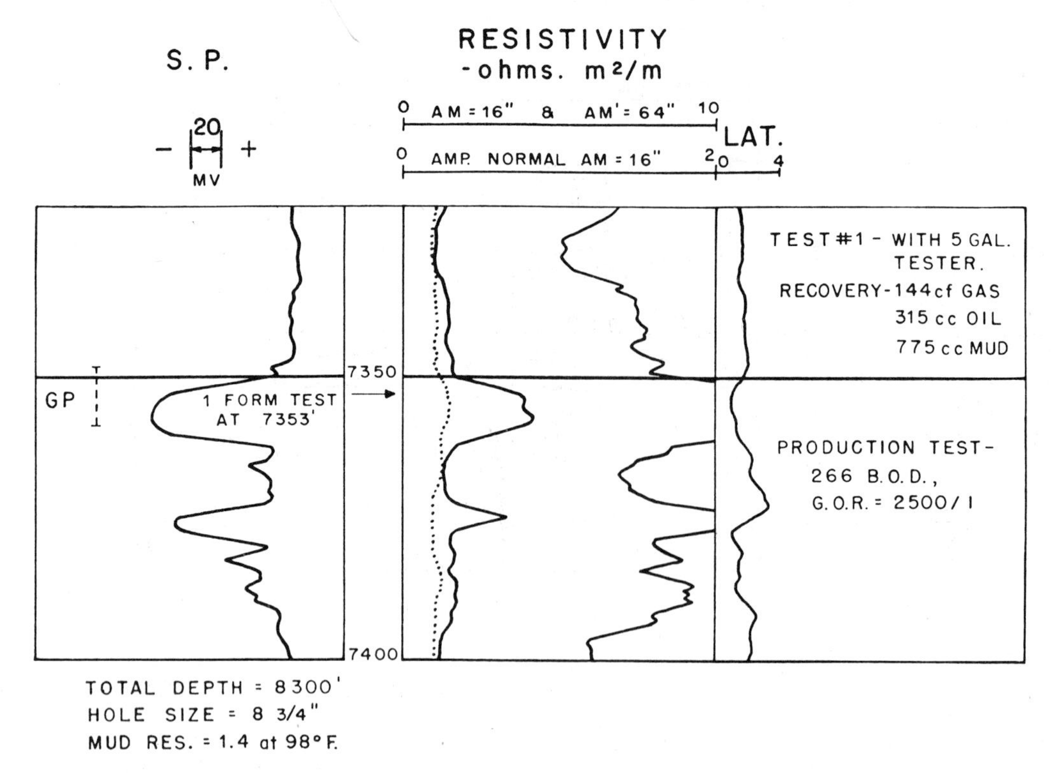

Fig. 13.21. Data for Example 13.4.

Other notations are as given on Figure 13.12.

Show that this is equivalent to Eq. 13.7 under certain conditions, and specify the conditions.

2. Following are DST data in a common abbreviated form. DST: 4990–5010 ft, tool open 30 min, shut-in 30 min. Recovered 1000 ft clean 35°API oil, 100 ft mud. ICIP = 2400, FFP = 250, FCIP = 2350.

(a) Estimate formation permeability, transmissibility factor, and productivity ratio for the interval. Make any assumptions necessary. *Ans.* (a) k_o = 16 md, PR = 0.4.

(b) Estimate the productivity index, J.

(c) Do you think this well will flow naturally?

(d) What steady state flowing production rate would you estimate for this well at its current PR?

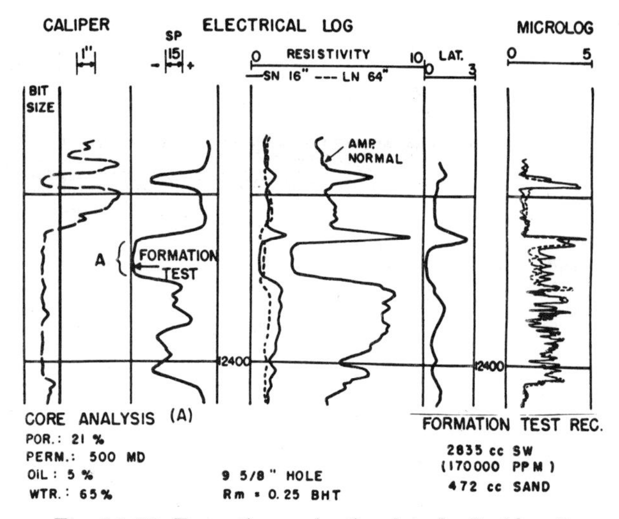

Fig. 13.22. Formation evaluation data for Problem 3.

3. Figure 13.22 shows electric log, core analysis, and wire line formation test data. What is your opinion concerning the productive potential of sand A? (For answer, see Reference 6, Example 4).

REFERENCES

1. Kirkpatrick, C. V., "Formation Testing," *The Petroleum Engineer*, Oct. 1954, p. B–139.

2. Black, W. M., "A Review of Drill-Stem Testing Techniques and Analysis," *Journal of Petroleum Technology*, June 1956, p. 21.

3. Dolan, J. P., C. A. Einarsen, G. A. Hill, "Special Applications of Drill-Stem Test Pressure Data," *Journal of Petroleum Technology*, Nov. 1957, p. 318.

4. Zak, A. J., Jr., and P. Griffin, 3rd, "Evaluating D.S.T. Data," *Oil and Gas Journal*, Apr. 15, 1957, p. 122.

5. Poettmann, F. H., and P. G. Carpenter, "The Multiphase Flow of Gas, Oil, and Water Through Vertical Flow Strings, with Application to the Design of Gas-lift Installations," *API Drilling and Production Practices*, 1952, p. 257.

6. Lebourg, M., R. Q. Fields, and C. A. Doh, "A Method of Formation Testing on Logging Cable," *Journal of Petroleum Technology*, Sept. 1957, p. 260.

7. Kirkpatrick, C. V., "An Integrated Summary of Formation Evaluation Criteria," AIME T.P. 595-G, presented at Formation Evaluation Symposium, Houston Univ., Oct. 1955.

8. van Poolen, H. K., "Damage Ratio Determined by Drill-Stem Test Data," *World Oil*, Nov. 1957, p. 139.

Chapter 14 . . .

Oil Well Cementing and Casing Practices

14.1 Introduction

During the course of drilling, it is necessary to run casing at various depth intervals, i.e., to lower the desired length of casing or pipe into the hole and cement it in place. The number and size of the casing strings used vary with the area, depth, anticipated producing characteristics of the well, and the choice of the operator. In addition to the casing, a smaller diameter string called tubing is used as the actual flow conduit for the produced fluids. The final appearance of a typical completed well is shown in Figure 14.1. Note that three separate casing sizes are indicated: the surface pipe, the intermediate string, and the oil string. The oil string has been set at the top of the oil sand in what is called an open hole completion. Note also that the tubing–oil string annulus is segregated from the pay sand by a packer. This is a common arrangement in wells which flow naturally. We will consider well completions in the next chapter, and further consideration of the subject is postponed until then.

Each casing string is cemented in place by a slurry pumped down the pipe and up the annulus between the casing and the open hole. The amount of slurry used is predetermined for the particular annular volume and fill-up height desired. The cement is then allowed to set for several hours before drilling or other operations are recommenced. The general functions of all casing strings are:

(1) To furnish a permanent borehole of precisely known diameter through which subsequent drilling, completion, and producing operations may be conducted.
(2) To allow segregation of formations behind the pipe, which prevents interformational flow and permits production from a specific zone.
(3) To afford a means of attaching the necessary surface valves and connections to control and handle the produced fluids.

Our discussion of casing programs will be enhanced by a consideration of the principles of cementing.

14.2 Primary Oil Well Cementing Techniques

The main function of primary cementing was given above as the second of the three general casing functions. Primary cementing pertains to the initial cementing jobs performed in conjunction with setting the various casing strings. Other purposes are:

(1) To afford additional support for the casing, either by physical bracing or prevention of formation pressures being imposed on the pipe. Although the latter is generally not considered in casing selection, it is nevertheless a beneficial aspect.
(2) To retard corrosion by minimizing contact between the pipe and corrosive formation waters.

Figure 14.2 illustrates the typical procedure for a single stage primary cement job. This is the conventional two-plug method. Neat cement (no sand or gravel) is introduced in a hopper where it is intimately mixed with water by a high velocity jet mixer. The resulting slurry is then pumped down the casing between two rubber plugs with wiping fins which are placed in the system at the proper time via a cementing head. When the bottom plug reaches the float collar, it stops; a pressure builds up which quickly ruptures the plug's diaphragm and allows the slurry to continue.

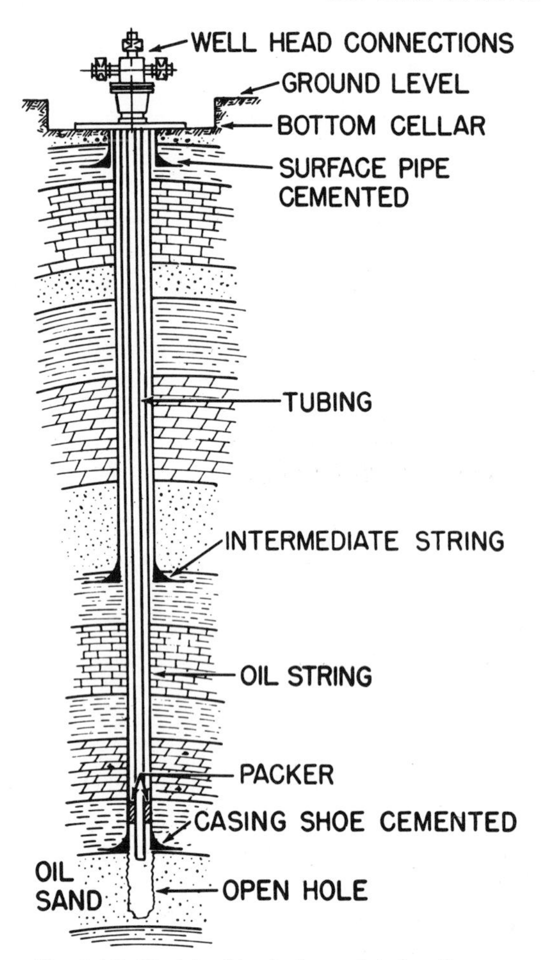

Fig. 14.1. Sketch of typical completed well.

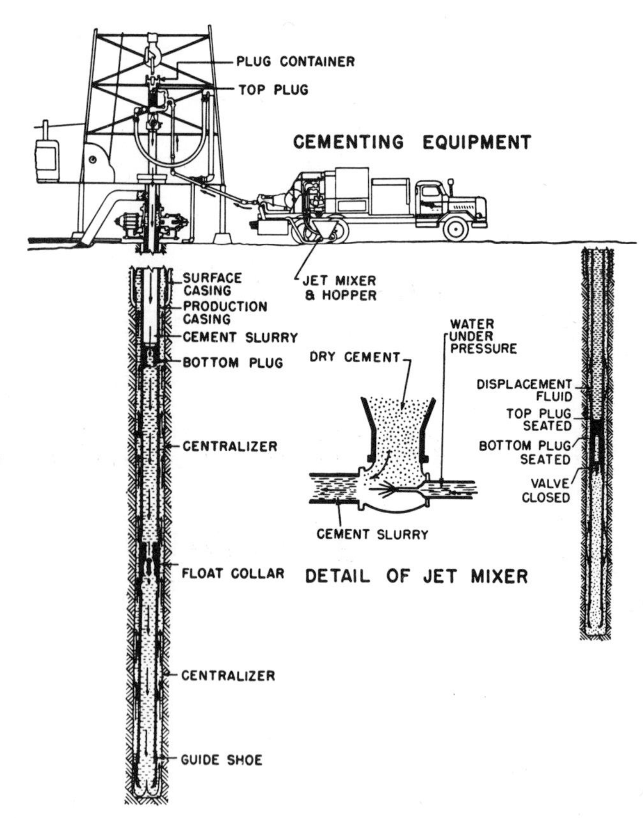

Fig. 14.2. Diagram of single-stage primary cementing job by conventional two-plug method. Courtesy Halliburton Oil Well Cementing Company.

The top plug, however, has a solid core, so that when it seats in the float collar, the surface pump pressure builds up sharply, thereby signaling the pump operator that the job is complete. The position of the top plug may also be checked either by metering the displacing fluid (since the casing volume is known), or by following the plug with a wire measuring line. The casing below the float collar is left full of cement which can be drilled out if necessary. This latter procedure is commonly called "drilling the plug."

Numerous variations of primary cementing techniques are in use. One or both of the plugs are sometimes omitted. Practices also depend on the depth and particular string being cemented. A study of surface pipe cementing practices has been made which illustrates well the wide diversity of opinion on proper procedure.[1]

It should be apparent that, if the cement slurry density is different from that of the mud and/or the displacing fluid, a considerable pressure unbalance will occur between the fluid columns in the casing and annulus. This situation has numerous possible consequences:

(1) Excessive pump pressure may be required to obtain a high fill-up.

(2) If cement density is higher than the mud, this greater pressure may break down formation(s) behind the pipe, and circulation may be lost during the job. This is of particular importance in a well which has already experienced lost circulation.

(3) If cement density is less than the mud, the displacing fluid should not be mud, since the pressure unbalance may move the slurry completely around the shoe and up the annulus.

Example 14.1

Casing is being cemented in a 10,000 ft well containing 10 lb/gal mud. The slurry density is 14 lb/gal and the plug is to be chased with 0.85 sp. gr. oil. If the anticipated fill-up is 1000 ft, what surface pressure must be held on the casing to prevent backflow at the end of the job in case the floating equipment fails to hold?

Solution:

Since the bottom hole pressure must be the same when calculated down both columns, we have (ignoring the cement plug inside the pipe and any thermal expansion effects):

$$p_s + 10{,}000(0.85)(0.433) = 1000\left(\frac{14}{8.33} \times 0.433\right) + 9000\left(\frac{10}{8.33} \times 0.433\right)$$

or,

$$p_s = 1700 \text{ psi (surface pressure)}$$

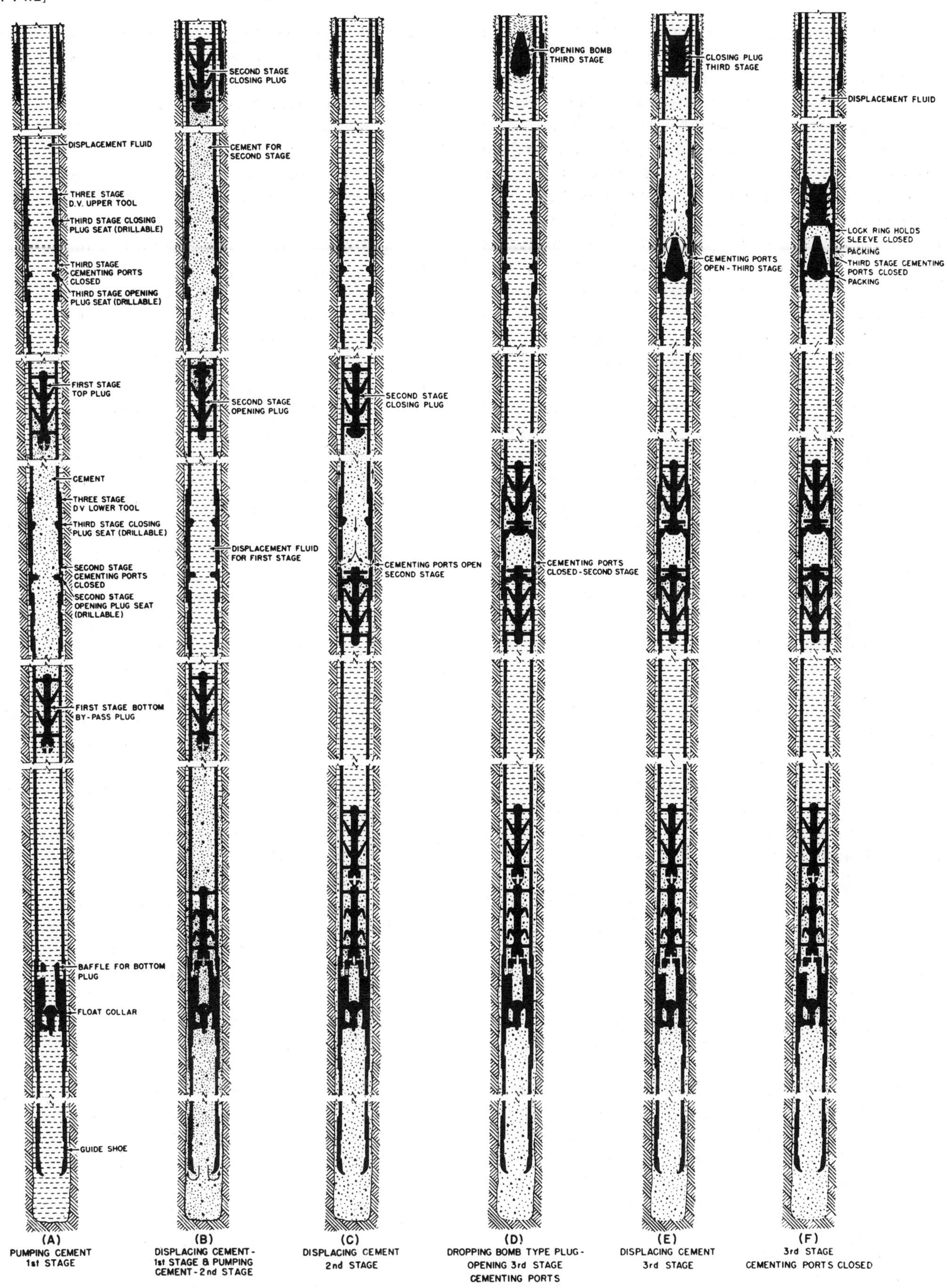

Fig. 14.3. Illustration of three-stage cementing technique. Note that cement is displaced out the casing shoe and at two separate points above bottom. Courtesy Halliburton Oil Well Cementing Company.

If, in this example, the drilling mud had been used instead of oil as the displacing fluid, the static pressure unbalance would have been only slightly above 200 psi. Suppose that for some reason it was desirable to cement the string over its entire length. This would require a final pressure of approximately 3600 psi using oil or 2100 psi using mud. The actual maximum pumping pressure would, of course, have to be considerably higher, to compensate for hydraulic losses. Another very important factor opposing such a procedure is that the leading cement segment must keep moving throughout the entire job. Since occasions do arise when long interval cementing is desirable, it is often necessary to find another method, such as that known as stage cementing. In our 10,000 ft case, for example, we might proceed as follows:

(1) Perform a normal cement job as a first stage, using enough cement to obtain a calculated fill-up of perhaps 3000 ft.
(2) Determine the actual fill-up by temperature survey, as 2500 ft (cement from 7500 ft to 10,000 ft).
(3) Run a perforating gun to perforate the casing at 7500 ft depth.
(4) Pump a second stage down the casing where it will pass out through the perforations and rise in the annular space above the now-solid first stage.
(5) Continue the procedure until the column has reached the desired height.

The above method may be altered to accomplish other jobs such as placing cement in discontinuous segments along the hole. The above method of stage cementing is obviously a long and costly procedure. To avoid it, stage tools have been developed which facilitate the placing of stages in a continuous process. A three-stage operation by this method is illustrated in Figure 14.3.[2] Any of the lower stages may be allowed to set before the ensuing stage(s) is injected. Two-stage cementing is naturally more common than three, and merely involves omission of the third stage collar from the casing string.

14.21 Cement Types, Specifications and Additives

Oil well cements are basically the same as those used in construction work. Common portland cement, for example, is widely used, although various additives may be required in a specific instance. The chemistry of cements is quite complicated and the reader is referred to a comprehensive discussion by Ludwig for such a coverage.[3] The API has set forth specifications for oil well cements in its Standard 10A[4] as well as recommended testing practices in RP 10B.[5] The six API cement classifications are as follows:

Class A: Intended for use to 6000-ft depth*, when special properties are not required. Similar to ASTM C 150, Type 1 Cement.

Class B: Intended for use to 6000-ft depth*, when sulfate resistance is required. Similar to ASTM C 150, Type II Cement.

Class C: Intended for use to 6000-ft depth*, when high early strength is required. Similar to ASTM C 150, Type III Cement.

Class D: Intended for use to 12,000-ft depth*, when moderately high temperatures and high pressures are encountered.

Class E: Intended for use to 14,000-ft depth*, when high temperatures and high pressures are encountered.

Class F: Intended for use to 16,000-ft depth*, when extremely high temperatures and extremely high pressures are encountered.

Numerous special cement types are also in common use. Some of the more important of these are:

1. Pozzolan Cements:[6] Pozzolans are siliceous materials which will react with lime to form a cementitious material. As such, they may be used as an additive to ordinary cement or prepared as a lime-pozzolan blend without portland-type cement. Both natural (volcanic origin) and synthetic pozzolans are in use. A common form of the latter is called fly ash and is the combustion product of pulverized coal as obtained from steam generating plants. Lime-pozzolan cement has proved to be a satisfactory deep well cement. Considerable variance in properties of pozzolans has been observed, and individual testing is required.
2. Diacel Cement Systems:[7–9] This designation refers to cement systems modified by one or more of the additives listed as Diacel D, LWL, and A in Table 14.1. Such cements have a large range of densities and thickening times, which gives them a wide scope of applicability. Fine sand (95% through 200 mesh) is sometimes added to increase early strength of the mixture.
3. Latex-Cement:[10] This is a special cement composed of latex, cement, a surface active agent, and water. It has proved useful in such special applications as plug-back jobs for water exclusion. It is especially resistant to contamination with oil and/or mud and exhibits a high strength bond with other materials (casing, rocks, etc.).
4. Diesel-Oil-Cements (DOC): Mixtures of portland cement, diesel oil (or kerosene), and a chemical dispersant have been found useful in well repair (remedial) work to seal off water-bearing strata. This material does not set until brought into contact with water and has, therefore, an unlimited pumping time. It has also been used to prevent lost circulation.

---

*The depth limits are based on the conditions imposed by the casing well-simulation tests, API RP 10B, schedules 1–10, inclusive, and should be considered as approximate values.

5. Oil-in-Water Emulsion Cements:[11] Low water loss, low density cements of adequate strength and thickening time have been prepared from kerosene, water, cement, and 2 to 4% bentonite. Calcium lignosulfonate is used as emulsifying agent and retarder. Such cements have applicability in both primary and remedial cementing.
6. Resin Cements:[12,13] Proper combination of synthetic resins, water, and portland cement are often used to provide an improved formation–cement bond in certain remedial operations. Cost prohibits use of this material for routine cementing of casing.
7. Gypsum Cements:[12,13] These are special mixtures which have high early strength and easily controlled setting times. Gypsum is the basic ingredient. Their principal use is to provide temporary plugs during testing and remedial work.

The selection of the proper cement for a specific job is a matter of conjecture, and differences of opinion exist among oil operators. In very deep, high pressure wells a special mixture may be custom tailored for the particular job. For shallower, moderate temperature wells a number of cement types may be satisfactory and the choice will depend on cost and personal preference based on experience. The following discussion lists the usual factors on which selection of cement type is based. Table 14.1 lists a number of additives, some of which are referred to in the discussion.

1. Slurry Density: Normally, the slurry density should be the same as that of the mud in the hole at the time of cementing. This minimizes chances of either blowouts or lost circulation occurring during cementing. The most common material used to decrease density is bentonite, although the reduction is primarily due to the increased water content of the slurry made possible by the presence of the clay. A special diatomaceous earth (Diacel D) of lower density than bentonite has also been developed and works on the same principle. Expanded perlite is often used for the same purpose when the addition of a bridging agent is desirable. This is a volcanic ore material which has been expanded by heating to its fusion point and subsequent cooling. It has a cellular, thin-walled texture which breaks down under high pressures, allowing water to be forced into the pores. Again, it is the additional water capacity which yields the bulk of the density reduction. Two to 6% bentonite is used with perlite slurries to prevent buoyant segregation of the perlite particles.

Coffer, Reynolds, and Clark[17] have proposed the use of small, hollow clay spheres (bubbles) to reduce slurry density. Laboratory tests indicated slurry densities as low as 10 lb/gal could be obtained while satisfactory strength was still maintained. This is an interesting departure from standard density reduction measures and offers some promise.

As a rule, the amount of a particular additive used in

TABLE 14.1

CEMENT ADDITIVES

Common Name	*Description*	*Function(s)*
Gel [14–16]	Bentonite	Reduce density, improve suspending qualities (gel strength), reduce filtration loss, improve perforating characteristics
Clay Bubbles [17]		Reduce density
Diacel D [7–9]	Special diatomaceous earth	Reduce density
Expanded Perlite [18]	Volcanic ore expanded (made cellular) by heating to fusion point	Reduce density, act as bridging material for lost circulation
Pozzolans[6]	(See special cements)	Reduce density, improve perforating characteristics, increase thickening time, furnish cementitious properties by combining with lime
Hydrocarbons[11]	Diesel oil, kerosene	Reduce density, also used in special cements
Barite[14]	$BaSO_4$	Increase density
Calcium Chloride	$CaCl_2$	Accelerate setting (reduce thickening time), decrease freezing temperature of mixing water*
Sodium Chloride [19]	NaCl	Same as above
Diacel A[7–9]	Special sodium silicate	Accelerate setting
Diacel LWL[7–9]	Carboxymethyl hydroxyethyl cellulose	Reduce filtration loss, retard setting
	Calcium lignosulfonates	Retarder, dispersant, emulsifier
Nut shells, cellophane flakes, etc.		Prevent lost circulation

*This is of particular interest in surface pipe jobs in the winter.

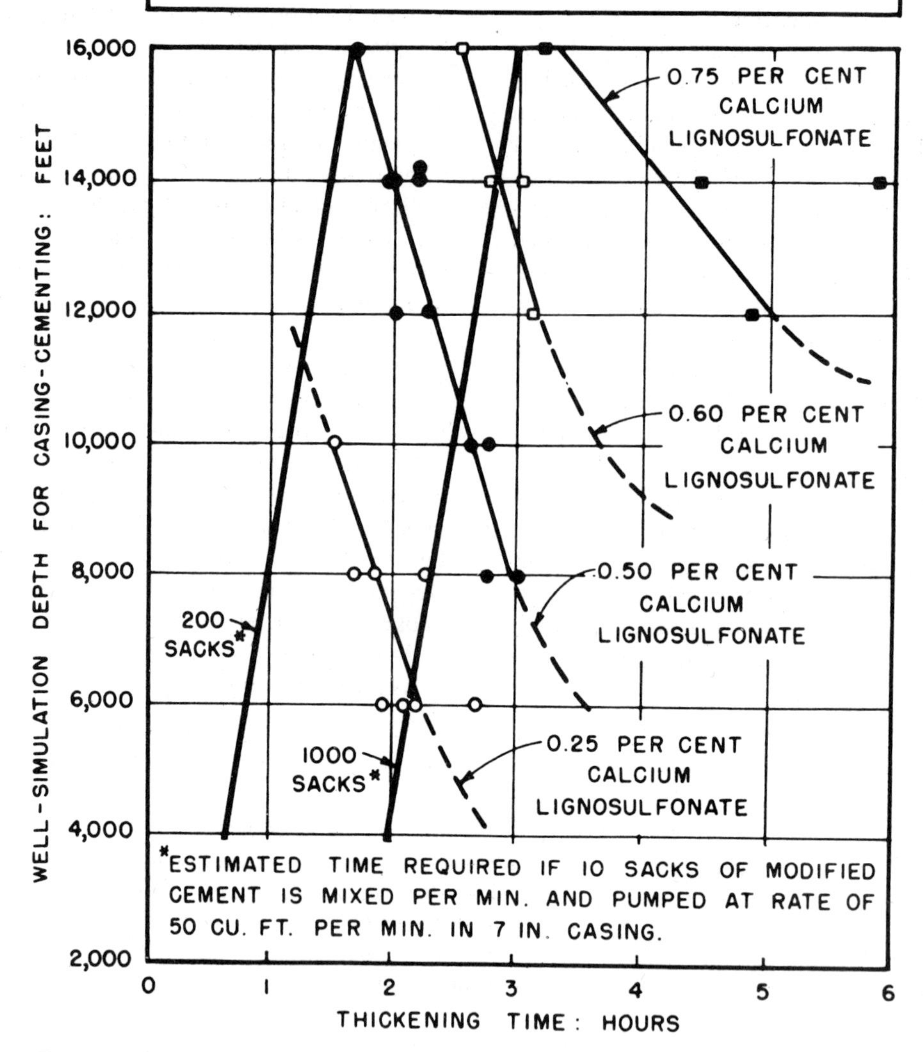

Fig. 14.4. Thickening times of portland cement slurries containing bentonite and calcium lignosulfonate. After Morgan and Dumbauld,[16] courtesy API.

a slurry is expressed as a wt. % of the dry cement. The following illustrates the practical units used in cement calculations.

Despite the current usage of bulk cement it is still customary to specify cement volume in sacks, where 1 sack = 94 lbs = 1 ft³ bulk. The absolute specific gravity of portland cement is 3.14. Hence a sack of cement contains only $94/(3.14 \times 62.4) = 0.48$ net ft³ of cement. In other words, its porosity is 52%. Slurry volume is computed by assuming the volumes additive. Water-cement ratios are usually expressed in gallons per sack. For example, if 5.5 gallons per sack is used, the slurry volume is

$$\frac{5.5}{7.48} + 0.48 = 1.22 \text{ ft}^3 \text{ per sack of cement}$$

Similarly, the slurry density is

$$\frac{1}{1.22}\left[\frac{5.5}{7.48}(62.4) + 94\right]\frac{1}{7.48} = 15.4 \text{ lb/gal}$$

If 6% bentonite (by weight of cement) of sp.gr. = 2.7 is added to the slurry, and the water-cement ratio is increased to 8.8 gal/sk, the resulting slurry volume is

$$\frac{(0.06)(94)}{(2.7)(62.4)} + \frac{8.8}{7.48} + 0.48 = 1.69 \text{ ft}^3\text{/sack of cement}$$

and the density is

$$\frac{(1.06)(94) + (8.8)(8.33)}{(1.69)(7.48)} = 13.7 \text{ lb/gal}$$

These calculations are identical to those for the mud system problems of Chapter 6, and the same formulae are applicable. Slurry density may be increased to approximately 19 lb/gal by use of barite, as in drilling muds. Perlite additions are usually expressed in terms of bulk volume ratio, i.e., a 1:1 mix = 1 bulk cu ft cement (94 lb) + 1 bulk cu ft perlite (13 lb). The absolute specific gravity of perlite is approximately 2.2, hence the total porosity of a bulk cubic foot is about 90%.

2. Thickening Time: A cement slurry must remain fluid sufficiently long for it to be pumped in place, with some margin of time allowed for possible trouble. Thickening time is the time necessary for the slurry consistency to reach 100 poises (measured with either the Stanolind or Haliburton Thickening Time Tester) as specified in API RP 10B. Well depth pressures and temperatures are simulated by standard test procedures which approximate field conditions. The testers resemble mud viscosimeters in that the torque needed to rotate a spindle through the slurry provides the consistency value.

Calcium lignosulfonate is the most common retarder. Pozzolan cements also increase thickening times, the pozzolan-lime type of cement (no portland cement) being particularly applicable to deep, hot wells.[5] CMHEC (Diacel LWL) retarded cements have been developed which are suitable for depths greater than 18,000 ft and have been used for special purposes to 21,600 ft and 326°F.[9] The availability of cementing materials which provide satisfactory thickening times for 20,000 ft drilling is limited and further development is needed. The general effect of calcium lignosulfonate on gel–cement slurries is shown in Figure 14.4.

Calcium chloride is the most common accelerator and has been in satisfactory use for years. The effect of sodium chloride has generally been considered more erratic; however, relatively recent work has shown it to be a satisfactory accelerator if properly used.[19] This is of interest in areas where sea water can be utilized as mixing water for the slurry. Acceleration of setting time is of interest only in shallow wells and surface pipe jobs.

3. Strength-Time Requirements: There is no universal agreement on strength requirements for oil well cements; however, figures of 200 to 500 psi compressive strength after 24-hr periods are often quoted. The value of 500 psi is the more widely accepted. Portland cements have far more strength than is generally needed; however, many additives reduce this to normal levels. Density reduction always results in considerable strength loss, this being the limiting factor. Retarders may also reduce cement strength. The time required for hardening to a desired strength is important, since rig

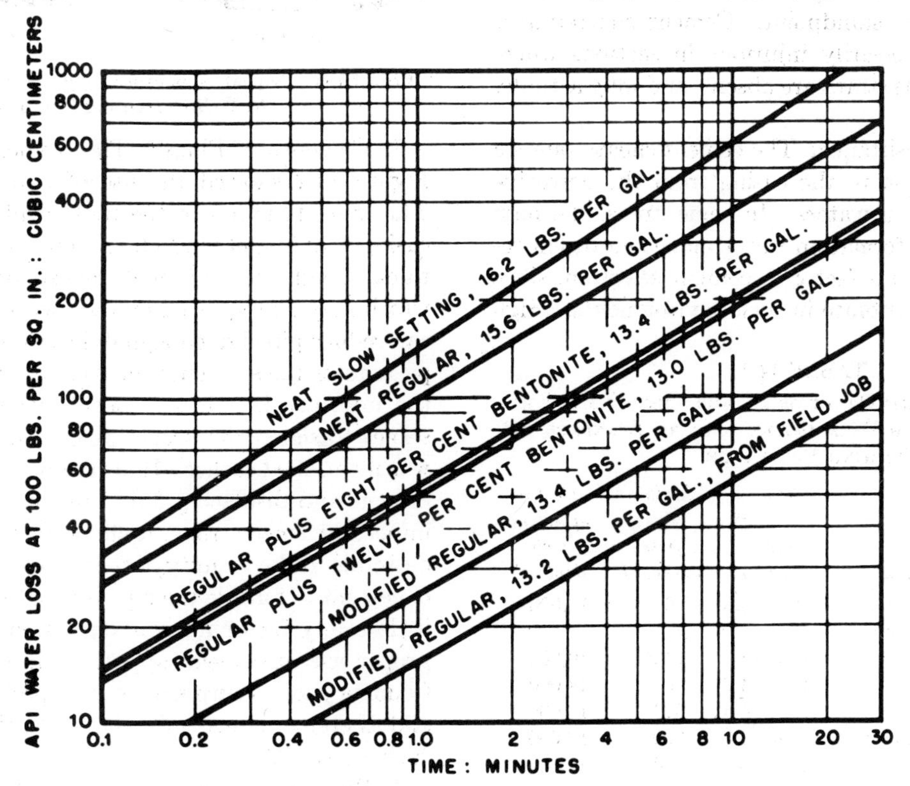

Fig. 14.5. Filtration properties of various cement slurries. Modified cement contains calcium lignosulfonate as dispersant. After Morgan and Dumbauld,[16] courtesy API.

time is lost during the waiting-on-cement (WOC) period. These delays are often excessive, and can many times be reduced by using proper cement mixtures.[20,21]

4. Filtration Properties: The water loss of neat cement is extremely high. Consequently, when a slurry contacts a porous formation from which the mud cake has been removed, it may quickly become dehydrated and undergo a so-called *flash set*. This may cause pipe to stick and prevent the rotation or reciprocation desirable when wall scratchers are used. Bentonite is of value in this regard, as may be seen in Figure 14.5. Water losses comparable to those of drilling fluids may be obtained with CMHEC additions of 0.25 to 0.9%. Water loss may also be of some interest from a formation damage standpoint.

5. Permeability of Set Cement: Since the main function of cementing is segregation, the lowest possible permeability is desirable. The permeabilities of many mixtures have been reported by Morgan and Dumbauld[16] and it appears that they are usually adequate if other properties are satisfactory. The permeability of atmospheric test samples is not always indicative of the down hole value, owing to the compacting effect of pressure.

6. Perforating Qualities: Completely hardened cement is apt to shatter and fracture excessively when gun perforated. For this reason, it is desirable to perforate while the cement is reasonably *green*. The low-strength cements have less tendency to shatter and are therefore desirable from this standpoint. Cement shattering is not, however, necessarily injurious in sections where nearby water or gas strata are absent, and may actually be beneficial.

7. Corrosion Resistance: The cement sheath offers a degree of protection to the casing from the corrosive actions of formation waters. In some areas, the high sulfate content of formation water makes corrosion resistance an important factor in cement selection, since many cements deteriorate in such environments. High temperatures alleviate this hazard, and hence it is in shallow wells that most problems arise. Cements having low tricalcium aluminate content are less susceptible to sulfate attack.

TABLE 14.2

General Applicability Range of Various Cement Mixtures. Temperature – depth Data Correspond to Standard API Testing Procedure.

Cement type or mix	*Common density range, lb/gal*	*Maximum static fmtn. temp., °F*	*Corresponding well depth, ft*
(0–14%) Bentonite-cement	12.3–16.8*	260	12,000
(2–6%) Bentonite-perlite-cement	12.3–14	230†	10,000†
Pozzolan-cement	13.5–14.2*	290	14,000+
Pozzolan-lime	13.8–14.3*	320+	16 000+
Oil-in-water emulsion-cement	11.4–12.8	320	16,000
Diacel systems	11–13.6	320+	16,000+

*Higher densities by using barite or illmenite ore.
†These data not obtainable with standard testing equipment.

Consideration of the above factors shows clearly that the selection of proper cement type may be a weighty problem. Certainly, no specific rules will fit all situations. Individual investigation of slurry properties of the particular cement brand (as well as grade) to be used should always be made in problem wells or other unusual circumstances. Consultation with cementing experts is always worthwhile in such cases. Table 14.2 is a summary of some commonly accepted limitations of various cementing mixtures.

14.22 Auxiliary Cementing Equipment

A discussion of certain auxiliary equipment items (Figures 14.6, 14.7) will serve to illustrate other factors influencing attainment of a successful primary cement job.

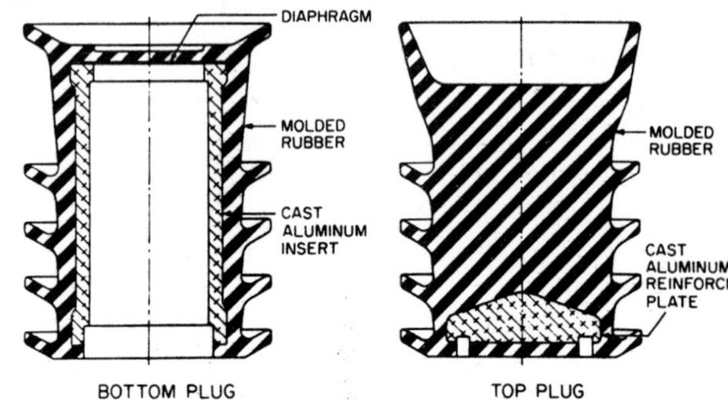

Fig. 14.6. Tyipcal cementing plugs. Courtesy Halliburton Oil Well Cementing Company.

1. Cementing Plugs: The cementing process in Figure 14.2 showed the use of two cementing plugs. The main function of the bottom plug is to wipe the casing wall free of mud film and prevent buildup of any thick, tough, cement–mud deposit on the casing wall which may be bypassed by the top plug. Such deposits can reduce pipe i.d. enough to interfere with subsequent passage of tools. Contamination of cement by mud at the interface is a lesser consideration. The top plug serves to signal the proper placement of the slurry, as was mentioned earlier. It also prevents commingling of cement and displacing fluid; in this case, this is quite important since strong, undiluted cement is especially desirable in the vicinity of the casing shoe. The volume of cement left inside the pipe is usually small and leaves little safety factor for contamination or dilution. The proper use of cementing plugs will greatly reduce the frequency of occurrence of the problems mentioned and thereby reduce primary cementing failures from such causes.[22]

2. Wall Scratchers: It is evident that a strong, positive bond between cement and formation is desirable

Obviously, the cement must contact the formation and hence, the mud cake must have been removed. Howard and Clark[23] have shown that the erosive or scouring action of the cement does not adequately perform this function, and that other means are required. The most common practice is use of mechanical wall scratchers of either the rotating (see Figure 14.7) or reciprocating type. These are tack-welded onto the string at the desired level(s) as determined from well logs. The spring loaded steel fingers literally dig into and loosen the wall cake as the pipe is rotated, allowing intimate contact between cement and the scratched area. Early use of these devices often resulted in sticking of the pipe (indicated by excessive rotating torque), which was generally due to rapid dehydration (flash set) of the cement as it contacted porous formations. This can be largely eliminated by reduction of the water loss from the slurry. Since the pipe is rotated at the desired setting depth, there is actually no cause for alarm if it does stick, provided, of course, that reasonable care is exercised in applying torque. Scratchers are generally used on oil strings only, as it is on these strings that precise segregation of zones is necessary. Cannon[24] has cited a study in which remedial cementing frequency due to primary cementing failure was reduced from 0.58 to 0.16 remedial jobs per oil string by use of rotation-type wall scratchers. Scratcher use may be advantageous on other strings, depending on the circumstances. Selection of intervals to be scratched depends on the pay section characteristics. Segregation of water zones, high gas-oil ratio or gas zones, and strata within the oil zone having greatly different permeabilities are the usual goals. Common sense along with well log inspection will show the best scratcher spacing pattern. There is, of course, no point in scratching dense impermeable zones where no mud cake exists. Caliper logs are also helpful in some instances for showing enlarged zones in which the scratcher fingers cannot reach the wall and are therefore useless. Temperature surveys consistently show that lower fill-up height is obtained when scratchers are used. This clearly implies that mud cake removal occurs.

3. Casing Centralizers: These devices are designed to provide a reasonably uniform cement layer around the pipe. This is the source of their name: they centralize the pipe in the hole and prevent it from lying against

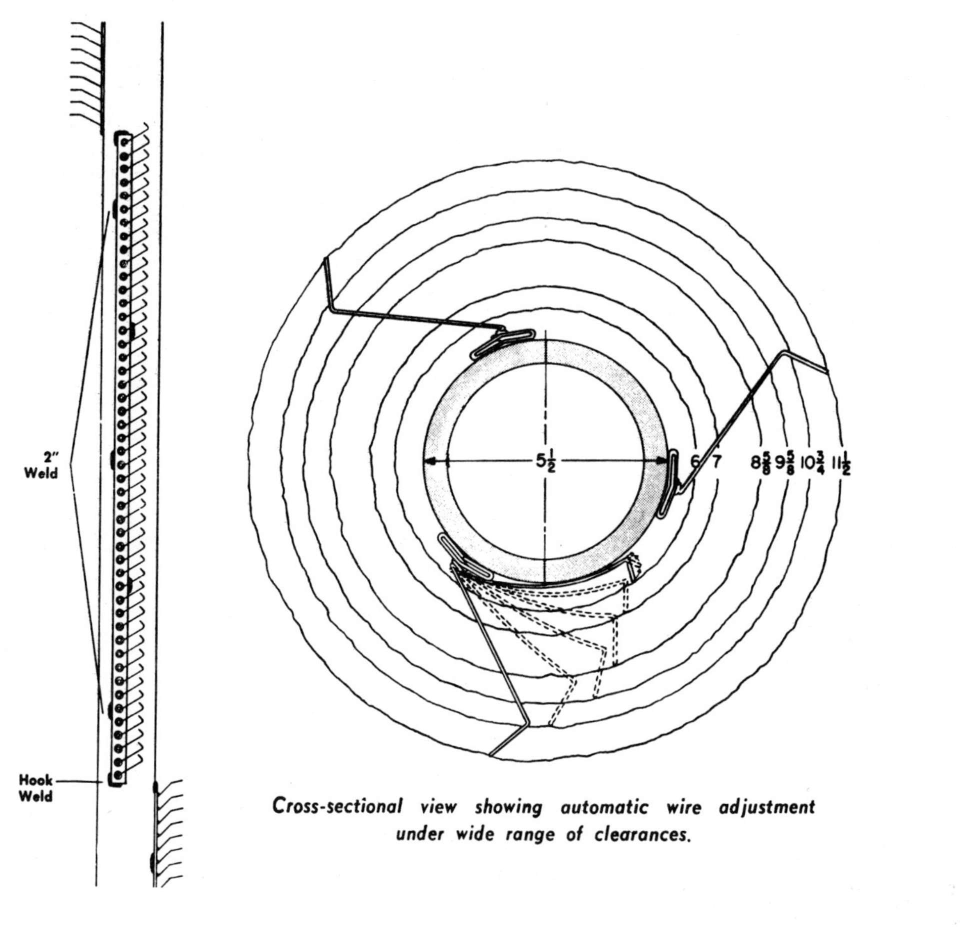

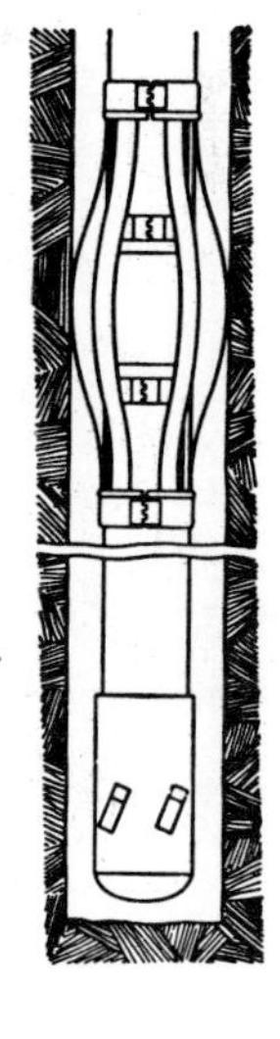

(B)

Fig. 14.7. Auxiliary cementing equipment. **(A)** Rotation type wall scratchers. **(B)** Function of casing centralizer. Courtesy Baker Oil Tools, Inc.

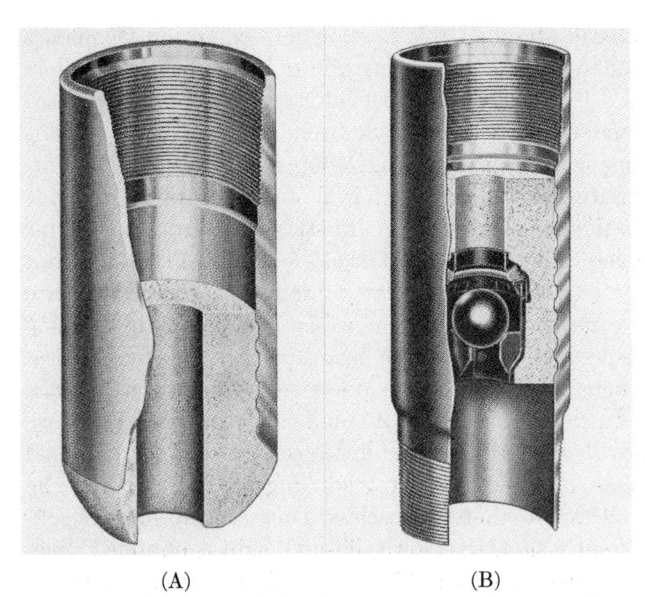
(A)
(B)

Fig. 14.8. Conventional floating equipment. Courtesy Baker Oil Tools, Inc. **(A)** Guide shoe. **(B)** Float collar.

the wall. Several designs are in use although the hoop skirt type (Figure 14.7) is quite common. These have hinged end bands which allow them to be readily slipped around the pipe. Sufficient clearance is allowed so that the pipe may rotate within the centralizer. Either casing couplings or various types of attachable stops are used to prevent excessive vertical travel of the centralizer. Centralizer spacing should be sufficiently close to keep pipe-wall clearance at some acceptable minimum value. A few, properly spaced centralizers are a valuable aid in obtaining a uniform cement sheath over critical hole intervals. Again, the caliper log may be used as an additional aid to centralizer location.

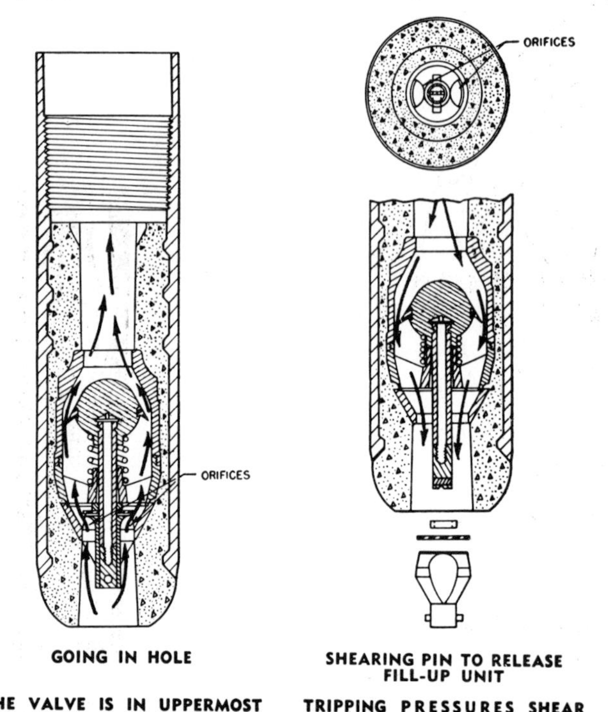
ORIFICES
ORIFICES
GOING IN HOLE
SHEARING PIN TO RELEASE FILL-UP UNIT
THE VALVE IS IN UPPERMOST POSITION. FLOW OF FLUID IS GOING UP THROUGH THE ORIFICES.
TRIPPING PRESSURES SHEAR THE PIN TO RELEASE FILL-UP UNIT AND PUMP IT AWAY INTO HOLE.

Fig. 14.9. Operating diagram for one type of automatic fill-up float shoe. Courtesy Halliburton Oil Well Cementing Company.

4. Floating Equipment: This commonly consists of a guide shoe and float collar (Figure 14.8). The round nosed guide shoe helps prevent the casing from catching or hanging up as it is lowered into the well. The float

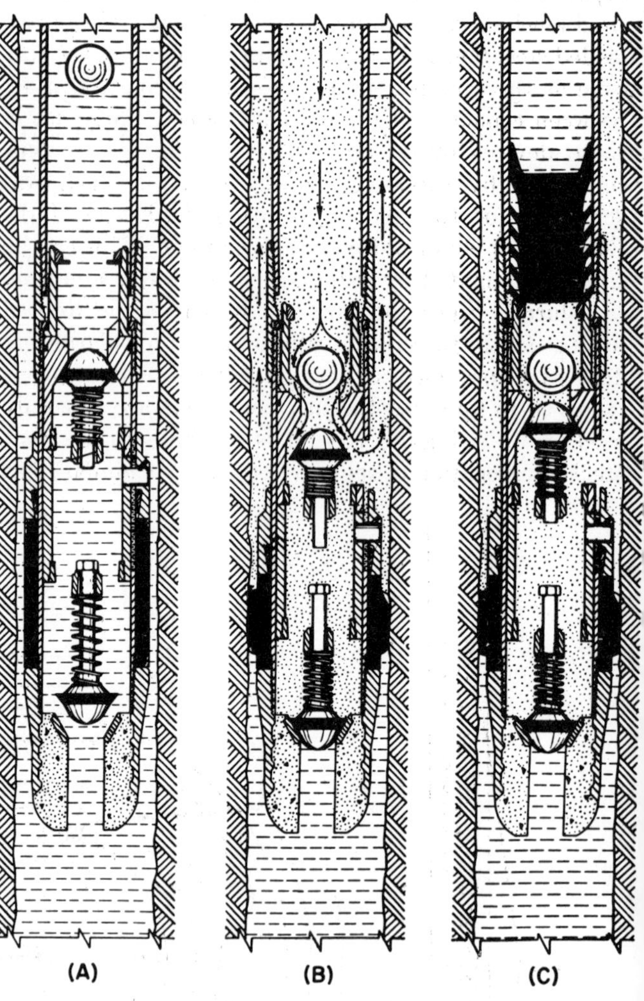
(A)
(B)
(C)

Fig. 14.10. Operating diagram of typical packer type cementing shoe. Courtesy Halliburton Oil Well Cementing Company. **(A)** Casing on bottom. Setting ball falling to the setting cage. **(B)** Packer set. Cementing job underway. **(C)** Job completed. Top plug at shut-off position in top of packer.

collar is generally inserted one to three joints above bottom, where it serves as a back pressure valve preventing backflow of cement after placement. Float valves are also incorporated into the guide shoe, in which case it is called a float shoe. Conventional floating equipment prohibits mud entry into the casing as it is run, hence the term, *floating*. While this lessens the hook load, it also increases down hole pressure surges and causes frequent delays in the casing job for filling the pipe. These disadvantages are overcome by use of automatic fill-up equipment which allows either partial or complete mud entry into the casing during running, while still retaining the back pressure valve feature after cementing. One such device is shown in Figure 14.9. The internal parts of all floating equipment are drillable.

5. Packer Type Cementing Shoes: In instances where casing is to be set some distance off bottom, it is often desirable to segregate the open hole zone(s) from the hydrostatic pressure of the cement column. Packer shoes allow this, as is shown by the operating diagram of Figure 14.10.[25] A casing shoe swivel run directly above the packer shoe permits pipe rotation if desired.

14.23 Cementing Volumes

The volume of cement needed for a particular case depends, of course, on the annular volume to be filled. Precise computation of this volume is difficult even when caliper logs are available, and some excess factor is

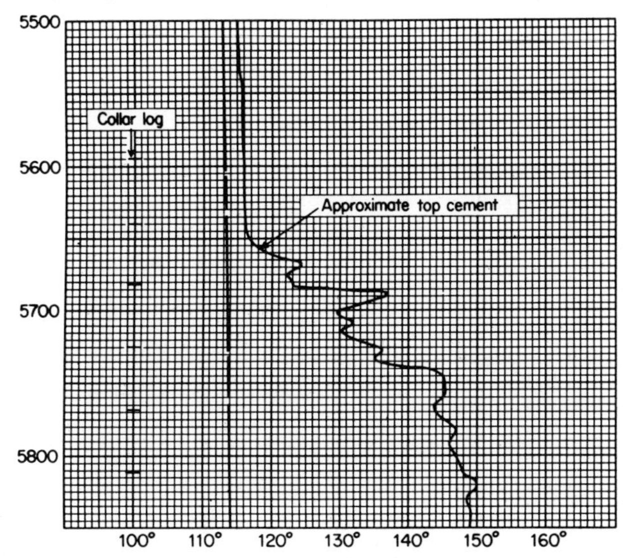

Fig. 14.11. Temperature survey showing cement top. Courtesy Halliburton Oil Well Cementing Company.

always applied. Previous experience in the same or similar areas is the best guide as to what this excess should be. Washing-out of surface or other loosely consolidated beds and salt sections must be accounted for if calculated fill-up is to approach the actual fill-up height. Filtration losses during placement reduce slurry volume by a generally incalculable amount, and reduction of slurry water loss results in higher fill-up per sack. Proper selection of an excess factor depends on the above considerations, and on reliability of hole volume data and past experience as well. Use of 15 to 25% excess is common when caliper logs are available, this being extended to as much as 50% or more when a gage hole is assumed.

The actual cement level or *top* is determined by running a temperature survey a few hours (commonly 6 to 12) after the job. The sudden increase in temperature opposite the top of the cemented zone is clearly detectable as a departure from the normal thermal gradient (Figure 14.11).

14.24 Other Considerations

Strength Retrogression at High Temperatures:[26–28] Limited testing has shown that many deep well cementing mixtures exhibit rather drastic reductions in strength with time at high temperatures. Partial results of one detailed study are given in Table 14.3. Note that the high bentonite and diatomaceous earth mixtures (also highest water content) showed the greatest strength losses. A critical temperature region of 230 to 290°F is the range in which severe retrogression takes place in most cases.

The only cement which did not lose strength was the pozzolan-lime type. Further data of this nature and, perhaps, data taken over longer periods of time are needed to evaluate fully the possibilities of these effects. It is clearly evident that 24-hr strength data are not always sufficient to evaluate oil well cementing materials.

TABLE 14.3

EFFECT OF HIGH TEMPERATURE ON SOME DEEP WELL CEMENTING MIXTURES[28]

Composition	*Slurry density, lb/gal*	*Average compressive strength, after 180 days, psi* 200°F	260°F	320°F	*Fraction of 24-hr strength retained after 180 days, %* 200°F	260°F	320°F
Retarded cements*	16.25	9200	5400	2040	145	87	39
Retarded cement-4% gel	15.2	5200	965	600	136	25	18
Pozzolan-API Class A cement-retarder	15.6	5300	5400	2100	240	195	54
API Class A cement-12% gel-retarder	12.6	1120	40	65	105	4	28
Pozzolan-lime-retarder	14.4	4000	4800	4700	140	140	153
API Class A-diatomaceous earth-CMHEC retarder	12.4	800	400	50	91	31	5

*Average of 5 brands

Effect of Drilling Mud Additives on Cement Properties: An extensive investigation of the effect of numerous common mud additives on cement properties has been conducted by an API committee.[29] Briefly, it was found that small degrees of contamination could cause appreciable injurious effects to cement properties.

These effects were erratic and difficult to predict in many instances. CMC, quebracho, and caustic soda were among the additives studied and were shown to be capable of seriously impairing thickening time and strength. This study lends further emphasis to the importance of minimizing mud-cement contamination and taking measures to remove the mud cake. Additions of decontaminants such as activated charcoal[30] or a paraformaldehyde-sodium chromate tetrahydrate mixture[31] have been proposed for eliminating effects of mud contamination. Some success in setting open hole plugs has been attributed to these mixtures.

Selection of Casing Point: The setting depth (casing point) for the oil string is generally dictated by the pay section characteristics. In the case of surface and intermediate strings, however, some degree of choice exists and it is worthwhile to give this matter some thought. The lower joints of these strings are subjected to shock loads during subsequent drilling, as well as erosive washing around the shoe by the mud. Cement strength is of little value, if the rocks at these points are themselves weak and unable to withstand imposed loads. Numerous situations have been observed in which several joints of surface or intermediate pipe have been knocked or whipped off during subsequent drilling. It is always advisable to *land* these strings in a dense formation where true hole gage exists. Well logs (including drilling time) are useful in selecting proper casing points.

14.3 Squeeze Cementing

This process is a secondary cementing method and as such will receive limited attention here. Many applications for squeeze cementing exist; however, most of these arise owing to some functional failure of the primary cement job. Squeeze cementing normally involves the injection under pressure of a slurry into a porous formation. The pumping rate is slow enough to allow for dehydration and/or initial setting of the slurry. This effect is indicated by increased pump pressure, and injection is continued until the desired final or squeeze pressure is attained. Example 14.2 illustrates the general procedure.

Example 14.2

A well was perforated for production from 7000 to 7015 ft following the primary cement job. Water and mud were produced as a result of communication of the 7000 ft zone with an upper strata. The following squeeze cementing procedure was carried out:

1. A retrievable squeeze cementing tool was run on 2-in. tubing to 7020 ft. Water was then circulated until it extended well above the perforated zone.
2. The squeeze packer was then raised and set at 6920 ft; tubing and casing were tested for leaks. Fifteen barrels of additional water were injected to break down the formation and insure that it would accept fluid.
3. A 15.8 lb/gal neat portland cement slurry was pumped down the tubing, the following factors having been considered:

(a) Slurry volume $= 1.14\ \text{ft}^3/\text{sack} \times 150 = 171\ \text{ft}^3$
(b) $5\frac{1}{2}$-in., 17 lb/ft casing capacity $= 0.1305\ \text{ft}^3/\text{ft}$
(c) Working casing volume below tool $= (7015 - 6920)(0.1305) = 12.4\ \text{ft}^3$, or the equivalent of 11 sacks.

4. The first 100 sacks were injected into the formation at a relatively high rate (2 to 3 bbl/min). No pressure build-up was noted and the pumping rate was reduced to $\frac{1}{2}$ bbl/min for the next 20 sacks and then to $\frac{1}{4}$ bbl/min. When 125 sacks were injected, pressure began increasing and reached 5000 psi (surface) when 135 sacks were in place. Pumping was stopped and this pressure was held for two minutes. The tool was then unseated and excess cement was rapidly reverse circulated out.

   Steps 2 to 4 may be visualized by noting the sequence of Figure 14.12.
5. The plug was drilled and the zone reperforated; it yielded clean oil, hence the job was successful.

It is obvious that such success may not always be obtained. Squeeze cementing is at best a ticklish business requiring close supervision. Numerous variations of equipment, cementing materials, and technique are used as dictated by well conditions and experience. Howard and Fast[32] have presented a detailed analysis of squeeze cementing in which the following were concluded:

(1) Proper choice of breakdown fluid (water, in our example) enhances the chance of success. Water is superior to mud, but an HCl-HF acid mixture is still more successful, owing to the cleaner exposed rock surface resulting, which allows better cement–rock bonding. Various breakdown fluids commonly called mud acids or mud cleanout agents are now available as breakdown fluids.

(2) A relationship exists between the final or squeeze pressure and success of the job. Bottom hole squeeze pressures of the following magnitudes were found to be generally successful:

$$\text{Bottom hole squeeze pressure} = 1.0 \times \text{depth} + 2000\ \text{psi}$$

For shallow wells the following was sometimes necessary:

$$\text{Bottom hole squeeze pressure} = 0.6 \times \text{depth} + 5000\ \text{psi}$$

We will not attempt to discuss the precise mechanism of squeeze cementing from the point of view of final cement distribution. This, of course, depends on the orientation of the fracture(s) formed, and the same principle is involved as that which governs well stimulation by hydraulic fracturing. This is a controversial

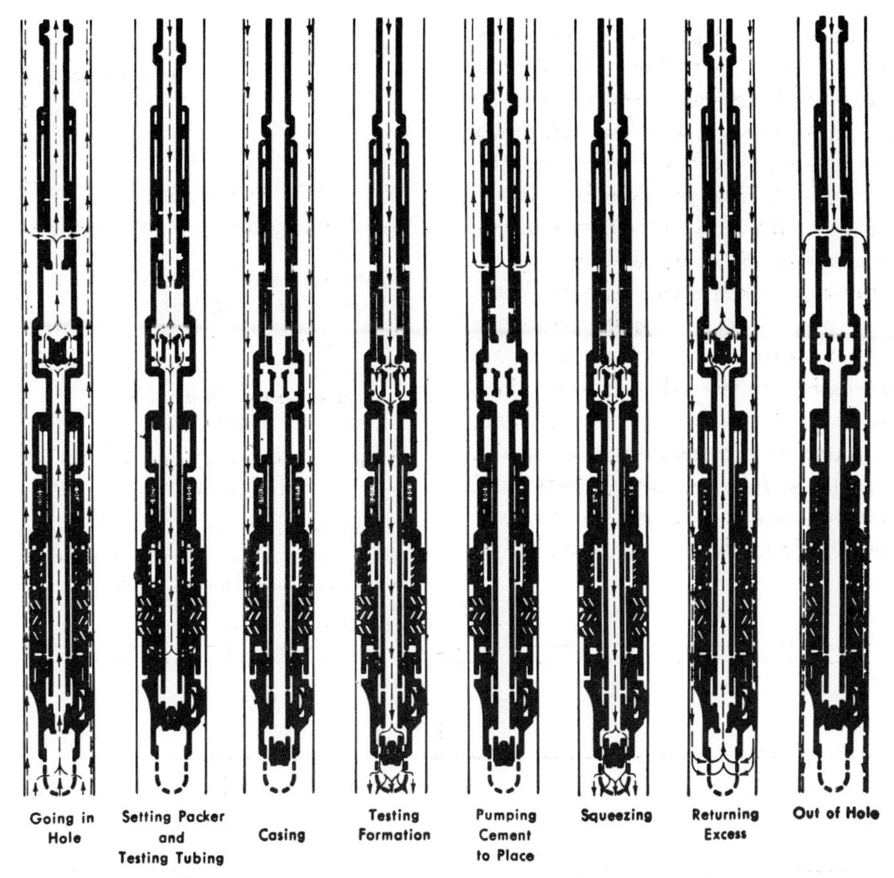

Fig. 14.12. Typical squeeze cementing procedure utilizing retrievable cementing tool. Courtesy Halliburton Oil Well Cementing Company.

topic of considerable importance which will not be discussed in this volume, owing to space restrictions. Papers by Walker,[33] Hubbert and Willis,[34] Scott, Bearden, and Howard,[35] and Reynolds, Bocquet, and Clark[36] are suggested as additional references on the subject.

14.4 Casing Types and Specifications

Some expansion of the general casing functions as applied to specific strings is in order.

Surface Strings: In hard rock areas, a single string of surface pipe set a few hundred feet deep is usually adequate to:

(1) Control caving and washing out of poorly consolidated surface beds
(2) Furnish a means of handling the return flow of drilling mud
(3) Protect fresh water sands from possible contamination by drilling mud, or oil, gas, and/or salt water from lower zones
(4) Allow attachment of blowout preventors

In soft rock areas, it is often necessary to set both a conductor string and a surface string to perform these functions. In such instances, the conductor string may be two or three hundred feet long with the surface pipe extending to a depth of two or three thousand feet. Some surface pipe is used on all wells of appreciable depth and is in fact required by law in many states. The main purpose of such legislation is to insure protection of fresh water supplies, as stated in function (3).

Intermediate Strings: The decision to use intermediate strings is largely dependent on well depth and geologic conditions in a specific area. Wells of moderate depth may use no intermediate pipe, while deep wells may require one or more protective strings. The principal function of intermediate casing is to seal off troublesome zones which

(1) contaminate the drilling fluid, making mud control difficult and expensive (salt, gypsum, heaving shales, etc.),
(2) jeopardize drilling progress with possible pipe sticking, excessive hole enlargement, or other fishing hazards.

In deep drilling, there is often a reluctance to drill with excessive intervals of open hole even though no particularly troublesome zones are exposed. In such instances, an intermediate string may be run as a precautionary measure so that any ensuing problems may be solved more easily. The intermediate pipe also affords greater safety in case of blowouts.

Oil Strings: This is the last and deepest casing string run. It is set above, (Figure 14.1), part way through, or completely through the lowermost pay zone, depending on the type of completion to be performed. The oil string furnishes the means of segregating the pay section from all other zones and is the work shaft through which access to the producing zone is gained. Normally, the produced fluids flow through tubing; sometimes, however, the oil string may actually serve as the flow conduit.

Oil well casing is specified by range length, type of construction, coupling type, steel grade, outside diameter, and weight per foot. Detailed specifications are set forth in the API Standard 5A[37] and Bulletin 5C2.[38] In addition to API casing types, other modifications and steel grades for special applications are available from various steel companies. The range lengths of API casing are shown in Table 14.4. Although butt weld and electric weld pipe are used to a limited extent, only seamless construction casing will be considered in our design calculations.

TABLE 14.4

RANGE LENGTH OF API CASING

	Range, ft		
	1	2	3
Total range length (incl.)	16–25	25–34	34 or more
Range length for 95% or more of carload			
Permissible variation (max.)	6	5	6
Permissible length (min.)	18	28	36
Range length for 5% or less of carload			
Permissible variation (max.)	9	9	—
Permissible length (min.)	16	25	34
*Jointers: Minimum length of shortest piece	5	5	5

*Jointers are two lengths connected by coupling, and may be shipped to a maximum of 5 per cent of the order; jointers not permissible on drill pipe and tubing.

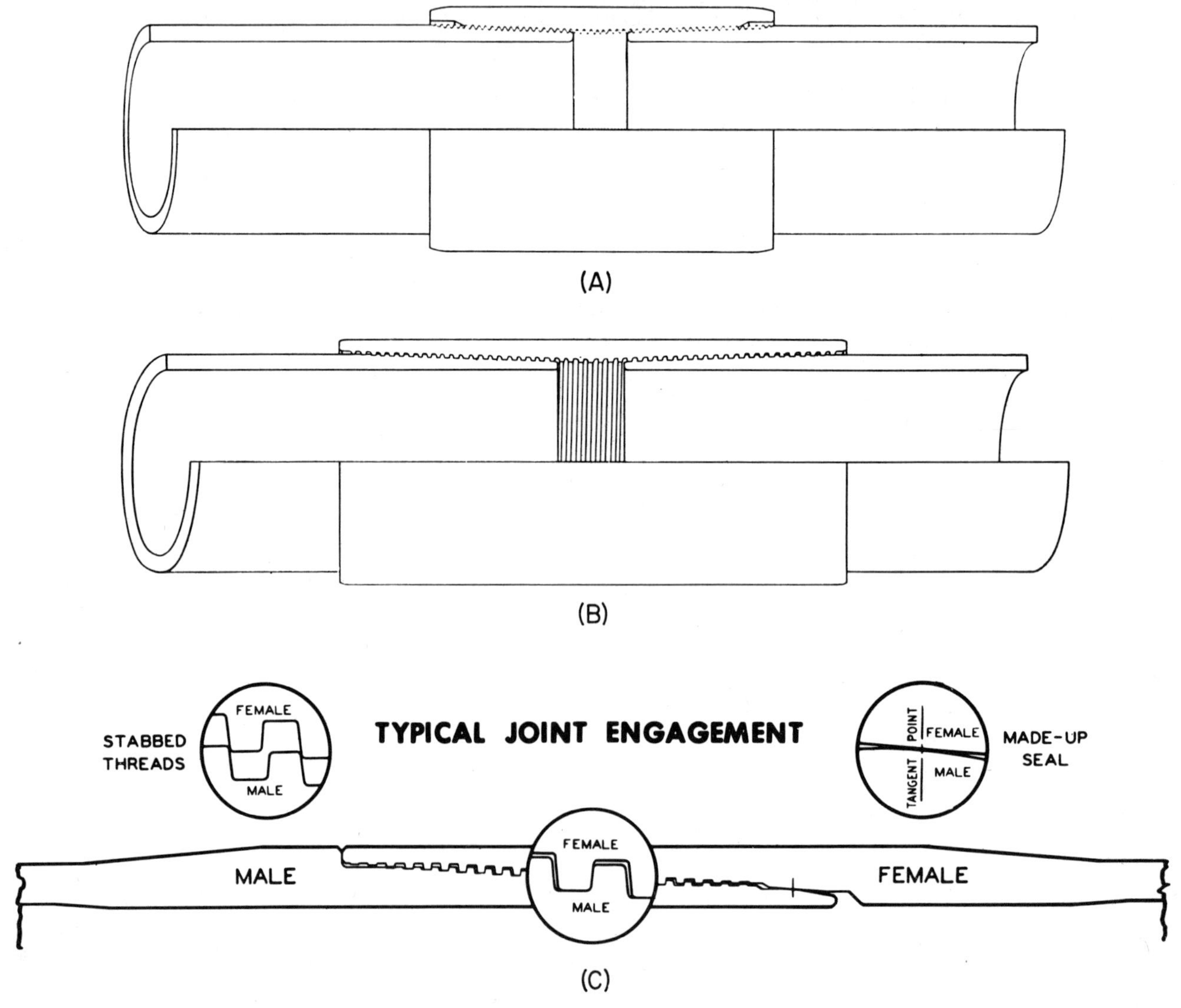

Fig. 14.13. Typical casing couplings. **(A)** API long thread. **(B)** Buttress-thread (National Tube Division, U.S. Steel Corp.) **(C)** Extreme line coupling (Spang-Chalfant Division, National Supply Company.)

The most common coupling is the API type shown in Figure 14.13. Some other available and widely used types are also shown.

Five grades of steel for oil well casing and tubing have been standardized by the API. These are designated by both a letter and number (J-55, N-80, etc.) as listed in Table 14.5. Note that the number represents the minimum yield stress in psi. This so-called minimum stress is defined as 80% of the average value from test data. Design calculations are normally based on minimum values.

TABLE 14.5
API STEEL GRADES FOR OIL WELL CASING AND TUBING (STANDARD 5A)

	API Grade Symbols				
	F-25	*H-40*	*J-55*	*N-80*	*P-110*
Yield strength, min., psi	25,000	40,000	55,000	80,000	110,000
Ultimate tensile strength, min., psi	40,000	60,000	75,000	100,000	125,000
Elongation in 2 in. strip specimens, min., %	40	27	20	16	15

Casing of higher strength than P-110 are available in limited sizes and/or special order from some of the steel companies. One such grade is the V-150 listed in the setting tables of section 14.6. This does not, however, fall into any API classification at this time. V-150 casing has a minimum yield strength of 150,000 psi at 0.7% total strain.[39]

14.5 Design Considerations

Casing strings are designed to withstand three principal types of loading:

(1) Tensile load: Each joint supports all the weight below it, hence the greatest tension occurs at the top of the string. Buoyancy is commonly neglected and design calculations are made as though the casing were freely hanging in air. Consideration of tension only would dictate that the strongest pipe should be placed near the surface.

(2) Collapse pressure: This is the unbalanced external pressure imposed on the pipe. The worst condition normally conceivable is for the pipe to be empty with the hydrostatic pressure of the mud column exerted upon it from the outside. This is a commonly assumed design basis. Therefore, consideration only of collapse pressure would dictate that the strongest pipe should be placed at the bottom of the string.

(3) Burst pressure: This, of course, refers to a condition of unbalanced internal pressure. A common design criterion is to assume that the formation pressure is exerted on the entire length of the string. This condition is approached in gas wells, owing to the low pressure gradient in a gas column. Burst considerations are most critical near the surface, where the opposing external pressure on the casing is low.

Qualitative consideration of these factors immediately suggests that casing strings need not be composed of the same grade and weight pipe throughout their entire length. Considerable economy is realized through the use of tapered strings composed of sections of several weights and/or grades which are adequate to withstand the imposed stresses at the depths they occupy. Consequently, it is standard practice to run tapered strings in virtually all wells of appreciable depth. In many instances, the above simple assumptions regarding the magnitude of tensile, collapse, and bursting forces are adequate and furnish a well-designed string, when used in conjunction with standard design or safety factors. Situations do arise, however, when simplified standard practices should be discarded in favor of a detailed analysis of the specific well(s) in question.

14.51 API Performance Formulae

As a result of detailed study, the API has adopted standard formulae which specify the allowable collapse resistance, burst pressure, and tensile loads for API casing. These are reproduced below from API Bulletin 5C2, 7th ed., May 1957.

1. Collapse pressures for F-25 casing* are calculated by means of Eqs. (14.1) and (14.2). For failure in the elastic range (for D/t values less than 43.5):

$$p_c = 0.75\left[\frac{86670}{(D/t)} - 1386\right] \tag{14.1}$$

For failure in the plastic range (for D/t values greater than 43.5):

$$p_c = 0.75\left[\frac{50{,}210{,}000}{(D/t)^3}\right] \tag{14.2}$$

2. For Grades H-40, J-55, N-80, and P-110: For failure in the plastic range and for D/t values less than 14, approximately:

$$p_c = 0.75 \times 2Y_a\left[\frac{(D/t) - 1}{(D/t)^2}\right] \tag{14.3}$$

For failure in the plastic range and for D/t values of 14, approximately, and up to the point at which the curve for plastic failure intersects the elastic curve:

$$p_c = 0.75 \times Y_a\left[\frac{2.503}{(D/t)} - 0.046\right] \tag{14.4}$$

---

*The elastic equation for grade F-25 casing is not related to that for the other grades.

For failure in the elastic range:

$$p_c = 0.75\left[\frac{62.6 \times 10^6}{(D/t)[(D/t) - 1]^2}\right] \tag{14.5}$$

In the above equations,

p_c = minimum collapse pressure, lb/in.²
D = nominal outside diameter, in.
t = nominal wall thickness, in.
Y_a = *average* yield stress in collapse, lb/in.², as follows:

Grade H-40	50,000
Grade J-55	65,000
Grade N-80	85,000
Grade P-110	123,000

3. Internal yield pressures are calculated for all grades of casing by means of the following equation.

$$p_i = 0.875(2Y_m t/D) \tag{14.6}$$

wherein p_i = minimum internal yield pressure, in lb/in.²
Y_m = specified minimum yield strength, lb/in.², as given in Table 14.5.
t = nominal wall thickness, in.
D = nominal outside diameter, in.

4. Joint strengths are, in all cases, calculated from the following equations. For casing with short threads and couplings:

$$L_j = 0.80C(33.71 - D)\left[\frac{1}{t - 0.07125} + 24.45\right]A_j \tag{14.7}$$

For casing with long threads and couplings:

$$L_j = 0.80C(25.58 - D)\left[\frac{1}{t - 0.07125} + 24.45\right]A_j \tag{14.8}$$

wherein L_j = minimum joint strength, lb
D = nominal outside diameter, in.
t = nominal wall thickness, in.
A_j = cross sectional area of the pipe wall under the last perfect thread, in.²
= $0.7854[(D - 0.1425)^2 - d^2]$
C = a constant for the grade of steel, as shown in the following tabulation.

Grade	*Short T and C*	*Long T and C*
F-25	53.5	—
H-40	72.5	—
J-55	96.5	159.0
N-80	112.3	185.0
P-110	146.9	242.0

The above equations are, in general, empirical adaptations of standard equations. In particular, the origin of the collapse and burst equations requires special attention. For a thin walled cylindrical tube, the tangential stress (hoop tension or compression) σ_t caused by either external or internal pressure is closely approximated (within the elastic limit of the material) by assuming that σ_t is uniformly distributed over the well thickness t (Figure 14.14). Summing the forces shown in Figure 14.14 in the vertical direction gives:

$$p_i ld - p_o ld - 2\sigma_t tl = 0$$

Note that the pipe length l cancels.
For a relatively small wall thickness t one can assume that $d \cong D$, obtaining

$$p_i - p_o = \frac{2\sigma_t t}{D} \tag{14.9}$$

wherein p_i, p_o = internal and external pressures, psi
σ_t = average tangential or hoop stress of the wall, psi. Note that for $p_i > p_o$, σ_t is a tension, while for $p_o > p_i$, σ_t is a compression, as denoted by the resulting negative sign in Eq. (14.9).
t = wall thickness of the pipe, in.
d = inside pipe diameter, in.
D = outside pipe diameter, in.

Equation 14.9 is commonly known as Barlow's formula and is sufficiently accurate for large values of D/t, i.e., thin walled tubes. Note that if $p_o = 0$ (internal pressure only) and $\sigma_t = Y_m$, Eq. (14.9) becomes the API formula Eq. (14.6), in which the .875 is essentially a safety factor. Actually, Eq. (14.9) is but a simplified case of a more general equation commonly referred to as the thick walled cylinder or the Lamé equation. Since this equation is often neglected in elementary mechanics courses, it will be reviewed here. Consider Figure 14.15(A), which shows the cross-section of a thick walled cylinder. Intuitively, one would guess that in this case the stress due to internal or external pressure will not be uniformly distributed over the wall thickness, but will vary with the radial distance r. Visualize a thin half-loop [Figure 14.15(B)] of unit length in the longitudinal or axial direction. Since Barlow's formula is applicable to the thin element, we use the thin wall tube procedure to sum forces in the vertical direction, hence

$$2\sigma_t dr - 2\sigma_r r + 2(\sigma_r + d\sigma_r)(r + dr) = 0 \tag{1}$$

Considering that $d\sigma_r\, dr$ is negligible compared to the other terms, we obtain:

$$\sigma_t + \sigma_r + r\frac{d\sigma_r}{dr} = 0 \tag{2}$$

Equation (2) is the basic differential equation relating tangential stress σ_t, radial stress σ_r, and radial position r. Its solution requires, however, an expression relating σ_t and σ_r. This may be obtained by assuming that the cylinder undergoes uniform deformation in the axial direction as σ_t and σ_r develop. This is equivalent to assuming that a plane cross-section remains plane, which is a satisfactory assumption, except for regions

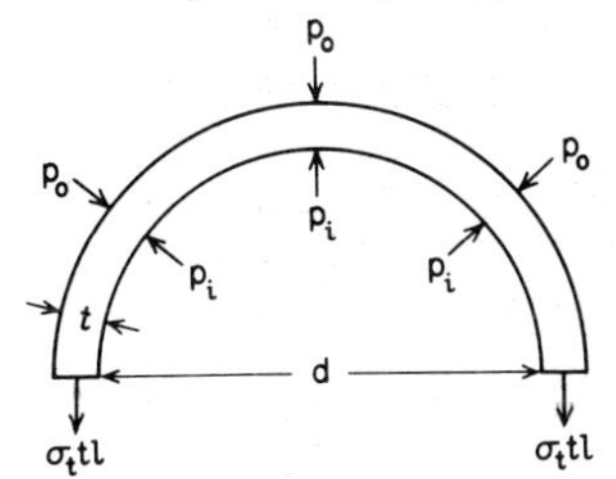

Fig. 14.14. Illustration of simple hoop tension in thin walled pipe.

close to the ends. The longitudinal strain δ_a is then constant for all values of r, hence:

$$(14.10)\quad \delta_a = \frac{\sigma_a}{E} - m\frac{\sigma_t}{E} + m\frac{\sigma_r}{E} = \frac{1}{E}[\sigma_a - m(\sigma_t - \sigma_r)]$$

wherein σ_a = longitudinal or axial stress, psi

E = Young's modulus = σ/δ = 30×10^6, psi, for steel

m = Poisson's ratio $\cong 0.3$ for steel.

Note that our sign convention assumed σ_t as a tension and σ_r as a compression. Since all quantities except σ_t and σ_r in Eq. (14.10) are constant with respect to r, the term $(\sigma_t - \sigma_r)$ is also constant, therefore

$$(3)\quad \sigma_t = \sigma_r + 2c, \qquad \text{where } 2c \text{ is a convenient constant}$$

Substituting Eq. (3) in Eq. (2), we get a readily solvable differential equation:

$$(4)\quad -2\frac{dr}{r} = \frac{d\sigma_r}{\sigma_r + c}$$

from which

$$(5)\quad \ln(\sigma_r + c) = -\ln r^2 + \ln K$$

wherein $\ln K$ = constant of integration

Rewriting Eq. (5) gives

$$(14.11)\quad \sigma_r = \frac{K}{r^2} - c$$

and from Eq. (3) and Eq. (14.11)

$$(14.12)\quad \sigma_t = \frac{K}{r^2} + c$$

The constants K and c are evaluated by noting the boundary conditions

$$r = r_i,\ \sigma_r = p_i$$
$$r = r_o,\ \sigma_r = p_o$$

from which

$$(6)\quad c = \frac{p_i r_i^2 - p_o r_o^2}{r_o^2 - r_i^2}$$

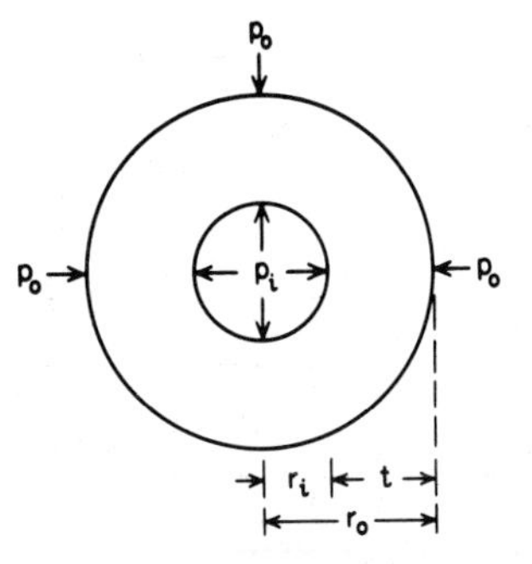

(A) Thick walled cylinder subjected to external pressure, p_o, and internal pressure, p_i

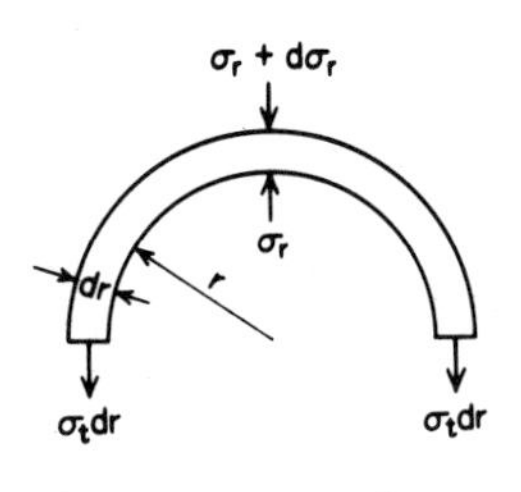

(B) Half loop element from cylinder of (A)

Fig. 14.15. Thick walled cylinder subjected to collapse or burst pressures.

$$(7)\quad K = \frac{(p_i - p_o)r_i^2 r_o^2}{r_o^2 - r_i^2}$$

Equations (6) and (7) substituted in Eqs. (14.11) and (14.12) give the general Lamé equations; however, our particular interest at this time lies in the collapse pressure solution for $p_i = 0$. From Eq. (14.12), we see that the maximum σ_t occurs when $r = r_i$, hence

$$(14.12a)\quad \sigma_t(\max) = \frac{K}{r_i^2} + c$$

Substituting for K and c in Eq. (14.12a) and setting $p_i = 0$, we obtain, finally,

$$(14.13)\quad p_o(\max) = -\frac{\sigma_t(\max)(r_o^2 - r_i^2)}{2r_o^2} = -\frac{\sigma_t(\max)}{2}\left(1 - \frac{r_i^2}{r_o^2}\right)$$

Actually Eq. (14.13) is the equivalent of the API Eq. (14.3). This may be verified by noting that $r_i = \frac{1}{2}(D - 2t)$, and $D = 2r_o$, and that for design purposes $\sigma_t(\max)$ is fixed at the minimum yield stress $0.75\ Y_a$.

The other API equations for collapse strength are essentially empirical variations of Eq. (14.13) which have been derived from extensive test data. In Eq. (14.5) it should be noted that the strength of the material does not alter the allowable collapse pressure, i.e., failure in this range of D/t values is dependent only on geometry. This may be seen from Figure 14.16, which shows plots of minimum collapse pressure p_c vs D/t. The upper limits for elastic failure are the intersections of the plastic failure curves with the elastic curve. For example, a tube of $D/t = 24$ made of either N-80 or P-110 casing exhibits a collapse resistance of about 3800 psi. Consequently, there would be no point in manufacturing such casing from steel of a grade higher than N-80, as far as collapse strength is concerned. The slight break in the plastic failure curves is due to the changeover from Eq. (14.4) to Eq. (14.3) for $D/t < 14$, as specified by the API.

The tensile load which API casing will support is

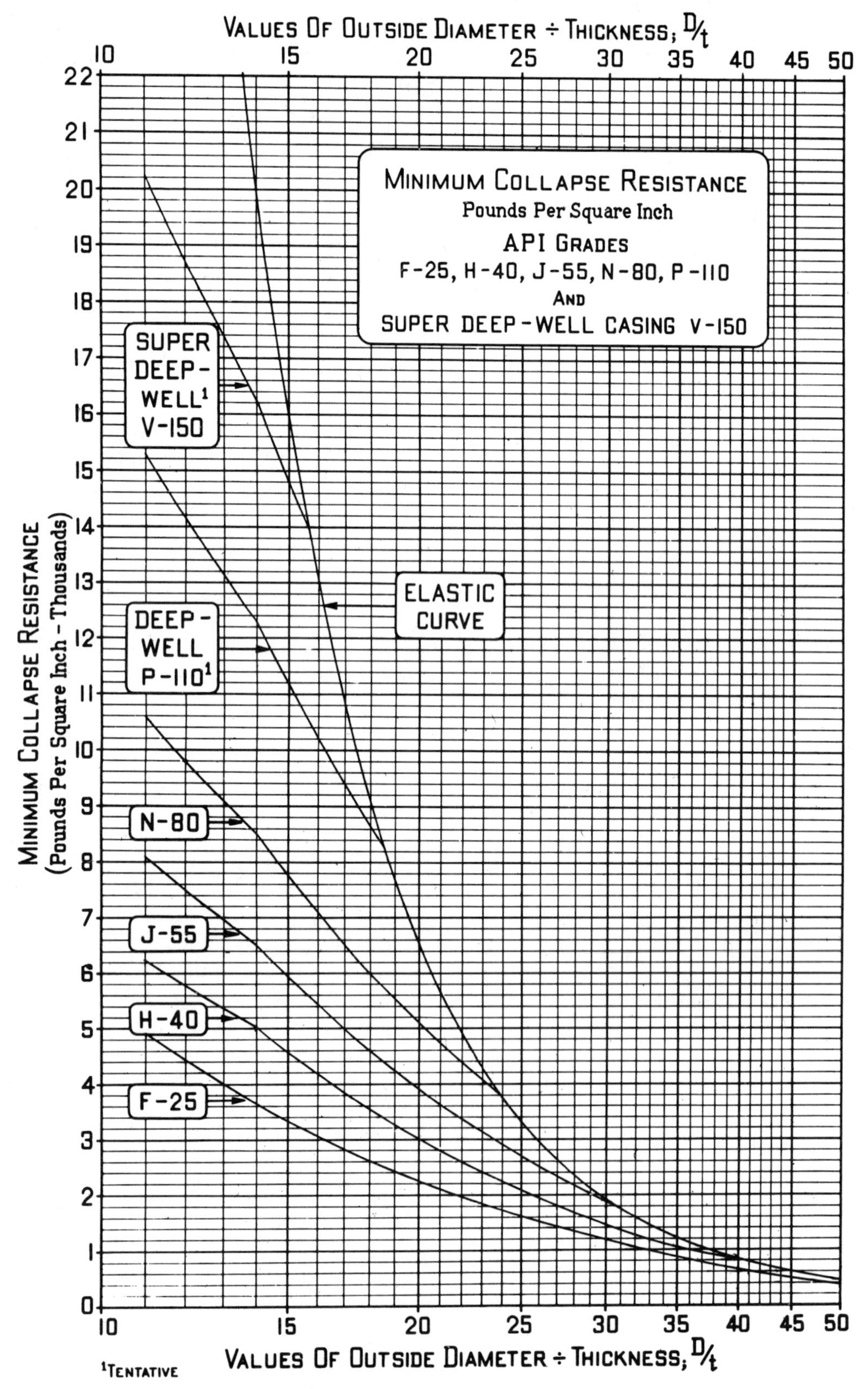

Fig. 14.16. Minimum collapse pressures for various casing grades as functions of D/t.[39] Courtesy National Tube Division, U. S. Steel Corp.

dependent on both the steel grade and the joint efficiency, where

$$\text{joint efficiency} = \frac{\text{Strength of joint, lb}}{\text{Strength of full pipe section, lb}}$$

The API joint strength equations are empirical and apply to the standard long and short API couplings. Various steel manufacturers offer improved joints, some

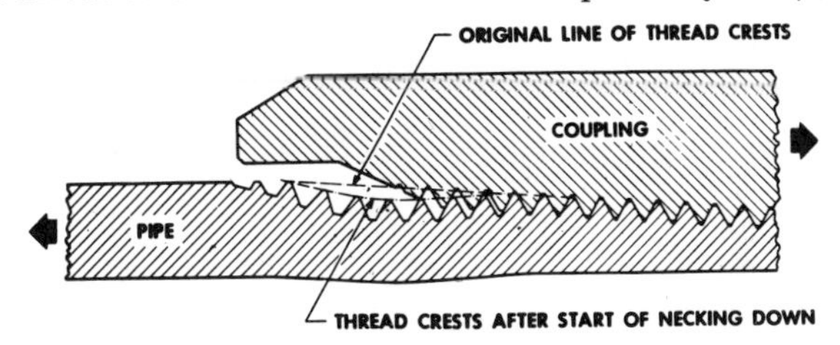

Fig. 14.17. Thread failure in tension by "unzippering." After Texter,[44] courtesy API.

of which exhibit essentially 100% efficiency. As a general rule, tension failures of API casing do not involve parting of the pipe body, but rather a jumping out of the coupling by the male end of the pipe.[40] This has been described as an unzippering effect, in which the last engaged thread slips inward as the pipe necks down (see Figure 14.17). This transfers excessive load to the next thread, which reacts similarly, and so on until the pin slips from the coupling.

Fortunately, it is not necessary to apply the API formulae each time a string is designed. Extensive precomputed tables and charts are available which facilitate design calculations. The use of these will be demonstrated later.

14.52 Biaxial Stress Considerations

The stresses imposed on casing occur in combination. Burst or collapse pressures occur simultaneously with tension or compression, and their combined effect must be recognized. These effects are best shown by the biaxial stress ellipse of Figure 14.18, which is based primarily on the theoretical work of Holmquist and Nadai[41] with experimental verification and refinement by Edwards and Miller.[42] Note that collapse strength is reduced by tension and increased by compression. In general, only the effect of tension on collapse is considered, although the other stress combinations may be of importance in unusually deep, high pressure wells. Since Figure 14.18 is plotted in general percentage terms, it is applicable to all casing. Special tension–collapse curves for specific sizes and grades are available in handbooks[39,43,44] and are actually replots of the tension–collapse quadrant of Figure 14.18, with more restricted coordinate scales.

As stated earlier, a common procedure for designing casing is to select pipe capable of withstanding the full unbalanced external pressure of the mud column. Such a pressure situation can arise during certain completion or testing operations. If such a possibility is remote, this procedure is often revised and either the normal hydrostatic gradient of the area or a salt water gradient

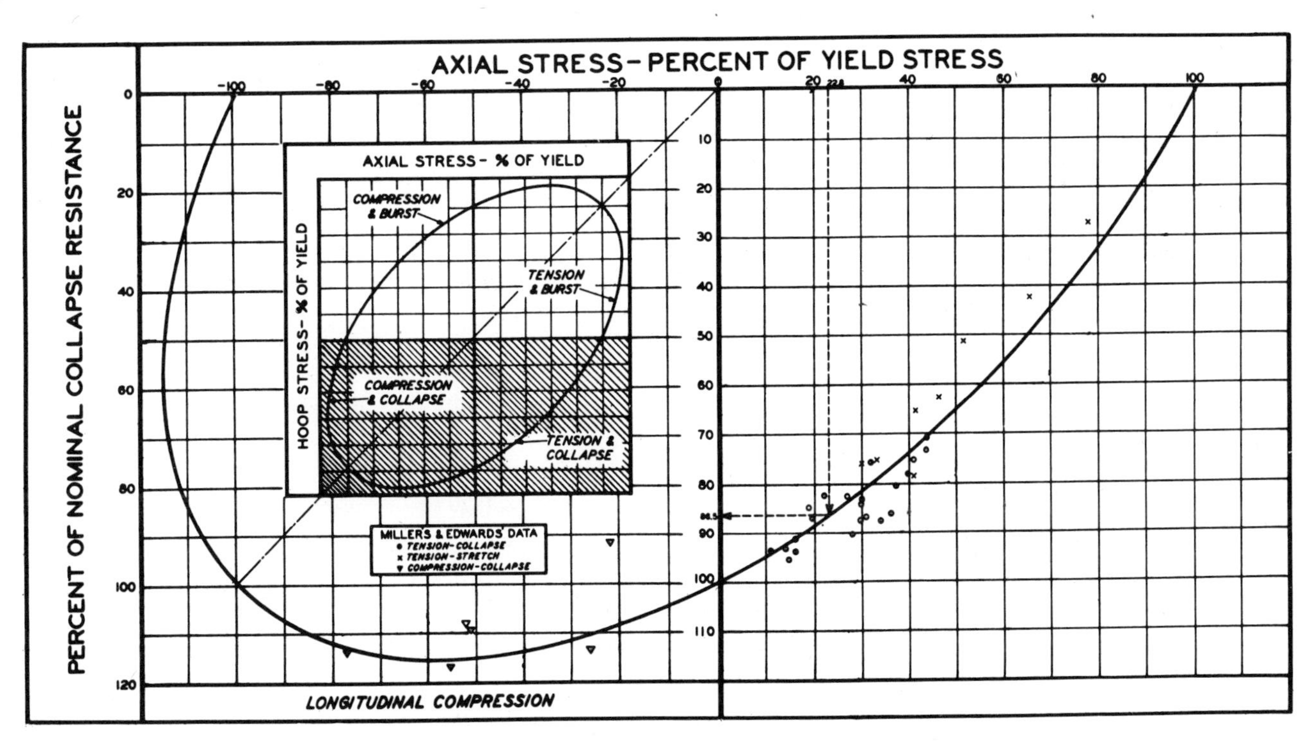

Fig. 14.18. Ellipse of biaxial yield stress. After Holmquist and Nadai,[41] courtesy API. Example shows 86.5% collapse resistance reduction when pipe is subjected to 22.8% tensile load. Reproduced from Spang-Chalfant Data Book.[44]

of 0.5 psi/ft is substituted for the mud gradient as a design basis. In either case, buoyancy is commonly neglected and the pipe is assumed to be in tension from the bottom up. This assumed set of conditions is, however, quite unrealistic, as shown by Lubinski,[45] Klinkenberg,[46] Hawkins and Lamont,[47] Holmquist,[48] and Payne.[49] The importance of this point makes it worthy of detailed consideration. The following analysis is essentially that presented by Klinkenberg. Consider the open-ended casing string depicted in Figure 14.19(A) which is hanging in a fluid and is therefore subjected to pressure at the lower end. This is the situation existing before the casing is cemented. The upward force due to this pressure is, of course, the buoyancy which equals the weight of fluid displaced by the casing.

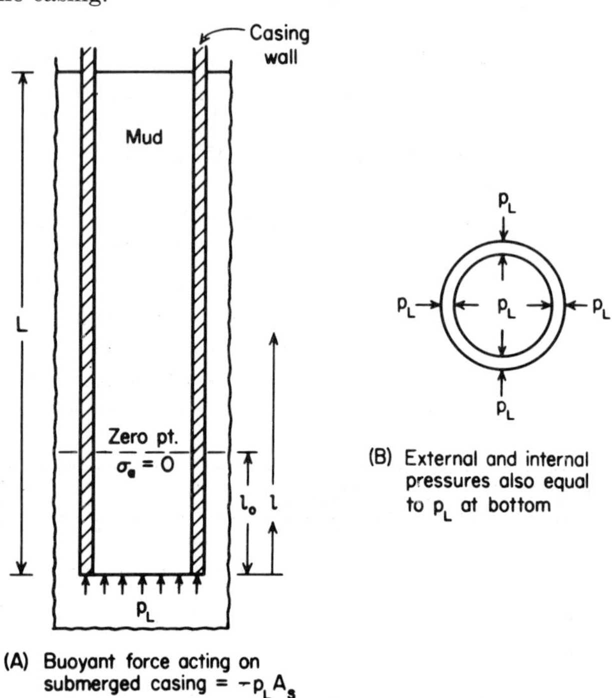

Fig. 14.19. Fluid pressure exerted on submerged, open ended casing. This is equivalent to situation prior to cementing.

$$\text{(14.14)}\quad \text{Buoyant force} = -\rho_f V_s = -\rho_f \frac{A_s}{144} L = -p_L A_s$$

wherein ρ_f = fluid density, lb/ft³

V_s = steel volume displaced, ft³

A_s = cross sectional area of steel, in.²

p_L = pressure in fluid column at depth L, psi, $= \rho_f L/144$

L = submerged length of casing, ft

It is then apparent that the bottommost element is under an axial tension of $-p_L$, i.e., it is under compression. Further, the radial and tangential stresses [Figure 14.19(B)] also equal $-p_L$. Thus, the very bottom of the string is under isotropic stress distribution, or, in other words, the three principal stresses are equal. While we did not mention this point in Chapter 9, this is also the definition of the neutral point used by Lubinski in his analysis of hole deviation problems. The location of the point in the string where the axial stress is zero—no tension or compression—is readily located by equation of the weight of pipe below this zero point and the buoyant force [Figure 14.19(A)].

$$\rho_s \frac{A_s}{144} l_o = \rho_f \frac{A_s L}{144}$$

or,

$$\text{(14.15)}\quad l_o = \frac{\rho_f}{\rho_s} L,$$

where l_o = ft from bottom to zero point

It should be noted that this point moves up as depth increases, while the true neutral zone is always at the bottom of the string.

Example 14.3

10,000 ft of 5½-in., 20 lb casing is freely suspended in 12 lb/gal mud. Compute: (a) The maximum compressive stress in the string, (b) The location of the neutral zone, and (c) The location of the zero axial stress plane.

Solution:

(a) The axial stress σ_a at any point in the string is defined as $\sigma_a = (wl/A_s) - p_L$, where w = pipe weight in lb/ft; hence, the maximum compressive stress is the greatest negative value of σ_a; clearly, this occurs at $l = 0$, or $-\sigma_a(\max)$ =

$$-p_L = \frac{-(10{,}000)(12)(0.433)}{8.33} = -6230 \text{ psi compression}$$

(b) The neutral zone is, by definition, the point at which the three principal stresses σ_a, σ_t, and σ_r are equal. This is possible only at the bottom, where

$$\sigma_a = \sigma_t = \sigma_r = -p_L$$

(c) The point of zero axial stress occurs where

$$\frac{wl_o}{A_s} = p_L$$

A_s = 5.828 in.² for 5½-in., 20 lb/ft pipe

$$l_o = \frac{(5.828)(6230)}{20}$$

= 1820 ft* off bottom, or a depth of 8180 ft

Equation (14.13) could have been used:

$$l_o = \frac{\rho_f}{\rho_s} L = \frac{(12)(7.48)}{490}(10{,}000) = 1830 \text{ ft*}$$

where ρ_s = 0.2833 lb/in.³ = 490 lb/ft³

*The slight discrepancy is due to use of the nominal 20 lb/ft which includes the coupling, while the value of A_s = 5.828 in.² pertains to the pipe body only. This difference is, of course, insignificant for our purposes.

Thus it is apparent that prior to cementing, the lower part of the casing is in axial compression. Suppose, now, that the string is cemented in such a manner that the stresses are unchanged during the process. This amounts to assuming either that the cement slurry density equals the mud density, or that fill-up height is negligible. After the cement sets, the casing is landed in the casing head without slacking off any weight. If, at this time, the mud is completely removed from the pipe, what will be the resulting change in axial stress distribution? This set of conditions represents quite closely the normal sequence of events which might lead to maximum collapse loading at the time of the well's completion.

The solution for this case is rather tedious and is obtained by solving Eq. (14.10) twice: once for the conditions $p_i = p_o = 0$ to $p_i = p_o \neq 0$, and again for $p_i = 0$, $p_o \neq 0$. This requires values of σ_t and σ_r which are obtained from the general Lamé Eqs. (14.11) and (14.12), realizing that the total casing stretch remains unchanged. When this is done, the change in axial stress $\Delta\sigma_a$ for these conditions becomes

$$(14.16)\quad \Delta\sigma_a = -\frac{\rho_f L m r_i^2}{144(r_o^2 - r_i^2)} = \frac{-p_L m r_i^2}{r_o^2 - r_i^2} \cong \frac{-0.804 p_L d^2}{w}$$

wherein w = pipe weight per foot, lb

d = inside pipe diameter, in.

$m = 0.3$

This is a further compression. Note that $\Delta\sigma_a$ is not dependent on depth but is a constant added equally to every point in the string. Hence the zero point $\sigma_a = 0$ has moved even higher by an amount

$$(14.17)\quad \Delta l_o = \frac{(\Delta\sigma_a)(A_s)}{w} = \frac{\rho_f L m r_i^2}{\rho_s(r_o^2 - r_i^2)} = \frac{0.236 p_L d^2}{w}$$

wherein $\rho_s = 490\ \text{lb/ft}^3$

The zero point is now located by

$$(14.18)\quad l_o' = l_o + \Delta l_o = \frac{\rho_f}{\rho_s} L\left(1 + \frac{m r_i^2}{r_o^2 - r_i^2}\right)$$

wherein l_o' = distance from bottom to zero point, $\sigma_a = 0$, for assumed conditions, ft

Example 14.4

Assume the casing of Example 14.3 was cemented and then emptied of mud according to the previous discussion. Compute: (a) The change in axial stress, and (b) The new location of the zero point.

Solution:

(a) Applying Eq. (14.16):

$$\Delta\sigma_a = -\frac{(0.804)(6230)[5.50 - 2(0.361)]^2}{20} = -5710\ \text{psi}$$

Note: 0.361 in. = wall thickness t of $5\frac{1}{2}$-in., 20 lb pipe

(b) Applying Eq. (14.17):

$$\Delta l_o = \frac{(0.236)(6230)(4.778)^2}{20} = 1680\ \text{ft}$$

Hence, $l_o' = 1830 + 1680 = 3510$ ft off bottom, or at 6490 ft depth.

There remains one more case of present interest, namely, the effect of evacuating a completely cemented or otherwise frozen casing, i.e., a casing unable to move at every point. Assuming the same original steps as before, it has been shown that in this case the change in axial stress due to the evacuation is:[40]

$$(14.19)\quad \Delta\sigma_a = \frac{-2\rho_f(L - l) m r_i^2}{144(r_o^2 - r_i^2)} = \frac{-2 p m r_i^2}{r_o^2 - r_i^2}$$

wherein p = pressure at any depth, $L - l$, psi

The additional axial stress $\Delta\sigma_a$ for this case is again a compression, but is now dependent on position. It varies from a maximum at the bottom where $p = p_L$, to zero at the surface, where $p = 0$. We shall not consider this quantitatively, but only note that the plane of $\sigma_a = 0$ has moved still further up the hole and a still greater length of pipe is under axial compression.

14.53 Design or Safety Factors

Once the design conditions are fixed, there remains the choice of design factors. In routine cases, these are generally standardized within a given company as dictated by its experience. Those who have had frequent failures are apt to apply more rigorous factors than those whose experience has been better. Such differences of opinion are evident in an API survey on the subject.[50] Considerable savings may be made by designing for specific field conditions once the latter are known with confidence. In addition to the standard stresses, other factors less subject to calculation, such as erosional wear and vibrational shock imposed by drill pipe, effects of perforating, and corrosion, must be considered. Consideration of these often dictates wall thickness requirements in excess of those called for by standard stresses alone.[50,51]

The most common range of design factors and assumed conditions are given below.

(1) Tension: 1.6–2.0 based on API minimum joint strength, with string freely hanging in air (no buoyancy).

(2) Collapse: 1.0–1.125 based on API minimum collapse pressures. The string is assumed empty, with either mud, salt water, or actual area pressure applied to the annulus. Biaxial stress corrections are usually applied with buoyancy neglected, although the latter is considered by some. Reinforcement by cement is usually ignored.

(3) Burst: 1.0–1.33 based on API minimum values. A gas column at formation pressure is generally assumed to be exerted on all depths within casing.

Surface and intermediate strings are often designed to withstand pressures equal to estimated formation breakdown pressures at their respective shoes, from blowout considerations. Reinforcement by cement is often considered in surface pipe selection.

Design factors are influenced by the possibility of exceeding the assumed conditions, and by the consequences of failure. Casings are therefore designed to sustain high tension, since acceleration during running and the possibility of pipe sticking make precise load calculations impossible. Personnel hazard is also involved. Collapse loads are more readily calculable and no personnel hazard exists; these factors lead to lower design factors. Burst pressures exceeding common design assumptions do occur; however, the API value is based on yield rather than ultimate strength, which permits the lower factors. It should also be realized that, in very deep wells, design factors are often low owing to the lack of available tubular goods which will withstand the assumed loads. In such cases caution is exercised to insure that actual loads are held within the casing strength limits.

Actual field experiments conducted by the Shell Oil Company to test tension and collapse design factors in the Elk City Field have been reported by Saye and Richardson.[52] Tensile tests at design factors as low as 1.4 (8 tests) based on string weight in air were conducted with no failures. Collapse tests were run, using a casing-type drill stem test tool to relieve internal pressure from the pipe. The following are partial data from these tests.

Collapse test number		*Design factor tested*	*Design factor at failure*
1.	uncemented	0.95, 0.90, 0.85, 0.80	0.80
2.	uncemented	0.95	0.95
3.	uncemented	0.95, 0.90, 0.85	0.85
4.	uncemented	0.95, 0.90	no failure
	cemented	0.85, 0.80, 0.75	no failure
5.	uncemented	0.90	no failure
	cemented	0.75, 0.70, 0.60, 0.50	no failure

The above design factors were calculated from the imposed mud pressure, with buoyancy being neglected. Biaxial stress effects were therefore not considered, as the test section was at the bottom where tension was assumed zero. Following these tests Shell revised casing design factors in the Elk City Field to 1.40 in tension, and 1.00 and 0.85 for collapse in uncemented and cemented casing, respectively. Field experience had shown no failures up to the time the results were published. Care was exercised, however, to insure a good cement job over the necessary interval.

Reduction in design factors must be considered as a calculated risk. A company which has had no failures in many years can certainly afford to investigate the possibility of economizing by reducing design requirements. In all cases, the severity of corrosion must be evaluated separately; practices may be allowed to vary from area to area, with considerable economies resulting.

TABLE 14.6

RELATIVE COSTS OF API CASING

Casing grade	*API coupling*	*Relative cost*
J-55	short	1.0 (base figure)
J-55	long	1.05
N-80	long	1.24
P-110	long	1.555
H-40	short	0.975
Tubing grade		
J-55	EUE	1.0
N-80	EUE	1.23
J-55	non-upset	0.94
N-80	non-upset	1.16

14.6 Casing String Design Procedures

Before working specific examples, we must consider the first step in any casing program, namely the selection of casing size. This decision must be made before the well is drilled, since it automatically governs bit size. In hard rock areas smaller hole–casing clearances are allowable than in soft rock regions, since competent beds are not so apt to develop tight spots due to slough-

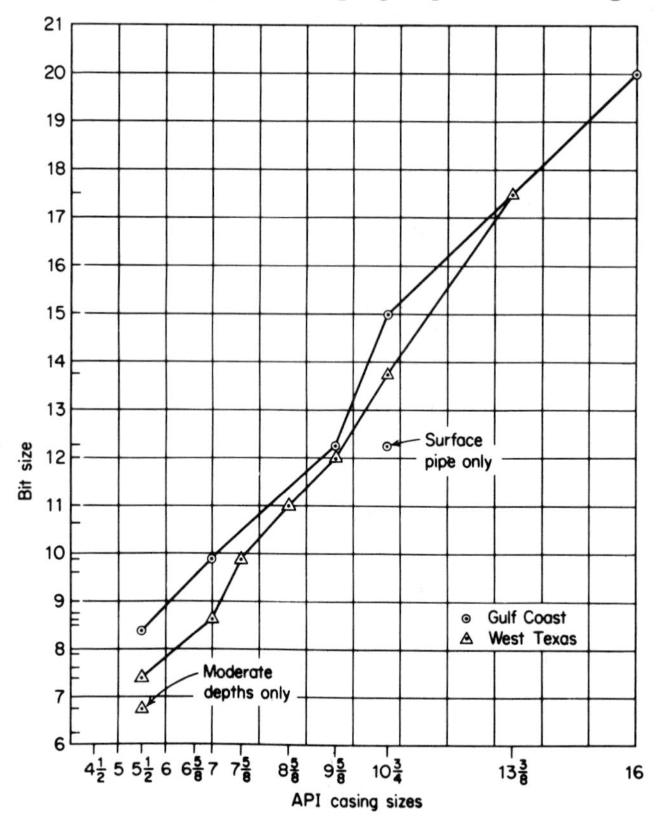

Fig. 14.20. Casing-bit programs for Gulf Coast and West Texas areas. Data plotted from Spang-Chalfant Data Book.[44] *Note*: Popular bit sizes shown as horizontal dashes to right of ordinate. For other available sizes consult bit supplier.

TABLE 14.7

DIMENSIONAL AND INTERNAL PRESSURE DATA FOR API SEAMLESS CASING.
(Courtesy National Tube Division, U.S. Steel Corp.[39])

Size: outside diameter, in.	Weight per foot, lb		Dimensions, in.								Steel grade	Internal pressures, psi				
			Casing			Coupling						Test			Ultimate[1]	
						Outside diam.		Length								
	Nominal: threads and coupling	Plain end	Wall thickness	Inside diameter	Drift diameter	API or buttress	[1, 3] Spec. clearance API or buttress	API short	API long	[1] Buttress		[3, 4] Mill	At fiber stress equal to 80% of min. yield strg.	[6] Yield (min.)	Min.	Avg.
5½	14.00	13.70	.244	5.012	4.887	6.050	5.875	6¾	..	..	H	2800	2800	3110	5990	6650
	14.00	13.70	.244	5.012	4.887	6.050	5.875	6¾	..	..	J	3000	3900	4270	7980	8870
	15.50	15.35	.275	4.950	4.825	6.050	5.875	6¾	8	9¼	J	3000	4400	4810	9000	10000
	17.00	16.87	.304	4.892	4.767	6.050	5.875	6¾	8	9¼	J	3000	4900	5320	9940	11050
	17.00	16.87	.304	4.892	4.767	6.050	5.875	6¾	8	9¼	N	7100	7100	7740	10940	12160
	20.00	19.81	.361	4.778	4.653	6.050	5.875	6¾	8	9¼	N	8400	8400	9190	13000	14440
	23.00	22.54	.415	4.670	4.545	6.050	5.875	6¾	8	9¼	N	9700	9700	10560	14940	16600
	17.00	16.87	.304	4.892	4.767	6.050	5.875	..	8	9¼	P	9700	9700	10640	12710	14040
	20.00	19.81	.361	4.778	4.653	6.050	5.875	..	8	9¼	P	11600	11600	12640	15100	16670
	23.00	22.54	.415	4.670	4.545	6.050	5.875	..	8	9¼	P	12000	13300	14520	17360	19170
7	17.00	16.70	.231	6.538	6.413	7.656	7.375	7¼	..	..	H	2100	2100	2310	4460	4950
	20.00	19.54	.272	6.456	6.331	7.656	7.375	7¼	..	..	H	2500	2500	2720	5250	5830
	20.00	19.54	.272	6.456	6.331	7.656	7.375	7¼	..	..	J	3000	3400	3740	6990	7770
	23.00	22.63	.317	6.366	6.241	7.656	7.375	7¼	9	10	J	3000	4000	4360	8150	9060
	26.00	25.66	.362	6.276	6.151	7.656	7.375	7¼	9	10	J	3000	4600	4980	9310	10340
	23.00	22.63	.317	6.366	6.241	7.656	7.375	7¼	9	10	N	5800	5800	6340	8960	9960
	26.00	25.66	.362	6.276	6.151	7.656	7.375	7¼	9	10	N	6600	6600	7240	10240	11380
	29.00	28.72	.408	6.184	6.059	7.656	7.375	7¼	9	10	N	7500	7500	8160	11540	12820
	32.00	31.68	.453	6.094	5.969	7.656	7.375	7¼	9	10	N	8300	8300	9060	12820	14240
	35.00	34.58	.498	6.004	5.879	7.656	7.375	7¼	9	10	N	9100	9100	9960	14080	15650
	38.00	37.26	.540	5.920	5.795	7.656	7.375	7¼	9	10	N	9900	9900	10800	15270	16970
	26.00	25.66	.362	6.276	6.151	7.656	7.375	..	9	10	P	9100	9100	9960	11890	13140
	29.00	28.72	.408	6.184	6.059	7.656	7.375	..	9	10	P	10300	10300	11220	13410	14800
	32.00	31.68	.453	6.094	5.969	7.656	7.375	..	9	10	P	11400	11400	12460	14880	16440
	35.00	34.58	.498	6.004	5.879	7.656	7.375	..	9	10	P	12000	12500	13690	16360	18070
	38.00	37.26	.540	5.920	5.795	7.656	7.375	..	9	10	P	12000	13600	14850	17740	19590
8⅝	28.00	27.02	.304	8.017	7.892	9.625	9.125	7¾	..	..	H	2300	2300	2470	4760	5290
	32.00	31.10	.352	7.921	7.796	9.625	9.125	7¾	..	..	H	2600	2600	2860	5510	6120
	24.00	23.57	.264	8.097	7.972	9.625	9.125	7¾	..	..	J	2700	2700	2950	5510	6120
	32.00	31.10	.352	7.921	7.796	9.625	9.125	7¾	10	10⅝	J	3000	3600	3930	7340	8160
	36.00	35.14	.400	7.825	7.700	9.625	9.125	7¾	10	10⅝	J	3000	4100	4460	8340	9270
	36.00	35.14	.400	7.825	7.700	9.625	9.125	7¾	10	10⅝	N	5900	5900	6490	9180	10200
	40.00	39.29	.450	7.725	7.600	9.625	9.125	7¾	10	10⅝	N	6700	6700	7300	10330	11480
	44.00	43.39	.500	7.625	7.500	9.625	9.125	7¾	10	10⅝	N	7400	7400	8120	11480	12750
	49.00	48.00	.557	7.511	7.386	9.625	9.125	7¾	10	10⅝	N	8300	8300	9040	12790	14210
	40.00	39.29	.450	7.725	7.600	9.625	9.125	..	10	10⅝	P	9200	9200	10040	12000	13250
	44.00	43.39	.500	7.625	7.500	9.625	9.125	..	10	10⅝	P	10200	10200	11160	13330	14720
	49.00	48.00	.557	7.511	7.386	9.625	9.125	..	10	10⅝	P	11400	11400	12430	14850	16400

[1] Sizes or values indicated are non-API.

[2] Since salt water is practically always encountered in drilling, the length of string is based upon 2 feet of water column to each pound of collapsing pressure.

[3] High test pressures for casing furnished with special clearance couplings are approximately 85 per cent of the listed pressures.

[4] Present mill equipment limits test pressure to 12,000 psi. API test pressures 10,000 psi maximum.

[5] Tension strengths for casing furnished with special clearance couplings are approximately 90 per cent of the listed strengths.

[6] Based on 87½ per cent for internal pressure at minimum yield strength.

TABLE 14.7 (Cont.)

Size: outside diameter, in.	Weight per foot, lb		Dimensions, in.								Steel grade	Internal pressures, psi				
			Casing			Coupling						Test			Ultimate[1]	
						Outside diam.		Length								
	Nominal: threads and coupling	Plain end	Wall thickness	Inside diameter	Drift diameter	API or buttress	[1,3] Spec. clearance API or buttress	API short	API long	[1] Buttress		[3,4] Mill	At fiber stress equal to 80% of min. yield strg.	[6] Yield (min.)	Min.	Avg.
9⅝	32.30	31.03	.312	9.001	8.845	10.625	10.125	7¾	..	..	H	2100	2100	2270	4370	4860
	36.00	34.86	.352	8.921	8.765	10.625	10.125	7¾	..	..	H	2300	2300	2560	4940	5490
	36.00	34.86	.352	8.921	8.765	10.625	10.125	7¾	10½	10⅝	J	3000	3200	3520	6580	7310
	40.00	38.94	.395	8.835	8.679	10.625	10.125	7¾	10½	10⅝	J	3000	3600	3950	7390	8210
	40.00	38.94	.395	8.835	8.679	10.625	10.125	7¾	10½	10⅝	N	5300	5300	5750	8130	9030
	43.50	42.70	.435	8.755	8.599	10.625	10.125	7¾	10½	10⅝	N	5800	5800	6330	8950	9940
	47.00	46.14	.472	8.681	8.525	10.625	10.125	7¾	10½	10⅝	N	6300	6300	6870	9710	10790
	53.50	52.85	.545	8.535	8.379	10.625	10.125	7¾	10½	10⅝	N	7200	7200	7930	11210	12460
	43.50	42.70	.435	8.755	8.599	10.625	10.125	..	10½	10⅝	P	8000	8000	8700	10400	11480
	47.00	46.14	.472	8.681	8.525	10.625	10.125	..	10½	10⅝	P	8600	8600	9440	11280	12460
	53.50	52.85	.545	8.535	8.379	10.625	10.125	..	10½	10⅝	P	10000	10000	10900	13020	14380
10¾	32.75	31.20	.279	10.192	10.036	11.750	11.250	8	..	..	H	1200	1700	1820	3500	3890
	40.50	38.88	.350	10.050	9.894	11.750	11.250	8	..	..	H	1600	2100	2280	4390	4880
	40.50	38.88	.350	10.050	9.894	11.750	11.250	8	..	10⅝	J	2100	2900	3130	5860	6510
	45.50	44.22	.400	9.950	9.794	11.750	11.250	8	..	10⅝	J	2500	3300	3580	6700	7440
	51.00	49.50	.450	9.850	9.694	11.750	11.250	8	..	10⅝	J	2800	3700	4030	7530	8370
	51.00	49.50	.450	9.850	9.694	11.750	11.250	8	..	10⅝	N	5400	5400	5860	8290	9210
	55.50	54.21	.495	9.760	9.604	11.750	11.250	8	..	10⅝	N	5900	5900	6450	9120	10130
	51.00	49.50	.450	9.850	9.694	11.750	11.250	8	..	10⅝	P	7400	7400	8060	9630	10630
	55.50	54.21	.495	9.760	9.604	11.750	11.250	8	..	10⅝	P	8100	8100	8860	10590	11700
	60.70	59.40	.545	9.660	9.504	11.750	11.250	8	..	10⅝	P	8900	8900	9760	11660	12880
	65.70	64.53	.595	9.560	9.404	11.750	11.250	8	..	10⅝	P	9700	9700	10660	12730	14060
	[1]71.10	70.12	.650	9.450	9.294	11.750	11.250	8	..	10⅝	P	10600	10600	11640	13910	15360
	[1]76.00	75.13	.700	9.350	9.194	11.750	11.250	8	..	10⅝	P	11500	11500	12540	14980	16540
13⅜	48.00	45.98	.330	12.715	12.559	14.375	.....	8	..	..	H	1200	1600	1730	3330	3700
	54.50	52.74	.380	12.615	12.459	14.375	.....	8	..	..	J	1900	2500	2730	5110	5680
	61.00	59.45	.430	12.515	12.359	14.375	.....	8	..	..	J	2100	2800	3090	5790	6430
	68.00	66.11	.480	12.415	12.259	14.375	.....	8	..	..	J	2400	3200	3450	6460	7180
	72.00	70.60	.514	12.347	12.191	14.375	.....	8	..	..	N	4900	4900	5380	7600	8450

The weight per foot of casing with threads and coupling is based on a length of 20 ft, including the coupling.

ing or caving. Well depth and crookedness are also factors to consider; however, it must be remembered that crookedness is defined in terms of abrupt changes in deviation and/or direction, and not deviation alone (see Chapter 9). In exploratory wells, the hole size is often larger than in field development wells of similar depth. This practice allows for the possibility that an extra protective string may be required owing to some unforeseen complication. Figure 14.20 shows some common bit-casing programs in the West Texas and Gulf Coast areas. This also shows the contrast between hard and soft rock areas. Note that, in general, larger bit sizes are used in the Gulf Coast region.

As we shall see, there is often more than one choice of pipe grade and/or weight which will meet design requirements. In most cases cost is the deciding factor, although in deep wells total string weight may be more important. The cost of pipe depends on tonnage, grade, coupling type, and freight charges; the latter is assumed to be proportional to tonnage and distance from the steel mills. In our examples, we will of necessity ignore freight charges and assume that, for a given grade and coupling, cost is proportional to weight only. Relative steel grade and coupling costs will be calculated from the data of Table 14.6; these should remain reasonably accurate despite price fluctuations.

TABLE 14.8

Collapse and Tension Properties of API Seamless Casing.
(Courtesy National Tube Division, U.S. Steel Corp.[39])

Size: outside diameter, in.	Nominal: weight per foot threads and coupling, lb	Steel grade	Minimum properties										Plain end strength[1]	
			Tension – API				Collapse		Tension – Buttress[1]					
			Short		Long						Spec. clearance			
			Equiv. length S.F. 2, ft	Ult. joint strength,[5] 1000 lb	Equiv. length S.F, 2, ft	Ult. joint strength,[5] 1000 lb	Setting depth[2] S.F. $1\frac{1}{8}$, ft	Pressure psi	Equiv. length S.F. 2, ft	Ult. joint strength, 1000 lb	Equiv. length S.F. 2, ft	Ult. joint strength, 1000 lb	Yield (min.) 1000 lb	Ultimate (avg.) 1000 lb
$5\frac{1}{2}$	14.00	H	4960	139	..	..	4340	2440	..	..	..	..	..	..
	14.00	J	6640	186	..	..	5640	3170	..	..	..	..	..	..
	15.50	J	6810	211	7970	247	6860	3860	14290	443	12650	392	248	483
	17.00	J	6880	234	8090	275	8000	4500	14320	487	11530	392	273	531
	17.00	N	8030	273	9410	320	10470	5890	15380	523	13850	471	397	571
	20.00	N	8150	326	9550	382	13480	7580	15350	614	11800	472	466	670
	23.00	N	8150	375	9570	440	15820	8900	15200	699	10260	472	530	763
	17.00	P	..	..	12290	418	15150	8520	18590	632	16740	569	546	690
	20.00	P	..	..	12500	500	19400	10910	18550	742	14520	581	641	810
	23.00	P	...	..	12520	576	22880	12870	18370	845	12630	581	729	922
7	17.00	H	4710	160	..	..	2440	1370	..	..	..	..	..	..
	20.00	H	4780	191	..	..	3410	1920	..	..	..	..	..	..
	20.00	J	6350	254	..	..	4440	2500	..	..	..	..	..	..
	23.00	J	6520	300	7480	344	5850	3290	13980	643	11110	511	366	712
	26.00	J	6630	345	7600	395	7220	4060	14020	729	9830	511	415	808
	23.00	N	7610	350	8700	400	7640	4300	15020	691	13430	618	532	765
	26.00	N	7730	402	8850	460	9460	5320	15060	783	11880	618	604	868
	29.00	N	7830	454	8970	520	11320	6370	15120	877	10660	618	676	972
	32.00	N	7890	505	9030	578	13160	7400	15110	967	9660	618	745	1071
	35.00	N	7910	554	9070	635	14970	8420	15090	1056	8830	618	814	1170
	38.00	N	7890	600	9050	688	16140	9080	14510	1103	8130	618	877	1260
	26.00	P	..	..	11580	602	12840	7220	18210	947	14620	760	831	1049
	29.00	P	..	..	11740	681	16390	9220	18280	1060	13100	760	930	1174
	32.00	P	..	..	11810	756	19020	10700	18270	1169	11880	760	1024	1295
	35.00	P	..	..	11870	831	21650	12180	18230	1276	10860	760	1119	1414
	38.00	P	..	..	11840	900	23340	13130	17860	1357	10000	760	1206	1523
$8\frac{5}{8}$	28.00	H	4500	252	..	..	2810	1580	..	..	..	..	..	..
	32.00	H	4610	295	..	..	3750	2110	..	..	..	..	..	..
	24.00	J	6000	288	..	..	2540	1430	..	..	..	..	..	..
	32.00	J	6140	393	6830	437	4870	2740	13580	869	12220	782	503	979
	36.00	J	6220	448	6930	499	6080	3420	13620	981	11180	805	568	1106
	36.00	N	7250	522	8070	581	7950	4470	14650	1055	13190	950	827	1189
	40.00	N	7360	589	8190	655	9580	5390	14740	1179	12150	972	925	1329
	44.00	N	7430	654	8280	729	11240	6320	14800	1302	11050	972	1021	1468
	49.00	N	7440	729	8290	812	13100	7370	14700	1441	9920	972	1129	1624
	40.00	P	..	..	10720	858	13190	7420	17810	1425	14950	1196	1272	1606
	44.00	P	..	..	10840	954	16250	9140	17890	1574	13590	1196	1404	1774
	49.00	P	..	..	10840	1062	18950	10660	17770	1741	12200	1196	1552	1962

[1]Sizes or values indicated are non-API.
[2]Since salt water is practically always encountered in drilling, the length of string is based upon 2 feet of water column to each pound of collapsing pressure.
[3]High test pressures for casing furnished with special clearance couplings are approximately 85 per cent of the listed pressures.
[4]Present mill equipment limits test pressure to 12,000 psi. API test pressures 10,000 psi maximum.
[5]Tension strengths for casing furnished with special clearance couplings are approximately 90 per cent of the listed strengths.
[6]Based on $87\frac{1}{2}$ per cent for internal pressure at minimum yield strength.

TABLE 14.8 (Cont.)

Size: outside diameter, in.	*Nominal: weight per foot threads and coupling, lb*	*Steel grade*	*Minimum properties*										*Plain end strength*[1]	
			Tension				*Collapse*		*Tension*					
			API						*Buttress*[1]					
			Short		*Long*						*Spec. clearance*			
			Equiv. length S.F. 2, ft	[5] *Ult. joint strength, 1000 lb*	*Equiv. length S.F, 2, ft*	[5] *Ult. joint strength, 1000 lb*	[2] *Setting depth S.F. 1⅛, ft*	*Pressure psi*	*Equiv. length S.F. 2, ft*	*Ult. joint strength, 1000 lb*	*Equiv. length S.F. 2, ft*	*Ult. joint strength, 1000 lb*	*Yield (min.)* 1000 lb	*Ultimate (avg.)* 1000 lb
9⅝	32.30	H	4320	279	..	..	2350	1320	..	..	..	..	..	..
	36.00	H	4420	318	..	..	3040	1710	..	..	..	..	..	..
	36.00	J	5860	422	6420	462	3950	2220	13380	963	12040	867	564	1097
	40.00	J	5960	477	6510	521	4920	2770	13450	1076	11080	886	630	1226
	40.00	N	6940	555	7580	606	6280	3530	14460	1157	13010	1041	916	1317
	43.50	N	7060	614	7700	670	7610	4280	14570	1268	12310	1071	1005	1444
	47.00	N	7090	666	7730	727	8710	4900	14570	1370	11390	1071	1086	1561
	53.50	N	7200	770	7860	841	10860	6110	14670	1570	10010	1071	1244	1788
	43.50	P	..	..	10060	875	8460	4760	17620	1533	15150	1318	1381	1746
	47.00	P	..	..	10130	952	10880	6120	17620	1656	14020	1318	1493	1887
	53.50	P	..	..	10280	1100	15700	8830	17730	1897	12320	1318	1711	1261
10¾	32.75	H	4050	265	..	..	1480	830	..	..	..	..	..	..
	40.50	H	4170	338	..	..	2380	1340	..	..	..	..	..	..
	40.50	J	5560	450	..	..	3080	1730	13100	1061	11790	955	629	1224
	45.50	J	5690	518	..	..	4090	2300	13260	1207	10730	976	715	1392
	51.00	J	5740	585	..	..	5100	2870	13250	1352	9570	976	801	1558
	51.00	N	6670	680	..	..	6670	3750	14250	1453	11570	1180	1165	1675
	55.50	N	6760	750	..	..	7860	4420	14330	1591	10630	1180	1276	1834
	51.00	P	8730	890	..	..	6670	3750	17220	1756	14240	1452	1602	2024
	55.50	P	8840	981	..	..	8960	5040	17320	1923	13080	1452	1754	2217
	60.70	P	8900	1081	..	..	12070	6790	17360	2107	11960	1452	1922	2429
	65.70	P	8980	1180	..	..	15180	8540	17420	2289	11050	1452	2088	2638
	[1]71.10	P	9060	1288	..	..	17280	9720	17490	2487	10210	1452	2269	2867
	[1]76.00	P	9110	1385	..	..	19180	10790	17530	2665	9550	1452	2431	3072
13⅜	48.00	H	3670	352	..	..	1320	740	..	..	..	..	..	..
	54.50	J	5000	545	..	..	2030	1140	..	..	..	..	..	..
	61.00	J	5020	613	..	..	2970	1670	..	..	..	..	..	..
	68.00	J	5110	695	..	..	3800	2140	..	..	..	..	..	..
	72.00	N	6030	868	..	..	5120	2880	..	..	..	..	..	..

TABLE 14.9

DIMENSIONAL AND STRENGTH DATA FOR V-150 (NON-API) SEAMLESS CASING.
(Courtesy National Tube Division, U.S. Steel Corp.[39])[a]

Size: outside diameter, in.	Weight per foot, lb		Dimensions, in.							Steel grade	Internal pressures, psi				
						Coupling					Test			Ultimate[1]	
						Outside dia.		Length							
	Nominal: threads and coupling	Plain end	Wall thickness	Inside diameter	Drift diameter	API or buttress	[1, 3] Spec. clearance API or buttress	API	[1] Buttress		[3, 4] Mill	At fiber stress equal to 80% of min. yield strg.	[6] Yield (min.)	Min.	Avg.
$5\frac{1}{2}$	20.00	19.81	.361	4.778	4.653	6.050	5.875	8	$9\frac{1}{4}$	Super	12000	15800	17230	18900	21000
	23.00	22.54	.415	4.670	4.545	6.050	5.875	8	$9\frac{1}{4}$	Deep-Well	12000	18100	19810	21730	24150
7	29.00	28.72	.408	6.184	6.059	7.656	7.375	9	10		12000	14000	15300	16790	18650
	32.00	31.68	.453	6.094	5.969	7.656	7.375	9	10	Super	12000	15500	16980	18640	20710
	35.00	34.58	.498	6.004	5.879	7.656	7.375	9	10	Deep-	12000	17100	18670	20490	22770
	38.00	37.26	.540	5.920	5.795	7.656	7.375	9	10	Well	12000	18500	20250	22220	24690
$8\frac{5}{8}$	44.00	43.39	.500	7.625	7.500	9.625	9.125	10	$10\frac{5}{8}$	Super	12000	13900	15220	16700	18550
	49.00	48.00	.557	7.511	7.386	9.625	9.125	10	$10\frac{5}{8}$	Deep-Well	12000	15500	16950	18600	20670
$9\frac{5}{8}$	53.50	52.85	.545	8.535	8.379	10.625	10.125	$10\frac{1}{2}$	$10\frac{5}{8}$	Super	12000	13600	14870	16310	18120
	[1]58.40	57.38	.595	8.435	8.279	10.625	10.125	$10\frac{1}{2}$	$10\frac{5}{8}$	Deep-	12000	14800	16230	17800	19780
	[1]61.10	60.08	.625	8.375	8.219	10.625	10.125	$10\frac{1}{2}$	$10\frac{5}{8}$	Well	12000	15600	17040	18700	20780
$10\frac{3}{4}$	65.70	64.53	.595	9.560	9.404	11.750	11.250	8	$10\frac{5}{8}$	Super	12000	13300	14520	15940	17710
	[1]71.10	70.12	.650	9.450	9.294	11.750	11.250	8	$10\frac{5}{8}$	Deep-	12000	14500	15870	17410	19350
	[1]76.00	75.13	.700	9.350	9.194	11.750	11.250	8	$10\frac{5}{8}$	Well	12000	15600	17090	18750	20840

[a]Footnotes given with Tables 14.7 and 14.8.

TABLE 14.9 (Cont.)

SUPER DEEP-WELL GRADE V-150
(FURNISHED WITH LONG THREADS AND COUPLINGS)
COLLAPSE AND TENSION PROPERTIES[a]

Size: outside diameter, in.	Nominal: weight per foot threads and coupling, lb	Steel grade	Steel grade	Minimum properties								Plain end strength[1]	
				Tension		Collapse		Tension					
				API				Buttress[1]					
										Spec. clearance			
				Equiv. length S.F. 2, ft	Ult. joint strength, 1000 lb [5]	Setting depth S.F. $1\frac{1}{8}$, ft [2]	Pressure psi	Equiv. length S.F. 2, ft	Ult. joint strength 1000 lb	Equiv. length S.F. 2, ft	Ult. joint strength, 1000 lb	Yield (min.)	Ultimate (avg.)
$5\frac{1}{2}$	20.00		Super Deep-Well	14880	595	25710	14460	22580	903	14520	581	874	..
	23.00			14910	686	30330	17060	19630	903	12630	581	994	..
7	29.00		Super Deep-Well	13970	810	18630	10480	22480	1304	13100	760	1267	..
	32.00			14060	900	25210	14180	21200	1357	11880	760	1398	..
	35.00			14130	989	28710	16150	19390	1357	10860	760	1526	..
	38.00			14080	1070	30950	17410	17860	1357	10000	760	1644	..
$8\frac{5}{8}$	44.00		Super Deep-Well	12900	1135	18310	10300	22010	1937	13590	1196	1914	..
	49.00			12900	1264	25140	14140	21860	2142	12200	1196	2118	..
$9\frac{5}{8}$	53.50		Super Deep-Well	12230	1309	17030	9580	21810	2334	12320	1318	2332	..
	[1]58.40			12230	1428	22400	12600	21700	2534	11280	1318	2532	..
	[1]61.10			12270	1499	25320	14240	21710	2653	10790	1318	2651	..
$10\frac{3}{4}$	65.70		Super Deep-Well	10690	1405	15860	8920	21430	2816	11050	1452	2847	..
	[1]71.10			10790	1534	20910	11760	20770	2953	10210	1452	3094	..
	[1]76.00			10850	1649	25420	14300	19430	2953	9550	1452	3315	..

[a]Footnotes given with Tables 14.7 and 14.8.

Necessary dimensional and strength data for some sizes of seamless casing are given in Tables 14.7 to 14.10. Use of these will be illustrated in the following examples.

TABLE 14.10

PLAIN END AREAS OF VARIOUS CASING SIZES AND WEIGHTS

Size OD, in.	Weight per ft, lb	Plain end (steel) area, in.^{2}
	14	4.029
	15.5	4.514
$5\frac{1}{2}$	17	4.962
	20	5.828
	23	6.630
	17	4.912
	20	5.749
	23	6.656
7	26	7.549
	29	8.449
	32	9.317
	35	10.173
	38	10.959
	24	6.934
	28	7.947
	32	9.149
$8\frac{5}{8}$	36	10.336
	40	11.557
	44	12.763
	49	14.118
	32.3	9.128
	36	10.254
	40	11.454
$9\frac{5}{8}$	43.5	12.559
	47	13.572
	53.5	15.547
	58.4	16.879
	61.1	17.672
	32.75	9.178
	40.5	11.435
	45.5	13.006
	51.0	14.561
$10\frac{3}{4}$	55.5	15.947
	60.7	17.473
	65.7	18.982
	71.1	20.625
	76.0	22.101

Example 14.5

Outline the casing-bit size program for both a West Texas and Gulf Coast field development well as follows:

Strings Required	*Depth, ft*
Surface	1,000
Intermediate	8,000
$5\frac{1}{2}$-in. oil string	14,000

1. Oil String = $5\frac{1}{2}$-in. From Figure 14.20, the necessary hole (bit) size is:

   W. Tex. = $7\frac{7}{8}$-in.

   G. Coast = $8\frac{1}{2}$-in. (nearest popular bit size)

2. Intermediate string must allow passage of $7\frac{7}{8}$-in. and $8\frac{1}{2}$-in. bits. From Table 14.7, using drift diameters:

   W. Tex. = $8\frac{5}{8}$-in. intermediate string

   G. Coast = $9\frac{5}{8}$-in. intermediate string

3. From Figure 14.20, necessary hole sizes for these are:

   W. Tex. = 11-in. bit

   G. Coast = $12\frac{1}{4}$-in. bits

4. Surface pipe must allow passage of 11″ and $12\frac{1}{4}$″ bits. From Table 14.7:

   W. Tex. = $13\frac{3}{8}$-in. surface pipe

   G. Coast = $13\frac{3}{8}$-in. surface pipe

5. Bit sizes for surface hole, from Figure 14.20.

   W. Tex. = $17\frac{1}{2}$-in.

   G. Coast = $17\frac{1}{2}$-in.

6. Summary:

	Bit Size, in.			*Casing Program Size, in.*	
Interval, ft	*W. Tex.*	*G. Coast*	*Interval, ft*	*W. Tex.*	*G. Coast*
0–1000	$17\frac{1}{2}$	$17\frac{1}{2}$	0–1000	$13\frac{3}{8}$	$13\frac{3}{8}$
1000–8000	11	$12\frac{1}{4}$	0–8000	$8\frac{5}{8}$	$9\frac{5}{8}$
8000–14,000	$7\frac{7}{8}$	$8\frac{1}{2}$	0–14,000	$5\frac{1}{2}$	$5\frac{1}{2}$

Note: Bit sizes are subject to slight changes when actual casing weights are selected.

Example 14.6

As a continuation of the above example, we will now select the surface pipe. Design factors are to be:

Tension: 2.0 on API joint strength

Collapse: $1\frac{1}{8}$ on API collapse strength

Burst: 1.33 on API internal yield pressure

Buoyancy is to be neglected in all cases. For a maximum collapse loading, we will assume the pipe empty with a 0.50 psi/ft (9.625 lb/gal density) gradient imposed externally. This condition could be approached if circulation were lost during intermediate hole drilling, allowing the annular mud level to drop below 1000 ft. No protection by the cement strength is assumed. The maximum burst condition is the possibility of blowout occurrence during drilling to 8000 ft. Since, however, this pipe is set at only 1000 ft, formation breakdown at the shoe could be anticipated at pressures on the order of 1000 psi (approximate overburden load). Therefore, we stipulate 1000 psi burst strength as the design figure. Again, no numerical value is placed on the cement strength.

From Table 14.8 it is noted that H-40, 48-lb, ST and C (short threads and coupling) is apparently adequate on all counts. Note that the table setting depth (collapse) and equivalent length (tension) values are based on our exact conditions, so that we may use them with no correction. From Table 14.7, allowable burst pressure is 1730 psi. We require

$$1000 \times 1.33 = 1330 \text{ psi} < 1730 \text{ psi.}$$

Therefore the burst condition is satisfied. Wall thickness is 0.330 in., which is commonly specified as a minimum for protection in case of damage during drilling. Our selection is:

		Actual Design Factors		
Casing	*Interval, ft*	*Tension*	*Collapse*	*Burst*
$13\frac{3}{8}$-in. H-40, 48-lb, ST and C pipe	0–1000	7.3*	1.32*	1.73*

*Tension = 3670/1000 × 2 = 7.3
*Collapse = 1320/1000 = 1.32
*Burst = 1730/1000 = 1.73

Example 14.7

We will now design a 14,000 ft oil string under the following assumed conditions.

Design Factors:

Tension: 1.80

Collapse: 1.00 (mud density = 10 lb/gal)

Burst: 1.00 (based on known reservoir pressure p_e = 6300 psi)

Buoyancy is to be neglected. Effect of tension on collapse to be calculated on the assumption that pipe is in tension at all points above bottom.

Tapered string is to be used, but design is arbitrarily restricted to 3 API weights and/or grades with either short (ST and C) or long (LT and C) API couplings.

1. We begin at the bottom, making a selection based on collapse consideration. It is first necessary to develop a factor F which changes table setting depths to our conditions.

$$F = \frac{1.125}{1.0} \times \frac{9.625}{10.0} = 1.08$$

where 1.125/1.0 = design factor (S.F.) correction

$$\frac{9.625}{10.0} = \text{mud density correction.}$$

2. From Table 14.8, N, 20-lb pipe has a setting depth of

$$D_c = 13{,}480 \times 1.08 = 14{,}558 \text{ ft} > 14{,}000$$

   and is therefore acceptable. Since, according to our table, both short and long couplings are available, we select short couplings for economy.

3. Progressing upward we find N, 17-lb as the next possible choice, which at zero tensile load will set to a depth of

$$D_o = 10{,}470 \times 1.08 = 11{,}300 \text{ ft}$$

   However, according to our design assumptions this must be corrected for axial tension. The weight below 11,300 ft = W_1 = (14,000 − 11,300) 20 = 54,000 lb. The axial stress at 11,300 ft in the 17-lb pipe is

$$\sigma_{a1} = \frac{W_1}{A_{17}} = \frac{54{,}000}{4.962} = 10{,}900 \text{ psi}$$

The minimum yield stress of N-80 steel is 80,000 psi, hence the per cent load based on minimum stress is 10,900/80,000 = 13.6%. From Figure 14.17, the collapse resistance is reduced to 92% of its tension-free value. The indicated change point is thus elevated to

$D_c' = 11{,}300 \times 0.92 = 10{,}400$ ft, an elevation of 900 ft

However, we note that the additional 900 ft of 20-lb casing will further decrease the collapse resistance, necessitating a further change. We anticipate this by multiplying the original reduction by approximately 3/2 (rule of thumb),

$$100\% - (3/2)(8\%) = 88\%$$

Checking the above:

$$\sigma_{ac} = \frac{(14{,}000 - 9950)20}{4.962} = 16{,}300 \text{ psi}$$

$$\text{Per cent load} = \frac{16{,}300}{80{,}000} = 20.4\%$$

From Figure 14.18, per cent collapse strength is approximately 88%, which checks our guess. This trial and error approach may be eliminated, as shown by Kastor.[53] We have now tentatively developed the following:

Interval, ft	*Casing*	*Coupling*
?–9,950	N-80, 17-lb	S
9,950–14,000	N-80, 20-lb	S

Note: Pertinent data are transferred to summary table in step 11 as they are obtained.

4. Returning to Table 14.8, we select J-55, 17-lb as the next possible choice.

$$D_o = 8000 \times 1.08 = 8640 \text{ ft}$$

Proceeding as in step 3:

$$\sigma_{a1} = \frac{81{,}000 + (9{,}950 - 8640)17}{4.962} = 20{,}800 \text{ psi}$$

$$\text{Per cent load} = \frac{20{,}800}{55{,}000} = 37.8\%$$

From Figure 14.18, collapse strength reduction = 24%. Applying our 3/2 rule, 3/2 × 24% = 36%, hence

$$D_c' = 8640 \times 0.64 = 5530 \text{ ft}$$

Checking:

$$\sigma_{ac} = \frac{81{,}000 + (9950 - 5530)17}{4.962} = 31{,}400 \text{ psi}$$

$$\text{Per cent load} = \frac{31{,}400}{55{,}000} = 57.2\%$$

From Figure 14.17, % collapse strength = 58%, which does not check our assumed 64% value.

5. From Table 14.8, the equivalent length of its own weight at S.F. = 2 which J, 17-lb, LT and C casing will support is 6880 ft. Correcting this to our tension design factor of 1.8,

$$l_e = \frac{2.0}{1.8} \times 8090 = 8990 \text{ ft}$$

The total casing weight below our tentative, and as yet incorrect, change point of 5530 ft is 156,000 lb. Therefore, from tension considerations, the allowable length of J, 17-lb, ST and C which can be run is

$$l_a = 8990 - \frac{156{,}000}{17} = -190 \text{ ft}$$

This, of course, means it cannot be used; thus further collapse consideration is unnecessary.

6. We will now investigate the tensile strength of N-80, 17-lb, ST and C casing to determine the length which can be run. From Table 14.8:

$$l_e = \frac{2.0}{1.8} \times 8030 = 8920 \text{ ft}$$

$$l_a = 8920 - \frac{81{,}000}{17} = 4150 \text{ ft}$$

Therefore we select N-80, 17-lb, ST and C for the interval 5800–9950 ft.

7. At this point we will change to long threads and couplings. For N, 17-lb, LT and C casing,

$$l_e = \frac{2.0}{1.8} \times 9410 = 10{,}450 \text{ ft}$$

$$l_a = 10{,}450 - \frac{151{,}600}{17} = 1530 \text{ ft}$$

Therefore N-80, 17-lb, LT and C will be used from 4270 to 5800 ft. This change in threads will require that either the top short coupling be replaced by a long, or that a special change nipple and collar be inserted. A long pin (male thread) will not make-up (screw completely into) in a short collar owing to lack of room; however, a short pin will make-up in a long collar.

8. N-80, 20-lb, LT and C is the next possibility.

$$l_e = \frac{2.0}{1.8} \times 9550 = 10{,}600 \text{ ft}$$

$$l_a = 10{,}600 - \frac{177{,}600}{20} = 1720 \text{ ft}$$

Therefore N, 20-lb, LT and C is to be used from 2550 to 4270 ft.

9. The next possibility is P-110, 17-lb, LT and C. However, a quick trial shows it will not extend to the surface. Since we were arbitrarily limited to 3 types of casing, we now investigate P-110, 20-lb, LT and C.

$$l_e = \frac{2.0}{1.8} \times 12{,}500 = 13{,}890 \text{ ft}$$

$$l_a = 13{,}890 - \frac{212{,}000}{20} = 3290 \text{ ft}$$

Since only 2620 ft are required, this will suffice for the top interval.

10. Burst is now checked. Maximum anticipated pressure was 6300 psi at S.F. = 1.0. Inspection of Table 14.7 shows all casing selected have API minimum yield pressures above 6300 psi, and no changes are necessary.
11. Design Summary:

Casing data					Weight, lb			S.F. in
Grade and wt., lb	T and C	Plain end area, in.²	Depth interval, ft	Length, ft	Interval	Cumulative	Governing stress	governing stress
N,20	S	5.828	9,950–14,000	4,050	81,000	81,000	col.	high
N,17	S	4.962	5,800–9,950	4,150	70,600	151,600	col. tens.	1.0 1.8
N,17	L	—	4,270–5,800	1,530	26,000	177,600	tens.	1.8
N,20	L	—	2,550–4,270	1,720	34,400	212,000	tens.	1.8
P,20	L	—	0–2,550	2,550	51,000	263,000	tens.	high
				14,000	263,000			

Example 14.7 points out the necessity for development of high strength joints. Had we, for example, allowed ourselves to use the buttress joint, no P-110 would have been needed. Had our well been slightly deeper, we would have been forced to use an improved joint to maintain our design factor.

The above calculation was quite tedious; as might be expected, a solution can be found graphically.

Example 14.8

The problem of Example 14.7 will now be solved graphically. The design chart (Figure 14.21) to be used was developed by W. C. Main and is readily available from a casing supplier.[54] Note that the chart shown is for 5½-in. casing; separate charts must be used for each casing size. This particular chart has the advantage that each operator may select the design factors and conditions which he prefers. Other design charts with fixed conditions are also in common use.[44]

Procedure:

1. Draw a vertical program line, *P-P*, through the desired mud wt. × safety factor, on the collapse scale. For our case, this is 10 lb/gal × 1.0 = 10 lb/gal. All depth values will be located by intersection of the sloping depth lines (left side of chart) with this line.
2. Locate slope point *S* by drawing a horizontal line to the right from the intersection of the program line and the 10,000 ft depth line. This is the procedure when buoyancy is neglected. If buoyancy is considered, a different procedure is followed (see reference 51).
3. Locate point *A* by drawing a horizontal line from *J*, the intersection of well depth and the program line. This line is extended to the left as shown, and is labeled 14,000 ft.
4. Point *A* falls above the N-80, 20-lb line and is therefore within its capacity. Thus the bottom section is selected as N, 20-lb.
5. Line *AB* is constructed parallel to *S*-20, where the 20 is located among the casing weights at the top of the right hand side of the chart (denoted by small circles).
   AB is actually a graph of tensile load vs depth, and has a slope of 20 lb/ft in the bottom section. *B* is the point at which use of N, 17-lb casing becomes permissible.
6. A horizontal line is extended from *B* to the left as shown. Intersection of this with the program line, point *K*, denotes the depth at which N, 17 lb may be used: 9950 ft.
7. Since at *B* the casing weight is changed, the slope must be altered to 17 lb/ft. This is done by constructing a line upward from *B* parallel to *S*-17. When this is done, it is noted that no further changes are permissible, and tension must be considered.
8. The tension limits based on joint strength with SF = 2 are indicated on the individual casing curves by circles with enclosed vertical marks (see legend). Our design factor is, however, 1.8, so those limits cannot be used. From Table 14.8 the API joint strength of N, 17-lb, ST and C pipe is 273,000 lb. We alter this to 273,000/1.8 = 151,600 lb, which is located as point *G* on the right hand abscissa. The intersection of a vertical line from *G* with the line from *B* of slope *S*-17 is shown as *C*. This is the limit for N, 17 lb, ST and C pipe. A horizontal line is drawn from *C* to the left which intersects the program line at point *L*, a depth of 5850 ft.
9. Point *D* is the upper limit for N, 17-lb, LT and C pipe located as the intersection of a vertical line from *H* (the tensile limit = 320,000/1.8 = 178,000 lb) and the extension of *BC*. This corresponds to a depth of 4350 ft, located at point *M* as before.
10. At point *D* the pipe weight is changed back to 20 lb, necessitating a change in slope. The new slope is, of course, 20 lb/ft, or parallel to *S*-20. Point *E* is located as the intersection of the load limit line from *I*, where *I* = 382,000/1.8 = 212,000 lb, the tensile limit of N, 20-lb, LT and C at SF = 1.8.
11. Extension of *DE* intersects zero depth at *F*, which corresponds to about 265,000 lb total string weight. Since this is within the load limit for P-110, 20-lb, LT and C, the design is finished.

The program is then quickly summarized in the column at the left. Comparison of this program with that of Example 14.7 will show that, for practical purposes, they are the same. The expedience of graphical solution is obvious.

We will now consider the same design except that the effect of buoyancy on both tension and collapse will be taken into account. Design factors will remain the same, but are to be applied to the buoyant load in 10-lb/gal mud.

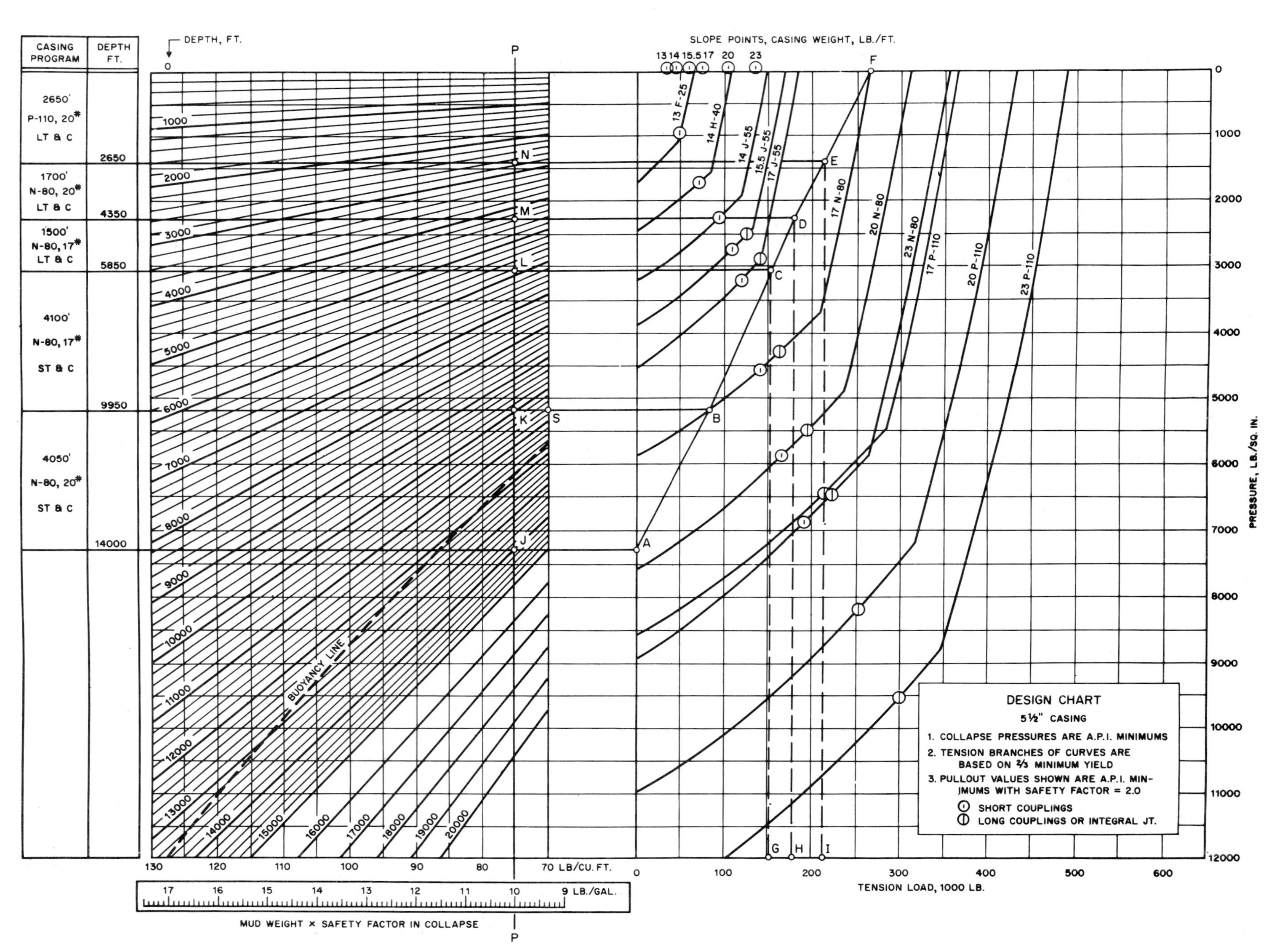

Fig. 14.21. Design chart for $5\frac{1}{2}$-inch casing strings. Courtesy Youngstown Sheet and Tube Company.[54]

Example 14.9

1. N-80, 20-lb, ST and C casing is again selected for the bottom section.
2. The tensile stress at the uncorrected change point, 11,300 ft, must be computed. The following relationship is evident from inspection of Figure 14.22.

$$\sigma_{a1} = \frac{20(14{,}000-11{,}300) + (5.828-4.962)p_1 - 5.828p_L}{4.962}$$

$$= 3360 \text{ psi}$$

where $20(14{,}000 - 11{,}300) = 54{,}000$ lb = pipe weight below 11,300 ft

$(5.828 - 4.962)[11{,}300 \times (10.0/8.33) \times 0.433] = 5100$ lb = downward force at 11,300 ft due to reduction of pipe area

$5.828\ [14{,}000 \times (10.0/8.33) \times 0.433] = 42{,}400$ lb = upward force at bottom.

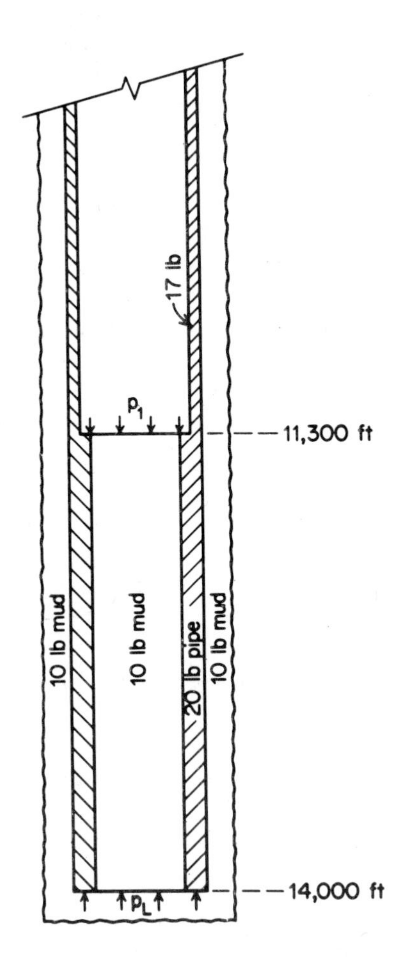

Fig. 14.22. Pressure forces affecting axial stress at uncorrected change point.

$$p_L = (14{,}000)\left(\frac{10}{8.33}\right)(0.433)$$

$$p_1 = (11{,}300)\left(\frac{10}{8.33}\right)(0.433)$$

3. Per cent load in 17-lb pipe is $3360/80{,}000 = 4.2\%$. From Figure 14.18, this amounts to a reduction of almost 3%, when corrected by our 3/2 rule. Actually, this reduction is practically negligible; nevertheless, the corrected setting depth is

$$D_c = 0.97 \times 11{,}300 = 11{,}000 \text{ ft}$$

The corrected stress is

$$\sigma_{ac} = 3340 + \frac{300 \times 20}{4.962} = 4550 \text{ psi}$$

which is a loading of $4550/80{,}000 = 5.7\%$, causing a 3% reduction of collapse resistance, which checks our guess. Therefore, for the interval 11,000–14,000 ft, N, 20-lb, ST and C casing is indicated.

4. J-55, 17-lb casing is still not satisfactory, so we turn to tensile consideration for the upper limit of the N, 17-lb, ST and C.

$$l_a = 8030 \times \frac{2.0}{1.8} - \frac{60{,}000}{17} = 5390 \text{ ft}$$

Therefore, N, 17-lb, ST and C will be used from 5610 to 11,000 ft.

5. Similarly, the N, 17-lb, LT and C has an allowable length of

$$l_a = 10{,}450 - \frac{151{,}600}{17} = 1530 \text{ ft (as in Example 14.7)}$$

The N, 17-lb, L interval is then 4080–5610 ft.

6. Switching now to N, 20-lb, LT and C casing, we note that this change in cross sectional area results in a load reduction, since the heavier walled pipe is on top.

$$\text{Upward force} = (5.828-4.962)(4080 \times 0.52) = 1840 \text{ lb}$$

Hence,

$$l_a = 10{,}600 - \frac{177{,}600 - 1840}{20} = 1800 \text{ ft}$$

and N, 20-lb, L casing may be used from 2280 to 4080 ft.

7. P-110 is, of course, satisfactory for the top and it is unnecessary to recheck burst. Again we summarize:

Casing	*Interval, ft*	*Length, ft*	*Casing, lb* *Interval*	*Cumulative*
N,20-lb, ST and C	11,000–14,000	3,000	60,000	60,000
N,17-lb, ST and C	5,610–11,000	5,390	91,600	151,600
N,17-lb, LT and C	4,080–5,610	1,530	26,000	177,600
N,20-lb, LT and C	2,280–4,080	1,800	36,000	213,600
P,20-lb, LT and C	0–2,280	2,280	45,600	259,200
		14,000	259,200	

8. Using the relative price data of Table 14.6, and assuming price proportional to weight for identical grades and couplings, the following comparison is obtained.

Program of Example 14.7

$$4050' \times \frac{20}{17} \times 1.24 = \quad 5{,}910$$
$$4150' \times 1.24 = \quad 5{,}150$$
$$1530' \times 1.24 \times 1.05 = \quad 1{,}990$$
$$1720' \times \frac{20}{17} \times 1.24 \times 1.05 = \quad 2{,}640$$
$$2550' \times \frac{20}{17} \times 1.555 = \quad 4{,}670$$
$$20{,}360$$

Program of Example 14.9

$$3000 \times \frac{20}{17} \times 1.24 = \quad 4{,}380$$
$$5390 \times 1.24 = \quad 6{,}680$$
$$1530 \times 1.24 \times 1.05 = \quad 1{,}990$$
$$1800 \times \frac{20}{17} \times 1.24 \times 1.05 = \quad 2{,}760$$
$$2280 \times \frac{20}{17} \times 1.555 = \quad 4{,}170$$
$$19{,}980$$

The cost reduction in Example 14.9 is 380/20,350, or about 2%. While not large, this may still represent a sizeable sum over a year's time and would no doubt amount to an average engineer's yearly salary. Other situations may show a higher contrast.

The decrease in casing cost was not obtained free, in that the second design is, of course, weaker than the first. The question is whether or not such reductions can be made on a sound basis. Current thinking seems to indicate that casing string design factors are often excessive and that substantial savings may be realized by relaxing requirements. A notable example is the Shell Oil Company experience cited earlier. It might be pointed out that the further increase in axial compression, which occurs if the pipe is evacuated after cementing, moves the zero point still higher. This, of course, further diminishes the importance of correcting collapse for tension. Consideration of reinforcement by cement may also allow further, sizeable reduction of collapse design factors in many cases. Example 14.9 may be worked graphically using an altered version of the previously shown design chart.[51]

14.7 Special Considerations

Strict adherence to standard design procedures is not feasible in all cases owing to unusual well or field conditions. Severely corrosive areas, for example, must always receive special consideration.[55,56] In other instances, high loadings imposed by either operating practices or geologic conditions must be recognized or accounted for in the design procedure.

High collapse pressures may arise during squeeze cementing, as depicted in Figure 14.23. This problem is

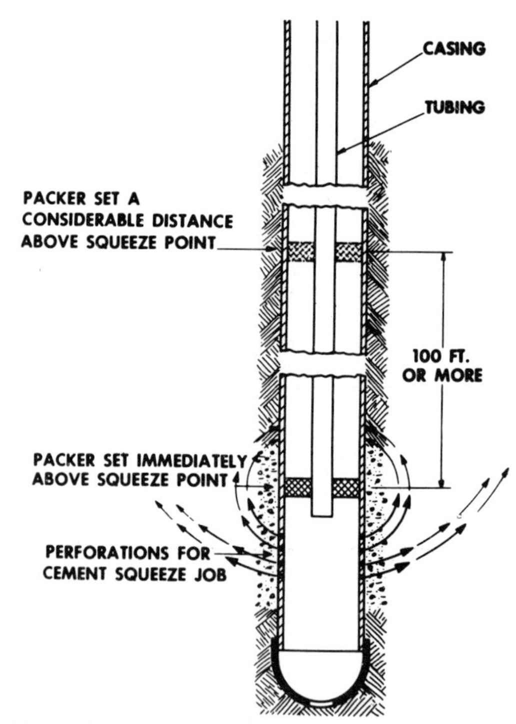

Fig. 14.23. Example of high collapse pressure imposed during squeeze cementing. After Texter,[40] courtesy API.

largely eliminated by either applying a balancing external casing pressure to the annulus above the retainer, or setting the retainer a substantial distance above the perforations, as shown.

In general there is no point in trying to design casing to withstand earth movements, as the forces involved are too great. One exception to this statement, however, is the case of the collapse failures caused by plastic flow of salt in a number of North Dakota wells.[40] Casing strings were redesigned for collapse pressures equal to overburden pressure (1 psi/ft) at the salt depth. Care was also taken in subsequent wells to prevent wash-out of the salt zone during drilling, in the hope of obtaining more uniform and complete cement coverage of the zone. Insufficient time has elapsed for complete evaluation of these measures, but to date they are apparently successful.

Changes in casing stresses may be caused by downhole temperature changes. For example, consider the elongation (or contraction) of a casing string caused by a uniform temperature change over its entire length, L (ft):

$$e_T = K_T 12 L \Delta T \tag{1}$$

where e_T = total stretch due to temperature change, in.

K_T = thermal expansion coefficient of steel casing, 6.7×10^{-6}/°F

ΔT = temperature change, °F

Axial stress is related to axial strain by Young's Modulus

(2) $$\sigma_a = E\delta_a$$

where δ_a = axial strain, ft/ft, etc. Since $e = 12\ \delta_a L$, the change in axial stress in a casing string fixed at both ends due to temperature change is

$$\Delta\sigma_{aT} = \frac{K_T 12 L \Delta T E}{12L} = K_T E \Delta T = 201\Delta T \tag{14.20}$$

Hence, a temperature change of 50°F will cause an axial stress change of approximately 10,000 psi. Such considerations may be of importance, in some instances.

The problem of correct casing landing procedure has received much attention.[57–59] API Bulletin D7 cites the results of a study on the subject and recommends in general that the pipe be landed at the same position it occupied when cemented, i.e., no weight should be either slacked-off or picked-up. The study committee recognized, however, that more complicated methods might be desirable in special cases. The principal concern in proper landing practice is the selection of a condition which will keep loading within desirable limits. This automatically requires some estimate of the producing conditions under which the casing will function. Cox[59] has presented a number of equations which may be used to determine the proper wellhead landing load under various conditions. Consideration of landing practices is, of course, most important in deep wells.

Wellhead equipment may consist of anything from simple slip bowl hangars screwed directly onto the top casing collars to elaborate, high pressure, bolted flange type *Christmas trees.* Specifications for these items are found in API Standard 6E as well as in manufacturers' catalogs. Wellhead equipment is in general not a weak link in well control, since adequate room is available for increasing size and strength as necessary.

14.8 Tubing Selection

Selection of the tubing string is based on the same considerations as selection of casing, though a few additional factors are involved. The same API steel grade designations are used, except that P-110 is replaced by P-105. Although regular or non-upset tubing is available, the majority of tubing used has the externally upset end shown in Figure 14.24. Sizes from 1.05-in. to 4½-in. O.D. are available; however, the dimensions and properties of only the common 2⅜- and 2⅞-in. sizes (usually called 2- and 2½-in.) are given in Table 14.11.

Additional factors to be considered in tubing selection are:

(1) In low fluid level pumping wells the tubing must support its own weight plus the weight of contained liquid.

(2) Tubing is frequently removed from wells. To avoid possible mixup, changes of steel grade are generally avoided.

(3) Variations of load in pumping wells plus factor (2)

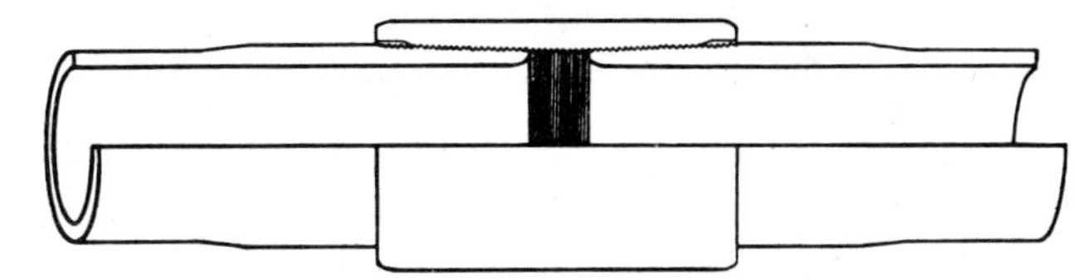

Fig. 14.24. External-upset tubing joint.

above make the use of upset joints attractive, even in shallow areas, since the greater metal area is subject to lower stress and has increased resistance to fatigue.

(4) Alternate buckling and straightening of tubing occurs in pumping wells as the fluid load transfers between rods and tubing on the up and down strokes. This causes excessive internal wear and hastens tubing failure. This difficulty is usually remedied by fixing the lower end with a tension anchor and landing the tubing with sufficient extra tension to prevent buckling.[60]

The design factors applied to tubing strings are about the same as those used for casing. Calculations are simplified by the general use of the same weight and grade of tubing for the entire string. Collapse and burst strengths are improved by the decreased D/t ratio. Sizes are sometimes tapered in deep pumping wells to accommodate large sucker rods in the top section. Size tapering also allows greater tension setting depths.

The true value of stresses imposed on oil well tubular goods are difficult if not impossible to predict. Therefore any design procedure must be a compromise between overly simplified and hopelessly complicated situations. Where possible, design procedures should be based on actual conditions with the design or safety factors being altered to fit the uncertainties which exist in the given area.

PROBLEMS

1. What surface pressure must be held to prevent cement backflow under the following conditions?

   mud = 9 lb/gal

   cement = 15 lb/gal

   oil = 30°API (used to displace cement)

   fillup = 4000 ft

   well depth = 10,000 ft

2. A string of 13⅜-in., 48-lb casing is to be cemented at 2000 ft with a 15 lb/gal slurry. Complete fillup is desired. Show that if water is used as the displacing fluid, the pressure unbalance plus the buoyant force will be sufficient to lift the pipe out of the hole. (Steel area of pipe wall = 13.5 in.^{2})

TABLE 14.11

DIMENSIONAL AND STRENGTH DATA FOR API EXTERNALLY UPSET TUBING.
(Courtesy National Tube Division, U.S. Steel Corp.[39])

Size: outside diameter, in.	Nominal: weight per foot threads and coupling, lb	Steel grade	Internal pressures, psi				Minimum properties					
			Test				Tension		Collapse		Tension	
							Regular OD coup.				Spec. clear. coup.	
			3, 4 Mill	At fiber stress equal to 80% of min. yield strg.	6 Yield (min.)	1 Ultimate (min.)	5 Equiv. length S.F. 2, ft	Yield load, lb	2 Setting depth S.F. $1\frac{1}{8}$, ft	Pressure psi	5 Equiv. length S.F. 2, ft	8 Yield load, lb
$2\frac{3}{8}$	4.70	H	3000	5100	5600	10500	5550	52170	9810	5520	5550	52170
		J	3000	7000	7700	14000	7630	71730	12760	7180	7630	71730
		N	10200	10200	11200	15400	11100	104340	16680	9380	11100	104340
		‡P	13400	13400	14700	17360	14570	136940	22580	12700	..	..
	‡5.95	‡N	13700	13700	14970	20590	11380	135400	21650	12180	..	..
		‡P	15000	18000	19650	23210	14930	177710	29300	16480	..	..
$2\frac{7}{8}$	6.50	H	3000	4800	5280	9910	5580	72480	9320	5240	5580	72480
		J	3000	6600	7260	13210	7670	99660	12090	6800	7670	99660
		N	9700	9700	10570	14530	11150	144960	15820	8900	11150	144960
		‡P	12700	12700	13870	16380	14640	190260	21400	12040	..	..
	‡8.70	‡N	13700	13700	15000	20620	11420	198710	21690	12200	..	..
		‡P	15000	18000	19690	23250	14990	260810	29330	16500	..	..

Size: outside diameter, in.	Nominal: weight per foot threads and coupling, lb	Weight per foot, lb	Dimensions, in.								Calculated weight of coupling, lb		Areas, in.[2]		
			Tubing					Coupling							
						External upset		Outside diam.,							
		Plain end	Wall thickness	Inside diameter	Drift diameter	Outside diameter	Min. length	Regular	1, 3, 8 Spec. clearance	Length	Regular O.D.	1, 8 Spec. clearance	Plain end	Pipe wall under last perfect thread	7, 8 Spec. clearance coupling
$2\frac{3}{8}$	4.70	4.43	.190	1.995	1.901	2.594	$3\frac{1}{2}$	3.063	2.910	$4\frac{7}{8}$	3.42	2.38	1.304	1.594	1.663
	‡5.95	5.75	.254	1.867	1.773	2.594	$3\frac{1}{2}$	3.063	..	$4\frac{7}{8}$	3.42	..	1.692	1.982	..
$2\frac{7}{8}$	6.50	6.16	.217	2.441	2.347	3.094	$3\frac{3}{4}$	3.668	3.460	$5\frac{1}{4}$	5.29	3.45	1.812	2.162	2.291
	‡8.70	8.44	.308	2.259	2.165	3.094	$3\frac{3}{4}$	3.668	..	$5\frac{1}{4}$	5.29	..	2.484	2.834	..

‡API.

[1]Sizes or values indicated are non-API.

[2]Since salt water is practically always encountered in drilling, the length of string is based upon 2 feet of water column to each pound of collapsing pressure.

[3]High-test pressures for tubing furnished with special clearance couplings are approximately 85% of the listed pressures.

[4]Present mill equipment limits test pressure to 15,000 psi. API test pressures 10,000 psi maximum.

[5]Tension setting depths for tubing are not related to those for casing which are based on joint pull-out strength. Tension setting depths shown are determined as the product of the minimum yield strength for the grade and the area of the section under the root of the last perfect thread or the body of the pipe, whichever is smaller.

[6]Based on $87\frac{1}{2}$% for internal pressure at minimum yield strength.

[7]Root of the special clearance coupling thread at the first perfect thread on the pipe when made up to power-tight position.

[8]On tubing furnished with special clearance couplings it is standard practice to furnish couplings in one steel grade higher than the tubing grade but P-105 is our present highest grade.

3. Ten thousand feet of $5\frac{1}{2}$-in. casing is to be cemented in an 8-in. hole. Desired fillup is 2500 ft. The cement to be used is API Class A with 4% gel. Desired slurry density is 13 lb/gal.

(a) How many sacks are required, if an excess factor of 1.20 is used?

(b) How much water will be needed for mixing and displacing the slurry from the pipe?

4. Given the following recommended mixing data and assuming material costs as cement = \$1.75/sk, and bentonite = \$2/100 lb,

Gel, %	*Required mix water, gal/sack*
0	5.5
2	6.5
4	7.7
6	8.8
8	9.7
10	11.1
12	12.3

Compute:
(a) Slurry density and volume per sack of cement for each of the above.
(b) Cost per cubic foot of slurry for each (ignore water cost).

5. Check both the tension and collapse setting depths in Table 14.9 for any two of the listed casing sizes, weights, and/or grades.

6. Calculate the allowable internal pressure for 7-in., 35-lb, N-80 casing based on $87\frac{1}{2}\%$ of the minimum yield stress.

7. Prove that at 100% joint efficiency, the allowable length of its own weight which any given grade of casing will support is independent of wall thickness.

8. (a) Using Barlow's formula for the average hoop stress and Eq. (14.13) for the maximum collapse stress ($p_i = 0$), show that the ratio of average to maximum stress may be expressed as:

$$\frac{\overline{\sigma_t}}{\sigma_{t\,\max}} = 1 - \frac{t}{D}$$

where t = wall thickness, and D = outside diameter.
(b) What error exists when Barlow's equation is used for a pipe of $D/t = 10$? $D/t = 14$?

9. A 10,000 ft string of 7-in., 32-lb casing is freely hanging in 14 lb/gal mud.
(a) What upward force is exerted on the casing?
(b) Where is the neutral point (isotropic stress distribution) and what stress exists there?
(c) At what depth does the axial or longitudinal stress vanish?

10. The string of Problem 9 is cemented and landed in its exact, free-hanging position. If the mud is then removed from inside the casing, what is the new location of the zero axial stress plane? What is the new axial stress at 10,000 ft?

11. Redesign the casing string of Example 14.5 using design factors of 0.90 in collapse and 1.5 in tension. Make a cost comparison. Would you consider this design feasible? Under what conditions?

12. Outline a complete casing and bit size program for a typical 16,000 ft Gulf Coast well for the following conditions:

Strings required	*Depth, ft*	*Mud density in hole at time string is run, lb/gal*
Surface	1,500	9
Intermediate	10,000	11
Oil String	16,000	14

Design Factors: Tension = 1.8 (on API joint strength)
Collapse = 1.0 (on actual mud density, pipe dry)
Burst = 1.1 (standard assumption, p_e = 9,000 psi)

Buoyancy is to be considered. Biaxial stress corrections are to be made based on the buoyancy-corrected axial stress. Use tapered strings on intermediate and oil strings; however, restrict design to only 4 weights and/or grades in each string. Change couplings at will. Make any necessary assumptions.

13. Select a tubing string for the well of Problem 12. Assume flowing production. Again, make any necessary assumptions.

14. Show that Eq. (14.13) is the equivalent of the API Eq. (14.3).

15. Derive Eq. (14.16). See reference 46 for help, if necessary.

REFERENCES

1. API Mid-Continent District Study Committee on Cementing Practices, "A Study of Surface Casing and Open-Hole Plug-Back Cementing Practices in the Mid-Continent District," *API Drilling and Production Practices*, 1955, p. 312.
2. Montgomery, P., "Multiple-Stage Cementing Uses 2 and 3-Stage Methods," *World Oil*, Dec. 1954.
3. Ludwig, N. C., "Portland Cements and Their Application in the Oil Industry," *API Drilling and Production Practices*, 1953, p. 183.
4. *API Specifications for Oil-Well Cements*, API Standard 10A, 2nd ed., Mar. 1955.
5. *API Recommended Practices for Testing Oil-Well Cements*, API RP 10B, 4th ed., May, 1955.
6. Smith, D. K., "A New Material for Deep Well Cementing," *Trans. AIME*, Vol. 207, (1956), p. 59.
7. *Diacel Cement Systems Handbook*, Bulletin D-11. Drilling Specialties Co., Bartlesville, Oklahoma, Nov. 1955.
8. Porter, E. W., "Special Additives Cut Cement Costs," *World Oil*, Feb. 1, 1958, p. 63.
9. Dunlap, I. R., and F. D. Patchen, "A High-Temperature Oil Well Cement," *The Petroleum Engineer*, Nov. 1957, p. B–60.
10. Rollins, J. T., and R. D. Davidson, "New Latex-Cement Solves Special Well Problems," *The Petroleum Engineer*, Feb. 1957.
11. Dumbauld, G. K., F. A. Brooks, Jr., B. E. Morgan, and G. W. Binkley, "A Lightweight, Low Water-Loss, Oil-Emulsion Cement for Use in Oil Wells," *Trans. AIME*, Vol. 207, (1956), p. 99.
12. Smith, D. K., "Cementing Procedures and Materials," *The Petroleum Engineer*, Apr. 1955.
13. Jahns, D. F., "Principles and Practices of Cementing," *The Petroleum Engineer*, Mar. 1957, p. B–64.
14. Mallinger, M. A., "Some Controlling Factors Regarding Variable Weighting of Cement Slurries," *Trans. AIME*, Vol. 189, (1950), p. 374.
15. Morgan, B. E., and G. K. Dumbauld, "A Modified Low-Strength Cement," *Trans. AIME*, Vol. 192, (1951), p. 165.

16. Morgan, B. E., and G. K. Dumbauld, "Recent Developments in the Use of Bentonite Cement," *API Drilling and Production Practices*, 1953, p. 163.

17. Coffer, H. F., J. J. Reynolds, and R. C. Clark, Jr., "A Ten-Pound Cement Slurry for Oil Wells," *Trans. AIME*, Vol. 201, (1954), p. 146.

18. Saunders, C. D., and F. W. Nussbaumer, "Oil Well Cementing Materials," *Oil and Gas Journal*, July 14, 1952, p. 133 and July 21, 1952, p. 115.

19. Ludwig, N. C., "Effects of Sodium Chloride on Setting Properties of Oil Well Cements," *Oil and Gas Journal*, May 24, 1951, p. 125.

20. Farris, R. F., "Method for Determining Minimum Waiting-on-Cement Time," *Trans. AIME*, Vol. 165, (1946), p. 175.

21. Davis, S. H., and J. H. Faulk, "Are We Waiting Too Long on Cement?" *Oil and Gas Journal*, Apr. 8, 1957, p. 99.

22. Owsley, W. D., "Improved Casing Cementing Practices in the U. S.," AIME Paper, presented San Antonio, Tex., Oct. 1949.

23. Howard, G. C., and J. B. Clark, "Factors to be Considered in Obtaining Proper Cementing of Casing," *API Drilling and Production Practices*, 1948, p. 257.

24. Cannon, G. E., "Improvements in Cementing Practices and the Need for Uniform Cementing Regulations," *API Drilling and Production Practices*, 1948, p. 126.

25. Montgomery, P., "Equipment, Materials, Tools for Cementing Deep Wells," *The Petroleum Engineer*, Jan., 1953.

26. Swayze, M. A., "Effect of High Temperatures and Pressures on Cements," *Oil and Gas Journal*, Aug. 2, 1954, p. 103.

27. Saunders, C. D., and W. A. Walker, "Strength of Oil Well Cements and Additives Under High Temperature Well Conditions," AIME Tech. Paper No. 390-G, presented San Antonio, Tex., Oct. 1954.

28. Carter, G., and D. K. Smith, "Properties of Cementing Compositions at Elevated Temperatures and Pressures," AIME Tech. Paper 892-G, presented Dallas, Texas, Oct. 1957.

29. "The Effects of Drilling-Mud Additives on Oil-Well Cements," API Bulletin D-4, Sept. 1951.

30. Morgan, B. E., and G. K. Dumbauld, "Use of Activated Charcoal in Cement to Combat Effects of Contamination by Drilling Muds," *Trans. AIME*, Vol. 195, (1952), p.225.

31. Beach, H. J., and W. C. Goins, Jr., "A Method of Protecting Cements Against the Harmful Effects of Mud Contamination," *Trans. AIME*, Vol. 210, (1957), p. 148.

32. Howard, G. C., and C. R. Fast, "Squeeze Cementing Operations," *Trans. AIME*, Vol. 189, (1950), p. 53.

33. Walker, A. W., "Squeeze Cementing," *World Oil*. Sept. 1949.

34. Hubbert, M. K., and D. G. Willis, "Mechanics of Hydraulic Fracturing," *Trans. AIME*, Vol. 210, (1957), p. 153.

35. Scott, P. P., Jr., W. G. Bearden, and G. C. Howard, "Rock Rupture as Affected by Fluid Properties," *Trans. AIME*, Vol. 198, (1953), p. 111.

36. Reynolds, J. J., P. E. Bocquet, and R. C. Clark, Jr., "A Method of Creating Vertical Hydraulic Fractures," *API Drilling and Production Practices*, 1954, p. 206.

37. *Specification for Casing, Tubing, and Drill Pipe*, 22nd ed., API Standard 5A. Mar., 1958.

38. *Performance Properties of Casing and Tubing*, 7th ed., API Bulletin 5C2. May 1957.

39. *Seamless Drill Pipe, Casing, and Tubing*, Bulletin 15. National Tube Division, U. S. Steel Corp., 1958.

40. Texter, H. G., "Oil-Well Casing and Tubing Troubles," *API Drilling and Production Practices*, 1955, p. 7.

41. Holmquist, J. L., and A. Nadai, "A Theoretical and Experimental Approach to the Problem of Collapse of Deep Well Casing," *API Drilling and Production Practices*, 1939, p. 392.

42. Edwards, S. H., and C. P. Miller, "Discussion on the Effect of Combined Longitudinal Loading and External Pressure on the Strength of Oil Well Casing," *API Drilling and Production Practices*, 1939, p. 483.

43. Zaba, J., and W. T. Doherty, *Practical Petroleum Engineers Handbook*, 4th ed., Houston, Texas: Gulf Publishing Company, 1956.

44. *Engineering Data*, Spang-Chalfant Division, The National Supply Company, Pittsburgh, Pa., 1956.

45. Lubinski, A., "Influence of Tension and Compression on Straightness and Buckling of Tubular Goods in Oil Wells," *Proc. API*, Sect. IV (Prod. Bull. 237), p. 31, (1951).

46. Klinkenberg, A., "The Neutral Zones in Drill Pipe and Casing and Their Significance in Relation to Buckling and Collapse," *API Drilling and Production Practices*, 1951, p. 64.

47. Hawkins, M. F., and N. Lamont, "The Analysis of Axial Stresses in Drill Stems," *API Drilling and Production Practices*, 1949, p. 358.

48. Holmquist, J. L., Discussion of Hawkins and Lamont paper, ibid.

49. Payne, J. M., "A Study-group Investigation of Equipment and Techniques for 20,000-ft Drilling," *API Drilling and Production Practices*, 1949, p. 123.

50. Moody, W. C., "Survey Report on Casing-string Design Factors," *API Drilling and Production Practices*, 1955, p. 154.

51. Hills, J. O., "A Review of Casing-string Design Principles and Practice," *API Drilling and Production Practices*, 1951, p. 91.

52. Saye, J. E., and T. W. G. Richardson, "Field Testing of Casing String Design Factors," *API Drilling and Production Practices*, 1954, p. 23.

53. Kastor, R. L., "Collapse Change Points in Casing Design," *The Petroleum Engineer*, Nov. 1953, p. B–121.

54. *Oil Country Tubular Goods*, Booklet No. 60. The Youngstown Sheet and Tube Company.

55. Villagrana, R. J., and Wm. W. Messick, "Economics of Oil-Well Corrosion Control," *API Drilling and Production Practices*, 1949, p. 391.

56. Goodnight, R. H., and J. P. Barrett, "Oil-Well Casing Corrosion," *API Drilling and Production Practices*, 1956, p. 343.

57. DeHetre, J. P., "Casing-Landing Practice," *API Drilling and Production Practices*, 1946, p. 34.

58. *Casing Landing Recommendations*, API Bulletin D7, June 1955.

59. Cox, W. R., "Down-Hole Factors That Influence Casing-Landing Procedures," *Oil and Gas Journal*, July 22, 1957, p. 86.

60. Lubinski, A., and K. A. Blenkarn, "Buckling of Tubing in Pumping Well, Its Effects and Means for Controlling It," *Trans. AIME*, Vol. 210, (1957), p. 73.

Chapter 15 . . .

The Well Completion

Much of the material in preceding chapters could be included under the general topic of well completions. Formation evaluation methods such as well logging, drill stem testing, and coring were, for example, cited as the means of determining whether or not a well could be completed for commercial production. These methods are also useful in defining certain individual characteristics of the pay section which dictate the completion method.

Various schemes are used to classify well completions, and some overlapping always occurs. For our purposes, we will consider the four major categories listed below and will discuss each in detail.

1. Open Hole Completions
2. Conventional Perforated Completions
3. Sand Exclusion Types
4. Permanent-Type Completions

15.1 Open Hole Completions

This category refers to cases in which the oil string is set on top of the pay zone, as was shown by Figure 14.1 of the last chapter. Such a method is only applicable to highly competent formations which will not slough or cave. Completions of this type are common in low pressure limestone areas (such as Kansas) where cable tools are used for the drilling-in operation. Rotary tools are used until the oil string is set, at which time the cable tool rig moves in, bails the mud from the hole, and drills the desired pay interval. This method permits testing of the zone as it is drilled, elimination of formation damage by drilling mud and cement, and incremental deepening as necessary to avoid drilling into water. This latter factor is quite important in thin, water-drive pay sections where no more than a few feet of oil zone penetration is desired.

Prior to the advent of hydraulic fracturing, a common method of well stimulation was nitroglycerine shooting. This, of course, had to be performed in an open hole interval and consequently restricted many wells to completions of this basic type. At the present time, nitro-shooting has been given a relatively minor role, making it, in general, an insignificant factor in selection of well completion methods, although it is still of importance in some of the shallow areas.

Intuitively, one would suspect that the productivity of an open hole completion should exceed that of a conventional perforated completion in which the fluids must enter the borehole through a few, small diameter holes in the casing. This advantage should be magnified in thin, laminated strata or other formations where vertical permeability is either low or discontinuous. Whether or not the advantage is appreciable is subject to some doubt at the present time. We shall consider it further in the light of statistical data which will be presented following our discussion of perforated completions.

15.2 Conventional Perforated Completions

This category of completions is restricted to wells in which the oil string is set through the pay section, cemented, and subsequently perforated at the desired interval. Such completions are extremely common and are feasible in all formations except those in which sand exclusion is a problem. The vital factor involved is, of course, the perforating process, to which we shall give special attention.

Prior to the early 1930's, casing could be perforated in place by mechanical perforators. These tools consisted of either a single blade or wheel-type knife which could be opened at the desired level to cut vertical slots in the

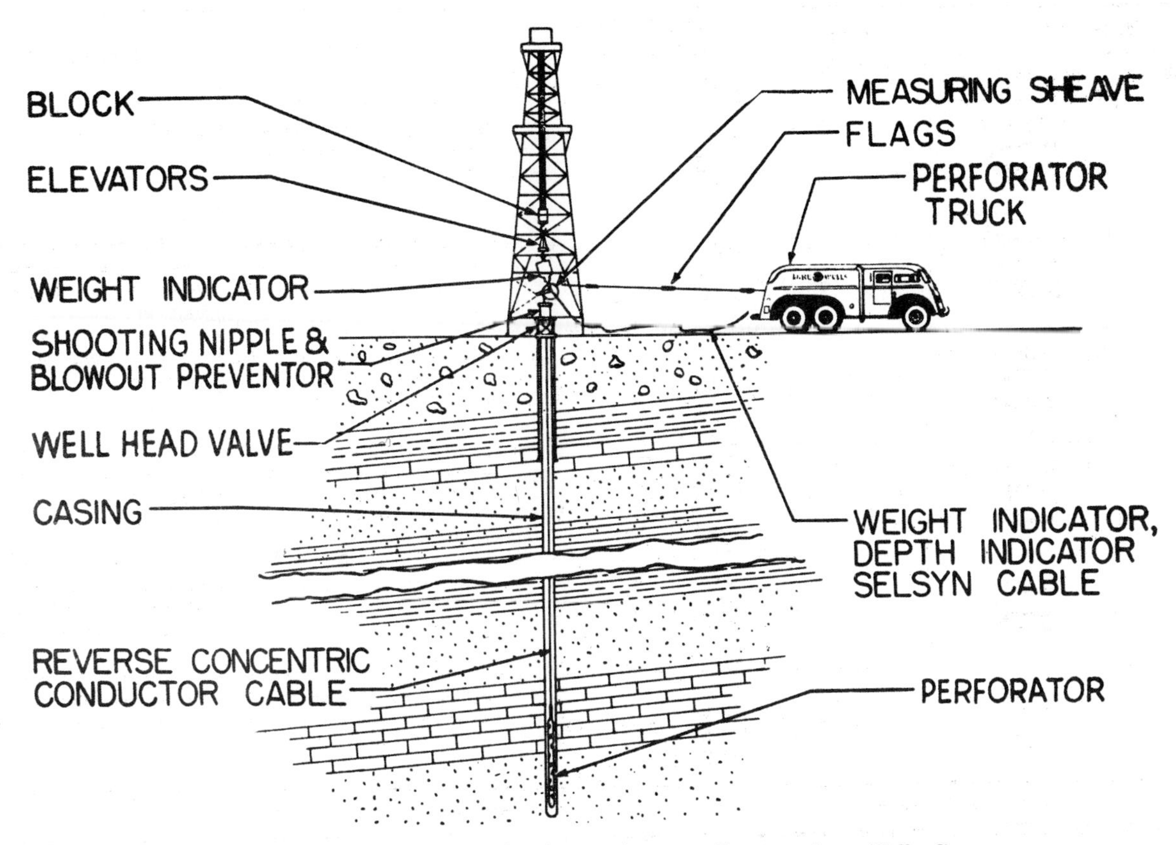

Fig. 15.1. Typical layout of perforating equipment. Courtesy Lane-Wells Company.

casing. The general unreliability of these devices was a natural deterrent to any type of *set-through* (casing set through pay zone) completion. Bullet perforating equipment was developed in the early 1930's and has been in continuous and widespread use since that time. Shortly after World War II the Monroe, or shaped-charge, principle was adapted to oil well work, and the resulting practise is now commonly referred to as jet perforating. This method was rapidly accepted by the industry and is widely used at this time. We will discuss the general features of each technique before proceeding to general perforating considerations.

Bullet Perforating: The bullet perforator is essentially a multi-barreled firearm designed for being lowered into a well, positioned at the desired interval, and electrically fired at will from surface controls. A schematic operating diagram is shown in Figure 15.1. Penetration of the casing, cement, and formation is accomplished by high velocity projectiles or bullets. Current equipment permits the selective firing of one bullet at a time, selective firing of independent groups of bullets, or simultaneous firing of all bullets, depending on the operator's need.

A number of bullet types are available, each being tailor-made for a particular purpose. A number of these and their individual characteristics are shown in Figure 15.2. Bullet guns designed for use in virtually all sizes of casing are readily available from a number of service companies.

An interesting application of projectile equipment is the formation fracturing tool or *bear gun* shown in Figure 15.3. This device fires a large ($1\frac{1}{2}$-in. diam.) projectile vertically downward. The missile passes through a vertical barrel and is deflected 90° as it leaves the muzzle. Horizontal penetrations of 2 to 5 ft as well as considerable fracturing of the formation are estimated to be obtainable with this device. The same assembly without the deflection channel has been successfully used as a powerful down-hole junk (items inadvertently left in the hole) shot. This gun is a special purpose tool generally applicable to dense, tight sections.

Jet Perforating: Numerous articles have described the theory and mechanism of this phenomenon.[1–6] The sequence of events in Figure 15.4 shows a generally accepted pattern of the jet's development during succeeding stages of the explosion. Penetration of the target is obtained from the jet stream's high velocity impact developed by the liner's inward collapse and partial disintegration. The velocity of the jet is on the order of 30,000 ft/sec, which causes it to exert an impact pressure of some 4 million psi on the target. The crumpled portion of the liner, called the carrot, is an undesirable feature of jet perforating as it may some-

BULLET TYPE		SIZES AVAILABLE	TYPICAL PERFORATION CHARACTERISTICS (1)		GIVES	USED FOR
NUMBER	NAME		BURR	HOLE DIA.		
1	STEELFLO 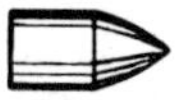	1/4" 3/8" 15/32" 9/16"		1/4" 3/8" 15/32" 9/16"	Maximum penetration. Improved rolled bead burr. Strengthened point structure. Flight stability.	Deep Formation penetration.
2	BURRFREE 	1/4" 3/8" 15/32" 9/16"		1/4" 3/8" 15/32" 9/16"	Elimination of burr. Less penetration than Steelflo. Otherwise possesses all advantages of Steelflo above.	All jobs where smooth inner surface of pipe is necessary and penetration is of secondary importance.
3	SEMI-MUSHROOM 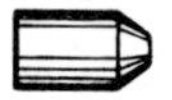	15/32" 9/16"		9/16" 5/8" (2)	Hole in casing larger than bullet. Considerable neat cement fracturing. Reduced penetration.	Casing removal. Cement fracturing.
4	FULL MUSHROOM 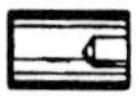	15/32" 9/16"		5/8" 7/8" (2)	Hole in casing considerably larger than bullet. Extensive neat cement fracturing. Considerably reduced penetration.	Casing removal. Cement fracturing.
5	NEEDLE 	1/16" 1/8"		1/16" 1/8"	Single string tubing penetration.	Establishing circulation in stuck tubing or any other purpose where small orifice is desired.
6	PUNCH 	15/32" 9/16"		15/32" 9/16"	Negligible burr. Controlled penetration. Hole same size as bullet.	Perforation of the inner of two strings without damage to outer.

(1) Based on observation of tests in J55 casing. (2) Dependent upon casing thickness and cement support.

Note: Although 15/32" steelflo type bullet is most widely used, other types and sizes are generally available on special order if specified by operator.

Fig. 15.2. Various bullet types and applications. Courtesy Lane-Wells Company.

times lodge in the perforation and obstruct flow. Improved charge designs have, however, been developed which eliminate this problem and produce an essentially carrot-free jet.[6,7]

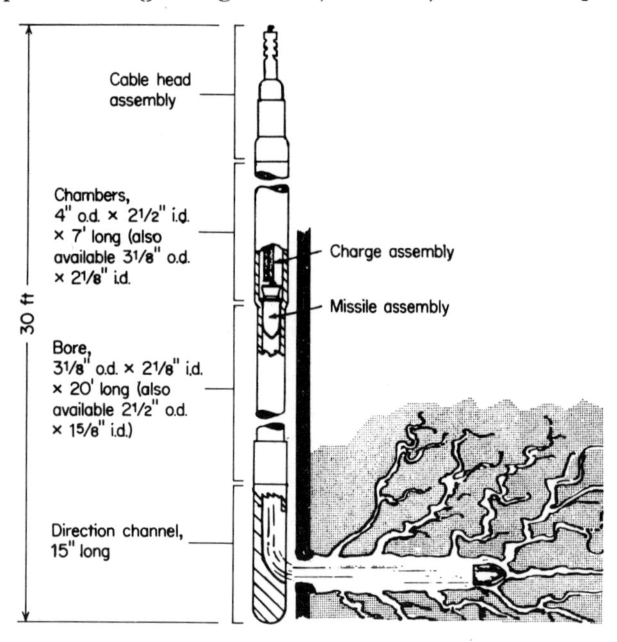

Fig. 15.3. Formation fracturing tool. Courtesy Welex Jet Services, Inc.

Two basic types of jet perforating equipment, retrievable and expendable guns, are available. As implied, a retrievable gun is composed of a cylindrical steel carrier with the charges opposite ports facing radially from the vertical axis of the carrier. Expendable guns are composed of materials which disintegrate into small particles when the gun is fired. The materials commonly used for the carrier are aluminum or cast iron while the cases housing the charge are constructed from glass, aluminum, plastic, cast iron, or ceramic material. A small diameter jet gun which can be run inside tubing and fired in the casing below is shown in figure 15.5.[8] This was designed primarily for use in permanent-type completions which will be discussed later.

Comparison of the Bullet and Jet Methods: There is always some argument as to which of these techniques should be used in a given instance. Comparisons of penetration data have generally shown that jet penetration is superior to bullets in dense formations, and for penetration of multiple casing strings. In softer targets, bullet penetration may equal or even exceed that of jets. Interesting data presented by Lebus[9] on this topic are shown in Table 15.1. The targets in these tests were composed of a $\frac{3}{8}$-in. mild steel plate in series with 16 lb/gal neat cement of various ages. Comparable formation hardnesses were determined by comparing

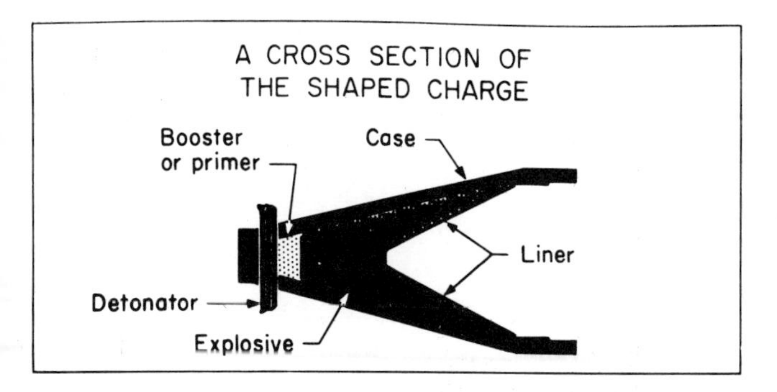

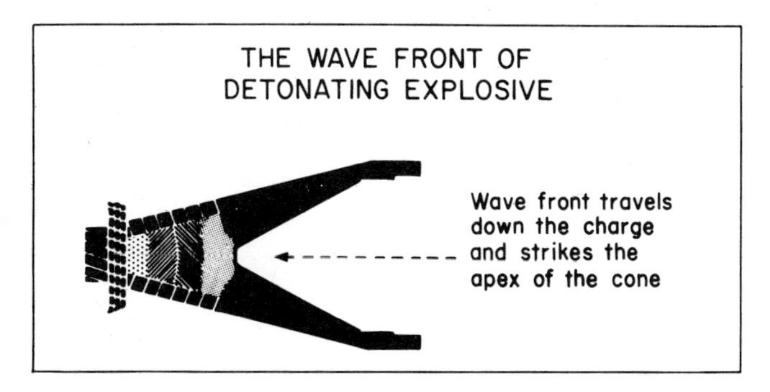

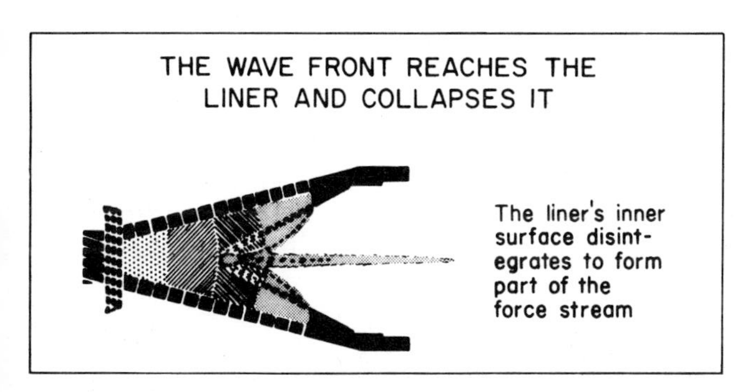

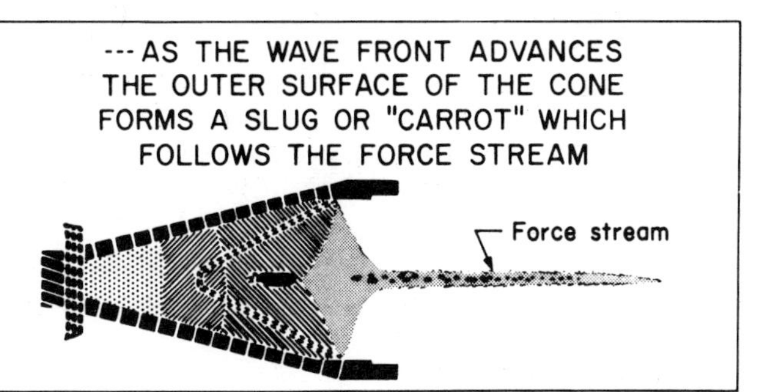

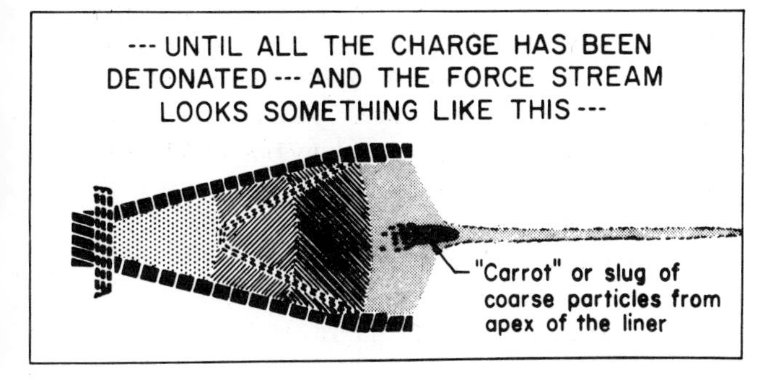

Fig. 15.4. Various stages in jet development. After Forsyth,[3] courtesy API.

penetration of side wall coring devices in the various aged targets with their known penetrations in actual formations. As in all laboratory data, there are always questions concerning differences between downhole and laboratory conditions and the following data are presented as merely indicative of a trend.

TABLE 15.1

Comparison of Bullet and Jet Penetration

Cement target age, hr	*Comparative formation type*	*Depth of penetration, in.* Bullet	Jet
24	very soft	15[a]	15[a]
48	soft	15[a]	15[a]
72	medium soft to medium	15[a]	$11\frac{7}{8}$
5 days	medium hard	$12\frac{7}{8}$	$12\frac{3}{16}$
7 days	medium hard to hard	$10\frac{1}{2}$	$11\frac{3}{16}$
10 days	hard	$9\frac{3}{8}$	$10\frac{1}{2}$
Other targets			
Berea sand core	very hard	$4\frac{1}{2}$	$6\frac{3}{4}$
Steel plate	?	2	$3\frac{1}{2}$

[a]Shot completely through 15-in. target.

Another factor often cited in bullet vs jet discussions is the flow capacity of the perforations. Many tests have indicated that bullets produce greater fracturing around and beyond the perforation which results in greater permeability, particularly in soft and/or shaly formations. Contradictory evidence has been cited from tests conducted on limestone cores in which jet perforating produced the greater permeability.[10] Tests reported by Allen and Worzel[11] and Krueger[12] showed substantially equal results on a productivity basis despite shallower penetration by bullets. Earlier jet equipment produced perforations which were highly tapered; however, this disadvantage also has been largely eliminated by improved shaped-charge design.[6]

Both types of equipment have functioned satisfactorily in thousands of wells and in all kinds of formations. Bullet gun usage is generally restricted to hole temperatures less than 275°F, while jet equipment is available for applications to 400°F. The generally accepted characteristics of both types are summarized in Table 15.2[13,14]

Operational Aspects: It is often tempting to choose a perforating gun considerably smaller in diameter than that of the casing in which it is to be run, rather than to use the size designed for the job. In this situation, the bullet or jet must expend considerable energy in penetrating the drilling fluid prior to reaching the target. This effect is more critical in jet than bullet equipment, and obviously will have an adverse effect on penetration. It is termed *standoff*, i.e., the distance the jet or bullet

TABLE 15.2

EQUIPMENT COMPARISONS

Jet

1. Deeper penetration in hard rocks and multiple casing strings.
2. Minimum burring of casing wall (of interest with regard to subsequent tool passage).
3. Minimum cement shattering.
4. Available for use in permanent-type completions.
5. Higher temperature range.

Steel Guns	*Expendable Guns*
1. Minimum junk left in hole.	1. Maximum junk left in hole
2. Semi-selective firing.	2. No swabbing action when retrieving line.
3. Less subject to damage while running in hole.	3. Large charges available for open-hole shooting.
	4. Angle shooting possible.

Bullet

1. Equal to superior penetration in soft to medium rocks.
2. Maximum fracturing of cement and soft rocks.
3. Completely selective firing available.
4. Controlled penetration by bullet selection where needed.
5. High powered, large diameter gun available if needed (Figure 15.4).
6. Generally cheaper, due to lower per shot cost and consideration (3) above.

must travel before encountering the target. The effect of standoff on jet penetration under some conditions has been evaluated by Reed and Carr[15] as noted in Figure 15.6.

Accurate depth measurements are essential to any perforating job. The most accurate placement of shots is obtained through the combined use of the collar and radioactivity logs. In this technique, the zone to be perforated is picked on the radioactivity log, and all measurements are made relative to casing collars which are located by wire detectors attached to the perforating gun.

A very critical factor in successful perforating is the type of fluid present in the hole when the operation takes place. Considerable restriction to flow can occur due to the dehydration of some drilling fluids which may form a complete or partial plug over the perforation. Other fluids may damage the formation simply by contaminating the pay section exposed by the perforations. Oil, oil base mud, or salt water spotted opposite the desired interval will in most cases be desirable, with general preference in the order listed.

It is also advantageous to perforate with the well bore pressure lower than formation pressure. This practice is necessarily restricted to low pressure wells or permanent-type completions. Numerous tests have shown this to result in less permeability restriction in the perforations themselves.

Theoretical Considerations of Perforated Completions: The prime objective in making a perforated completion is to obtain productivity as near to that of an uncased hole as possible, while still enjoying the advantages of

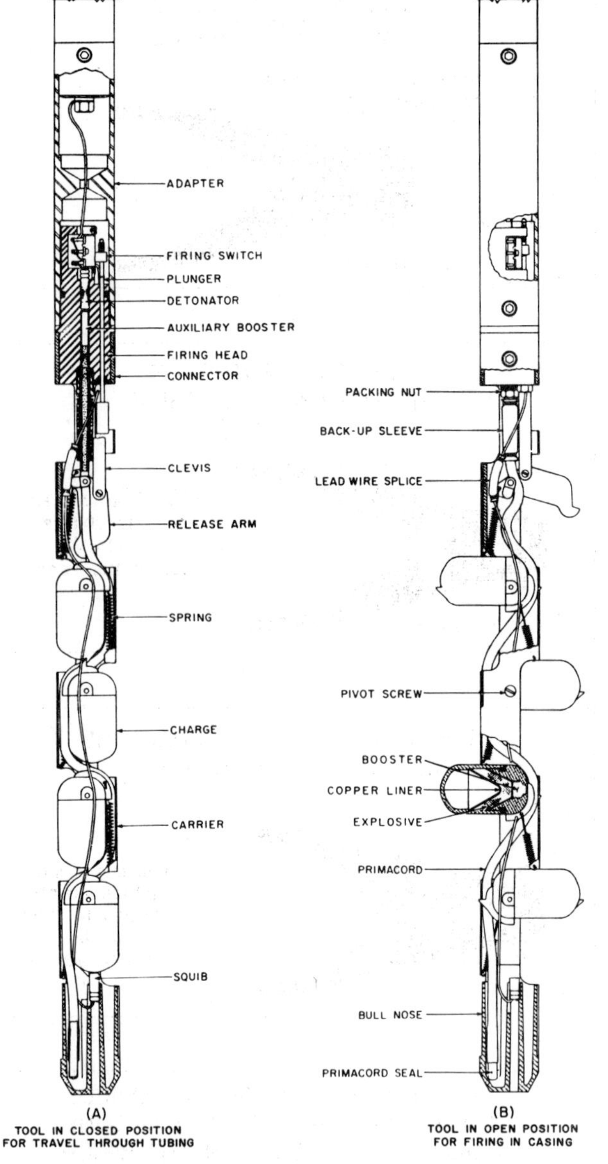

Fig. 15.5. Tubing type jet perforating gun for perforating casing below tubing. After Caldwell and Owen,[8] courtesy AIME.

casing. The principal factors to consider in this respect are:

1. Perforation diameter
2. Perforation density (number of holes per foot)
3. Depth of penetration (radial distance perforated)

The effects of these variables on well productivity have been the subject of considerable mathematical[16] and experimental[10,11,17–19] study. Partial results of the McDowell and Muskat experiments conducted with electrolytic models are reproduced in Figure 15.7. The effect of perforation diameter is not shown; however, this was found not critical if the diameter was greater than $\frac{1}{4}$ in. at normal perforating densities. Other conclusions apparent from Figure 15.7 are as follows:

(a) For penetration depths of $\frac{1}{2}$ to 1 well diameter, little increase in productivity is gained from increasing perforating densities above four $\frac{1}{2}$-in. holes per ft.

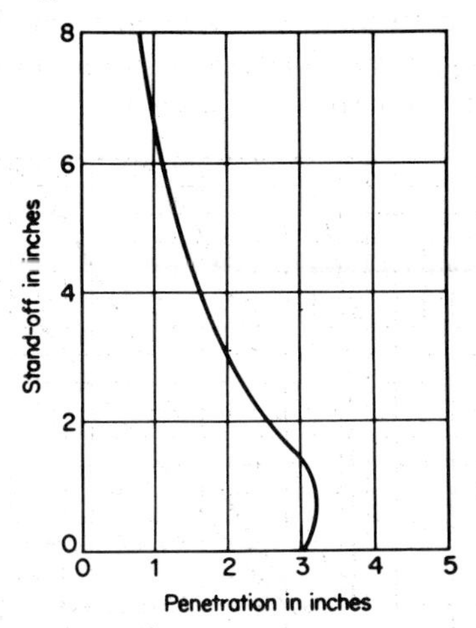

Fig. 15.6. Effect of standoff on jet penetration. After Reed and Carr,[15] courtesy *Oil and Gas Journal.*

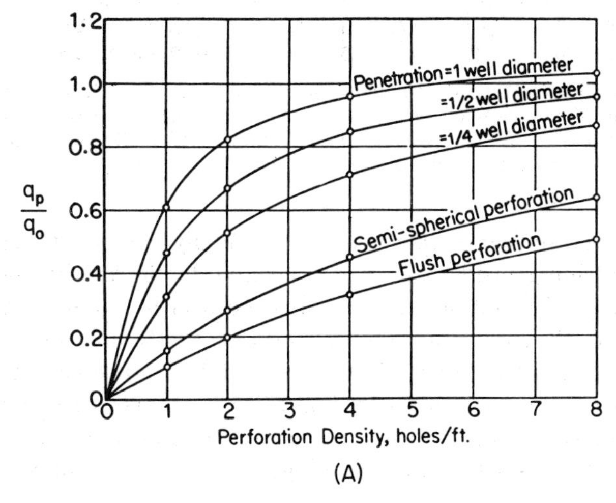

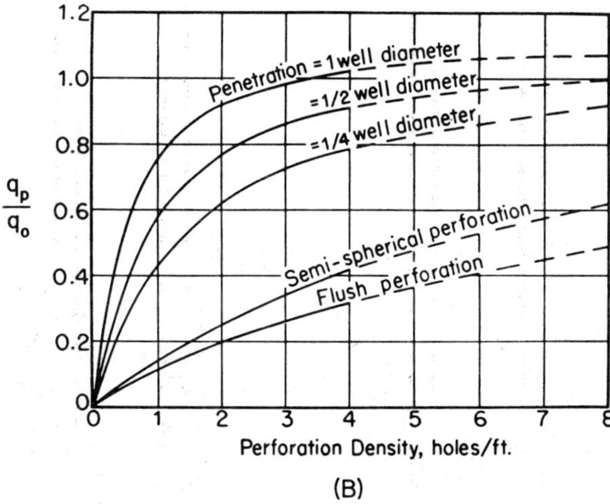

Fig. 15.7. Effects of perforation density and depth of penetration on relative well productivity for 6-in. **(A)** and 12-in. **(B)** casing. q_p/q_o = productivity of perforated well/productivity of open hole. Perforation diameter = $\frac{1}{2}$ inch. After McDowell and Muskat,[18] courtesy AIME.

(b) The productivity of perforated completions equals that of an uncased hole at penetration depths of one well diameter and normal shot densities.

The above data assume a homogeneous section and no formation damage either prior to, or during, perforating. For formations of low or irregular vertical permeability, higher perforating densities may be desirable, to insure that all strata have access to the borehole. However, this is generally taken care of by cement shattering and fracturing during either perforating or stimulation treatments. The effect of penetration depth may be greatly magnified if a zone of impaired permeability exists, since the perforation may extend beyond the most highly damaged region. Hence perforating is itself a stimulation method whereby permeability in the vicinity of the borehole may be increased. This comparison is not, however, strictly valid since the open hole may also be perforated with a similar effect. This is, in fact, practiced in some areas.

Open Hole vs Perforated Completions: It is worthwhile, perhaps, to consider briefly the relative merits of these two basic completion methods. The open hole method is initially cheaper, since perforating costs are eliminated. Contamination by cement is, or can be, avoided. On the other hand, the perforated completion offers a much higher degree of control over the pay section, since the interval can be perforated and tested as desired. Individual sections may, in general, be isolated and selectively stimulated much more easily and satisfactorily through perforations than in an open hole. There is also considerable evidence that hydraulic fracturing is more successful in perforated completions. This is clearly indicated by Figure 15.8 which is taken from API Bulletin D6. Productivity ratios of perforated wells were about 50% higher than those of similar open hole completions. This superiority is apparently

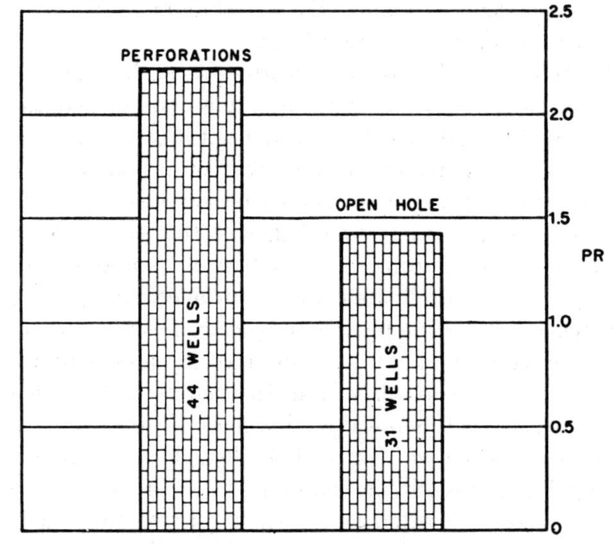

Fig. 15.8. Productivity ratios of perforated vs. open hole completions in stimulated sand wells. Courtesy API.[13]

due to more uniform treatment over the entire pay section plus the stimulation benefit gained from penetration of the perforations themselves. The improved zonal control is also of considerable value when remedial measures, such as water or gas exclusion, are undertaken.

With perhaps a few exceptions in low pressure or thin water-drive pay areas, the benefits of the perforated completion overshadow those of the open hole type, making it generally preferable. This priority has been made possible by modern perforating and stimulation techniques as well as advances in drilling muds, cementing materials and methods, and numerous other phases of petroleum technology.

15.3 Sand Exclusion Problems

The completion of a well in an unconsolidated sand is not as simple as the types of completion discussed above, in that the additional problem of excluding any sand produced with the oil must be solved. Sand production, if unchecked, can cause erosion of equipment, and well bore and flow string plugging to the extent that well operation becomes uneconomical. Sand production is, in general, sensitive to the rate of fluid production. At very low rates, little or no sand may be produced, while at high rates large quantities will be carried along in the production stream. In the early days of the oil industry, sand production was tolerated in flowing wells, some means of preventing its accumulation being the only control measure initiated. When it became necessary to pump such wells, exclusion methods were required to prevent pumping equipment erosion. Many wells now produce which, without sand control measures, would be uneconomical.

The most common methods of excluding sand employ some means of screening. These techniques include:

(1) Use of slotted or screen liners,

(2) Packing of the hole with aggregates such as gravel.

The basic requirement of these methods is that the openings through which the produced fluids flow must be of the proper size to cause the formation sand to form a stable bridge and thereby be excluded. This, of course, requires some knowledge of the sand size in question. It is difficult to obtain a representative sample of unconsolidated sand; however, special core barrels are available for this purpose.[20] The maximum opening size which will exclude a given sand is determined from screen analysis, in the manner determined by Coberly.[21] Reference to this method was made in Chapter 6 when we discussed the bridging phenomena as applied to lost circulation materials. The sand's size distribution is plotted as shown in Figure 15.9. The appropriate liner slot or screen width is then taken as twice the 10-percentile size indicated from the screen analysis. For gravel packs, the gravel size is generally selected as 4 to 6 times the 10 percentile sand size.[22–24] As is evident in Figure 15.9, highly variable sands may have the same 10 percentile size, and smaller slots or screen openings may be advisable if an unusually high percentage of fine grains are present.

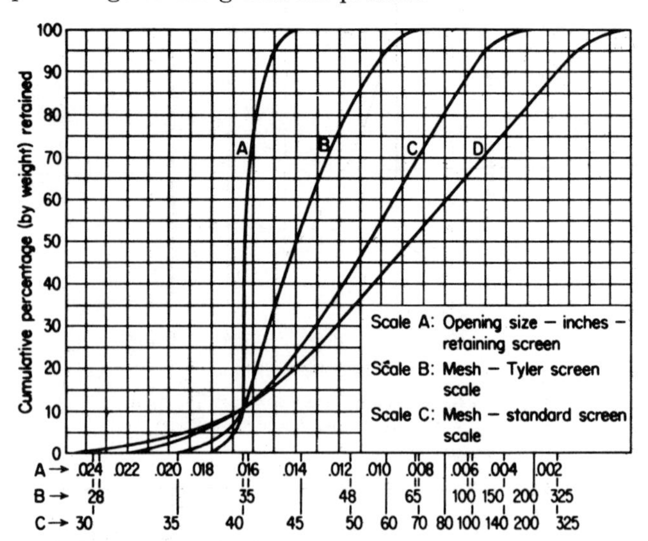

Fig. 15.9. Method of plotting screen analysis data. Note that four sands shown have essentially same 10 percentile point of 0.0165-in. despite size range variations. After Tausch and Corley,[25] courtesy *Petroleum Engineer.*

The slotted or screen liner is normally run on tubing and hung inside the oil string opposite the producing zone. The oil string may have been either cemented through the section and perforated, or set on top, as indicated by Figure 15.10. Typical slot and screen arrangements are shown in Figure 15.11. It is extremely important that the sand face be free of mud cake before the liner is set, to prevent plugging. This is commonly

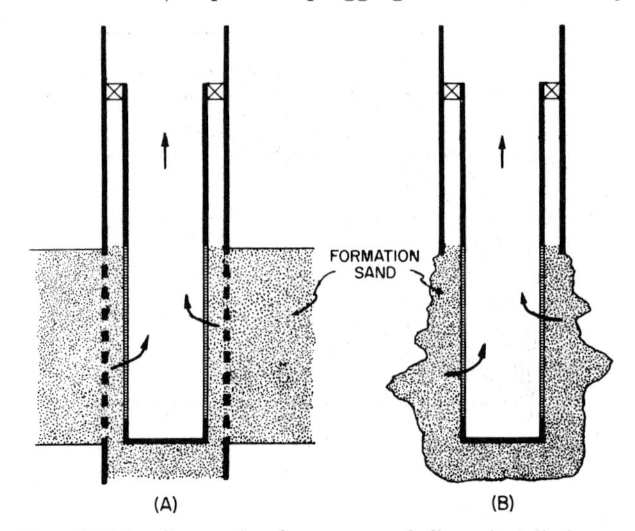

Fig. 15.10. Conventional screen and liner installations: **(A)** Screen set inside casing opposite perforations, and **(B)** screen set in open hole section. After Tausch and Corley,[25] courtesy *Petroleum Engineer.*

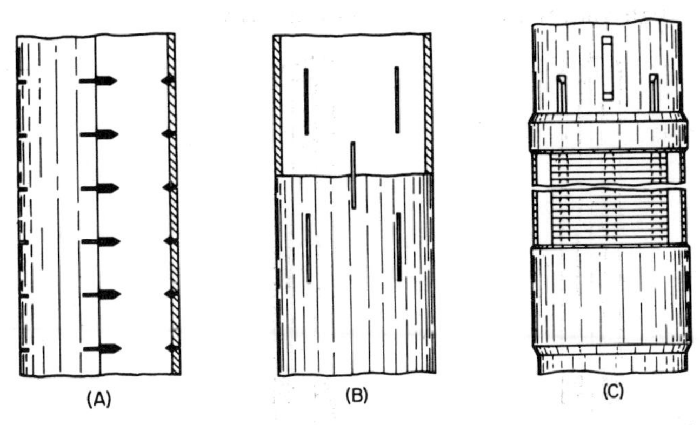

Fig. 15.11. Types of screens for sand control. **(A)** Horizontal slotted screen. **(B)** Vertical slotted screen. **(C)** Wire-wrapped screen. After Tausch and Corley,[25] courtesy *Petroleum Engineer*.

accomplished by either using clay-free completion fluids or washing the well with salt water before hanging the liner.

Gravel packing may be performed in several ways in either perforated or open hole intervals. Figure 15.12 shows one placement method commonly practiced in the Gulf Coast. Since the screen is used to exclude gravel only, the slots may be larger than in the previous case and are usually only slightly smaller than the gravel. The required thickness of the gravel pack is only four or five gravel diameters. The formation sand then bridges within the pores of the gravel pack while gravel entry is prevented by the screened liner. Uniform placement of gravel is facilitated by using a penetrating (high fluid loss) fluid such as oil or salt water. Perforation washing is also necessary to insure that all perforations are open, and to allow proper gravel placement.

The methods discussed above may be applied at the time of completion or later in the well's life when an unexpected sand problem occurs. It is, of course, good to anticipate sand problems if possible; however, these preventive measures cost money and are best avoided, if conditions allow it.

Sand may also be excluded by the use of a consolidating plastic material which actually cements the sand grains together, thereby preventing caving. This obviously results in some reduction in permeability and is therefore not desirable in low permeability, dirty sands. This technique is not so widely used as the sand-bridging methods, but has been applied successfully in some instances.

The engineer who works in hard rock country may never encounter sand problems; however, in California and the Gulf Coast, sand is an everyday problem and is handled almost routinely. The series by Tausch and Corley[25] is an excellent and complete discussion for those wishing to pursue the subject.

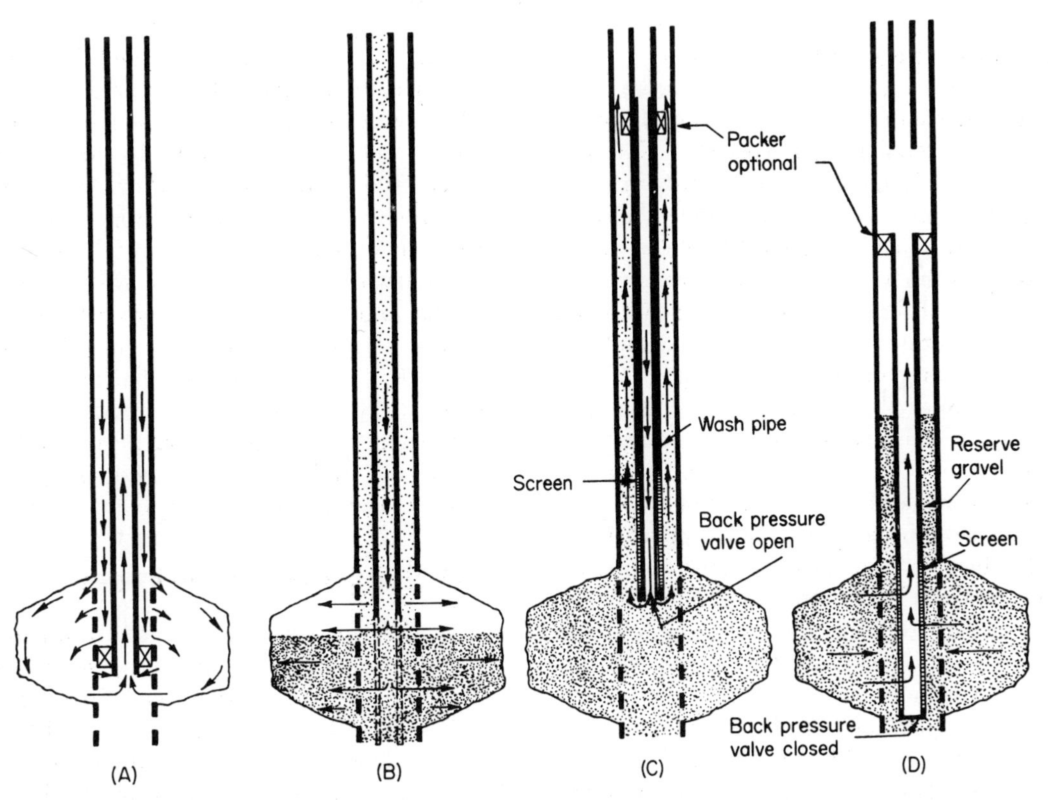

Fig. 15.12. Simplified method of gravel packing commonly used in the Gulf Coast: **(A)** Perforations are washed; **(B)** gravel is squeezed through perforations; **(C)** cavity is filled and screen is washed through gravel; **(D)** after screen placement, wash pipe is removed. After Tausch and Corley,[25] courtesy *Petroleum Engineer*.

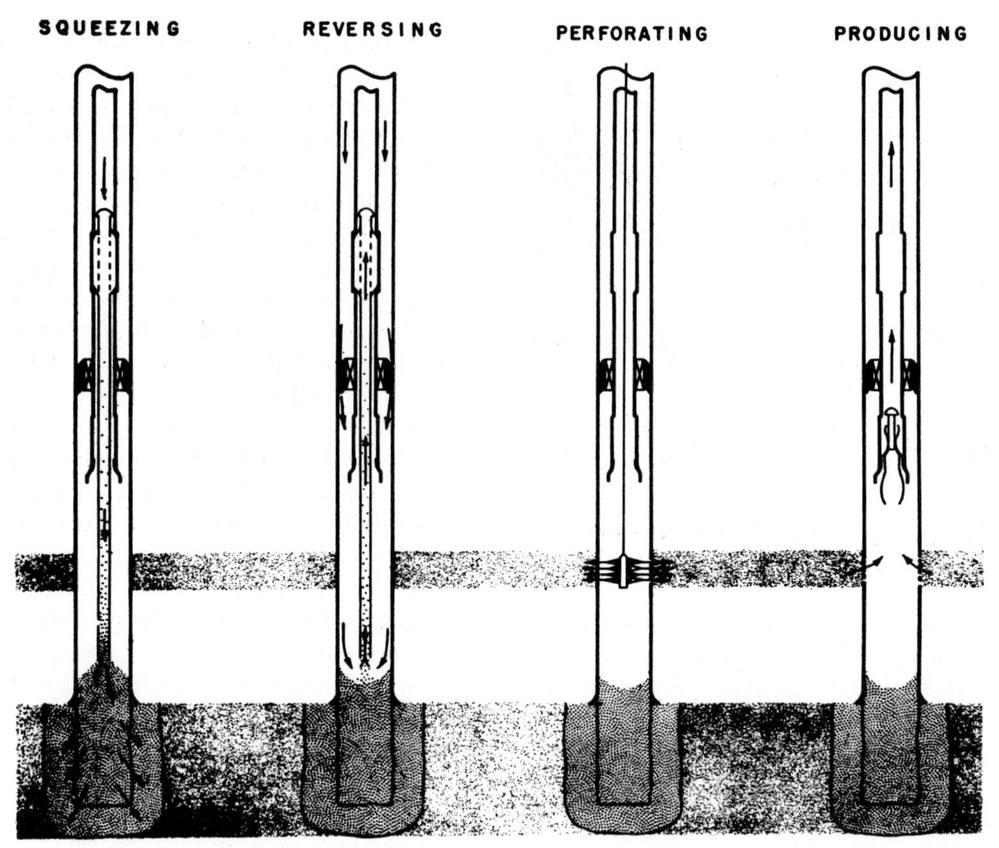

Fig. 15.13. Operational techniques, permanent-type well completions. After Huber and Tausch,[27] courtesy AIME.

15.4 Permanent-Type Well Completions

This type of well completion was developed by the Humble Oil and Refining Company.[26,27] No new principles are involved; however, the manner in which various completion operations are performed is somewhat novel. A permanent-type completion is one in which the tubing is run and the well head is assembled only once in the life of the well. Perforating, swabbing, squeeze cementing, gravel packing or other completion or remedial work is performed with special, small diameter tools capable of being run inside the tubing. A jet perforating gun of this type was shown earlier. The primary advantage of this technique is economy; savings of 75% on some types of completion and workover jobs have been reported. The main application of the method has been in the Gulf Coast area where sand exclusion and multiple zone perforating and testing in high pressure flowing wells make it particularly attractive. As familiarity is gained, the usage will undoubtedly become more extensive in other areas.

A number of the operations which may be performed are shown in the sequence of Figure 15.13. This shows an open hole section being squeeze cemented through a wire-line tubing extension. The extension is then removed and the desired upper zone is perforated and placed on production. The tubing head is not touched during the entire operation and it is therefore not necessary to kill the well with mud to control pressure. The new zone may therefore be perforated under oil or salt water with a favorably unbalanced pressure differential, and consequently, formation damage or perforation plugging during the operation seldom occurs. In a conventional recompletion of this type it would be necessary to (1) kill the well with mud, (2) pull the tubing and rerun with squeeze retainer, (3) pull the tubing, (4) run a perforating gun and perforate, (5) rerun tubing and place the well back on production. The saving in time, labor, and equipment is apparent. The improvement and development of special tools and other materials for this method have progressed rapidly, and improvement is certain to continue.[8,28–31]

15.5 Multiple Zone Completions

A multiple zone completion is one in which two or more separate pay zones are produced simultaneously from the same well bore without commingling of the fluids. This segregation is commonly required for purposes of reservoir control, and is generally made compulsory by state regulatory bodies. Two-zone or dual completions are the more common, although triple and quadruple zone completions have been performed. As might be expected, the principal factor opposing this practice is the complexity of downhole equipment necessary to maintain zonal segregation. These complications become more serious when one or both zones require artificial lifting (pumping, gas lift, etc.). Some typical downhole arrangements of packers and tubing strings for both flowing and artificially lifted production are shown in Figures 15.14 and 15.15.

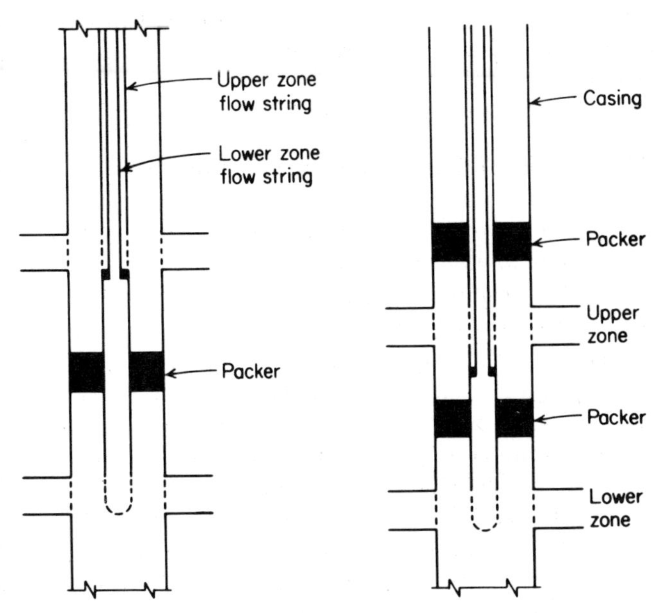

Fig. 15.14. Single and double packer arrangements with concentric tubing strings for dually completed flowing wells. After Turner and Morgan,[32] courtesy *Petroleum Engineer.*

In the past, operators have been reluctant, in general, to apply dual completion techniques. While the initial savings were high, due to elimination of one well for each dual completion, production and remedial work problems were multiplied to the point that the initial economy did not always provide a profit. In recent years, the improvement in dual completion equipment, particularly for artificial lift,[33] has brought these completions into more widespread use. Equipment is also available for permanent-type dual completions, which is a further economic incentive in some cases.[34] Advanced planning is always required to insure that the drilling–casing program fits the completion requirements. The applicability of multiple zone completions depends entirely upon the economic saving as opposed to separate wells. The following considerations are favorable:[33a,35]

1. Two or more marginal zones which do not warrant separate wells may be economically produced.
2. Lease obligations (offset clauses) may be fulfilled for marginal zones.
3. Saving of steel tonnage may be extremely important in times of shortage.
4. Field development may be speeded.
5. Capital may be released for development of other, more profitable properties.

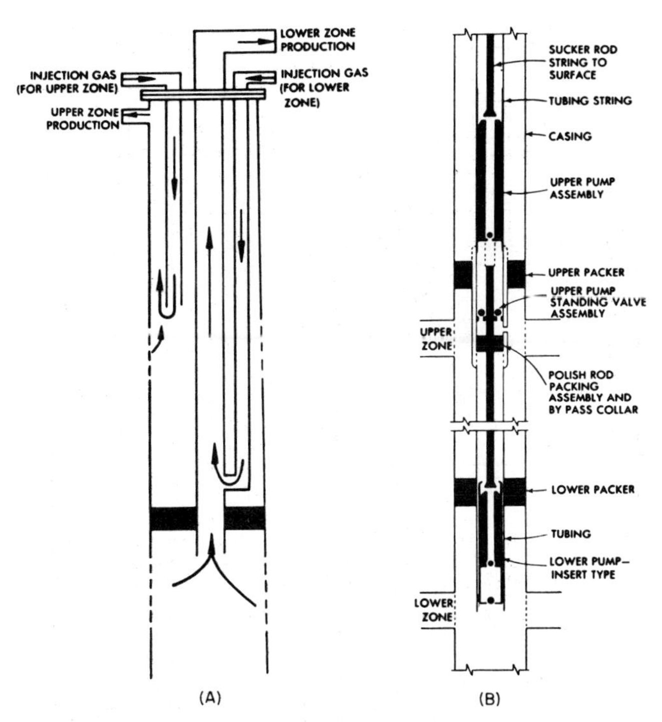

Fig. 15.15. Typical dual completion installation where both zones are artificially lifted. **(A)** Gas lift. **(B)** Pump. After Turner and Morgan,[32] courtesy *Petroleum Engineer.*

As was stated with regard to permanent-type completions, the future will undoubtedly show an increase in multiple zone completions. The continued development and improvement of equipment and techniques, plus increased industry familiarity, will furnish an increasing economic incentive which will overcome much of the reluctance caused by past uncertainties.

15.6 Drainhole Drilling

A number of wells have been completed by drilling one or more lateral offshoots from the main borehole in an attempt to increase productivity. This is accomplished by the use of special directional drilling techniques and equipment.[36] These offshoots, or auxiliary wellbores, are called drain holes. The soundness of such a practice depends upon the balance between the additional cost and the improvement obtained. Considerable data on the subject as based on electrolytic model studies have been presented.[37–39] The effect of drainholes on well productivity may be considered in terms of a productivity ratio, i.e., the ratio of productivity with drainholes to productivity without. This is quite analogous to our previous definition of productivity ratios for formation damage and well stimulation. Partial results of one study are given in Figure 15.16. The curves apply to horizontal drainholes of the dimensions shown in a homogeneous formation. While perfectly horizontal drainholes are never obtained in practice, the results of the simplification are apparently not serious, as reason-

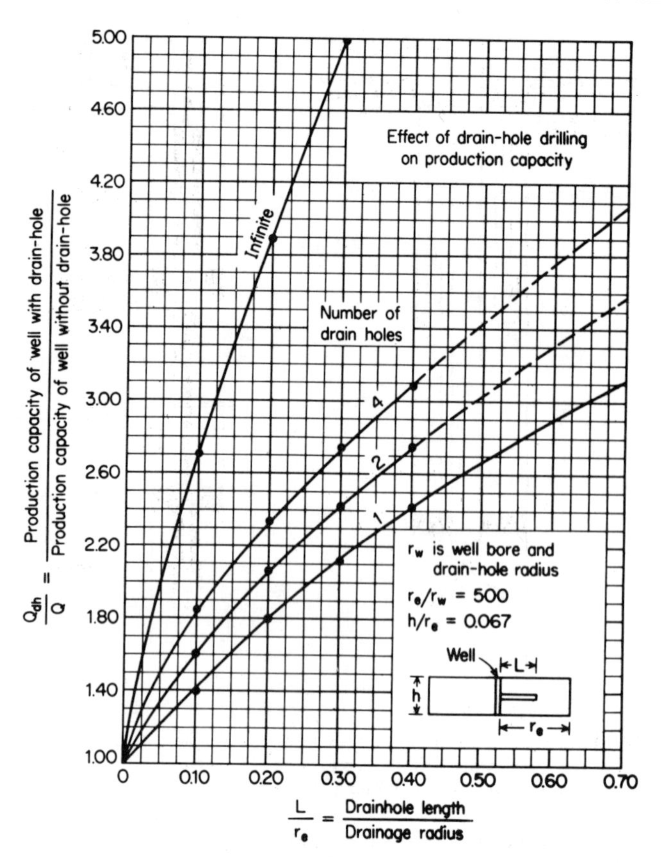

Fig. 15.16. Effect of drainholes on productive capacity. After Landrum and Crawford,[38] courtesy AIME.

able checks with field data have been obtained. Note that for drainhole lengths of $L/r_e = 0.10$ to 0.20, the productivity will be approximately doubled if 2 to 4 drainholes are drilled. This is in line with field experience and practice.

This type of well completion has had limited use in some California and Venezuela fields, and in the Spraberry field of West Texas. With continued improvement and cost reduction, applications may increase; however, drainhole completions must compete with normal stimulation techniques which are capable of obtaining similar productivity increases in most areas (Figure 15.8). That drainholes can be oriented in any desired direction is a distinct advantage, which may result in economical improvement of sweep efficiencies, and hence total oil recovery, in secondary recovery projects (water-floods, etc.).

The preceding sections have indicated the principal features of various well completion methods. We now turn to general topic of considerable interest in any well completion, namely, the exclusion of unwanted gas or water.

15.7 Water and Gas Exclusion — Coning

In many reservoirs, the oil zone occurs with an overlying gas zone or an underlying water zone, or both. In these cases, it is desirable to complete the producing wells in such a manner that free gas and/or water production is avoided. Production of large quantities of gas per unit of oil (high gas/oil ratio) is undesirable, on account of the waste of reservoir energy and consequent rapid decline of pressure. A further incentive is the reduction of oil allowable which is the common penalty imposed by regulatory bodies when excessive gas is produced. Production of water is undesirable for many obvious reasons which stem, mainly, from increased operating costs. The question immediately arises as to the proper selection of the producing interval within such zones, either by selective perforating or by the depth to which the well should penetrate the oil zone. To answer this we will consider the approach of Muskat and Wyckoff[40,41] which was the first basic appraisal of the problem.

Consider Figure 15.17 which depicts an idealized, homogeneous sand body partially penetrated by a producing well. In order to produce the well, a pressure gradient must be established between the well and its drainage radius. The latter extends both vertically and horizontally, and acts on the water as well as the oil. Consequently, both fluids tend to flow toward the well.

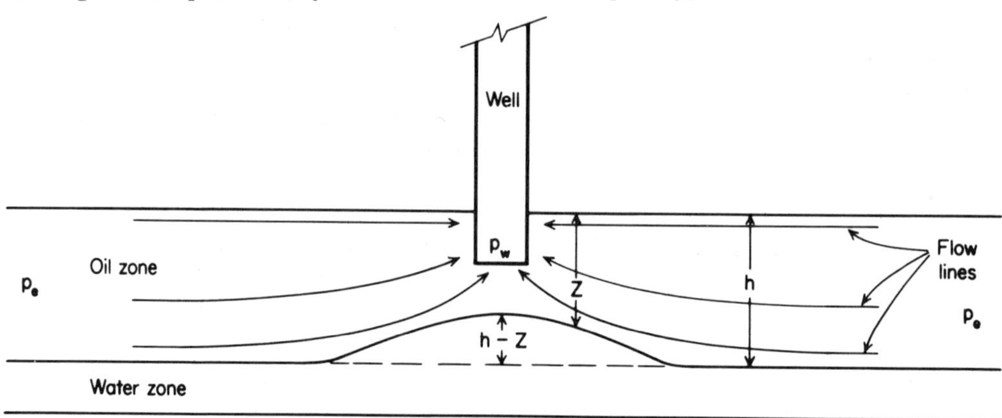

Fig. 15.17. Oil zone underlain by water. Elevation of oil-water interface (cone) is due to producing pressure gradient. After Muskat and Wyckoff,[40] courtesy AIME.

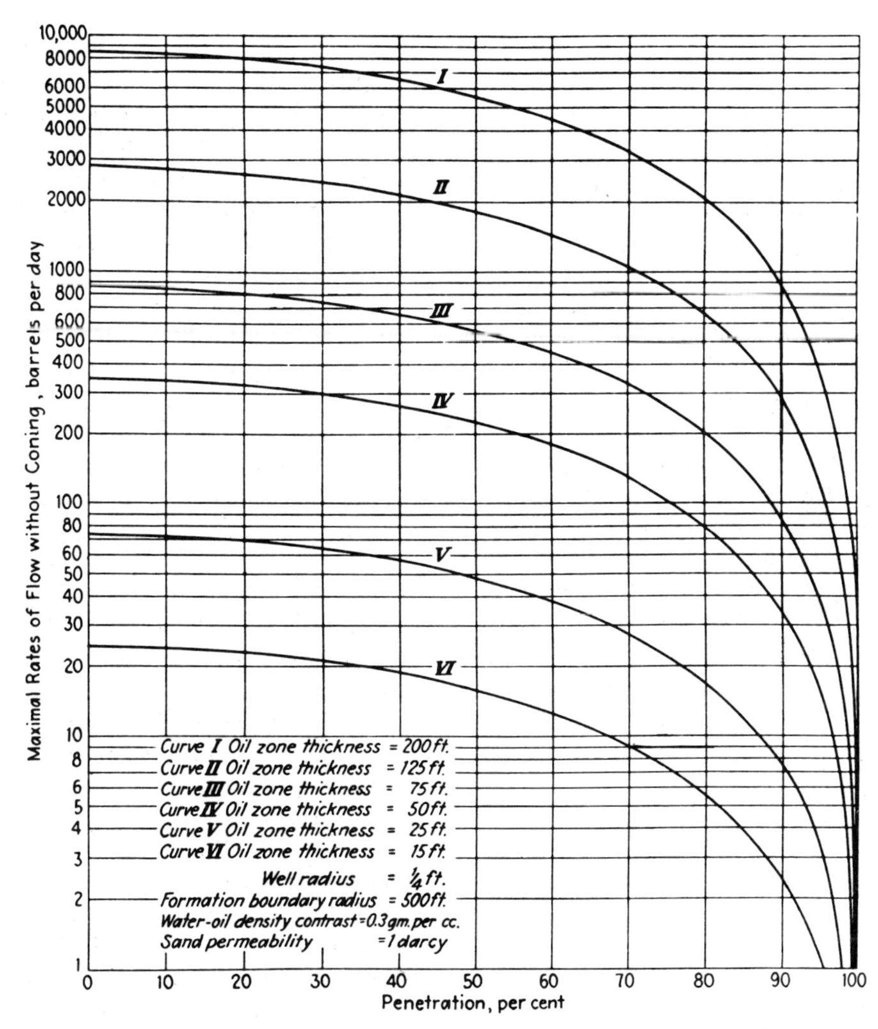

Fig. 15.18. Maximal flow rates without coning as functions of penetration and pay thickness. Other conditions:

$$\frac{k_o}{\mu_o B_o} = \frac{1 \text{ darcy}}{\text{cp}},$$
$$r_e = 500 \text{ ft},$$
$$r_w = \tfrac{1}{4} \text{ ft},$$
$$\Delta\rho_{wo} = 0.3 \text{ gm/cc}.$$

After Muskat and Wyckoff,[40] courtesy AIME.

Fortunately, water is more dense than oil and this density difference furnishes an opposing gradient tending to prevent the water's rising above its static level. If the producing rate is not too high, the oil-water interface will merely become elevated until it reaches an equilibrium position (Figure 15.17) such that the two opposing gradients are equal. If, however, the producing rate is too great, the cone becomes unstable and cusps into the well. This results in water production and its attendant difficulties.

The solution for the pressure distribution above the interface is quite complex and has been treated in detail by Muskat.[42] The general mode of solution is, however, quite obvious: namely, to find solutions for the maximum cone height, h-z, at which the gradients above and below the interface can be equal. Figure 15.18 shows the solution in terms of maximum oil production rate for certain specified conditions. Example 15.1 illustrates the use of these curves.

Example 15.1

A well is producing from an oil zone having the following characteristics:

$h = 100$ ft	$B_o = 1.30$ (formation volume factor)
$k_o = 100$ md	oil density $= 0.70$ gm/cc
$\mu_o = 2$ cp	water density $= 1.10$ gm/cc
$r_e/r_w = 2000$	

If the oil zone is underlain by water, what will be the maximum water-free producing rate if the following intervals are perforated? (a) Upper 50 ft, (b) Upper 25 ft.

Solution:

Note that the curves are constructed for the following conditions:

$$\frac{k_o}{\mu_o B_o} = 1 \frac{\text{darcy}}{\text{cp}}$$

$r_e = 500$ ft

$r_w = \frac{1}{4}$ ft

$\Delta\rho_{wo} = 0.30$ gm/cc (oil-water density difference)

We must change these to conform to our conditions:

$$\frac{k_o}{\mu_o B_o} = \frac{0.1}{(2)(1.3)} = 0.038$$

$r_e/r_w = 2000$ (no correction necessary)

$\Delta\rho_{wo} = 1.10 - 0.70 = 0.40$

From Darcy's law we know that the steady state pressure gradient (and the flow rate) is proportional to $k_o/\mu_o B_o$.

In coning, the allowable flowing gradient also depends on the water-oil density contrast and may be increased or decreased depending on the existing $\Delta\rho_{wo}$. Hence the values from Figure 15.18 may be applied to other situations by:

$$q_{ot} = \frac{(k_o)(\Delta\rho_{wo})}{(\mu_o B_o)(0.30)} q_c = \frac{3.3 k_o \Delta\rho_{wo} q_c}{\mu_o B_o} \tag{15.1}$$

where q_{ot} = maximum tank oil producing rate, bbl/day

k_o = effective permeability to oil, darcys

$\Delta\rho_{wo}$ = density contrast, water-oil, gm/cc

μ_o = reservoir oil viscosity, cp

B_o = formation volume factor, vol/vol

q_c = maximum rate from appropriate curve of Figure 15.18

For our example:

(a) Penetration = 50 ft/100 ft = 50%

$q_c = 1000$ bbl/day (interpolated)

$q_{ot} = (3.3)(0.038)(0.40)(1000) = 50$ bbl/day

(b) Penetration = 25 ft/100 ft = 25%

$q_c = 1500$ bbl/day

$q_{ot} = 75$ bbl/day

The same procedure may be used to estimate maximum rates without gas coning, the only change being substitution of the oil-gas density difference $\Delta\rho_{o\text{-}g}$ for $\Delta\rho_{wo}$. Such a system may be visualized by mentally inverting Figure 15.17. Gas coning prevention is less critical since $\Delta\rho_{og} > \Delta\rho_{wo}$. Corrections for other values of ln (r_e/r_w) are generally negligible, due to its relative insensitivity. It is worthwhile, however, to emphasize the other assumptions involved and the possible effect they may have on the correctness of the estimated q_{ot}.

(1) Homogeneous fluid: The relatively low pressure around the well bore may cause gas evolution and a resultant decrease of B_o, increase of μ_o, and decrease in k_o. Since it is this region which governs coning, these factors may cause the estimated value of q_{ot} to be high.

(2) Homogeneous sand: It is actually unusual for vertical permeability to be as great as horizontal permeability. If any vertical permeability barrier exists, such as a continuous shaly or dense streak, the actual water-free (or gas-free) tank oil producing rate may be much higher than the predicted value. For example, if even one inch of impermeable shale separates the oil-water zones, the entire oil section could be perforated, providing the shale barrier remained intact. From this standpoint, the theoretical values can represent a minimum. If, however, vertical exceeds horizontal permeability, the reverse occurs.

It should be noted that the curves of Figure 15.18 become quite flat at low penetrations, and that little benefit is gained by restricting penetration to less than 20%. This limit is of extreme importance in remedial work where bottom plugging and reperforating is attempted. If the section is homogeneous, it is fùtile to attempt to exclude bottom water by such practices if the penetration is already low. This does not rule out those heterogeneous cases where an impermeable streak exists which can act as a shield. In the latter cases, the success of water shut-off depends on the cementing job. Such barriers do not have to be of large areal extent, since the pressure gradient dissipates rapidly with distance from the well.

It must be realized that it is the pressure gradient which governs the occurrence of coning. Hence a damaged zone will aggravate the problem by increasing the gradient around the well. Stimulation, or an improved productivity ratio, will then be beneficial in preventing coning — providing that the permeability is improved equally in both horizontal and vertical directions. An increase of permeability in only the horizontal direction will further diminish coning, while an increase in vertical permeability may worsen the situation. Constant q_o was assumed in the above discussion.

Many water production problems attributed to coning are actually due to faulty cement jobs or casing leaks. True coning is inherently rate-sensitive and should dissipate if the well is shut in until the cone subsides, and then reopened at a lower rate. It is doubtful that complete subsidence can occur within a reasonable time in moderate to low permeability sands, and the increased water saturation immediately below the well may reduce the critical oil production rate below its pre-coning value. This is, however, conjectural. Water analysis data are the best guide in defining the source of produced water, and will often show whether or not coning is the true problem.

While the subject of reservoir performance is beyond our scope, it is perhaps worthwhile to mention the water drive behavior briefly at this time. Oil sands underlain by water are apt to perform as water drive reservoirs. This means that as oil is produced, the pressure drop in the oil column will allow adjacent edge or bottom water to expand and encroach into the oil zone. This expanding water is then a source of pressure maintenance — hence the name, water drive. The extent to which the water encroaches is governed by the amount of water in the aquifer, and if this is sufficiently large it will eventually invade the entire oil zone. If bottom water is of this nature, the fight against coning is a losing battle, since the *static* level is continually rising, and water production must eventually be tolerated. This is of more interest in post-completional problems than for our present purpose, but is a point worth mentioning. Gas coning is an entirely analogous problem, although it is less serious from an operating standpoint.

Water or gas exclusion problems other than coning are often due to irregular permeability and saturation distributions over the producing interval. In sandwich-type reservoirs (composed of alternate and essentially separate layers) exclusion problems may be solvable by squeeze cementing and reperforating. The success of this procedure is always dependent on the cement job's providing the necessary inter-layer seals behind the casing. Lateral fingering, which is similar to coning except that the water level is both horizontally and vertically displaced from the well, has been treated by Arthur,[43] as have other aspects of coning. M. King Hubbert's classic geological paper on hydrodynamic oil entrapment also covers coning and presents definitive equations.[44] Chaney et al[45] have prepared curves based on the Muskat and Arthur papers which simplify the calculations of perforation positions for maximum water- and/or gas-free rates. A further treatment may also be found in Pirson's text.[46]

15.8 Stimulation Methods

In Chapter 12, well stimulation was mentioned as a means of increasing well productivity. Several methods may be applied, depending on the individual situation. Each of the basic methods is in turn subject to numerous variations with regard to techniques and materials. We will limit our discussion to the basic principles involved and the general applicability of each method. The three principal stimulation methods in their chronological order of development are:

1. Nitro-shooting
2. Acidizing
3. Hydraulic Fracturing

These have been classified as large area penetrators because their effect extends an appreciable distance from the well, as compared to other techniques which will be mentioned later.

15.81 Nitro-shooting:[47–49]

The use of explosives to improve productivity is practically as old as the oil industry. This process involves the placing and detonating of an explosive adjacent to the producing strata. The explosion shatters and fractures the rock, which enlarges the borehole and increases permeability, thereby increasing productive capacity. Solidified or gelatin type nitroglycerin is commonly used. The explosive is placed in suitable containers (often called torpedoes) and lowered to the desired open hole interval. The upper casing is protected by placing a temporary plug, tamped with cement, plastic, and/or gravel above the shot (see Figure 15.19). The shot is detonated with a time bomb. The well must then be cleaned of debris prior to being placed on production.

Nitro-shooting is much less widely applied since the development of hydraulic fracturing. High powered, expendable shaped-charge guns have also been substituted for nitro-shooting in many instances. The latter is, however, still used to a limited extent and no doubt has advantages in some areas.

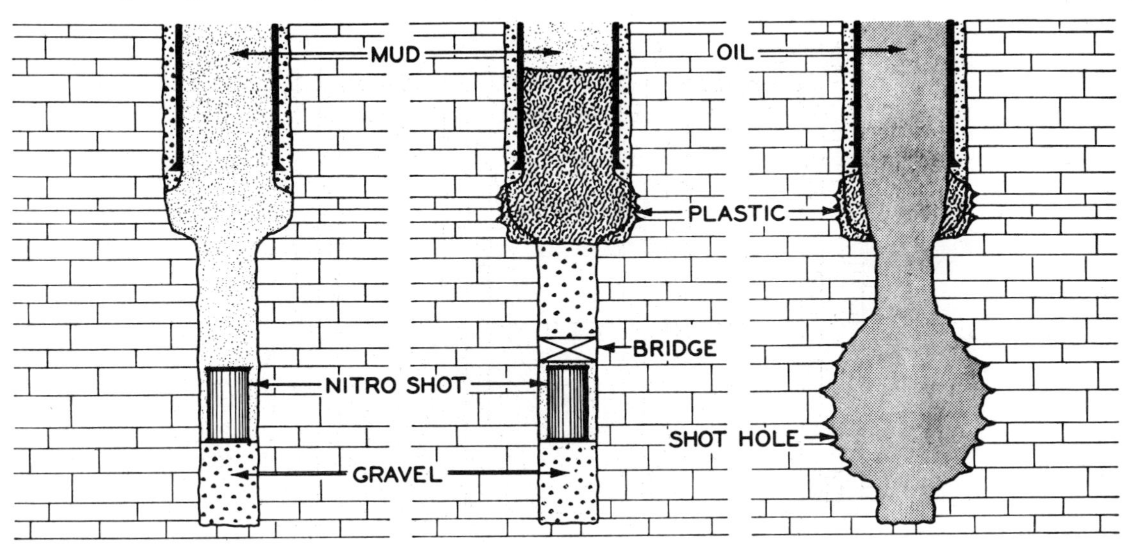

Fig. 15.19. Typical nitro-shooting procedure. Courtesy Dowell, Inc.

15.82 Acidizing:

Acid was used to enhance productivity before 1900; however, it was not until the 1930's that its usage became widespread. Acidizing involves the injection of acid into an acid-soluble pay zone where its dissolving action enlarges existing voids and thereby increases the permeability of the zone. The acid commonly used is 15% hydrochloric (by weight) which reacts with limestone or other carbonates according to the following reaction:

$$2HCl + CaCO_3 \rightarrow CaCl_2 + H_2O + CO_2$$

Only the carbonate rocks are generally susceptible to acid treatment; however, some sands have sufficient calcareous content (usually cementing material) to warrant acidization. Numerous additives are used in the acid, including inhibitors to retard corrosion of casing and tubing. Non-emulsifying agents are often added to prevent formation of an oil-acid emulsion during the stimulation treatment. Such emulsions, if formed, are often highly viscous and cause permeability damage which can largely cancel the benefits of the treatment. This emulsifying tendency varies with the crude oil, and selection of the proper non-emulsifying agent is best determined from tests with the field crude oil.

Since HCl does not react with silicates, it will not dissolve mud cake. Special solutions called mud acids have been developed for this purpose and are often used, in relatively small volumes, either to prepare the well bore for a conventional treatment, or to serve as the sole means of stimulation. The chemical nature of mud acid varies among service companies; however, a common type is a mixture of HCl + HF (hydrofluoric acid), the latter being a silicate solvent.

Acid treatments are performed in many ways, depending on the particular characteristics and existing equipment of the well in question. Figure 15.20 shows the Carr method which is often used. The hole is initially filled with oil or another fluid, then acid is pumped down the tubing while the casing annulus valve at the well head is left open to permit discharge of the displaced oil at the surface. When sufficient acid volume has been injected to displace the entire tubing string and annular section opposite the pay zone, the annulus valve is closed. Continued pumping forces the acid into the formation. Oil is then used to displace the last of the acid. Afterwards, the pressure is released and the well either is allowed to backflow or is swabbed to remove the spent acid and residue, and is then placed on production. In wells completed with tubing-casing packers, slightly altered, but basically similar, methods are used.

It is apparent that the above procedure offers little control over the interval of pay section to be acid-treated. In general, the most permeable spots receive the bulk of the treatment. To prevent this, the injection pressure is generally maintained at the highest possible level in an attempt to obtain more uniform treatment of the entire pay section. Since this practice is not entirely satisfactory, many methods of selective treating have been developed whereby more uniform coverage is obtained. These include the use of temporary blocking agents, as well as the use of multiple packer arrangements to isolate specified intervals. The idealized application of a temporary plugging material is shown in Figure 15.21. In the first stage, the treatment affects the high permeability section. A thickened hydrocarbon gel is then injected which enters and seals this zone, allowing the second acid treatment to break down a second region. The procedure is repeated for a third interval. Techniques of this and similar nature have proved beneficial in both acidizing and hydraulic fracturing.

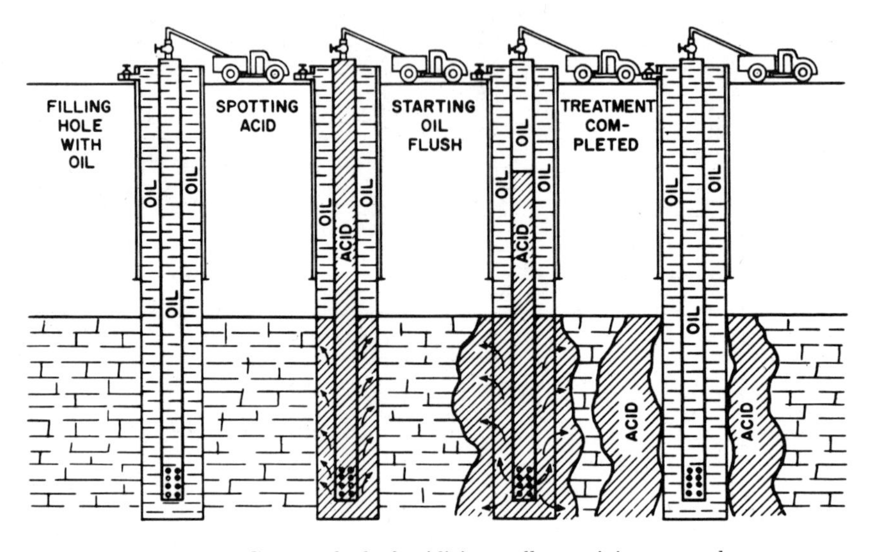

Fig. 15.20. Carr method of acidizing well containing no packer.

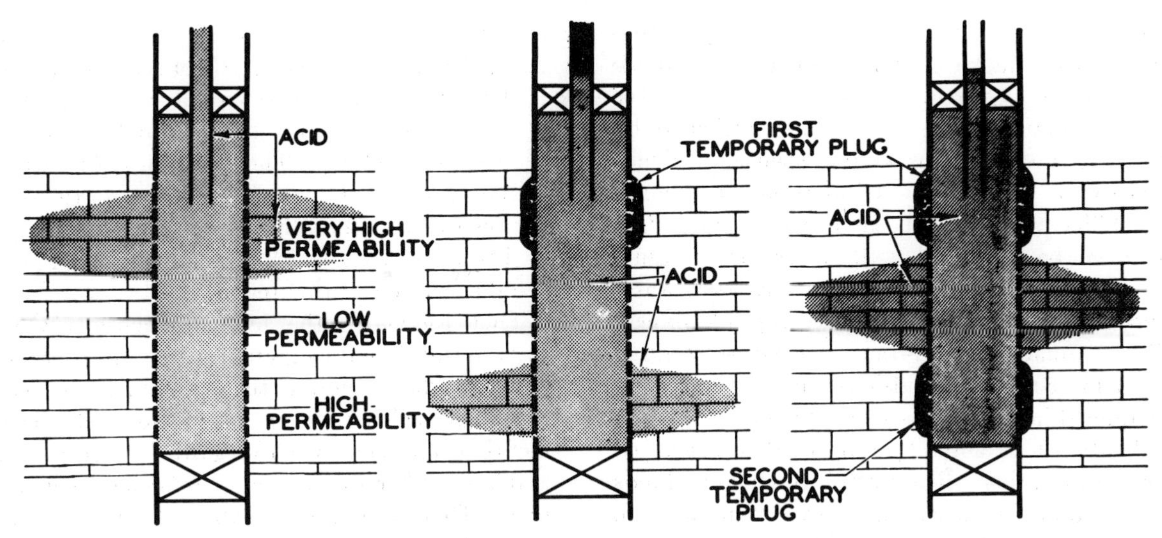

Fig. 15.21. Idealized illustration of three stage acid treatment utilizing temporary plugging materials. Courtesy Dowell, Inc.

A large number of temporary plugging materials are in use. Such materials must, of course, be easily removable after the treatment. Many of these contain reactive solids which dissolve within a short time and destroy the gel, allowing it to backflow from the formation. Others require injection of a breaker solution which breaks down the plugs in a similar manner.

There is always some question as to the quantity of acid to be used in a particular case. Generally, a conventional acid job does not create fractures but merely enlarges existing voids. In nearly all cases, it is the secondary voids and not the intergranular ones which govern the flow capacity of carbonate reservoirs. In highly permeable sections where acidization is required only because of damage, a small 500 gallon mud acid treatment may be more than adequate. In other cases several thousand gallons of HCl may be required to obtain a reasonable increase in productivity. In unfractured limestone sections, acidization may yield little if any improvement. Experience usually dictates the amount needed, although little agreement exists between companies in different areas. Kingston's handbook is the most complete single treatment of acidizing and contains an extensive bibliography.[50]

15.83 Hydraulic Fracturing:

This method of well stimulation, developed by the Pan American Oil Company (then the Stanolind Oil and Gas Company), was first introduced to the industry by J. B. Clark[51] in 1948, and received immediate and widespread acceptance.[52] The basic procedure involves the injection of a fracturing fluid and propping agent into the pay zone under sufficient pressure to open existing fractures and/or create new ones. These are extended some distance around the well by continued high pressure injection after the initial breakdown or rock rupture has occurred. Upon cessation of pumping (as pressure is reduced) the fractures remain open, being held in place by the propping agent, a carefully sized, silica sand. This process is applicable to virtually all reservoir rocks and may be combined with acid treatments in limestone areas.

Hydraulic fracturing of reservoir rocks was not a new idea; it had long been experienced (though not always recognized) in acidizing, lost circulation, squeeze cementing, and pressure parting of water injection wells in waterflood projects. The idea of using a propping agent to prevent fracture closing was the key to the new method's success. The sand most commonly used as a propping agent is 20-40 mesh, (.0328–.0164 in.) well rounded, silica sand which has a packed permeability of about 300 darcys. Multiple sand sizes also are often applied in fracture treatments. Relatively small sand is used at first, a larger size being applied next to prop the greater fracture width near the well.

Field procedure in fracturing requires the presence of the necessary pumping, mixing, and fracture fluid storage space as dictated by the size and type of job. This equipment is mobile and is normally furnished by the service company performing the job. The desired proportion of sand is mixed either continuously or in batches with the fracture fluid in special blending tanks. From these, the sand–fluid mixture is pumped down the well through either the tubing or casing. If treatment is carried out through tubing, a packer is generally used to prevent imposition of excessive internal casing pressure. As in acidizing, selective treating methods involving packers and temporary plugging agents are used. A large amount of research and field experience has shown the success of fracturing to be dependent on a number of variables. Some of these will be discussed briefly.

1. Fracture fluid:[53–55] Early fracture techniques generally utilized thickened gels made from kerosene and diesel oil. Currently, lease crude oil is the principal fracture fluid; it may be thickened by additives if necessary for sand suspension. Fluids native to the formation are less prone to damage permeability and should be used if available. Gas wells have been treated with water-base fracture fluids, with superior results; however, fresh water should not be used if the sand is susceptible to clay swelling. Combined acid-fracture treatments using gelled acid or acid-oil emulsions have been successfully applied in various carbonate areas.

The filtration loss of the fracture fluid has received considerable attention. Penetrating (high loss) fluids reduce breakdown pressures and are often used to spearhead the treatment. Low fluid loss is generally desirable as it promotes extension of fractures, by virtue of the higher injection pressures made possible by the small leak-off of filtrate as new rock surface is exposed to the fracture fluid.

2. Sand-fluid ratio:[56] Sand concentrations of $\frac{1}{2}$ to 4 lb/gal have been frequently used in fracturing. It is difficult to define any universally applicable optimum concentration and quite possibly such a figure may vary with the area. From field experience, it appears that 1 to 2 lb/gal is the most commonly applied range of concentration. The proposed injection rate and the fracture fluid's filtration loss, as well as formation characteristics, generally govern field practice. Screening-out, due to sand bridging at the well bore face, has occurred and may generally be eliminated by reducing sand size, fluid loss, and/or sand concentration.
3. Injection rate during treatment:[57] Injection rates are controlled by the fracture fluid flow properties, available pump horsepower, and the size of the injection string (tubing or casing). Treatment through tubing under packer(s) was once standard; however, treatment through casing has definite advantages and is now applied where possible. High injection rates and pressures are more apt to create multiple and deeply extending fractures; this in turn brings about greater productivity. High friction losses restrict down-tubing injection rates to about 3 to 5 bbl/min while casing treatments often average injection rates of 40 to 70 bbl/min. In thin pay zones with bottom water, excessive fracture extension may be undesirable, and small, low rate treatments may give superior results.
4. Size of treatment: In moderate to high permeability zones which have been badly damaged during completion, small treatments may be completely adequate. In tight zones, the large volume treatment may give optimum results. The pay thickness is also a governing factor. Current practices seem to average from 500 to 1000 gal/ft of section. In the early days, fracturing treatments averaged 2000 to 4000 gal; however, this figure rapidly increased to around 10,000 gal. The economics of treatment size requires careful analysis; it is certain that considerable overtreatment has been applied to many wells. The paper by Howard, Flickinger, and Fast should be consulted in this regard.[58]

There are many other factors to consider with regard to hydraulic fracturing application and procedure. In addition, there remains the question of vertical, horizontal, or inclined fracture orientation. This aspect was mentioned previously with regard to squeeze cementing, and no further discussion or references will be given here. There are cases in which vertical fractures are desirable, e.g., sections of poor vertical permeability where a fracture across bedding planes would promote more complete drainage. In other cases, a vertical fracture across such boundaries might promote excessive gas or water production. It would indeed be desirable if fracture orientation could be controlled by the operator; procedures facilitating such control have been proposed. Whether or not control of fracture orientation is practically feasible is the subject of much controversy; all claims should be carefully examined. As space does not permit presentation of these arguments, we will restrict ourselves to this statement of caution and suggest that the reader explore the literature cited in the squeeze cementing section of the last chapter.

15.84 Miscellaneous Stimulation Methods

A number of minor, or small area penetrator, stimulation techniques are in use. Among these are perforation and mud acid washes which have already been mentioned. Removal of sand face deposits is sometimes accomplished by washing the sand face with high velocity streams (jets) of acid. This is known as jet acidization. Various surface active agents are also used to remove flow restrictions resulting from emulsification and/or clay swelling. Marble-shooting is a small charge variation of nitro-shooting which utilizes the abrasive or scouring action of glass marbles placed around the charge. All of these techniques have restricted penetration and are not applicable in many cases. Table 15.3 lists the common stimulation techniques and the general considerations involved in their selection.[13]

15.9 Benefits and Limitations of Well Stimulation

Occasionally, the statement is made that stimulation methods merely increase producing rate and do not affect ultimate oil recovery. Such a statement is

TABLE 15.3

CONSIDERATIONS INVOLVED IN THE SELECTION OF WELL STIMULATION METHODS. COURTESY API.[13]

LARGE AREA PENETRATORS

NITRO-SHOOTING

Benefits:
1. Bore-hole enlargement combined with fracturing.
2. Not selective to single fracture at weakest bedding plane.
3. No hydrostatic or fluid effect on permeability.
4. Stimulant itself relatively inexpensive.

Limitations:
1. Cleanout problems and expense.
2. Hazard to personnel, well, equipment.
3. Limited to open-hole completions.

ACIDIZING

Benefits:
1. Moderate bore-hole enlargement.
2. Primarily adapted to formations of appreciable calcareous content (not generally adaptable to sandstone).
3. Cleans out, enlarges, interconnects fractures, vugs, other channels.
4. Stimulant relatively inexpensive.
5. Adaptable to both open-hole and set-through completions.

Limitations:
1. May require residue cleanout.
2. Somewhat hazardous and corrosive.

HYDRAULIC FRACTURING

Benefits:
1. No bore-hole enlargement.
2. Highly flexible procedure:
   a. Permits multiple or single fracture.
   b. Can combine advantages of fracturing and acidizing.
   c. Wide latitude of sand-carrier agent.
3. Maximum effective area of stimulation.
4. Maximum extension of inherent or induced fractures.
5. Propping agent maintains high permeability.
6. Permits relatively localized fracture level if desired (in approximately horizontal bedding planes).
7. Adapted to either open-hole or set-through completions.

Limitations:
1. May involve cleanout of propping sand.
2. Somewhat hazardous with some carriers.
3. Relatively expensive.
4. High pressures may damage tubing or casing.
5. Intricate down-hole operations requiring packer manipulations.

SKIN BREAKERS

GUN SHOOTING (Open Hole)

Benefits:
1. Improves permeability adjacent to bore holes* blocked by: *a*, mud solids; *b*, mud filtrates; *c*, emulsions; *d*, formation clay swelling.
2. Relatively inexpensive.
3. Minimum cleanout required.
4. Highly selective (location).
5. Fracture starting at given level (if followed by hydraulic fracturing).

Limitations:
1. Limited penetration.
2. Ineffective in large bore holes.

*If penetration sufficient.

MUD ACID

Benefits:
1. Improves permeability adjacent to bore holes blocked by: *a*, mud solids; *b*, mud filtrates; *c*, emulsions; *d*, formation clay swelling.
2. Limited benefits from usual acidizing effects.

Limitations:
1. May require residue cleanout.
2. Somewhat hazardous and corrosive.
3. More expensive than usual acidizing.
4. Limited selectivity to level.

SURFACE-ACTIVE AGENTS (Fluid Squeezes)

Benefits:
1. Improves permeability adjacent to bore holes blocked by: *a*, emulsions; *b*, mud filtrates.
2. In certain combinations with other agents may remove clay particle blocks.
3. May reduce breakdown pressure for hydraulic fracturing.

Limitations:
1. Relatively expensive (some commercial types).
2. Sensitive to chemistry of formation and fluids.
3. May require residue cleanout.
4. Limited selectivity to level.

JET ACIDIZATION

Benefits:
1. Especially adapted to remove this sand face deposits (not soluble by usual methods).
2. Minor cleanout required.
3. Limits bore-hole enlargement (where not needed).
4. Suitable for recurring treatment.
5. Highly selective (location).

Limitations:
1. Low penetration.
2. Comparatively expensive.

MARBLE SHOOTING

Benefits:
1. Especially adapted to remove thin sand face deposits (not soluble).
2. Minor cleanout required.
3. Limits bore-hole enlargement (where not needed).
4. Suitable for recurring treatments.

Limitations:
1. Least penetration of listed stimulants.

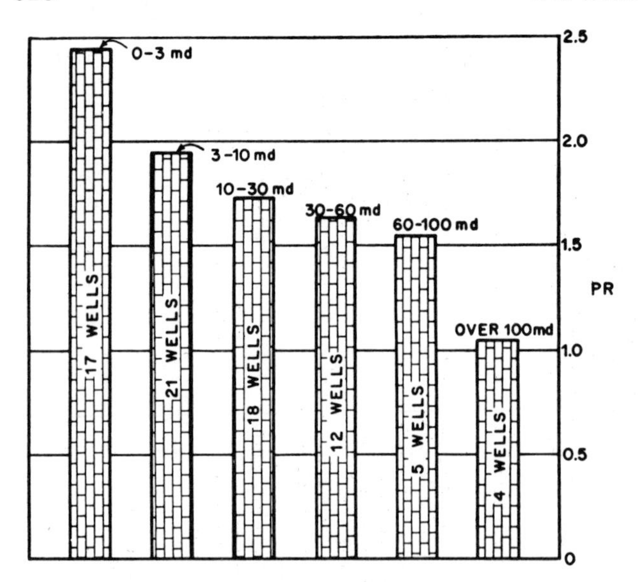

Fig. 15.22. Productivity ratio vs. permeability of stimulated sand wells. Courtesy API.[13]

unsound, since a minimum producing rate or economic limit always exists below which the operator cannot afford to produce. Many wells which produce virtually nothing before stimulation become prolific, flowing producers after acid or fracture treatments. In such cases, the total production from the well has been obtained as a direct consequence of stimulation. Many old wells, which had declined to bare stripper operations, have been transformed to highly profitable operations by application of newly developed techniques which allowed much additional oil to be economically obtained. Ultimate recovery is therefore closely allied with maintaining economic producing rates.

There have been, of course, numerous occasions when the benefits of stimulation have been overestimated. For example, no treatment is applicable if the reservoir pressure is depleted, for fluids will not flow without some driving force, regardless of permeability.

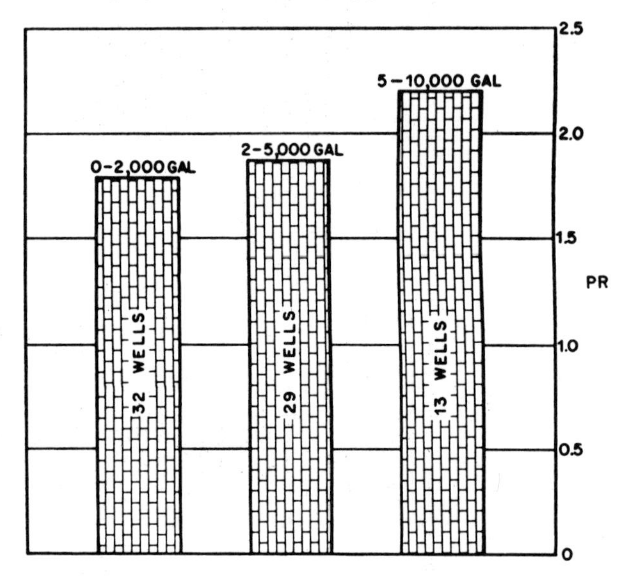

Fig. 15.23. Productivity ratio vs. size of fracture treatment in sand wells. Courtesy API.[13]

Obviously, wells which swab dry on completion need stimulation before commercial productivity can be obtained. In other cases the need, or lack of need, is not so apparent and analysis is required. The pressure build-up method of analysis is the best means of evaluating both the need for and feasibility of stimulation. This technique was demonstrated in Chapter 12. It is well, however, to note the general range of productivity improvement which can be obtained by stimulation. API Bulletin D6[13] contains a large amount of statistical data based on actual well performance. The average effect of fracture treatments on sand wells is shown in Figure 15.8. Figure 15.22 shows the same data distributed according to formation permeability. Figure 15.23 shows a further breakdown with respect to treatment size.

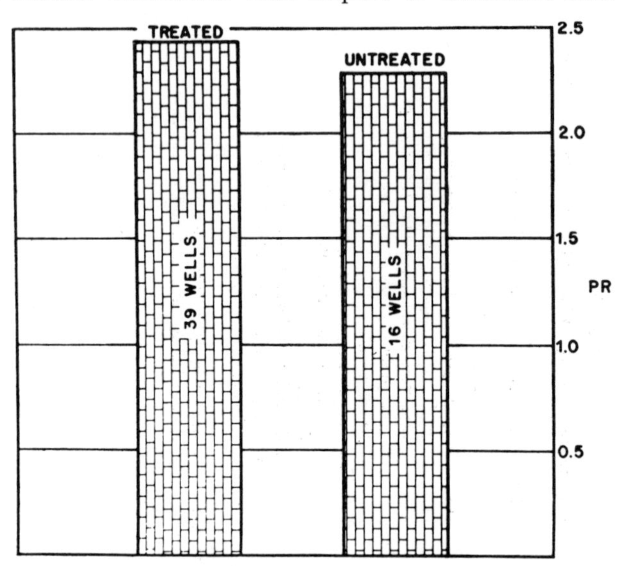

Fig. 15.24. Productivity ratio of treated and untreated limestone wells. Courtesy API.[13]

Observed average productivity ratios for 55 limestone wells, both treated and untreated, are shown in Figure 15.24. It appears that limestone wells are less susceptible to damage than sands; however, the committee did find *PR*'s less than one for some limestone wells. The existence of initial *PR*'s above 1 in limestone wells is believed to be due to fracturing during drilling plus effective well bore enlargement by existing vugs and fracture systems.

These data allow important generalizations to be made concerning the potential benefit of stimulation.

1. A typical fracture treatment in a low permeability sand generally results in a productivity ratio of approximately 2–2.5. This diminishes with permeability, approaching 1 in highly permeable sands. This means that improvement in production rate will be, in general, restricted to the ratio of these improved

values to the PR before treatment. Hence a damaged well whose untreated $PR = 0.2$ might be expected to undergo a ten-fold increase, while a well with an initial $PR = 1$ will respond less sensationally, although still appreciably. This is an extremely important concept. Treatment of a low capacity well producing 10 bbl/day from a 5 md sand at a PR of 1 is apt to yield a two-fold increase. Stimulation does not appreciably alter the average properties of the bulk rock, and cannot be expected to transform dry holes into gushers.

2. Treatment of highly permeable zones results in lower PR's; however, these walls have adequate native flow capacity and require stimulation only if damaged. Hence small treatments may be completely effective in restoring normal productivity.

These generalizations, like any, are subject to error in specific cases where some fortuitous circumstance may provide additional benefits; however, it is not wise to count too heavily on such occurrences. Stimulation is an individual well problem and should be approached on as sound an analytical and economic basis as possible.

Thus the cure of formation damage is accomplished by stimulation. Whether this best involves fracturing, acidizing, shooting, or combinations of these and other methods depends on the formation and the choice of the operator. Careful analysis of early field wells will permit application of more efficient methods to subsequent completions. Large treatments, small treatments, or no treatment may result in the most economic operation. Proper selection of drilling and completion fluids may also be determined for a given formation. Premium drilling fluids are justified on a formation damage basis only if their use results in sufficient completion improvement to warrant the additional cost.

This discussion of well completions has been quite qualitative, due to the complexity of mathematical treatment of the subject. Many individual well problems which apply directly to well completions are, however, presented in other texts commonly used in petroleum reservoir engineering courses.

PROBLEMS

1. At what perforation density will the theoretical productivity of a perforated completion equal that of an open hole for 12-in. casing and penetration of 12 in.?
2. Rearrange Figure 15.7(A) as a plot of q_p/q_o vs penetration depth for shot densities of 2, 4, and 6 holes/ft. From this, estimate the relative productivity of 6-in. casing with 4 holes/ft at a penetration of 12 in.
3. Given the following screen analysis of an unconsolidated sand, determine:

(a) The proper slot width for sand exclusion
(b) The gravel size to be used for sand exclusion by gravel packing.

Screen size, in.	*Cumulative % retained*
0.03	1
.02	5
.01	20
.005	80
.003	90
.001	100

Note: Show this as a semi-log plot, % vs log screen size.

4. A well is completed with 3 approximately horizontal drainholes as follows:

$r_e/r_w \simeq 500$, $h = 200$ ft Drainhole length $= 50$ ft

What productivity increase would you expect from this technique?

5. An oil well is to be completed in a 125-ft section of homogeneous sand which is underlain by water. The following data are pertinent:

Spacing = 40 acres $B_o = 1.40$
Hole Size = 8 in. $\rho_o = 0.75$
$k_o = 200$ md $\rho_w = 1.05$
$\mu_o = 1.5$ cp

Estimate the maximum water-free producing rates if the following intervals are perforated: (a) Upper 100 ft, (b) Upper 50 ft, (c) Upper 25 ft.

6. Assume that instead of having bottom water, the pay zone of Problem 5 is overlain by a gas cap:

Gas gravity = 0.80
$\mu_g = 0.02$ cp
$T = 200°$F (Reservoir temp.)
$p_e = 4000$ psia in gas zone

Estimate the maximum producing rate without gas coning if the following intervals are perforated: (a) lower 100 ft, (b) lower 50 ft, (c) lower 25 ft.

7. Pressure build-up analysis indicates the following characteristics for a newly completed sand well. The current producing rate is 50 bbl/day.

$k_o = 50$ md $PR = 0.5$

(a) What productivity would you expect after stimulation? Assume an average result as based on the API study (Figures 15.24 to 15.26) cited in this chapter.
(b) Suppose $k_o = 5$ md and $PR = 0.20$. What productivity increase would you estimate was obtainable?

REFERENCES

1. McLemore, R. H., "Application of the Shaped-Charge Process to Petroleum Production," *The Petroleum Engineer*, Aug. 1948.
2. Forsyth, V. L., "Koneshot Perforating — A Study in the Effective Use of the Shaped-Charge Process," *Tomorrow Tools Today*, 2nd Quarter, 1949.
3. Forsyth, V. L., "A Review of Gun-Perforating Methods and Equipment," *API Drilling and Production Practices*, 1950.
4. Box, W. T., and R. F. Meiklejohn, "The Jet Perforation Story," *World Oil*, Mar. 1950.

5. Painter, T. W., "Jet Perforating in the West Texas Area," *The Petroleum Engineer*, July 1952.

6. Poulter, T. C., and B. M. Caldwell, "The Development of Shaped Charges for Oil Well Completion," *Trans. AIME*, Vol. 210, (1957), p. 11.

7. Delacour, J., M. P. Lebourg, and W. T. Bell, "A New Approach to Elimination of Slug in Shaped Charge Perforating," AIME Tech. Paper No. 941-G, presented Dallas, Texas, Oct. 1957.

8. Caldwell B. M., and H. D. Owen, "A New Tool for Perforating Casing Below Tubing," *Trans. AIME*, Vol. 204, (1954), p. 29.

9. Lebus, J., "Should We Use Bullets or Jets?" *World Oil*, Mar. 1957.

10. Lewelling, H., "Experimental Evaluation of Well Perforation Methods As Applied to Hard Limestone," *Trans. AIME*, Vol. 195, (1952), p. 163.

11. Allen, T. O., and H. C. Worzel, "Productivity Method of Evaluating Gun Perforating," *API Drilling and Production Practices*, 1956, p. 112.

12. Krueger, R. F., "Joint Bullet and Jet Perforation Tests," *API Drilling and Production Practices*, 1956, p. 126.

13. *Selection and Evaluation of Well Completion Methods*, API Bulletin D6, July, 1955. Also in *API Drilling and Production Practices*, 1955, p. 421.

14. Gatlin, C., "Perforating of All Types," in *Fundamentals of Rotary Drilling*, Dallas, Texas: The Petroleum Engineer Publishing Co., p. 98.

15. Reed, J. E., and F. F. Carr, "Effect of Temperature, Pressure, Standoff Upon Perfo-Jet Penetration," *Oil and Gas Journal*, Aug. 3, 1950.

16. Muskat, M., *Physical Principles of Oil Production*, 1st ed. New York: McGraw-Hill Book Co., Inc., 1949, pp. 215–220.

17. Howard, R. A., and M. S. Watson, Jr., "Relative Productivity of Perforated Casing — I and II," *Trans. AIME*, Vol. 189, (1950), pp. 179 and 323.

18. McDowell J. N., and M. Muskat, "The Effect of Well Productivity of Formation Penetration Beyond Perforated Casing," *Trans. AIME*, Vol. 189, (1950), p. 309.

19. Allen, T. O., and J. H. Atterbury, Jr., "Effectiveness of Gun Perforating," *Trans. AIME*, Vol. 201, (1954), p. 8.

20. Hildebrandt, A. B., H. C. Bridwell, and J. M. Kellner, "Development and Field Testing of a Core Barrel for Recovery of Unconsolidated Oil Sands," AIME Tech. Paper No. 880-G, presented Dallas, Texas, Oct. 1957.

21. Coberly, C. J., "Selection of Screen Openings for Unconsolidated Sands," *API Drilling and Production Practices*, 1937, p. 189.

22. Hill, K. E., "Factors Affecting the Use of Gravel in Oil Wells," *API Drilling and Production Practices*, 1941, p. 134.

23. Gumpertz, B., "Screening Effect of Gravel on Unconsolidated Sands," *Trans. AIME*, Vol. 142, (1941), p. 76.

24. Rogers, C. J., "Some Practical Aspects of Gravel Packing," *Trans. AIME*, Vol. 201, (1954), p. 15.

25. Tausch, G. H., and C. B. Corley, Jr., "Sand Exclusion in Oil and Gas Wells," *The Petroleum Engineer*, June and July 1958, pp. B–38 and B–58, respectively.

26. Huber, T. A., T. O. Allen, and G. F. Abendroth, "Well Completion Practices," Presented API Annual Meeting, Los Angeles, Calif., Nov. 1950.

27. Huber, T. A., and G. H. Tausch, "Permanent Type Well Completion," *Trans. AIME*, Vol. 198, (1953), p. 11.

28. Lebourg, M. P., and G. R. Hodgson, "A Method of Perforating Casing Below Tubing," *Trans. AIME*, Vol. 195, (1952), p. 303.

29. McGhee, E., "Here's the Equipment You'll Need for That Permanent Type Well Completion," *Oil and Gas Journal*, Dec. 7, 1953.

30. "Permanent-Type Well Completions," Special Section, *World Oil*, Mar. 1954.

31. Huber, T. A., and G. H. Tausch, "Permanent-Type Well Completion Developments," *API Drilling and Production Practices*, 1955, p. 103.

32. Turner, M. C., and F. C. Morgan, "Multiple Zone Completions," *The Petroleum Engineer*, June 1956, p. B-38.

33. Zaba, J., H. Schaefer, and G. E. O'Neal, "Producing Dually Completed Wells," *API Drilling and Production Practices*, 1956, p. 26.

33a. Mallander, R. G., Discussion of Zaba, Schaefer and O'Neal paper (Ref. 33), *API Drilling and Production Practices*, 1956, p. 36.

34. Tausch, G. H., and J. W. Kenneday, "Permanent-Type Dual Completions," *API Drilling and Production Practices*, 1956, p. 208.

35. Turner, M. C., "When Do Dual Completions Pay?" *Oil and Gas Journal*, May 17, 1954, p. 111.

36. Eastman, H. J., "Lateral Drain Hole Drilling," *The Petroleum Engineer*, Nov. and Dec. 1954, pp. B–57 and B–44, respectively.

37. Roemershauser, A. E., and M. F. Hawkins, Jr., "The Effect of Slant Hole, Drainhole, and Lateral Hole Drilling on Well Productivity," *Journal of Petroleum Technology*, Feb. 1955, p. 11.

38. Landrum, B. L., and P. B. Crawford, "Effect of Drainhole Drilling on Production Capacity," *Journal of Petroleum Technology*, Feb. 1955, p. 45.

39. Perrine, R. L., "Well Productivity Increase from Drain Holes as Measured by Model Studies," *Trans. AIME*, Vol. 204, (1955), p. 30.

40. Muskat, M., and R. D. Wyckoff, "An Approximate Theory of Water-Coning in Oil Production," *Trans. AIME*, Vol. 114, (1935), p. 144.

41. Muskat, M., *Physical Principles of Oil Production*, 1st ed. New York: McGraw-Hill Book Co., 1949, pp. 226–236.

42. Muskat, M., The Flow of Homogeneous Fluids Through Porous Media, 1st ed. Ann Arbor, Michigan: J. W. Edwards, Inc., 1946, pp. 263–276.

43. Arthur, M. G., "Fingering and Coning of Water and Gas in Homogeneous Oil Sand," *Trans. AIME*, Vol. 155, (1944), p. 184.

44. Hubbert, M. K., "Entrapment of Petroleum under Hydrodynamic Conditions," in Bull. 37, American Association of Petroleum Geologists, (1953), p. 1954.

45. Chaney, P. E., M. D. Noble, W. L. Henson, and J. D. Rice, "How to Perforate Your Well to Prevent Water and Gas Coning," *Oil and Gas Journal*, May 7, 1956, p. 108.

46. Pirson, S. J., *Oil Reservoir Engineering*, 2nd ed. New York: McGraw-Hill Book Co., Inc., 1958, pp. 431–437.

47. Rison, C. D., "Manufacture of Nitroglycerin and Use of High Explosives in Oil and Gas Wells," *Trans. AIME*, Vol. 82, (1929), p. 240.

48. Grant, B. F., W. I. Duvall, L. Obert, R. L. Rough, and T. C. Atchison, "Research on Shooting Oil and Gas Wells," *API Drilling and Production Practices*, 1950, p. 303.

49. Atchison, T. C., Jr., B. F. Grant, and W. I. Duvall, "Progress Report on Well-Shooting Research," *API Drilling and Production Practices*, 1952, p. 63.

50. Kingston, B. M., *Acidizing Handbook*. Houston, Texas: The Gulf Publishing Co., 1947.

51. Clark, J. B., "A Hydraulic Process for Increasing the Productivity of Wells," *Trans. AIME*, Vol. 186, (1949), p. 1.

52. Roberts, G., Jr., "Hydraulic Fracturing," in *Technical Manual on Hydraulic Fracturing*. Tulsa, Oklahoma: Oil and Gas Journal.

53. Scott, P. P., Jr., Wm. G. Bearden, and G. C. Howard, "Rock Rupture as Affected by Fluid Properties," *Trans. AIME*, Vol. 198, (1953), p. 111.

54. Hurst, R. E., J. M. Moore, and D. E. Ramsay, "Development and Application of Frac Treatments in the Permian Basin," *Trans. AIME*, Vol. 204, (1955), p. 58.

55. Hendrickson, A. R., G. L. Foster, and R. B. Rosene, "Investigation of Various Refined Oils for Formation Fracturing," *Trans. AIME*, Vol. 204, (1955), p. 285.

56. Dehlinger, P., W. H. Browne, and C. O. Bundrant, "Optimum Sand Concentrations in Well Treatments," *Trans. AIME*, Vol. 201, (1954), p. 273.

57. Brown, R. W., G. H. Neill, "Hydraulic Horsepower Requirements for Well Treatments," *The Petroleum Engineer*, Mar. 1957, p. B–53.

58. Howard, G. C., D. H. Flickinger, and C. R. Fast, "How to Calculate Maximum Net Profit from Hydraulic Fracturing," *Oil and Gas Journal*, May 26, 1958.

APPENDIX A

CONVERSION FACTORS

Multiply →	*By* →	*To Obtain*
To Obtain ←	*By* ←	*Divide*
Acres	43,560	Square feet
Acres	4,047	Square meters
Acre-feet	43,560	Cubic feet
Acre-feet	7,758	Barrels
Atmospheres	14.70	Lb/square inch
Atmospheres	29.92	Inches of mercury
Atmospheres	33.90	Feet of water
Atmospheres	76.0	Cm of mercury
Barrels	42	Gallons*
Barrels	5.6146	Cubic feet
Barrels	0.159	Cubic meters
Barrels	0.159×10^6	Cubic centimeters
Barrels/hour	0.0936	Cu ft/minute
Barrels/hour	0.70	Gallons/minute
Barrels/hour	44.2	Cu centimeters/second
Barrels/day	0.0292	Gallons/minute
Barrels/day	1.84	Cu centimeters/second
Centipoise	0.01	Poise
Centipoise	0.01	Grams/cm-sec
Centipoise	6.72×10^{-4}	Pounds/ft-sec
Cubic feet	7.48	Gallons
Feet	30.48	Centimeters
Feet	0.360	Vara (Texas)
Feet/minute	0.5080	Cm/sec
Gallons (U.S.)	0.833	Gallons (Imperial)
Gallons	3785	Cubic centimeters
Grams	980.7	Dynes
Grams	0.0353	Ounces
Grams	2.205×10^{-3}	Pounds
Grams	15.43	Grains
Hectares	2.471	Acres
Horsepower	33,000	Ft-lb per minute
Horsepower	550	Ft-lb per second
Inches	2.54	Centimeters
Inches of mercury	1.134	Feet of water
Inches of mercury	0.4912	Lb per sq inch
Kilograms	2.2046	Pounds
Kilowatts	1.341	Horsepower
Meters	3.281	Feet
Meters	39.37	Inches
Miles	1.609	Kilometers
Miles	1,900.8	Varas (Texas)
Parts per million	0.05835	Grains per gallon
Pounds per Cu Ft	0.1337	Lb per gallon
Pounds per Cu Ft	0.01602	Grams per cc
Pounds per gallon	0.1198	Grams per cc
Sacks cement	94	Pounds
Sections	640	Acres
Specific gravity	8.33	Lb per gallon
Specific gravity	350.5	Lb per barrel
Townships	23,040	Acres

*All gallons are U.S.

APPENDIX B
GRAPHICAL SLOPE DETERMINATIONS

In many engineering problems, it is necessary to determine graphically the slope of some particular function. If the plotted curve is not a straight line, a tangent to the desired point is constructed, and the slope at the point is then determined as that of the tangent line. For some reason, these operations are a common source of errors, probably because plots on different types of graph paper require slightly different

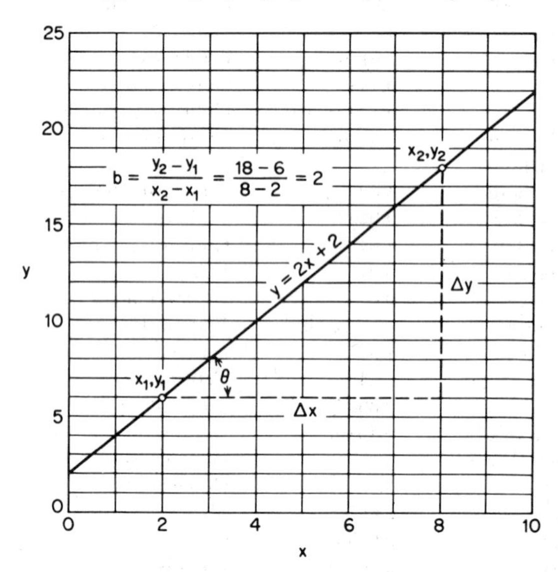

Fig. B-1. Cartesian plot.

interpretation procedures. Following are examples of proper procedure for cartesian, semi-log, and log-log graph paper.

B-1. Cartesian Plots:

All functions of the type $y = a + bx$ are linear on this paper. The slope b is obtained as

$$b = \frac{y_2 - y_1}{x_2 - x_1}$$

Physical measurement, i.e., protractor measurement, of $\tan \theta$ as it exists on the paper has no significance except for the very special case when the x and y scales happen to be equal. Figure B-1 shows a plot of the function $y = 2 + 2x$. Obviously, the slope $= dy/dx = 2$, which is verified graphically on the figure.

B-2. Semi-Log Plots:

All functions of type: $y = B^{a+bx}$ are linear on this type of paper since, taking the log of both sides, $\log y = a + bx$; hence, $\log y$ (base 10 assumed) is linear with x. Rather than plotting $\log y$ vs x on Cartesian paper, we may plot y vs x on semi-log paper with y on the logarithmic scale. However, in taking the slope it must be remembered that $\log y$ is the variable and not y, i.e.,

$$b = \frac{\log y_2 - \log y_1}{x_2 - x_1}$$

Note that it is most convenient to choose Δy as one cycle, giving $\Delta y = \pm 1$. Again, physical measurement

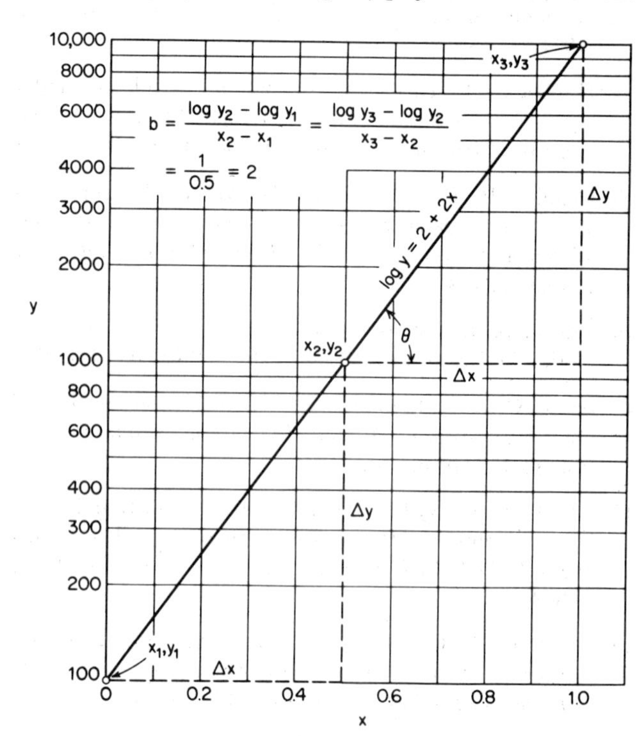

Fig. B-2. Semi-log plot.

of $\tan \theta$ is insignificant, since changes in scale distort the true functional value of θ. If the function portrayed is to another base (such as e), the plot is, of course, still linear; however, the slope becomes:

$$b = \frac{\ln y_2 - \ln y_1}{x_2 - x_1} = \frac{2.3(\log y_2 - \log y_1)}{x_2 - x_1}$$

since $\ln y = 2.3 \log y$. Consequently, the difference in the ln of two numbers one cycle (power of 10) apart is always $2.3 \times 1 = 2.3$. Figure B-2 is a plot of $y = (10)^{2+2x}$, or $\log y = 2 + 2x$. The slope,

$$b = \frac{d(\log y)}{dx} = 2,$$

as can be verified graphically.

B-3. Log-Log Plots:

All functions of type $y = ax^b$ are linear on this paper since $\log y = \log a + b \log x$. The slope,

$$b = \frac{\log y_2 - \log y_1}{\log x_2 - \log x_1}$$

Here, the physical value of $\tan \theta$ is the slope and may be measured in in., cm, etc. if desired. This is true, since changes in scale merely result in parallel lines. For convenience choose Δx as one cycle; thus:

$$b = \frac{\log y_2 - \log y_1}{\pm 1}$$

Also note that the base does not affect this computation, i.e.,

$$b = \frac{\cancel{2.3}(\ln y_2 - \ln y_1)}{\cancel{2.3}(\ln x_2 - \ln x_1)}, \quad \text{etc.}$$

Shown to illustrate this is the equation $y = 2x^2$ or $\log y = \log 2 + 2 \log x$, which is plotted as Figure B-3. Differentiating, $d(\log y)/d(\log x) = 2$; this is the slope, as verified both graphically and by direct measurement of $\tan \theta$.

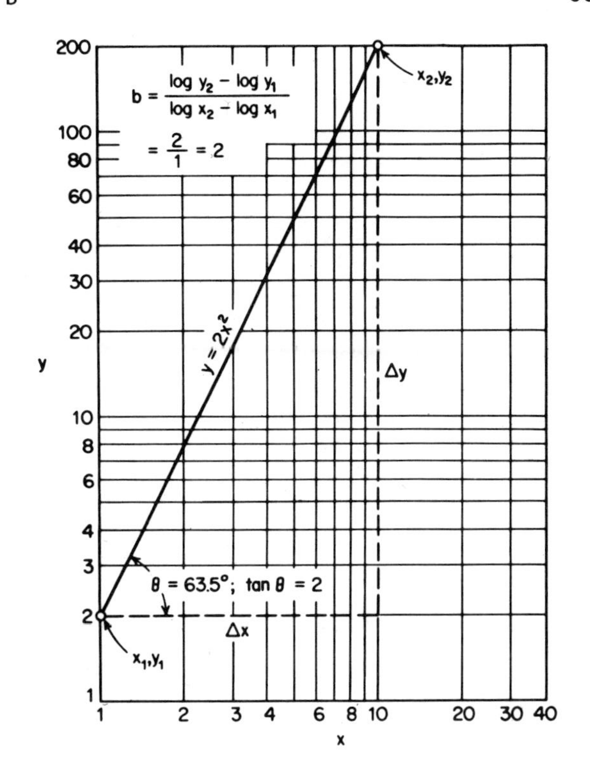

Fig. B-3. Log-log plot.

Index

D

E

F

C